Developments in Economic Geology, 1

GEOCHEMICAL EXPLORATION 1974

Association of Exploration Geochemists
Special Publication No. 2

Developments in Economic Geology, 1

GEOCHEMICAL EXPLORATION 1974

Proceedings of the Fifth International Geochemical Exploration Symposium held in Vancouver, B.C., Canada, April 1—4, 1974, sponsored and organized by the Association of Exploration Geochemists

Edited by

I.L. ELLIOTT

Chief Geochemist, Falconbridge Nickel Mines Limited, Vancouver, B.C.

and

W.K. FLETCHER

Assistant Professor, University of British Columbia, Vancouver, B.C.

Association of Exploration Geochemists
Special Publication No. 2

ELSEVIER SCIENTIFIC PUBLISHING COMPANY
Amsterdam — Oxford — New York 1975

ELSEVIER SCIENTIFIC PUBLISHING COMPANY
335 Jan van Galenstraat
P.O. Box 211, Amsterdam, The Netherlands

AMERICAN ELSEVIER PUBLISHING COMPANY, INC.
52 Vanderbilt Avenue
New York, New York 10017

Library of Congress Card Number: 74-21855

ISBN 0-444-41280-8

With 317 illustrations and 102 tables

Printed in The Netherlands

FOREWORD

In the first week of April 1974, over 550 people from thirty countries came
to Vancouver to attend the Fifth International Geochemical Exploration
Symposium. This meeting, sponsored and organized by the Association of
Exploration Geochemists, heard speakers from Universities, Government
Agencies and the Mining Industry describe approaches and solutions to
practical problems of geochemical prospecting. Following the technical
sessions 130 of the delegates visited the porphyry copper deposits of the
Lornex, Bethlehem and Utah mining companies.

The Association was honoured by the presence of Dr. D.J. McLaren,
Director of the Geological Survey of Canada, who welcomed the delegates to
the meeting, and Dr. J.S. Webb, Professor of Applied Geochemistry at the
Royal School of Mines, and President of the Institution of Mining and Metallurgy of London, who gave the opening address.

The technical programme was largely concerned with exploration for
mineral deposits concealed beneath various types of overburden, and with
the use of primary dispersion patterns as guides to mineralization. Other
sessions dealt with analytical and sampling techniques, statistical interpretation of geochemical data, and other topics of a more general nature.

Forty-nine papers were scheduled for presentation during the four-day
meeting, and forty-seven were delivered. We thank all those who offered
synopses; due to constraints of time and space it was not possible to include
many excellent contributions. It is a matter of particular regret that our
Russian colleagues Dr. R.I. Dubov and Dr. E.M. Kvatakovskiy were unable
to attend the meeting. The gap in the programme caused by their absence
was fortunately filled with a film on the use of dogs in prospecting made
available to us by Dr. Bölviken of the Norwegian Geological Survey. We are
also grateful to Dr. H.V. Warren for presenting his paper on "Barium—
strontium relationships -- a possible geochemical tool in the search for ore
bodies", at short notice to further fill up the programme.

On behalf of the Association we warmly thank all the authors for taking
the trouble to prepare and deliver their papers. The successful outcome of
the meeting rests upon their efforts. To the chairmen of the sessions must
go the credit for the smooth organization and progression of the technical
sessions. We would also like to thank all those delegates who contributed to
the discussion of the papers.

Forty-five of the papers read at the meeting are presented here. One paper
has been published elsewhere and appears here in summary form and one
manuscript was not available at the time of going to press. The manuscripts
were reviewed by numerous people; and we thank all who gave their time and

energy to this task, particularly J. Barakso, M.A. Chaffee, R.F. Horsnail and A.J. Sinclair. The editors are most grateful to those authors who generously reduced the length of their papers to enable us to meet the page limitations imposed by the publisher.

As chairman of the Organizing Committee, and on behalf of the Association, I am pleased to be able to place on record our thanks to the members of the committee for the hard work, patience and enthusiasm with which they carried out their responsibilities. It is fitting that recognition should also be made of the essential contributions of Mr. W. Young, our accountant and Miss Julia Robinson and various members of the staff of the University of British Columbia for their secretarial work.

This meeting would not have been possible without substantial donations of time and facilities from the employers of the members of the organizing committee. To all these organizations the Association gratefully acknowledges its debt.

The Sixth International Geochemical Exploration Symposium will be held in Sydney, Australia; we can look forward to an enjoyable and stimulating meeting "Down Under" in 1976.

I.L. ELLIOTT
Chairman, Symposium Organizing Committee

W.K. FLETCHER
Chairman, Technical Session Committee

SYMPOSIUM COMMITTEE

I.L. Elliott (Chairman)	*Falconbridge Nickel Mines Limited*
J.A. Barakso	*Min-En Laboratories Limited*
R.W. Boyle	*Geological Survey of Canada*
A.A. Burgoyne	*Union Miniere Mining and Exploration Limited*
A. Burton	*Phelps Dodge Limited*
E.M. Cameron	*Geological Survey of Canada*
W.K. Fletcher	*University of British Columbia*
F.D. Forgeron	*Bondar-Clegg Limited*
R.F. Horsnail	*AMAX Exploration Incorporated*
J.D. Knauer	*Noranda Exploration Limited*
B.M. Smee	*Barringer Research Limited*

LIST OF CONTRIBUTORS

CONTENTS

X

WELCOMING ADDRESS

D.J. McLAREN

Director, Geological Survey of Canada, Ottawa, Ont. (Canada)

It is a great pleasure for me to welcome all of you to this Fifth International Geochemical Exploration Symposium on behalf of the Canadian Government. I would, in particular, like to welcome those from overseas to our continent and country. We hope that your visit here, both during the meetings and elsewhere, will be enjoyable.

Those of you who are not familiar with this part of the world may not know that you are meeting in a city whose climate and scenery is envied by many of us who live further east. But please do not let this discourage you from visiting other parts of Canada. This country has much to interest the majority of you who I know are still geologists at heart.

At the opening of this Fifth International Geochemical Exploration Symposium, it gives me pleasure to recall that the Geological Survey of Canada hosted the first in Ottawa in 1966. At that time we were pleasantly surprised to find that more than two hundred people were interested in attending the meeting. As you know this was followed by biennial meetings in Denver and then in Toronto. Two years ago the symposium moved overseas for the first time to London and I understand that the meeting to be held in 1976 will be in Australia. Perhaps I should add that I am well aware of this latter fact since my geochemical staff mention it on all suitable occasions.

These successive symposia have clearly filled a need for communication among applied geochemists. Out of these meetings has come the Association of Exploration Geochemists, which at this time has 500 members in 50 countries of the world. The Association has, in its turn, created the first international journal in the field of exploration geochemistry.

I am particularly pleased that by hosting the first meeting, the Geological Survey of Canada has played some part in this development of intranational and international communication. This is because the discovery and management of resources are now playing a far greater role in the life of Man than most of us could have guessed in 1966.

The political events that have caused a shortage of petroleum in many countries have only accelerated the inevitable restriction on the supply of this commodity from conventional sources. This has caused for the first time a serious consideration of the profligate use by the developed world of irreplaceable energy resources. Great efforts are now being applied to discovering new sources of energy and to the proper evaluation and management of what resources there are. Perhaps, in the next century, this sudden crisis will be seen as a blessing. However, to be so, we must apply the same determination to the conservation and proper development of the world's non-energy sources.

This brings one to a point that I would particularly like to make to this audience of applied geochemists. I know that I need give you no encouragement to continue your vital work on finding conventional types of mineral resources. I would suggest to you, however, two areas in which the geosciences must greatly increase their efforts. The first is in developing methods of evaluating resource potential. The second, and related area, is in discovering unconventional sources of metals and minerals.

In both these fields geochemistry has a critical role to play. Perhaps I can best show this by analogy with petroleum. The "unconventional" sources of petroleum that will see most development during the next few years are oil sands, oil shales, and coal. These sources are fortunately not difficult to find and the impediments to their development have been largely economic and technological. In contrast, most of the unconventional or low-grade sources of metals are much less obvious.

I have no need to tell you that while economists may speak glibly of extracting metals from granites or shales the amount of information on the distribution of most metals in the crust of the earth is woefully lacking. Most of the deposits which are easy to discover have been found, and we must look to more expensive frontier areas and greater depths for new ore. The message for the geologist is clear, and we hope that it is fully understood by the economist and politician. Assurance of resource adequacy does not follow simply from the crustal abundance of the elements. It can only come about through detailed and exhaustive data collecting and interpretation, combined with research into new methods and systems of exploration and extraction. The basic requirement is for maps: geological, geophysical, and geochemical maps of every kind, including, and supported by surface and subsurface three dimensional representations of the structural configurations of the materials of the crust, their composition, and geological history.

As an example, although the present situation in the marketing of gold is an artificial one, it perhaps foretells the future patterns for other metals vital to our way of life. At present prices gold may possibly be extracted profitably at a level of one part per million — not as a by-product but as a primary product. Yet our knowledge of the world's reserves of ore at this level of abundance is negligible. For low-grade ores, geochemistry will be the principal tool of discovery.

The Geological Survey's continuing program of airborne geophysical and ground geochemical surveys is designed to assist in the evaluation of this country's mineral potential and to develop new methods of data collecting and interpretation. As well as providing data that may be used immediately by the exploration industry for the discovery of deposits with presently workable grades of metal, it will allow our scientists to gain some knowledge of the metal contained in bodies of much lower grade and which may not be exploited for many years.

This brings me to another important point to draw to the attention of this meeting. Governments can provide reconnaissance data in the geosciences that are difficult for individual companies to obtain. Conversely, mining companies can provide much of the information that governments now require for the proper management of resources. Increasing costs of exploration should discourage duplication of effort in collecting basic information, and prevent misdirected effort through lack of information. Larger amounts of non sensitive or term sensitive information should be reported, marshalled, pooled, and used, thereby increasing the efficiency of industry. These remarks apply particularly to the mining industry, in this country, which might still, with profit, learn from the benefits that have accrued to the petroleum industry of freer exchange of information without serious loss of competitive advantage.

It is unfortunate, but perhaps inevitable, that in the western world we are now entering a period of some conflict between resource industries and governments that are responding to or anticipating the demands of their people. It is associations such as yours, made up of men working in industry, university and government, and with members drawn from societies with differing political beliefs, that must ensure continuity of responsibility during this time of change. I would hope that you will play your part in developing the understanding and cooperation that is necessary if the world's resources are to be used in the best interests of its peoples.

And we are faced with another conflict: two schools of thought as to what estimates of mineral potential show, which necessarily governs the mineral policy to be adopted by the nations of the world. One school believes that the supply of economically obtainable minerals is indeed limited and that already we are in danger of running out of certain commodities. The other school looks at the general geochemical abundance of elements in the crust and asserts that future technologies will be able to develop "ores" many times leaner than those being developed today. They argue that the supply is therefore almost limitless. We are faced, therefore, with a very important philosophical issue, which may, in fact, be one of the most important to face mankind, at least in physical terms, in the future. For on its resolution must depend rational planning for the future. This, surely, is a sufficient argument for rapid development of research into methodology of mineral assessment which must be closely tied with economic and social considerations. A major confrontation must come between the optimists and pessimists in these two schools and the solution may well lie in your hands and in the hands of those who work with you. We may claim, therefore, that the meeting held in Ottawa in 1966 and the one beginning here today may have significance far greater than perhaps we even now appreciate. Good luck with your meeting.

ENVIRONMENTAL PROBLEMS AND THE EXPLORATION GEOCHEMIST

JOHN S WEBB*

President, Institution of Mining and Metallurgy, London (Great Britain)

Variation in the geographical distribution of the elements is a central
factor in a wide spectrum of environmental problems. The ability to detect
and interpret such variations in prospecting for mineral deposits is of course
the essence of exploration geochemistry. By virtue of the job he has to do,
the exploration geochemist has acquired a unique breadth of experience in
developing and using cost-effective methods for mapping the distribution of
trace elements. This expertise can also be employed or adapted to advantage
in the environmental field. In order to achieve this to greatest effect, it is
clearly desirable that exploration geochemists in university, industry or government agencies should be generally aware of the problem areas (other than
mineral exploration) in which their perspectives, techniques and results can
be used to the benefit of the community as a whole. In principle this has long
been apparent but the present social and political climate underlines the need
for more widespread positive action than we have seen hitherto. I would like,
therefore, to take the opportunity afforded by this Symposium to illustrate
briefly the overall scope for geochemical mapping and to make some suggestions towards extending the inter-disciplinary co-operation that is essential
in the best interests of all concerned.

Starting with agriculture: the importance of trace elements in crop and
animal nutrition has been recognized for many years. The optimum ranges of
concentration in soil or herbage may often be narrow, and deficiency and/or
excess of such metals as copper, zinc, molybdenum, cobalt, selenium, iron,
manganese and chromium can lead to crop failure or animal death. The onset
of such extreme conditions is generally readily recognized and treated
accordingly.

Of far greater significance are those areas where marginal imbalance is
responsible for sub-clinical conditions resulting in infertility, loss of production and unthriftiness. In these circumstances the problem may not be recognized and, therefore, remain untreated, with serious consequences not only
to individual farmers but also to agricultural economy on the provincial or
national scale depending on the size of the affected areas. In addition there

---

*Present address: Department of Geology, Royal School of Mines, Prince Consort Road,
London SW7.

are the potential or latent problems where the trace element status of the soil is such that clinical or sub-clinical disorders may only become apparent following changes in farm management or agricultural practice. Furthermore, the importance of multi-element interaction in relation to agricultural productivity is also receiving increasing attention. The value of producing multi-element maps for agricultural purposes alone is self-evident.

In fisheries, information concerning the effect of variations in trace elements on edible species is meagre although the evidence of toxic symptoms in trout and oyster larvae subjected to elevated concentrations of such metals as zinc, cadmium, copper and lead and the ability of some species to accumulate heavy metals, indicates that further research will increase the economic significance to be attached to trace elements in this field. Again, all too little is known of possible sub-clinical damage at marginal concentrations or, indeed, of the effects of abnormal metal contents on lower organisms in the food chain. Because of massive dilution by seawater offshore, adverse conditions related to variations in the chemical composition of the aquatic environment are largely limited to rivers, estuaries and inshore waters. In short, the problems lie primarily in those environments where the trace element status of the adjoining inland catchment areas has a direct impact.

In addition to their actual or potential bearing on productivity in agriculture and fisheries, there is the equally important question of the influence of trace elements on the quality of the produce obtained for human consumption; similar considerations apply to potable water supplies. Increasing awareness of these problems has led to the need for more intensive systematic assessment and monitoring of trace elements in food and water, and prior knowledge of the regional distribution of the elements provided by geochemical maps would materially assist in the selection of the most suitable sites for these activities.

The importance of trace elements in human health and nutrition is well known and there is no doubt as to the existence of geographical patterns in the incidence of many diseases. There are, however, very few instances where a direct relationship with the geochemical environment has been established, these include iodine deficiency and goitre, fluoride intoxication and fluorosis, and fluorine deficiency and dental caries. Several diseases are known to be caused by excessive or deficient intake of one or more elements, amongst which are inadvertent excess of cadmium, mercury and lead and deficiency of iron and zinc, but these are in the main due to abnormal exposure resulting from certain industrial activities or to dietary shortcomings that are not related to the geochemical environment. Apparent correlations with no proven causal relationship are both more numerous and controversial; in this category are, water hardness and cardio-vascular diseases, lead and multiple sclerosis, cadmium and hypertension and atherosclerosis, and a range of trace elements with various types of cancer, to mention but a few. Quite apart from identifiable disease, there is a very real possibility of sub-clinical debilitation due to trace element imbalances which overall could have far more serious consequences and of which, as in the case of agriculture and fisheries, we are very ignorant indeed. The link between the distribution of elements and human health inevitably becomes more tenuous the greater the extent to which a community relies on bulked and processed foods accompanied by movement of populations. Nevertheless, while the greatest correlation can be expected in those relatively primitive communities that live closest to the ground, it is surprising the degree to which rural communities in the more advanced societies may in fact rely on local produce of one sort or another; in parts of the United Kingdom, for example, over 40% of the population receive locally-produced milk and, on average, considerable proportions of the vegetables consumed are home-grown. There is, therefore, every reason for producing geochemical maps in the interests of human health which in appropriate circumstances could be of value to the epidemiologist, medical geographer and health authorities in general.

The geochemist's involvement in pollution has been reviewed elsewhere by Helen Cannon and Barbara Anderson who rightly emphasize the importance of the need to accumulate data concerning the background contents of the elements that can be expected to occur in natural geological settings against which to compare the effects of man-made contamination, and to increase our understanding of the factors influencing metal dispersion and concentration in rocks, soils, vegetation and the atmosphere. I would like only to add the obvious value of geochemical maps in this connection and to stress two equally obvious facts that do not seem to be generally recognized in some quarters. First, many of the environmental problems outlined in the preceding paragraphs are often the result of entirely natural variations in metal concentration which can cover extremely wide ranges even without including those related to undisturbed mineralization. Secondly, wherever man has been, the environment by definition will have been polluted to a greater or lesser degree; in places this has been both extensive and intensive as in parts of the United Kingdom, for example, where mining and smelting have been widespread and date back to Roman times and beyond. To be fully effective as aids in evaluating the contaminating influence of present-day industrial activities (either actual or proposed) geochemical maps should preferably show not only the patterns of natural metal variations but also those resulting from past human activity. One last point on this subject: pollution of the environment will inevitably increase with the growth of populations, their need for food, metals and other resources necessary for survival, and the worldwide demand for higher standards of living. Food can only be grown economically under certain conditions and metals and other mineral resources occur where nature has placed them. If a nation wants to have these commodities it has only two options: either to develop its own resources, technologies and monitoring systems as best it can or to pay for importing them, thereby passing responsibility for dealing with the consequential pollution onto someone else — a short-sighted policy with which I, for one, have little sympathy. The problem is not one of stopping pollution — which is manifestly impossible —

but of determining the best means and *cost* of ameliorating its impact in
relation to the damage to the environment that can be tolerated in the
interests of world society as a whole. In these circumstances, emotive
approaches though understandable can all too easily lead to ill-advised
political and social actions. Objective decisions can only be based on full
knowledge of the facts relating to any particular problem. Where metals are
involved geochemistry must play its part in arriving at equitable solutions.

In addition to their many practical applications for environmental purposes
and mineral exploration, multi-element geochemical maps also have fundamental scientific significance not only in geology but in any field of research
where the geographical distribution of the elements is a relevant factor. The
question is not, therefore, whether multi-purpose regional geochemical atlases
are required but how they may be compiled and used to best effect.

Before taking a brief look at these aspects, I would first like to indicate
the environmental potential of such atlases with some examples from a
recently completed experimental reconnaissance of England and Wales. This
reconnaissance, which formed part of an extended research programme on
regional geochemical mapping carried out by the Applied Geochemistry
Research Group at Imperial College, was based on sampling stream sediment
at tributary/road intersections at a mean density of one sample per square
mile. The preliminary computer plots for a selection of the 22 elements
determined are shown in Figs.1—5.

The distribution of molybdenum (Fig.1) is largely related to molybdeniferous
marine shales ranging in age from Cambrian to Recent. These formations
include rocks of similar lithology that are not molybdeniferous. Consequently,
neither the geological map nor detailed knowledge of the lithology give
adequate information. The patterns are further complicated by the patchy
occurrence of glacial and alluvial drifts of various origins which may or may
not be molybdeniferous and the development of molybdeniferous soils of
recent origin in some coastal environments, the reason for which is obscure.
Clinical molybdenum toxicity and molybdenum-induced copper deficiency
in cattle have long been recognized in parts of these geochemically anomalous
areas. Subsequent investigations have demonstrated the more widespread
existence of sub-clinical conditions where copper supplementation can substantially improve live-weight gain. Not all land within the anomalous patterns
disclosed by the stream sediment reconnaissance is molybdeniferous but the
map serves the very useful purpose of focussing attention on the most likely
suspect areas. Similar considerations apply to a number of other elements of
agricultural interest, and regional liaison committees have been set up by the
U.K. Agricultural Development and Advisory Service in conjunction with the
Imperial College Group, in order to examine the geochemical information as
a whole in the light of local problems and to initiate follow-up programmes
as required.

In contrast to molybdenum, the anomalous patterns for lead, zinc, cadmium
and arsenic (Figs.2—5) are mainly related to mineralization and in most cases

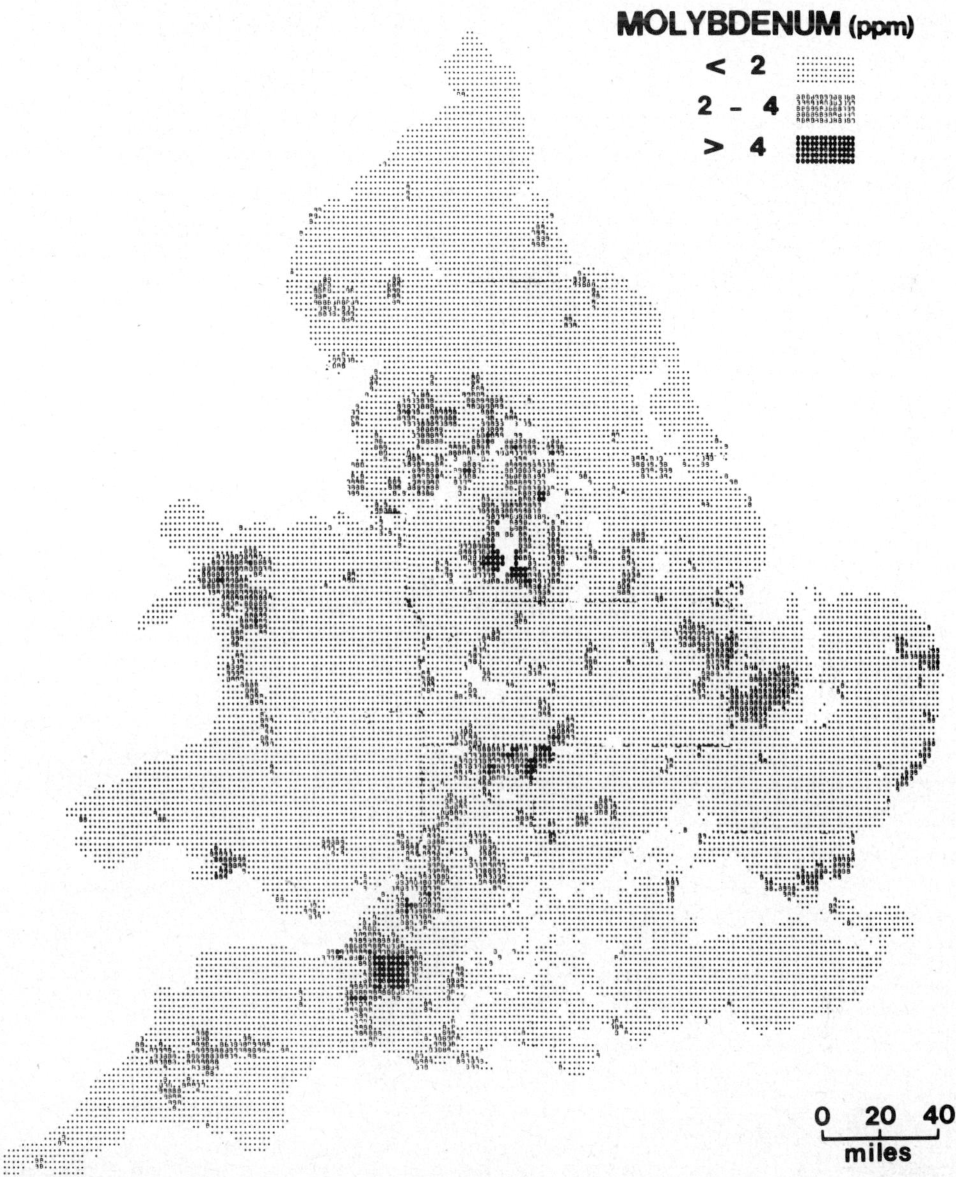

Fig.1. Regional geochemical map for molybdenum in England and Wales.

have been profoundly modified by past mining and smelting, while others are due solely to industrial contamination. The maps for these and other potentially toxic elements are being used in selecting areas for investigating metal distribution in relation to food quality.

Abnormal concentrations of zinc have been implicated in erratic loss of

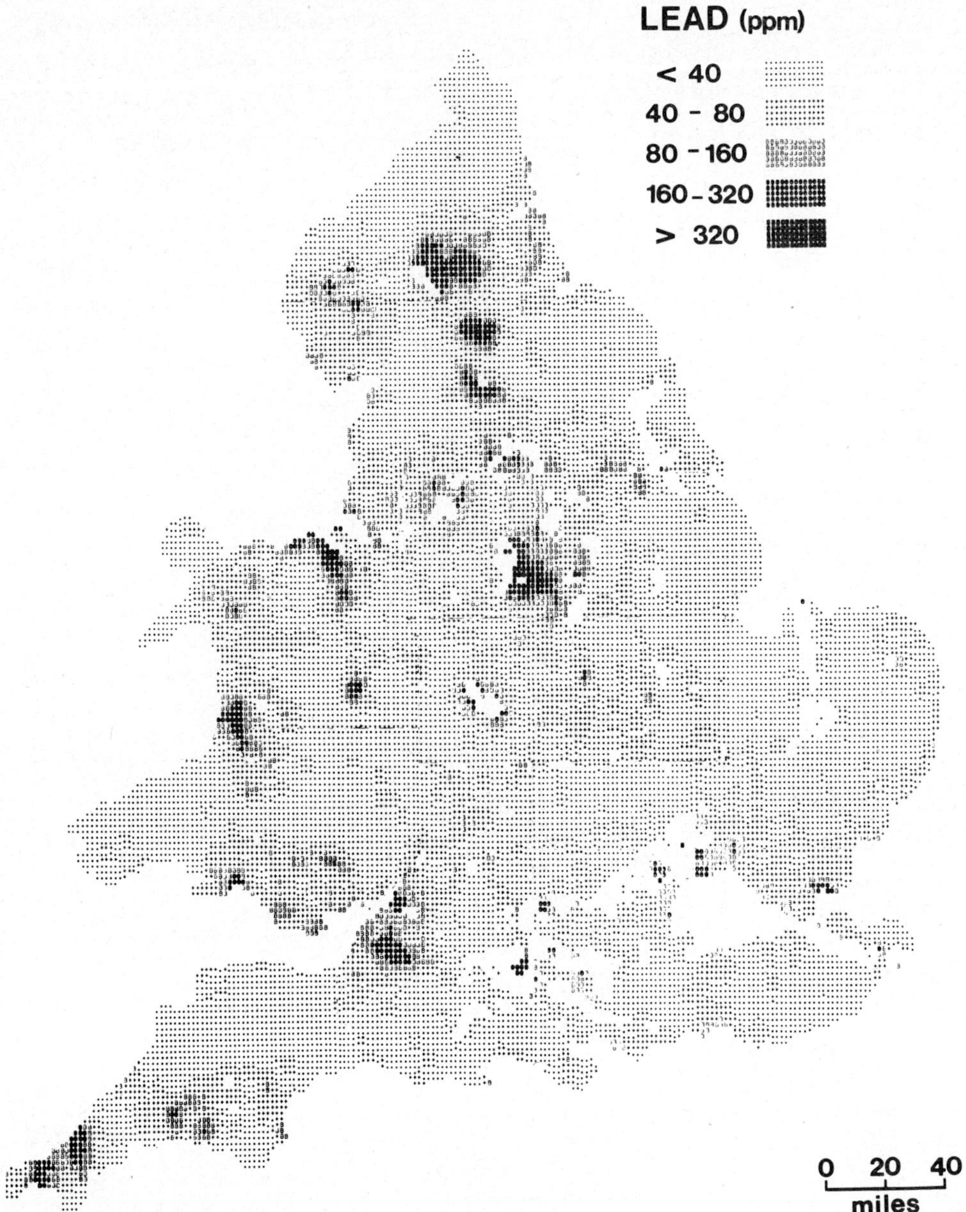

Fig.2. Regional geochemical map for lead in England and Wales.

productivity experienced in a pilot oyster hatchery in northwest Wales belonging to the White Fish Authority. The hatchery is situated on an estuary which receives drainage from an old Pb—Zn mining district. Analysis of the hatchery water showed fortnightly fluctuations ranging from 10 to 500 μg/l

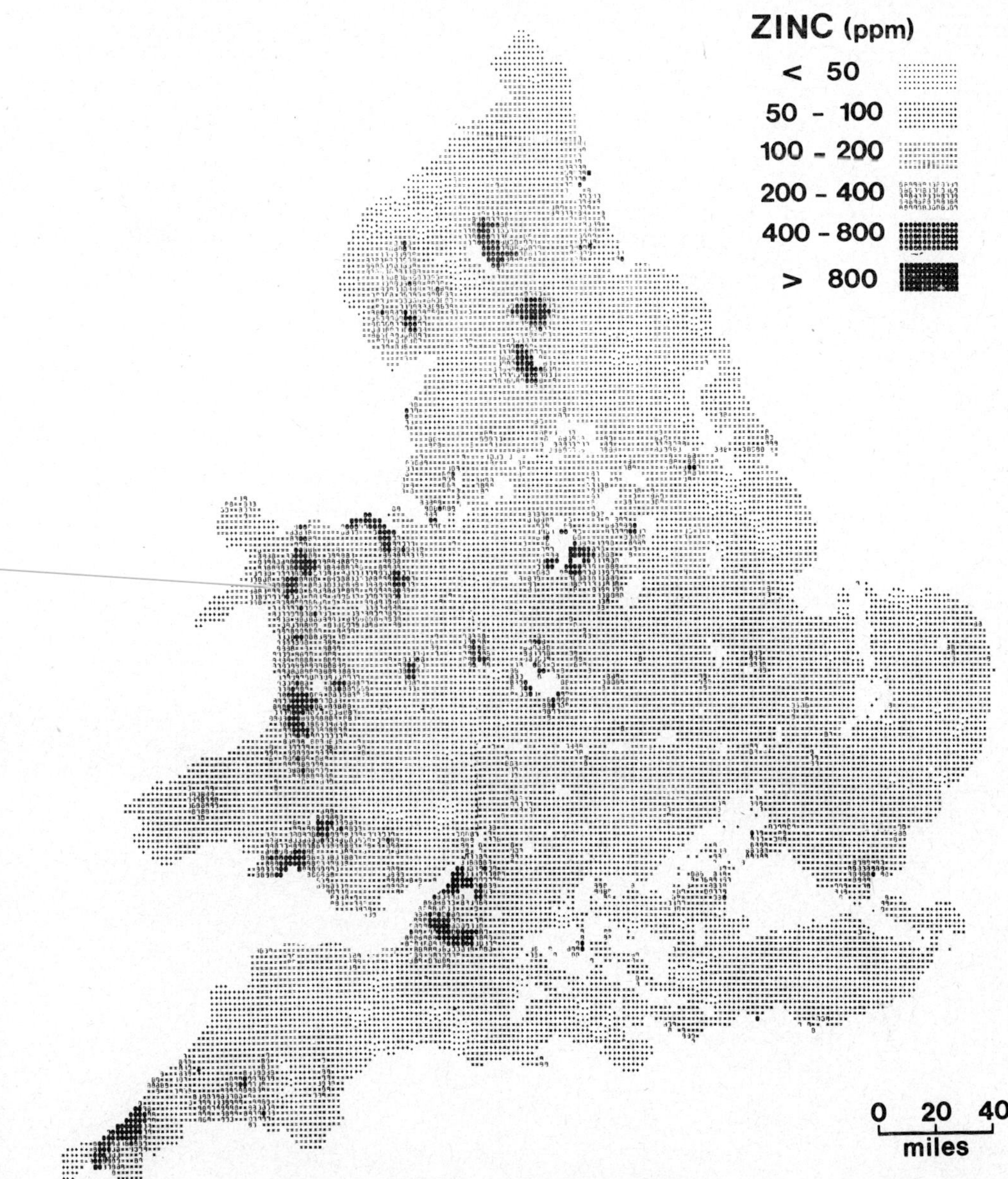

Fig.3. Regional geochemical map for zinc in England and Wales.

Zn related to the tidal cycle and time of year. Subsequent experiments confirmed the toxic effect of zinc on the oyster larvae at concentrations in
excess of 100 μg/l. The regional geochemical maps for zinc and other metals
have since been used in the selection of a number of estuaries with different
trace element characteristics for further studies of the effect of geochemical

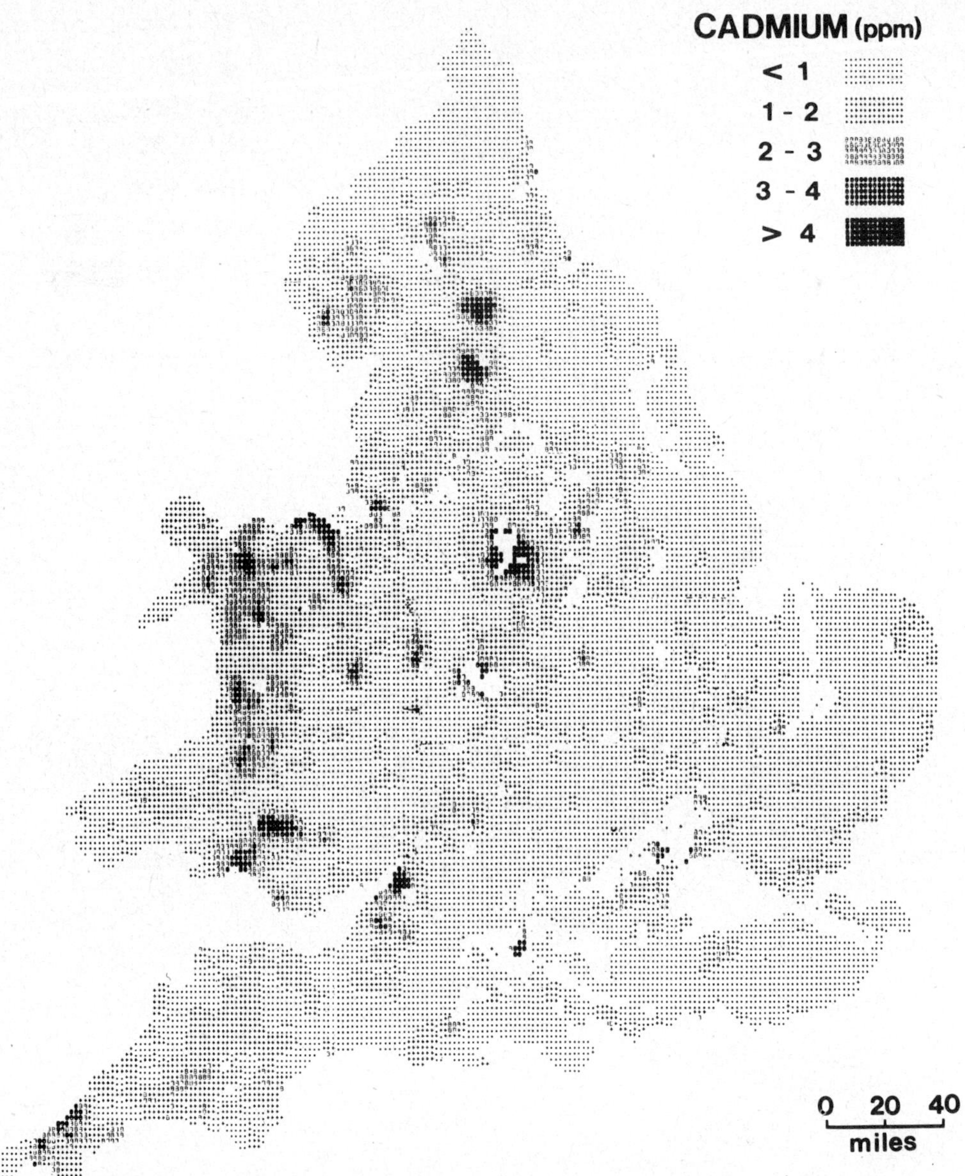

Fig.4. Regional geochemical map for cadmium in England and Wales.

environment on the composition and productivity of shellfish.

Preliminary studies have indicated that stream sediment data can be a more stable index of the trace element status of river catchment areas than analysis of the water, the composition of which is liable to marked diurnal and seasonal fluctuations. Further investigation of the usefulness of stream sediment in-

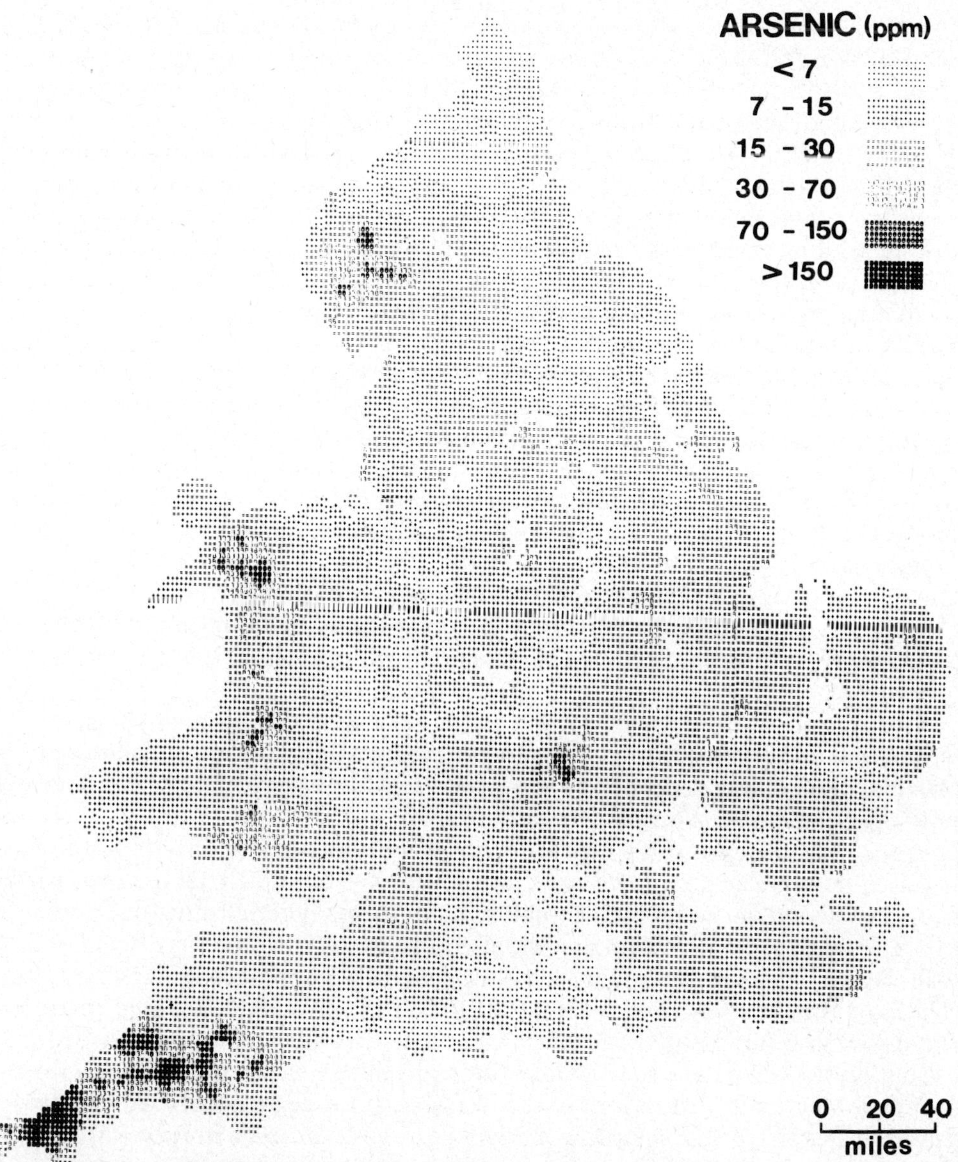

Fig.5. Regional geochemical map for arsenic in England and Wales.

formation as an aid in selecting sites for water monitoring and the location
of reservoirs is now underway in collaboration with the Cornish River Authority and could eventually be extended to other regions.

As a final example, the geochemical reconnaissance data combined with
the national geological map provided one of the bases for area-selection in a
study of childhood lead burdens and the metal content of garden soils carried

out in collaboration with the Paediatric Unit of St Mary's Hospital Medical School, London. The results showed no correlation between lead levels in blood and in soils which contained up to 1000 ppm. A significant correlation was obtained, however, in the results from two groups of villages, one of which included dwellings in close proximity to old mineworkings where garden-soil lead contents up to 3% were reported. Even so, all the observed blood lead levels were well below the presently accepted "safe" level of 80 μg/100 ml, although a few were in the region of the upper limit of the normal range (40 μg/100 ml).

The reconnaissance survey to which the foregoing examples refer was carried out primarily for applied research purposes concerned with examining the multi-purpose scope of broad-scale regional geochemical mapping and the considerable problems of handling, processing and presenting the large volumes of multi-component data involved. The resulting maps are now being used for a wide range of applied research in agriculture, fisheries and pollution, while at the same time providing information of immediate practical value. The cost-effectiveness of the reconnaissance may be judged from the total expenditure of less than $250,000 spread over five years. Clearly there is much room for improvement before arriving at a definitive national atlas, and the more rigorous regional surveys currently being undertaken by the Institute of Geological Sciences are the next important steps towards this objective.

Stream sediment has been chosen as the preferred sampling medium (supplemented by rock, soil and water samples where required) because it provides the closest approach to a *composite* sample of soil and rock weathering products required for primary regional reconnaissance of a country where the bedrock, overburden, soil types and parent materials are markedly heterogeneous. These factors, coupled with a long and varied industrial history, presuppose the existence of many small-scale patterns which individually may be of local extent but which taken overall could be of national significance. A well-developed surface drainage system, generally good accessibility and the relatively small survey area (ca. 50,000 sq. miles) provide additional reasons for systematic sampling at relatively high densities for the initial surveys.

In different circumstances where there are broad expanses of essentially homogeneous rock formations and soil types, poor accessibility or a sparse drainage system, other sampling media or procedures may well be preferable. This could be particularly true where the national area is so large as to necessitate a highly selective initial approach. Here the answers may lie (a) in the extremely low-density stream sediment or soil sampling used by Ian Nichol and his co-workers in Africa and H. Shacklette and his colleagues in the U.S.A., respectively, (b) in the phased (hierarchical) sampling of rock, soil, vegetation and water carried out by A.T. Miesch, Jon Connor and co-workers in Missouri, or (c) in the even broader concept of biogeochemical provinces proposed by V.V. Kovalsky and others in the U.S.S.R. Regional geochemical mapping is in its infancy and we need far more experience of the technical problems and economics of the many different ways of acquiring, handling, processing and presenting the data, before it will be possible to determine the most appropriate procedures to adopt in any given set of conditions. All that can safely be said at the present time is that broad-scale geochemical atlases are now a national cartographic requirement analogous to the geological, topographical and other maps relating to the environment in which we live.

The responsibility for compiling national multi-element multi-purpose geochemical atlases clearly lies with government and the need is urgent. In many parts of the world, even a preliminary broad-scale coverage will take a considerable time to accomplish. In the interim, geochemical surveys of very extensive areas have been or are being undertaken for mineral exploration by national and international agencies and industry. These, even at the reconnaissance level, are normally at a higher degree of detail than is required for a primary atlas and are mostly based on stream sediment, soil or rock sampling. New methods are constantly being developed and in the future there could well be growing application of regional lake-water and lake-sediment sampling and, even, airborne geochemical surveys. Much of these data would be invaluable for environmental purposes both now and in the future. The idea of national geochemical data banks for general use is not new — in fact, the creation of such a bank initiated by the Institute of Geological Sciences is imminent. Extending this concept to include geochemical samples collected during the course of such surveys, particularly reconnaissance surveys, would also provide a permanent source of material for any further analyses that may be required for purposes other than those for which the samples were originally collected.

The compilation of national geochemical maps and archives from material gathered by national and international agencies should present no insuperable problems. For industry, the constraints imposed as a consequence of operating in a competitive climate are readily appreciated. Nevertheless, it would clearly be advantageous to all concerned if a formula could be found (such as a guaranteed period of confidentiality) whereby geochemical data and samples obtained by industry could be made available for deposit with the appropriate government agency to ensure the widest possible use of the results in the national interest.

However the basic data may be acquired, geochemical atlases solve no problems in themselves. Their prime function lies in delineating areas wherein to concentrate more detailed investigations. The degree to which they are useful is dependent on the extent to which they can be interpreted for the widest possible range of practical and fundamental purposes. The exploration geochemist, geologist and groundwater hydrologist are primarily concerned with the metal content of the bedrock; the interests of the agriculturalist, epidemiologist and others in the environmental field mostly start with the metal content of the soil and surface waters (irrespective of the relationship with the underlying bedrock) and the biological pathways by which the

elements may enter the food chain or otherwise influence the ecosystem.
Despite the magnitude of the range of these different interests, there is a great
deal of common ground in the problems of determining the factors influencing
metal distribution, mode of occurrence, origin and bio-availability. The
apparent divergence in objectives underlines the necessity for a multi-disciplinary approach in arriving at the best methods of compiling and presenting
regional geochemical information and in the multi-disciplinary research
necessary to develop interpretational criteria to meet the requirements of all
the eventual users.

Even assuming that all these praiseworthy goals are attained, the full return
for the effort invested will not be forthcoming unless effective steps are also
taken to overcome the problems of mutual education and communication —
not only between the professionals concerned but also the public at large.

Finally, as the programme for this Symposium shows, the exploration geochemist continues to exercise ingenuity in solving the problems of metal dispersion and in the development of new techniques for finding mineral deposits.
There is no doubt as to his ability also to contribute in the environmental
field. The extent to which he does so will depend on a more general recognition
of his achievements and potential, and on the encouragement he receives to
become involved in the broader issues.

POSTSCRIPT

All too few applied geochemists have been actively concerned with both
mineral exploration and environmental problems. Professor Harry Warren,
recently retired from the University of British Columbia, is one of these few.
It is particularly appropriate on this occasion in Vancouver that we should
pay tribute to his efforts pursued with characteristic enthusiasm and tenacity
over many years, which have stimulated so much thought and activity (and no
little controversy!) in the geomedical field.

SELECTED REFERENCES

Armour-Brown, A. and Nichol, I., 1970. Regional geochemical reconnaissance and the
location of metallogenic provinces. Econ. Geol., 65: 312—330
Cannon, H. and Hopps, H. (Editors), 1971. Environmental geochemistry in health and
disease. Geol. Soc. Am., Mem. 123
Hemphill, D. (Editor), 1967—1971. Trace substances in environmental health. Proc. 1st—
5th Ann. Conf., Univ. of Missouri Environmental Health Center and Extension Division
Kubota, J. and Allaway, W.H., 1972. Geographic distribution of trace element problems.
In: Micronutrient in Agriculture. Soil Science Society of America, Madison, Wisc., chapter 21
Mills, C.F. (Editor), 1969. Trace element metabolism in animals. Proc. WAAP/IBP Int.
Symp., Aberdeen. Livingstone, Edinburgh and London
Shacklette, H.T. et al., 1971. Elemental composition of surficial materials in the conter-
minous United States. U.S. Geol. Survey Prof. Paper 574-D
Thornton, I., 1975. Applied geochemistry in relation to mining and the environment. In:

Proc. Symp. Minerals and the Environment. Institution of Mining and Metallurgy, London (in press)
Underwood, E.J., 1971. Trace Elements in Human and Animal Nutrition. Academic Press, London, 3rd ed.
U.S. Geological Survey, 1971—1973. Branch of Regional Geochemistry. Open file reports
Webb, J.S. et al., 1973. Provisional geochemical atlas of Northern Ireland. Appl. Geochem. Res. Group., Imperial College, London, Comm. No. 60

Part 1
GENERAL TECHNICAL SESSION

Chairmen:

P.K. THEOBALD
United States Geological Survey, Denver, Colo., U.S.A.
H.V. WARREN
University of British Columbia, Vancouver, B.C., Canada

GEOCHEMICAL METHODS OF EXPLORATION FOR MASSIVE SULPHIDE MINERALIZATION IN THE CANADIAN SHIELD

E.M. CAMERON

Geological Survey of Canada, Ottawa, Ont. (Canada)

ABSTRACT

The massive sulphide (volcanic-exhalative) deposits of the Canadian Shield produce a large part of Canada's output of zinc, copper, lead, silver and gold. Since they are a relatively small target and are often covered by thick glacial sediments in the southern Shield, geochemistry has not played an important part in their discovery. However, the majority of the Shield has only a thin cover of overburden and here conventional geochemical methods based on primary and secondary dispersion may be readily applied. Much of this area is in the zones of continuous or discontinuous permafrost.

Major and trace element data are compared for volcanic sequences, both barren and mineralized. Most of these data are from the Slave Province of the Shield, but analyses are also given for the highly productive Noranda belt. All rocks are of Archean ($>2.5 \times 10^9$ years) age.

While Archean volcanism produced cycles of both tholeiitic and calc-alkaline composition, it is confirmed that massive sulphides occur only in calc-alkaline cycles. Massive sulphides of the Pb—Zn(Cu) type occur in more siliceous volcanic sequences than those of the Zn—Cu type. For the ore-associated elements, sulphur in basic volcanics is shown to be relatively constant from belt to belt, whether mineralized or not. In volcanics of intermediate or acidic composition this element is significantly higher in sequences that host massive sulphides, than in barren sequences. For zinc the background level is slightly to moderately elevated in volcanic rocks hosting Zn—Cu mineralization, compared to barren volcanics or those containing Pb—Zn(Cu) deposits.

Alteration zones in the footwall volcanics around massive sulphides may be quite extensive. On the basis of the irregular loss of calcium and sodium and gain of magnesium and iron in these zones, the ratio (Mg + Fe)/(Ca + Na) is suggested as a useful geochemical indicator of hydrothermal alteration.

Lake sediment sampling has been shown to be an effective method of reconnaissance and semi-detailed exploration in permafrost areas of the Shield. This is because of the active post-glacial oxidation of massive sulphides, with the consequent dispersion of mobile elements, such as zinc and copper, in drainage waters. The relatively high relief of volcanic rocks, compared to other lithologies, assists in this dispersal. Massive sulphide deposits contain a variety of elements, with differing geochemical characteristics. As well as highly mobile elements they contain elements of intermediate (As) to low mobility (Pb, Ag, Au). This allows a multi-stage exploration program based on lake sediment and soil analysis, successively reducing the size of the target area.

INTRODUCTION

Massive sulphide deposits are *the* typical ore of the Canadian Shield. These

22

bodies produce more than half of Canada's output of zinc, copper and silver, one-third of its lead, and significant gold. Applied geochemistry has played a relatively minor role in exploration for these deposits. There are, perhaps, two important reasons for this. The first is that the ore bodies are not large and mineralization of the host rocks is not extensive. Thus the exploration target is relatively small. Secondly, the traditional exploration country of the southern Shield is covered in many places by thick and impermeable glacial sediments; elsewhere in the south, weathering of sulphides and dispersion of metals in the drainage systems is often minimal. By contrast, the high sulphide content of the bodies and their sharp contact with wall rocks provides an excellent target for geophysical methods. These methods, plus conventional geological work, have and continue to dominate exploration for massive sulphides in the Shield.

Over the last decade geochemical methods that may be applied in areas of thick overburden have been developed. These are finding increasing use in the southern Shield. It is not the purpose of this paper to discuss these methods; the reader is referred to the work of Fortescue and Hornbrook (1969), Garrett (1971), Gleeson and Cormier (1971), Hornbrook and Gleeson (1972), Skinner (1973), and Gleeson and Hornbrook (1975).

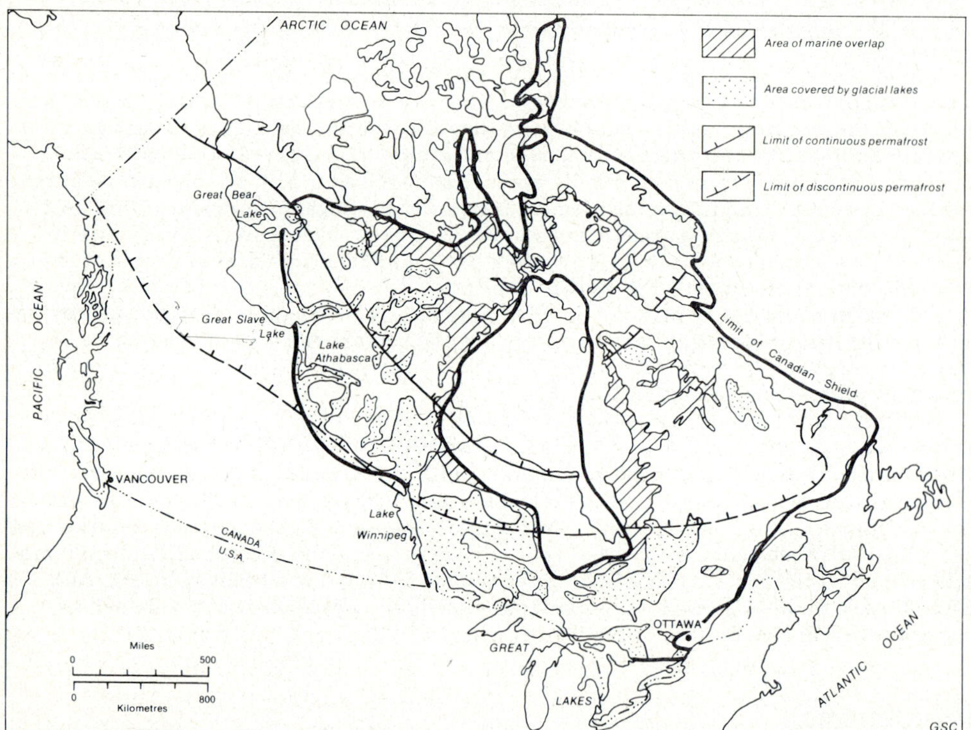

Fig.1. Areas of the Canadian Shield that may contain thick overburden deposited during marine overlap or within glacial lakes; permafrost distribution.

Study of the glacial geology of the Shield shows that the majority of its land surface is free of the thick overburden that was deposited in areas of glacial lakes or of marine overlap (Fig.1). For these areas more conventional methods of geochemical exploration are applicable. Much of the area of the Shield covered by thin overburden is within the permafrost zone or the zone of discontinuous permafrost (Fig.1). It is the principal purpose of this paper to discuss geochemical methods that may be applied to the search for massive sulphides within these zones.

Before developing methods based on secondary dispersion, it was considered necessary to discover the nature and extent of primary metal dispersion and hydrothermal alteration around massive sulphide deposits. Apart from providing information essential for developing methods based on secondary dispersion, these results suggest lithogeochemical approaches to exploration.

MASSIVE SULPHIDES OF THE CANADIAN SHIELD

Following from the pioneer ideas of Oftedahl (1958), most Canadian economic geologists now support the volcanic-exhalative nature of these deposits. That is, the mineralization was precipitated on the sea floor around submarine hot springs present near volcanic centres. The features of these deposits and their environment of deposition has recently been summarized by Sangster (1972). A number of the features that are of particular significance to exploration geochemists will be briefly reviewed here.

Massive sulphide mineralization

The form of the typical deposit is shown in Fig.2. It resembles a mushroom. The stem is the pipe in the footwall volcanics up which the mineralizing solutions passed. These hypogene solutions leached some elements from the rocks forming the pipe and deposited others. Stringer-type mineralization occurs in the upper part of the pipe. According to Sangster (1972) the mineable grade seldom extends more than 200 ft down the pipe, although iron sulphides may continue down for several thousand feet. The cap of the mushroom consists of massive ore in a body that seldom exceeds 100 ft in thickness but a diameter several times this dimension.

Massive sulphide bodies lie along stratigraphic planes in the volcanic rocks. These represent intervals in volcanism during which precipitates from hot springs had an opportunity to accumulate. In addition to the mineralization that was precipitated near to the hot springs, generally thinner chemical sediments were deposited along these horizons for miles out from the springs. The convenient term "exhalite" has been applied to these chemical sediments of volcanic origin (Ridler and Shilts, 1973). They are most commonly composed of chert and iron sulphides, carbonates or oxides. Limestones of volcanic origin are often present.

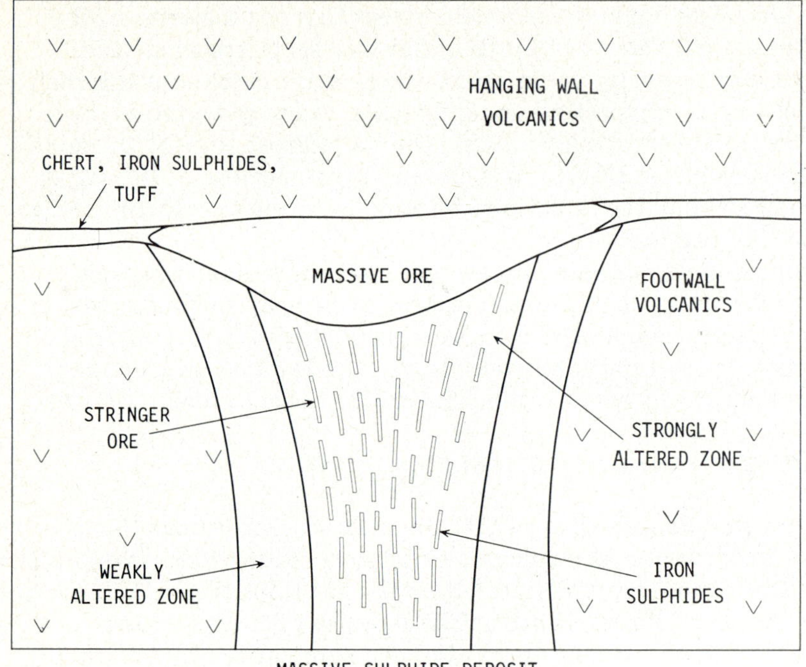

Fig.2. Typical massive sulphide deposit, Canadian Shield.

Environment of deposition

Archean volcanism in the Shield comprised, very generally, cycles of basic to acidic volcanism, that are capped by flysch-type sediments. The sediments are derived, in part at least, from the underlying volcanic rocks. Massive sulphide deposits are predominantly associated with the siliceous phase of the cycle, commonly during the time interval at the end of a cycle. Since the hot springs developed around the volcanic centres, the deposits are associated with the coarser silicious pyroclastic rocks. Intrusives, commonly acidic, are intruded into the volcanic pile.

Numbers of deposits may be found along the same stratigraphic plane and massive sulphides generally occur in clusters 10—20 miles in diameter.

Composition of deposits

Pyrite and pyrrhotite typically make up 50% or more of the sulphide content, and sphalerite, chalcopyrite and galena the remainder. Silver and gold are economically important constituents of the ore in the range 10—150 ppm Ag and 0.3—15 ppm Au. For geochemical exploration these are important indicator elements.

There are three main compositional types of deposit. The Zn—Cu type is the most common in the Shield, and in Archean rocks generally. The Pb—Zn-(Cu) type is also commonly found in the Shield. The cuprous-pyrite type or Cu(Zn) is chiefly found in Phanerozoic rocks.

Zoning of the deposits may be quite marked. Pyrite and sphalerite are most common in the massive ore and pyrrhotite and chalcopyrite in the stringer ore. Galena occurs only in the massive ore.

SAMPLING LOCATIONS

The data given in this report are all from volcanic areas of Archean ($>2.5 \times 10^9$ years) age. All but one of the locations — Noranda, Quebec — are in the Slave Province of the Canadian Shield (Fig.3).

Barren volcanics

Indin Lake, N.W.T. 109 samples of volcanic rock collected at 100-ft intervals (see Cameron, 1974, fig.4). There are no known massive sulphide deposits in the area but there are abandoned gold workings that commonly occur in shear zones in volcanic and sedimentary rocks.

Yellowknife, N.W.T. 194 volcanic rock samples collected by Dr. W.R.A. Baragar from the immediate vicinity of Yellowknife and from the nearby Cameron River volcanic belt. Sampling traverses are shown in Baragar (1966)

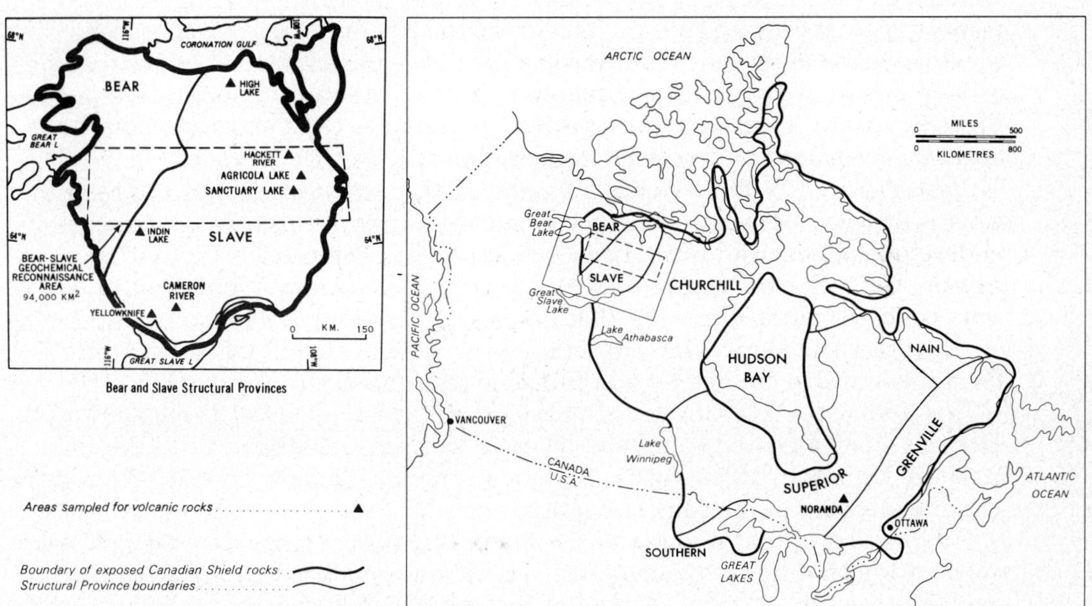

Fig.3. Sampling locations, Canadian Shield.

and in Cameron and Baragar (1971). There are no known massive sulphide deposits in these areas, but important gold deposits, associated with volcanic and sedimentary rocks, occur at Yellowknife.

Mineralized volcanics

High Lake, N.W.T. 94 samples of volcanic rocks were collected from three traverses in the vicinity of an undeveloped deposit containing 5.2 M tons of 3.5% Cu and 2.5% Zn. Traverses C1 and C2, closest to the deposit, were sampled at 100-ft intervals; traverse C3, two miles to the south, was sampled at 200-ft intervals.

Agricola Lake, N.W.T. The reconnaissance lake sediment sampling of the Bear-Slave Operation (Allan et al., 1973b) showed anomalous values for copper, zinc and a number of other metals in this area. Follow-up studies by Cameron and Durham (1974a) located a volcanic sequence 7000 ft thick that contains two exhalite horizons. One of these horizons contains Zn-, Cu-, Pb-, Ag- and Au-bearing mineralization. This area has since been staked and may be drilled in 1974.*

Many of the volcanic rock samples collected from this area are in the immediate vicinity of exhalite or are in hydrothermally altered rocks. In order to avoid a biased sample of the volcanics, these samples have been excluded from the set of 40 samples described in this report.

Sanctuary Lake, N.W.T. This area is 25 miles to the southwest of Agricola Lake in the same volcanic belt, which here is 25,000 ft thick. Again, highly anomalous values for zinc, copper and other metals were found in lake sediments during follow-up studies. For this area also, only samples of volcanic rocks collected along the main traverse lines have been included in the data set of 117 samples, to avoid biased sampling around exhalite and alteration pipes. Figures showing rock sampling sites for the Agricola Lake and Sanctuary Lake areas are given in Cameron and Durham (1974b).

Hackett River, N.W.T. This is 25 miles north-northwest of Agricola Lake in what is believed to be the same belt of volcanic rocks. This area contains one undeveloped deposit of estimated 10 M tons 8% combined Pb—Zn and 9 oz. Ag/ton, but other bodies are under exploration in the area. Here 37 samples were collected along a traverse that lies 3—4 miles along strike from the main ore deposit (Allan et al., 1973a). In contrast to the other areas, the samples from this traverse are mainly volcanogenic sedimentary rocks.

Noranda, Que. 701 samples of volcanic rock were collected at regular 100-ft intervals along a north—south and an east—west traverse through the Noranda area (see Cameron, 1974, fig.2). This area contains a number of Zn—Cu massive sulphide deposits and also gold deposits.

For the subsequent discussion it is important to recognize that the volcanic rocks of the Noranda and High Lake areas are associated with massive sulphides of the *Zn—Cu type*, while the Hackett River-Agricola Lake-Sanctuary Lake belt of volcanics are associated with massive sulphide mineralization of the *Pb—Zn(Cu)* type.

* The first drill hole by the Yava Syndicate shows 134 ft of massive sulphide mineralization averaging 3.7% Zn, 0.7% Pb, 1.1% Cu, 2.7 oz. Ag/ton and 0.04 oz. Au/ton.

SAMPLE PREPARATION AND ANALYSIS

Sample preparation

The fist-sized rock samples were reduced to ¼-inch chips in a Chipmunk crusher, then further reduced to about 20-mesh by a Braun pulverizer with ceramic plates. 10 g of this material was sampled, then ground to approximately minus 150-mesh in an alumina ceramic mill.

The lake sediments, soils and gossans were air-dried, then sieved to minus 250-mesh.

Analytical methods

The major elements in the Indin Lake, High Lake and Agricola Lake samples have been determined by direct-reading emission spectrometry (see Allan et al., 1973a). Silica in the Yellowknife, Agricola Lake, Sanctuary Lake, and Noranda samples was measured by X-ray fluorescence on unfused powders. Other major elements in the Sanctuary Lake and Agricola Lake samples were determined by atomic absorption spectrometry, following a lithium tetraborate fusion.

Zn, Cu, Pb, Ni, Co and Ag were determined in all sample types by atomic absorption spectrometry, following a hot HNO_3/HCl leach (see Allan et al., 1973a). Arsenic was extracted by a hot HCl leach, then measured colorimetrically using silver diethyldithiocarbamate. Gold in soil and gossan samples was measured by a combined fire assay—atomic absorption procedure. Sulphur was analysed by the method of Bouvier et al. (1972). Some samples were below the detection limit for Pb (3 ppm), Ag (0.5 ppm), Au (5 ppb), or S (0.01%). These samples have been assigned values of 2 ppm, 0.2 ppm, 2 ppb or 0.005% respectively.

In order to better interpret the trace element content of the volcanic rocks they have been classified into basic (40.0—54.9% SiO_2), intermediate (55.0—67.9% SiO_2) or acidic ($\geqslant$ 68% SiO_2). Since a few samples may analyse less than 40% SiO_2 the sum of acidic plus intermediate plus basic may not equal the total number of samples for a given area.

MAJOR ELEMENT COMPOSITION OF HOST VOLCANICS

For massive sulphides of Phanerozoic age a good correlation has been established between the type of deposit and the tectonic setting and chemical composition of the host rocks. Sillitoe (1973) notes that massive sulphides of the cuprous pyrite type are found with tholeiitic basalts in an ocean-ridge environment (e.g. Cyprus) or in marginal ocean basins (e.g. Notre Dame Bay, Newfoundland). In island-arc environments over converging crustal plates Pb—Zn(Cu) deposits are associated with calc-alkaline volcanics (e.g. Kuroko deposits). It is possible, from this correlation between the composition of volcanic rocks and the composition of the ores, that the major element con-

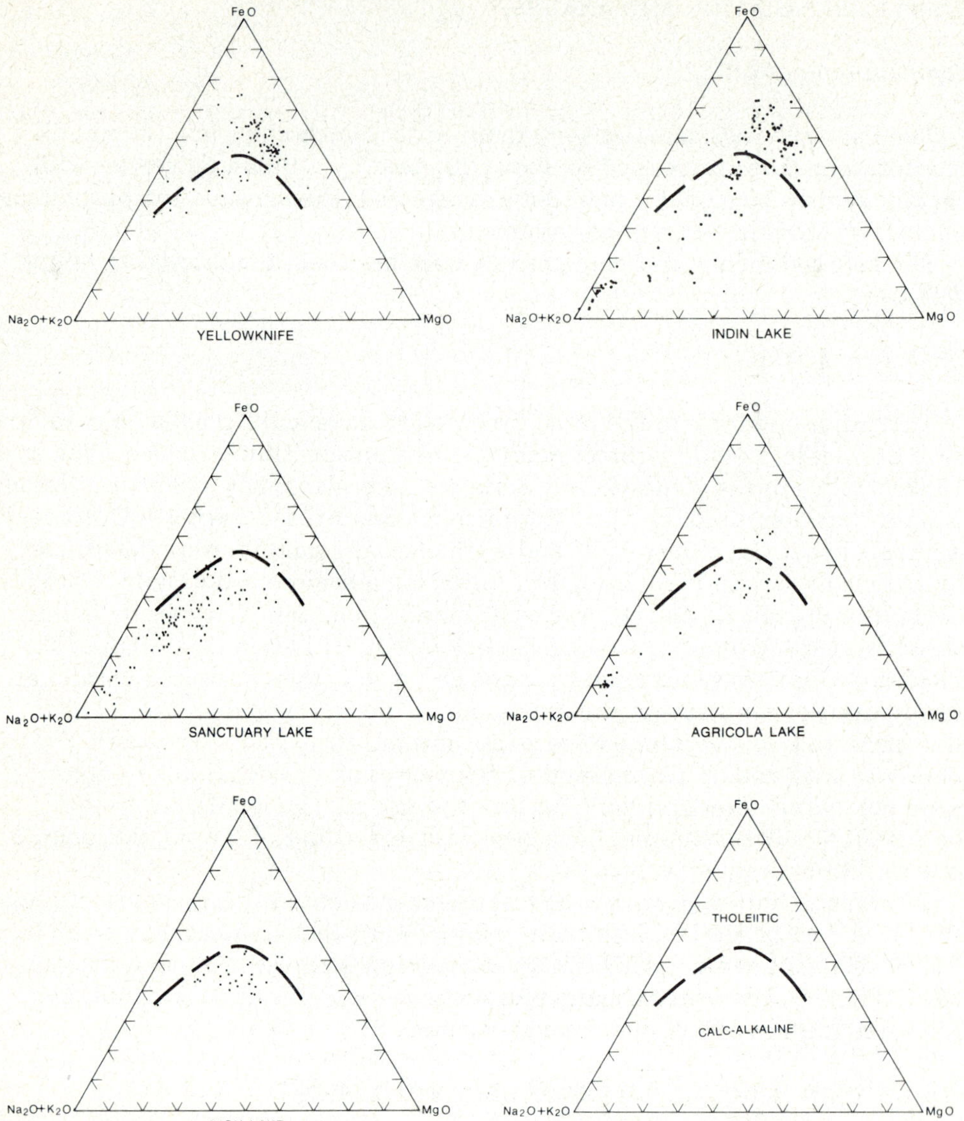

Fig.4. AFM plots, Slave Province, Canadian Shield. Yellowknife data from Baragar (1966). Total Fe as FeO.

tent of volcanic rocks may assist in identifying mineralized volcanic belts.

Archean volcanic belts in the Canadian Shield contain rocks of tholeiitic and calc-alkaline character. Both are commonly found in the same volcanic belt. The mixed character of the volcanic rocks from Indin Lake and from Yellowknife is shown in the AFM plots of Fig.4. Descarreaux (1973) has shown that while the Abitibi belt contains both types of volcanic rocks, the massive sulphides are found in rocks of dominantly calc-alkaline character. The AFM plots given in Fig.4 for the mineralized Agricola Lake, Sanctuary Lake, and High Lake areas confirm that these strata are dominantly calc-alkaline. Only traverse C3 is given for High Lake, because the rocks from the other two traverses have been, in part, metasomatized.

There is also a clear difference between the character of the volcanic rocks associated with Pb-poor deposits at Noranda and High Lake and the Pb-bearing mineralization found in the Hackett River-Agricola Lake-Sanctuary Lake belt. The volcanic rocks from these latter areas are much more siliceous than those from Noranda (Fig.5). This is in accord with the observation of Hutchison (1973).

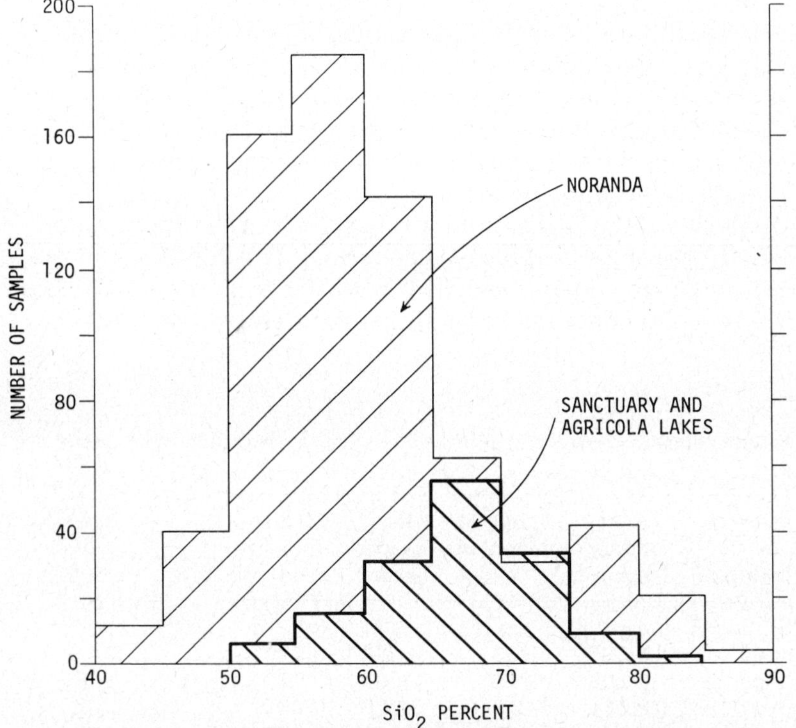

Fig.5. Histograms for SiO_2, Agricola and Sanctuary Lakes, N.W.T.; Noranda, Que.

Baragar and Goodwin (1969) provide some of the most complete data on the composition of Archean volcanic rocks in the Shield, based on samples from the Timmins-Noranda, Birch-Uchi, Lake of the Woods-Wabigoon and Yellowknife belts. Comparing the highly productive Timmins-Noranda belt with the others, the most notable differences are found for MgO and FeO in salic volcanic rocks. An average of 2.6% MgO for Timmins-Noranda compares with 1.4%, 1.7% and 1.3% MgO respectively for the other belts; and 4.0% FeO with 3.0%, 2.5% and 2.5% FeO. These differences may be related to hydro-thermal alteration effects in the volcanic rocks that will be discussed later in this report.

It seems probable that other differences will be found between the non-ore element content of barren and mineralized volcanic sequences. Ideally, this will be studied by multivariate discriminant analysis. However, a fairly large number of cycles of volcanic rocks, both barren and productive, are required for comparison, in order that spurious discriminant functions are not obtained.

ORE ELEMENT COMPOSITION OF HOST VOLCANICS

Sulphur

The distribution of sulphur in Archean volcanic rocks of the Canadian Shield has recently been discussed by Cameron (1974). Its geochemistry is more complex than that of most other elements, since it may be lost as a vapour or in solution during eruption and cooling of the lavas. Also, sulphur may be introduced into permeable lavas by sulphur-bearing waters or during the formation of massive sulphide ores.

Cumulative frequency distribution curves for sulphur in the Indin Lake volcanic rocks (Fig.6) are typical of barren sequences. The curves for basic, intermediate, and acidic are widely separated, with the basic variety containing most sulphur (Table I). The data for basic samples closely approximates a

TABLE I

Sulphur content (%) of volcanic rocks from Indin Lake, N.W.T.; Noranda, Que.; and High Lake, N.W.T.

Sample location	Number of samples	Arith-metic mean	Standard deviation	Geo-metric mean	Median
Indin Lake					
all samples	109	0.067	0.093	0.036	0.04
basic	55	0.103	0.117	0.065	0.07
intermediate	21	0.034	0.043	0.020	0.03
acidic	31	0.025	0.019	0.017	0.03
Noranda					
all samples	701	0.119	0.318	0.043	0.04
basic	214	0.091	0.182	0.043	0.05
intermediate	372	0.127	0.357	0.045	0.04
acidic	114	0.124	0.295	0.037	0.02
High Lake					
all samples	94	0.205	0.542	0.044	0.04
basic	22	0.144	0.421	0.031	0.04
intermediate	54	0.177	0.555	0.046	0.04
acidic	16	0.249	0.490	0.045	0.04

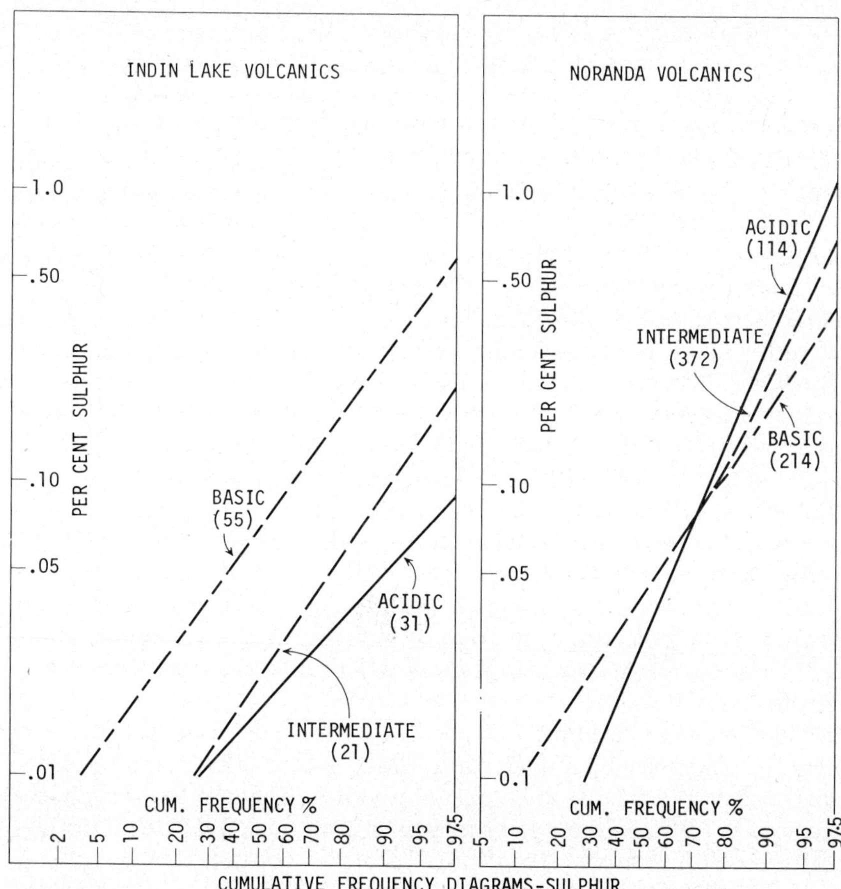

Fig.6. Cumulative frequency distributions for sulphur, Noranda, Que., and Indin Lake, N.W.T.

lognormal distribution. The sulphur content of these basic samples is close to the limit of solubility of sulphur in basaltic melts of ~0.1% S. The sulphur present in the original melt was probably largely retained in the lavas by rapid submarine quenching (see data for modern lavas by Moore and Fabbi, 1971).

The intermediate and acidic rocks at Indin Lake contain less sulphur and their distribution about the frequency curve is more scattered than for the basic group (this is probably, in part, a function of the analytically low levels of sulphur present, and the small number of samples in the acidic and intermediate groups). Acidic volcanic sequences from the Shield, including those from Indin Lake, contain a high proportion of pyroclastic material. This indicates that degassing, with probable sulphur loss, was taking place during eruption. Also, pyroclastic rocks are frequently permeable, allowing further loss — or gain — of sulphur. A further factor is that silica-rich melts may have a lower solubility for sulphur than basaltic melts.

The cumulative frequency curves for sulphur in the Noranda samples are believed to be typical of sequences containing massive sulphides. There is little change in the position of the frequency curve for basic samples, compared to that for Indin Lake. In fact, the basic samples from Noranda have a slightly lower mean sulphur content. This indicates that there was probably little difference in the sulphur content of the basaltic melts between the two areas.

The major difference in the Noranda data is the considerable increase in the sulphur content of the acidic and intermediate groups (Fig.6, Table I). In fact, the right-hand side of the frequency distributions are in reverse order from that at Indin Lake, with the acidic rocks containing the most sulphur. In view of the well-known association between massive sulphide deposits and acidic volcanic rocks, this finding is hardly surprising. It is believed to have been caused by the introduction of sulphur into rocks near massive sulphide deposits at the time of their formation. Note from the median contents of sulphur in Table I that there is no general difference in the sulphur content of the acidic rocks between Noranda and Indin Lake.

Data for sulphur at High Lake are given in Table I. Note that sulphur in these samples is greater than for Noranda and even the basic rocks contain increased amounts. The explanation for this is that two of the sampling traverses are relatively closer to the massive sulphide body than is the case with the Noranda samples.

It appears that in certain situations the sulphur content of acidic and intermediate volcanic rocks may be a useful exploration tool. The cumulative frequency curves for Indin Lake and for Yellowknife indicate that sulphur levels in excess of 0.1% in acidic volcanics are anomalous. Only a fractional percentage of the acidic samples from these localities contain this level or more, compared to 25% of those from Noranda and High Lake. It is of interest that analyses coming to hand as this paper was being prepared show that a good many of the acidic and intermediate samples from the Agricola Lake area contain more than 0.1% S, including samples well removed from exhalite or hydrothermal alteration zones.

Zinc

Zinc must be given consideration in searching for an element to discriminate between ore-bearing and barren volcanic rocks. It is a universal constituent of the massive sulphides of the Shield and, unlike copper, shows only moderate variation from basic to acidic rocks. This latter feature is an important consideration in practical exploration, since it dispenses with the need for a precise estimation of the silica content of samples.

Table II contains data on the zinc content of volcanic rocks from various areas across the Shield. The Noranda and High Lake samples discussed above are associated with mineralization of the Zn—Cu type. The Confederation and Woman Lake volcanic cycles of the Uchi Lake belt are also associated with this

TABLE II

Zinc (ppm) in volcanic rocks from the Canadian Shield

Sample location	Number of samples	Arithmetic mean	Standard deviation	Geometric mean	Median	90 Percentile	95 Percentile
Indin Lake							
all samples	109	44.4	31.5	33.4	38	93	103
basic	55	48.2	31.4	39.8	42	98	105
intermediate	21	50.4	33.4	38.4	50	96	100
acidic	31	32.3	27.7	21.4	32	89	94
Noranda							
all samples	701	70.9	106.9	51.9	50	110	154
basic	214	70.0	92.4	53.9	51	115	153
intermediate	372	69.7	113.0	51.1	49	98	134
acidic	114	67.4	57.4	49.5	51	152	179
Agricola and Sanctuary Lakes							
all samples	157	40.7	30.7	32.8	36	69	83
basic	9	53.3	72.2	35.4	31	—	—
intermediate	82	47.3	30.4	39.2	43	79	105
acidic	66	30.8	17.1	26.0	30	68	54
High Lake, all samples (traverse C3)	37	67.6	19.6	64.2	68	92	101
Hackett River, all samples	37	67.7	83.3	42.1	43	146	370
Lake St. Joseph[1], all samples	2055	54.2	40.6	43.5	50	90	105
Confederation Lake[2]							
basic	60	105					
intermediate	29	165					
acidic	151	126					
Woman Lake[2]							
basic	17	79					
intermediate	17	77					
acidic	25	58					

[1] From R.H.C. Holman, unpublished data.
[2] From Davenport (1972).

type of deposit (Davenport, 1972). The Hackett River, Agricola Lake, and Sanctuary Lake samples are associated with mineralization of the Pb—Zn(Cu) type. To represent presumably barren volcanics are the data from Indin Lake (no zinc analyses are available for the Yellowknife samples) and previously unpublished data by R.H.C. Holman from the Lake St. Joseph map sheet in western Ontario (Emslie and Holman, 1966). This map sheet is to the east of the Uchi Lake area.

In all of these cases it is the background content of zinc that is being examined. Sampling close to sulphide bodies has been avoided, or is confined to a few samples. Thus for High Lake only traverse C3 is included. The Hackett River data should be interpreted with caution, since the samples are largely volcanogenic sediments, rather than true volcanics.

As noted earlier, the data for Noranda, Indin Lake, Agricola Lake, Sanctuary Lake and High Lake were obtained using a HNO_3/HCl leach. This is thought to extract approximately 80% of the zinc of volcanic rocks on average (Allan et al., 1973a). The Lake St. Joseph data are by a dithizone colorimetric method, following a pyrosulphate fusion. The Uchi Lake data are by atomic absorption, following a $HClO_4/HNO_3$ leach. For both these latter two methods of decomposition, extraction is believed to be near 80% for zinc (J.J. Lynch, personal communication, 1974).

There appears to be some indication in the data that modestly greater amounts of zinc are present in volcanic rocks associated with Zn—Cu deposits, compared to barren volcanics. The differences are not large and clearly require to be confirmed by sampling of other areas. The greatest contrast is found by comparing siliceous samples. These data confirm the observation of Davenport and Nichol (1973) that increased amounts of zinc are associated with the ore-bearing Confederation Lake Cycle, rather than the barren Woman Lake Cycle at Uchi Lake. Examining the data for Hackett River, Agricola Lake and Sanctuary Lake, it appears that volcanic rocks associated with the Pb—Zn(Cu) type of massive sulphide mineralization are not enriched in zinc. The somewhat higher arithmetic mean for Hackett River is caused by the inclusion of a few mineralized samples associated with the Cleaver Lake massive sulphide body at that camp.

The differences noted above for the volcanics hosting the Zn—Cu mineralization, compared to barren volcanics, are not large and require careful sampling and precise analysis to be defined. The fact that massive sulphide deposits are not enclosed in host rocks that contain distinctly higher levels of zinc is of considerable significance in designing methods of reconnaissance sampling utilizing surficial materials. One must deal with secondary dispersion from a very small target of the massive sulphide deposit itself and the mineralized wall rocks, rather than a much larger area of metal-rich volcanic rocks.

Interflow shales

As was discussed earlier, exhalative-type sulphide deposits formed during an interval in active volcanism. During this interval, shales had an opportunity to accumulate along the given stratigraphic plane. Pyritic, carbonaceous shales, clearly deposited in an euxinic environment, are typically found interbedded in volcanic sequences in the Canadian Shield, The euxinic environment is ideal for the precipitation of heavy metals that may be dispersed in seawater from nearby hot springs.

A reconnaissance sampling of shales from the southern Shield was carried out by Dr. A. Baumann and the author. Sampling locations and a discussion of the major element chemistry of these shales are given in Cameron and Baumann (1972). In the course of this work it became apparent that some shale samples from the Timmins area of Ontario were anomalous in heavy metals, most notably zinc. The 265 shales of Archean age from areas other than Timmins had a relatively uniform distribution of zinc (Table III), with a mean of 123 ppm. In contrast, the 141 samples from Timmins, which were mainly obtained from drill core, gave a mean of 632 ppm Zn. This high value

TABLE III

Zinc content (ppm) for shales from the southern Canadian Shield

Sample location	Number of samples	Arith-metic mean	Standard deviation	Geo-metric mean	Median	90 Per-centile	95 Per-centile
Timmins	141	632	3700	165	110	500	1276
Other areas							
Southern Shield	265	123	177	104	98	154	233

is determined by a number of anomalous to highly anomalous samples. Note that the median values for the two groups are similar, but the 90 and 95 percentile values are widely dissimilar.

At Timmins, metasedimentary rocks overlie a thick volcanic sequence. Shales were deposited with this metasedimentary unit and also as thin horizons within the volcanic rocks. Fig.7 shows the main portion of the area sampled and the zinc content of the shales. There are four massive sulphide deposits in the area, including the rich Kidd Creek mine. It may be seen that the shales from the vicinity of this mine are anomalous, or highly anomalous, in zinc. There are anomalous values elsewhere in the area, but mainly from the area underlain by volcanic rocks.

Pyritic, carbonaceous shales are frequently drilled on the basis of the geophysical response that they give. It would seem wise to routinely analyse these samples. Anomalous values would require to be followed up by careful stratigraphic mapping, since it is possible that massive sulphide mineralization may lie several miles along strike from an anomalous shale sample. From Table III it may be seen that only 5% of the samples from outside the Timmins area contain more than 233 ppm Zn. This would be an appropriate value to consider anomalous. It should be noted that many of the samples from outside the Timmins area are also pyritic and carbonaceous, but mostly do not have anomalous zinc values.

Anomalies caused by migration of compaction fluids

Since the hanging wall volcanics are necessarily younger than the associated massive sulphide body, they will not contain a primary dispersion halo. As a chemical precipitate on the sea floor, massive sulphides presumably once contained a high water content. As other volcanics or sediments were deposited on this exhalite horizon, gradual loss of water by compaction would have taken place. By analogy with other sediments that have been drilled in oil fields, the loss of compaction fluid might have continued after burial to several thousand feet. Although the author knows of no studies on the likely composition of fluids expelled from massive sulphides, it is probable that

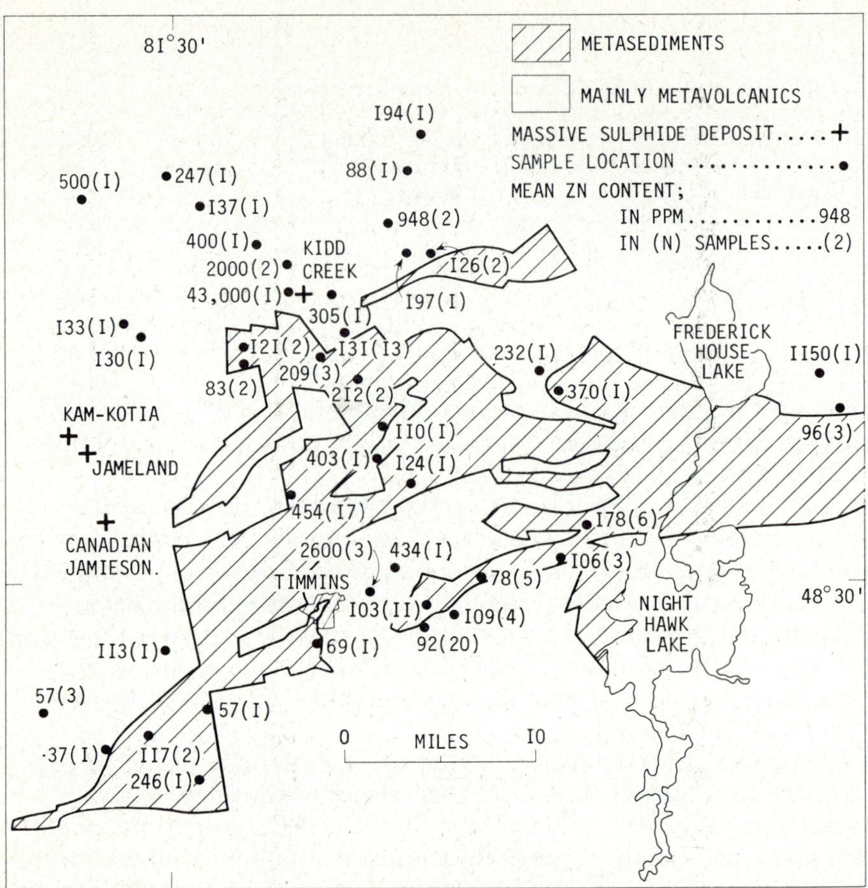

Fig.7. Distribution of zinc in Archean shales, Timmins area, Ont.

they contain indicator elements which migrated through the hanging wall volcanics.

It is possible that such a mechanism might explain the leakage halo for mercury, outlined by J. Boldy (in: Sakrison, 1971) for a massive sulphide deposit in the Noranda area. There is no evidence, however, to prefer this explanation to any other. Boldy was careful to sample fractures in the volcanic rocks. Migration presumably took place after initial fracturing of the rocks. Boldy's work is a fine example of applied geochemistry that deserves to be followed-up by detailed research on this type of anomaly. The dispersion by compaction fluids of a highly mobile element, such as iodine, that is not retained within the sulphide mass (Safronov et al., 1970) should be examined.

Composition of sulphides

Faulkner (1964) and Sangameshwar (1972) have studied the trace element composition of sulphide minerals associated with "economic" and "barren" massive sulphides of the Snow Lake and Flin Flon districts of Saskatchewan and Manitoba. The "economic" group is composed of pyrrhotite and pyrite with mineable quantities of sphalerite and chalcopyrite. The "barren" group contains only small quantities of the latter two minerals. Sulphide minerals, including pyrite and pyrrhotite, from the "economic" group contain more copper than nickel, while the reverse is true for the "barren" group. Further, the "barren" sulphides have a wide spread of $\delta\,^{34}S^0/_{oo}$ values (+10 to -1) compared to the "economic" group (+4 to -1).

If these criteria could be shown to have more general application, they might be quite valuable to mineral exploration. The common iron sulphides are much more readily found than sphalerite, chalcopyrite and galena, particularly in the northern permafrost weathering environment. Wide-ranging studies of sulphide composition such as that of Jonasson and Sangster (1975, this volume) are likely to provide valuable information.

ALTERATION AS A GUIDE TO ORE

Alteration of the footwall volcanics is a notable feature of most massive sulphide deposits. The hypogene solutions that passed up the feeder pipe to form a hot spring were strongly reactive with the wall rock; they have generally produced a pipe-like alteration zone.

One of the most detailed studies of wall-rock alteration about a massive sulphide deposit is for the Boliden deposit, Sweden, by Nilsson (1968). He found a central zone which had lost Fe and Mg and gained Al, K, Ti, Si and H_2O. An outer zone showed gain of Fe, Mg, H_2O and S. Both zones had lost a substantial part of their Ca and Na. In geological studies, the feature most commonly commented upon is magnesium metasomatism which has chloritized the rocks (e.g. Gilmour, 1965; Boldy, 1968). A distinctive mineralogy and chemistry remains after metamorphism. Often, cordierite-anthophyllite rocks are formed, but high magnesium chlorites are stable under high grades of metamorphism (Sangster, 1972). Nilsson believes that the high magnesium in the outer part of the alteration zone at Boliden was derived by leaching from the lower part of the pipe. For Zn—Cu ores of the Abitibi belt, Descarreaux (1973) finds that the dominating feature of alteration is loss of sodium and gain of magnesium within acid volcanics. Potassium increases close to the ore bodies.

The rock samples from traverses C1 and C2 at High Lake show extensive and intensive alteration. Calcium and sodium are reduced to 0.1% or less along substantial lengths of these traverses, while Mg, Fe, Mn and S have been added. For the other major components measured in the samples — Si, Al, Ti and K — it is more difficult to estimate possible changes. It is of interest that loss of

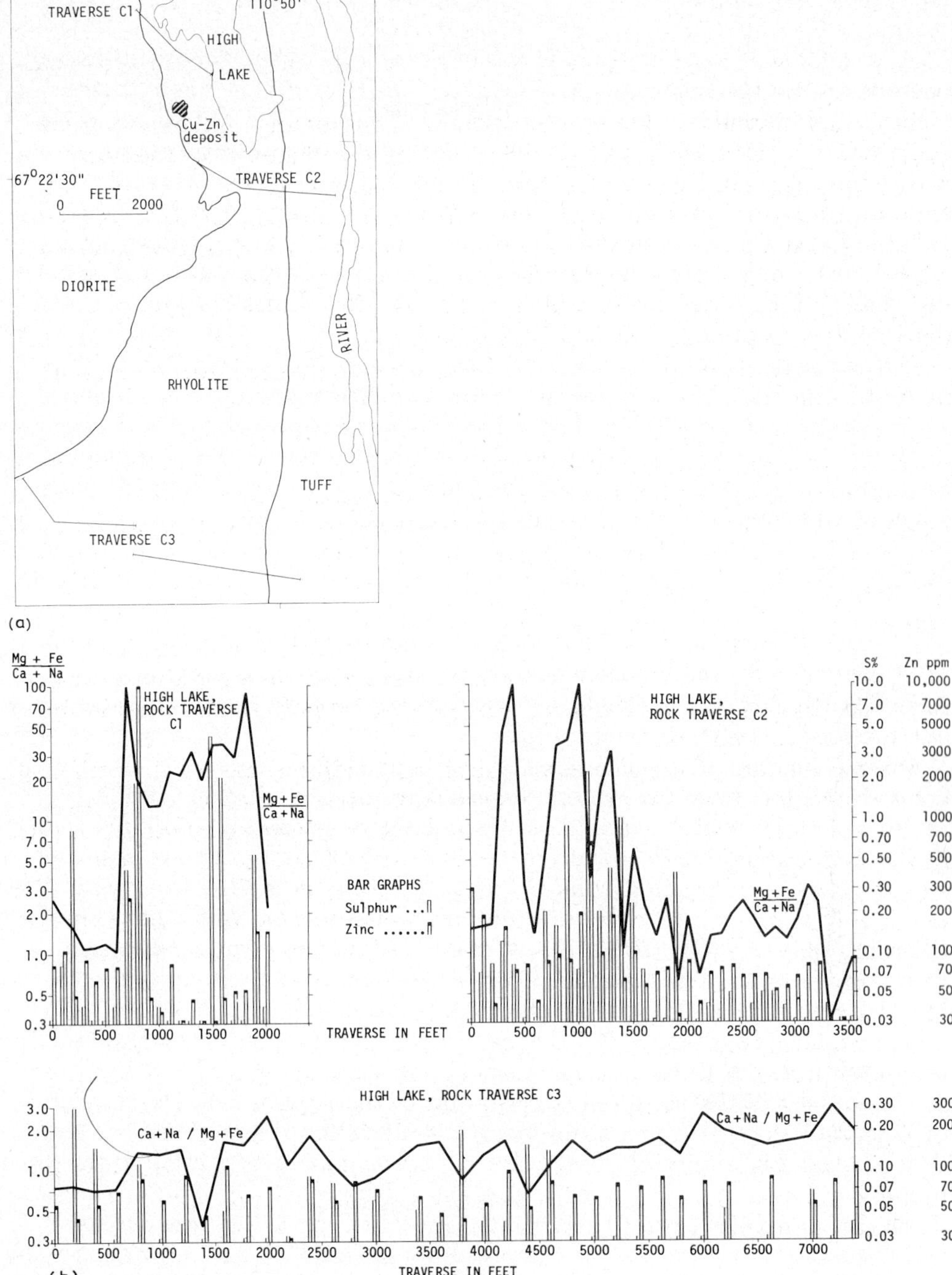

Fig.8. (a) Sampling traverses, High Lake, N.W.T. (b) Chemical index of hydrothermal alteration, and distribution of zinc and sulphur; High Lake, N.W.T.

Ca and/or Na is more extensive than is gain of Mg, Fe, Mn or S. Some samples appear to have lost Ca and/or Na and gained little if any of these other five elements. Conversely, a few samples have gained Mg, Fe or S, but have apparently lost little Ca or Na.

In using a chemical guide to alteration as a prospecting method it is more appropriate to measure *any* chemical change in the character of the host rock, rather than rely on just one effect, such as magnesium metasomatism. It is therefore suggested that the ratio (Mg + Fe)/(Ca + Na) be used as an index of hydrothermal alteration of volcanic rocks. This ratio will change with change in any of the four components. In unaltered rocks this ratio decreases only moderately during the main sequence of calc-alkaline differentiation. Of course, for some types of deposit, it may be necessary to change some of the elements comprising this ratio to reflect different types of hydrothermal alteration.

In Fig.8 is plotted the (Mg + Fe)/(Ca + Na) ratio for the three rock traverses at High Lake. For comparison, sulphur and zinc are also shown. Along traverse C1, approximately 1000 ft north of the deposit, there is a 1400-ft zone with highly elevated ratios. It is of interest that the ratios are at their maximum at the margins of the zone. This is due to maxima for magnesium (both margins) and for iron (west margin). Along traverse C2, 1200 ft or more south of the deposit, the alteration is in the form of a number of zones or strata with unaltered or partly altered rock between. The total width of the zone is, however, similar to that in traverse C1. In traverse C3, two miles south of the deposit, no obvious sign of alteration is shown by the ratio, with values in the range 0.65—3.0, with one sample at 0.37.

What is the fate of the elements calcium and sodium, that are the principal components lost from the altered pipe rock of massive sulphide deposits? Sodium is most likely to have passed into solution in the sea of the time. The limestones that are found near many massive sulphide deposits are very possibly derived from the calcium removed from the rocks forming the pipe. At High Lake there are a number of limestone horizons of probable exhalative origin. At Hackett River similar limestone horizons occur near the deposit. Samples of this limestone have a very high manganese content (to 1.23% MnO), which is quite atypical of normal marine limestones. Plagioclase may also form in the outer part of the alteration zone. The ratio (Mg + Fe)/(Ca + Na), if determined outside the immediate limits of the alteration pipe may, therefore, show lower values due to the precipitation of $CaCO_3$ or formation of plagioclase.

The detection of alteration haloes is not restricted to the analyses of rock samples. Allan et al. (1973a) show a well-developed anomaly for MgO in lake sediments around the High Lake deposit.

GEOCHEMICAL METHODS BASED ON SECONDARY DISPERSION IN PERMAFROST
AREAS

Principles

At first sight, permafrost areas may appear unpromising for geochemical
exploration methods based on secondary dispersion. It is widely thought that
chemical weathering is minimal in cold regions, thus restricting the flux of
dissolved species necessary to develop extensive anomalies in soils and drainage
systems. In the Canadian Shield this problem appears to be further com-
pounded by the predominantly flat terrain that hinders the dispersion of particulate matter.

Work over the past two decades, particularly in the permafrost areas of the
U.S.S.R. has, however, shown that chemical weathering takes place in permafrost regions. In particular, the oxidation and leaching of sulphide bodies is
active with the resulting dispersion of elements in solution.

The critical element to an understanding of how chemical weathering takes
place was the discovery that in frozen ground thin films of water are present
along mineral/ice interfaces and along ice/air interfaces (Tyutyunov, 1960,
1961). Ionic mobility within these films is only slightly less than in normal
aqueous solution (Anderson and Morgenstern, 1973), allowing chemical reactions to proceed along grain boundaries. The water in these films moves in
response to gradients, such as pressure, gravitational, temperature, compositional, and electrical. However, apparently the most important influence is thermal
gradients which cause the movement of the water films and their contained
ions from warmer to cooler regions (Ferguson et al., 1964).

During the long, cold winters experienced in permafrost areas, the soil surface is the cold front, and water and dissolved salts move upwards towards
this. As a result of changes in pH or formation of ice, salts may precipitate
near the surface. Some water and salts may continue to move up into the
overlying snow (Jonasson and Allan, 1973). In the spring thaw the snow and
active layer of soil melts and the runoff waters carry the soluble salts into the
drainage system.

The first stages of sulphide oxidation, that is to sulphate, are exothermic
reactions that can proceed readily in the permafrost environment. Many of
the succeeding stages of oxidation are endothermic and these changes are,
therefore, restricted (Bugel'skii, 1962). As a result, sulphates are a commonly
found product of sulphide oxidation in permafrost regions and the solubility
of metal sulphates is likely to have an important bearing on the differential
dispersion of metals.

In the U.S.S.R. oxidation of sulphide bodies in permafrost terrain may be
active at depths of 100 m or more. There is some dispute as to the dominant
age of this oxidation — preglacial, interglacial or postglacial — but it seems
virtually certain that some of the oxidation is contemporary (Shvartsev and
Lukin, 1965). In Canada, Boyle et al. (1971) discuss the deep oxidation of a

number of sulphide deposits and conclude that, although some of the oxidation is pre-Pleistocene in age, oxidation and migration of elements is continuing today. During the course of flights in light aircraft and helicopters across the tundra of northern Canada, the present author has been impressed by the extensive oxidation shown by many hundreds of gossans. Since exploration in this region is in its infancy, there is little information available on the depths of oxidation in these gossans.

The exothermic nature of sulphide oxidation, plus the high ionic concentration of the resulting solutions, often causes deeper thawing over sulphide bodies and the development of taliks, or thawed channels in permafrost (Shvartsev and Lukin, 1965).

Reconnaissance geochemical exploration using lake sediments

A large part of the Canadian Shield is covered by lakes that range in size from ponds to bodies the size of inland seas. Provided that there is movement of indicator elements into these lakes in solution, analysis of lake waters and/or sediments should provide a useful method of reconnaissance. Because of the very low content of most indicator metals in lake waters of the region (Boyle et al., 1971; Allan et al., 1973a) attention has been focussed on lake sediments. Early work using lake sediments has been reviewed by Allan (1971).

In 1972, a large-scale reconnaissance of 36,000 square miles of the Bear and Slave Provinces of the Shield (Fig.3) was carried out using lake sediments at a sample density of one per 10 square miles (Allan et al., 1973b). In 1973 follow-up studies were carried out in the eastern part of this survey area to determine the significance of the data in evaluating mineral potential (Cameron and Durham, 1974a, b). In the course of this study, data were obtained which illustrate well secondary dispersion in soils and the drainage system from massive sulphide mineralization. This will be related in the section below.

GEOCHEMICAL DISPERSION FROM MASSIVE SULPHIDE MINERALIZATION, AGRICOLA LAKE AREA, N.W.T.

The study area is shown on Fig.9. In the 1972 reconnaissance survey, anomalous results for Zn, Cu, Ni, Co, Mn and As were obtained from a lake sediment sample collected from Agricola Lake. Follow-up lake sediment and soil sampling showed that these metals are probably derived from one or both of two gossan zones in volcanic rocks that lie immediately to the west of Agricola Lake (Fig.9). The volcanic rocks are of relatively higher relief than the metasedimentary rocks, so that drainage is eastwards along the Tooshort Lake-Agricola Lake chain. The metasedimentary rocks overlie the volcanic rocks; both are near vertical dipping. The eastern gossan zone (the A horizon) is at the volcanic-metasedimentary contact. It is relatively thin and may be followed along strike for many miles. The western zone (the B horizon) is thicker, up to 100 ft, and is more restricted in its lateral extent. Its thickest

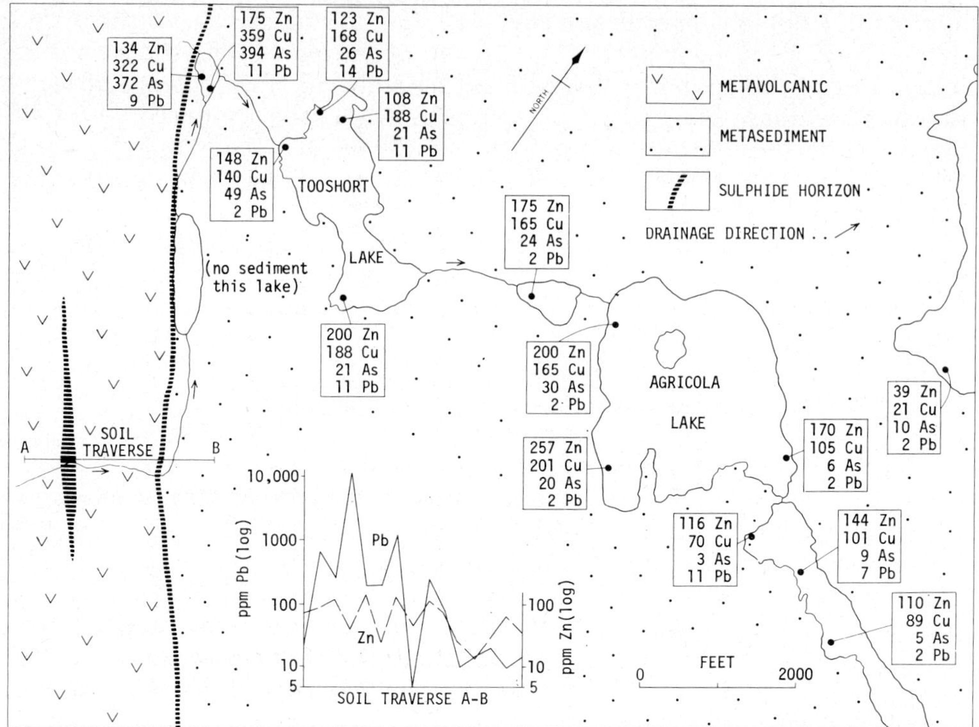

Fig.9. Secondary dispersion from volcanic exhalative zones, Agricola Lake area, N.W.T.

portion is underlain by hydrothermally altered rock. This gossan horizon, therefore, has the morphology and associations of typical massive sulphide mineralization, although of undetermined economic value.

Soil traverse *A—B* (Fig.9 and Table IV) was made over both gossan zones. The A horizon is near samples 646 and 647 and the B horizon near samples 654 and 656. There is a gentle downslope from the western end of the line to sample 646; eastwards from this the ground is flat. There are sharp peaks for Pb, Ag, As, and Au over the B horizon and lesser peaks over the A horizon.

For comparison, median values for these elements in another soil traverse in this area were computed. These may be considered as background values: Zn 62 ppm, Cu 45 ppm, As 23 ppm, Ag less than detection limit, arbitrarily 0.2 ppm, Pb 8 ppm, Au 5 ppb. The anomaly peaks along soil traverse *A—B* are greater than these background values by the following multiples: Zn × 2, Cu × 19, As × 39, Ag >× 205, Pb × 1575, Au × 162.

The interpretation of the above data is as follows. Massive sulphide mineralization within the B horizon contains Zn, Cu, Pb, Ag, Au, and As. Oxidation and leaching of this mineralization has carried considerable amounts of the mobile elements zinc and copper into the streams and lakes. The immobile elements lead, silver and gold have been largely retained in the soils, where

TABLE IV

Zn, Cu, Pb, Ag, As (all as ppm) and Au (as ppb) in soils from traverse A—B (see Fig.9)

Soil sample	Zn	Cu	Pb	Ag	As	Au
641	37	25	15	0.2	44	<5
642	67	44	10	0.2	309	5
643	31	31	20	0.2	239	<5
644	14	10	15	0.7	92	5
645	26	20	10	0.2	72	5
646	79	297	83	3.4	317	<5
647	112	206	271	3.1	92	10
648	47	116	2	0.2	7	5
650	136	856	1370	23.0	145	80
651	26	56	206	4.7	13	20
653	141	163	189	0.2	28	5
654	42	450	12,600	41.0	890	810
656	126	189	259	3.1	7	25
657	93	158	644	11.0	22	60
658	75	38	25	1.0	12	<5

they may have undergone some transport in particulate form downslope from
the B horizon. Arsenic is of intermediate mobility (Boyle and Jonasson, 1973)
and has, in part, been retained on the soils, but has also migrated into the lakes.

Lead in soil sample 654 is present as the sulphates plumbojarosite and
anglesite. No galena is present. It is probable that the dispersion of the metals
Zn, Cu, Pb, and Ag from the mineralized zone has been largely governed by
the solubility of their sulphates. In Table V these solubilities are shown. They
decrease in the order: Zn, Cu, Ag and Pb — similar to the order of anomaly
contrast given above. From the data given above in Table IV it is apparent
that the most mobile element, zinc, has been largely removed from the soils
around the gossans.

In the Tooshort Lake-Agricola Lake drainage system, zinc also appears to
be the most mobile element. It is anomalous over the entire drainage system
in the range 108—257 ppm Zn. By comparison, a sample from the lake north-

TABLE V

Solubilities of heavy-metal sulphates

Sulphate	Solubility (g/100 ml at $x°$C)
$ZnSO_4$	86.5[80]
$CuSO_4$	14.3[0]
Ag_2SO_4	0.57[0]
$PbSO_4$	0.004[25]

east of Agricola Lake in a different drainage system contains 39 ppm Zn
(Fig.9). Note that the maximum contents of zinc are not found near the
presumed western source end of the system, but in Agricola Lake. Copper,
however, does show a maximum near the western source and decreases to the
east. Arsenic shows this tendency to a much greater degree, but its distribution
is influenced by a high content of this element in the slates immediately above
the volcanic content (note the high arsenic values along the eastern part of
soil traverse $A{-}B$). Lead (and Ag) shows virtually no dispersion in the drainage system.

The varying mobility of the elements typically found in massive sulphide
deposits allows a three-stage geochemical exploration program for this type of
deposit in northern Canada. At the first stage, the highly mobile elements, zinc
and copper, are measured in lake sediments and/or waters sampled at a reconnaissance scale. This is followed by analysis for zinc, copper and arsenic in lake
sediments and waters at a follow-up scale. The program is completed by
analysis of Cu, As, Pb, Ag and Au in soils.

It also follows from the above results that the common practice of gossan
sampling and analysis for zinc and copper in exploration for massive sulphides
may give quite misleading results. It is the immobile elements that should be
determined. In Fig.10 are shown histograms for Zn, Cu, As, Pb and Ag in 83
samples of gossan material collected from various localities in the eastern
part of the Slave Province. Since the data are plotted on a logarithmic scale
the dispersion of all elements are directly comparable. It may be seen that

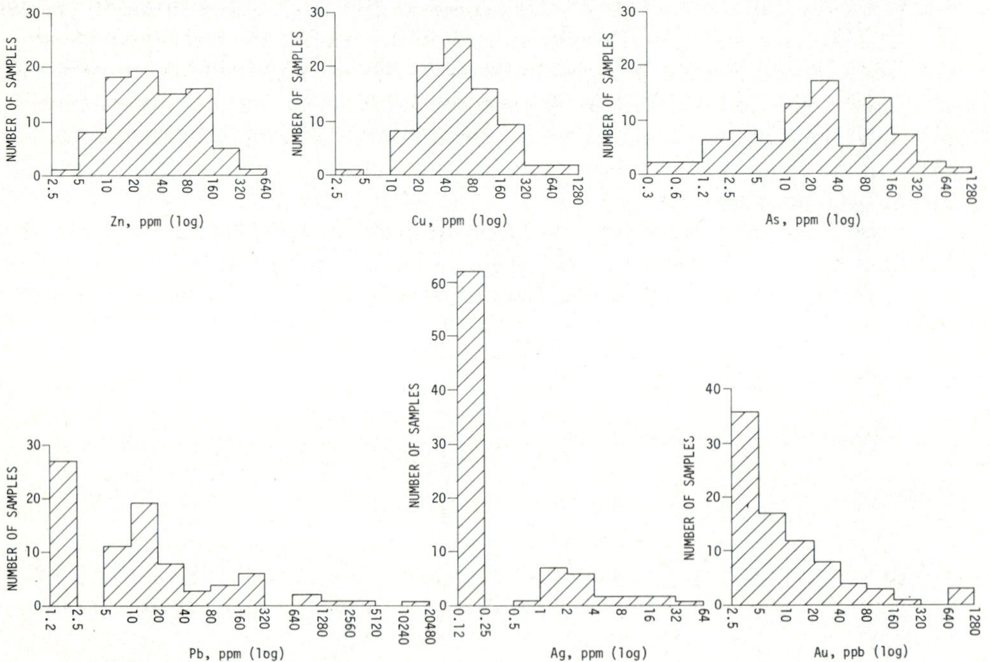

Fig.10. Distribution of zinc, copper, arsenic, lead, silver and gold in gossan samples,
eastern part Slave Province, N.W.T.

the range displayed by the mobile elements, zinc and copper, are limited, presumably because of leaching of these metals from gossans. The range shown by the other, less mobile, elements are much greater, although that for silver, gold and perhaps lead, is artificially truncated by detection limits of 0.5 ppm, 5 ppb and 3 ppm, respectively.

Till sampling

An alternative method of geochemical exploration in northern Canada has been suggested by Shilts (1973) and by Ridler and Shilts (1973). It involves sampling till that is exposed in the frost-boils that are developed in permafrost terrain.

The method is based on the assumption that during glacial action, indicator trains of sulphides were formed down-ice from massive sulphide bodies. These sulphides have since been oxidized in the active zone and some of the derived metal ion has been sorbed onto the clay fraction of the tills or soils. The authors suggest separating the minus 2-μ fraction and analysing this for zinc, copper and other metals. At a reconnaissance scale this requires a sample interval of at least one sample per square mile. This and the cost of separating the minus 2-μ fraction makes the method costly, ~$50 per square mile (Shilts, 1974), compared to $7.50 per square mile for reconnaissance lake sampling at one sample per ten square miles (Allan et al., 1973). This method will, therefore, probably be confined to areas of known high potential where the dispersion of metals in the drainage system is inadequately developed. Analysis of the minus 2-μ fraction of soils for zinc and copper may also be a useful method of following-up lake sampling, since the resulting dispersion halo may be broader than that of the less mobile elements.

CONCLUSIONS

The majority of the Canadian Shield is overlain by only thin glacial drift. Much of this area is within the zones of continuous or discontinuous permafrost. For these regions of the Shield the most effective and economical geochemical method of reconnaissance and semi-detailed exploration for massive sulphides is lake sediment sampling. In the Slave Province, where the bulk of lake sediment geochemistry has been carried out, the following conditions are probably critical to the success of this technique:

(1) Post-glacial weathering of massive sulphide bodies is active, with dispersion of mobile elements, such as zinc and copper, in solution in drainage waters.

(2) Volcanic rocks are of relatively greater relief than the other two main rock types, metasedimentary rocks and granites. The dispersion of elements in solution in drainage waters therefore is relatively wide.

Before this type of survey is extended widely from the bounds of the Slave Province, it is essential to ascertain that these criteria exist within other regions. Also, much work remains to be carried out on the controls on the dispersion of mobile elements; in particular, the effect of organic material.

Massive sulphide deposits generally occur in clusters. It is therefore not necessary to use sampling intervals in reconnaissance surveys that will locate every deposit and body. As long as sufficient anomalous values are discovered to define a mineralized belt or cluster, follow-up methods — geochemical, geophysical, or geological — can locate the other deposits.

The varying geochemical mobility of elements contained in massive sulphide deposits allows certain of these elements to be used for reconnaissance surveys, and others for follow-up lake sediment and soil sampling. Thus, by choosing the proper combination of elements, the target areas can be gradually reduced in size.

Within the mixed tholeiitic and calc alkaline volcanic sequences of the Archean of the Shield, massive sulphides appear to be confined to calc-alkaline cycles. The cycles that host Pb—Zn(Cu) mineralization are distinctly more siliceous than those with the Zn—Cu type. The sulphur content of basic volcanic rocks is constant from area to area. Its content in acidic and intermediate volcanics associated with massive sulphides is, however, significantly higher than for barren sequences. There is also a slight to moderate increase in the Zn content of volcanics hosting the Zn—Cu type of massive sulphides, compared to barren volcanics or those containing Pb—Zn(Cu) mineralization. The alteration zone around massive sulphide deposits may be rather extensive. The (Mg + Fe)/(Ca + Na) ratio is suggested as a geochemical index of alteration.

All of the above characteristics of the volcanic rocks may be of value in exploration for massive sulphide deposits. However, if conditions are suitable for lake sediment sampling, it is doubtful if such lithogeochemical techniques will be used at the reconnaissance or semi-detailed stages of exploration. In detailed exploration and the search for buried deposits, methods based on alteration zones or on the limited dispersion of ore and associated elements from the deposit will play an important role.

ACKNOWLEDGEMENTS

I am indebted to a number of analysts in the Geological Survey of Canada who have supplied the data given in this report: Mr. R. Crook, Mr. W. Nelson, and Mr. R.E. Horton (major elements); Mr. G. Lachance and Mr. J. Gravel (X-ray fluorescence); Mr. W. Alexander, Mrs. G. Aslin, Mr. G. Gauthier, Mr. J.J. Lynch, Mrs. A. MacLaurin, and Miss E. Ruzgaitis (trace elements); Mr. J.L. Bouvier and Dr. J.G. Sen Gupta (sulphur); Mr. P. Lavergne (sample preparation); Bondar Clegg and Company (gold). Dr. A. Dass and Mr. G. Thomas provided some excellent assistance in some of the field operations. The manuscript was typed by Mrs. R. Chaffey and Mr. R.F. Daugherty supervised the preparation of figures.

Some of the work reported here was carried out on joint programs with Dr. R.J. Allan; my appreciation is due for his stimulating company both in the field and in Ottawa. Dr. I.R. Jonasson has also provided much useful comment on secondary dispersion processes. Finally, my sincere thanks to Mr. C.C. Durham who has shared with the author all of the pleasures and pains of field work in northern Canada. Without his efforts this paper could not have been written.

REFERENCES

Allan, R.J., 1971. Lake sediment: a medium for regional geochemical exploration of the Canadian Shield. Bull. Can. Inst. Min. Metall., 64(715): 43—59

Allan, R.J., Cameron, E.M. and Durham, C.C., 1973a. Lake geochemistry — a low sample density technique for reconnaissance geochemical exploration and mapping of the Canadian Shield. In: M.J. Jones (Editor), Geochemical Exploration 1972. Institution of Mining and Metallurgy, London, pp.131—160

Allan, R.J., Cameron, E.M. and Durham, C.C., 1973b. Reconnaissance geochemistry using lake sediments of a 36,000 square mile area of the northwest Canadian Shield (Operation Bear-Slave, 1972). Geol. Survey Can. Paper 72-50, 70 pp.

Anderson, D.W. and Morgenstern, N.R., 1973. Physics, chemistry, and mechanics of frozen ground: a review. In: Permafrost. North American Contrib., Sec. Int. Conf., Nat. Acad. Sci., Washington, D.C., pp.257—288

Baragar, W.R.A., 1966. Geochemistry of the Yellowknife volcanic rocks. Can. J. Earth Sci., 3: 9—30

Baragar, W.R.A. and Goodwin, A.M., 1969. Andesites and Archean volcanics of the Canadian Shield. Proc. Andesite Conf., Int. Upper Mantle Project, Sci. Rep., 16: 121—142

Boldy, J., 1968. Geological observations on the Delbridge massive sulphide deposit. Trans. Can. Inst. Min. Metall., 71: 247—256

Bouvier, J.L., Sen Gupta, J.G. and Abbey, S., 1972. Use of an "automatic sulphur titrator" in rock and mineral analysis: determination of sulphur, total carbon, carbonate and ferrous iron. Geol. Survey Can., Paper 72-31, 22 pp.

Boyle, R.W. and Jonasson, I.R., 1973. The geochemistry of arsenic and its use as an indicator element in geochemical prospecting. J. Geochem. Explor., 2: 251—296

Boyle, R.W., Hornbrook, E.H.W., Allan, R.J., Dyck, W. and Smith, A.Y., 1971. Hydrogeochemical methods — application in the Canadian Shield. Bull. Can. Inst. Min. Metall., 64(715): 60—71

Bugel'skii, Yu. Yu., 1962. Supergene migration of ore components in different climatic regions. In: Weathering Crusts, 4. Izd-vu. Acad. Nauk U.S.S.R., pp.261—287

Cameron, E.M., 1974. Sulphur in Archean volcanic rocks of the Canadian Shield. Geol. Survey Can., Paper 74-18, 9 pp.

Cameron, E.M. and Baragar, W.R.A., 1971. Distribution of ore elements in rocks for evaluating ore potential: frequency distribution of copper in the Coppermine River Group and Yellowknife Group volcanic rocks, N.W.T., Canada. In: R.W. Boyle and J.I. McGerrigle (Editors), Geochemical Exploration — Can. Inst. Min. Metall., Spec. Vol., 11: 570—576

Cameron, E.M. and Baumann, A., 1972. Carbonate sedimentation during the Archean. Chem. Geol., 10: 17—30

Cameron, E.M. and Durham, C.C., 1974a. Follow-up investigations on the Bear-Slave Geochemical Operation. Geol. Survey Can., Paper 74-1, Part A: 53—60

Cameron, E.M. and Durham, C.C., 1974b. Geochemical studies on the eastern part of the Slave Province, 1973. Geol. Survey Can., Paper 74-27, 20 pp.

Davenport, P.H., 1972. The application of geochemistry in base-metal exploration in the Birch-Uchi Lakes volcano-sedimentary belt, northwestern Ontario. Thesis, Queen's Univ., Kingston, Ont.

Davenport, P.H. and Nichol, I., 1973. Bedrock geochemistry as a guide to areas of base-metal potential in volcano-sedimentary belts of the Canadian Shield: In: M.J. Jones (Editor), Geochemical Exploration 1972. Institution of Mining and Metallurgy, London, pp.45—57

Descarreaux, J., 1973. A petrochemical study of the Abitibi volcanic belt and its bearing on the occurrences of massive sulphide ores. Bull. Can. Inst. Min. Metall., 66: 61—69

Emslie, R.F. and Holman, R.H.C., 1966. The copper content of Canadian Shield rocks, Red Lake-Landsdowne House Area, northwestern Ontario. Geol. Survey Can. Bull., 130: 31 pp.

Faulkner, E.L., 1964. The distribution of cobalt and nickel in some sulphide deposits of the Flin Flon area, Saskatchewan. Ph.D. Thesis, Univ. of Saskatchewan, Saskatoon, Sask.

Ferguson, H., Brown, P.L. and Dickey, D.D., 1964. Water movement and loss under frozen soil conditions. Proc. Soil Sci. Soc. Am., 28: 700—702

Fortescue, J.A.C. and Hornbrook, E.H.W., 1969. Progress report on biogeochemical research at the Geological Survey of Canada, 1963—1966. Geol. Survey Can., Paper 67-23, Part II: 39—63

Garrett, R.G., 1971. The dispersion of copper and zinc in glacial overburden at the Louvem deposit, Val d'Or, Quebec. In: R.W. Boyle and J.I. McGerrigle (Editors), Geochemical Exploration — Can. Inst. Min. Metall., Spec. Vol., 11: 157—158

Gilmour, P., 1965. The origin of massive sulphide mineralization in the Noranda district, northwestern Quebec. Geol. Assoc. Can. Proc., 16: 63—81

Gleeson, C.F. and Cormier, R., 1971. Evaluation by geochemistry of geophysical anomalies and geological targets using overburden sampling at depth. In: R.W. Boyle and J.I. McGerrigle (Editors), Geochemical Exploration — Can. Inst. Min. Metall., Spec. Vol., 11: 159—165

Gleeson, C.F. and Hornbrook, E.H.W., 1975. Semi-regional geochemical studies demonstrating the effectiveness of till sampling at depth. In: I.L. Elliott and W.K. Fletcher (Editors), Geochemical Exploration 1974. Elsevier, Amsterdam, pp.611—630

Hornbrook, E.H.W. and Gleeson, C.F., 1972. Regional geochemical lake bottom sediment and till sampling in the Timmins-Val d'Or region of Ontario and Quebec. Geol. Survey Can., Open File Report, 112

Hutchison, R.W., 1973. Volcanogenic sulphide deposits and their metallogenic significance. Econ. Geol., 68: 1223—1325

Jonasson, I.R. and Allan, R.J., 1973. Snow: a sampling medium in hydrogeochemical prospecting in temperate and permafrost regions. In: M.J. Jones (Editor), Geochemical Exploration 1972. Institution of Mining and Metallurgy, London, pp.161—176

Jonasson, I.R. and Sangster, D.F., 1975. Variations in the mercury content of sphalerites from some Canadian sulphide deposits. In: I.L. Elliott and W.K. Fletcher (Editors), Geochemical Exploration 1974. Elsevier, Amsterdam, pp.313—332

Moore, J.G. and Fabbi, B.P., 1971. An estimate of the juvenile sulfur content of basalt. Contrib. Mineral. Petrol., 33: 118—127

Nilsson, C.A., 1968. Wall rock alteration at the Boliden Deposit, Sweden. Econ. Geol., 63: 472—494

Oftedahl, C., 1958. A theory of exhalative-sedimentary ores. Geol. Foren. Stockholm Forh., 80: 1—19

Ridler, R.H. and Shilts, W.W., 1973. Exploration for Archean polymetallic sulphide deposits in permafrost terrains: an integrated geological/geochemical technique; District of Keewatin, Kaminak Lake area. Geol. Survey Can., Open File Report, 146: 45 pp.

Safranov, M.L., Lapp, M.A. and Meshcheryakov, S.S., 1970. Scientific principles of prospecting for deep-seated ore deposits. In: Scientific Principles of Geochemical Methods for Prospecting for Deep-seated Ore Deposits. Acad. Sci. U.S.S.R., Irkutsk, pp.37—78 (in Russian)

Sakrison, H.C., 1971. Rock geochemistry — its current usefulness on the Canadian Shield. Bull. Can. Inst. Min. Metall., 64(715): 28—31

Sangameshwar, S., 1972. Trace element and sulphur isotope geochemistry of sulphide deposits from the Flin Flon and Snow Lake areas of Saskatchewan and Manitoba. Ph.D. Thesis, Univ. of Saskatchewan, Saskatoon, Sask.

Sangster, D.F., 1972. Precambrian volcanogenic massive sulphide deposits in Canada: a review. Geol. Survey Can., Paper 72-22, 44 pp.

Shilts, W.W., 1973. Drift prospecting, geochemistry of eskers and till in permanently frozen terrain, District of Keewatin, Northwest Territories. Geol. Survey Can., Paper 72-45, 34 pp.

Shilts, W.W., 1974. Drift prospecting in the Ennadai-Rankin Inlet greenstone belt, District of Keewatin. Geol. Survey Can., Paper 74-1, Part A: 259—261

Shvartsev, S.L. and Lukin, A.A., 1965. Hydrogeochemical zonality of groundwaters of some sulphide ore deposits in deep frozen ground. In: Cryogenic Processes in Soils and Rocks. Nauka, Moscow, pp.141—148

Sillitoe, R.H., 1973. Environments of formation of volcanogenic massive sulphide deposits. Econ. Geol., 68: 1321—1336

Skinner, R.G., 1973. Drift prospecting in the Abitibi Clay Belt. Geol. Survey Can., Open File Report, 116

Tyutyunov, I.A., 1960. The processes leading to alteration and reconstruction of soils and rocks at negative temperatures. Izd-vu Acad. Nauk U.S.S.R., Moscow (in Russian)

Tyutyunov, I.A., 1961. Introduction to the theory of the formation of cryogenic rocks. Izd-vu Acad. Nauk U.S.S.R., Moscow (in Russian)

THE STATUS OF EXPLORATION GEOCHEMISTRY IN SOUTHERN AFRICA

E. BUHLMANN, D.E. PHILPOTT, M.J. SCOTT and R.N. SANDERS

Johannesburg Consolidated Investment Company Limited, Johannesburg (South Africa)

ABSTRACT

Extensive application of geochemical techniques to "grassroot exploration" in southern Africa has resulted in a number of significant mineral discoveries. The success of geochemical exploration under local conditions depends largely on the rapid turnover of large volumes of samples. Optimization of the routine procedures of sample collection, processing, interpretation and follow-up work permits exploration of large areas at low cost.

Geological as well as geophysical programs become more effective if geochemical techniques are incorporated because the latter tend to eliminate unmineralized targets while enhancing those due to mineralization. Effective and cheap conventional geochemical prospecting techniques are applicable in a wide variety of geological and geomorphological environments in southern Africa.

INTRODUCTION

Large-scale geochemical exploration, as practiced by some companies in southern Africa today, is founded on the work of Webb (1953, 1958), Tooms (1955), Williams (1956) and Webb and Tooms (1959). They demonstrated the applicability of geochemical drainage and soil surveys to base-metal exploration under subtropical conditions in Zambia. These methods were adopted by several exploration companies and successfully applied under a wide variety of conditions in Angola, Botswana, Rhodesia, South Africa and South West Africa.

The initiation of the intensive search for new base metal deposits in southern Africa dates back only 10—15 years. It is said that in many countries the number of exposed ore deposits which can be found by conventional geological and geochemical techniques is rapidly decreasing. This does not apply to southern Africa where large areas of exposed bedrock are virtually unexplored. The reason for this may be found in the strong orientation towards gold- , platinum- and diamond-mining. Until recently, base-metal exploration ventures appeared to be marginally attractive and of secondary importance.

Table I lists a number of mineral deposits which were discovered in southern Africa over the last 20 years. From visits and scanty literature it can be inferred that nearly all of them had surface exposures ranging from a few square metres to many thousands of square metres. In several instances the deposits were known as mineral occurrences for many years and the eventual discovery of their economic potential resulted invariably from the thorough application of

52

TABLE I

Discoveries made in southern Africa during the last 20 years[*]

Deposit	Metals/minerals	Exposure
Aggeneys	Cu, Pb, Zn	(+)
Annex, Prieska	Cu, Zn	(—)
Areachap	Cu, Zn	(+)
Buffalo Mine	CaF_2	(+)
Damba	Ni, Cu	(+)
Ell Mine	Cu	(+)
Empress Mine	Ni, Cu	(+)
Gams	Zn, Pb	(+)
Gorob & Hope	Cu, Zn	(+)
Madziwa	Ni, Cu	(+)
Matchless	Cu	(+)
Matsitamma	Cu	(+)
Otjihase	Cu, Zn	(+)
Palaborhwa	Cu	(+)
Prieska	Cu, Zn	(+)
Rosh Pinah	Zn, Pb	(+)
Roessing	U	(+)
Rozyn Bos	Pb, Zn	(+)
Selebi/Pikwe	Ni, Cu	(—)
Shangani	Ni, Cu	(+)
Trojan	Ni, Cu	(+)
Vergenoeg	CaF_2	(+)

[*] Deposits with surface expression are marked (+).

established prospecting methods and economic concepts. The observation that perhaps not more than two out of twenty-two discoveries represent concealed orebodies indicates that the chances for the discovery of further exposed ore bodies are by no means exhausted.

A review of the deposits listed and their settings shows that the surface expression of nearly all of them could have been detected by soil geochemical techniques making, of course, the assumption of adequate sampling density per unit area. In addition, each deposit displays a variety of geological and physical characteristics which are detectable with suitable techniques. The point to be made here is that the deposits have only one characteristic in common which is their anomalous metal/mineral content. Aspects of concentration factors leading from primary abundances to ore grades were discussed by Sullivan (1970). From a theoretical as well as a practical viewpoint it is obvious that chemical exploration methods may offer the only direct approach towards the detection of anomalous metal concentrations. Our conclusion is that any geomorphological environment that is likely to respond to chemical techniques should be subjected to geochemical exploration methods at some stage of the exploration program.

VOLUME OF GEOCHEMICAL WORK IN SOUTHERN AFRICA IN 1973

Table II shows the approximate number of samples collected and processed by a number of companies in southern Africa in 1973. The table lists only some of the samples as figures from several large and active exploration concerns could not be obtained. Figures from Zambia are not included. More than 95% of the samples listed are soils taken on grid or strip traverse patterns. The remainder is made up of stream sediment, percussion drill and rock samples. Few vegetation, water and dust samples are being taken. The table does not include the considerable number of heavy-mineral samples collected in diamond exploration.

TABLE II

Exploration geochemical samples collected by some companies in the territories of Angola, Botswana, Moçambique, Rhodesia, South Africa and South West Africa in 1973

Company	Number of samples	Number of determinations
Johannesburg Consolidated Investment Co.	901,000	3,100,000
Anglo American Corporation	300,000	900,000
Phelps Dodge of S.A.	30,000	100,000
Cominco	20,600	61,800
U.S. Steel	20,000	60,000
Total	1,271,000	4,221,800

The figures in Table II suggest a distinct difference of opinion among exploration companies with regard to the applicability of large-scale geochemical exploration in southern Africa. Southern African companies appear to make greater use of geochemical methods than foreign companies.

Johannesburg Consolidated Investment Company which accounts for a large number of the geochemical samples applies geochemistry as a routine blanket or saturation technique, whereby over preselected areas a high density of geochemical information is gathered in a short period of time. The stage of saturation sampling is looked upon merely as a production process aiming at high output, speed, low cost and effective coverage.

GEOCHEMICAL SURVEY ORGANIZATION

General organization

Our base-metal exploration is carried out by four regional exploration units based in Angola, Rhodesia, South Africa and South West Africa. Each unit is staffed with five to seven geologists, one geophysicist, six to twelve geological and geophysical assistants and 80—150 labourers. In addition each unit has its geochemical laboratory which is operated by one technician and eight to twelve labourers.

Geochemical laboratory

The capital cost of each laboratory, excluding buildings, is of the order of U.S. \$30,000. The running cost is US. \$6000 to \$8000 per month. Cost per metal determination is U.S. \$0.05 to \$0.09. Combined monthly capacity of the laboratories is in excess of 160,000 samples or 400,000 metal determinations. The four metals determined most frequently are copper, zinc, lead and nickel. The substantial excess capacity of the laboratories guarantees rapid processing of any likely number of samples.

Sample collection

Individual sampling teams consist of four to five men. In typical regional exploration programs preselected areas of 20—600 km^2 are subjected to sampling grids of, for example, 200 $\times$ 50 m or 100 $\times$ 25 m. The teams move along traverses and take samples at 50-m, 25-m or even 5-m intervals. The amount of sample taken is 50—100 g per site from a depth of between 5 and 30 cm. Traverse bearing is maintained with a prismatic compass and sample intervals are controlled with tape measures. Base lines are cut or demarcated, but traverse line cutting is avoided. Over traverse distances of 20 km length sampling teams usually do not deviate from the projected line by more than 100 m. Samples are placed in small numbered paper envelopes and carried in rucksacks. Sampling teams may traverse between 0.5 and 25 km per team-shift. Traversing starts early in the morning and the teams return to camp by 12 noon.

Sample processing

Samples are sorted at camp and analysis sheets filled in. The analysis sheets show the sample numbers and provide space for results. Samples and sheets are packed into boxes and despatched to the regional laboratory on a weekly basis by truck or air. On arrival at the laboratory, both samples and assay sheets are checked. Samples are screened through 80-mesh sieves and the minus 80-mesh fraction returned into the original sample envelope while the coarse fraction is discarded. The boxes containing the samples in their original order and accompanied by the assay sheets are passed to the weighing section where 100 mg of sample are weighed into test tubes at a rate of 600—900 samples per man shift. The tubes are stored in metal racks holding 100 tubes each. The original sample order is maintained in the racks to avoid labelling tubes. Each rack contains two check samples.

Analysis

With tilting burettes 1 ml of perchloric-nitric acid mixture is added into each test tube. The mixture contains 19 parts of 76% perchloric acid and one part of concentrated nitric acid. The racks are placed in sandbaths on hotplates, brought to boiling temperature and after 5—20 minutes of boiling allowed to cool. 10 ml of 5% hydrochloric acid are added to each test tube bringing the volume of sample solution to between 10 and 11 ml. The mixture is thoroughly agitated and allowed to settle overnight. The metal content of the supernatant solution is determined by atomic absorption spectrophotometry. The instruments are calibrated to permit direct reading of metal content in parts per million and the readings are entered into the analysis sheets immediately after reading.

Data processing

Assay sheets are posted to the field camps where they usually arrive within six to fourteen days after sample despatch. Results are plotted as profiles along traverse lines on a suitable scale.

For reference and reassessment purposes some results are stored and handled by computer methods. Attempts are being made to computerize representation and the statistical analysis of results. To date conventional processing remains superior to computer processing in terms of speed and quality of interpretation.

Interpretation

Generally all soil geochemical values exceeding the regional background by a factor of two and more are considered as anomalous. This definition is arbitrary but experience shows that all anomalies related to significant mineralization will fall within the anomalous category. In stream sediment surveys the threshold value may be set at an even lower level, e.g., 1.2 times background. Similarly in areas of mixed soils containing a large proportion of windblown sand a very low threshold value may be chosen. High analytical precision is required under such conditions.

It was found that the geochemical response of soils can vary widely from area to area and large areas must invariably be sampled before a feel for the interpretation of results is acquired.

The main factors to be considered in interpretation are: response of known mineralization to the technique under local conditions, type of mineralization to be expected (e.g., Cu, Pb, Zn, Cu—Ni), mono-element or multi-element anomalies, prevailing rock types and their response, prevailing soil profile, weathering characteristics, rainfall, topography, and geometry of anomaly.

Experiments are underway to quantify these factors in a way that will permit computerized interpretation along lines proposed by DeGeoffroy and Wignall (1970). Although it is doubted that a computerized interpretation

56

technique will replace field interpretation in the foreseeable future, a valuable
check and reference system may emerge.

In our experience the best interpretive results are produced by persons
with a good background in exploration geochemistry who, in addition, have
an intimate knowledge of the respective area. These people are invariably the
ones who actually supervise the local sampling programs. Although all results
are reviewed at the regional centres, great importance is placed on the interpretation by the field staff directing the individual programs.

Anomalies are classified into A-, B-, C-, and D-priorities, of which A- and
B-priorities are followed up routinely. It is common that the number of samples taken during follow-up exploration is much larger than that taken during
regional, area selection and area exploration stages.

ROLE OF GEOCHEMISTRY IN THE VARIOUS EXPLORATION STAGES

According to Mackenzie (1970), "In the exploration environment, search
targets are successively narrowed and the level of detail of information increased by proceeding sequentially through a number of information gathering
investment stages". In general five sequential exploration stages may be distin-

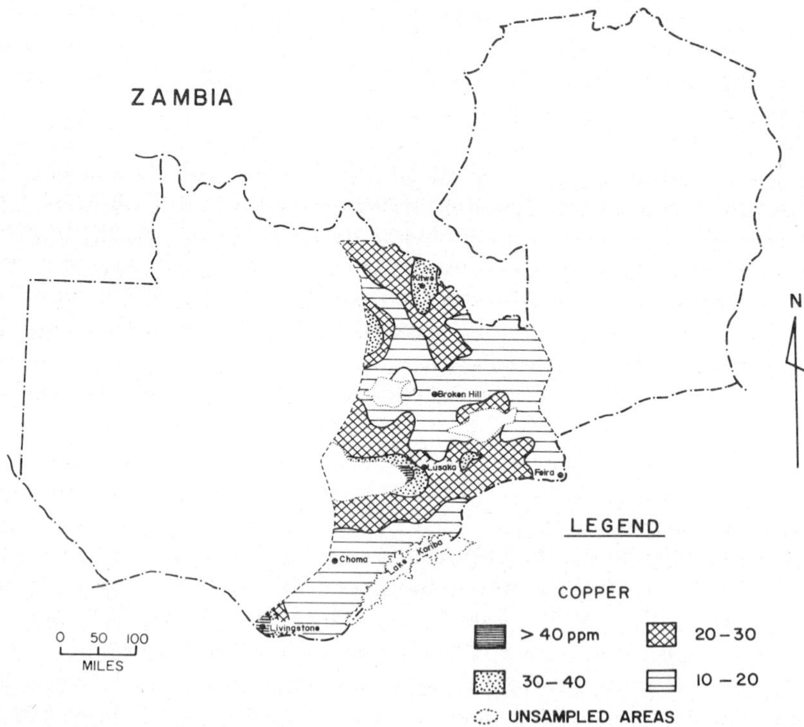

Fig.1. Reconnaissance stream sediment survey in Zambia (after Armour-Brown and Nichol,
1970).

guished, namely: (1) regional selection, (2) area selection, (3) area exploration, (4) follow-up exploration, and (5) detailed exploration.

Under local conditions geochemical techniques can make important contributions to every exploration stage and a number of examples is given below.

Regional and area selection

The necessity for greater venture selectivity has been stressed by Brant (1968). DeGeoffroy et al. (1968) and DeGeoffroy and Wignall (1970) have pointed to statistical selection techniques and it is clear from their work that the quality of venture selection depends on the amount and detail of geological, geochemical and geophysical information per unit area. Garrett and Nichol (1967) and Armour-Brown and Nichol (1970) have shown how low-density stream sediment sampling techniques can produce meaningful results for regional selection. Fig.1 shows some of the results obtained by Armour-Brown and Nichol (1970) in a regional survey of parts of Zambia, covering approximately 200,000 km^2 and involving about 1100 stream sediment samples. The principal areas of known copper mineralization were clearly indicated at this scale. Although few companies are willing to commit themselves to the expenditure of such regional work, government institutions have recognized and accepted its value and play an increasing part in the compilation of regional geochemical maps.

Area exploration

Fig.2 shows an area of approximately 580 km^2 in western Rhodesia which was subjected to a blanket soil geochemical survey on a 300 × 30 m grid. The area is generally flat and bush covered, with little outcrop. Soils are residual and annual rainfall is about 800 mm per year. Approximately 60,000 soil samples were taken during the area exploration stage and analysed for copper, nickel and cobalt. The work which involved about 2000 km of traversing was completed in 3.5 months. Fig.2 shows the contoured results for copper. The high copper background in the west is due to intermediate and basic volcanics of Precambrian greenstone-type environment. The sporadic anomalies in the centre of the area are due to pyritic carbonaceous shale containing trace amounts of chalcopyrite. Fig.3 shows the nickel contours for the same area. It can be seen that a zone of probably ultrabasic rocks exists in the central portion of the area extending as an arc in a northerly direction through the entire area. In the centre high nickel values are associated with high copper values and close scrutiny reveals that serpentinite and carbonaceous shale are in contact here. Detailed geochemical work indicates that in small parts of the area sympathetic Ni—Cu values occur over serpentinite. These values are derived from disseminated nickel- and copper-sulphide mineralization. The mineralization was known prior to the survey. The survey failed to delineate further zones of mineralization.

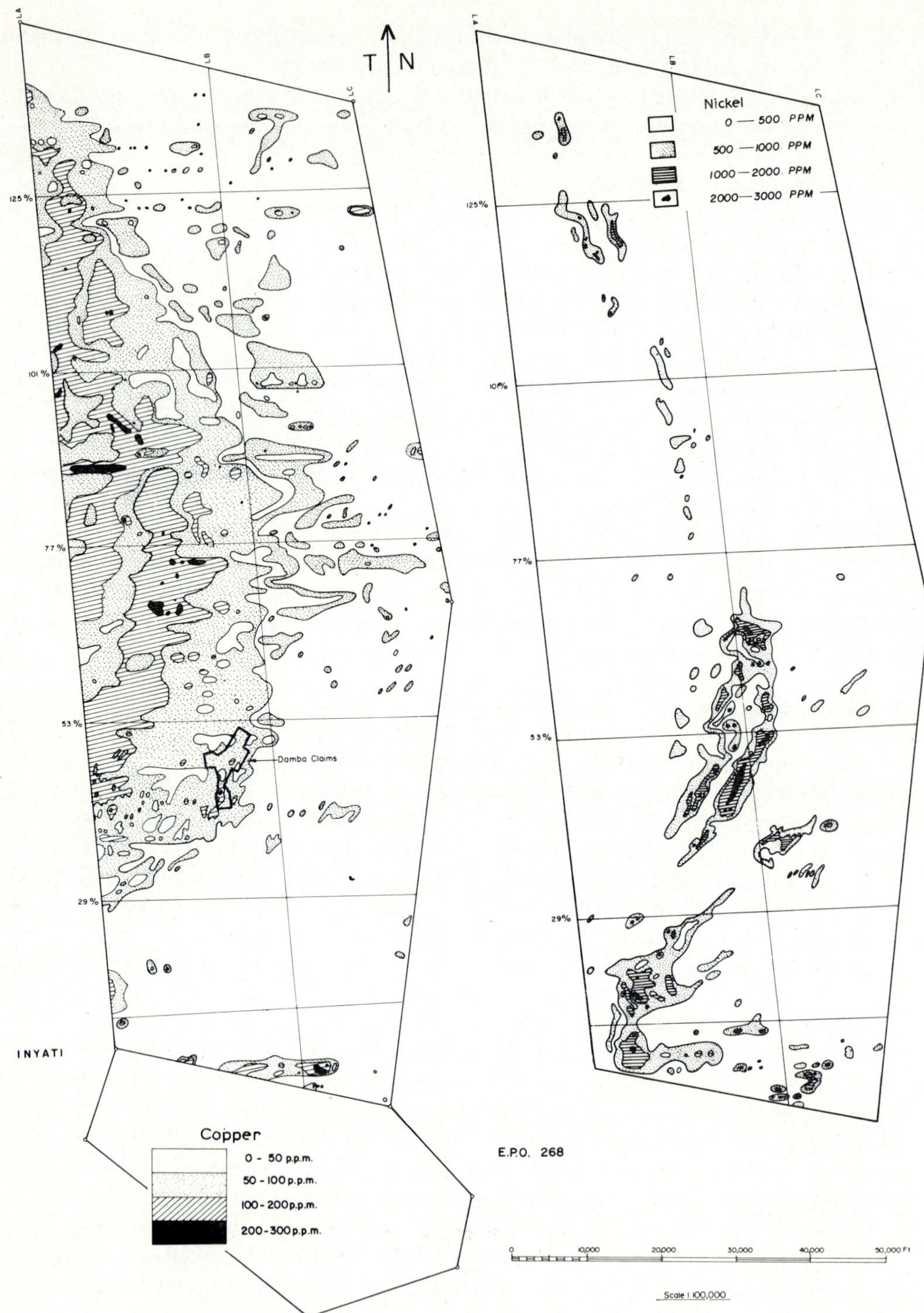

Fig. 2. Soil copper contours of Inyati area, Rhodesia.

Fig. 3. Soil nickel contours of Inyati area, Rhodesia.

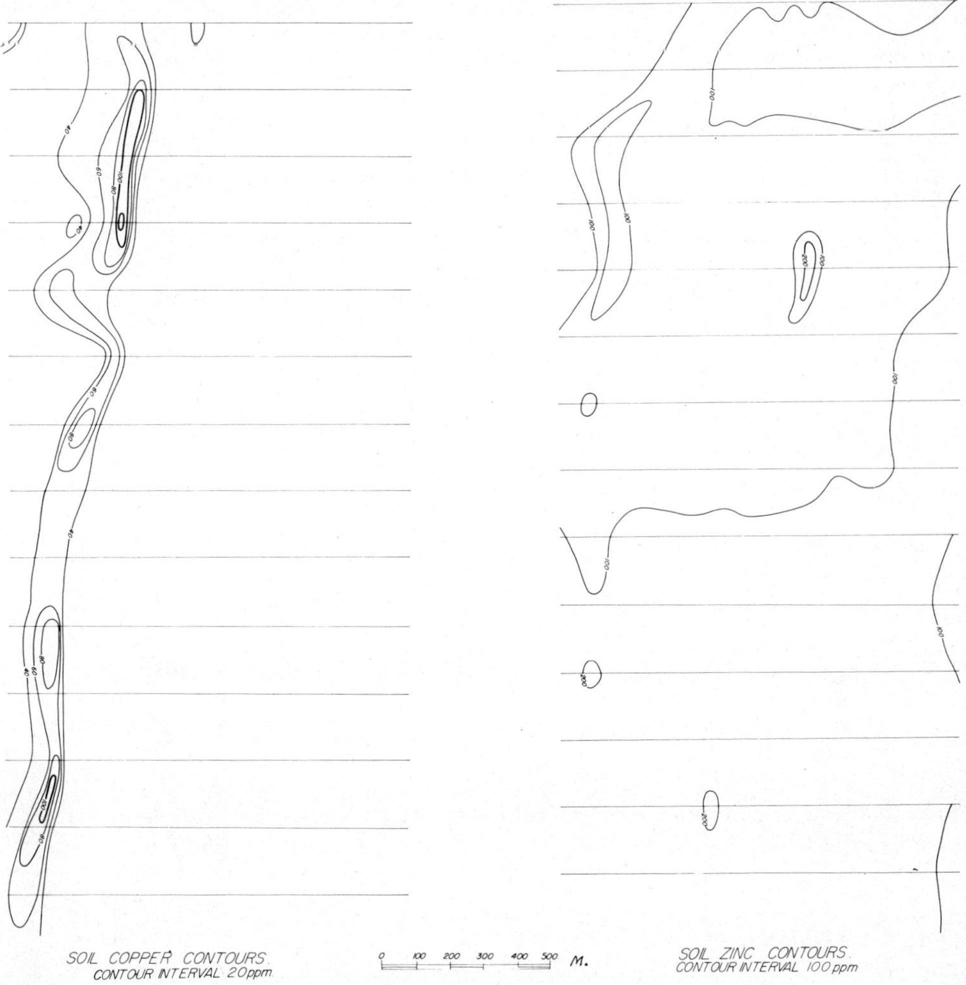

Fig.4. Soil copper and zinc contours in sand- and calcrete-covered area.

Fig.4 shows soil-geochemical results for copper and zinc from an area which is characterized by low rainfall, sparse outcrop and extensive sand-, calcrete- and rubble-cover. Such areas which are classified as sand/rock deserts cover considerable parts of Angola, Botswana, South West Africa and South Africa. The area in Fig.4 was sampled on a 200 × 50 m² grid and sympathetic relationships between copper and zinc can be recognized. Fig.5 shows part of the same area subsequently subjected to closer sampling and the copper contours now indicate a much better defined anomalous zone. The anomaly was shown to be related to underlying Cu—Zn mineralization.

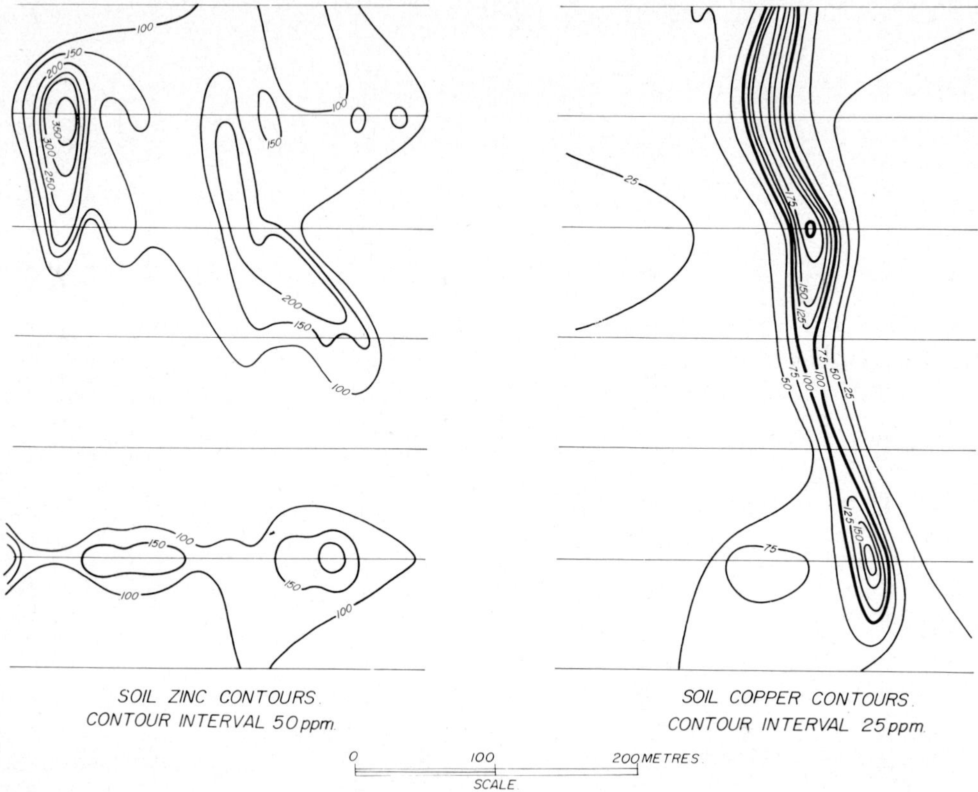

Fig.5. Soil copper and zinc contours resulting from follow-up survey over part of area shown in Fig.4.

Detailed exploration

Geochemical techniques used commonly in detailed exploration are trench sampling, percussion drilling and systematic assaying of all exploration-type diamond drill core for relevant trace metals. Percussion drilling has proved particularly useful in the investigation of mineralization hidden under sand- and calcrete-cover. Low water tables permit the application of this technique on a large scale at a cost of less than U.S. $3.00 per metre.

Integration of geochemical and other techniques

It is obvious that in many instances the combination of several techniques will lead to better results than one technique on its own. Limits are imposed by available funds.

In many of our programs geochemistry is applied hand in hand with geophysics. Massive sulphide exploration is dominated by target selection using

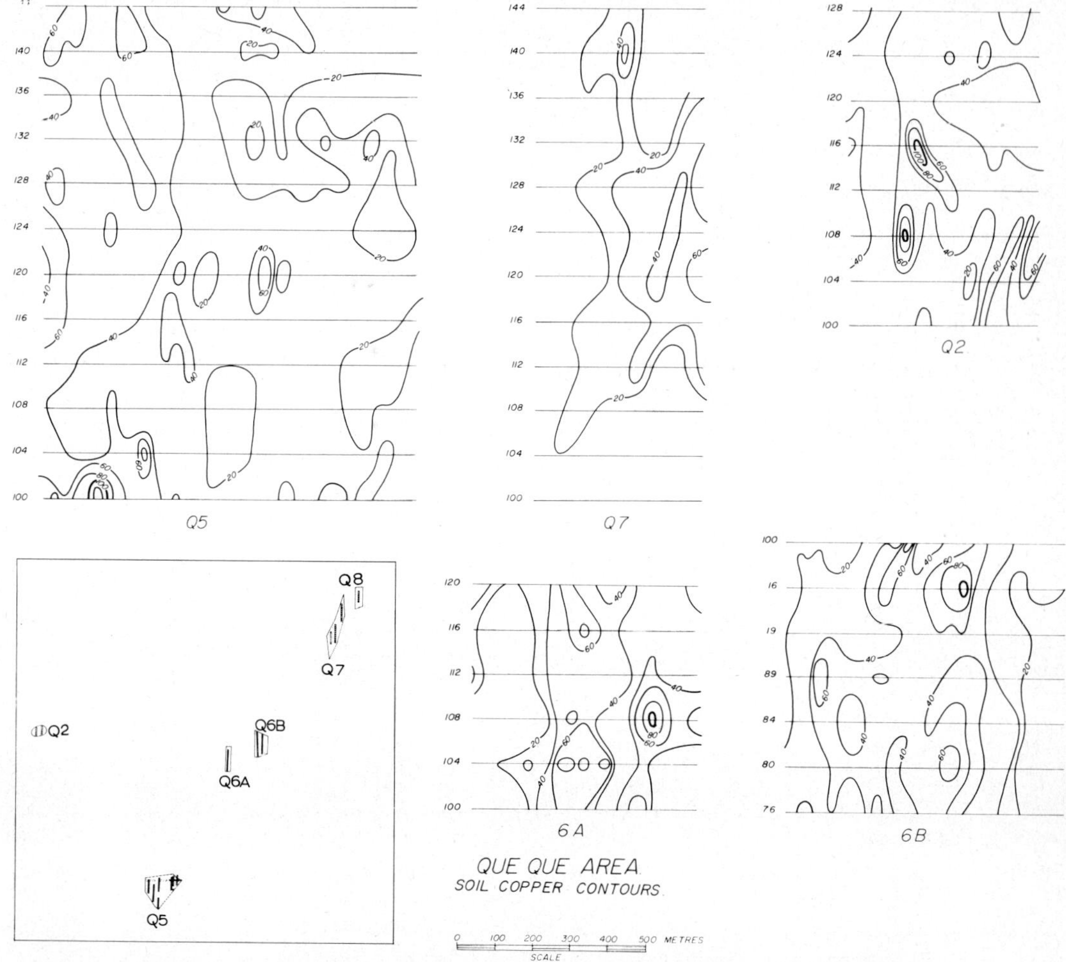

Fig.6. Soil copper contours resulting from follow-up survey over bedrock conductors. Inset shows position of bedrock conductors resulting from airborne electromagnetic survey, Que Que area, Rhodesia.

airborne electromagnetic techniques which were introduced to southern Africa on a large scale only about 8 years ago, mainly by Canadian companies. Fig.6 (inset) shows an area in Rhodesia underlain mainly by intermediate to acid volcanics and associated sediments. The area was blanket covered by an airborne electromagnetic survey, delineating 24 conductors. Of these, six were recommended for follow-up work as potentially significant bedrock conductors. Follow-up work employed electromagnetic and induced polarization techniques and all the conductors were located on the ground. Diamond drilling confirmed the presence of massive sulphide or graphitic sediment. No significant copper, lead or zinc mineralization was encountered. Only small

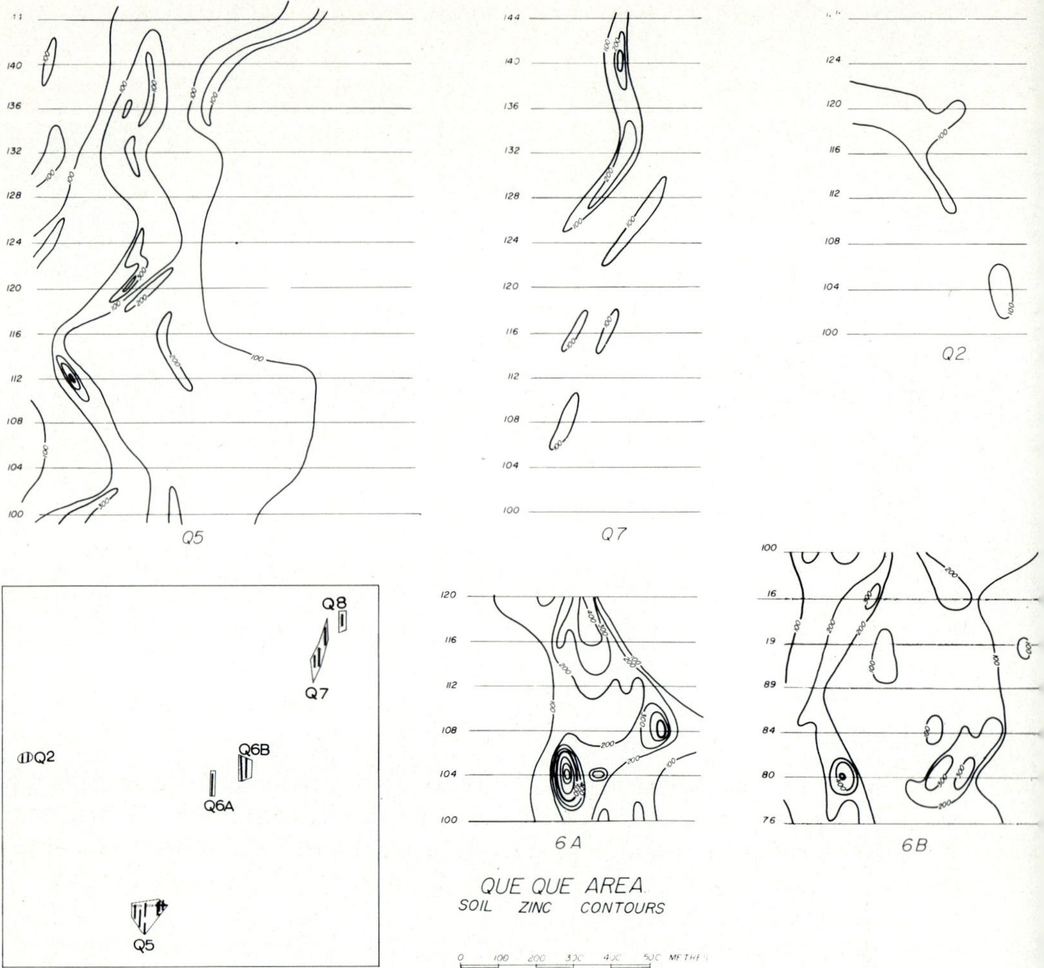

Fig.7. Soil zinc contours resulting from follow-up survey over bedrock conductors shown. Inset shows position of bedrock conductors resulting from airborne electromagnetic survey, Que Que area, Rhodesia.

amounts of zinc were found in one of the areas drilled. For study purposes geochemical soil grid sampling was applied. Fig.6 shows the soil copper contours for a number of the target areas. The values do not suggest the presence of anomalous copper concentrations. Similarly the zinc values in Fig.7 are considered to be low with the exception of targets Q5 and 6A where further follow-up work would be considered. The above examples indicate how soil-geochemical methods can aid in the elimination of conductors without resorting to diamond drilling in areas of shallow soil cover.

Under local conditions most ore deposits are oxidized and leached to a varying extent and the resultant gossans may show widely varying mineralogy.

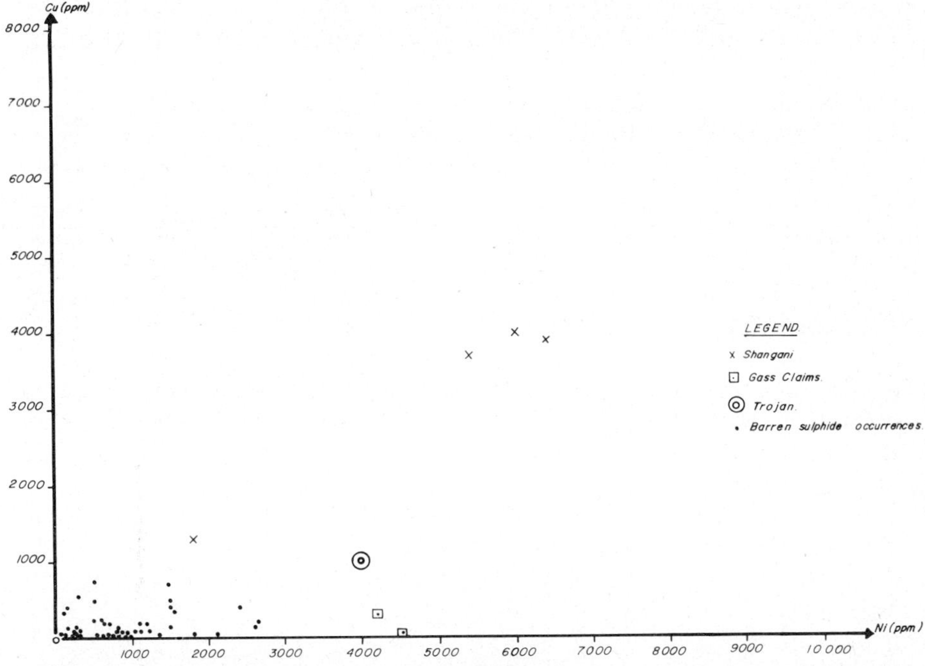

Fig.8. Copper and nickel content of gossans from several localities in southern Africa.

Despite leaching the gossans retain, by and large, characteristic amounts of metal giving a clear indication of the type of mineralization from which they were derived. Fig.8 shows the plot of the copper and zinc content of a number of gossan samples. The ones marked as open circles are from localities with significant Cu—Zn mineralization. The remainder, shown as black dots, were collected over barren sulphide occurrences. Although the samples from mineralized localities show a broad scatter of Cu—Zn contents it is clearly seen that the average content of these two metals is much higher here than in samples from barren localities.

DISCUSSION

Our experience is that the scope for the application of chemical prospecting techniques is growing in southern Africa. Soil and stream sediment sampling techniques are dominant and they have been sufficiently refined to compare favourably with other saturation techniques in widely differing geological and geographic environments. At present geochemical techniques are more widely applied than other blanket techniques and their use is increasing from year to year, Developments in exploration geochemistry show a trend towards improvement of established techniques with regard to quantity, cost and wider application.

The main use of geochemistry is at present found in area exploration and follow-up exploration while its application to regional and area selection is still secondary.

Research into ore-forming processes as well as weathering and dispersion processes is followed with much interest as it continues to provide aids towards an improved understanding of the applicability of geochemistry in exploration. The development of airborne geochemical techniques is being watched closely. The large sand-, tillite-, and calcrete-covered areas of southern Africa continue to offer a challenge to improved geochemical techniques.

ACKNOWLEDGEMENTS

The authors gratefully acknowledge the assistance of many members of the exploration teams in gathering the data presented in this paper. Jim Sutcliffe drafted the illustrations and Cederic Cousins read the manuscript critically.

Thanks are due to the management of Johannesburg Consolidated Investment Company Limited for the permission to publish this paper.

REFERENCES

Armour-Brown, A. and Nichol, I., 1970. Regional geochemical reconnaissance and the location of metallogenic provinces. Econ. Geol., 65: 312—330

Brant, A.A., 1968. The pre-evaluation of the possible profitability of exploration prospects. (3rd Hugh Exton McKinstry Memorial Lecture, Harvard Univ., April 25, 1968.) Mineral. Deposita (Berl.), 3: 1—17

DeGeoffroy, J. and Wignall, T.K., 1970. Application of statistical decision techniques to the selection of prospecting areas and drilling targets in regional exploration. CIM Trans., LXXIII: 196—202

DeGeoffroy, J., Wu, S.M. and Heins, R.W., 1968. Selection of drilling targets from geochemical data in southwest Wisconsin zinc area. Econ. Geol., 63: 787—795

Garrett, R.G. and Nichol, I., 1967. Regional geochemical reconnaissance in Eastern Sierra Leone. Trans. Inst. Min. Metall., Sec. B, Appl. Earth Sci., 76: B97—B112

Mackenzie, B.W., 1970. Corporate Exploration strategy. Apcom 2, 8 pp.

Sullivan, C.J., 1970. Relative discovery potential of the principal economic metals. CIM Trans., LXXIII: 163—173

Tooms, J.S., 1955. Geochemical dispersion related to mineralisation in Northern Rhodesia. Ph.D. Thesis, Univ. of London

Webb, J.S., 1953. A review of American progress in geochemical prospecting. Inst. Min. Metall. Trans., 62: 321—348

Webb, J.S., 1958. Observations on geochemical exploration in tropical terrains. Int. Geol. Cong., XX Sess., pp.143—173

Webb, J.S. and Tooms, J.S., 1959. Geochemical drainage reconnaissance for copper in Northern Rhodesia. Inst. Min. Metall. Trans., 68: 125—144

Williams, D., 1956. Research in applied geochemistry at Imperial College, London (abstract). Int. Geol. Cong., XX Sess., p.380

GEOCHEMICAL EXPLORATION FOR KUROKO DEPOSITS IN NORTH-EAST HONSHU, JAPAN

MAKOTO SHIIKAWA, KISAKU WAKASA and NORIO TONO

Department of Geology, Akita University, Akita (Japan)
Akita Holy Spirit High School, Akita (Japan)
Geological Survey of Japan, Tokyo (Japan)

ABSTRACT

High-grade strata-bound Pb—Zn—Cu sulphide deposits occurring in the Miocene rocks of Japan have been mined since the middle of the nineteenth century.

Exploration for these "Kuroko" deposits is guided by concepts of their mode of formation; principally that they are of submarine exhalative sedimentary origin.

Soil sampling surveys have been found to be an effective prospecting tool even for blind orebodies occurring at depths of 200 m below surface.

Statistical analysis of geochemical data from drill holes in selected areas around known orebodies can be used to predict the presence of additional mineralization.

INTRODUCTION

In the past, exploration for Kuroko deposits has been based on prospecting, geological mapping, geophysics and drilling. More recently, the gradual exhaustion of known ore reserves led to intensified exploration efforts in the course of which geochemical methods were found to be powerful prospecting tools.

Two of the most characteristic features of this type of deposit are, first, a submarine, exhalative, sedimentary origin and, second, the orebodies are concealed beneath thick post-mineralization rocks.

Our preliminary studies using geochemical techniques were aimed at delineating anomalies in soils overlying the deposits; examining the distribution of elements in the post-mineralization cover and developing statistical methods to interpret multi-element data.

GENERAL GEOLOGY OF KUROKO DEPOSITS

All Kuroko deposits are found in the so-called Green Tuff region of Japan (Fig.1) (Watanabe, 1973). The Green Tuff region is named after a distinctive unit of Cenozoic age (Tatsumi et al., 1970). In Cenozoic time the region was characterized by explosive volcanic activity and the development of sedimentary basins caused by severe subsidence. In the early Miocene volcanic products were of basaltic to andesitic composition and were followed by

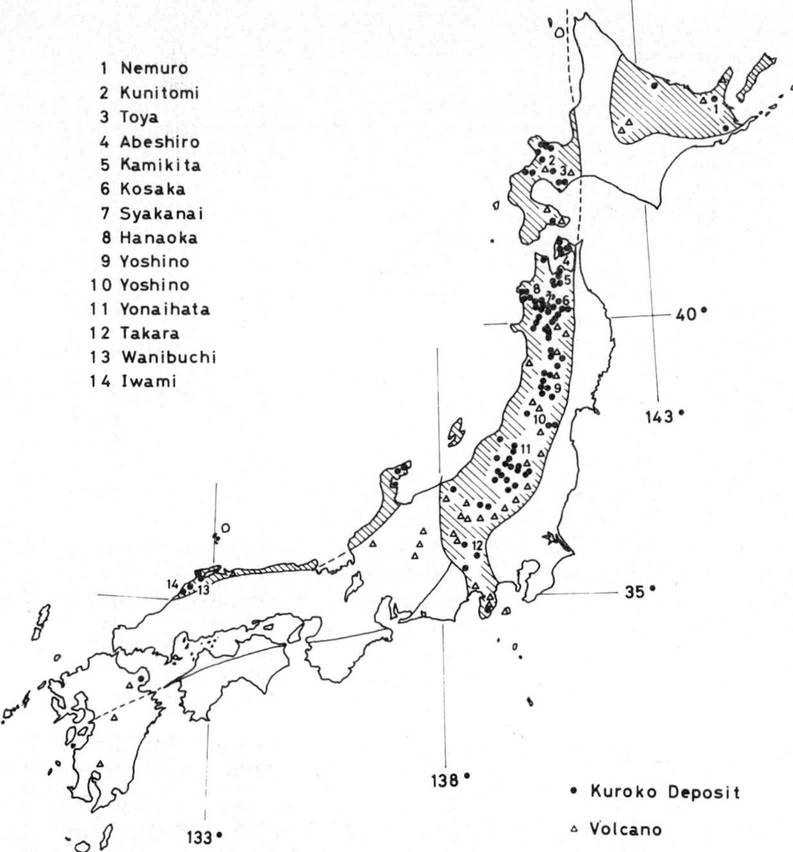

Fig.1. Distribution of Kuroko mines in Japan (after Watanabe, 1973).

the eruption of an enormous amount of andesitic material. The middle Miocene was a period of intense felsic submarine volcanism. Progressive subsidence of the region reached a maximum in late Miocene time and was accompanied by a reduction in volcanic activity.

Kuroko mineralization is related to felsic volcanic events of early geosynclinal development (Kajiwara, 1970; Suga et al., 1972). Most of the presently known deposits occur in the Nishikurosawa stage which is considered to be equivalent to the middle Miocene Serravillian of Europe (Table I) (Chinzei, 1967; Suzuki et al., 1971).

The general stratigraphic succession of Kuroko ore zones from top to bottom is shown in Fig.2 (Matsukuma and Horikoshi, 1970). Matsukuma and Horikoshi described the major constituent minerals of the ore zones as follows:

Oxide zone: quartz haematite, minor pyrite.

Barite zone: barite.

Black ore zone (Kuroko): sphalerite, galena, barite.

Yellow ore zone (Oko): chalcopyrite, pyrite.

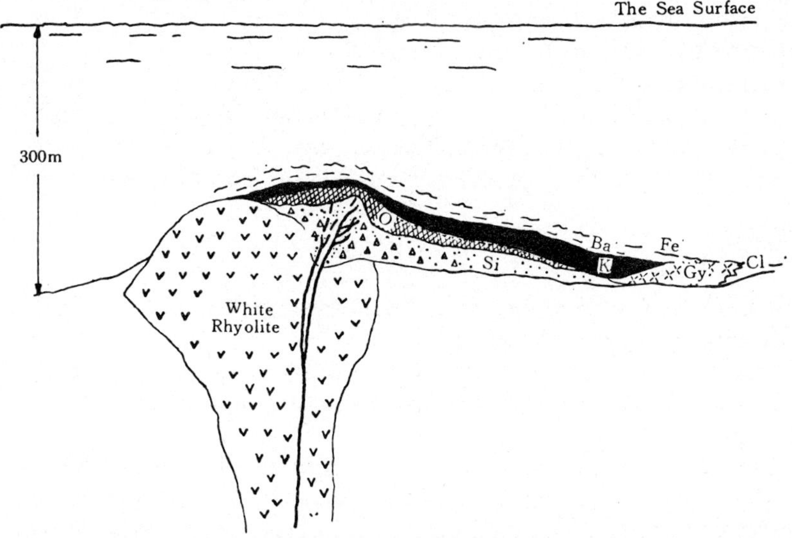

Fig.2. Idealized section of Kuroko deposit. (After the Dowa Mining Co. Ltd.)
Fe = ferruginous bed, Ba = barite bed, K = black ore (Kuroko), O = yellow ore (Oko), Si = siliceous ore (Keiko), V = ore vein, Gr = gypsum ore, Cl = clay.

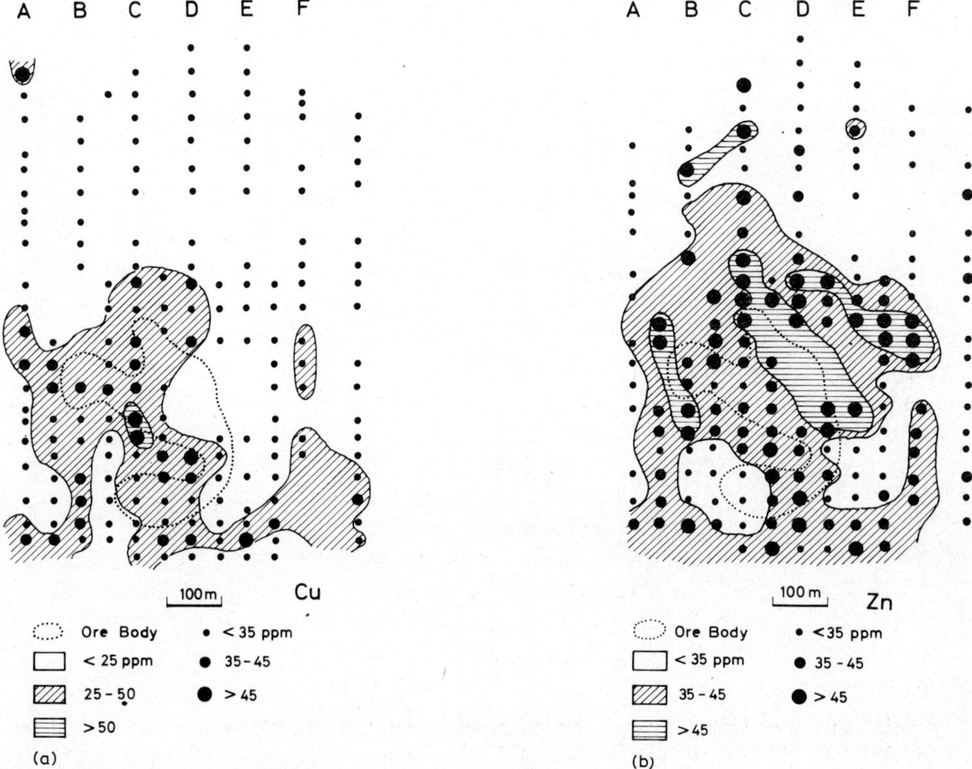

Fig.3. Copper (a) and zinc (b) in soils, Tashiro mine.

Siliceous zone (Keiko): pyrite, chalcopyrite.

Gypsum zone (Sekko): anhydrite, gypsum, pyrite.

Kuroko deposits are characteristically associated with an extensive suite of trace elements such as As, Bi, Cd, Ga, Ge, Hg, In, Mo, Sb, Sn and Ti. In addition, silver and gold are present in commercial amounts. Vanadium often occurs in high concentrations in the uppermost part of the Kuroko zone.

SOIL GEOCHEMISTRY

An investigation of the usefulness of soil geochemistry was made in the vicinity of the Tashiro mine, Fukushima Prefecture.

The Tashiro deposit consisting of oxide, barite, kuroko, yellow and stockwork ore zones lies within the upper horizons of the middle Miocene Takizawagawa rhyolites.

Soil samples were taken from an area where the mineralization, buried deeply under post-mineralization sedimentary rocks is covered by a thin (10—30 cm) residual soil cover. The dark brown B horizon underlying 2—5 cm of organic material was sampled at 50-m intervals along lines 50 m apart.

The minus 120-mesh fraction of oven dried soils (110°C) was analysed for copper and zinc by atomic absorption following a hot nitric acid extraction.

Both copper and zinc, which exceed 45 ppm in anomalous samples compared to less than 35 ppm in background samples, outline the area of the concealed ore deposit (Fig.3).

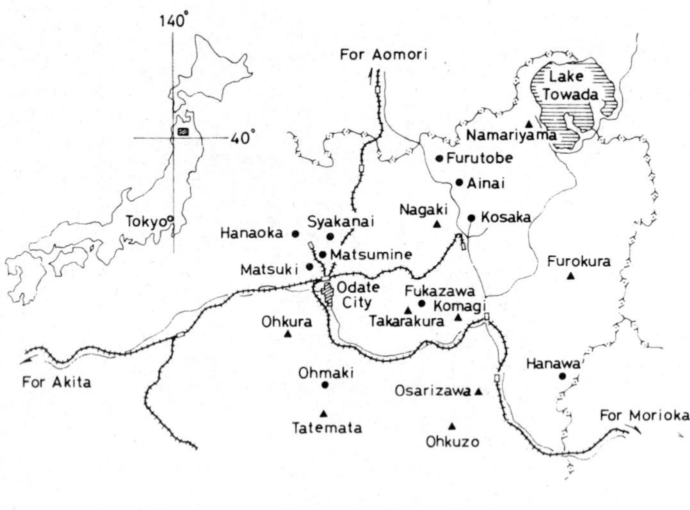

Fig.4. Distribution of Kuroko mines in Hokuroku district, Akita Prefecture (after Tatsumi et al., 1970).

ROCK GEOCHEMISTRY

The distribution of elements in the wallrocks of Kuroko deposits was
examined in the Matsumine and Fukazawa mines in Akita Prefecture (Fig.4).
The Matsumine mine lies in the upper part of the rhyolitic Hanoaka formation
and consists of kuroko, yellow, siliceous and gypsum ore zones (Fig.5).
(Toraiwa and Hashiguchi, 1967). The mineralization at the Fukazawa deposit
consists of kuroko, siliceous and gypsum ore occurring in the dacitic Yukisawa
formation (Fig.6). (Tanimura et al., 1972). The comparative stratigraphy of
these deposits is illustrated in Table I.

Rock samples about 10 cm long were collected every 2—4 m along the length
of selected drill cores. After grinding and pulverizing, the samples were
analysed by semiquantitative spectrographic methods and, for copper and zinc,
by atomic absorption.

TABLE I

Correlation table of Matsumine and Fukazawa mining fields (after Suzuki et al., 1971)

Age	Standard of northeast Honshu Stage	Matsumine		Fukazawa	
		formation	lithology	formation	lithology
Late Miocene	Tentokuji				
	Funakawa				
	Onnagawa	Shishigamori	pumice tuff tuff breccia	Shigenai	pumice tuff mudstone tuff breccia
		Tsutsumizawa	mudstone pumice tuff mudstone	Kagoya	mudstone pumice tuff mudstone basalt
Middle Miocene	Nishikurozawa	Hanaoka	ore pumice tuff rhyolite	Yukisawa	ore tuff breccia dacite
		Hotakizawa	mudstone basalt lava		
	Daijima				
Early Miocene	Monzen	Menaichizawa	sandstone altered andesite	Menaichizawa	altered andesite tuff
Pre-Tertiary					

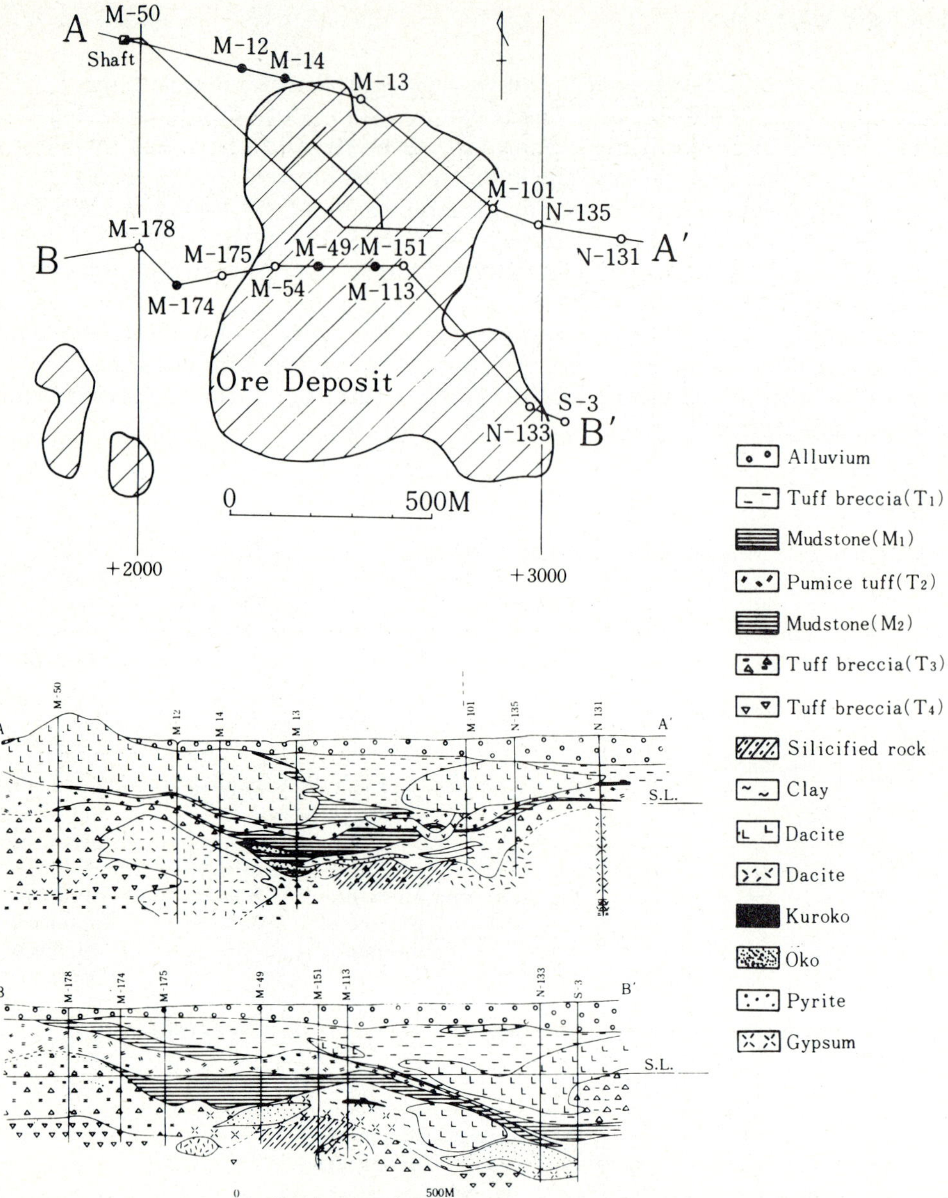

Fig.5. Stratigraphic column of the Matsumine mine. (After the Dowa Mining Co. Ltd.)

The distribution of iron, manganese (Pb + Cu + Zn), silver, molybdenum, and barium for holes passing through the Matsumine and Fukazawa orebodies is shown in Figs.7 and 8, respectively. Whilst iron is very anomalous in the tuffs overlying the mineralization, there is distinct depletion in manganese.

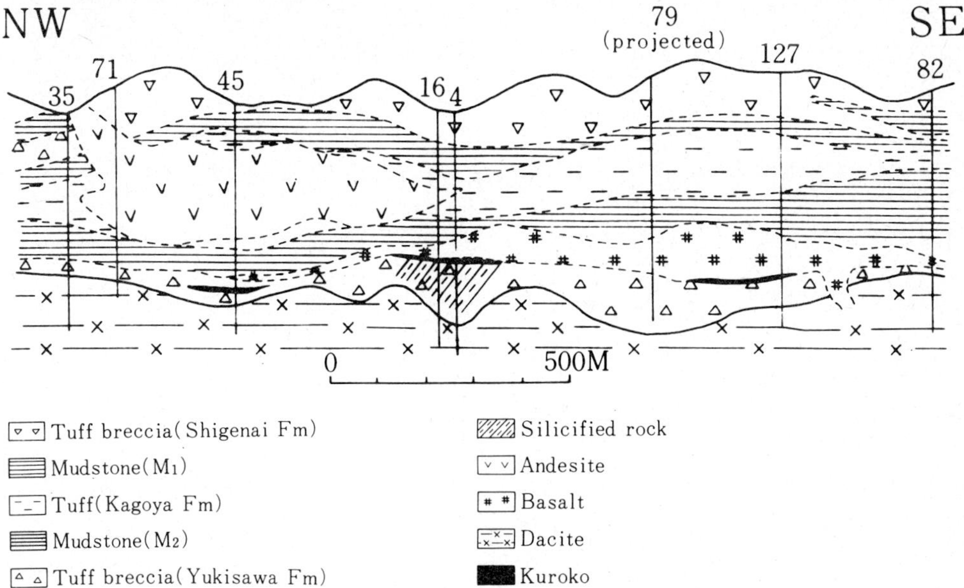

Tuff breccia(Shigenai Fm) Silicified rock

Mudstone(M₁) Andesite

Tuff(Kagoya Fm) Basalt

Mudstone(M₂) Dacite

Tuff breccia(Yukisawa Fm) Kuroko

Fig.6. Stratigraphic column of the Fukuzawa mine. (After the Dowa Mining Co. Ltd.)

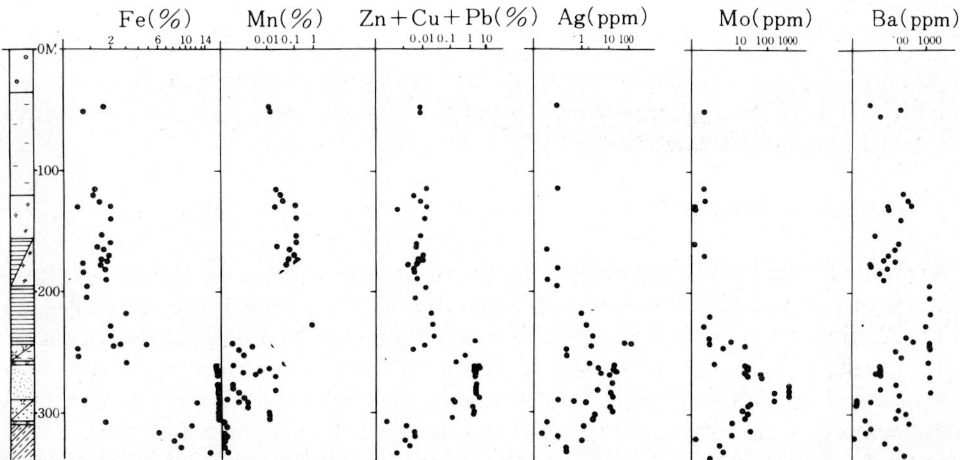

Fig.7. Distribution of Fe, Mn, Zn, Cu, Pb, Ag, Mo and Ba in drill core through ore deposit, hole M49, Matsumine mine.

The high iron is, of course, due to pyrite mineralization in the ferruginous siliceous zone underlying the yellow ore. Silver and molybdenum decrease gradually with increasing distance from the ore zone. Barium is not only high in the main ore zone but also in the overlying lithic tuff. The lateral extent of

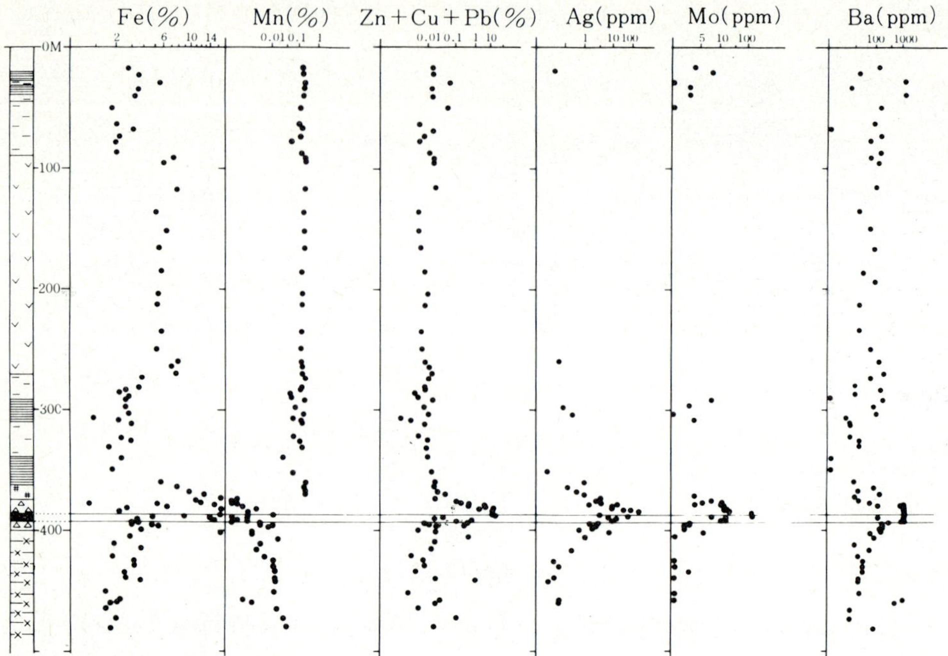

Fig.8. Distribution of Fe, Mn, Zn, Cu, Pb, Ag, Mo and Ba in drill core through ore deposit, hole TK45, Fukuzawa mine.

these geochemical patterns is demonstrated by the fact that anomalous concentrations of the ore elements are found in adjacent barren beds corresponding stratigraphically to the ore horizons.

STATISTICAL INVESTIGATIONS

Samples for statistical investigations were obtained from drill holes in the vicinity of the Yoshino and Kumanosawa deposits in Akita Prefecture (Fig.9). In the Yoshino district the mineralization is found in Nishikurosawa formation consisting of mudstones with intercalated felsic and bentonitic tuffs.

Statistical methods which have been examined are factor analysis, trend surface analysis and discriminant analysis. One of the objectives of the investigation was to find some method which would aid in deciding whether to stop or to continue exploration drilling.

One successful method has been to make trend surface maps of significant factor scores. Factor analysis based on 884 drill core samples led to the isolation of three major factors. Factor I (Co—V—Cu—Ni) accounting for 24.5% of the data variability is related to the mudstone underlying the mineralization and has high negative scores over the mineralization. Factor II (Pb—Ag—Zn—Cu) is considered to be related to mineralization. High positive scores correlate with the ore bearing Nishikurosawa formation whereas negative scores are associated

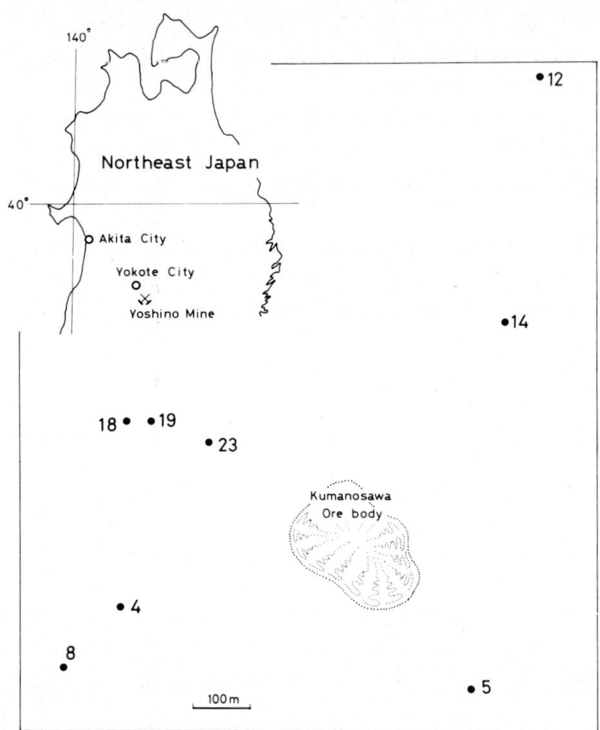

Fig.9. Location of Yoshino mine, Akita Prefecture, in relation to investigated drill cores.

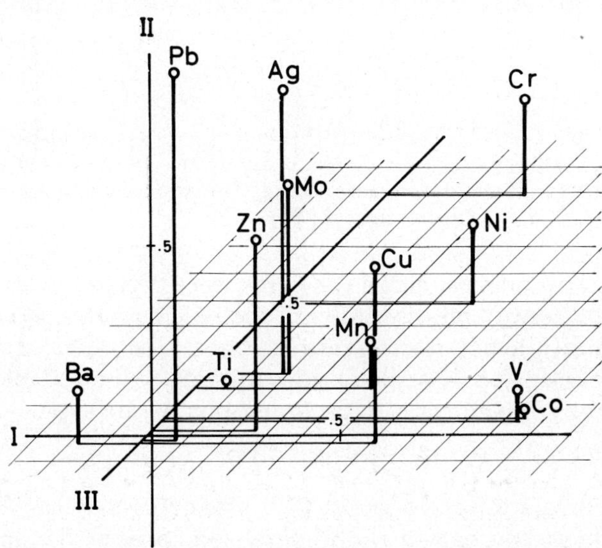

Fig.10. Configuration of elements in common factor space.

with the barren Onnagawa formation. Factor III (Cr—Ni) is essentially a chromium factor and seems to be associated with mafic minerals; in this case it correlates well with mudstone intercalations in the predominantly rhyolitic sequence. Fig.10 shows the configuration of the elements in common factor space. The anomalous position of barium is of interest since it suggests that this element has a different origin to the other elements associated with Kuroko mineralization.

The fourth-degree trend surface of the mineralization factor of drill core data from the mined out Kumanosawa deposit is shown in Fig.11. The high factor score occurring between drill holes 23 and 5 disclosed the position of already known mineralization.

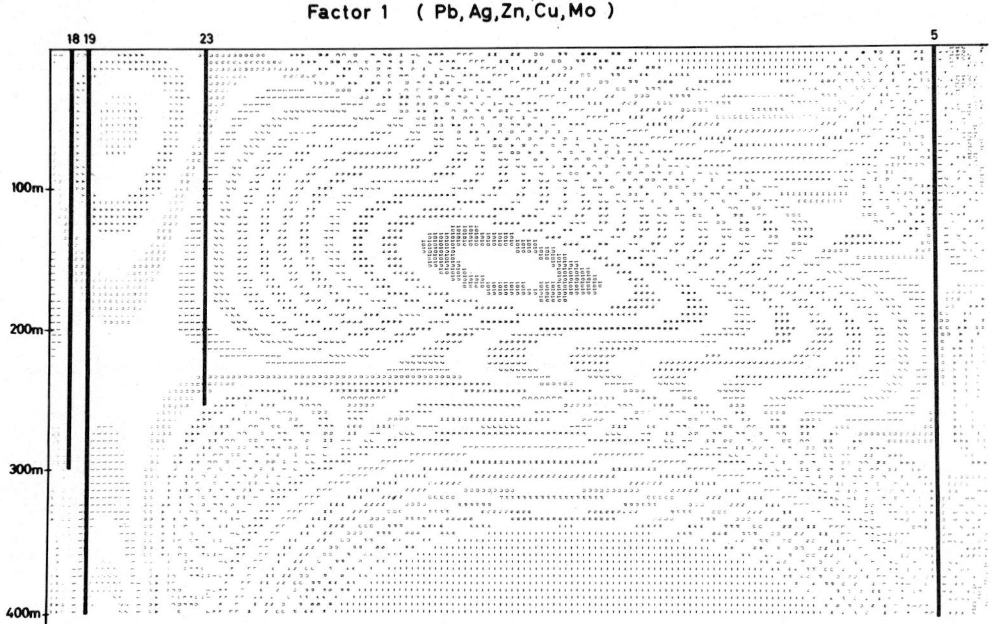

Fig.11. Fourth-degree trend surface of mineralization factor score, Kumanosawa deposit. Counter interval = 1.0; increasing: 1, 2, 3, . . .; decreasing: A, B, C . . .

To test the usefulness of discriminant analysis the cores were divided into mineralized and unmineralized groups and discriminant scores calculated from the analytical data using the multiple correlation ratio method (Shiba, 1972). Initially, twelve elements were used for this calculation, but later, due to lack of data, scores were derived from five element data. Equations of the type:

$$Y = 0.228 \text{ Ti} + 0.232 \text{ V} + 0.588 \text{ Zn} - 0.765 \text{ Ag} + 0.396 \text{ Pb}$$

were obtained. The multiple correlation ratio was 0.445. The efficiency of this method in discriminating between barren and mineralized cores is seen in Fig.12.

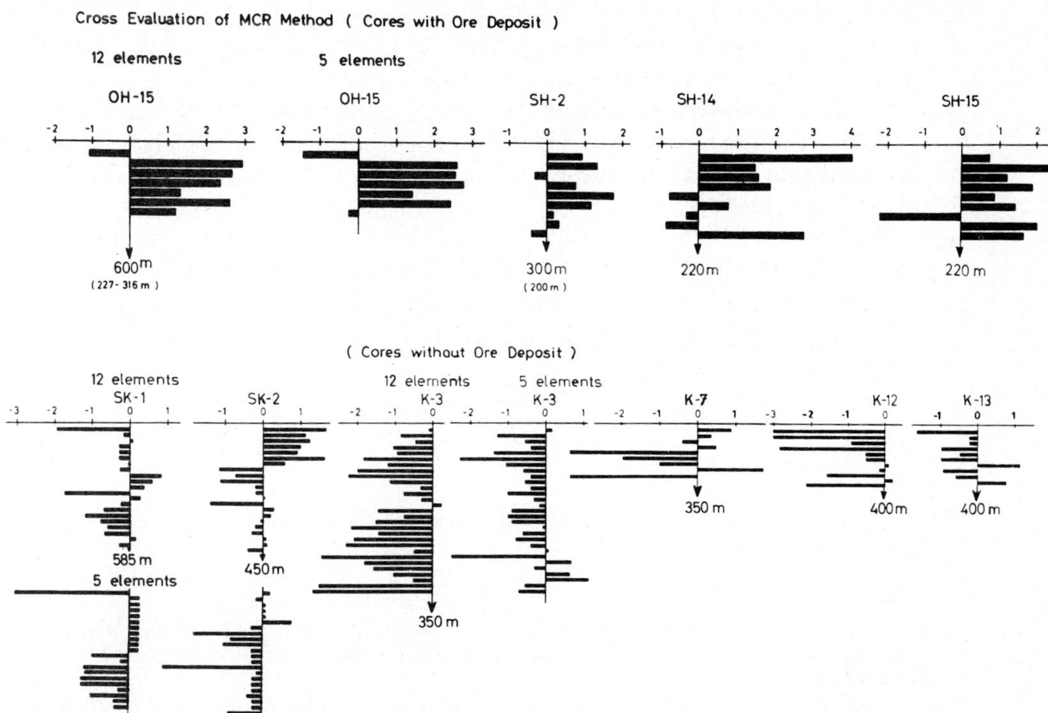

Fig.12. Evaluation of discriminant functions.

Although much more experience is needed with these methods, these preliminary tests indicate their great potential in assisting in exploration for Kuroko deposits.

SUMMARY AND CONCLUSIONS

Kuroko deposits are characterized by a submarine exhalative sedimentary origin and burial beneath thick post-mineralization cover.

This study was carried out to establish the usefulness of geochemical techniques, such as soil and rock sampling, in detecting concealed mineralization.

Although the soil anomalies detected are of low intensity and contrast, when taken together with the results of other techniques such as geological mapping and geophysics, they add to the effectiveness of exploration.

The application of standard statistical techniques such as factor, trend and discriminant analysis to the interpretation of drill core data has proved to be useful in defining mineralization trends.

ACKNOWLEDGEMENTS

The authors wish to express their thanks to the Sumitomo Metal Co. Ltd., the Dowa Mining Co. Ltd. and the Nippon Mining Co. Ltd. for providing drill core and soil samples at the Yoshino, Matsumine, Fukazawa and Tashiro mines.

Sincere thanks are offered to Dr. Sakakibara of Sumitomo Metal Mining Co. Ltd. for his co-operation during the drill core sampling at the Yoshino mine.

Particular thanks are due to Assistant Professor S. Shiba, Tokyo University, for his initiation of our programme of investigations and to President T. Watanabe, Akita University, for his useful suggestions during the course of our studies.

The statistical analyses were carried out by use of HTAC 8800/8700 system at the computer centre, Tokyo University, and partial financial support by Grant-in-Aid for Fundamental Scientific Research from the Ministry of Education, Japan, is acknowledged.

REFERENCES

Chinzei, K., 1967. Absolute ge in the Neogene sediments in Japan. J. Geol. Soc. Japan, 73: 220—221 (in Japanese)

Horikishi, E. and Sato, T., 1970. Volcanic activity and ore deposition. In: T. Tatsumi (Editor), Vulcanism and Ore Genesis. University of Tokyo Press, Tokyo, pp.181—196

Kajiwara, Y., 1970. Syngenetic features of the Kuroko Ore from the Shakanai Mine. In: T. Tatsumi (Editor), Vulcanism and Ore Genesis. University of Tokyo Press, Tokyo, pp.197—206

Matsukuma, T. and Horikoshi, E., 1970. Kuroko deposits in Japan, a review. In: T. Tatsumi (Editor), Vulcanism and Ore Genesis. University of Tokyo Press, Tokyo, pp.153—180

Shiba, S., 1972. Factor Analysis Methods. University of Tokyo Press, Tokyo, 424 pp.

Suga, K., Ito, T. and Takahashi, T., 1972. Volcanic sedimentary features in the Matsumine "Kuroko" deposits, Hanaoka Mine, Japan. Min. Geol. (Tokyo), 22: 225—249 (in Japanese)

Suzuki, Y., Tanimura, S. and Hashiguchi, H., 1971. Geology and geological structure in the Hokuroku district, Akita Prefecture. Min. Geol. (Tokyo), 21: 1—21 (in Japanese).

Tanimura, S., Shimoda, T. and Sawaguchi, T., 1972. On the Fukuzawa "black-ore"-type deposits in the Hokuroku district — a history of its discovery and geological setting. Min. Geol. (Tokyo), 22: 108—120 (in Japanese)

Tatsumi, T., Sekine, Y. and Kanchira, K., 1970. Mineral deposits of volcanic affinity in Japan: metallogeny. In: T. Tatsumi (Editor), Vulcanism and Ore Genesis. University of Tokyo Press, Tokyo, pp.3—51

Toraiwa, T. and Hashiguchi, H., 1967. Exploration of the Matsumine ore deposit in the Hanaoka mine. Min. Geol. (Tokyo), 17: 179—189 (in Japanese)

Watanabe, T., 1973. Some Geochemical Problems on Kuroko Deposits. Geochemical Society of Japan, Tokyo, 86 pp. (in Japanese)

GEOCHEMICAL PROSPECTING FOR ORES IN THE BOHEMIAN MASSIF, CZECHOSLOVAKIA

J. POKORNÝ

Geological Survey, Prague (Czechoslovakia)

ABSTRACT

Since 1952, geochemical methods have been applied to ore prospecting in Czechoslovakia. The main natural conditions controlling the working methods are a temperate climate, deep overburden connected with a long geological development of the Bohemian Massif, and a strong influence of anthropogenic effects on the landscape geochemistry (old mines, recent industry, dense population and agriculture).

Four stages of development can be distinguished in the short history of geochemical prospecting in Czechoslovakia: a period of method development and prospecting in the neighbourhood of known ore deposits (1952—1958), a period of broad field applications (1958—1964), a period of stagnation (1964—1969), and a recent period of a new activity, mostly an exploration for buried and hidden ore deposits (since 1969).

Successful geochemical exploration methods used in Czechoslovakia are as follows: soil surveying, base-of-slope sampling, stream sediment surveys, hydrogeochemical and biogeochemical prospecting and rock-sampling surveys. These methods are very often used in combination with geophysical measurements or heavy-mineral prospecting (panning).

Of the chemical techniques, emission spectrometry, atomic absorption spectrometry and radiometric methods are of special importance. Colorimetry, polarography and neutron activation analysis are uncommon. The use of selective ion electrodes, e.g., the fluoride electrode, seems to be promising.

In the last decade, methods for locating tin ores, including hidden deposits were developed. Soil surveying is a suitable method used in the search for greisen zones in granites, for chlorite-cassiterite sulphide ores and tactite tin ores in phyllite complexes. Prognostic zones of hidden deposits in granites were predicted on the basis of detailed geochemical maps, e.g., trend maps of lithium, fluorine, rubidium, gallium, beryllium and tin. Primary dispersion patterns in overlying quartz porphyries characterized by anomalous concentrations of tin, rubidium, molybdenum and tungsten indicate hidden tin ores in granitic intrusives.

Most geochemical exploration activity is concerned with the search for Pb—Zn deposits of the vein and stratiform types. A great deal of knowledge was gained in the Příbram ore district and in the Českomoravská Vysočina Highland. In the Jeseniky Mountains primary dispersion patterns in rocks were investigated on a broad regional scale in combination with hydrogeochemical exploration.

Ground- and stream-water fluorine anomalies were found by using a selective fluoride electrode in direct field measurements resulting in the discovery of new fluorite veins.

INTRODUCTION

Since 1952, geochemical methods have been applied to ore prospecting in

Czechoslovakia. The main natural conditions controlling the working methods are a temperate climate, deep overburden connected with the long geological development of the Bohemian Massif, and the strong influence of anthropogenic effects on the landscape (ancient mines, modern industry, dense population and agriculture). Successful geochemical exploration methods used in Czechoslovakia are as follows: soil surveying, base-of-slope sampling, stream sediment surveys, hydrogeochemical and biogeochemical sampling, and rock-sampling surveys. The best results have been obtained in geochemical exploration for tin deposits, base-metal vein- and stratiform deposits and fluorite-barite deposits.

GEOLOGICAL AND METALLOGENIC VIEW OF THE BOHEMIAN MASSIF

The Czechoslovakian part of the Bohemian Massif consists of a consolidated basement belonging to the western sector of the West European and Variscan platform. This basement is built up mostly of Precambrian and Paleozoic folded rock complexes, both strongly metamorphosed as well as unmetamorphosed, intruded by granitic and mafic rocks. The platform cover mainly consists of Carboniferous, Permian, Cretaceous and Neogene sediments and volcanic rocks.

Numerous mineral deposits ranging from ore bodies to insignificant occurrences occur within the 65,000-km^2 of the Bohemian Massif in Czechoslovakian territory.

The Proterozoic basement contains sedimentary ores with an accumulation of iron in oxides, e.g., magnetite ores of the Sydvaranger type, in sulphides usually in pyritic slates or as carbonates associated with manganese.

Oölitic iron ores of sedimentary origin typically occur in Lower Paleozoic rocks. In the Jeseniky mountains important stratiform base-metal deposits are mined in the volcano-sedimentary complexes of Devonian age. The Ransko gabbro-peridotite Massif contains pentlandite-pyrrhotite and sphalerite ores.

Metallogenic processes of Variscan age are very important in producing a varied scale of postmagmatic element associations such as Sn, W, Au, Cu, Pb, Zn, Ag, U, Ba, F, Hg, and Sb. The succession of associations was spread over a long period ranging from Paleozoic to Tertiary time. Postmagmatic ore deposits are developed as veins, impregnation zones and metasomatic bodies. Their positions are due to numerous faults and other tectonic effects.

GENERAL CONDITIONS OF GEOCHEMICAL EXPLORATION

From Carboniferous to Lower Cretaceous time the area occupied by the Bohemian Massif was characterized by a rather low-lying landscape having only limited topographic relief. Its present geomorphology is due to tectonic activity which reached its peak in Neogene times. Prominent fault scarps were formed and the landscape was deeply dissected by the drainage network.

The greater part of the area is covered by residual sandy clay soils formed under tropical conditions during the Mesozoic and Tertiary periods. During Quaternary and Recent times these soils were modified by a more temperate climatic regime. The thickness of the superficial cover is generally 10—20 m in the lowlands but thins considerably in hilly areas.

Chernozems, podsols, brownsoils and meadow soils passing into chernozems are the characteristic soils of the region.

Although the Bohemian Massif lies in the temperate zone local climatic conditions vary markedly with altitude. Mean annual temperature is 6—8°C falling to 0—4°C in the highest mountains. Annual precipitation ranges from 500 to 800 mm with as much as 1600 mm in exceptional years. Runoff rates vary from 3—10 l/sec/km^2 in the lowlands to 10—30 l/sec/km^2 in the upland areas. A dense and regular drainage system has been developed.

Sufficient moisture, deep overburden and mature soils are all factors which create favourable conditions for the development of vegetation. The length of the vegetation period (i.e., the annual number of days with an average daily temperature over 5°C) ranges from 230 days in the lowlands to 140 days in mountain regions. As a result the Bohemian Massif is covered by well-cultivated agricultural soils and forests.

One of the most important factors affecting the application of geochemical surveys is the effect of human settlement. Exploration for ore deposits has been carried on in the region since the time of Celtic colonization in the 5th century B.C. This is evidenced by hundreds of mines and workings for non-ferrous metals. In areas adjacent to such workings the original geochemical environment of the landscape was changed by the effects of mine drainage, ore haulage and natural leaching of mine and smelter dumps. For instance, during a geochemical survey in the Příbram district a dispersion halo of over 100 ppm Pb in rocks extended over more than 10 km^2 in the vicinity of a metallurgical plant.

At the present time, human settlement in the area is comparatively dense; averaging 125 persons per square kilometre. Zinc anomalies were often discovered in stream sediments in the vicinity of small villages.

The development of industry, agriculture and the road network are all contributing factors to both extensive and intensive pollution of the landscape. These factors must be considered in the interpretation of geochemical surveys carried out in the area.

HISTORICAL DEVELOPMENT OF GEOCHEMICAL EXPLORATION IN THE BOHEMIAN MASSIF

In the first period (1952—1958), geochemical methods were tested and applied to the search for copper, lead, zinc, antimony and tin ore deposits. Soil surveys were the most generally adopted method using emission spectro-chemistry and, more rarely, colorimetry and polarography. About 5000—20,000 soil samples were analysed annually. They were collected from dense grids covering limited areas. The methods used were developed by the Ore

Research Institute and the Czech Geological Survey based in Prague (Pácal, 1961, 1967; Pokorný and Litomiský, 1960; Pokorný, 1965). Shortly after 1958, regional exploration programmes commonly used geochemical techniques and generated up to 100,000 samples per year. Geochemical exploration in the general vicinity of known ore districts was often supplemented by geophysical measurements. Soil surveys and base-of-slope sampling were also used. Emission spectrochemistry was still the dominant analytical method. The low degree of accuracy and imperfect interpretation of the results gave rise to strong criticism of geochemical techniques and led to a reduction in the use of these methods.

A revival of geochemical exploration based on new analytical methods which permitted more precise determination of element concentrations, such as quantitative emission spectrochemistry, atomic absorption spectrochemistry, and X-ray spectrometry, took place in the latter half of the sixties. Since then, soil surveys have been supplemented by hydrogeochemical, lithogeochemical and stream sediment sampling, whose application is critically evaluated by geochemists (Hettler and Urbánek, 1967; Bubeníček, 1971).

Statistical evaluation of such data is an essential part of interpretations. In the next six years it is estimated that 120,000—130,000 samples will be collected annually and analysed. 80% of those samples will be rocks and soils. The main problem to be solved in the coming period is to develop techniques to search for hidden ore deposits using primary dispersion halos.

RESULTS AND DISCUSSION

In the following section some interesting results obtained from a number of current geochemical exploration projects in the Bohemian Massif will be described.

The geochemistry of intrusive rocks occurring in the entire area has carefully been studied since 1964. The most important group of calc-alkali intrusive rocks is divided into the following petrometallogenic series: ultramafic series (Cr), gabbro series (Fe—Ti; Ni—Cu), intermediate series including mafic intrusions (Au), and transitional series of granites compositionally approaching intermediate igneous rocks (Mo—Au; Mo—W). The alkaline members of this group partly coincide with rocks classified as subvolcanic granites of the Mediterranean potassium series and represent a very important group that contains tin ores (Sattran and Klomínský, 1970). This general geochemical study of granitic rocks will, it is hoped, facilitate broad exploration planning in the Bohemian Massif.

In 1974 the first steps were taken to put into practice a planned regional geochemical exploration for the whole territory on a scale of 1:500,000. In the first three-year period, general geochemical maps will be produced based on the analysis of 3000 rock samples for 25 elements. Thus geochemical provinces will be delimited and the statistical evaluation of lithologically different rock types used as a basis for further, more detailed rock-sampling surveys.

Primary hydrothermal dispersion patterns are being studied in great detail at the present time. Wall-rock anomalies associated with hydrothermal veins are generally narrow. One-dimensional trends in the cross sections of such veins are characterized by the use of Spearman's rank correlation coefficients. Very often the dispersion patterns are discontinuous depending on the frequency of joints and dislocations. Changes in concentration values are very irregular and highly variable and difficult to predict.

Geochemical aureoles surrounding greisen-type tin ores have proved to be very useful in search for hidden tin deposits (Štemprok, 1970). In the area of the well-known Cinovec deposit, quartz porphyry overlying an ore-bearing granite cupola showed anomalies up to 150 ppm Sn (background values about 10 ppm Sn) and 1110 ppm Rb (background values 100 ppm Rb) at places where the upper contact of the tin-bearing granite was at a depth of 100 m and the greisen ore deposit at a depth of more than 200 m. In addition to tin and rubidium, tungsten and molybdenum were found to be useful pathfinder elements (Janečka, 1973).

In the same district, some prognostic zones containing more than 20 ppm Sn were differentiated on a map of the tin distribution in the overlying quartz porphyry. By using a computer technique, two three-dimensional sections showing tin distribution trends were produced which showed the general trend of mineralization and its longitudinal variation (Janečka, 1973). In the Slavkovský les Mountains area, sampling distinguished several anomalous zones in granites containing 50—200 ppm Sn or more.

Both the three-dimensional exploration of primary dispersion halos and the construction of distribution maps at different levels seem to be a very promising method of prospecting for all base-metal deposits.

The geochemical investigation of bore holes in the sediments of the Bohemian Cretaceous Basin (Čadek, in: Malkovský et al., 1974) provides an example of deep-reaching "geochemical introspection". Platform sediments were sampled at 5-m intervals in bore holes 20—50 m apart. The minimal weight of one sample was 250 g. By using an X-ray spectrochemical analysis, contents of such elements as Ag, As, Cr, Cu, Mo, Nb, Ni, Pb, Rb, Sn, Sr, U and Zn were determined. Geochemically anomalous rocks were then examined separately, i.e., rocks containing coal, organic compounds, or sulphides, etc.

As a result, numerous anomalies were found in these lithologically complex sediments. A number of them were produced by geochemical processes at the time of sedimentation, others are due to various epigenetic processes. The anomalies will be systematically investigated further.

Rock sampling methods have been applied on a broad scale, in combination with hydrogeochemical prospecting, in the promising areas of stratiform base-metal deposits in the Devonian volcano-sedimentary complexes and the Příbram ore vein district. On the resulting 1:10,000 and 1:25,000 geochemical maps prominent anomalous zones were found to have a very complicated geochemical history ranging from Paleozoic to recent time.

In the Příbram ore district, which covers about 300 km^2, rock samples were collected from exposures and scree material. Each sample weighed about 2 kg and was collected on an irregular grid resulting in a sample density of about 20 samples/km^2. The rock types in the area are Cambrian greywackes, arkoses and conglomerates and Proterozoic greywackes, slates and volcanic rocks. The concentration levels of Ag, Pb, Zn, Cu, Hg, Sb, As and Sn in the samples were determined by emission spectrometry. Background concentrations of copper and lead are less than 10 ppm and that of zinc about 10 ppm. Anomalous areas returned 50—150 ppm Zn, 50—235 ppm Pb, 50—165 ppm Cu and 0.1—0.3 ppm Ag. A supplementary hydrogeochemical exploration was made using water from springs, wells and streams on a grid pattern with 4—5 samples/km^2. Areas containing well-developed industrial operations were not sampled. Temperature, pH and Cu, Pb, Zn, Ba, SO_4^{2-}, Cl, F contents were determined on the water samples. Anomalies of lead and zinc showed concentrations ranging from 50 to 250 μg/l.

Secondary dispersion patterns are frequently used in geochemical exploration of ore deposits. The well-developed residual soils give rise to high and prominent anomalies. In the mountainous areas dispersion patterns are very often asymmetrically deformed on slopes. The intricate forms of hydromorphic lateral anomalies make the design of soil surveys rather complicated and led us to develop more efficient base-of-slope sampling methods (Maštera et al., 1965).

Soil surveys are the most general method of geochemical exploration in the Bohemian Massif (Veselá and Pokorný, 1964). Very often this method is used in combination with geophysical measurements. In the base-metal ore districts of the Českomoravská Vysočina Highland, several methods were applied on a 200 x 10 m^2 grid as follows: initially, graphitic zones, pyritized zones and polymetallic ore zones were mapped by SP measurements. Then the anomalies were classified by carrying out a soil survey. Copper, lead and zinc contents were determined by emission spectrometry. Resistivity measurements were used to detect vein zones in areas of broad geochemical anomalies.

A further example is in prospecting for tin-bearing skarn deposits. In the first stage the positions of tactite horizons in a phyllitic complex were found by a magnetic survey. Negative gravimetric anomalies were used to locate high-level granitic intrusives. Geochemical soil samples from 1 m depth taken on a 300 x 20 m^2 (locally 150 x 20 m^2) grid were analysed for tin, zinc, copper, arsenic and boron to locate stockwork zones of tin mineralization. Because the ores of economic value are developed as accumulations at the intersection of tin stockworks and tactite horizons when combined together, the anomalous zones indicated promising areas for further drilling (Strnad, in: Janečka, 1973).

Geochemical exploration has played an important role in a recently designed programme of prospecting for fluorite on a large scale. Of particular importance in this type of prospecting is the membrane fluoride electrodes "Crytur" made in Czechoslovakia for hydrogeochemical prospecting. (This method for fluorine was developed by the laboratory staff of the Geological Survey, Prague.) Water samples (250 ml) taken from springs and streams in

mountain regions were sampled after measuring temperature and pH. Fluorine
was determined in the laboratory by means of a potentiometer. In areas composed of phyllite and mica schist, granite and orthogneiss complexes, the content of fluorine was mostly less than 0.1 mg/l, 0.1—0.2 mg/l and 0.2—0.3 mg/l,
respectively. In anomalous zones, the concentration of fluorine varied between
0.5 and 1.0 mg/l. In comparison water from a fluorite mine contained more
than 7 mg/l F (Chrt and Woller, 1970).

REFERENCES

Bubeníček, J., 1971. Geochemical exploration in the Krásná Hora district. Geol. průzkum,
    XIII(1): 3—7 (in Czech.).
Chrt, J. and Woller, F., 1970. Hydrogeochemical prospecting — a new method of search for
    fluorite in the Bohemian Massif. Geol. pruzkum, XII(8): 255—258 (in Czech.).
Hettler, J. and Urbánek, J., 1967. Mercurometry in the search for base metal deposits in
    Nízký Jeseník Mountains. Geochem. Česk., Sb. Vys. skoly bánské, Ostrava, pp.381—391
    (in Czech.).
Janečka, J., 1973. Geological Research on Tin and Tungsten Ore Deposits in the Bohemian
    Massif. Geological Survey, Prague (in Czech.).
Malkovský, M. et al., 1974. Bohemian Cretaceous Basin and its Basement. Geological Survey,
    Prague (in Czech.).
Maštera, L., Pokorný, J. and Veselý, J., 1965. Regional metallometry of the Nasavrky
    pluton. Geol. průzkum, VIII(2): 43—45 (in Czech.).
Pácal, Z., 1961. Deep geochemical prospecting of greisen rocks. Sb. ÚÚG, XXVI, Odd.
    Geol., Prague, pp.373—396 (in Czech.)
Pácal, Z., 1967. Metallometry of Sn—W—Li ores in the area of Karlovy Vary granite massif.
    Geochem. Česk., Sb. Vys. školy bánské, Ostrava, pp.403—408 (in Czech.)
Pokorný, J., 1965. Tungsten metallometry. Geol. průzkum, VII(1): 9—11 (in Czech.)
Pokorný, J. and Litomiský, J., 1960. Geochemical exploration of the Hoffman vein near
    Vrančice. Čas. Miner. Geol., V(4): 414—423 (jn Czech.)
Sattran, V. and Klomínský, J., 1970. Petrometallogenic series of igneous rocks and
    endogenous ore deposits in the Czechoslovak part of the Bohemian Massif. Sb. Geol. Věd.
    Rada LG, Prague, 12: 65—154
Štemprok, M., 1970. The use of the lithogeochemical prospecting in the search for tin ores.
    Geol. průzkum, XII(2): 36—38 (in Czech.)
Veselá, M. and Pokorný, J., 1964. Metallometric surveys at the geological mapping. Geol.
    průzkum, 3: 81—82 (in Czech.)

DISCOVERY AND EXPLORATION OF ASHNOLA PORPHYRY COPPER DEPOSIT, NEAR KEREMEOS, B.C.: A GEOCHEMICAL CASE HISTORY

J.H. MONTGOMERY[1], D.R. COCHRANE[2] and A.J. SINCLAIR[1]

[1] *Department of Geological Sciences, University of British Columbia, Vancouver, B.C.*
(Canada)
[2] *Cochrane Consultants, Ladner, B.C. (Canada)*

ABSTRACT

The Ashnola copper prospect, a typical porphyry copper deposit, was discovered by regional stream sediment sampling. Subsequent geochemical studies included additional stream sediment sampling, soil sampling of A and B horizons, biogeochemical sampling, and rock sampling. The results of the geochemical studies are compared to geology and geophysical expression of the deposit.

Important results of the study are:

(1) B-horizon Cu and Mo provide a sound basis for a soil geochemical survey in the general area of Ashnola prospect because of close correlation of their sub-populations with geological features including those of economic importance.

(2) A-horizon Zn is more useful than B-horizon Zn but neither appears necessary in this particular case.

(3) Biogeochemical analyses for copper and zinc correlate best with A-horizon soil analyses.

(4) Zinc is concentrated preferentially relative to copper in the vegetation analyzed. The Zn/Cu ratio in A-horizon soils is about 2/1, whereas the ratio in ash of lodgepole pine needles is 10/1.

(5) The method of data analysis utilizing thresholds estimated from partitioned probability plots of all variables aided the interpretation immeasurably and appears a useful general procedure in the routine analysis of geochemical survey data.

INTRODUCTION

This geochemical case history deals with sequential exploration of a porphyry Cu—Mo deposit, the Ashnola prospect, in southern British Columbia. Sampling of stream sediments, soil horizons, vegetation and rock was undertaken over a period of years by several different groups. Consequently, there are some gaps in available data. The history encompasses a period from the discovery of the property in 1966 by a regional stream sediment sampling program up to diamond drilling in 1973.

The Ashnola Cu—Mo prospect is about 30 miles southeast of Princeton, British Columbia, in Osoyoos Mining Division (Fig.1). The area is characterized by rugged terrain with steep-walled glaciated valleys. Local relief is about

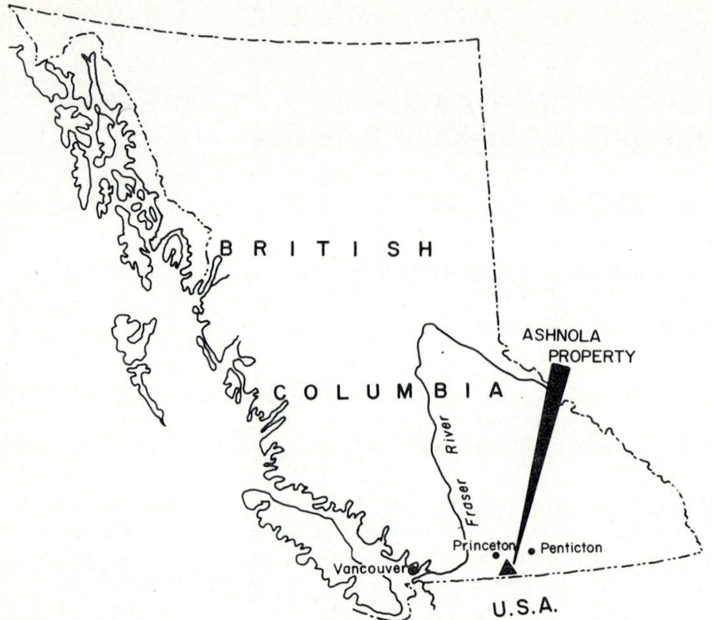

Fig.1. Location map, Ashnola porphyry Cu—Mo prospect.

1500 ft. The region is abundantly forested, particularly in valleys and lower slopes, with lodgepole pine (*Pinus contorta*), ponderosa pine (*Pinus ponderosa*) and sparse underbrush.

Soils have formed in three distinct but overlapping environments. On the upper part of the area, glacial deposits are absent and residual soils have developed with little or no transport. On the steep slopes, the parent material is composed of fluvio-glacial deposits and talus characterized by rapid transport which, in some places, has caused two levels of soil development. Apparently, an earlier soil profile has been buried by downward-moving material and the new profile developed above it. Valley bottoms are covered with fluvio-glacial deposits. The upper part of Cat Creek valley resembles a miniature flood plain and an alluvial fan has formed at the mouth of Cat Creek.

General geology of Ashnola prospect is shown in Fig.2. The area is underlain by a series of silicic porphyries and pyroclastic rocks with a small intrusive quartz monzonite boss and associated dykes.

Mineralization which accompanied the intrusion, consists of chalcopyrite, molybdenite and abundant pyrite. It occurs almost entirely in fracture-fillings in rhyolite but disseminated sulphides are common within the quartz monzonite. Leaching of sulphides has taken place in surface rocks to depths of about 100 ft. Oxidation is evident to 300 ft. Supergene minerals, such as chalcocite, cuprite and native copper are present in trace amounts but enrichment is only local and poorly developed.

Alteration is widespread and typical of that found associated with porphyry

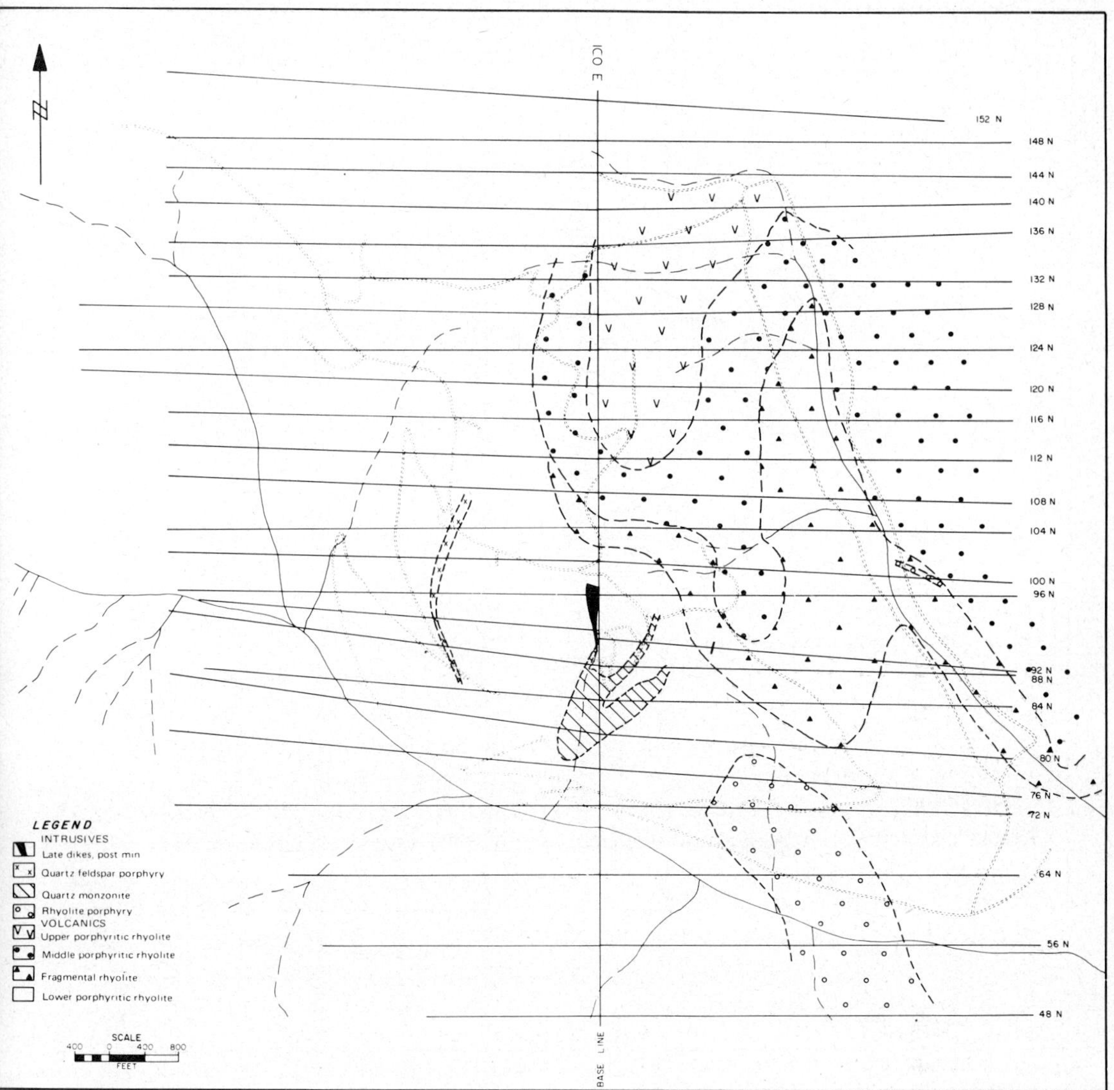

Fig. 2. General geology, Ashnola property.

copper deposits. Fig.3 is a composite map showing the relationship between alteration and chargeability as determined by an induced polarization survey. An alteration zone about 1.8 miles in diameter, is characterized by quartz veining with silification and sericitization. Pyrite mineralization accompanies this "phyllic zone" alteration and is distributed in a semi-circle as outlined by anomalous chargeability. A zone of argillic alteration is coincident with part of the quartz sericite alteration as shown in Fig.3. A small zone of potassic alteration (biotite and K-feldspar) is virtually coincident with the quartz diorite boss.

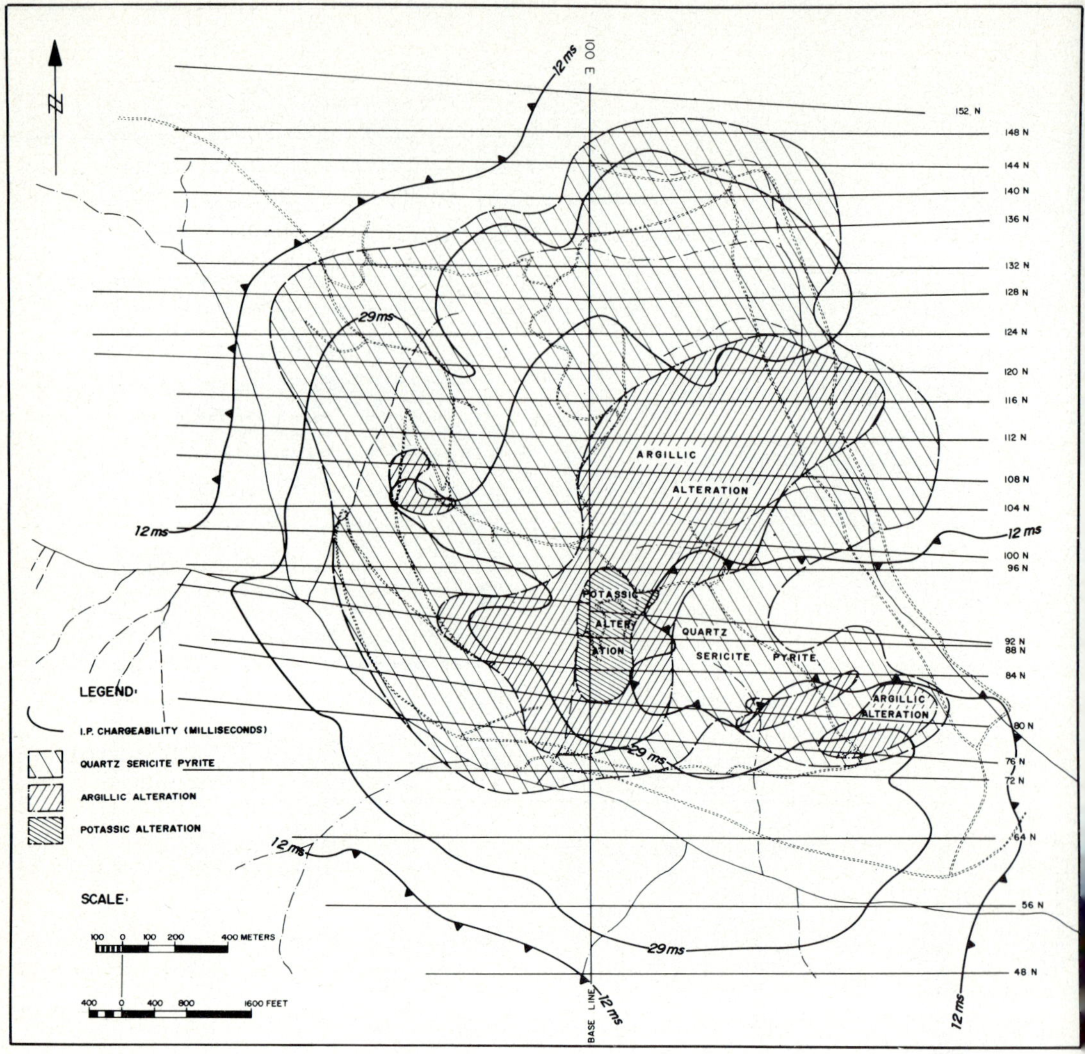

Fig.3. Relationship of alteration to an induced polarization (I.P.) chargeability high that coincides with a zone of abundant pyrite.

In summary, the geology, mineralization and alteration, closely resembles a high-level part of Lowell's porphyry model (Lowell and Guilbert, 1970).

HISTORY

Ashnola prospect was discovered many years ago but was rediscovered independently in 1966 by a regional stream sediment sampling program. Since that time, work on the project has included geological mapping, soil geochemistry, rock geochemistry, biogeochemistry, magnetometer and induced polarization surveys and drilling. In all, forty holes, including rotary, percussion and diamond-drill holes, have been completed. Approximately $300,000 has been expended to date on the deposit.

STREAM SEDIMENT SURVEY

In 1966, a regional stream sediment survey was conducted over parts of southern British Columbia, mainly south of Highway 3 from Princeton to Osoyoos. The only streams significantly anomalous in copper and molybdenum were McBride and Cat Creeks, both easterly flowing tributaries of Ashnola River. Sediment samples from these creeks contained Mo — 12 ppm, Cu — 550 ppm, and Zn — 150 ppm.

The anomalous creeks were sampled in detail with stream sediments taken at intervals of roughly 500 ft, and soil samples taken from several adjoining areas. Fig.4 shows sample locations and anomalous areas for copper in stream sediments. Results for molybdenum were comparable.

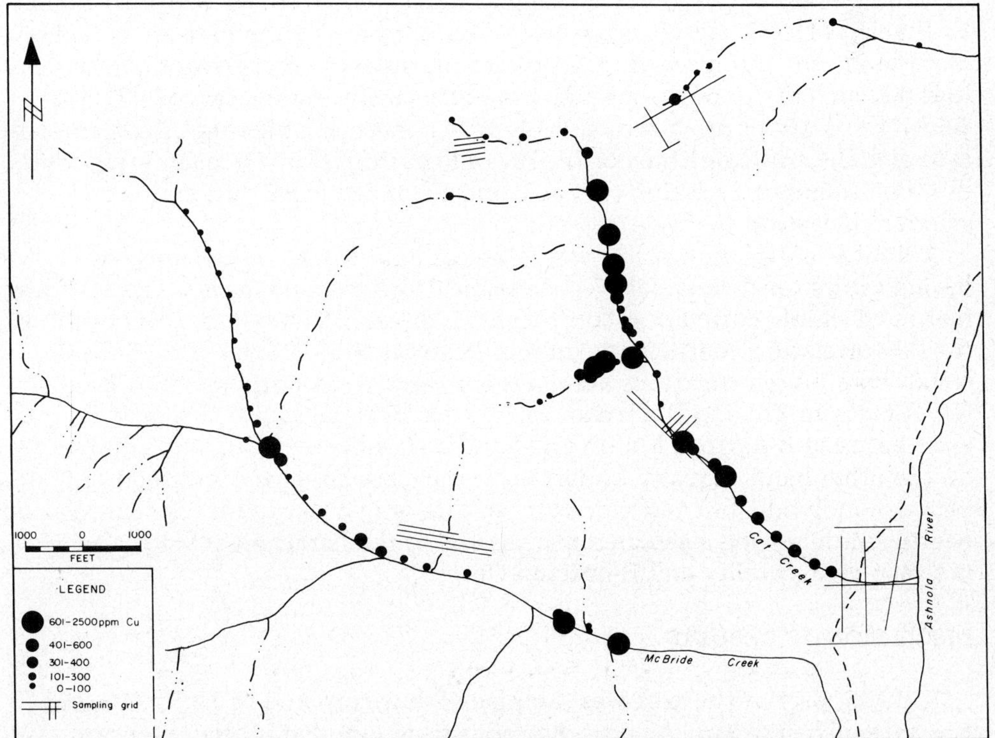

Fig.4. Copper data, detailed stream sediment survey, Ashnola property.

SOIL SURVEY

As a result of detailed stream sediment sampling, several small areas were soil sampled (B horizon). Numerous high values led to a more extensive grid-controlled B-horizon survey in 1966. A number of detailed profile studies show that copper and molybdenum are depleted in the upper few inches to 1 ft and increase in amount at depth by several fold or as much as an order of magnitude. Zinc commonly shows the reverse pattern, being much enriched near surface.

The grid was extended in 1970 and samples from both A and B horizons were analyzed, as were a limited number of biogeochemical samples.

DATA ANALYSIS — DETAILED SOIL SURVEY

Map plots with arbitrary contour intervals were prepared for all geochemical variables to permit intercomparison of patterns and relationships to geology, geophysical data, etc. A more informative procedure based on analysis of probability plots of each variable was also undertaken. In each case polymodal, lognormal models seemed applicable and each probability graph was partitioned on that basis (see examples in Figs.5 and 8). Thresholds separating the partitioned populations as efficiently as practicable were chosen, as described by Sinclair (1974). Resulting groups of data for each variable were coded on maps to aid the interpretation. Isopleths on these maps represent thresholds that discriminate populations relatively efficiently. Parameters of all partitioned populations are listed in Table I. The method of listing these parameters is to give the antilogarithms of the mean logarithm (b), the mean logarithm plus one standard deviation ($b + s_L$), and the mean logarithm minus one standard deviation ($b - s_L$).

A limited correlation study was done for data from 203 stations, each having values for 8 variables (5 geochemical and 3 geophysical). The resulting matrix of simple correlation coefficients is given in Table II. A total of 12 of the 28 correlation coefficients are significant at the 1% level. Six of these result from intercorrelations among the 4 variables A-horizon Cu, B-horizon Cu, A-horizon Zn, and B-horizon Zn. Two additional significant correlations exist between B-horizon Mo on one hand and A-horizon Cu, and B-horizon Cu on the other hand. Equally important is the clear absence of correlation between molybdenum and zinc in soils. The only other significant correlation coefficient involving a geochemical variable is the barely significant one between chargeability and B-horizon Cu.

DISCUSSION OF RESULTS

In 1966, part of the area was sampled (B horizon) and in 1970 the grid was extended and both A and B horizons were sampled in the extension. B-horizon Cu results in the original central area are considerably higher than

TABLE I

Parameters of partitioned populations, geochemical variables

Variable[*]	Population	Proportion (%)	Values (ppm)		
			b	$b + s_L$	$b - s_L$
ACU	I	5	62	123	31
	II	95	5.3	12.3	2.3
BCU (1966)	I	6	222	268	184
	II	6.5	120	142	103
	III	97.5	26	51	13.3
BCU (1970)	I	2.4	316	—	—
	II	33.6	26	37	20
	III	64	5.4	7.9	3.7
AMO	I	2.4	8.8	12.3	6.3
	II	97.6	0.69	1.6	0.34
BMO (1966)	I	3.5	62	77	50
	II	8.5	24.5	33.1	18.0
	III	88	2.6	4.9	1.4
BMO	I	7	20	43	9.6
	II	93	1.1	2.5	0.51
AZN	I	4.5	380	490	295
	II	85.5	58.9	123	28.9
	III	10	9.6	13.5	6.8
BZN	I	5	363	490	275
	II	65	74.1	123	43.6
	III	30	23.4	33.9	16.0
BIO-CU	I	15	182	199	166
	II	85	117	140	98
BIO-ZN	I	8	2090	2320	1850
	II	92	1080	1370	840

[*]ACU = A-horizon Cu, BCU = B-horizon Cu, AMO = A-horizon Mo, BMO = B-horizon Mo,
AZN = A-horizon Zn, BZN = B-horizon Zn, BIO-CU = biogeochemical copper, BIO-ZN =
biogeochemical zinc.

TABLE II

Correlation coefficients for arithmetic data[*]

Variable[**]	CH	MAG	ACU	AZN	BCU	BZN	BMO	RES
CH	1.0							
MAG	−0.279	1.0						
ACU	0.105	−0.081	1.0					
AZN	−0.023	−0.039	0.582	1.0				
BCU	*0.183*	−0.138	*0.863*	*0.490*	1.0			
BZN	0.118	−0.142	*0.740*	*0.779*	*0.806*	1.0		
BMO	0.138	−0.170	*0.332*	0.083	*0.374*	0.132	1.0	
RES	*−0.583*	*0.558*	−0.066	0.082	−0.084	−0.018	−0.091	1.0

[*]Absolute values greater than 0.181 are significant at the 1% level and are italicized.
[**]CH = chargeability, MAG = ground magnetometer values, RES = apparent resistivity,
ACU = A-horizon Cu, AZN = A-horizon Zn, BCU = B-horizon Cu, BZN = B-horizon Zn,
BMO = B-horizon Mo.

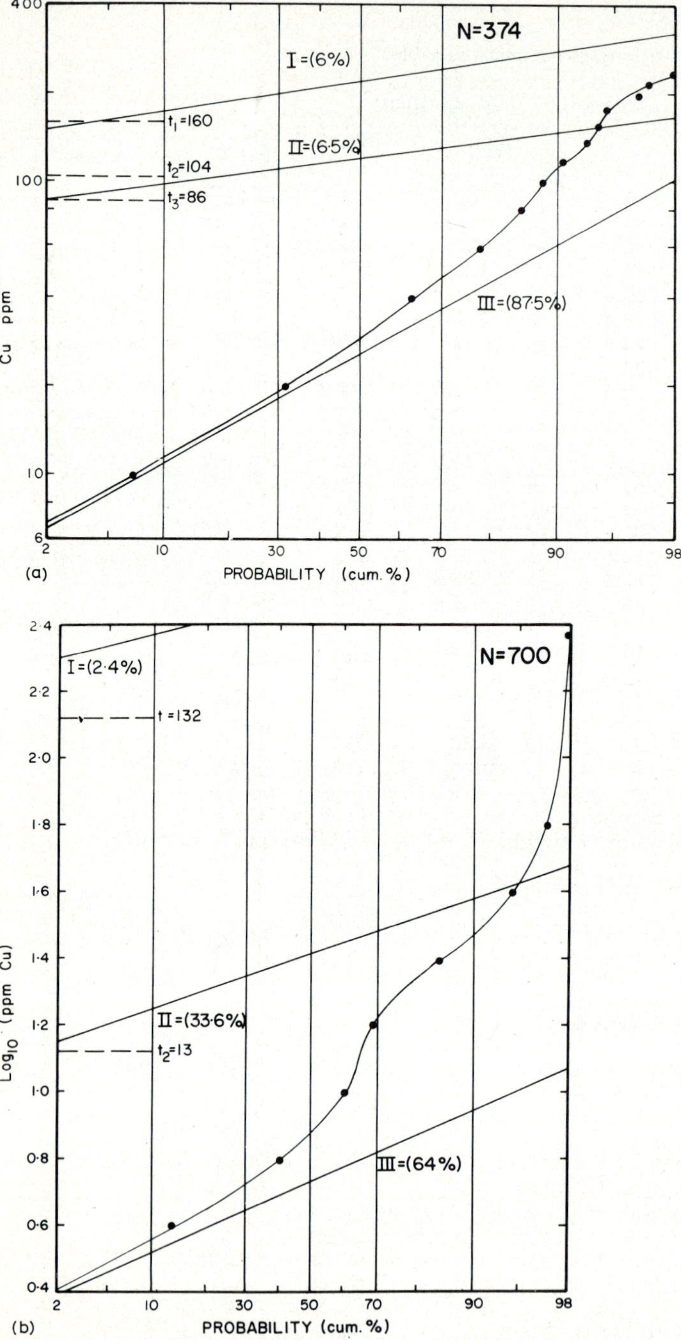

Fig.5. Probability plots: (a) B-horizon Cu data for 1966; (b) B-horizon Cu data for 1970. Each plot is partitioned into populations shown by straight lines. Thresholds (in ppm) separating populations are indicated by t_1, t_2, etc.

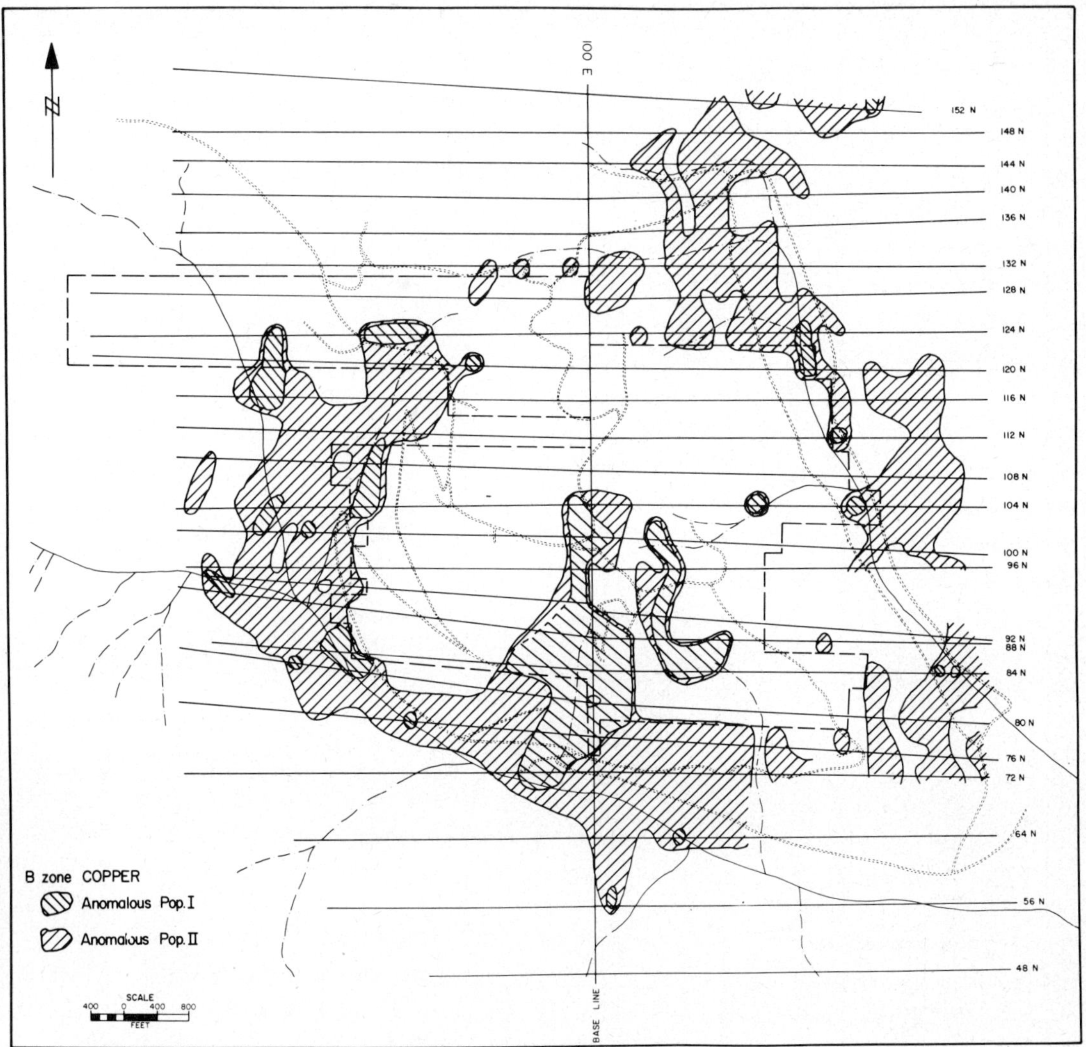

Fig.6. B-horizon Cu populations. Hachured areas show high and low anomalous populations. Isopleths are thresholds between populations (see Fig.5).

in the extension. This is probably the result of different analytical procedures. The two sets of data, however, are statistically similar. They are both made up of three populations: I — related to mineralization, mainly in the quartz monzonite boss; II — related to the pyrite halo; III — background. Fig.5a, b shows probability plots of B-horizon Cu data for original and extended areas respectively. Fig.6 is a geochemical plan combining both sets of data.

A-horizon Cu data is composed of two populations: I — a series of localized highs that correspond erratically to the pyrite halo; II — background. The third upper population corresponding to that for B-horizon Cu is missing,

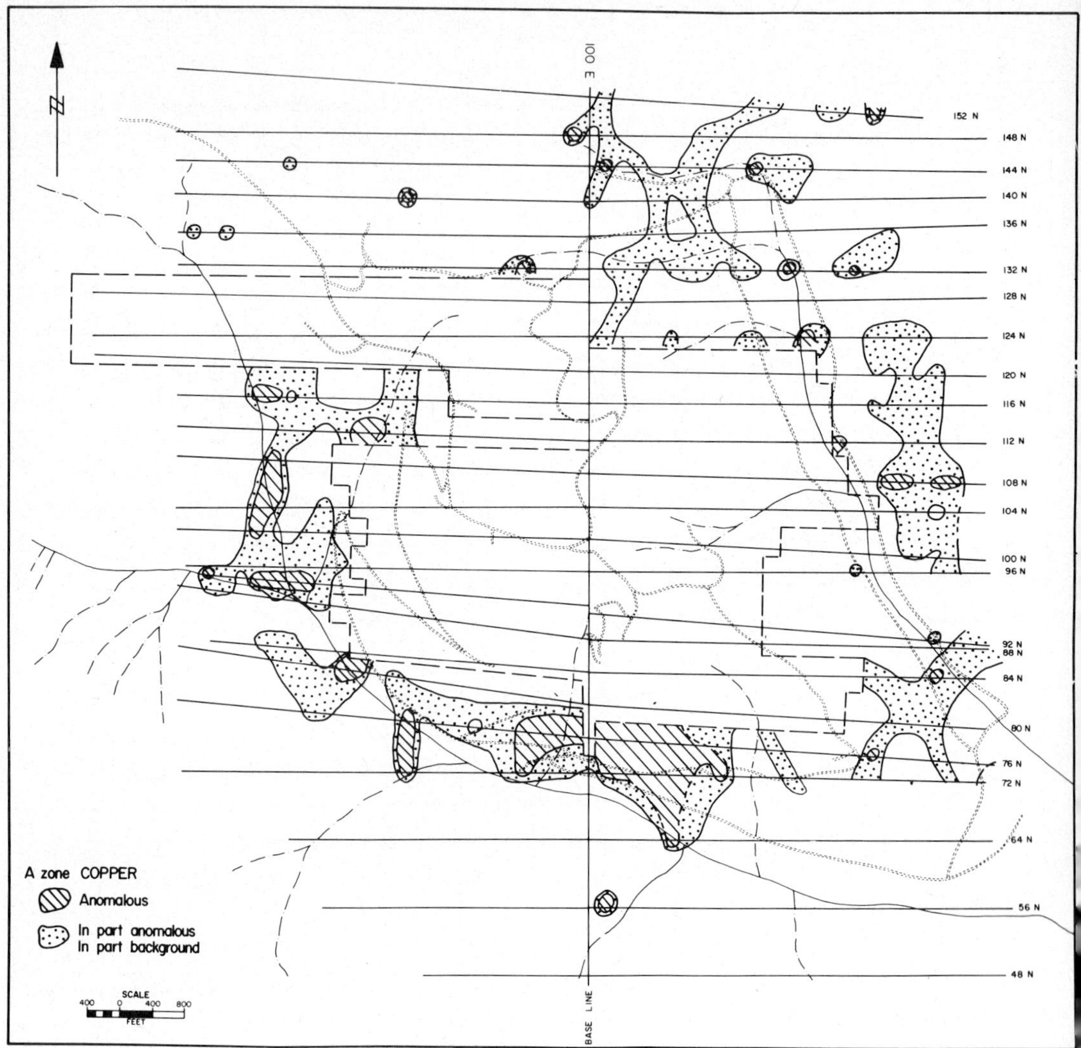

Fig.7. A-horizon Cu populations. Hachured area is anomalous. Dots show overlap of anomalous and background values. Isopleths are thresholds chosen from probability plots.

probably because the quartz monzonite boss is not exposed in the extended area for which A-horizon data are available. Fig.7 is a geochemical plan of A-horizon Cu populations. A comparison of A-horizon Cu with B-horizon Cu shows that good correlation exists between the two (see Table II) but that, in the type of environment found at Ashnola, the B-horizon Cu is diffused over a much larger area.

A limited area, mantled by overburden of unknown depth, was sampled for biogeochemical analysis. Second year growth from lodgepole pines were collected, ashed and analyzed for copper and zinc. Both data sets contain

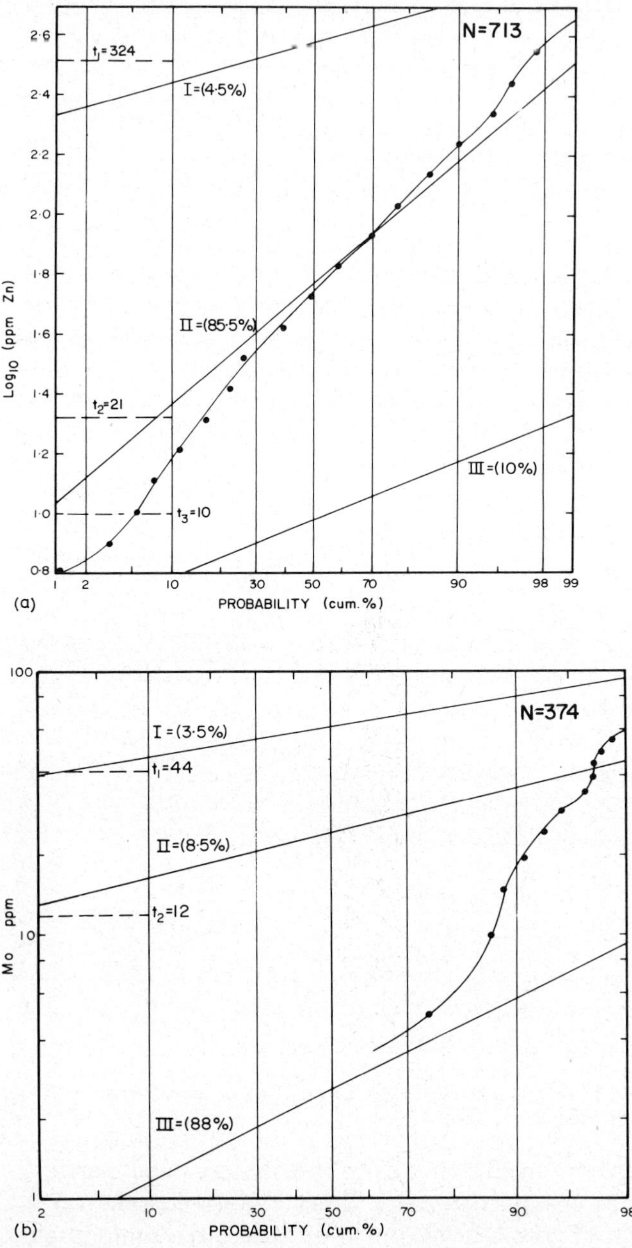

Fig.8. Probability plots: (a) A-horizon Zn; (b) B-horizon Mo, 1966. Each plot is partitioned into populations shown by straight lines. Thresholds (in ppm) separating populations are indicated by t_1, t_2, etc.

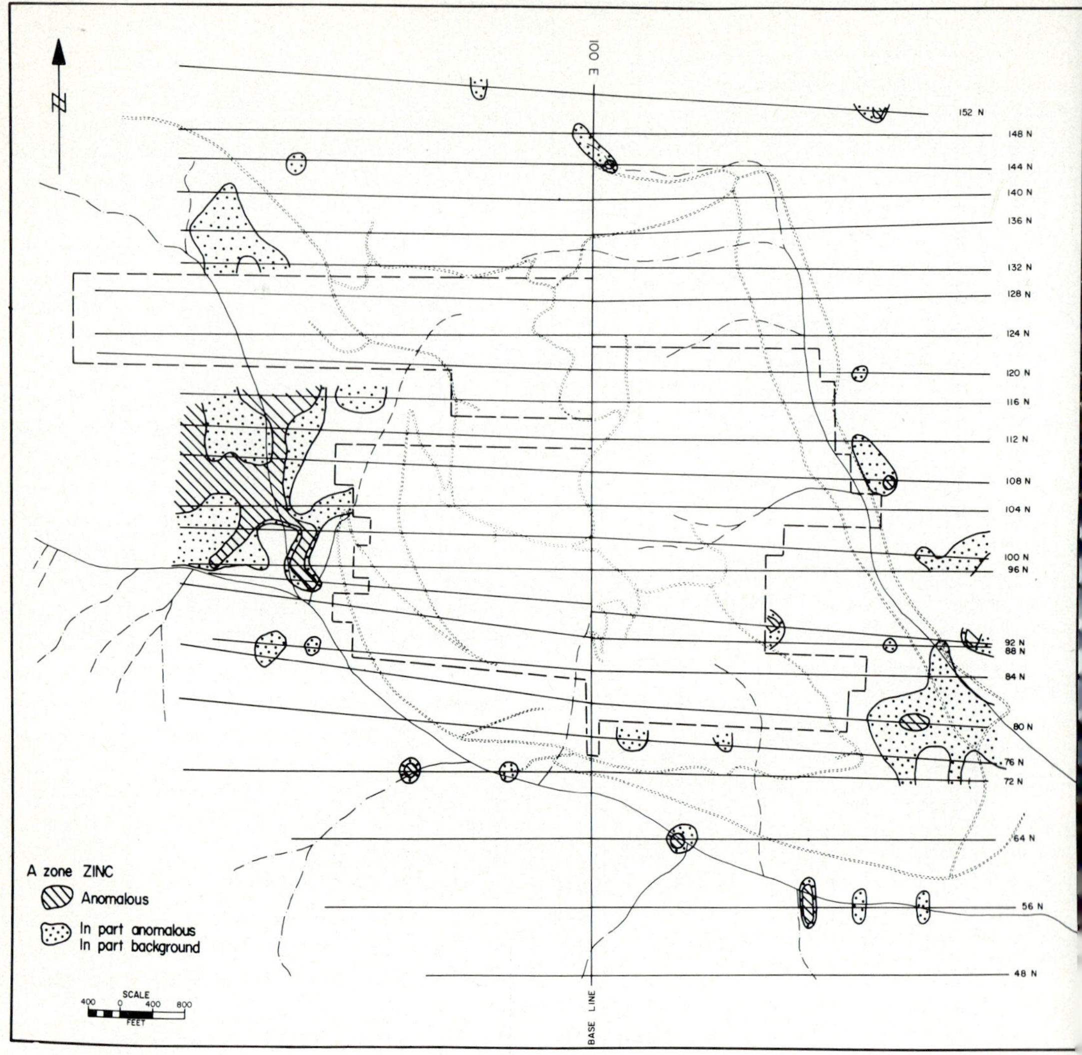

Fig.9. A-horizon Zn plan showing upper population (hachured) and range of overlap (dots) between upper and lower populations.

two populations that overlap somewhat; therefore threshold values, particularly those of copper, are not sharp. In general, there is good correlation between biogeochemical copper and A-horizon Cu but there are some unexplained differences. Figs.10 and 11 show the distribution of copper and zinc, respectively. There is no apparent correlation between biogeochemical copper and zinc.

A-horizon Zn, as shown in Fig.8a, is composed of three populations: I — peripheral to and partly overlapping (in some areas) the pyrite halo (population II in B-horizon Cu); II — background; III — also background but anoma-

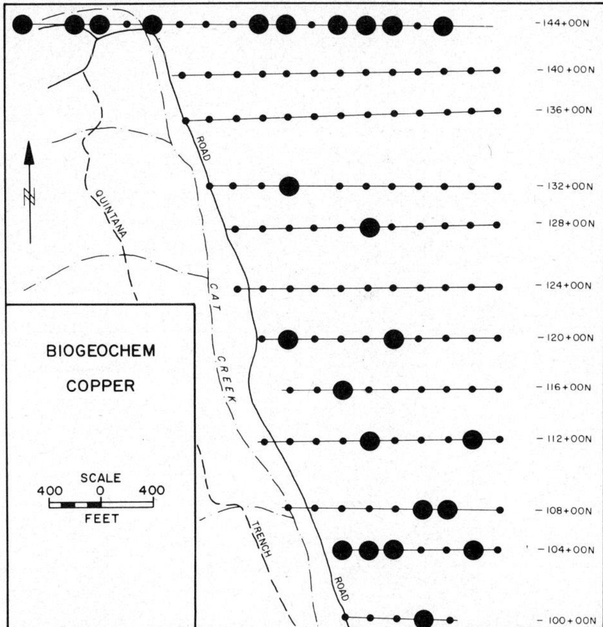

Fig.10. Biogeochemical copper. Large dots are upper population; small dots are lower population.

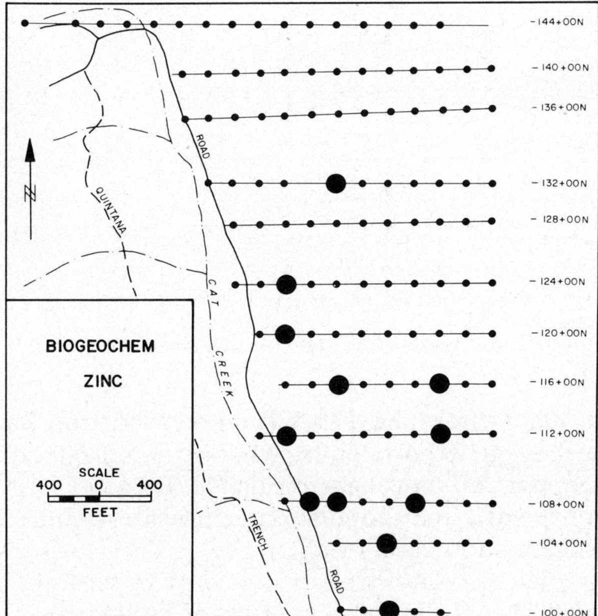

Fig.11. Biogeochemical zinc. Large dots are upper population; small dots are lower population.

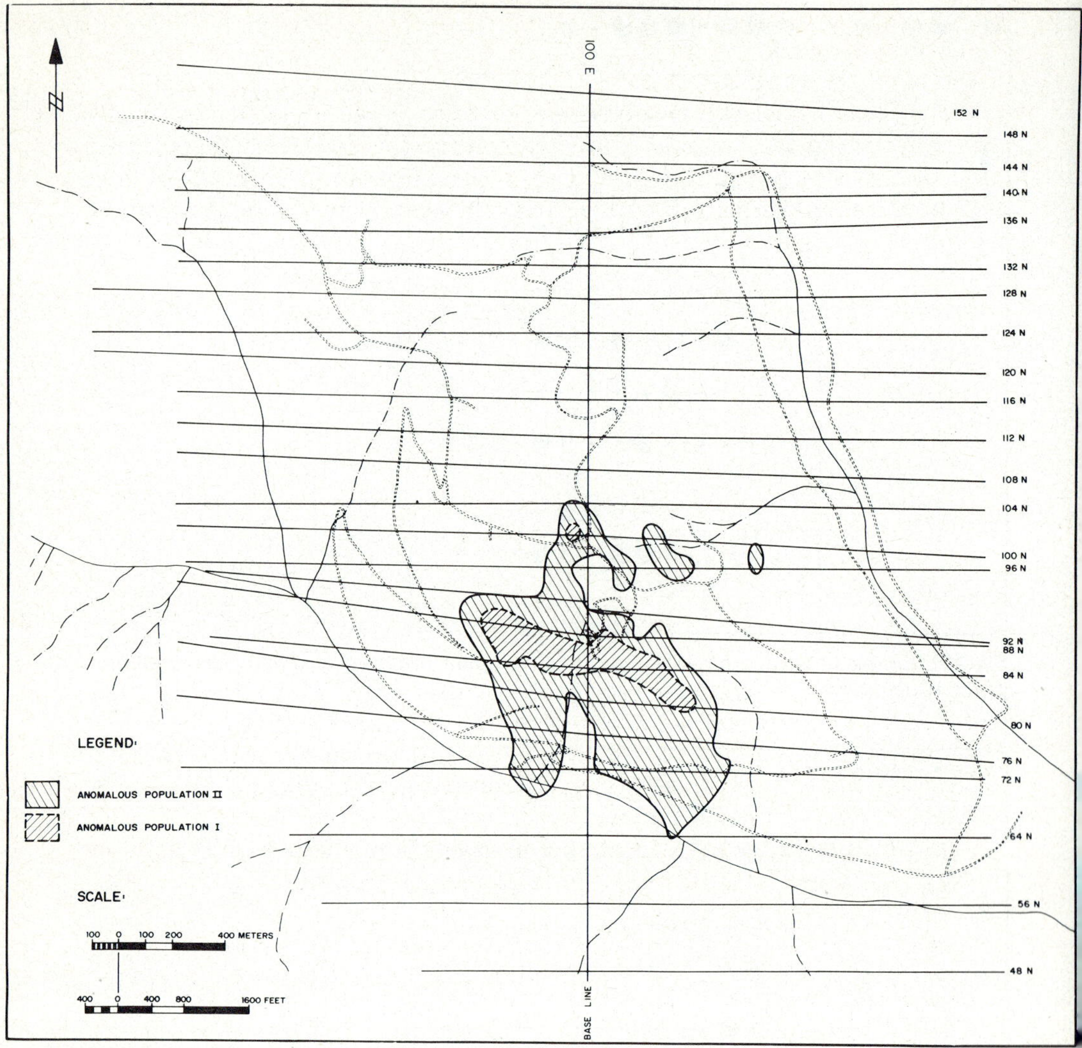

Fig.12. B-horizon Mo populations. Anomalous populations I and II (hachured) are surrounded by background values.

lously low for reasons unknown. Fig.9 shows the distribution of A-horizon Zn. A-horizon Zn and A-horizon Cu show good correlation, whereas no correlation exists between biogeochemical copper and biogeochemical zinc. Biogeochemical zinc and A-horizon Zn have different population compositions but their uppermost populations correlate moderately well.

B-horizon and A-horizon Zn values correlate strongly, not only in terms of calculated correlation coefficient (Table II), but also in terms of individual populations. A comparison of the geochemical plans of the two variables shows that good coincidence is present between A- and B-horizon Zn anomalies but that A-horizon anomalies are, in general, more abundant and larger than B-horizon anomalies.

B-horizon Mo values are composed of three populations (e.g. Fig.9b): I — related to in-place molybdenite mineralization; II — mechanical dispersion and dilution of I (active talus); III — background. A possible exception to the above is that several small scattered areas of population II could be related to molybdenite mineralization with an overburden cover. A geochemical plan for B-horizon Mo is shown in Fig.12. Visual examination suggests that A-horizon Mo correlates strongly with B-horizon Mo.

B-horizon Mo and B-horizon Cu also show good correlation. Comparison of Figs.6 and 12 shows that this correlation stems mainly from a high degree of coincidence of anomalous values of both variables on and near the mineralized quartz monzonite boss.

CONCLUSIONS

(1) The application of data analysis based on partitioning of cumulative probability plots has greatly facilitated the recognition of geochemical populations and their correlation with geological and geophysical information.

(2) A comparison of A- and B-horizon Cu analyses has shown that B-horizon data provided more complete information with respect to population definition with efficient thresholds and correlation with geological features. A horizon anomalies are more erratic and less dispersed than those of B horizon. Little of significance was added to B-horizon information by A-horizon data.

(3) Biogeochemical copper data contributed little additional information to that gained from B-horizon data.

(4) The concentration ratio of zinc in vegetation is very high relative to that of copper and there is no correlation between abundances of the two elements.

(5) A-horizon Zn anomalies are discontinuous and are peripheral to and partly overlapping the pyrite halo. B-horizon Zn data adds little to A-horizon Zn.

(6) B-horizon Mo values are significantly higher than those of A horizon and the threshold values are more useful.

ACKNOWLEDGEMENTS

The writers wish to thank Mr. A.L.J. MacDonald, President, Prism Resources Ltd., for permission to publish this case study of the Ashnola property. Costs of data analysis and manuscript preparation were offset partially by an operating grant (to A.J. Sinclair) from the National Research Council of Canada. Technical assistance from Mrs. Joan Mullen, Mr. M. Waskett-Myers, and Mrs. Sheila A. Montgomery is appreciated.

REFERENCES

Lowell, J.D. and Guilbert, J.M., 1970. Lateral and vertical alteration-mineralization zoning in porphyry ore deposits. Econ. Geol., 65: 373—408
Sinclair, A.J., 1974. Selection of threshold values in geochemical data using probability graphs. J. Geochem. Explor., 3: 129—149

SOIL CONDUCTIVITIES: ASSESSMENT OF AN ELECTROGEOCHEMICAL EXPLORATION TECHNIQUE

G.J.S. GOVETT

Department of Geology, University of New Brunswick, Fredericton, N.B. (Canada)

ABSTRACT

A study of the relation between geochemical and electrochemical processes is part of a
research programme at the University of New Brunswick to develop geochemical exploration techniques capable of locating deeply buried deposits. As a result of this work it has
been found that the electrical conductivity of water slurries of soil samples show anomalous
values in the vicinity of both sub-outcropping and buried mineralization.

The determination of soil conductivities is simple and rapid. 1 g of minus 80-mesh soil
is weighed into 100 ml deionized water and stirred for one minute; the conductivity is
then immediately measured by a dip-type conductivity cell and a conductivity bridge.

Soil slurry conductivity measurements and comparative trace element content are given
for B-horizon soils in the vicinity of a number of different sulphide deposits. These are
massive Zn—Pb deposits in northern New Brunswick and Spain, a massive vein-type antimony deposit in south-central New Brunswick, massive and disseminated Ni—Cu—Co
deposits in southern New Brunswick, and disseminated Zn—Pb deposits in northern New
Brunswick and Greece. The New Brunswick deposits have 3—120 ft of glacial debris overlying them; some of the Ni—Cu—Co deposits have a zone of barren rock above mineralization. The deposit in Spain has 3—5 ft of residual soil overlying it. Nevertheless, all these
varied deposits give an anomalous soil slurry conductivity response at least as good as and,
in some cases, better than the conventional geochemical trace element anomaly. Only the
deposit in Greece — which has a very thin layer of rocky residual soils over mineralization —
fails to give a marked response.

The form of a conductivity anomaly is not a simple peak. The form varies with depth
and type of mineralization (i.e., massive or disseminated); apparently it does not vary
with the mineralogical composition of the deposit. Thus, the response over near-surface
massive deposits is generally high-amplitude and short-wavelength peaks and troughs; over
disseminated and more deeply buried deposits the amplitude decreases and the wavelength
increases. There is an indication that anomalously high values persist for many thousands
of feet from mineralization.

Although these results represent only a preliminary investigation, they are sufficiently
encouraging to suggest that soil slurry conductivities could add another dimension to the
interpretation of geochemical data.

INTRODUCTION

During the course of research to develop geochemical techniques capable
of detecting deeply buried sulphide deposits it was found that quite marked
dispersion of elements occurs not only in transported overburden, but also in

102

post-mineralization rocks overlying ore deposits (Govett, 1972). It has been
proposed that the mechanism causing this type of secondary dispersion is at
least partly electrochemical and is influenced by electrical currents around a
sulphide deposit (Govett, 1973a); this view has received support from the
work of Logn and Bölviken (1973), and the concept is further discussed in
the paper presented at this meeting by Bölviken and Logn (1975, this volume).
Furthermore, it was reasoned that if dispersion of elements under the influence
of electrical currents is a continuing mechanism, there should be a greater
than usual proportion of free mobile ions of all types within the electrical
field around a sulphide body. It was concluded that a simple means of determining this would be to measure the conductivities of aqueous slurries of
samples taken from the overburden; measurements of soil slurry conductivities
from a single traverse over the Armstrong "A" deposit in northern New Brunswick did, in fact, show an intense anomaly associated with the orebody
(Govett, 1973a).

Subsequent extensive tank experiments have been undertaken in an
attempt to confirm the influence of electrochemical processes on secondary
dispersion (Govett and Whitehead, 1974). Conductivity measurements of
slurries of representative soils from the vicinity of a wide range of sulphide
deposits confirm that anomalous conductivities occur in the vicinity of sulphide deposits (Govett, 1974).

The form of the conductivity anomaly is not as simple as first reported for
Armstrong "A", as is illustrated in this paper which gives the results of conductivity measurements for a number of different sulphide deposits and
compares the responses with conventional trace element data. Tentative
empirical conclusions regarding the practical application of the technique to
mineral exploration are offered.

TECHNIQUES

All determinations were made on minus 80-mesh B-horizon soil samples.
A sample of 1.0 g was weighed into a tall-form 250 ml capacity beaker; 100 ml
of deionized water added and the sample stirred for one minute on a magnetic
stirrer. Measurement of the conductivity of the soil slurry was then made
immediately. The conductivity of each fresh batch of deionized water was
measured (normally this was about 100 micromhos), and the measurement
subtracted from the soil readings. Measurements were made with a Barnstead
Conductivity Bridge, Model PM-70CB, with a dip-type conductivity cell
having a cell constant of 0.1.

The conductivity increased with time, but compared with the one-minute
reading, measurements were only 10—20% greater after two hours and were
generally not much more than 50% higher after 24 hours. Insofar as the contrast between low and high readings did not systematically improve with
time, the great convenience of a one-minute stirring time and immediate
measurement was retained for all measurements.

Some hundreds of samples have been measured on two separate occasions; repeatability is regarded as good, and there is no significant difference between any two sets of readings at the 99% confidence level by the Kolmorgorov-Smirnov Nonparametric Test.

A series of 60 samples covering a conductivity range of 30×10^{-5} to 1000×10^{-5} mhos measured on 1.0 g sample in 100 ml deionized water was repeated using 1.0 g sample in 50 ml, 200 ml, 300 ml, and 500 ml water. There is a parallel straight line negative correlation between conductivity and amount of water used over the entire range; the contrast between low and high readings is therefore not affected by dilution.

The pH was determined on the soil slurry of 70 samples immediately after the conductivity reading was taken. Over a pH range of 4.5—5.6 and a conductivity range of 25×10^{-5} to 120×10^{-5} mhos, there was no discernable correlation between pH and conductivities.

DESCRIPTIONS OF AREAS AND RESULTS OF CONDUCTIVITY MEASUREMENTS

Conductivity measurements and corresponding trace element concentrations in B-horizon soils from seven different mineralized zones are shown in Figs. 1—6. Each of the seven mineralized areas is described briefly below. The

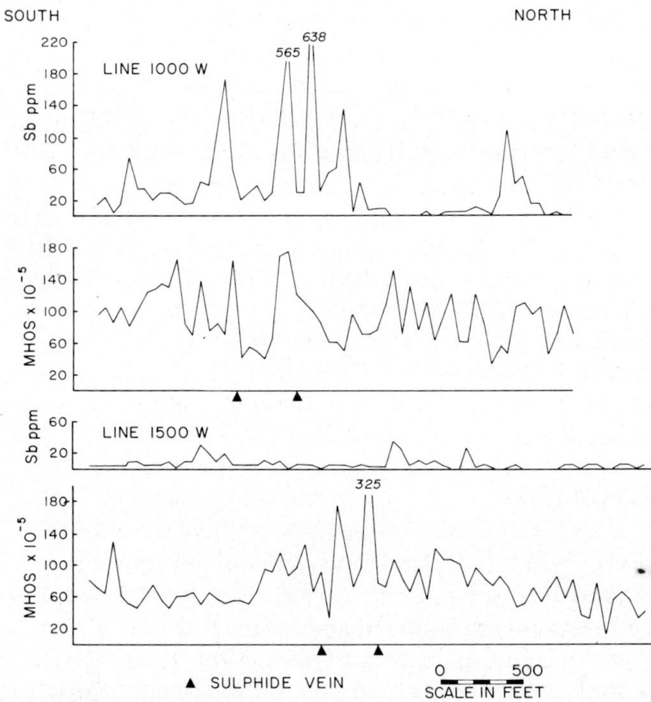

Fig.1. Distribution of antimony and conductivity in soils, Lake George (Sb values from Crosby, 1973).

104

sulphide deposits X and Y illustrated on Figs.8 and 9 are not described here
because details of their location and grade are confidential at present.

Lake George, south-central New Brunswick

Veins of massive stibnite and arsenopyrite dip about 45° north-northeast
within Silurian sandstone and shale. The grade of the deposit is not known,
but it is being actively mined at present (Crosby, 1973). The soils are podzols
and bedrock is 3—6 ft below the surface. The distribution of antimony and
soil conductivities are shown on Fig.1.

Armstrong "A", northern New Brunswick

The Armstrong "A" deposit consists of two massive sulphide lenses
separated by a narrow zone of disseminated pyrite. The lenses have a vertical
to steep easterly dip and lie within porphyroblastic quartz-feldspar chlorite
schist; there are green schists to the west and andesites to the east. These
rocks are pre-Silurian in age. The general grade of the deposit is 0.2—0.3% Cu,
0.6% Pb, and 2.2—3.2% Zn. The soils are podzols derived from glacial till and
are 3—5 ft thick.
The distribution of soil conductivities and lead are shown in Fig.2 (other
trace element data for the deposit are given in Pilch, 1970, and Govett, 1973b).

Rubiales, northwestern Spain

The Rubiales Zn—Pb deposit is in the Cantabrican Mountain range in
Galicia. The deposit appears to be essentially stratiform and occurs in a
siliceous zone within an alternating series of argillite, limestone, and quartzite
of Cambrian—Silurian age (Felder, 1972). There are no data available on the
grade of the deposit; however, it is reported that the content of zinc is four
to six times greater than that of lead, and chalcopyrite and pyrite are present
in minor amounts. The dominant soil type is a residual brown podzol (with
rendzina soils over limestones); the soils are 3—5 ft thick, and B-horizon samples
were taken from a depth of 1—2 ft. The distribution of soil conductivities and
of lead and zinc is shown in Fig.3.

St. Stephen, southern New Brunswick

Massive and vein-type Ni—Cu—Co mineralization occurs in gabbro and
norite, and disseminated mineralization occurs in periodotite. Argillaceous
and arenaceous sedimentary lenses of pre-Silurian age occur within the basic
intrusions. The grade of the more significant occurrences is about 0.9% Ni,
0.2% Cu, and 0.1% Co. The soils are generally podzols, although acid brown
forest soil profiles occur in places (Lahti, 1971). The parent material is un-

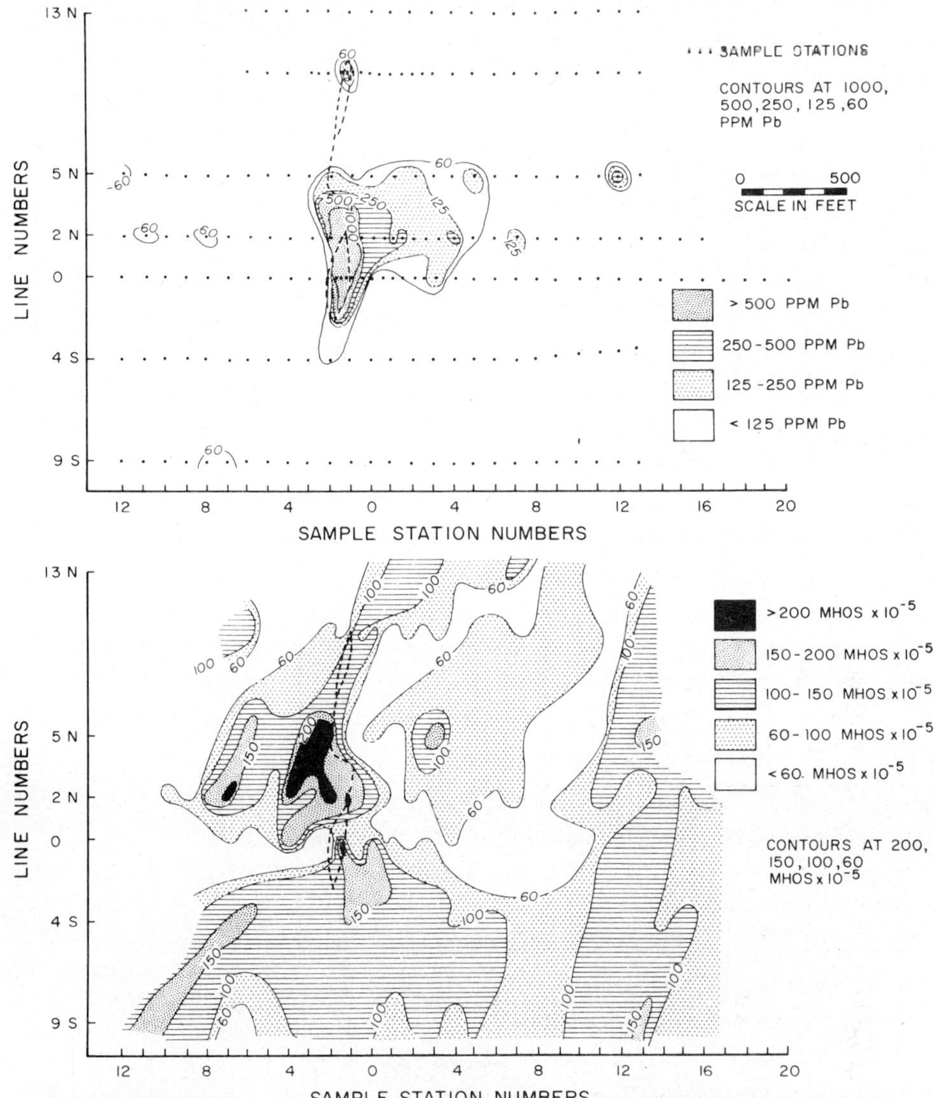

Fig.2. Distribution of lead and conductivity in soils, Armstrong "A" (Pb values from Pilch, 1970).

sorted glacial till of local origin, and there are beach and marine deposits interstratified with it at several localities.

The distribution of soil conductivities and the nickel content for three occurrences are shown in Figs.4 and 5. According to Mersereau (1969), the sulphides at the C-zone (Fig.4) have a total sulphide content of 33%, containing about 98,000 tons of 1% combined nickel and cobalt. The mineralization is vein-like and ranges from massive to disseminated. The eastern zone

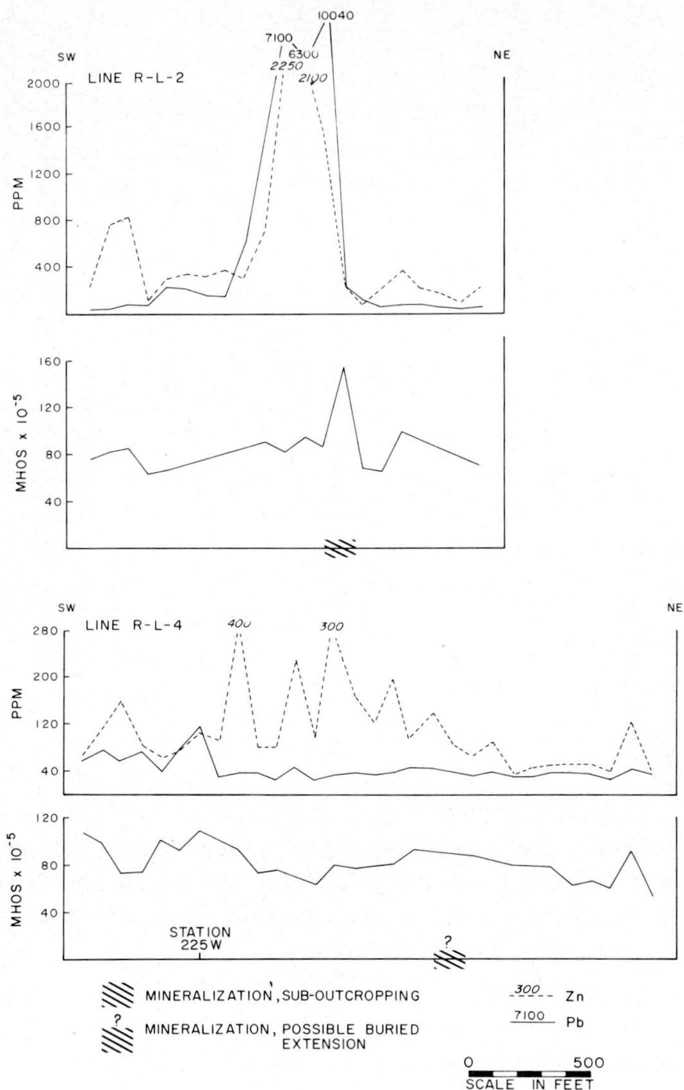

Fig.3. Distribution of lead, zinc and conductivity in soils, Rubiales (Pb and Zn values from Felder, 1972).

shown on the section on Fig.4 is partly sub-outcropping, whereas the top of the western zone is 150 ft below the surface. The thickness of overburden ranges from 3—12 ft.

The B-zone (Fig.5) is in anorthosite-peridotite and lies beneath 75 ft of glacial and fluvioglacial overburden; the sulphides are disseminated. There are reported to be 1400 tons per vertical foot of more than 0.5% combined nickel and copper.

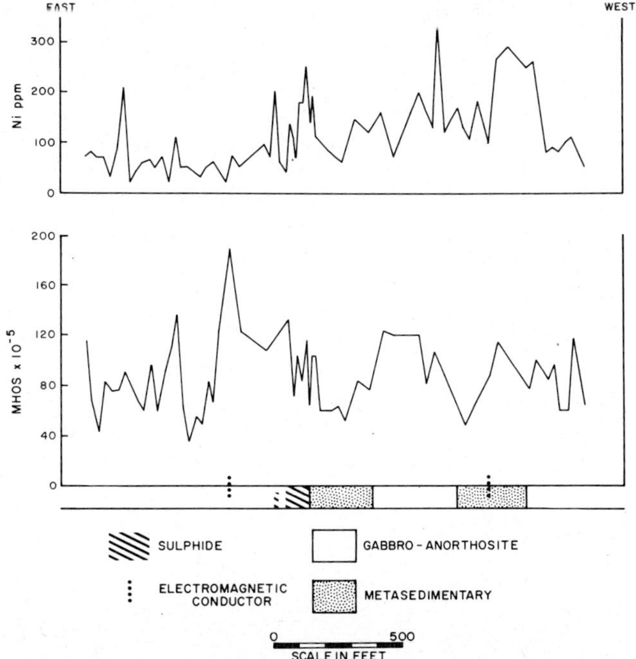

Fig.4. Distribution of nickel and conductivity in soils, St. Stephen C-zone (Ni values from Mersereau, 1969).

The K-zone (Fig.5) mineralization is very weak and not well defined by drilling. It is generally disseminated, containing 0.1—0.25% Ni and 0.1—0.5% Cu and is encountered in gabbro at depths of about 40—130 ft. The soils are podzols developed over glacial till which ranges from 10 to 20 ft to as much as 120 ft thick.

King Arthur, northeast Greece

The King Arthur mineralization in the Kirki area of northeastern Greece consists of disseminated sulphides along a tectonic contact between Tertiary andesite and Basement gneiss and pegmatite. There are about 30,000 tons of approximately 6.0% Zn, 4.0% Pb, and less than 1.0% Cu (Galanos, 1973). The area is a rugged upland region lying between about 300 and 2700 ft above sea level; the local relief at King Arthur is about 500 ft. The soils are residual podzolic, thin (8—12 inches), and stoney, with very poorly developed horizons. The distribution of lead, zinc, and soil conductivities along two traverses is shown on Fig.6.

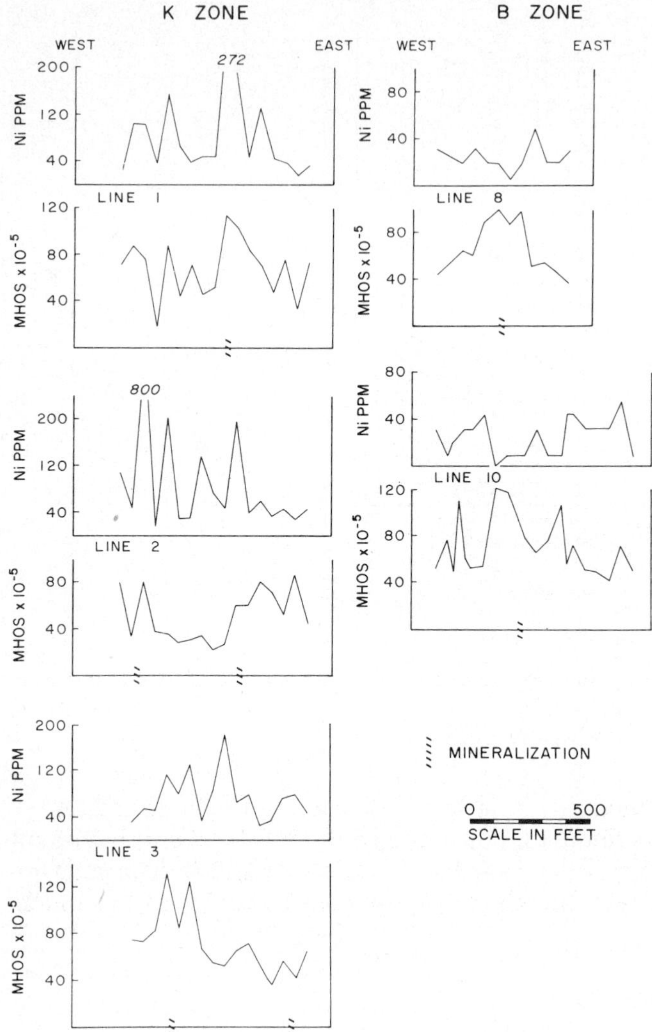

Fig.5. Distribution of nickel and conductivity in soils, St. Stephen K-zone and B-zone (Ni values from Lahti, 1971, and Mersereau, 1969).

DISCUSSION OF RESULTS

Interpretation of the data presented can only be empirical in the absence of a satisfactory explanation of the mechanisms responsible for the observed differences in conductivities. The ensuing discussion is an attempt to determine whether there are diagnostic variations of conductivity in soils which may be attributed to proximity to mineralization and, therefore, to determine whether there is at least prima facie evidence that soil conductivity measurements may form the basis of another mineral exploration technique.

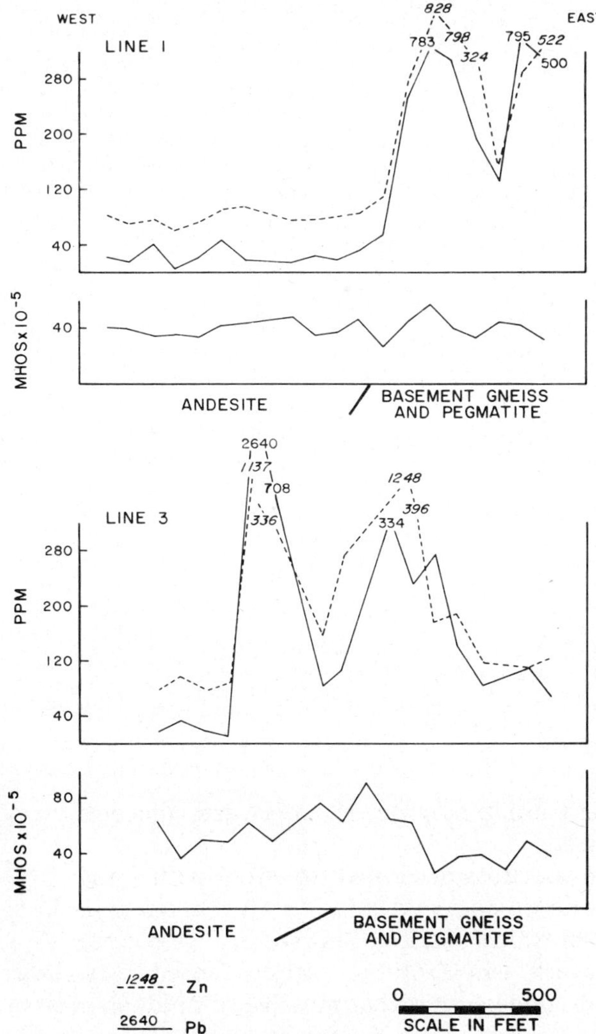

Fig.6. Distribution of lead, zinc and conductivity in soils, King Arthur (Pb and Zn values from Galanos, 1973).

The frequency distribution of conductivity measurements for all the data given here is shown in Fig.7. Despite the wide range of geological and soil conditions represented in the data, the spread of values is remarkably narrow; only 4.4% of the measurements exceed twice the upper limit of the modal class (Fig.7) and less than 6% are less than one-half the lower limit of the modal class. A cumulative percent plot of the frequency on arithmetic probability paper shows a distinct break in the distribution at about 70×10^{-5} mhos, with lesser breaks at 40×10^{-5} mhos and at 130×10^{-5} mhos. The relatively simple frequency distribution is in marked contrast with the

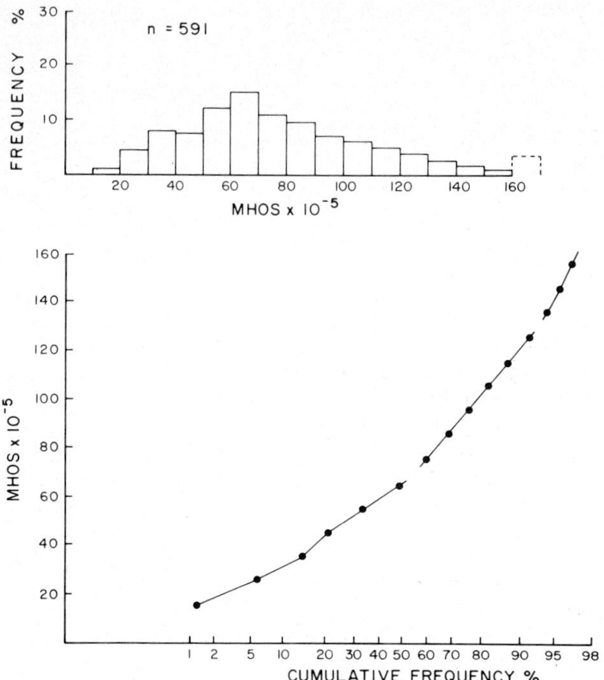

Fig.7. Frequency distribution and cumulative frequency of all soil conductivity measurements.

wide and apparently erratic variation in conductivity between adjacent samples along individual traverses.

Representative profiles from the areas described, together with some profiles from two other mineralized zones and some totally barren areas in northern New Brunswick, are shown as a series of curves in Figs.8 and 9. A base line of 70×10^{-5} mhos, rather than 0.0×10^{-5} mhos, is used to facilitate comparison of data. The salient conclusion is that massive sulphide deposits — Armstrong "A", Rubiales, Lake George, Zone X, and St. Stephen C-zone — are associated with narrow, well-defined peaks greater than 130×10^{-5} mhos, with adjacent well-defined troughs of less than 70×10^{-5} mhos and commonly less than 40×10^{-5} mhos (Fig.8). There is also an indication from Lake George and Rubiales that there is a very broad zone of greater than normal values in the region of specific anomalies; this is supported to some extent by the general level of conductivities along the three background traverses. The implication is that a regional conductivity anomaly exists of much greater dimension than the trace element anomaly, and most of the traverses illustrated here do not extend far enough away from mineralization to reveal this.

Difficulties of interpretation arise in the case of disseminated, low-grade, or deeply-buried mineralization (Fig.9). Thus, background traverse C is un-

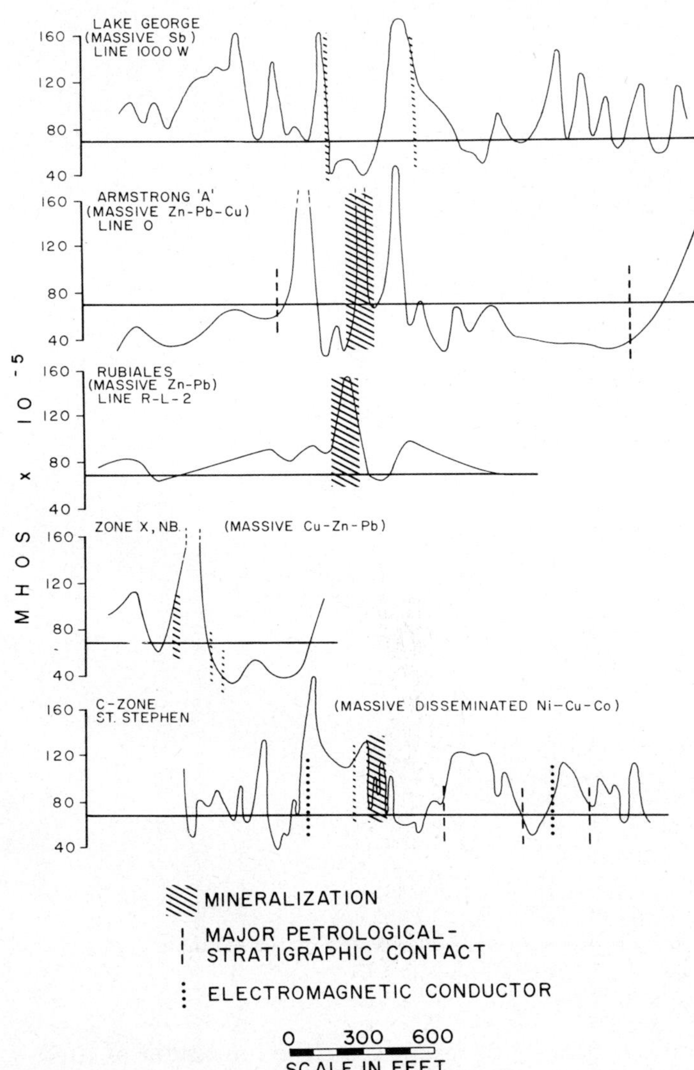

Fig.8. Comparative conductivity profiles for soils over massive sulphide deposits.

ambiguous, and even the fluctuations along background traverse B are unlikely
to be confused with any of the anomalous traverses shown (except King
Arthur, which had a very poor response). However, in background traverse A
the pattern of conductivities attributed to a major change in lithology and
even the variation in the vicinity of the river are remarkably similar to line 10
of the B-zone at St. Stephen and to line 3 of the K-zone at St. Stephen, as
well as to the disseminated and discontinuous mineralization of Zone Y in
northern New Brunswick. Regrettably, the traverses at both locations at St.
Stephen and at Zone Y are too short to determine whether the pattern in the

112

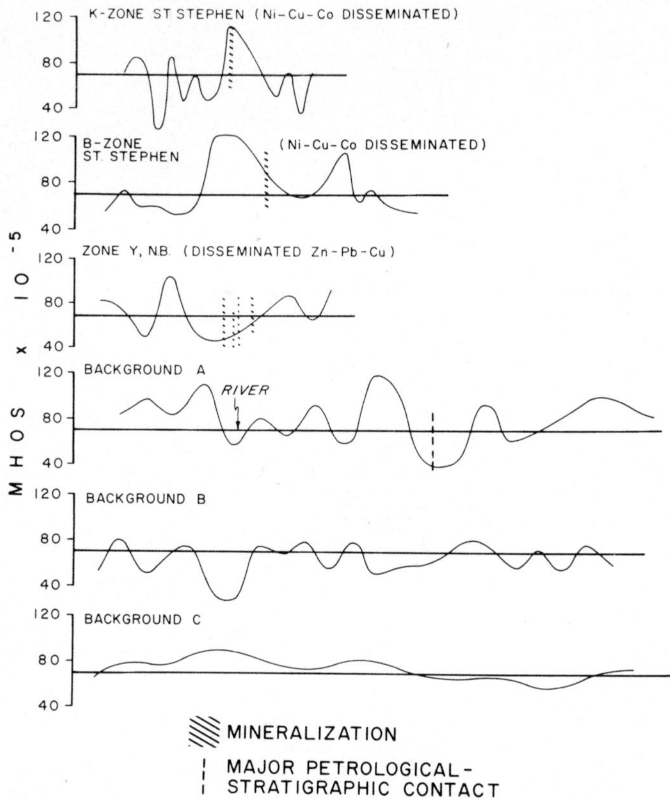

Fig.9. Comparative conductivity profiles for soils over disseminated sulphide deposits and background areas.

vicinity of mineralization is distinctly different from the remainder of the surrounding area.

DETAILED INTERPRETATIONS

Lake George

The conductivity measurements across the Lake George antimony deposit demarcate the mineralized zones (Fig.1). It is of interest that conductivities most clearly define the sulphide veins on traverse 1500 W where there is no response of Sb in the soils. Lesser peaks along the traverses may be related to changes in lithology, but there are no geological data to confirm this.

Armstrong "A"

The contoured conductivities shown in Fig.2 clearly indicate that there is no simple relation between conductivity and mineralization; there is, however, a fairly pronounced north—south lineation which is parallel to the main structure. The conductivities are more readily interpreted from profiles; a

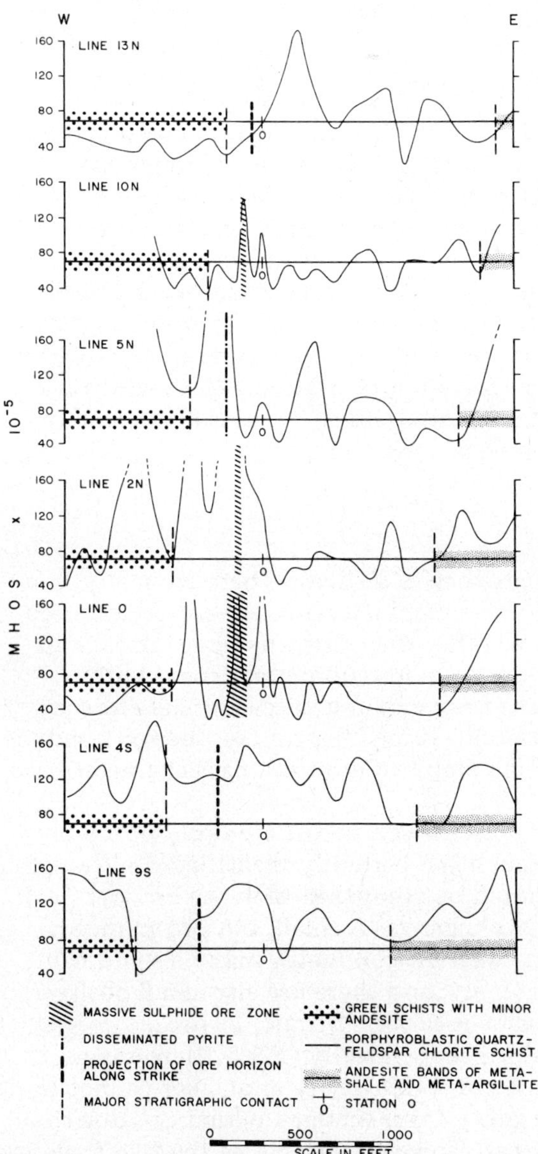

Fig.10. Comparative conductivity profiles in soils across Armstrong "A" deposit.

114

series of east—west profiles are shown in Fig.10. The massive sulphides are
demarcated by high contrast peaks and troughs, whereas most major stratigraphic breaks are marked by a broad trough and adjacent broad peaks. There
is an indication that the mineralized horizon may be identified by broad
positive anomalies both north and south of the deposit. There is inadequate
geological control available to the writer to substantiate this; moreover, soil
sampling should be extended at least 1000 ft in all directions to determine
whether values decrease to a lower background level.

Rubiales

Traverse R-L-2 over sub-outcropping mineralization is quite unambiguous;
there is a strong positive anomaly associated with an adjacent trough over
mineralization. Traverse R-L-4, which is believed to be over buried mineralization (Felder, 1972) does not show a recognizable anomaly over the presumed
position of mineralization (Fig.3). The distribution of lead and zinc also does
not support the presence of mineralization at the location indicated. Based
on the fact that on traverse R-L-2, lead exceeds zinc only over mineralization,
and that this is also true at Armstrong "A" and elsewhere (Govett, 1973b), it
is arguable that the location of mineralization is at Station 225W, which is
the only station where lead values exceed zinc values. There is also a small
conductivity anomaly here associated with a marked trough.

St. Stephen

Interpretation of results from St. Stephen is difficult: there are many thin
zones of mineralization; only a few of the many electromagnetic conductors
have been investigated to determine whether they carry mineralization; and
there are frequent occurrences of lenses of metasedimentary rocks within
ultrabasic intrusions. The C-zone is the only example discussed here of sub-
outcropping mineralization which contains some massive ore; the soil conductivity pattern of short-wavelength—high-amplitude variations is characteristically
anomalous.

The known sulphide zone and the western conductor are fairly positively
identified; the eastern conductor is much less certainly identified — although
the sharp peak and trough (Fig.4) should be compared with the broader peak
and trough associated with a lithological change. A small, but pronounced,
soil conductivity anomaly west of the western conductor may be significant;
it is associated with a small nickel anomaly, and there are also small positive
copper and cobalt anomalies at the same location (Govett, 1973b).

The buried B-zone is not identifiable geochemically (Fig.5, Mersereau,
1969). On the data available the small soil conductivity anomalies cannot be
discriminated with any degree of certainty from similar fluctuations due to
lithological changes in background areas. However, neither of the two traverses
illustrated in Fig.5 extend far enough from mineralization to properly assess

the results. It is possible that in an area of such thick drift cover background is low and that lithological changes in bedrock are not so positively reflected in surface soil conductivities; in this case, the anomalies would be clearly identifiable.

The mineralization at the K-zone appears to be insignificant, but it is of interest that soil conductivities respond at least as well as the distribution of nickel in soils. The eastern conductor on line 2 (Fig.5) is not reflected by a conductivity anomaly — but nickel does not respond either. Given the nature of the mineralization, the rock is quite possibly barren here. As with the B-zone, the samples do not extend far enough away from the mineralized zone to allow a precise assessment of the conductivity pattern.

King Arthur

There is remarkably little variation in soil conductivities at King Arthur which, if compared directly with conductivity data from New Brunswick, clearly indicate background conditions. It should be noted that at King Arthur the soil is extremely thin and poorly developed, with a high proportion of bedrock fragments indicating relatively little chemical weathering compared with physical erosion.

King Arthur is the only example so far found which does not give a soil conductivity response where there is a trace element response. Since the only unique feature of the King Arthur deposit compared with the others cited is the extremely thin soil cover, the lack of soil conductivity response may be attributed to the same factors responsible for the thin soil. It is of interest that the range of fluctuation of conductivity away from the mineralized zone is only about one-fifth that in the other examples; a five-fold expansion of the conductivity scale gives a profile quite similar to anomalous areas in New Brunswick (Fig.11).

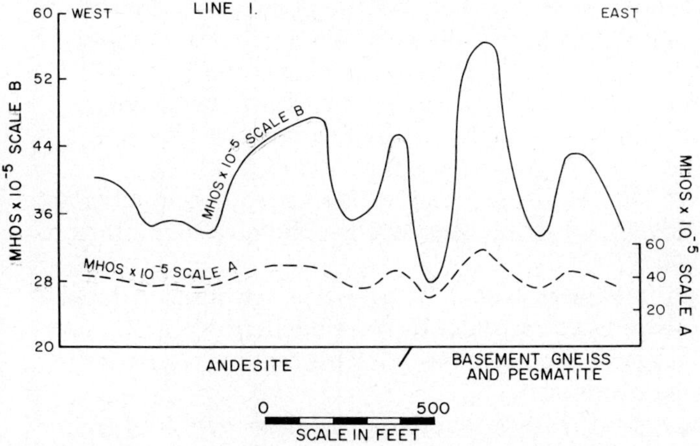

Fig.11. Effect of scale on conductivity profile in soils, King Arthur.

SUMMARY AND CONCLUSIONS

The main practical conclusions of this study are:

(1) Near-surface, massive sulphide deposits cause high-amplitude—short-wavelength soil conductivity responses.

(2) The amplitude tends to decrease and the wavelength tends to increase with increases in depth of the sulphides and also as the sulphides become less massive in character.

(3) Major lithological changes tend to give a response which, with present data, is difficult to discriminate from the response due to disseminated or deeply buried deposits.

Whereas highly suggestive trends may be discerned from the soil conductivity measurements presented here, there are not enough data to substantiate and classify the responses. Detailed studies must be made over "type" ore deposits and over large regions around them; a high degree of geological control is essential for such studies. Furthermore, concomitant detailed multi-element geochemical measurements and electrical geophysical measurements should be made. Similarly detailed studies are required in a variety of climatic, pedological, and physiographic situations.

The determination of conductivities on soil slurries is a measurement of *all* ions which are released into water under the conditions of the test. It is not surprising, therefore, that there is little consistent correlation between the conductivity measurements and trace element distribution. A fundamental problem which requires solution before significant progress can be made in the practical application of soil conductivity measurements to exploration is *why* there are greater than, and less than, normal quantities of ions which are water-soluble in the vicinity of mineralization and, to a lesser extent, in the vicinity of lithological contact zones.

Soil slurry conductivity measurements are in some sense analogous to the cold extractable metal tests. The advent of specific ion electrodes and flameless atomic absorption spectrophotometry allows the possibility of measuring particular water-soluble metal ions in soil slurries; this is being investigated as a possible exploration technique. (Since this paper was written in July, 1973, a joint research programme with Barringer Research Ltd., sponsored by a NRC-PRAI grant, has commenced on many of the matters cited here as requiring investigation.)

The data presented here indicate four possible uses for soil conductivity measurements which, if developed, could improve geochemical exploration. These are:

(1) Soil conductivity anomalies appear to be present over massive deposits even in situations where there is no response in trace elements (e.g., line 1500 W over the Lake George antimony deposit). This feature would be useful in tracing extensions of deposits.

(2) There is a different type of response in conductivities over low-grade disseminated deposits compared to massive deposits under similar circum-

stances. This difference is not always obvious from trace element data (see Govett, 1973b). Soil conductivity measurements can assist in the discrimination between these two types of deposits from surface data.

(3) There are indications that deposits which are quite deeply buried and which give a poor or no geochemical response are associated with anomalous soil conductivity patterns (e.g., B-zone at St. Stephen). Work needs to be done to develop the technique for locating such deposits.

(4) There are also indications that higher than normal conductivities are associated with sulphide deposits in soils many thousands of feet from near-surface mineralization. This feature is potentially useful for regional-scale soil exploration.

It is concluded that there is sufficient evidence to suggest that soil conductivity measurements may provide a useful interpretative adjunct to soil geochemical data and offer the possibility of detecting deeply buried deposits which have no conventional geochemical trace element response.

ACKNOWLEDGEMENTS

Acknowledgement is due to D.A. Galanos, F. Felder, H.R. Lahti, T.G. Mersereau, and P.G.H. Pilch for use of soil samples they collected as part of their M.Sc. studies at the University of New Brunswick. Acknowledgement is also made of the National Research Council of Canada Grant A5585 and the Geological Survey of Canada Grant 17-68. Heath Steele Mines are thanked for supplying some soil samples from northern New Brunswick.

Sincere thanks are given to Mrs. M.H. Govett for carrying out all the soil conductivity measurements and to Mr. R. Phillips of the University of New Brunswick for preparing the diagrams.

REFERENCES

Bölviken, B. and Logn, Ö., 1975. An electrochemical model for element distribution around sulphide bodies. In: I.L. Elliott and W.K. Fletcher (Editors), Geochemical Exploration 1974. Elsevier, Amsterdam, pp.631—648

Crosby, R.M., 1973. Distribution of some trace metals in the secondary environment in southwestern New Brunswick. M.Sc. Thesis, University of New Brunswick, Fredericton, N.B.

Felder, F., 1972. Trace element dispersion patterns associated with lead—zinc mineralization in northwestern Spain. M.Sc. Thesis, University of New Brunswick, Fredericton, N.B.

Galanos, D.A., 1973. The distribution of Cu, Pb, Zn, Ni, Co, Mn, Fe in soils, rocks and stream sediments at Kirki area in northern Greece. M.Sc. Thesis, University of New Brunswick, Fredericton, N.B.

Govett, G.J.S., 1972. Interpretation of a rock geochemical exploration survey in Cyprus — statistical and graphical techniques. J. Geochem. Explor., 1: 77—102

Govett, G.J.S., 1973a. Differential secondary dispersion in transported soils and post-mineralization rocks: an electrochemical interpretation. In: M.J. Jones (Editor), Geochemical Exploration 1972. Institution of Mining and Metallurgy, London, pp.81—91

Govett, G.J.S., 1973b. Geochemical exploration studies in glaciated terrain of New

Brunswick, Canada. In: M.J. Jones (Editor), Prospecting in Areas of Glaciated Terrain. Institution of Mining and Metallurgy, London, pp.11—24

Govett, G.J.S., 1974. Soil conductivity measurements: a technique for the detection of buried sulphide deposits. Trans. Inst. Min. Metall., 83: B29—B30

Govett, G.J.S. and Whitehead, R.E.S., 1974. Origin of metal zoning in stratiform sulfides: a hypothesis. Econ. Geol., 69: 551—556

Lahti, H.R., 1971. Factors contributing to secondary dispersion of trace elements in glacial soils, St. Stephen area, New Brunswick. M.Sc. Thesis, University of New Brunswick, Fredericton, N.B.

Logn, Ö. and Bölviken, B., 1973. Self potentials at the Joma pyrite deposit, Norway. Geoexploration, 12: 11—28

Mersereau, T.G., 1969. Secondary dispersion of the metals nickel, cobalt and copper near the St. Stephen gabbro, Charlotte County, New Brunswick. M.Sc. Thesis, University of New Brunswick, Fredericton, N.B.

Pilch, P.G.H., 1970. Dispersion of some trace elements in soils of glacial origin near the Armstrong "A" sulphide deposit. M.Sc. Thesis, University of New Brunswick, Fredericton, N.B.

GEOCHEMICAL EXPLORATION IN SARDINIA

H. HEETVELD and S. PRETTI

Bureau de Recherches Géologiques et Minières, Paris (France)
Ente Minerario Sardo, Cagliari, Sardinia (Italy)

ABSTRACT

Geochemical exploration is being carried out in Sardinia as part of a research programme
to determine the mineral potential of this island with a centuries-old mining tradition. This
paper deals with the field methods which had to be adapted to the particular conditions
of a heavily explored country. Analytical techniques are discussed, as well as the processing
by computer of all geochemical data. A short description is given of some interesting
results.

Application of a systematic geochemical and alluvial sample grid, covering all geological
formations appearing to present mineral potential, seems to lead to interesting discoveries
and opens new horizons for old mining districts.

INTRODUCTION

Sardinia is an island with a centuries-old mining tradition. In the past,
however, exploration has been concerned nearly exclusively with sulphide
minerals of lead, zinc, copper, silver and iron and their associated oxidation
products. Not until very recently has attention been given to other substances
like antimony, manganese, aluminium, coal, kaolin, clay, barite and fluorite.
Furthermore, although indications of tungsten, nickel, cobalt, molybdenum
and tin have been known for some time, related exploration has been only
sporadic. No research whatsoever has taken place on substances like titanium,
zirconium and rare-earth minerals, whose presence is quite compatible with
the metallogenetic context of the island.

The existence of all these unknown factors in the mining potential of
Sardinia has led the local government to create a public organization, Ente
Minerario Sardo (EMSa), charged with exploring and developing all potential
mineral resources of the island (Salvadori, 1973).

As can be seen on the flow-chart for this programme (Fig.1), the first stage,
now completed, consisted of assembling all available data on geology,
metallogeny, location and nature of mineral outcrops and, in particular, on
the exploration results for all types of mineral occurrences. All the information
obtained has been recorded, analysed, and prepared for computer processing.
A synthesis of the data, compiled on geo-mineral provinces, makes it possible

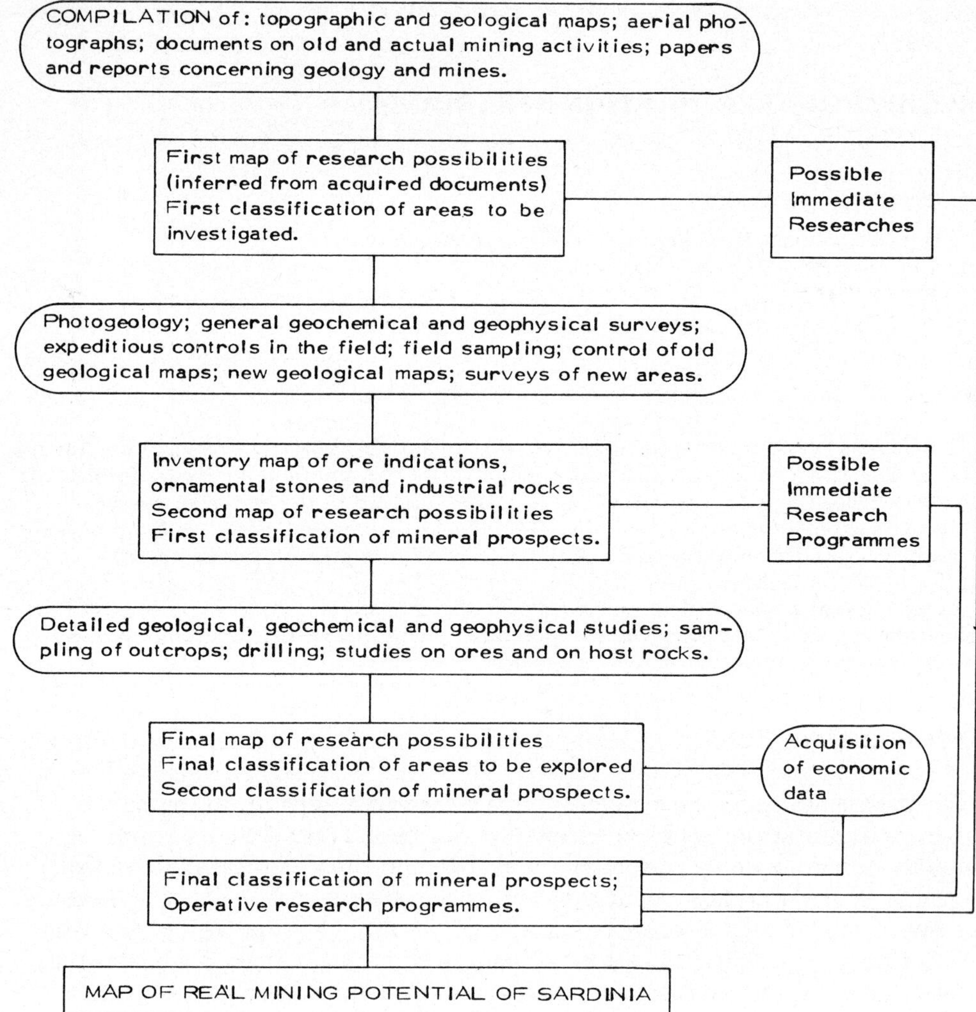

Fig.1. Outline of general research programme.

to outline with reasonable precision the various mineral provinces, to infer their respective metallogenetic characteristics and to estimate their potential for future development.

This exhaustive documentation clearly reveals extensive knowledge of certain types of mineral deposits, whereas for other occurrences information is lacking. It emphasizes also the disparity between extensively explored areas and other parts of the island, where for several reasons mineral research has been negligible.

This situation obviously requires application of an exploration technique capable of supplying in a reasonably short time adequate information on the distribution of a great number of elements and substances. Geochemical and alluvial prospecting have been selected for this, the second stage of the EMSa programme. This strategic geochemical programme is now being carried out by ITALCONSULT of Rome and the Bureau de Recherches Géologiques et Minières (B.R.G.M.) of France, under the supervision of the EMSa geological staff.

SELECTION AND GEOLOGICAL CONTEXT OF EXPLORATION AREAS

Selection of the areas to be included in the geochemical programme has been based on geological and metallogenetic considerations (Fig.2). The formations with a proven or a theoretical mineral potential can be grouped into four units (Cocozza et al., 1974):

(1) The pre-Permian sedimentary formations, metamorphosed to varying degrees: principal mineral occurrences Pb, Zn, Cu, Ag, Fe, W, Ba, F (Ni, Co, Mo).

(2) Igneous rocks of the Caledonian and Hercynian cycles: principal mineral occurrences Sn, W, Mo (Ba, F, Pb, Zn).

(3) The base of the Mesozoic formations: principal mineral occurrences: Pb, Zn (indications only).

(4) The volcanic series of the Alpine cycle, mainly of trachytic and andesitic nature: principal mineral occurrences Cu, Mn, Pb, Zn (Ba, F) (generally merely indications).

The remaining formations — Mesozoic calcareous rocks, Tertiary continental sediments and basalts, Quaternary sediments — have been excluded from this study, as they apparently lack mineralization of any importance. This selection reduces the exploration area from about 24,000 km² to approximately 13,500 km².

In organizing the programme the following factors had to be considered: (a) speed of operation (the whole study to be completed in approximately two years); (b) flexibility of methods depending on geology and the type of mineralization expected, and (c) a sample pattern combining accuracy and expediency.

In view of these considerations the exploration areas were divided into two main groups on the basis of geology (Fig.3). Areas underlain by formations 1, 3 and 4 and the extrusives of formation 2, covering approximately 6900 km², were investigated by stream sediment geochemistry. Sediments were analysed for 11 elements, of which ten are common to all samples (Pb, Zn, Cu, Ag, Co, Ni, Sb, Mo, Cr, Mn), while the eleventh is either mercury or fluorine depending on mineral associations. In areas of granitic intrusives (6600 km²) alluvial prospecting, involving mineralogic and chemical analysis of heavy-mineral concentrates, was employed.

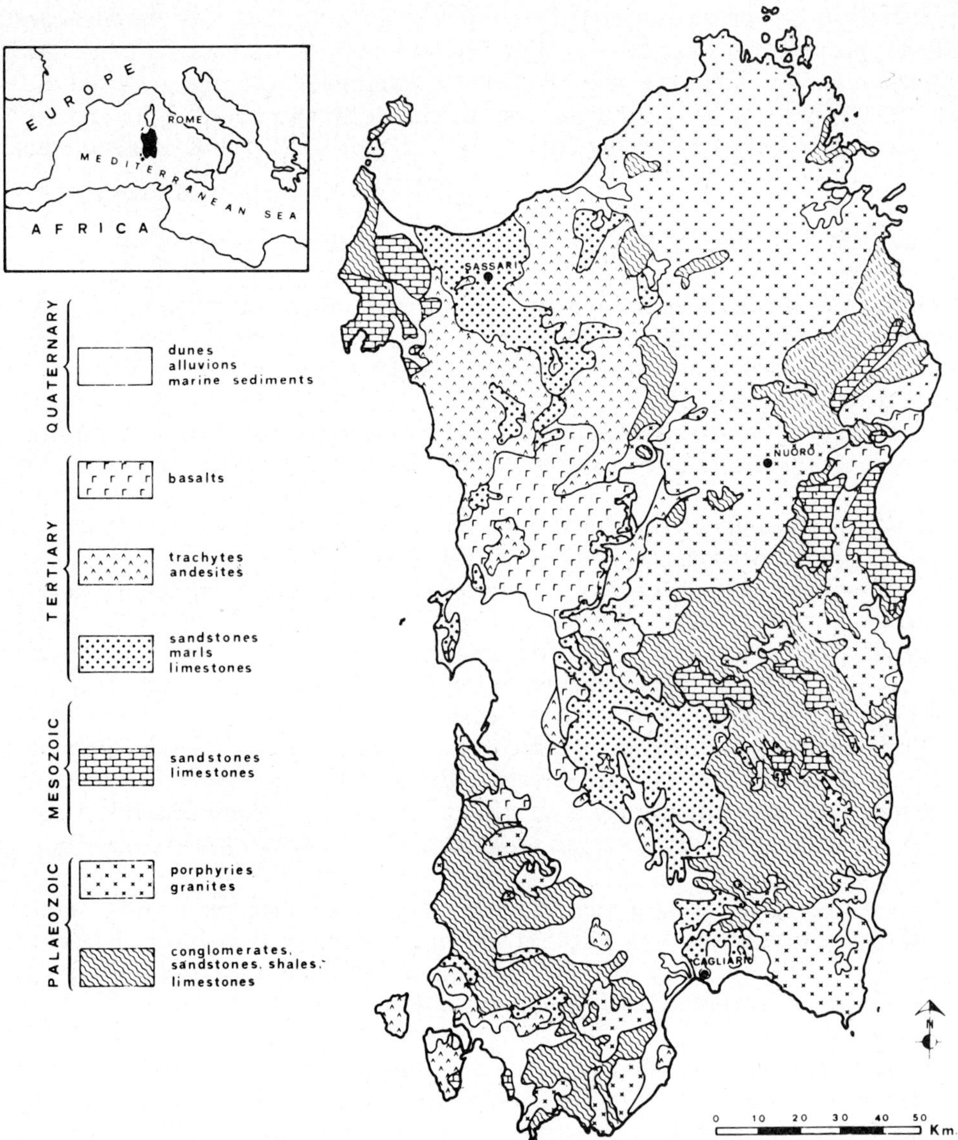

Fig.2. Geological map of Sardinia.

FIELD METHODS

Taking into consideration topography, vegetation and prevailing climatic conditions, it has been found that a density of 3—5 stream sediment samples per km^2 is adequate to obtain complete coverage of an area. Sample density is kept fairly flexible to allow the field teams to adapt the pattern to local

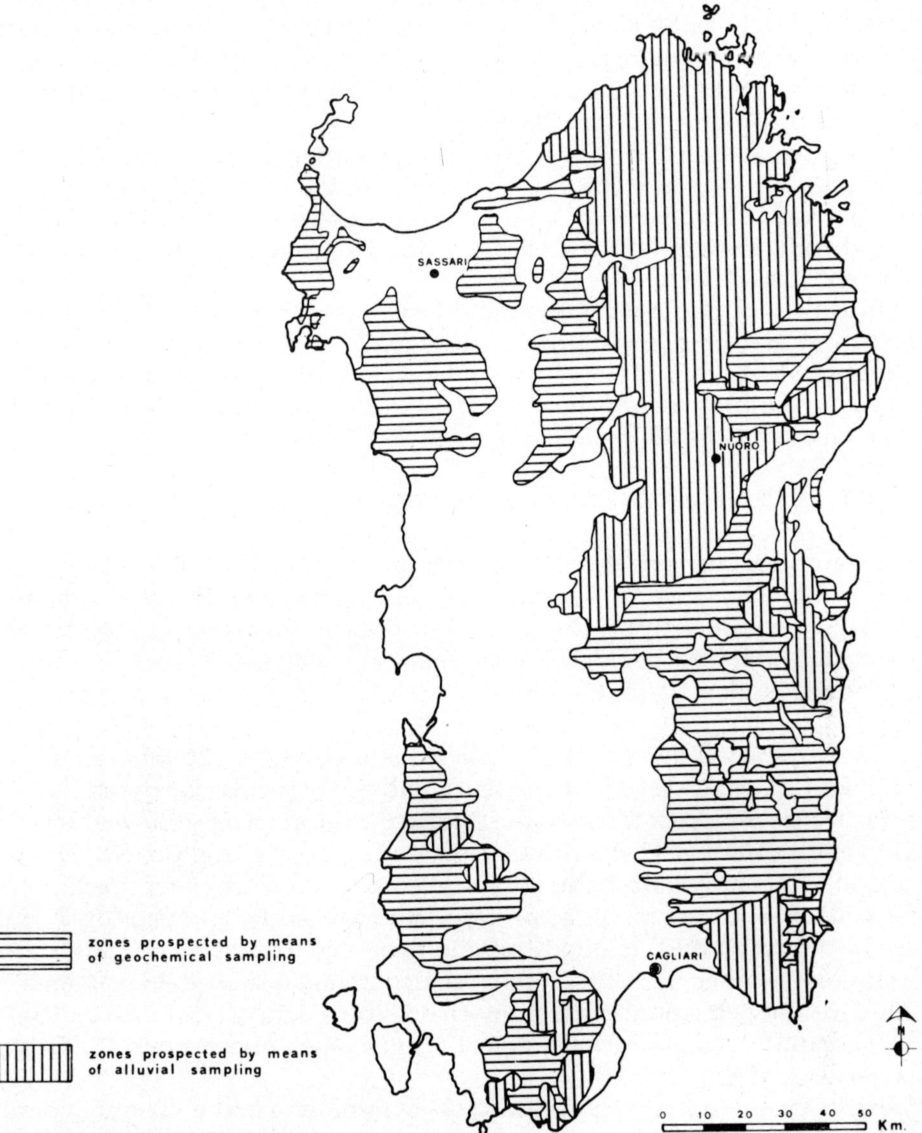

Fig.3. Geochemical and alluvial survey of Sardinia.

conditions: topography, structural features, geological phenomena, mineralizations, etc.

In a heavily explored country like Sardinia, pollution of samples by old tailings cannot always be completely avoided, although it is kept to a minimum. Field teams re-arrange sample locations when tailings are encountered that were not previously recorded. However, many small workings are very old and completely masked by erosion and vegetation. Related contaminated samples have to be eliminated at a later stage. Discovery of hitherto unknown and unmapped old workings by geochemical pollution updates and completes the knowledge on mineral occurrences in Sardinia, and forms a valuable addition to the documentation data bank.

The pattern chosen for alluvial prospecting is approximately one sample per km^2. Here also the density is flexible and can be adapted to local conditions and requirements. At all alluvial sites a geochemical sample is collected and generally all the alluvial prospecting areas have been covered by sediment sampling.

Fieldwork is carried out by two-men teams using four-wheel-drive vehicles. Efficiency varies according to local and seasonal conditions. The fieldteams have attained a rhythm of 25—30 geochemical or 7—10 alluvial samples per day. During the first year of the exploration programme three field teams covered an average of 580 km^2 per month.

ANALYTICAL TECHNIQUES AND DATA PROCESSING

Heavy-mineral concentrates, resulting from the panning of 10 litres of alluvial sediments, sieved to 5 mm, are subjected to a complete semi-quantitative mineralogical analysis. Composites of the concentrates, representative of a drainage basin or a geological formation, are analysed by X-ray fluorescence for Ni, Ti, Sc, Cr, Sn, Ce, Ba, Mo, Zr, Nb, Y, Th, Pb, Bi, Zn, W, Ta and Cu.

Stream sediments (dried at 35°C, crushed and sieved to 120 mesh) are divided with one-half stored for reference and the other half analysed. Samples are treated with warm hydrofluoric-perchloric acid, followed by dilution and buffering. They are analysed for Pb, Zn, Cu, Ag, Co, Ni, Cr, Sb, Mo and Mn by a fully automatic atomic absorption spectrometer (Perkin-Elmer 403). Results are recorded on tape and checked by an operator. F is determined by a classical method. As a control measure and to obtain information on other elements, some samples are analysed for 38 elements by direct-reading emission spectrometer; analytical data are fed into an IBM 360/40 computer, to produce listings of results which also contain the field characteristics of the samples.

Preliminary computer output includes the minimum and maximum contents for each metal in each geological group, the mean, the deviation and the coefficient of variation in both arithmetic and logarithmic form. Subsequently, other statistical parameters, for example correlation coefficients, are calculated and histograms plotted. Cumulative frequencies of the distributions are plotted on arithmetic or logarithmic probability paper and thresholds defining different classes of anomalies obtained. The final step in data processing is factor analysis. This enables multi-element associations to be determined and represented as factor score maps.

All data are processed in such a way that they can be introduced ultimately into the memory bank of the EMSa documentation system. In addition, a synthesis of the geochemical results will be represented on metallogenetic maps.

DISCUSSION

As the geochemical programme has only just passed its half-way mark, it is too early to make a complete assessment of the results. On a sketch map (Fig.4) the main anomalous areas that seem to emerge from the available

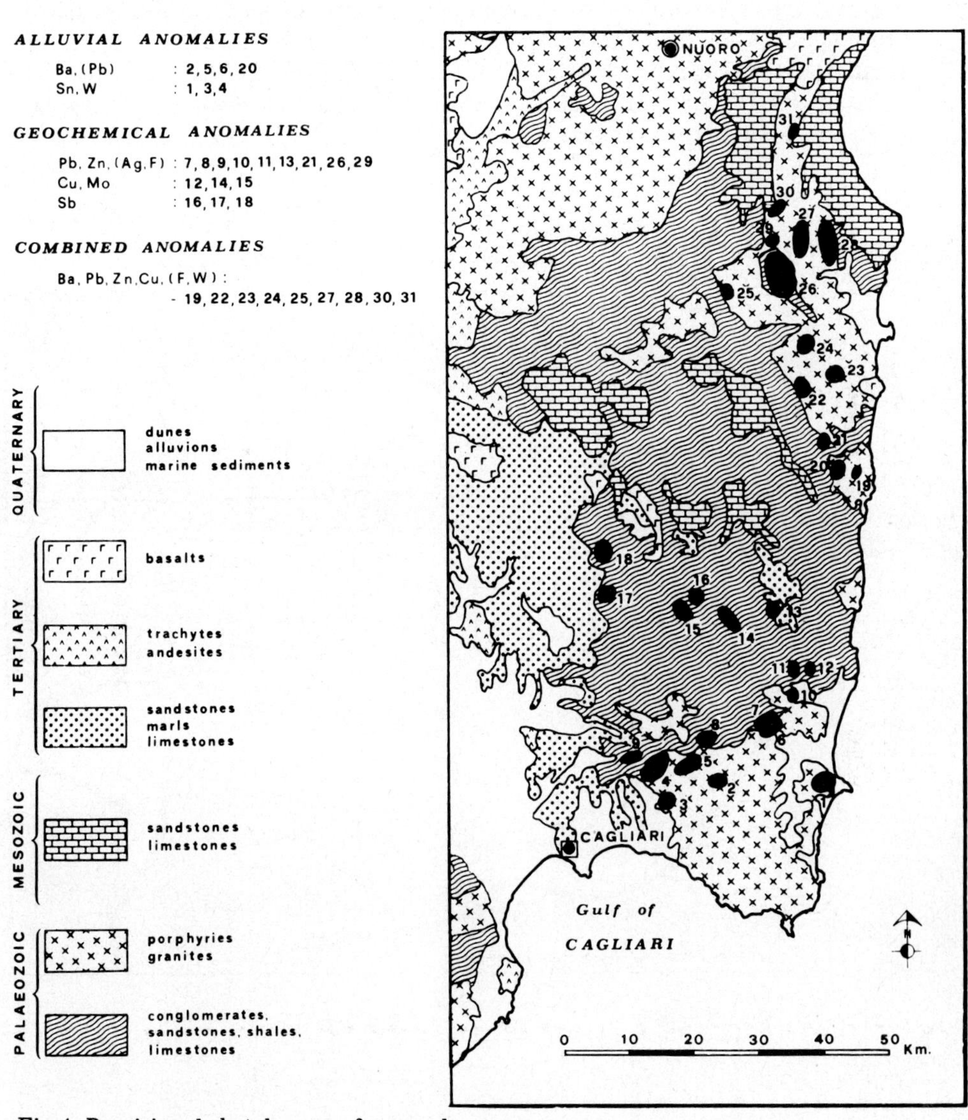

Fig.4. Provisional sketch map of anomalous areas.

data have been assembled. Approximately 10% of the explored surface is anomalous, but it is obvious that the numerous anomalies are of very unequal value. Polluted samples and areas will have to be eliminated. Field control of the remaining anomalies and their geological setting will permit their more precise definition and enable a provisional appraisal to be made of their value.

Here we propose to discuss some isolated examples which illustrate the main trends so far observed in the results, and prove the geochemical method to be efficient, even in previously explored areas.

The first case (Fig.5) shows an antimony anomaly in the vicinity of an abandoned mine. After detailed mapping of the tailings and old exploration pits, tunnels and trenches, samples obviously polluted by old workings have been eliminated. Field control shows that elsewhere on the anomaly no previous exploration has taken place and documents concerning the mining activities of the area confirm this conclusion. A non-polluted antimony

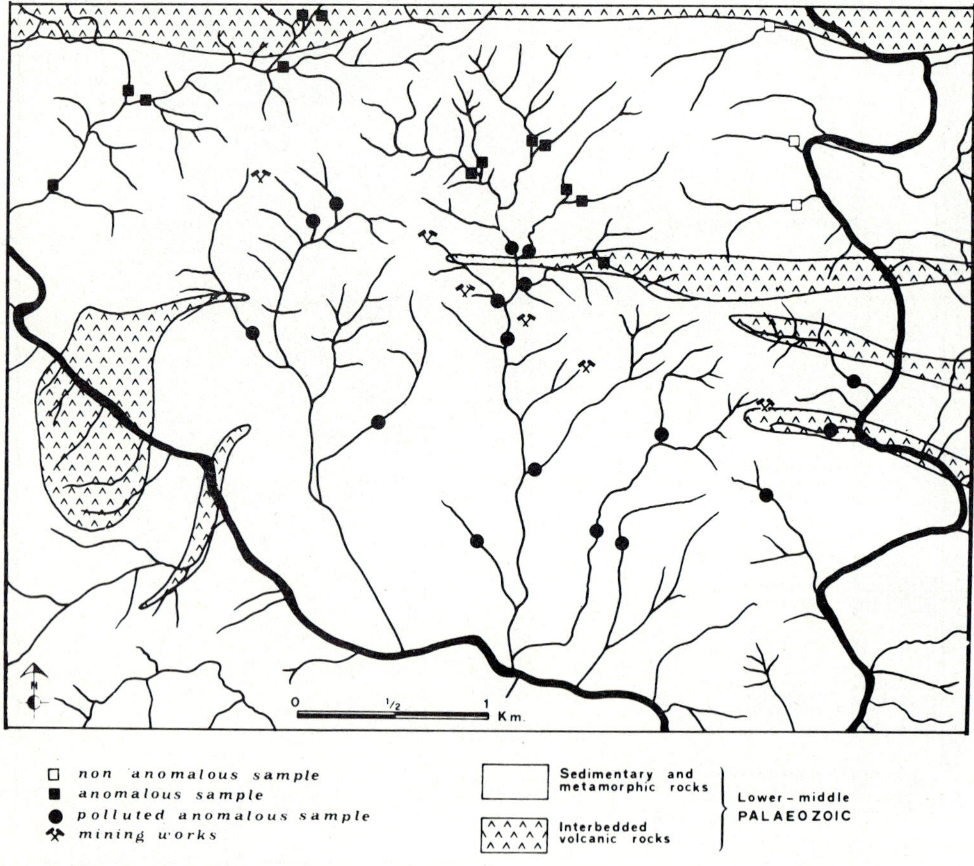

Fig.5. Geochemical anomaly: antimony.

anomaly seems to be present to the northwest of the orebody. This coincides with the prolongation of the same stratigraphic horizon that has been worked in the old mine. Thus, the geochemistry seems to indicate the presence of a hitherto unknown extension of the mineralized structure.

Sample density in this area was in the order of 4 per km². Use of a wider grid could easily have resulted in discovery of the polluted samples only, while the anomalous extension of the orebody would not have been detected.

The second example (Fig.6) shows a sulphide orebody of medium size trending northwest—southeast in a valley underlain by Paleozoic formations. The surrounding hills are covered by sub-horizontal Tertiary sediments and previous exploration had been restricted to the east of the Tertiary table mountain. To the northwest the Tertiary cover is also eroded but outcrops

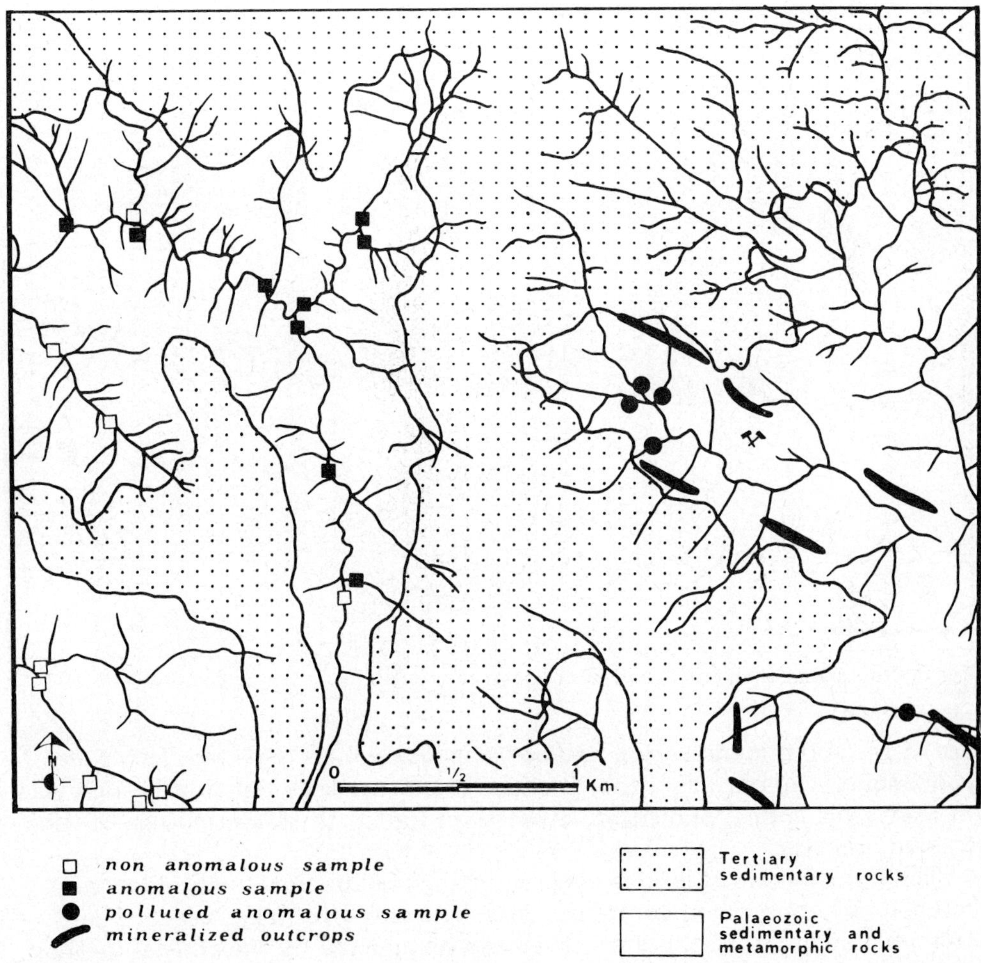

Fig.6. Geochemical anomaly: Pb—Zn—Cu—Ag.

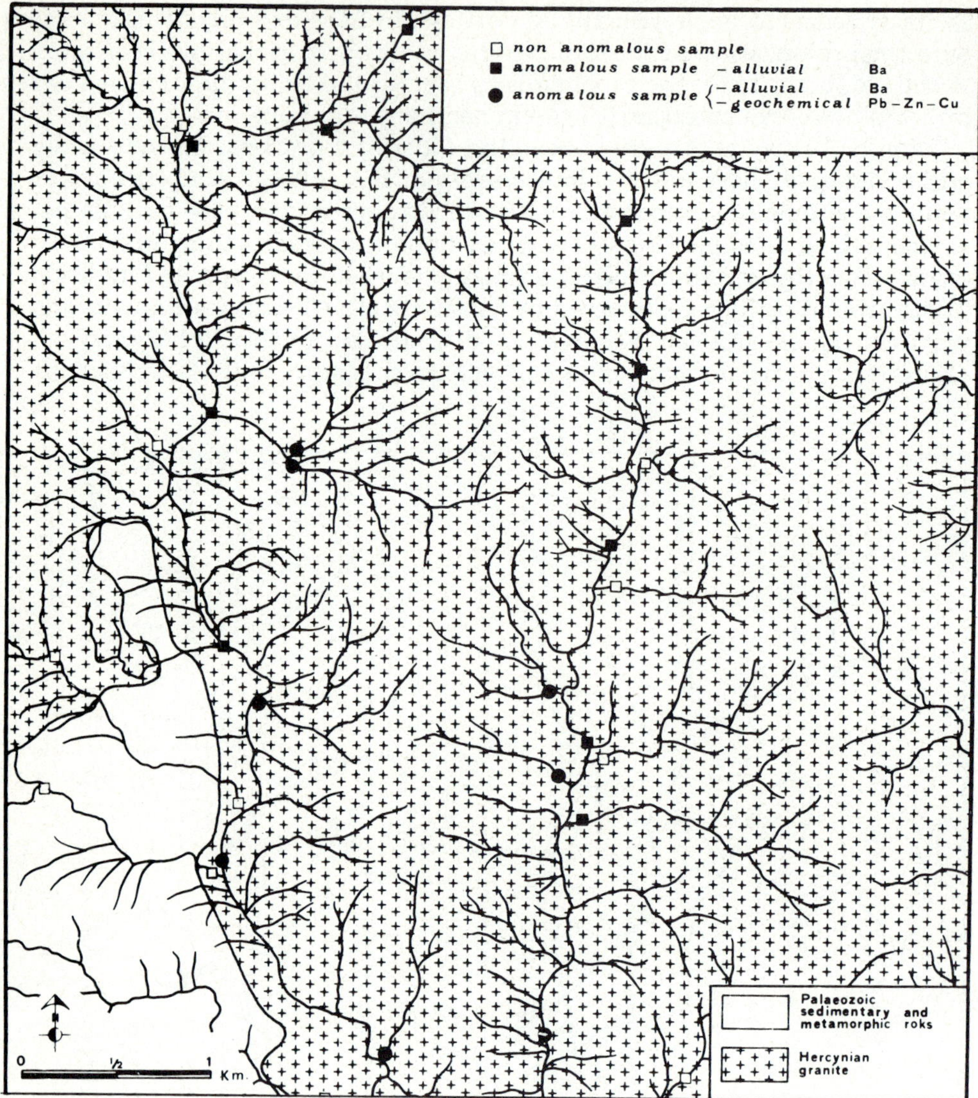

Fig.7. Combined geochemical and mineralogical anomaly.

are scarce and mineralization had not previously been detected. However, geochemical samples from this drainage basin contain anomalously high values of lead, zinc, copper and silver, situated on the northwest extension of the known orebody.

The last example (Fig.7) is situated in an area apparently devoid of any mineral outcrops, where no mining activities have ever taken place. The Hercynian granites of this area have been prospected by the alluvial method. A rather extensive zone was found to be anomalous with samples containing important quantities of barite, locally associated with noteworthy concentrations of scheelite and monazite. In nearby districts a paragenetic association of barite with fluorite and sulphides is known to exist.

It was, therefore, decided to verify the mineralogical anomaly by analysing geochemical samples collected during the alluvial exploration campaign. The geochemical samples outline a zone with high values of lead and zinc and some indications of fluorine within the barite anomaly (Fig.7). This seems to be a convincing example of the complementary nature of the alluvial and geochemical prospecting methods. The combination of the two techniques gives more information on the paragenesis of the mineralization and permits more precise definition of the most promising zones within anomalous areas.

CONCLUSIONS

It has been pointed out that at this stage no definite conclusions can be drawn. However, a preliminary interpretation of the available data suggests that the main objectives of the geochemical exploration are being attained. The principal results can be summarized as follows:

(1) Known mineral provinces are confirmed and their extensions and mineral associations are defined.

(2) New mineralized areas are outlined.

(3) New minerals are discovered in old mining districts.

(4) The distribution of the geochemical values provides valuable information on the position, extension and nature of mineral occurrences for which records are incomplete.

(5) On a larger scale the metallogenetic provinces of Sardinia are being defined by geochemistry, and litho-stratigraphical and structural features are outlined.

The following specific results have been obtained. Alluvial prospecting has discovered: (1) appreciable quantities of monazite, a mineral not previously mentioned in the geological literature of Sardinia, (2) concentrations of scheelite and cassiterite in areas where these minerals were unknown; and geochemical prospecting has discovered: (1) anomalies of Pb, Zn, Cu, Mo, and Sb in previously unexplored districts, (2) anomalies representing possible and as yet unknown extensions of existing ore structures situated in heavily explored and mined districts.

REFERENCES

Cocozza, T., Jacobacci, A., Nardi, R. and Salvadori, I., 1974. Schema geologico-strutturale del massicio Sardo-corso e minerogenesi della Sardegna. Mem. Soc. Geol. It. (in press)
Salvadori, I., 1973. Programma generale straordinario di ricerca: metodologia e finalita. Ente Minerario Sardo, Not. Tecnico-econ., 28 pp.

Part 2

APPLICATION OF STATISTICAL TECHNIQUES

Chairmen:

A.W. ROSE
Pennsylvania State University, University Park, Pa., U.S.A.
D. SAMPEY
Contech Pty. Ltd., Midland, W.A., Australia

SOME CONSIDERATIONS REGARDING GRID ORIENTATION AND SAMPLE SPACING

A.J. SINCLAIR

Department of Geological Sciences, University of British Columbia, Vancouver, B.C.
(Canada)

ABSTRACT

In planning a grid-controlled geochemical survey, it is important to consider relative geometries of anomalous zones (targets) and rectangular grid cells. For any desirable probability of success in locating targets of a particular type, there are optimal grid orientations and sample spacings. In some simple cases these probabilities of success can be estimated reasonably closely by simple geometric considerations. In other more general cases it is necessary to specify density distributions for such parameters as anomaly size, anomaly orientation and anomaly density (number of targets per unit area). Although such cases might be impossible to apply in detail to specific problems, they illustrate the importance of knowing something of the density distribution of targets. In addition, they show the advantage of increasing probability of success by distributing a fixed number of grid intersections through a large area rather than through a relatively small area, all other factors in the two areas being equal.

INTRODUCTION

Location of anomalous zones during grid-controlled geochemical exploration programs depends on: (1) the existence of clearcut thresholds between background and anomalous values of elements considered, and (2) the interrelations of shapes and sizes of anomalous zones relative to the geometry of the exploration grid.

Here we assume that a single, efficient threshold exists, a fairly common real situation, and confine our discussion to the second aspect involving relative geometries of anomalous zones and grid cells. These problems have been considered in the field of geophysics (e.g. Slichter, 1955, 1960; Ellis and Blackwell, 1959; Drew, 1967) and general approaches exist (e.g. Singer, 1972). A recent broaching of the topic in the field of geochemistry is that of Hodgson (1972).

From a theoretical point of view, a specific problem is easily stated and solved. Given anomalous areas of certain sizes, shapes and orientations, what are the optimal grid spacings and orientation for locating such targets? We here define optimal grid spacing as that spacing providing a specified probability of success, P_s, in locating targets with particular geometries. It is important to realize that this desired probability of success in locating such targets might be very different in one situation compared to another, due to restraints that cannot be considered here, such as topography, fixed maximum costs, etc.

IMPORTANCE OF GRID SPACING AND ORIENTATION

A few simple geometric examples are useful to illustrate the importance of grid spacing and orientation in relation to anomaly shape and size.

Case 1

A rectangular anomaly $2s \times s$ would be found using a $2s \times s$ rectangular grid whose unit cell were oriented in the same direction as the anomalous zone (Fig.1a). This would be true no matter where the origin of the grid system were located. Assume, however, that the long direction of the unit cell was inadvertently oriented at 90° to the long direction of anomalous zones. With random location of anomalous zones in the area of such a grid, the probability of locating such zones is 50%, a remarkably large reduction from the 100% probability of success in the case of the first orientation. Of course, real anomalies generally are not rectangular. The example is useful, however, because (1) it is not completely unrealistic and avoids complicated formulae that are sometimes necessary in calculating probabilities, and (2) it emphasizes the importance of grid orientation — an optimal grid spacing for one grid orientation need not be optimal for other grid orientations.

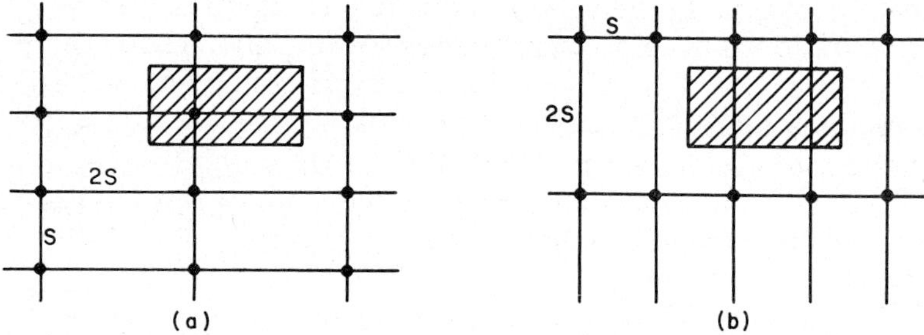

Fig.1. A simple example to illustrate the importance of grid orientation. The rectangular target (hachured) and grid cell are equivalent in size and shape. In (a) the probability P_s that the target will be located by the grid is 100%, whereas in (b) P_s is 50%.

Case 2

Consider an isotropic anomaly, i.e., one with circular outline (Fig.2). A square grid, regardless of orientation, will have a probability of locating the anomaly that depends only on grid dimensions relative to target diameter, assuming that targets are distributed randomly throughout the gridded area. In this case it is a simple matter to show that for certainty of locating all circular

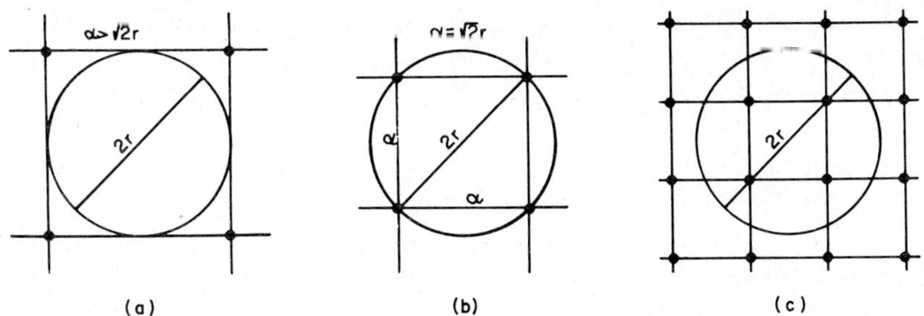

Fig.2. Grid orientation is unimportant for isotropic (circular) anomalies. In (a) $P_s < 1.0$, (b) optimal grid spacing $\alpha = \sqrt{2}\,r$ for $P_s = 1.0$, (c) grid spacing not optimal unless more than one "hit" per target is desired.

targets with specified diameter, the maximum possible grid spacing is $\sqrt{2}\,r$. This spacing is therefore optimal if 100% probability of success is desired at minimum cost.

Case 3

The vast majority of anomalous zones can be approximated by ellipses (cf. Slichter, 1960) of which the circle of the last example is a special case. Departure of an elliptical anomalous zone from a circle can be represented by the ratio α/β where α and β are semi-major and semi-minor axes respectively of the ellipse. This ratio is analogous to the fineness ratio (λ) of Slichter (1955). In Fig.3, a fineness ratio $(\lambda = \alpha/\beta)$ of 2 applies in three cases shown, i.e. one axis is twice as long as the other. For individual elongate targets of this nature, grid orientation is not important in cases where grid cells entirely contain anomalous zones ($P_s = A_{anom}/A_{cell}$, Fig.3a) or are contained entirely within an anomalous zone ($P_s = 1.0$, Fig.3c). For certainty that an anomalous zone of given fineness ratio is intersected by at least one grid point the grid must be oriented parallel to axes of the elliptical target with sample spacings *of* $\sqrt{2}\,\alpha$ parallel to the major axis (2α) of the ellipse and $\sqrt{2}\,\beta$ parallel to the

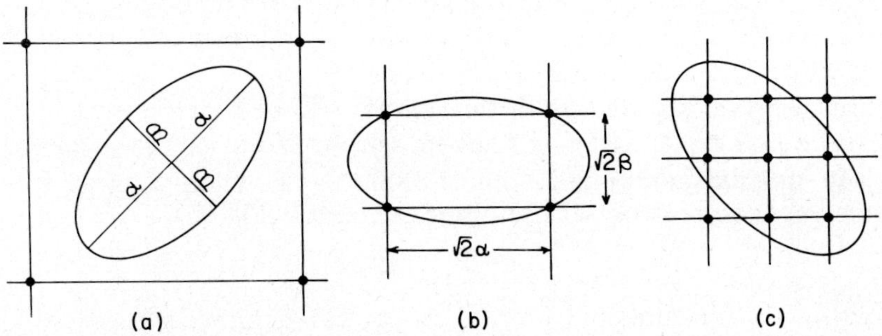

Fig.3. Elliptical anomaly with axes 2α (major) and 2β (minor): (a) illustrates $P_s < 1.0$, (b) optimal grid spacing and orientation, (c) grid spacing less than optimal unless more than one "hit" per target is desired.

minor axis cf the ellipse (2β), as shown in Fig.3b. With such an optimum grid orientation and sample spacing, it is of interest to note that there is a 73% probability that two grid intersections (samples sites) will intersect the anomalous zone!

These simple examples illustrate the following generalizations noted by various authors:

(1) Targets in mineral exploration commonly can be approximated closely by an ellipse (e.g. Slichter, 1960). This generalization applies reasonably well to many types of geochemical anomalies.

(2) If an anomaly is contained entirely within a unit cell of a sample grid, the probability of locating one such anomaly or target is given by the relation:

$$P_s = A_t/A_c$$

where P_s = probability of success, A_t = area of target, and A_c = area of grid cell.

In the general case of an elliptical target in a rectangular grid, the probability of success is, $P_s = \pi\,\alpha\beta/xy$, where α and β are semi-major and semi-minor axes respectively of the ellipse and x and y are sides of the rectangular grid cell (Fig.3a).

(3) If a grid cell can be contained entirely within a target of specified size, certainty (P_s = 1.0) exists that all such targets will be found by the sampling grid (Fig.3c).

(4) In cases intermediate to (2) and (3) above, the probability of success is a function of relative sizes, shapes and orientations of targets and grid cells. Calculating this probability is a relatively simple matter for an isotropic anomaly but is awkward for other shapes, even ellipses. Such probabilities have been tabulated by Savinskii (1965). A Fortran program has been published recently (Singer, 1972) that allows calculation of probabilities of success for elliptical targets of various sizes and orientations relative to rectangular and hexagonal grids.

(5) Grid orientation has no effect on the probability of success in locating isotropic (circular) anomalies.

(6) For an elliptical target, optimal grid orientation arises when one principal grid direction parallels the elongation direction of the target (Matheron and Marachel, 1969).

(7) For absolute certainty that an elliptical target will be located by a survey, the grid should be oriented as in (6) above, with optimal grid spacings of $\sqrt{2}\,\alpha$ along the long axis, and $\sqrt{2}\,\beta$ along the short axis, where α and β are major and minor axes respectively of the elliptical target (Fig.3c).

FURTHER COMMENTS ON GRID ORIENTATION

The foregoing simple examples suffice to illustrate the advantage of knowing anomaly size and shape prior to specifying either grid orientation or unit cell dimension. Of these two distinct problems, orientation and unit cell geometry, only the latter has received much attention in the literature. The question of cell orientation arises, of course, only if an anomaly is anisotropic, that is, elliptical in plan rather than circular. Routinely, an orientation is imposed on a grid by having one principal direction parallel to the regional trend in the area. In many cases, such a procedure has much to recommend it but one can imagine situations where anomalous zones are elongate in directions that bear no obvious relationship to regional trends. Furthermore, there are numerous cases where surveys are carried out in terrains in which no obvious regional trend is apparent, e.g. glaciated shield areas, areas of flat-lying sedimentary rocks, and large areas underlain by plutonic complexes. In such cases as these, no preferential direction for grid orientation need be obvious from regional features. In a blind survey of such an area where no a priori knowledge exists concerning anomaly sizes, shapes, orientations, distributions, etc., the best approach is to use a randomly oriented square grid. However, if knowledge exists regarding size distribution and preferred orientation of anomalies, this information should be used to obtain the best estimate of optimum grid orientation and sample spacing.

A GENERAL APPROACH

To this point, the problems we have been dealing with compare with the simplest situations in nature. They include estimation of the probability of success for isotropic anomalies and either: (1) assume that all anomalies are of the same size — leading to the acceptance of the probability of success as a strict figure, i.e. if one anomaly exists in the gridded area, the calculated P_s is the chance that one anomaly will be found by the survey, or (2) deal with a minimum sized anomaly that we might wish to locate and accept the probability of success as a minimum.

Either of these approaches might well be considered realistic because to go further requires mathematical approximations that cannot always be made with reasonable certainty. Furthermore, calculations even to this point have been based on serious restrictions imposed on size and shape of the anomalous target. However, there are cases where a more sophisticated approach might be desirable, in particular, where anomalies occur with a wide range of sizes and/ or shapes and/or orientations.

More sophisticated analyses, such as that of Matheron and Marachel (1969), involve: (1) mathematical approximations related to the assumed simplified geometry of targets, (2) a distribution function for the probability of success in locating a target, (3) a distribution function for the abundance of targets (e.g. in terms of number of targets per unit area as determined by experience in a neighbouring area of similar geology) and (4) a distribution function for size of targets (perhaps including a minimum cut-off size based on experience in nearby similar areas).

It should be apparent from the foregoing that a generalized approach re-

138

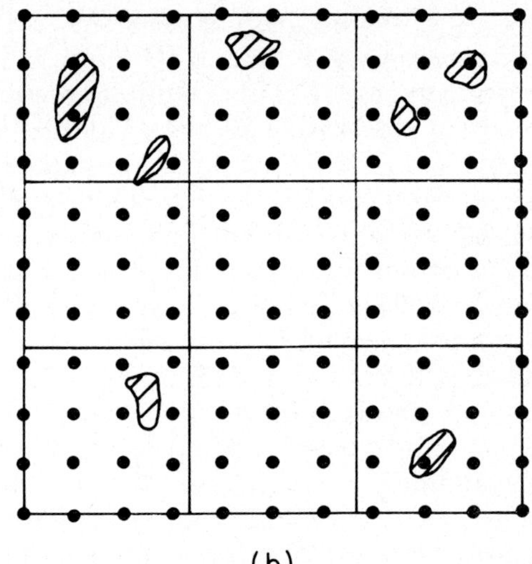

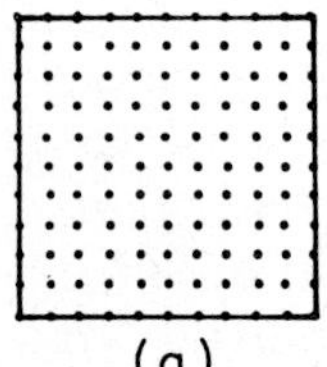

(a)   (b)

Fig.4. Diagram illustrating the generalization that the probability of failure of a survey is higher for a small area than for a large area, all other things (including number of sample sites) being equal.

quires a considerable range of assumptions that might be entirely or partly impossible to justify in any one particular area. In spite of this, certain general conclusions from sophisticated models have important bearings on some types of geochemical programs. For example, such studies emphasize the importance of the *number* of targets as well as their morphology. In addition, for a constant number of samples, the probability of *failure* is higher for a small area than for a larger area, all other things being equal. This arises primarily from the fact that as a survey area becomes smaller there is an increased likelihood that *no* target will occur within it! This latter conclusion is of particular importance in planning regional soil surveys where the grid cell is much larger than individual targets, and the probability exists that zero to several targets occur within each grid cell. In such a case, it is advisable to spread a fixed number of sample sites over as large an area as possible. assuming of course, that the large area exhibits the same characteristics as does the smaller area. The situation is illustrated diagrammatically in Fig.4 where, for example, it is apparent that a much higher probability exists of zero deposits in a single cell relative to a square area consisting of 4 cells.

SERIAL CORRELATION TECHNIQUES

Recently, Hodgson (1972) has suggested that the serial correlation method described by Agterberg (1965) could be used to determine optimum grid spacing for geochemical soil surveys. This method involves determining the simple coefficient for pairs of a variable at a number of different sample separations to produce a correlogram. The technique presupposes (1) that the grid orientation is defined by having one principal grid direction parallel to the regional strike, and (2) anomalous and background values are distinguishable such that correlograms can be determined for each separately. The detailed procedure recommended by Hodgson (1972) for correlogram analysis is as follows:

(1) From a close-spaced sampling grid of the type normally used for orientation surveys, select at random 100 sample pairs separated by constant distance h.

(2) Calculate the simple correlation coefficient for each element for the 100 pairs.

(3) Repeat steps 1 and 2 for sufficient h spacings that a correlogram can be constructed for each element involved in the survey.

(4) For each element define a *critical sampling interval*, the distance beyond which the correlation coefficient cannot be distinguished from zero. A "safe" sampling interval (grid cell dimension) for a given element in a regional soil survey is then taken at the nearest 100's of feet *less than* the smaller of the critical sampling intervals for both background and anomalous data.

There are several fundamental problems in the widespread application of correlograms in the evaluation of soil geochemical data. In the first place, the correlogram might not exist despite the fact that mathematical values can be calculated for it. Secondly, the co-variance (the numerator in the equation for the correlation coefficient) is always biased because the mean is used in its estimation. On these points alone one might consider the use of another somewhat comparable auto-correlation function, the semi-variogram (Matheron, 1971). Apart from this are the numerous procedural difficulties with the correlogram method, including: (1) assuming isotropy when in fact elongate anomalies are very common, (2) ignoring the presence of a background trend or drift, (3) practical problems in determining the correlogram for anomalous areas because of shortage of data, and (4) the fact that the critical sampling interval in background areas need bear no particular relationship to the critical sampling interval in anomalous areas and, what is more, neither of these critical sampling intervals tells us anything about (a) sizes or shapes of anomalous zones within a background field, or (b) the probability of success in locating anomalous targets.

This latter point is worth some amplification. Certainly correlograms provide us with useful information, but useful for what? The distances over which background values show significant correlation obviously relates to "average structure" within background areas and is completely unrelated to anomalous zones. In fact, it is somewhat presumptuous to expect to put limits on the sampling interval for locating anomalies by considering background data. Even to consider anomalous data alone, assuming sufficient were available, would, in general, lead to estimating the average length of structures *within* anomalous zones and result in too small a grid cell and, hence, too high costs for line-cutting, sampling and analysis. Of course, there are other, quite different problems for which the range of auto-correlation in geochemical data might be very important, such as contouring problems using kriging procedures that give linear interpolations *with variance estimates.*

DISCUSSION

Simple geometric considerations are enough to indicate the importance of anomaly shape and size in determining optimum grid orientation and sample spacing for a geochemical soil survey. The major difficulty is to define these shape and size parameters prior to an extensive survey. Two general procedures are available: (1) an orientation survey in a region thought to be representative of the survey area, or (2) analysis of other geochemical data from similar nearby areas. In either of the foregoing cases, the initial problem is to characterize the anomalies in terms of geometric pattern(s) and presence or absence of preferred orientation(s). Once this is done, an optimum grid can be designed to meet the required probability of success for the survey.

In reality of course, the problem of characterizing geochemical anomalies is a difficult one because of the many factors that affect metal dispersion, and the fact that these factors can and do vary drastically even locally. At the present time, research is being conducted into these problems utilizing techniques of pattern recognition and various auto-correlation functions, some of which will be exposed further in presentations that follow.

ACKNOWLEDGEMENTS

Drs. J.H. Montgomery and W.K. Fletcher offered constructive comments that improved the manuscript. This work has been supported financially by the National Research Council of Canada.

REFERENCES

Agterberg, F.P., 1965. The technique of serial correlation applied to continuous series of element concentration values in homogeneous rocks. J. Geol., 13: 142—154

Drew, L.J., 1967. Grid-drilling and its application to the search for petroleum. Econ. Geol., 62: 698—710

Ellis, R.M. and J.H. Blackwell, 1959. Optimum prospecting plans in mining exploration. Geophysics, 24: 344—358

Hodgson, W.A., 1972. Optimum spacing for soil sample traverses. In: 10th APCOM Symposium. South African Institution of Mining and Metallurgy, Johannesburg, pp.75—78

Matheron, G., 1971. The theory of regionalized variables and its applications. Les Cahiers du Centre de Morphologie Mathématique, Fontainebleau, No.5, 211 pp.

Matheron, G. and A. Marachel, 1969. Note sur la probabilité de succes dans une reconnaissance aveugle. Centre de Morphologie Mathématique, Fontainebleau, Rep. N-109

Savinskii, I.D., 1965. Probability tables for locating elliptical underground masses with a rectangular grid. Consultants Bureau, New York, N.Y., 100 pp.

Singer, D.A., 1972. Elipgrid, a Fortran IV program for calculating the probability of success in locating elliptical targets with square, rectangular and hexagonal grids. Geocom Programs, 4, 16 pp.

Slichter, L.B., 1955. Optimum prospecting plans. Econ. Geol., 15th Ann. Vol., pp.885—915

Slichter, L.B., 1960. The need of a new philosophy of prospecting. Min. Engr., 12: 570—576

AUTOCORRELATION STUDIES IN THE ANALYSIS OF STREAM SEDIMENT DATA

S. DIJKSTRA and K. KUBIK*

Mining Exploration Department, International Institute for Aerial Survey and Earth Sciences (ITC), Delft (The Netherlands)

ABSTRACT

Efficiency of large-scale medium-density geochemical surveys depends largely upon the development of appropriate methods of data analysis. The paper describes the experience obtained with the application of a number of statistical techniques for separating trends from local effects, including the practical use of autocorrelation studies. Examples are given of subjects including: selection of sampling densities, smoothing techniques, semi-quantitative assessment of map accuracies and of criteria for defining threshold values.

INTRODUCTION

The present paper documents the authors' experience in the application of various existing mathematical and statistical techniques to geochemical data analysis, techniques which until now appear to have received limited attention only. These techniques are being employed in several related disciplines concerned with the variation of natural quantities in space and their utility in other fields of geology has been amply demonstrated.

The authors' experience would indicate that many of these techniques and their underlying mathematical concepts are also applicable to study of spatial distribution of trace metal values. In agreement with a general model for the distribution of geological quantities as proposed by Agterberg (1970), they wish to convey the concept that geochemical surfaces can be described by a trend function $t(x,y)$ plus a homogeneous random function $z(x,y)$ with expectation $E(z_i) = 0$ and covariance function $Cov(i,j) = E[z_i - E(z_i)] [z_j - E(z_j)]$. This covariance function serves to measure the mutual relationship of values $z_i = z(x_i,y_i)$ and $z_j = z(x_j,y_j)$ where x and y are plane coordinates. The property of homogeneity entails that the covariance function $Cov(i,j)$ is independent of the absolute location of the pair of points i and j and depends only on their relative position. By measuring the relationship between these points the covariance function is a measure of the roughness of the surface $z(x,y)$.

In the following sections the above conceptual model is developed further,

---

* Also: Data Processing Department, Ministry of Transport and Waterstaat, The Hague, The Netherlands

specific examples of estimated covariance functions are given and techniques of geochemical data analysis, as applied by the authors, are described.

A CONCEPTUAL MODEL FOR TRACE METAL DISTRIBUTIONS

The model presented in this section is based on the generalization that trace metal distributions within a drainage system are controlled by a combination of several events of differing natures, including both trace metal dispersion from distant and nearby source materials and changes in environmental conditions. With regard to their contribution to the overall geochemical landscape, events may be divided according to their effects, for example, areal, local and point effects or, rather, point deviations.

The authors' reflections on a possible model are probably best demonstrated with the aid of a hypothetical drainage system as presented in Fig.1. This shows a series of primary catchment areas, each with three different types of source rocks, where adjacent catchment areas have two source rocks in common. The catchment areas are drained by three first-order streams, which join to form one second-order stream and ultimately form a drainage pattern as outlined. The effect of a single event on trace metal distributions in the drainage system will be considered first. Trace metal dispersion from source rocks occurring within the primary catchment areas has been selected for this purpose; subsequent trace metal contributions from sources within areas of second and higher-order streams are initially excluded.

It is assumed that the trace metal values c_i of rocks in primary catchment areas represent a series of uncorrelated random variables. Accepting that con-

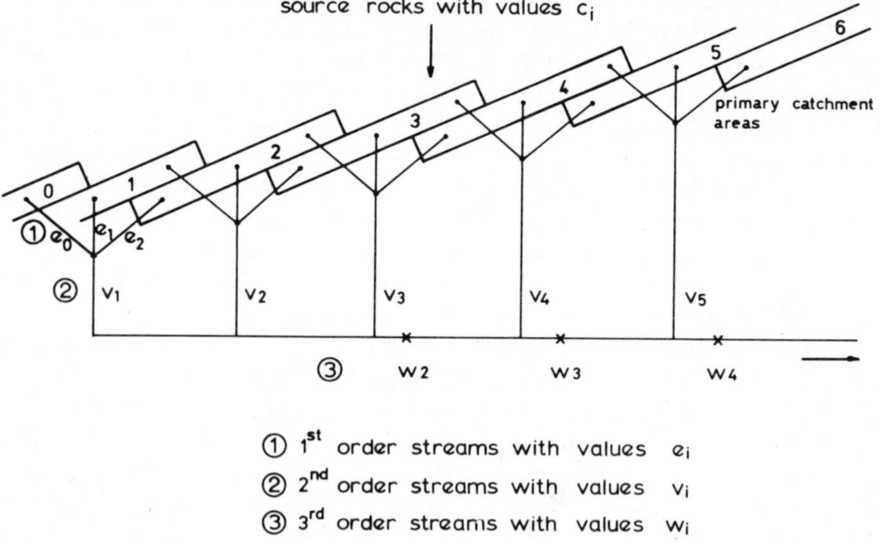

Fig.1. Generation of autocorrelated series of trace metal values in sediments of a hypothetical drainage system.

ditions for trace metal dispersion in all areas are comparable, this implies that values e_i of sediments of first-order streams are directly proportional to the trace metal values of the source rocks and are also uncorrelated random variables. Hence under the present assumptions and with reference to Fig.1 this latter series of random variables can be described by:

$$E(e_i) = \bar{e} \qquad\qquad i = 0, \dots N$$
$$Var\,(e_i) = E\,(e_i - \bar{e})^2 = \sigma_e^2$$
$$Cov\,(e_i, e_{i+d}) = E\,(e_i - \bar{e})\,(e_{i+d} - \bar{e}) = E\,(e_i' \cdot e_{i+d}') = 0$$

with $e' = e_i - \bar{e}$, E = expectation, Var = variance, Cov = covariance, and d = lag.

The various parameters for the series of values v_i and w_i of drainage sediments of higher-order streams can now be computed as follows:

$$v_i = 1/3\,p_1\,(e_{i-1} + e_i + e_{i+1}), \qquad\qquad i = 1, \dots N-1$$

or symbolically:

$$v = P_1\,\Sigma e, \qquad\qquad P_1 = 1/3\,p_1$$
$$E\,(v_i) = P_1\,[E\,(e_{i-1}) + E(e_i) + E(e_{i+1})]$$
$$Var\,(v_i) = 1/9\,p_1^2\,E\,(e_{i-1}' + e_i' + e_{i+1}')^2 = 3/9\,p_1^2\,\sigma_e^2$$

Note that the factor p_1 takes into account the phenomenological datum that values of dispersion trains decrease downstream; $0 < p_1 < 1$. For the covariances it holds that:

$$Cov\,(v_i, v_{i+1}) = 1/9\,p_1^2\,E\,(e_{i-1}' + e_i' + e_{i+1}')\,(e_i' + e_{i+1}' + e_{i+2}')$$
$$= 2/9\,p_1^2\,\sigma_e^2$$
$$Cov\,(v_i, v_{i+2}) = 1/9\,p_1^2\,E\,(e_{i-1}' + e_i' + e_{i+1}')\,(e_{i+1}' + e_{i+2}' + e_{i+3}')$$
$$= 1/9\,p_1^2\,\sigma_e^2$$
$$Cov\,(v_i, v_{i+3}) = 0$$

Similarly for the series of values w_i one finds:

$$w_i = 1/3\,p_2\,(v_{i-1} + v_i + v_{i+1}), \qquad\qquad \begin{array}{l} i = 2, \dots N-2 \\ 0 < p_2 < 1 \end{array}$$

or symbolically:

$$w = P_1 P_2\,\Sigma\Sigma e, \qquad\qquad P_2 = 1/3\,p_2$$
$$E\,(w_i) = P_1 P_2\,[E(e_{i-2}) + 2E(e_{i-1}) + 3E(e_i) + 2E(e_{i+1}) + E(e_{i+2})]$$
$$Var\,(w_i) = 19/81\,p_1^2\,p_2^2\,\sigma_e^2$$

$$Cov\,(w_i, w_{i+1}) = 16/81\,p_1^2\,p_2^2\,\sigma_e^2$$
$$Cov\,(w_i, w_{i+2}) = 10/81\,p_1^2\,p_2^2\,\sigma_e^2$$
$$Cov\,(w_i, w_{i+3}) = 4/81\,p_1^2\,p_2^2\,\sigma_e^2$$
$$Cov\,(w_i, w_{i+4}) = 1/81\,p_1^2\,p_2^2\,\sigma_e^2$$
$$Cov\,(w_i, w_{i+5}) = 0$$

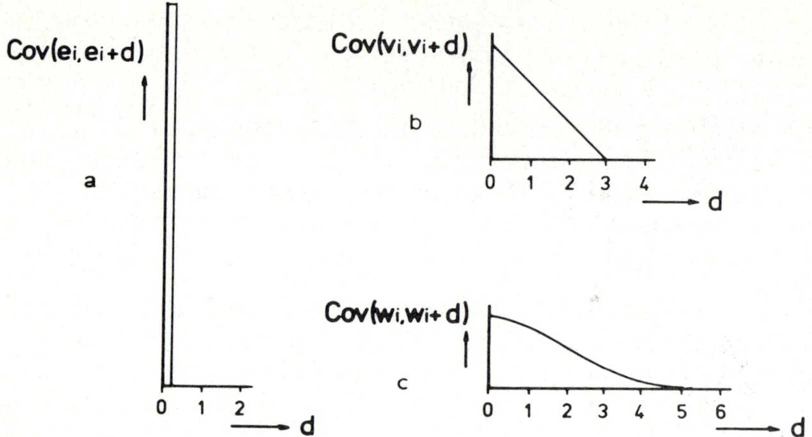

Fig.2. Covariance functions for the series of values e, v and w of the hypothetical drainage system of Fig.1.

The computations show that values of higher-order streams become increasingly autocorrelated. In instances where rock units of the primary catchment areas have regional extensions, it can be shown that this increase in autocorrelation will also hold for the series of values z_i of fourth-order streams.

In conclusion it may be stated that each of the series of values e, v, w and z constitutes separate series of increasingly correlated random variables, as also illustrated by the graphical presentation of the covariance functions Cov (e_i, e_{i+d}), Cov (v_i, v_{i+d}) and Cov (w_i, w_{i+d}) in Fig.2. These series of increasingly correlated random variables become more and more regular in their appearance and ultimately can be regarded as trend components of an areal or regional nature. The results of the above computations thus demonstrate a possible generation of autocorrelated series of trace metal values for stream sediments by a process of trace metal dispersion from rocks in primary catchment areas.

Referring to the earlier generalization that trace metal distributions in space are controlled by an accumulation of effects, resulting from a variety of events, the trace metal value of any point within the drainage pattern can be written symbolically as follows:

$$z_i = p_i + S_1 \Sigma h_j + R_1 R_2 \Sigma\Sigma g_k + Q_1 Q_2 Q_3 \Sigma\Sigma\Sigma f_l + \ldots \qquad (1)$$

where p_i is the effect of a point deviation while h_j, g_k, f_l, ..., represent the effect of different events in sequence of increasing order or regional significance; $S_1 \Sigma h_j$ stands for $1/3 \, S_1 \, (h_{j-1} + h_j + h_{j+1})$ and can be defined as a local effect; $R_1 R_2 \Sigma\Sigma g_k$ is an areal effect, $Q_1 Q_2 Q_3 \Sigma\Sigma\Sigma f_l$ is a regional effect, etc. The factors P, Q, R and S all relate to the fact that values of dispersion trains decrease in a downstream direction. The equation is applicable if the original assumption, that the effects of separate events can be represented by separate series of homogeneous random variables, remains valid.

The covariance function of the series of random variables z_i is obtained by summation of the corresponding covariance functions of all terms in cq.1. Bearing in mind the results of the previous covariance computations illustrated in Fig.2, this summation yields a covariance model as presented in Fig.3.

Owing to the great complexity of geochemical processes the authors do not pretend that the simple model holds true for every geochemical surface. The importance of the model, however, is that it has proven useful in the interpretation of covariograms of actual data. In this model all surface components, and most especially the trend components, can be described by the accumulation of random variables. They are thus accessible to unified treatment. The model indicates that knowledge of the respective covariance functions might assist in separating surface components and it suggests that such a separation will be particularly efficient if the number of effects (in terms of eq.1) is limited and each differs widely in regional significance. If values z_i reasonably satisfy the model:

$$z_i = p_i + R_1 R_2 \Sigma \Sigma u_j + P_1 P_2 P_3 P_4 \Sigma \Sigma \Sigma \Sigma t_k \tag{1a}$$

then the surface components can be separated into a regional trend t, areal undulations u, and point deviations p. It then follows that:

$$Cov(z,z) = Cov(p,p) + R_1^2 R_2^2 \, Cov(\Sigma \Sigma u, \Sigma \Sigma u)$$
$$+ P_1^2 P_2^2 P_3^2 P_4^2 \, Cov(\Sigma \Sigma \Sigma \Sigma t, \Sigma \Sigma \Sigma \Sigma t)$$
$$\sigma_z^2 = \sigma_p^2 + R_1^2 R_2^2 \sigma_u^2 + P_1^2 P_2^2 P_3^2 P_4^2 \sigma_t^2 \tag{1b}$$

What is regarded to be trend, and what local undulation, is subject to seemingly arbitrary definitions; in practice, however, these two concepts are fairly well

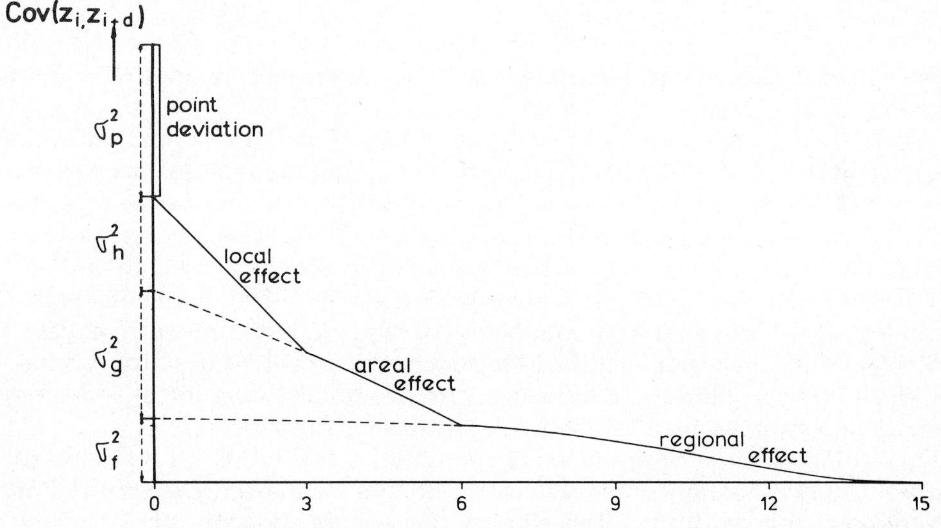

Fig.3. Graphical illustration of the covariance model of eq.1.

defined within a particular level of detail and the size of the area of a given survey. On the other hand it is certainly possible for a trend in a small area to be part of an undulation if the area is enlarged.

In the presence of large-scale trends within the limits of a surveyed area, the geochemical surface commonly can be described best in practice by the summation of a deterministic function and a homogeneous random function. After removal of the deterministic component, the covariance model of Fig.3 is again applicable.

EXAMPLES OF COVARIOGRAMS AND AUTOCORRELOGRAMS OF ACTUAL DATA

Covariance functions can be estimated from values at sample points; the following computational rule applies, for instance, for evenly spaced data along sample profiles:

$$C\,(z_i, z_{i+d}) = \sum_1^n \; (z_i - \bar{z})\,(z_{i+d} - \bar{z})/(n-1) \tag{2}$$

where $\bar{z} = E\,(z)$, d = distance in units of sampling intervals, and n = number of terms in the summation.

Plots of discrete estimates of $C\,(z_i, z_{i+d})$ against lags d will be referred to as (auto)covariograms. Autocovariograms are closely related to autocorrelograms, which are defined by:

$$r_d = C\,(z_i, z_{i+d})/V \tag{3}$$

$$V = \sum_1^n \; (z_i - \bar{z})^2/(n-1) \tag{4}$$

where r_d = autocorrelation coefficient, and V = estimated variance of random function.

Figs.4—7 show covariograms and correlograms of data for several different medium density stream sediment surveys. The examples demonstrate the practical usefulness of covariograms and correlograms for the purpose of separating various surface components, including point deviations. They also render information on the efficiency of sampling spacing in terms of correlation distance. The figures are in fair agreement with the theoretical model of Fig.3. Differences that are observed could be due to the limited number of values available within the given length of the geochemical profiles used for estimation of the covariograms. Differences might also point to the presence of trends that can be described by a low order deterministic function.

Fig.4 refers to an orientation survey conducted in the Central Belgian Ardennes. Duplicate samples were collected along a second-order stream at 250-m intervals and assayed for hot HCl-extractable lead; in addition, some sections of the stream were sampled at intervals of 30 m and 3 m, respectively. The

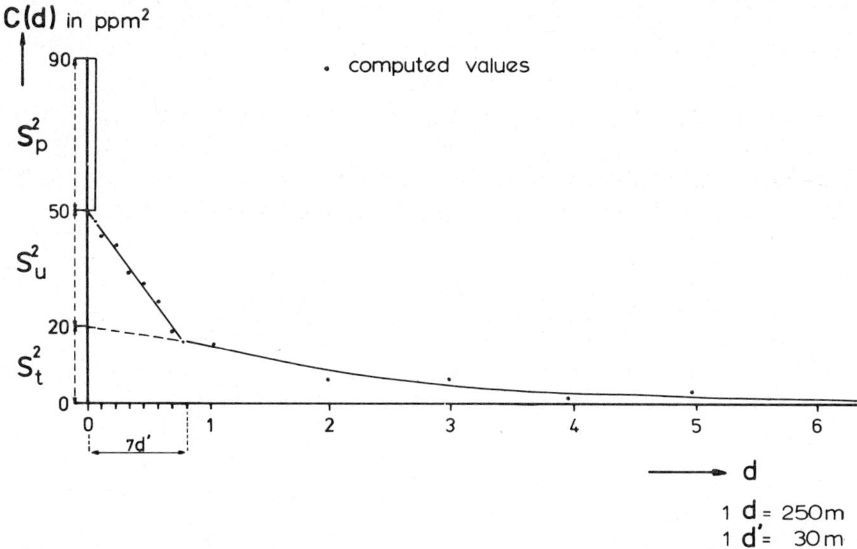

Fig.4. Covariogram of lead values from stream sediment samples of an orientation survey in the Belgian Ardennes. The covariogram is estimated from over 100 pairs of values of three different sets of samples, collected at intervals of 250 m, 30 m and 3 m, respectively, along a single stream profile; S_p^2, S_u^2 and S_t^2 are estimates for variances, relating to point deviations, undulations and trend, respectively (cf. eq.1b).

covariogram of Fig.4 agrees well with the simple covariance model of eq.1a and indicates the presence of point deviations p, local undulations u, over a distance of 200 m, and an areal trend t. From this covariogram it can be observed that sampling intervals of some 400 m will quantify the areal trend and that efficient delineation of local undulations demands sampling intervals in the order of 100 m. The overall variance of Pb values is largely due to point deviations; point deviations can possibly be reduced by revising sampling techniques.

Fig.5 presents a correlogram of copper values from a stream sediment survey in the Ingessana Hills area (Sudan). The correlogram is estimated from a series of profiles parallel to the general direction of the drainage system. It fits the model presented in Fig.3 fairly well and shows a trend extending over more than three sample lags. The dotted line gives an estimation of the covariance function of the trend. It shows that medium to large point deviations with an estimated variance $S_p^2 = (1.00 - 0.53) S_{total}^2$ are present.

Fig.6 displays a number of correlograms of stream sediment data from a survey undertaken in the Chongwe River area in the Central Province of Zambia; the sampling interval is 1.5 km. Correlograms of arsenic values in profiles parallel to the general direction of the given drainage system are not comparable to the correlograms of profiles perpendicular to this system (see Fig.6a, b); it follows that the arsenic surface is strongly anisotropic. In contrast correlograms of chromium values reveal isotropy (Fig.6c, d).

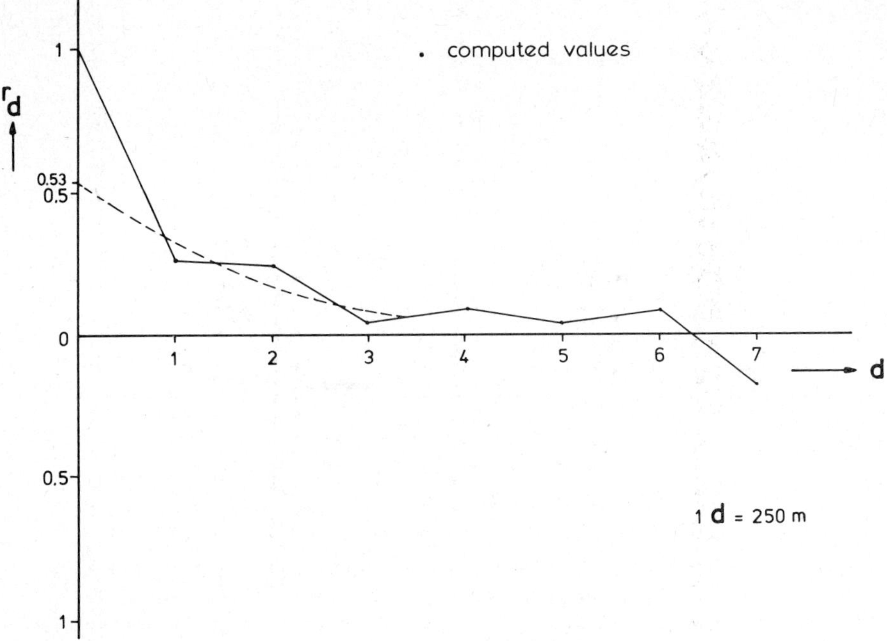

Fig. 5. Correlogram of copper values from a stream sediment survey in the Ingessana Hills area (Sudan). The correlogram is estimated from values of 100 samples, collected at 250-m intervals along four different profiles, all parallel to the general direction of the drainage system. $3d = 750$ m is a conservative estimate of the correlation distance.

Agreement between the above correlograms and the covariance model of Fig. 3 varies from fair to rather poor. The differences are mainly due to the presence of a deterministic trend component, and to a lesser extent to the presence of oscillations, and minor discontinuities.

Fig. 7 refers to a geochemical survey within the Biga Peninsula area in Turkey. The correlogram shows the estimated mutual relationship between zinc values in a profile oblique to the general direction of the drainage pattern; the correlogram deviates considerably from the model presented in Fig. 3. These extensive deviations are caused by the occurrence of discontinuities including a geographic cluster of highly anomalous values within the profile. The geochemical profile of the raw data requires review in such instances; the profile might have to be split into two or more subprofiles.

ANALYSIS OF SURFACE COMPONENTS

With regard to the separation of surface components, according to eq. 1, covariograms of raw geochemical data could provide a fair insight into data structure. The usefulness of autocovariograms for the purpose of separating surface components, however, is greatest if the information supplied by the covariograms is combined with information obtained from other surface analysis techniques. In this context the authors have been employing a combination of covariance studies, a particular cell average method and techniques of variate difference analysis with considerable success.

The cell average method referred to is described by DeGeoffroy et al. (1968). In application this method is simple. The geographic centre of a cluster of relatively high trace metal values is selected as the stationary midpoint of a series of square-shaped cells, which increase in size in a stepwise fashion. Means are computed for all cells in the series and the mean cell values are plotted against cell size. If the data structure is anisotropic the procedures have to be modified, but remain simple. An example of a cell average plot is given in Fig.8 for copper data from the geochemical survey of the Ingessana Hills area in Sudan which was discussed earlier. The plot shows the various components of the geochemical landscape and suggests that these components can be separated by moving average techniques using cell sizes of 5.0×5.0 km^2, 2.5×2.5 km^2 and 1.5×1.5 km^2, respectively. The correlation distance of the autocorrelogram (Fig.5) and the smallest cell size, being twice the correlation distance of 0.75 km, are in good theoretical agreement. A further comparison of Fig.5 and Fig.8 shows that the cell average method renders useful information on the occurrence of large-scale surface components, whereas autocorrelograms provide maximum information in terms of the small-scale components. Cell average plots reflect in one way or another the autocovariance properties of the geochemical landscape; the relationship, however, is complicated.

Methods of variate difference analysis proved to be useful for separating trends from local undulations, especially in those instances where the geochemical surfaces can be described by a summation of a deterministic trend function and uncorrelated random point deviations p with $E(p) = 0$ and $Var(p) = \sigma_p^2$. For a full description of this method reference is made to Kendall (1966). Briefly, the technique is based on the property that the $(n + 1)$th derivative of an nth degree polynomial is zero. It involves computation of the first differences Δz of adjacent equally spaced sample values and their variance $S^2_{\Delta 1}$, followed by the computation of the second differences $\Delta\Delta z$ and their $Var\ S^2_{\Delta 2}$, and so on. From those variances, σ_p^2 is estimated by dividing $S^2_{\Delta 1}, S^2_{\Delta 2}, \ldots S^2_{\Delta j} \ldots$ by appropriate coefficients C_j^*. The procedure is continued until the magnitudes $S_j^2 = S^2_{\Delta j}/C_j$ and $S^2_{j+1} = S^2_{\Delta j+1}/C_{j+1}$ are no longer significantly different. The value j then yields an estimate of the degree of a polynomial function suitable for trend elimination while S_j^2 provides an estimate of σ_p^2. The expression $(S^2_{total} - S_j^2)/S^2_{total}$ constitutes an a-priori estimate of the goodness of fit of the polynomial function.

In geochemical data analysis the assumption that residuals are not autocorrelated often does not hold. In such cases other coefficients C'_j must be employed to arrive at a reasonable estimate for goodness of fit. Estimates for the new factors can be computed with the aid of covariograms of residual

---

* The coefficients C_j for $j=1$ to $j=4$ are 2, 6, 20 and 70, respectively.

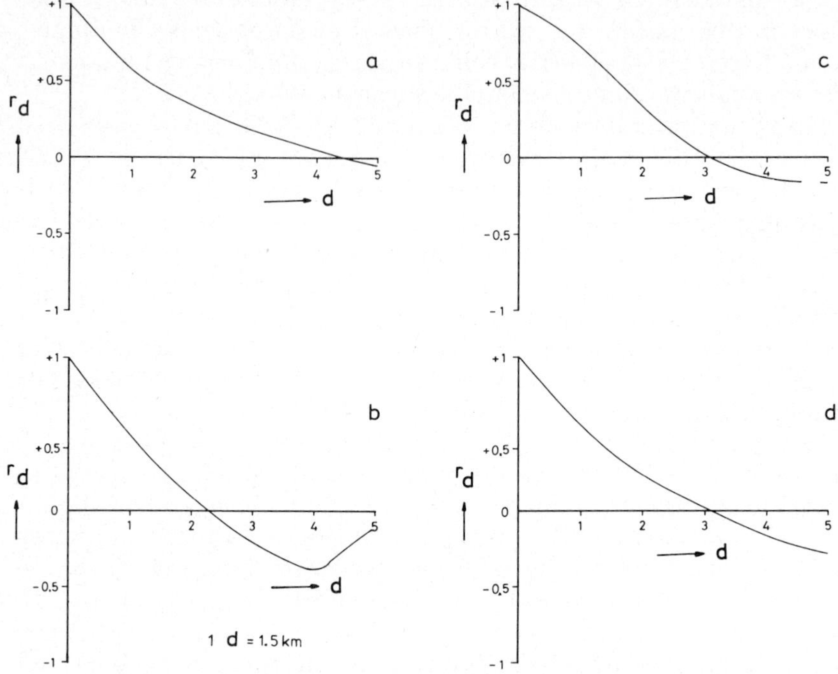

Fig.6. Correlograms from stream sediment data of a geochemical survey in the Chongwe River area (Zambia) for arsenic (a, b) and chromium (c, d), respectively. The correlograms of (a) and (c) refer to series of values parallel to the direction of the drainage system and the correlograms of (b) and (d) to series perpendicular to this direction. On average the correlograms are estimated from three separate series of 25 samples each. The mean sampling interval is 1.5 km.

values obtained after low degree deterministic trend functions have been fitted by trial and error. Fig.9 gives an example of an S_j^2 plot computed by using coefficients estimated in this manner. The plot refers to a profile of chromium data from the Chongwe River area (Zambia).

The plot indicates that the trend component can be eliminated by a polynomial function of degree 4 or 3; it is recommended that the polynomial of lowest degree be chosen. The goodness of fit of a 3rd degree polynomial estimated a priori is 76%; the goodness of fit of the 3rd degree polynomial computed a posteriori is 67%. Such agreements are criteria for the efficiency of the removal of the deterministic component. In the presence of anisotropy, separate S_j^2 plots should be computed for different directions; the results may lead to such decisions as to employ polynomes that are linear in x and cubic in y.

The combined information contained in covariograms, cell average plots, and S_j^2 plots provides a good starting point for the construction of empirical models yielding a maximum understanding of any geochemical surface.

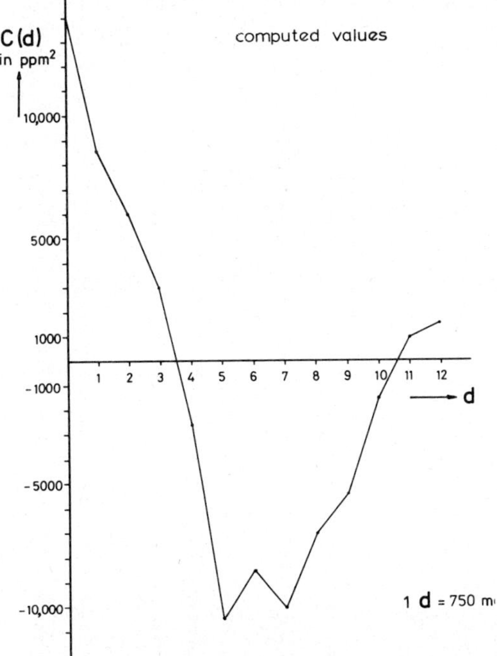

Fig.7. Covariogram of zinc values as estimated from a single series of 40 stream sediment samples along a profile oblique to the general direction of the drainage pattern within an area of the Biga Peninsula (Turkey); the unit d corresponds with sampling intervals of 750 m.

COMPUTATION OF REGIONAL TREND SURFACES

For computation of regional trend surfaces the authors have used both moving average techniques and techniques employing polynomial and piecewise polynomial functions. In the moving average technique cell sizes are chosen with the aid of cell average plots and covariograms. Results were generally satisfactory. Simple polynomial functions are generally insufficiently flexible for following a trend over extended distances, nor are such functions capable of avoiding large undulations of local character. For these reasons the authors prefer piecewise polynomial functions; a general review of the techniques employed is given below.

A piecewise polynomial function is that function $u(x,y)$ which is continuous, possesses continuous derivatives in the whole area under consideration, and equals polynomials of low degree in small subregions of the area as illustrated in Fig.10. The trend surface is thus defined by separate polynomials for the various subregions; these join smoothly with the polynomials of the adjacent subregions. Piecewise polynomials of degree 3 are currently being used. These functions are very flexible and are capable of describing the trend over extended areas. The parameters of the separate polynomials u for the individual subregions

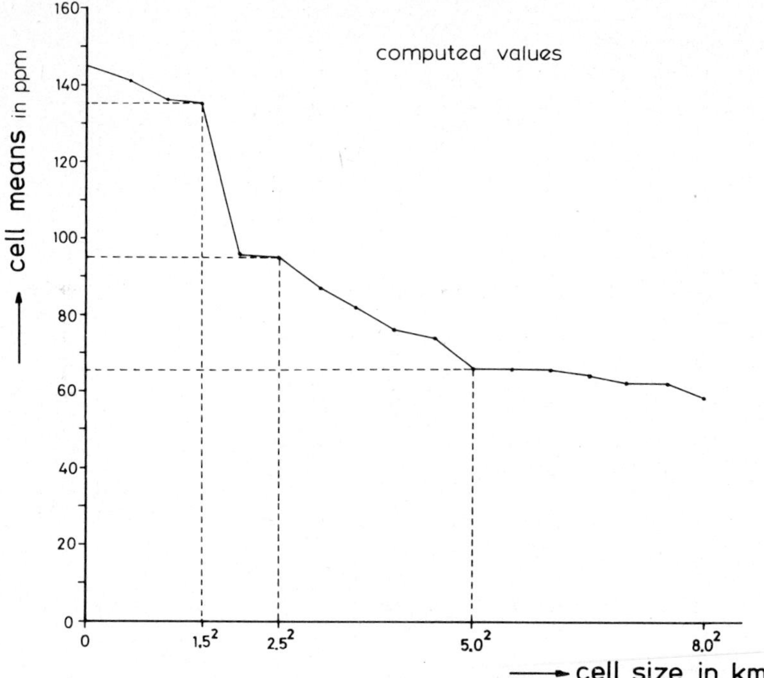

Fig.8. Example of a cell average plot with reference to copper data of a geochemical stream sediment survey in the Ingessana Hills area (Sudan) (cf. correlogram Fig.5).

are computed simultaneously from the expression:

$$P \sum_{\text{points}} [u(x,y) - z(x,y)]^2 + \iint_{\text{region}} (\partial u/\partial x)^2 + (\partial u/\partial y)^2 dx dy = \text{minimum}$$

The first term symbolizes the sum of squares of the residuals, the second term ensures a smooth trend surface. P is a weighting factor. Obviously additional conditions can be imposed, depending on requirements. Knowledge of the homogeneity or inhomogeneity of the trace metal distribution can also be incorporated in the model.

The magnitude of the weight factor, P, is chosen empirically in such a way that the average magnitude of the residuals $(u - Z)$ is equal to its pregiven value. Consequently, if after one run of the computation the values $(u - Z)$ are still too large, the factor P is changed and the computation is repeated with this new value. The pregiven values may either be derived from variate difference plots or be based on the experience of the geochemist regarding the properties of geochemical surfaces, or on both.

Agreement between the pregiven and computed magnitudes of the residuals ensures that the computed surface represents a reasonably defined trend surface. Covariograms of residuals can further determine how efficiently the trend component has been removed. An example of a third-order piecewise polynomial surface with a medium degree of smoothness is displayed in Fig.11.

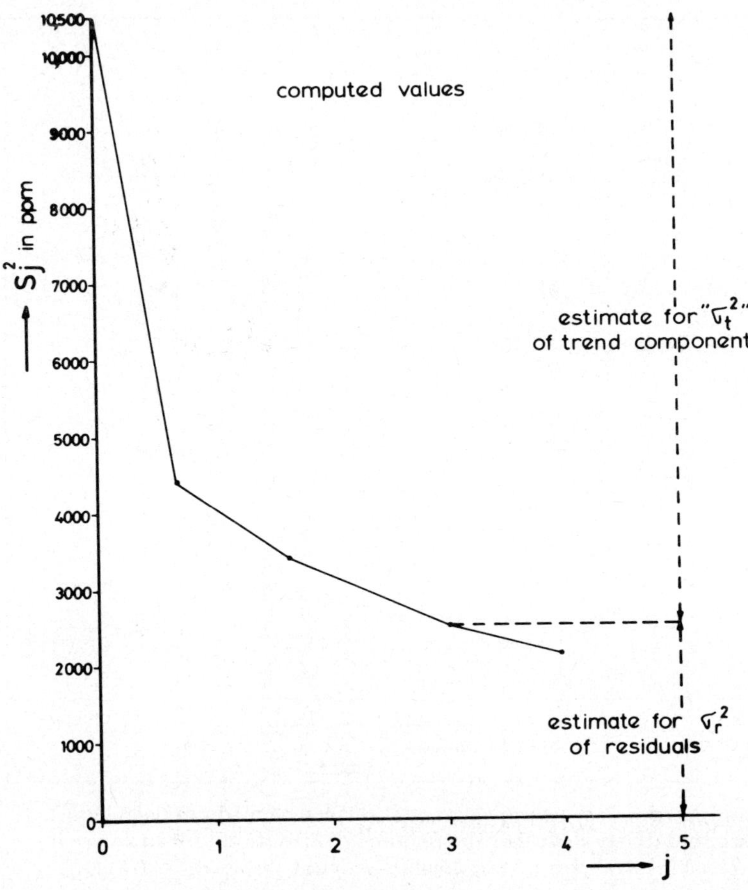

Fig.9. Example of an S_j^2 plot, computed from the same series of raw chromium values, as used for estimating the correlograms of Fig.6c, d. "σ_t^2" is a rough estimate for the fraction of the total variance that can be accounted for by the trend component.

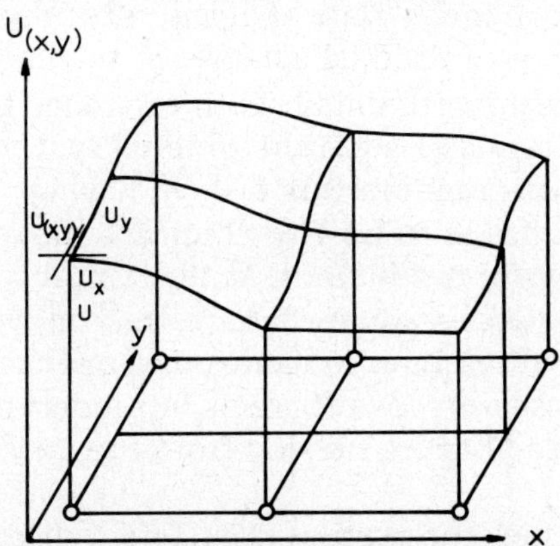

Fig.10. Diagrammatic illustration of a piecewise cubic function; the quantities u, u_x, u_y, u_{xy} at mesh points are the parameters of the function.

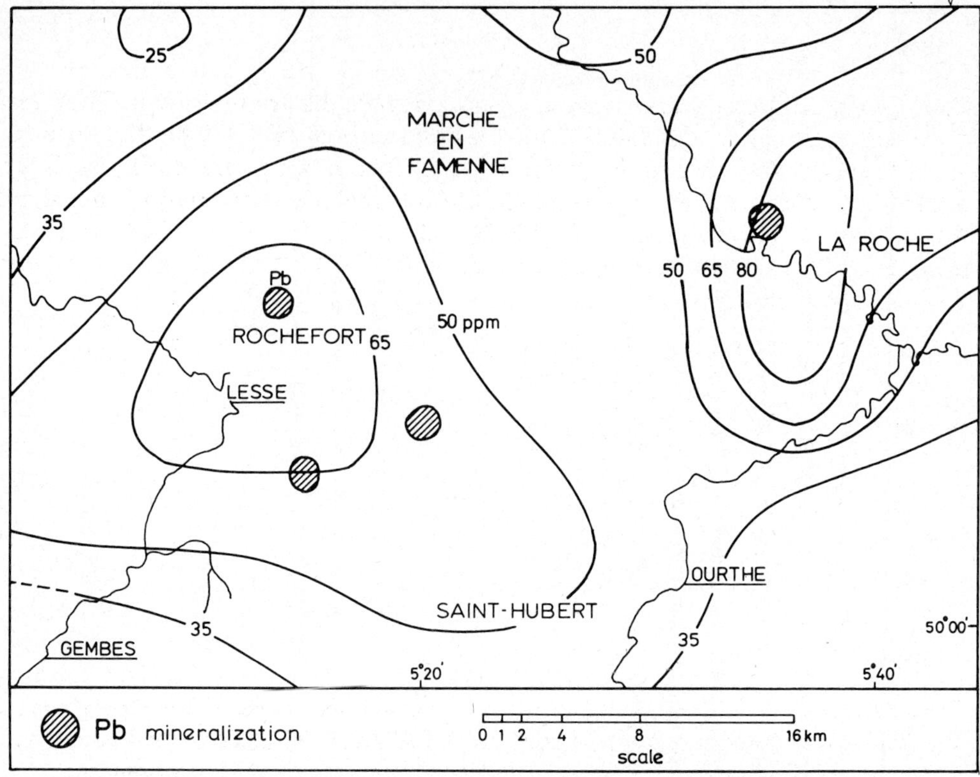

Fig.11. Example of a third-order piecewise polynomial surface of medium degree of smoothness. The surface tentatively describes the regional distribution of lead values in stream sediments within the central part of the Belgian Ardennes.

APPROXIMATION OF RESIDUAL SURFACES

One of the most crucial operations in geochemical data analysis, after the trend has been eliminated, is the computation of residual surfaces in such a way as to extract a maximum amount of meaningful detail. With reference to the covariance model of Fig.3 and the first terms of eq.1, the residual surface can be tentatively described as a homogeneous random function, h, having some autocorrelation. Problems inevitably arise as to how to eliminate the effect of point deviations, p, from the meaningful residuals, h, at the sampling points. In practice one may attempt to introduce a somewhat arbitrary moving average rule, but experience gained in other disciplines indicates that one can do much better if the rules employed are based on covariograms or models of covariance functions. Various approaches are possible ranging from the qualitative to the rather more sophisticated.

Rules can be computed for profiles and rules for profiles in different directions can be combined in moving average designs to provide two-dimensional patterns.

The possibilities of simple moving average rules are demonstrated with reference to copper data collected in the Ingessana Hills area (Sudan); for the sake of this example it is assumed that no large-scale regional trend is present. As no duplicate samples were collected, point deviations are unknown. They can, however, be roughly estimated from the covariogram by splitting the covariogram into a trend component and a point deviation component (Fig.5). If it is accepted that moving average rules should not include data beyond the correlation distance, then a number of rules for estimating the meaningful residuals, h_i, from actual residuals, r_i, appear feasible; for profiles of equally spaced sample points one may use:

$$g_i = 1/5\ (r_{i-2} + r_{i-1} + r_i + r_{i+1} + r_{i+2}) \tag{5a}$$
$$g_i = 1/7\ (r_{i-3} + r_{i-2} + r_{i-1} + r_i + r_{i+1} + r_{i+2} + r_{i+3}) \tag{5b}$$

or weighted moving average rules, e.g.:

$$g_i = 1/4\ (r_{i-1} + 2r_i + r_{i+1}) \tag{6a}$$
$$g_i = 1/9\ (r_{i-2} + 2r_{i-1} + 3r_i + 2r_{i+2} + r_{i+2}) \tag{6b}$$

where g stands for the estimator of h.

Experience shows that an increase in the number of residuals within the computation rule is more efficient in reducing the effect of the point deviation p than is weighting. The relative efficiency of the above rules can be tested by computing the covariance matrix of the errors in g, $e = g - h$ where h stands for the "true" meaningful residuals (cf. Appendix). The best of the above rules is the one which keeps the variances of the errors e as small as possible.

With respect to accuracy, optimum rules for (weighted) moving average schemes can be computed by the prediction theory (cf. Appendix), which was developed by Wiener (1949), empirically applied by Krige (1966) and utilized by Matheron (1965). A fair estimate of the covariance function of the functional values is required in the application of this theory. In the absence of a model this function may be tentatively approximated by the manually smoothed covariogram. Approximations can be improved if σ_p^2 is known from an orientation survey.

An appropriate model of a covariance function for many residual surfaces is given by:

$$Cov\ (h_i, h_{i+d}) = \sigma_0^2 \exp\ (-d/a) \tag{7}$$

It has been tacitly assumed that correlation distances of residuals exceed the sampling distance; otherwise contouring of individual residuals is obviously not meaningful. It emerges that covariograms of residuals constitute a powerful check on sampling density requirements. Pending determination of the magnitude of prevailing point deviations it is thought preferable that sampling intervals in direct exploration surveys be less than 1/2 to 1/3 the correlation distance of the residuals.

ACCURACY OF RESIDUAL SURFACES AND DEFINITION OF THRESHOLD

With regard to the selection of equivalue lines for delineation of meaningful anomalies on residual maps, the authors are inclined to advocate an approach that is based on the concept of surface accuracy. This approach demands that trend components as defined in section "Analysis of surface components" (p.148) have been satisfactorily removed and that the residuals can be described by homogeneous random functions with expectation equal to zero and appreciable autocorrelation. If point deviations have largely been eliminated and equivalue lines are then drawn by linear interpolation, map accuracy is essentially governed by the spacing of the sampling points in terms of the correlation length of the meaningful residuals, h. The accuracy of this interpolation process can be derived by the law of propagation of errors (Kubik and Clerici, 1973). For the particular covariance function of eq.7, this interpolation accuracy is given in the Appendix. In many instances covariograms of the meaningful residuals, h, reasonably satisfy the covariance function of eq.7 and mean surface accuracies, σ_{mean}, for given sampling spacings, D, can be read off from the accuracy graph of Fig.12, which summarizes the results of the Appendix. The parameters σ_0 and a needed to enter the graph must be estimated from the covariogram of the residuals h. The use of the graph is demonstrated employing the data found in Fig.13. If it is accepted that point deviations, p, are zero then σ_0^2 is 3600 ppm, σ_0 is 60 ppm, and $D/a = 0.75$. According to the accuracy graph, the average accuracy of the interpolated surface $\sigma_{mean} = 60 \times 9/100 = 5.4$ ppm. The influence of sampling spacing on the map accuracy is evident. Doubling the spacing in the given example increases the average interpolation error to $60 \times 37/100 = 22$ ppm, a further increase of the sampling distance to three times the original spacing gives a further increase in the map error to 43 ppm.
In the presence of point deviations the mean map accuracy will be inferior, and for sparse sample point arrangements it is approximately true that:

$$\sigma^2 = \sigma^2_{mean} + \sigma_p^2$$

where σ_{mean} stands for the interpolation accuracy as read off Fig.12 and σ_p is the standard deviation of the point error. This approximation formula is valid for surveys with sample intervals larger than 1/5 the correlation distance of residual values.
In choosing the contouring interval between equivalue lines, the authors suggest the use of intervals of 2 times the surface accuracy, the equivalue lines $\pm 2\sigma$ being the two threshold equivalue lines. This quantity 2σ may ultimately be decisive for the selection of sample patterns.

CONCLUSIONS

We have attempted to demonstrate the usefulness of the concept of autocovariance in solving a great variety of problems encountered in geochemical data

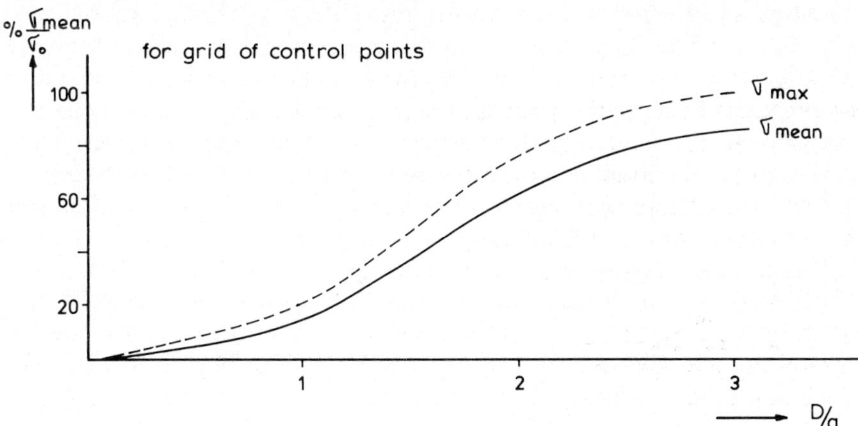

Fig.12. Relationship between sampling spacing and interpolation accuracy; this relationship holds if point deviations have been eliminated and eq.7 applies.

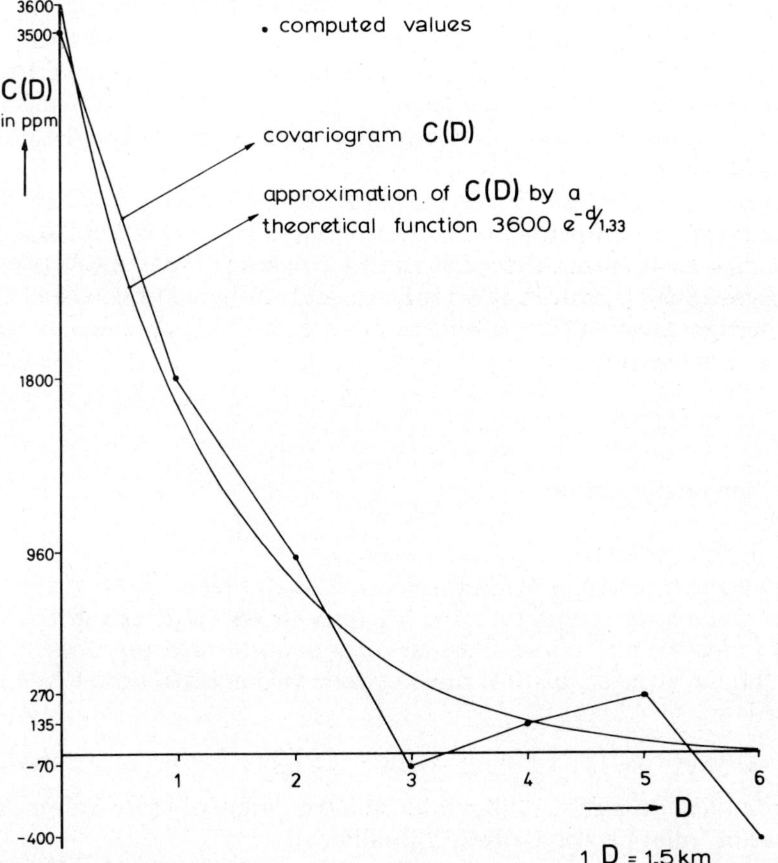

Fig.13. Covariogram $C(D)$ from residual stream sediment data of the Chongwe River area (Zambia) for chromium, after trend elimination. The covariogram is estimated from three series of 25 residual values each (compare Figs.6a, b and 9). The covariogram is fairly approximated by the function of eq.7.

analysis. Qualitative studies of autocovariograms already provide a great deal of information on data structure; maximum information, however, is obtained through quantitative approaches in which autocorrelation studies are combined with other available techniques of data analysis. It is strongly recommended that medium density geochemical surveys be preceded by orientation surveys. Early knowledge of the character of autocovariograms may help to define realistic objectives for a survey and favourably effect decisions regarding sampling techniques and sampling densities to be employed.

The authors are fully aware that their model, which in fact is a dynamic model, requires additional verification by surveys that are continuously analysed by techniques as outlined in the various sections of this paper. They hope that the model and the techniques proposed will contribute to optimizing the effectiveness of medium density geochemical surveys.

ACKNOWLEDGEMENTS

The authors wish to thank the Director of T.C. Maden Tetkik Ve Arama Enstitüsü, Genel Direktörlüğü, Ankara, the Director of the Geological and Mineral Resources Department of the Democratic Republic of Sudan and the Director of the Geological Survey Department of the Republic of Zambia for their kindness in providing a wealth of information with regard to the data of actual geochemical surveys.

Thanks are also due to Mr. A.G. Botman, Mr. D. Kovacs, Mr. A.A. Stamatiou and all former and present students of the International courses in Mining Exploration at ITC, for their substantial contributions towards writing this paper. The valuable suggestions of Prof. H.J. Roorda are gratefully acknowledged.

Mr. W.A. Hugens is thanked for drawing all figures and Miss A.H.J. Mallegrom for preparing the manuscript.

APPENDIX

Accuracy of surface interpolation

The problem of interpolation
In the interpolation problem a set of sample values $z_\alpha = z(x_\alpha, y_\alpha)$ $\alpha = 1, \ldots n$ is given after which an approximation g for the trace metal value at a given position (x_p, y_p) must be computed. Various interpolation rules are possible; one example is the linear interpolation rule between two control points in a profile (cf. Fig.A-1):

$$g = [1 - (x_p - x_1)/(x_2 - x_1)]z_1 + [(x_p - x_1)/(x_2 - x_1)]z_2 \tag{A1}$$

others include the local quadratic, cubic interpolation, etc. All these commonly known interpolation rules can be written formally as:

$$g(x_p, y_p) = A \cdot z_c \tag{A2}$$

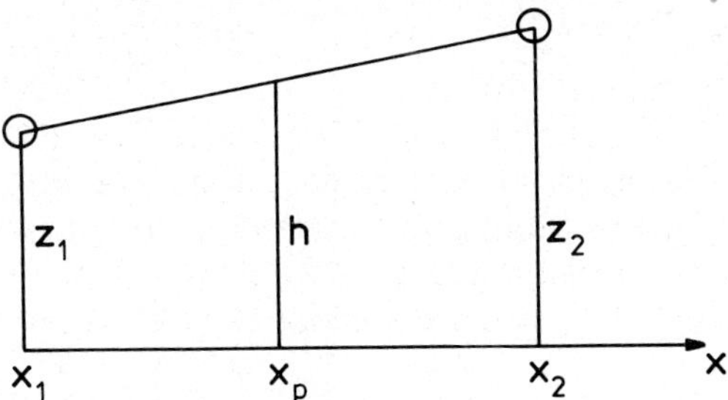

Fig.A-1. Linear interpolation.

where z_c denotes the array of sample values, $z_c = (z_\alpha)^T$, $\alpha = 1, \ldots n$, and A denotes an array with elements $[a_\alpha (x_p,y_p)]$ depending on the position (x_p,y_p) of the point to be interpolated. The error e of the interpolated value g will be equal to

$$e(x_p,y_p) = z(x_p,y_p) - g(x_p,y_p) = z(x_p,y_p) - A \cdot z_c \tag{A3}$$

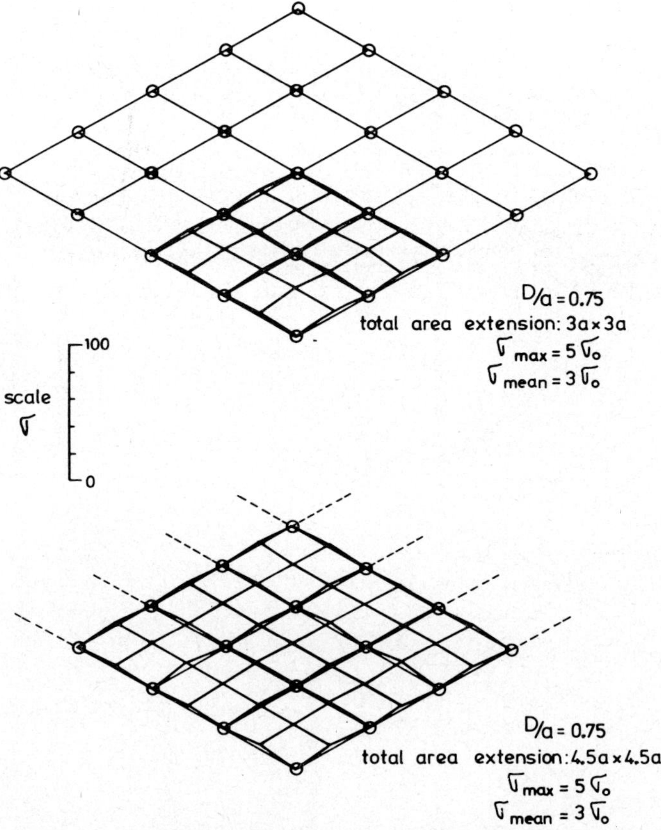

Fig.A-2. Prediction interpolation for grids with a spacing between control points of $0.75a$ (accuracies in percentages of σ_0).

160

and the variance of this error is given by:

$$Q_{ee} = Cov(p,p) + A \cdot Cov(\alpha,\beta)A^T - 2A \cdot Cov(\alpha,p), \qquad \alpha,\beta = 1, \ldots n \qquad (A4)$$

An interpolation method exists which minimizes this variance, namely the method of prediction (cf. Wiener, 1949). The coefficient matrix, A, for prediction interpolation is:

$$A = Cov\,(p,\alpha)\,Cov^{-1}\,(\alpha,\beta) \qquad (A5)$$

and the error variance is found by the operation:

$$Q_{ee} = Cov(p,p) - Cov(p,\alpha)Cov^{-1}(\alpha,\beta)Cov(\beta,p) \qquad (A6)$$

Accuracy formulae A4 and A6 are used for the accuracy computations in the following section.

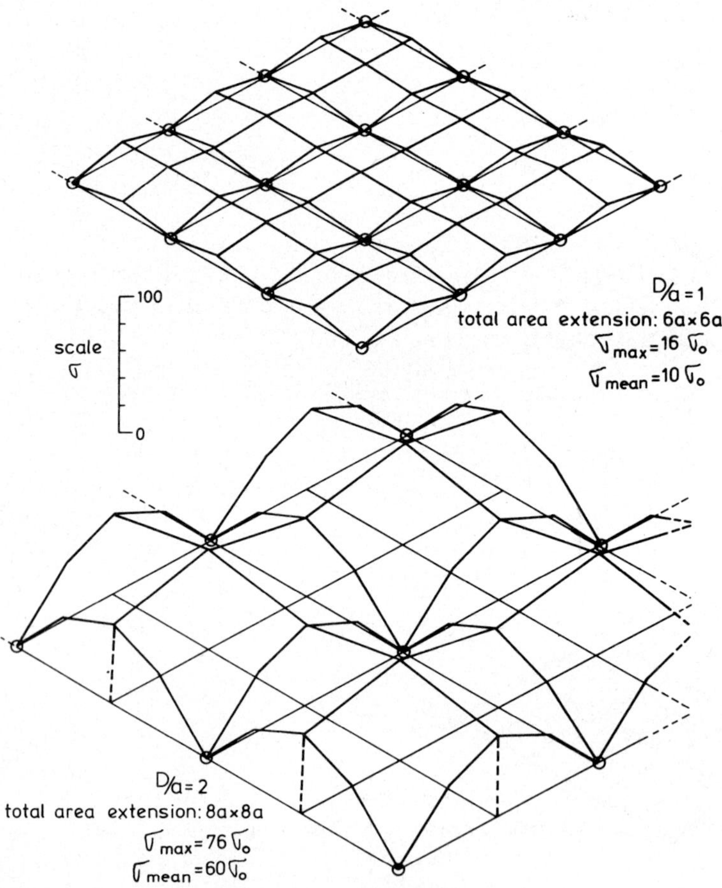

Fig.A-3. Prediction interpolation for grids with spacings between control points of 1a and 2a respectively (accuracies in percentages of σ_0).

The accuracy results indicated are based on the covariance function given by:

$$Cov\ (i,j) = \sigma_0^2 \exp\ (-\,d(i,j)/a) \tag{A7}$$

These theoretical results can be readily applied to practical cases. For studying the accuracy of interpolation the standard deviation $\sigma = Q_{ee}^{1/2}$ of the interpolation error, its maximum σ_{max} and its mean σ_{mean} (equal to the sum of the standard deviations of all interpolated points, divided by the number of interpolated points) are used. The standard deviation σ is expressed in units of σ_0.

Interpolation accuracy with grids of sample points

In this interpolation problem a grid of error free sample points with equal grid spacing both in x- and y-direction is given. The accuracies were computed for prediction interpolation. They apply approximately to all other meaningful interpolation methods. Figs.A-2 and A-3 show the accuracy results for prediction interpolation for various sample point spacings and area extensions.

The figures show that local maxima of σ appear in the centres of the meshes. These maxima are dependent on the extension of the grid and the position of the mesh in the grid. For prediction interpolation the local maxima of σ decrease to a limiting value with increasing grid extension. This limiting value is reached for grid extensions equal to $4a$ independently of control spacing.

The accuracy parameters σ_{mean} and σ_{max} increase with increasing control spacing. The increase in both σ_{mean} and σ_{max} is relatively low for very short control spacings up to $0.5a$, (cf. Figs.A-2, A-3 and Fig.12), and for control spacings above $3a$. Note that the interpolation accuracy becomes practically independent of the control spacing for spacings above $3.5a$.

REFERENCES

Agterberg, F.P., 1970. Autocorrelation functions in geology. Proceedings of a Colloquium on Geostatistics, University of Kansas, Lawrence, June 1970. In: D.F. Ed. Merriam (Editor), Series Computer Applications in the Earth Sciences, Geostatistics. Plenum Press, N.Y., 113—141

DeGeoffroy, J., Wu, S.M. and Heins, R.W., 1968. Selection of drilling targets from geochemical data of the southwest Wisconsin zinc area. Econ. Geol., 63: 787—795

Kendall, M.G., 1966. The Advanced Theory of Statistics, 3. Griffin, London, 552 pp.

Krige, D.G., 1966. Two-dimensional weighted moving average trend surfaces for ore valuation. In: Symp. on Mathematical Statistics and Computer Applications in Ore Valuation, Johannesburg, March, 1966 — J. S. Afr. Inst. Min. Metall., pp.13—38

Kubik, K. and Clerici, E., 1973. The theoretical accuracy of point interpolation on topographic surfaces. Rept. Ministry of Transport and Waterstaat, 26 pp. (unpublished)

Matheron, G., 1965. Les Variables régionalisées et leur Estimation. Masson, Paris, 306 pp.

Wiener, N., 1949. Extrapolation, Interpolation and Smoothing of Stationary Time Series. John Wiley and Sons, New York, N.Y., 163 pp.

LAKEVIEW REVISITED: VARIOGRAMS AND CORRESPONDENCE ANALYSIS — NEW TOOLS FOR THE UNDERSTANDING OF GEO-CHEMICAL DATA

MICHEL DAVID and MICHEL DAGBERT

Department of Mineral Engineering, Ecole Polytechnique, Montreal, Que. (Canada)

ABSTRACT

This paper intends to show how two new statistical techniques can help to answer several questions raised in geochemical data interpretation. The variogram will show the interdependence of samples and possible trends. It will thus help to answer questions concerning sample spacing, sample size, and level of variation. It also helps to point out anisotropies in an environment. Correspondence Analysis is a new distribution free factorial analysis technique where both sample-space and variable-space are considered at the same time. It produces diagrams which can be thought of as generalized petrographic diagrams, showing both samples and variables, and from which geological factors can be extracted. Several examples will be discussed, and particular attention will be given to the Lakeview Pluton of southern California, which was carefully sampled several years ago by Baird and McIntyre. Several examples of this technique applied to exporation geochemistry will be discussed, and it will be shown how factors can be mapped and anomalous areas detected.

INTRODUCTION

The intention of this paper is to introduce two statistical techniques originally developed in France but largely unknown to North American geologists and geochemists — variogram analysis and correspondence analysis. The first might be familiar to geologists dealing with ore reserves estimation; it was developed mainly by Matheron (1963). The second, attributed to Benzecri (1973), has recently been applied in petrology (David and Woussen, 1973; David and Beauchemin, 1974; David et al., 1974). The usefulness of these techniques in analyzing geochemical data is described with reference to the Lakeview Mountains Pluton, studied by Baird et al. (1967), Morton (1969) and Morton et al. (1969).

We will consider questions relating to size and variability of samples, the level of homogeneity in the distributions of chemical elements and their consequences for sampling patterns and mapping techniques. These questions will be answered through variogram studies and their relationship to the usually used, but sometimes wrongly applied, analysis of variance. A second group of questions relates to description of zoning within the batholith and possible inference regarding its origin. These aspects will be examined by correspondence analysis.

VARIOGRAM ANALYSIS

Spatial correlation: We will restrict ourselves to two-dimensional problems, because geochemical samples are usually taken in a two-dimensional field. The problem of sample size is the problem of resemblance between a sample of size v and its environment; in other words we should study the relationship between grades at one point and grades at other points. The same problem of resemblance appears when one wants to study levels of variations. In other words both problems are associated with spatial correlation. Correlograms have already been computed for geochemical samples but to our knowledge the full sequence of deductions which could have been drawn has not been exploited, e.g., computing the variance of other kinds (sizes) of samples, and estimation variances.

Rather than studying correlation, which can be shown in some cases to raise serious statistical problems (Matheron, 1972), we will study the variogram which, besides having interesting statistical properties, appears to be the simplest function to describe the relationship or difference between values of a variable at a given spacing.

The variogram: Considering two points x and $x + h$ where h is a given fixed vector, the simplest way to compare the two values $f(x)$ and $f(x + h)$ is to take their difference. However, we are not interested in the sign of this difference nor in the two particular points x and $x + h$. We therefore square the difference and do the same for all pairs of samples separated by a distance h and then average these values. This averaging means that we assume that the amplitude of the variations is the same everywhere. This is an assumption of stationarity, which can also be used for the reverse purpose: that is we can compute this average in different areas and draw conclusions about the homogeneity of the process.

This average value is the variogram, $2\gamma(h)$:

$$2\gamma(h) = 1/V \int_{\underline{V}} [(f(x) - f(x+h)]^2 \, dx$$

where $\underline{V}$ is the field in which variable $f(x)$ is studied and V the measure of that set $\underline{V}$.

Anisotropy: The preceding function varies with the direction of h, thus giving information about anisotropies in variations of the chemical element considered, and allowing precise definition of the shape of the zone of influence of a sample.

Continuity and structure: The rate of increase of $\gamma(h)$ allows us to characterize the continuity of the chemical elements. In many cases, after increasing to a given value, the variogram tends to level off at a sill. The curve can in many circumstances be represented by an equation like the following one, the spherical model:

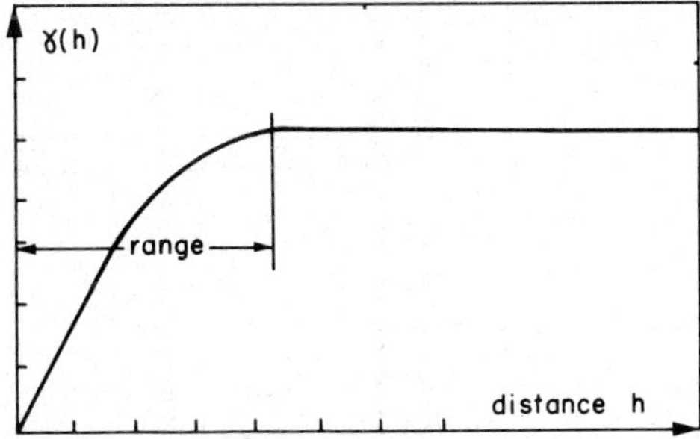

Fig.1. The spherical variogram.

$$\gamma(h) = C\,(1.5\,h/a - 0.5\,h^3/a^3) \qquad h \leqslant a$$
$$\gamma(h) = C \qquad\qquad\qquad\qquad\qquad h > a$$

Its shape is shown in Fig.1. Such a curve is in fact characteristic of a structure; it denotes the existence of units of size a within which grades are dependent on each other and independent of the grades of the next unit. This curve can also characterize the level of variations sought by people interested in sampling or genetic studies. It summarizes all the information needed to compute the variance of samples within a given field, a grid unit or an outcrop.

The variance of sample s within a larger area, S

If there is no structure to the spatial variations of an element, then the variance of samples should be the same, whatever the size of field in which they are taken, variability of the estimates being only a function of the number of samples (see analysis of variance). In fact, where a structure exists as is usually the case, the variance $\sigma^2\,(s/S)$ of a sample s within a field S, can be computed from the variogram (Matheron, 1963). The formula is very simple:

$$\sigma^2\,(s/S) = (1/S^2)\,\int_S dy \int_S \gamma\,|x - y|\,dx - (1/s^2)\,\int_s dy \int_s \gamma\,|x - y|\,dx$$

where γ is the variogram of the element under study. The function $(1/S^2)\,\int_S dy \int_S \gamma\,|x - y|\,dx$ is sometimes called the F function and it has been tabulated for simple shapes of S. An example is given in Fig.2 for a rectangle (a, b) and a spherical variogram $\gamma\,(h)$.

Influence of the size of the sample

The preceding formula also allows us to decide what size of sample should be taken to sample a given area. Thus, if rewritten as:

$$\sigma^2\,(s/S) = F(S) - F(s)$$

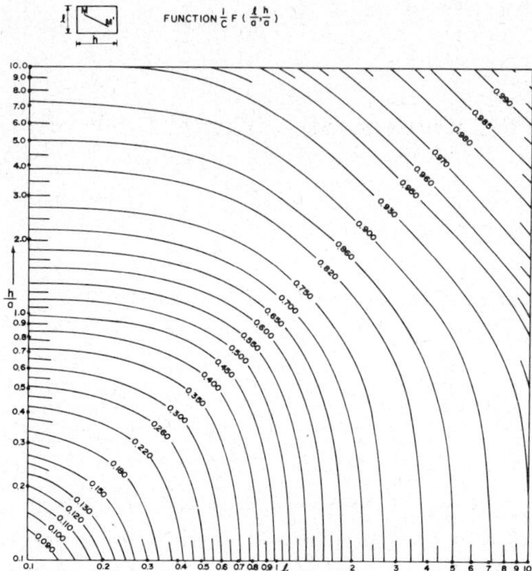

Fig.2. The F function for a rectangle and a spherical variogram.

we see that for a change in the size of a sample, s, the only value which changes is $F(s)$. Usually $F(s)$ is very small and $F(S)$ large, so that doubling or tripling the size of s, usually has very little effect on $\sigma^2 (s/S)$, except if the structure has a range a comparable to the size of the samples. Let us take an example: suppose a variable with a zone of influence of 2000 ft, a sample s_1, of 6 in. $\times$ 3 in., and an investigated zone of $(10{,}000 \text{ ft})^2$. Then the ratios $(l/a, h/a)$ for the sample, are virtually zero and $F(s_1)$ is zero, while $F(S)$ is equal to $F(10{,}000/2000, 10{,}000/2000) = 0.98C$ after Fig.2. Now if we take samples, $s_2 = 1 \text{ ft}^2$, $F(s_2)$ is almost unchanged and practically zero, so that there is absolutely no point in taking the trouble to carry such samples!

Now suppose a variable with a range of 30 in., the same samples as before and the same field under investigation. Then $F(s_1)$ becomes $F(6/30, 3/30) = F(0.2, 0.1) = 0.12C$ after Fig.2; $F(S)$ is equal to C after the same chart. Now for samples s_2, $F(s_2)$ is $F(12/30, 12/30) = 0.32C$ so that we have:

$$\sigma^2 (s_1/S) = C (1 - 0.12) = 0.88C$$

and

$$\sigma^2 (s_2/S) = C (1 - 0.32) = 0.68C$$

which makes a significant difference, i.e. a small range of influence makes it worthwhile to take large samples. Ranges of a few inches are not uncommon and would have an even larger effect.

Of course it is not possible to know the range before sampling, but once a series of samples has been taken, then it is possible to decide how the next series should be collected.

This phenomenon is of course well known to geologists. One usually says that if there are large small-scale variations, large samples should be collected. The variogram quantifies the process. In fact, most real situations are more complicated and there are usually nested structures with different levels of variations.

Examples of multiple levels of variations and nested structures

Serra (1970) has studied in great detail the iron ore deposit of eastern France. He reported the existence of 6 successive levels of variation, the ranges of which were as follows: 200 μ, 2 cm, 2 m, 300 m, 3 km and 15 km. This means that there exist intermeshed units like the one shown in Fig.3,, where grades inside one unit are correlated, but independent of the ones in the next unit, while the units of one size are correlated within a large envelope and so on. Obviously, the same kind of sample cannot demonstrate the existence of both a 2 cm and 3 km range. However, successive grouping will do so and simple variogram computations have shown the existence of levels of variation, especially the 300 m structure, which were sometimes unknown to experienced mine geologists. This shows that it is hard to make the right choice of sample size from intuition alone.

A similar type of nested structure was sought by Krumbein and Slack (1959) when they studied the radioactivity of Pennsylvanian black fissile shales in Illinois. They arbitrarily defined a priori structural units like super-township, township, mines "major units", "minor units" and samples, and constructed an hierarchical model of variance analysis. The observations, however, were probably not independent of each other and thus the statistical tests of significance are probably questionable. Results of Serra (1970) indicate that a variogram study would probably have allowed an objective definition of the real units, with their real size.

When several interlocked structures exist the situation can usually be described by one variogram which is the sum of several independent terms. Now in variance computations which are linear with respect to $\gamma(h)$, each term contributes its part. The calculation has only to be repeated several times. Usually, however, the relative size of the samples and field under investigation are such that only 2 or 3 levels of variation have to be considered at a time. In the iron ore example, if samples are 1 m long, the first two structures are irrelevant as are the last two. We now consider the common and simple case of only two structures.

The nugget effect and common case of two levels of variations

If for a given size of sample, there exist two ranges, one of 3 ft and one of 10,000 ft, then the first structure is represented by:

$$\gamma_1(h) = C_1 \left(1.5\, h/a_1 - h^3/2a_1^3 \right) \qquad\qquad h \leqslant a_1 = 3 \text{ ft}$$

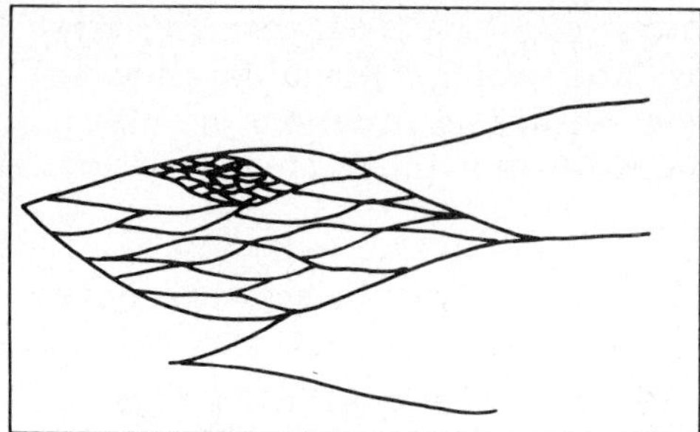

Fig.3. Idealized nested structures.

and the second by:

$$\gamma_2(h) = C_2 \ (1.5 \ h/a_2 - h^3/2a_2{}^3) \qquad\qquad h \leqslant a_2 = 10,000 \ \text{ft}$$

The resulting process has a variogram:

$$\gamma(h) = \gamma_1(h) + \gamma_2(h)$$

If plotted on graph paper, it is almost impossible to see $\gamma_1(h)$. It looks as if $\gamma(0) \neq 0$. This discontinuity at the origin is called the nugget effect because this phenomenon is very common in gold mines. If samples closer to each other than a_1 are never considered we can write the equation very simply:

$$\gamma(h) = C_0 + C_2 \ (1.5 \ h/a_2 - 0.5 \ h^3/a_2^3) \qquad\qquad h \leqslant 10,000 \ \text{ft}$$
$$\gamma(h) = C_0 + C_2 \qquad\qquad h \geqslant 10,000 \ \text{ft}$$

Effect of C_0 on variance computation (Matheron, 1972)

The simplified equation cannot be used to find the effect of C_0. Thus, the complete equation $\gamma = \gamma_1 + \gamma_2$ is needed to show the relationship between γ_1 and C_0. The practical result, however, is very simple and we find that given a variogram $\gamma(h) = C_0 + \gamma_2(h)$ computed from samples of size s, the variance $\sigma^2(s/S)$ is increased by C_0, provided a_1 is smaller than s. If a_1 is larger than s then one simply applies the variance algorithm twice to the two variograms γ_1 and γ_2 and adds the resulting contributions.

It is apparent that the effect of taking two samples a short distance apart does not necessarily halve the nugget effect. The samples have to be independent and this is not the case when there exists a range of say 10 ft and the samples are 3 ft apart. Similarly if the range is only 1 or 2 inches, a sample twice as large is equivalent to two samples a few feet apart and a sample nine times as large is equivalent to nine carefully chosen samples a few feet from each other.

In the case of the Lorraine iron ore deposit, 3 samples collected at 1-m intervals provide a sample with a nugget effect reduced by a factor of 3; if they were collected at 20-cm intervals they would not reduce the nugget effect at all! In the case of the Lakeview batholith, pairs of samples were collected at

5-ft intervals which is probably better than two samples at a 6-in. interval. The nugget effect is especially important in the problem of precision of the estimation of a mean.

Estimation variance of a mean value

For n independent samples with a variance σ^2 the average value of these n samples has a variance σ^2/n. This very simple formula, which indicates how many independent samples should be collected to achieve a given precision in terms of average value, is counterbalanced by the following one in the case of non-independent samples:

$$\sigma^2{}_N = (2/NS)\sum_i \int_S \gamma|x_i - x|\,dx - (1/S^2)\int_S dx \int_S \gamma|x - y|\,dy - (1/N^2)\sum_i\sum_j \gamma|x_i - x_j|$$

if n samples are collected to estimate the average value on set S. This is not solvable on a slide rule or even a desk top computer. Fortunately charts are available and, for instance, a good approximation for the case where area S is sampled regularly on a grid of side l can be obtained from Fig.4. This curve gives the estimation variance $\sigma^2{}_e$ of one grid estimated by one central sample. If n grid cells exist, the estimation variance of their average value is approximately $\sigma^2{}_e/n$. If a nugget effect is present it is added to $\sigma^2{}_e$. Other simple estimation variances have been charted by Matheron and Serra so that the question, "How many samples are needed to provide a given precision with a given geometric pattern?" can be answered readily. Matern (1960) showed how to collect samples in a forest. This is directly applicable to geochemistry. Again it is a function of the variogram and thus a first sampling campaign is necessary to define the variogram in the area under investigation.

The nugget effect and mapping

Sometimes there is a perfect nugget effect, the variogram is flat, meaning that samples are all independent of each other. This means then that no mapping should be attempted, since the expected value at all points other than sampling points is the same, namely the average value of all the samples.

Application to the Lakeview Mountains Pluton

We quote from the paper of Morton et al. (1969): "A part of the southern California batholith, it (the Lakeview Mountains Pluton) is a steep-walled body with a tear-shaped ground plan, exposed discontinuously over an area of 100 to 130 km^2 .... The Pluton is almost entirely coarse-grained hornblende biotite quartz diorite that lacks potassium-feldspar. The sampling was done in two successive phases, one to define the size of samples and scale of variation and a second systematic one, following a regular square grid of

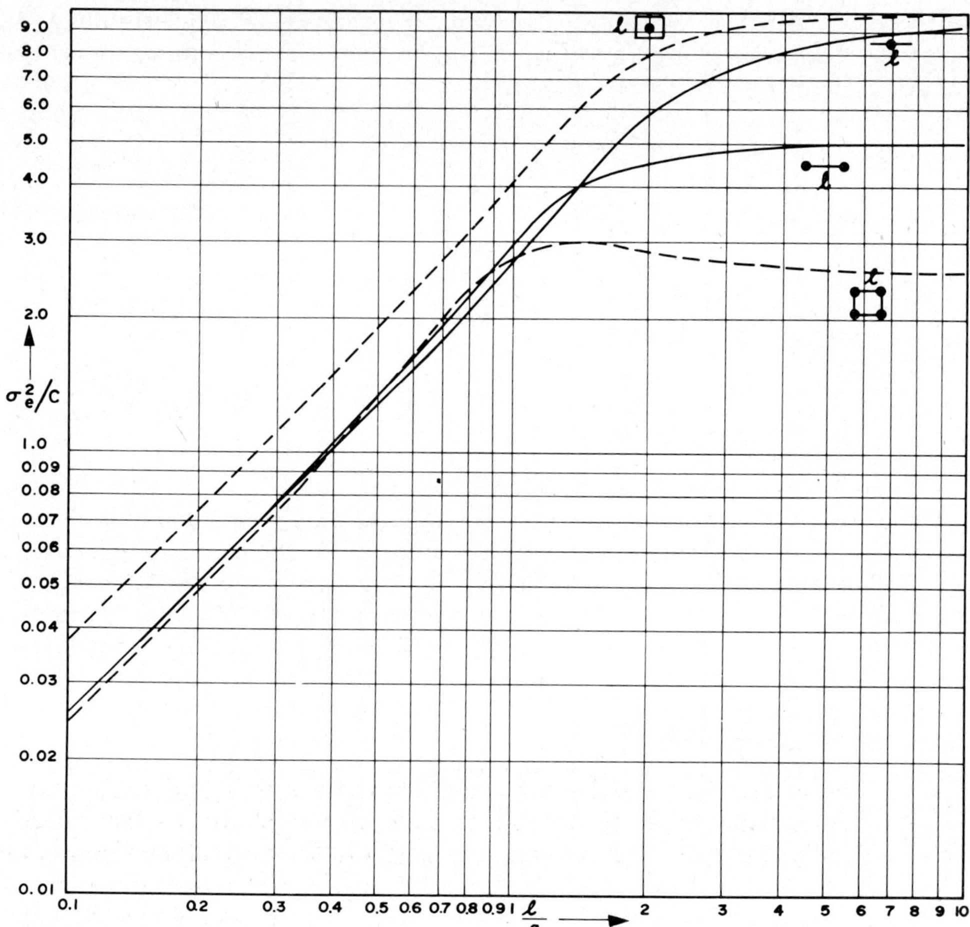

Fig.4. Elementary extension variances.

2000 ft × 2000 ft. This yielded 165 samples, which were analysed for Na, Mg, Al, Si, K, Ca, Fe and specific gravity." We obtained the data from the ASIS National Auxiliary Publication Service and deleted a few aberrant values, keeping a total of 147 samples. It should be made clear that we do not wish to criticize or even comment on Morton's excellent paper but we found his data set the most convenient to suit our purpose.

The variograms

The first general remark which can be made is that they exhibit no real anisotropy (Fig.5). The variations are approximately equally important in all the directions with a tendency to be higher in the direction U, reflecting the shape of the Pluton. The second point is that they can be separated into three groups. Silicon, calcium and potassium are similar with a low nugget effect,

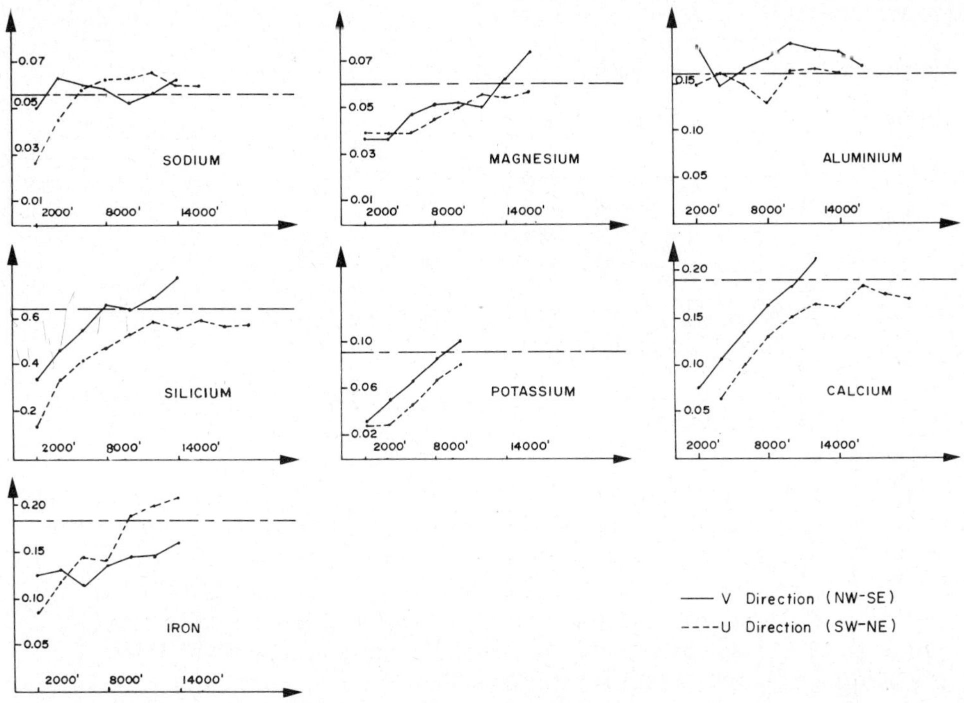

Fig.5. Variograms of elements for Lakeview Mountains Pluton.

while iron and magnesium show the same behaviour with a marked C_0, and sodium and aluminium constitute a third class with an almost perfect nugget effect. Finally, except for aluminium and sodium, they can all be represented acceptably by spherical models with a range of approximately 12,000 ft.

As an example of computation, we can determine the precision with which the average composition of the batholith is known with 162 samples, as defined in the paper (nine 0.400-kg samples averaged at each location). For a given element, x, having a variogram defined by a nugget effect, C_0, a variance (sill) $C + C_0$ and a range a, the variance σ^2_e will be read from the curve of Fig.4; the estimation variance of the whole area is then given by $\sigma^2_E = (C_0 + \sigma^2_e)/162$. An approximate calculation yields Table I.

The most important aspect is the small-scale structure of which little is known. The way samples are taken at each grid point influences the nugget effect, and unfortunately the range of this small-scale phenomenon is not known. We can, however, state that a smaller number of locations with more samples is much better than a large number of locations with a smaller number of samples per location.

CORRESPONDENCE ANALYSIS APPLIED TO LAKEVIEW MOUNTAINS PLUTON

Correspondence analysis is attributable to Benzecri (1973). It has been de-

TABLE I

Estimation variance of the average composition of the Lakeview Mountains Pluton

	(C_0+C)	C_0	C	σ_e^2/C	$\sigma^2 e$	$\sigma^2 e + C_0$	$\sigma^2 E$	$2\sigma E$
Na	0.0077	0.0077	0	0.06	0	0.007	0.00005	0.013
Mg	0.063	0.035	0.028	0.06	0.00168	0.036	0.00022	0.03
Al	0.16	0.16	0	0.06	0	0.16	0.001	0.06
Si	0.064	0.02	0.044	0.06	0.00264	0.022	0.00013	0.023
K	0.091	0.02	0.071	0.06	0.00426	0.024	0.00014	0.024
Ca	0.22	0.05	0.17	0.06	0.0102	0.06	0.00037	0.04
Fe	0.25	0.10	0.15	0.06	0.0090	0.11	0.00068	0.05

C_0 = nugget effect, $C+C_0$ = variance (sill), σ_e^2 = variance from Fig.4, σ_E^2 = estimation variance of the average composition.

scribed in detail in many papers and might be more attractive to the geologist than variogram methods, since the bulk of the computations are performed by a single program and the interpretation requires no special knowledge except that of standard factorial analysis. Recent papers describing the technique include David and Woussen (1973), David and Beauchemin (1974), David et al. (1974), Dagbert and David (1974); also see Cazes (1970).

We assume that the reader is familiar with R- and Q-mode analysis. The correspondence analysis method is a distribution free technique of factorial analysis which uses a special weighting function (equivalent to a χ^2 distance instead of a euclidean distance) and makes use of the relationship between factors of the sample space and those of the variable space to project, at the same time, variables and samples in the same space. The basis for this is that the two matrices usually diagonalized in R- and Q-mode analysis have in fact the same eigenvalues. This saves considerable computing time and makes possible analyses that so far have been claimed to be impossible. Weighting procedures and the definition of contributions are given in the appendices.

We will now proceed directly to the interpretation of results of correspondence analysis on the data for the Lakeview Mountains Pluton. Output from our program provides: (1) factors in variable space, (2) factors in sample space, (3) contributions of samples and variables to factors, (4) plot of variables and samples on factor planes, and (5) maps of factors.

Interpretation of results

The factors are listed in Table II. It includes the loading of each variable and its absolute contribution to the factor as well as the relative contribution. The percentage of variation explained by each factor is listed at the bottom of the table. The corresponding diagrams are shown in Figs.6 and 7. On these plots, the proximity of samples reflects their similarity, while the proximity of variable points denotes a similarity of behaviour. Proximity of a variable

TABLE II

J Factors (variables coordinates) — absolute and relative contributions

	Weight	FACI(1)	CA(1)	CR(1)	FACI(2)	CA(2)	CR(2)	FACI(3)	CA(3)	CR(3)	FACI(4)	CA(4)	CR(4)	FACI(5)	CA(5)	CR(5)
Na	0.02	0.02	0.75	59.45	−0.01	3.28	36.90	−0.00	0.39	1.56	0.00	0.70	2.00	0.00	0.08	0.10
Mg	0.02	0.07	6.89	59.82	0.04	18.43	22.69	0.03	33.01	14.58	−0.01	5.63	1.76	−0.00	8.93	1.15
Al	0.11	0.01	1.79	23.76	−0.02	27.05	50.80	−0.00	9.50	6.40	−0.01	38.90	18.56	−0.00	2.41	0.47
Si	0.34	−0.02	14.76	92.94	−0.00	0.68	0.61	−0.00	0.16	0.05	0.00	28.14	6.38	0.00	0.27	0.03
K	0.01	−0.24	44.78	94.54	0.04	9.82	2.94	0.00	0.31	0.03	−0.03	26.62	2.02	0.01	14.67	0.46
Ca	0.06	0.07	19.43	92.09	−0.00	1.94	1.30	0.01	9.53	2.30	−0.00	0.00	0.00	0.01	61.39	4.31
Fe	0.06	0.05	11.57	60.18	0.04	38.62	'28.49	−0.02	42.53	11.26	−0.00	0.00	0.00	0.00	0.95	0.07
SG	0.35	0.00	0.03	5.67	0.00	0.19	5.84	0.00	4.57	51.45	−0.00	0.01	0.05	−0.00	11.29	36.97
% Explained			79.17			11.23			4.03			2.85			1.17	

FACI (1) = loading of variable on factor 1, CA(1) = absolute contribution of variable to factor 1 (see Appendix), CR(1) = relative contribution of variable to factor 1 (see Appendix).

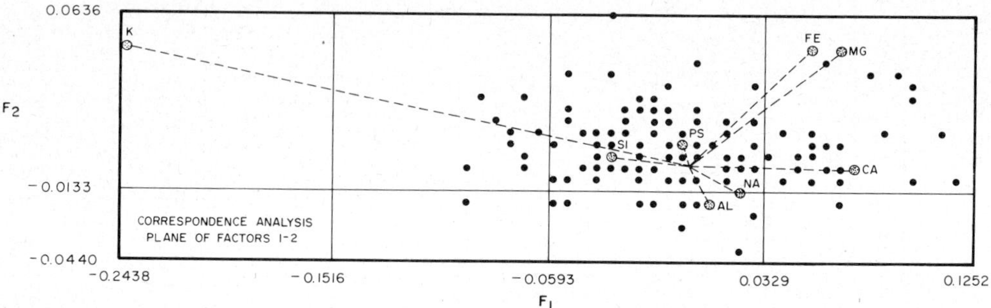

Fig.6. Lakeview Mountains Pluton: plots of samples and variable in the plane of factors 1 and 2.

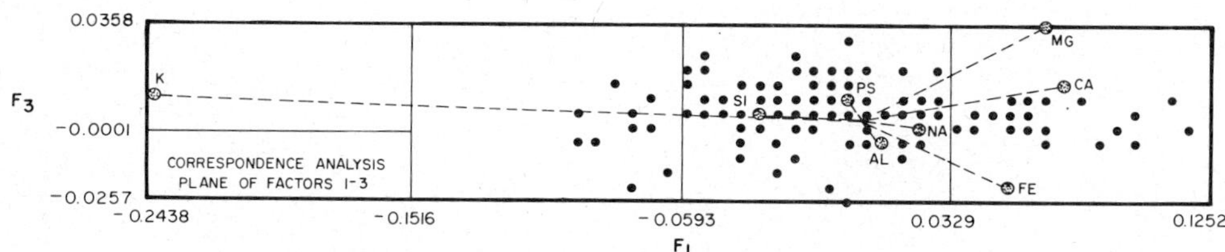

Fig.7. Lakeview Mountains Pluton: plots of samples and variables in the plane of factors 1 and 3.

and sample denotes a high relative concentration of the element in that sample. The lines drawn from the centre of the plane to the elements show the manner in which each element contributes to the differences between samples and an average rock which would plot at the centre.

The first factor, which explains 80% of the total variation in the batholith, shows the antipathetic relationship of the mafic elements calcium, magnesium and iron to silicon and potassium, the absolute contribution of each variable being, respectively, K (45%), Si (15%), Ca (19.5%), Fe (11.5%), Mg (7%). This distribution of elements is the usual one encountered when the main factor is magmatic differentiation. In a complementary map of factor scores (Fig.8) concentric zoning is strikingly obvious. Similar results were obtained by Morton (1969) and Morton et al. (1969) using trend surfaces on each element. Here it is easily mapped using only one variable: factor 1, which incorporates the effect of all elements. This shows a relatively basic centre and acid margins. A second basic centre also appears on the border. Also, since most potassium is contained in biotite and sodium in andesine, one can consider that factor 1 also distinguishes biotite- and quartz-rich areas from hornblende-rich areas.

The second factor explains 11% of the variability and subtly influences the degree of magmatic differentiation. For a given acidity it separates Al-rich samples from Fe-, Mg- and K-rich samples. The absolute contributions of these elements to the factor are, respectively, Al (27%), Fe (38%), Mg (18%),

Fig.8. Lakeview Mountains Pluton: map of factor 1.

K (9%). Taking into account that there are only four minerals in the rock, this factor might represent internal variation in the biotite as noticed and discussed by Morton et al. (1969). Mapping of this factor (Fig.9) shows that it

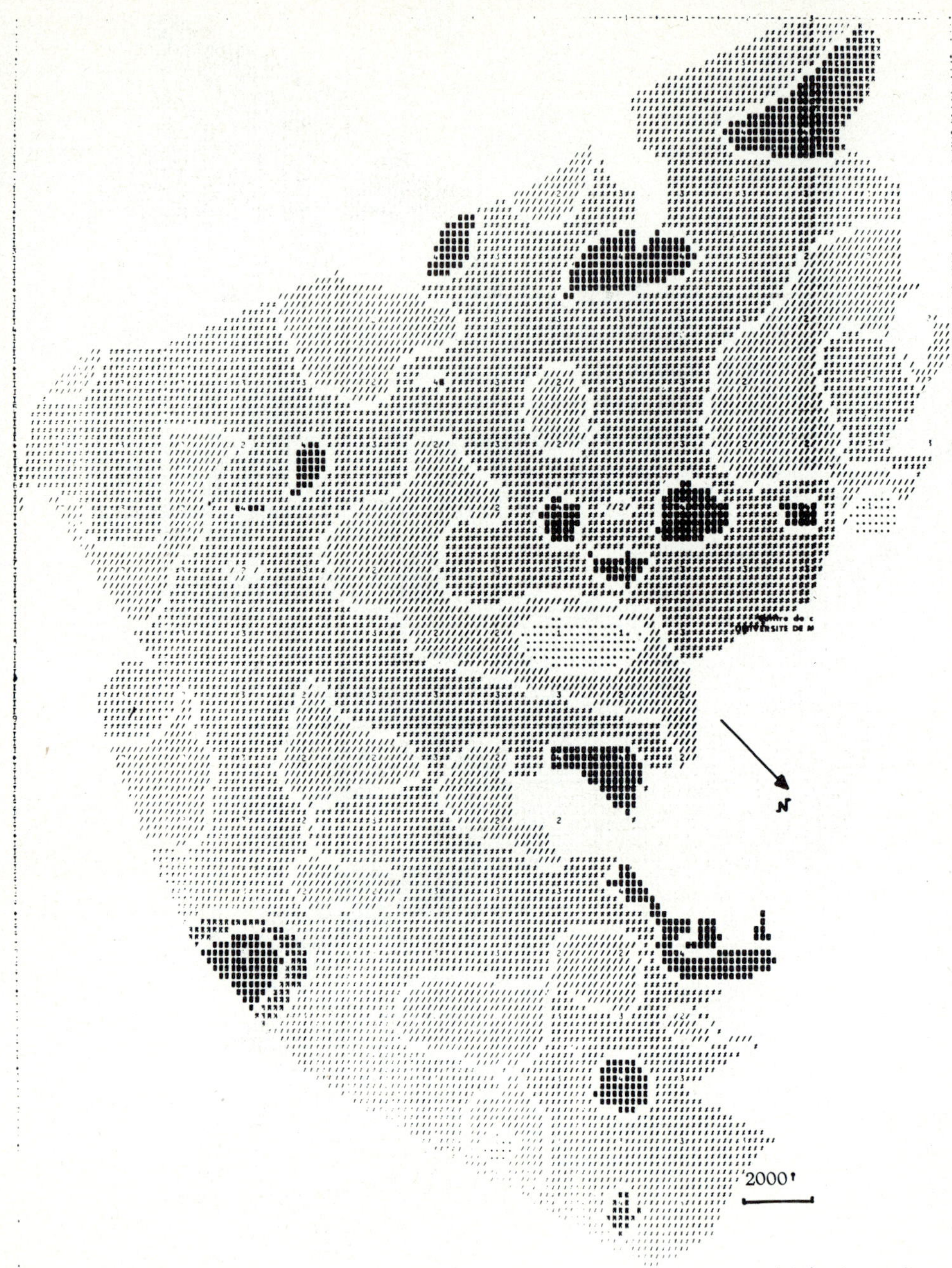

Fig.9. Lakeview Mountains Pluton: map of factor 2.

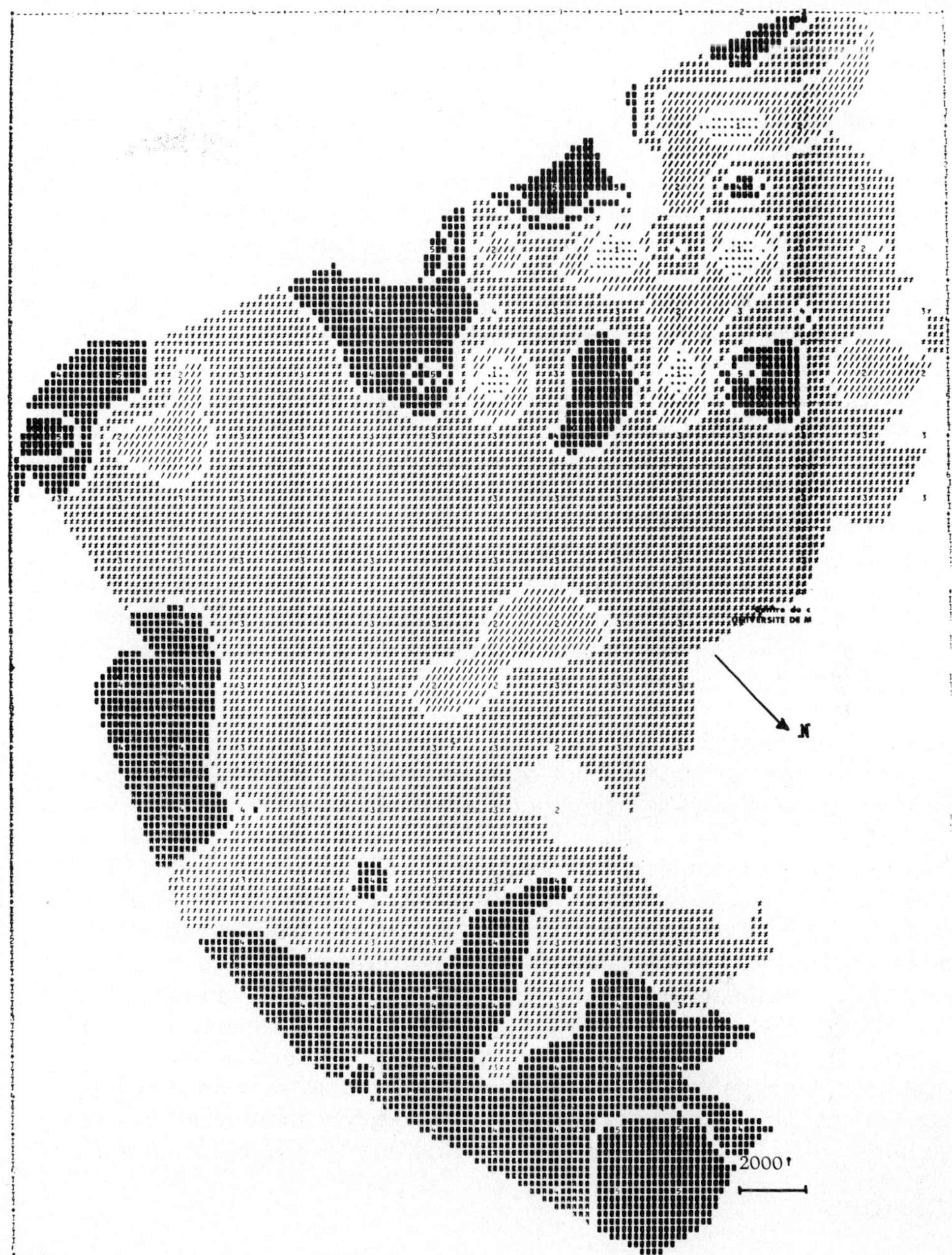

Fig.10. Lakeview Mountains Pluton: map of factor 3.

extends over the whole area, but that there is only patchy local variation without any trend.

The third factor accounts for only 4% of the variation and its interpretation might thus be questionable. Its variations (Fig.10) are only local and involve the acid margins. It separates Fe- and Al-rich samples from those rich in magnesium and calcium. This might reflect variations in hornblende composition, these variations being especially noticeable in acid rocks, rather poor in that mineral. Supporting evidence is that this factor makes a 51% contribution to variations of specific gravity.

The fourth factor explains less than 3% of the variation and might be totally void of significance. It shows local enrichment in aluminium and potassium versus silicon and might be related to the local presence of alkali feldspar.

Alternatively Table II can be read horizontally, rather than vertically, and the behaviour of the elements considered. Silicon, potassium and calcium have over 90% of their variations explained by the first factor which is regional. In the case of iron and magnesium 60% of their variation is explained by the first regional factor, while the remaining 40% is contributed by local factors 2 (25%) and 3 (15%). Aluminium is over 75% "local" since it is mainly explained by the second (50%) and fourth (25%) factor, whereas the regional components yield only 25%. Sodium is almost completely regional since 60% is explained by the first factor and 40% by the second. However, the amplitude of its variations is very low.

Link with variograms

In the first part we claimed that variograms could describe levels of variation. We will see that these two very different tools support the same interpretation.

Silicon, calcium and potassium were grouped by the shape of their variograms. The fact that they exhibit mainly regional rather than local variations was shown by the low nugget effect and long range; we might even dare to relate the ratio C_0/C to the ratio of the local over regional factors. The same interpretation holds for iron and magnesium which exhibit similar variograms with a high nugget effect, almost as important as the variance of the structured component, C. The ratio is almost 1/1. Aluminium which is mainly "local", will be expected to have an almost flat variogram — an almost pure nugget effect. This is the case, as Fig.5 shows. Finally sodium shows a very high nugget effect and a low sill, reflecting its very low overall variation.

CONCLUSION

Running the computer programmes for variogram and correspondence analysis, including all graphs, represents a total computing cost of $15.00. The cost of the four maps produced using Symap is $30.00. If one considers the cost of a single analysis, and the quantity of information contained in the variograms and correspondence analysis there is no financial justification for neglecting these tools. The less impressive results provided by variogram analysis should not be neglected and the basic ideas of computing sample variances or estimation variances in the case of dependent samples should be remembered. They are essential to a good understanding of sampling variations.

ACKNOWLEDGEMENT

The study was made possible through NRC Grant A 7035. Yves Beauchemin and Guy Daoust carried out most of the programming involved.

APPENDIX

Weighting procedures in Correspondence Analysis

Starting with a matrix X of n samples and p variables we will consider each entry as a probability and for this we divide all of them by the sum of all numbers in the matrix. (All the entries should be non-negative numbers.) So that we replace X_{ij} by:

$$P_{ij} = X_{ij} / \sum_i \sum_j X_{ij}$$

We also sum each line and each column as $P_{i.}$ and $P_{.j}$. Now to represent the matrix in both spaces, we will plot the samples as points having coordinates $(P_{ij})/P_{i.}(j = 1, \ldots .p)$ and the variables as $(P_{ij})/P_{.j}$ $(i = 1, \ldots .n)$.

In other words we consider the relative role of each variable or element. In order to take into account the fact that some points are more important than others, we give to each sample point a weight $P_{i.}$ and to each variable point a weight $P_{.j}$.

Let us note that in fact our points are in an $(n-1)$-dimensional space on one hand and a $(p-1)$-dimensional one, for the other set.

We have yet to introduce another weighting factor in distance computation since the one we have used so far does not remove the excessive weight which silica, for instance, would have.

Instead of using as distance between two samples l and k the usual quantity

$$d^2\,(l,k) = \sum_{j=1}^{p} \left[\frac{P_{lj}}{P_{l.}} - \frac{P_{kj}}{P_{k.}} \right]^2$$

we weight the contribution of each variable by its relative importance:

$$d^2\,(l,k) = \sum_{j=1}^{P} \frac{1}{P_{.j}} \left[\frac{P_{lj}}{P_{l.}} - \frac{P_{kj.}}{P_{k.}} \right]^2$$

Similarly in the other space, the distance between two variables will be

$$d^2 (j,j') = \sum_{i=1}^{n} \frac{1}{P_{i.}} \left[\frac{P_{ij}}{P_{.j}} - \frac{P_{ij'}}{P_{.j'}} \right]^2$$

Now that we have clouds of points where distances and masses have been defined, we can, by classical methods, find the axes of these clouds of points and treat them as we do in normal Principal Component Analysis. We will also take full advantage of the fact that we can project both clouds onto the same plane.

Definition of contributions

To help interpretation of factor k, two coefficients are calculated for every sample i and variable j.

The "absolute contribution", $ca_k(i)$ or $ca_k(j)$ expresses the part given by sample i or variable j to the dispersion along factor k.

As the variance of the loadings $f_k(i)$ or $g_k(j)$ along k is s_k:

$$ca_k(i) = \frac{pi. \, f_k^2(i)}{s_k} \; ; \; ca_k(j) = \frac{p.j \, g_k^2(j)}{s_k}$$

The absolute contribution allows determination of the samples or variables responsible for the appearance of factor k.

The "relative contribution", $cr_k(i)$ or $cr_k(j)$ expresses the part given by factor k to the dispersion of sample i or variable j.

The dispersion of sample i being evaluated by the distance from point i to the centre of gravity G of the cloud:

$$d^2 (i, G) = \sum_{k=1}^{p} f_k^2(i)$$

so:

$$cr_k(i) = \frac{f_k^2(i)}{d^2(i, G)} \; ; \; cr_k(j) = \frac{g_k^2(j)}{d^2(j, G)}$$

REFERENCES

Baird, A.K., McIntyre, D.B., Welday, E.E. and Morton, D.M., 1967. A test of chemical variability and field sampling methods, Lakeview Mountain tonalite, Lakeview Mountains southern California batholith. Calif. Div. Mines and Geology Spec. Rep. 92, pp. 11—19

Benzecri, J.P., 1973. L'Analyse des Données. Dunod, Paris, 600 pp.

Cazes, P., 1970. Application de l'analyse de données au traitement des problèmes géologiques. Thèse de 3e cycle, Paris

Dagbert, M. and David, M., 1974. Pattern recognition and geochemical data. An application to Monteregian Hills. Can. J. Earth Sci. (in press)

David, M. and Woussen, G., 1973. Correspondence Analysis, a new tool for geologists. Proc. Min. Pribram, 1: 41—65

David, M. and Beauchemin, Y., 1974. The Correspondence Analysis method and a Fortran IV program. Geocom Program 10

David, M., Campiglio, C. and Darling, R., 1974. Progresses in R- and Q-mode analysis: correspondence analysis and its application to the study of geological processes, Can. J. Earth Sci., 11: 131—146

Krumbein, W.C. and Slack, H.A., 1956. Statistical analysis of low-level radioactivity of Pennsylvanian black fissile shale in Illinois. Bull. Geol. Soc. Am., 67: 739—762

Matern, B., 1960. Spatial variation. Medd. Gran Statens Skogsforskningsinst., 49(5): 149 pp.

Matheron, G., 1963. Principles of geostatistics. Econ. Geol., 58: 1246—1266

Matheron, G., 1972. The theory of regionalized variables. Cahiers du Centre de Morphologie Mathématique de Fontainebleau, No.6, 211 pp.

Morton, D.M., 1969. The Lakeview Mountain Pluton, Southern California batholith, I. Petrology and Structure. Geol. Soc. Am., Bull., 80: 1539—1552

Morton, D.M., Baird, A.K. and Baird, K.W., 1969. The Lakeview Mountains Pluton, Southern California batholith, II. Chemical composition and variation. Geol. Soc. Am., Bull., 80: 1553—1564

Serra, J., 1970. Les structures gigognes, morphologie mathématique et interprétation métallogénique. Mineral. Deposita (Berl.), 3: 135—154

TOWARDS QUANTITATIVE UTILIZATION OF GEOCHEMICAL EXPLORATION: THE THRESHOLD AND ITS ESTIMATION IN SOIL SURVEYS

F. CACHAU-HERREILLAT et al.

Societe Nationale des Petroles d'Aquitaine, Pau (France)

ABSTRACT

Despite recent considerable development of geochemical exploration, a striking feature is that it continues to be devoted mainly to location and description of geochemical surfaces and not to quantitative estimation of anomalies. First, the oft-debated problem of the threshold level for anomalous values is reviewed: a theoretical discussion shows the arbitrariness of the usual rule of "two standard deviations" for the threshold level. A deconvolution solution is proposed to enable a probabilized estimation of cut-off values for regional surveys. The unresolvable nature of the problem in detailed surveys is demonstrated by the variable value of the ratio of the constant anomaly surface to the variable surface of the surveyed area. The problem of quantitative estimation in detailed soil surveys is examined using A.P. Solovov's (1965) determination of "K" ratios between estimated quantities of metal in soil, to estimated quantities of metal in bedrock. Two estimations of this ratio in areas of France with continental or Atlantic climates lead to consistent values of 1.2—1.5 for lead, zinc, copper, tungsten, titanium and other elements in landscapes with mechanically dispersed soils. These values are interpreted as showing a non-leaching process of dispersion, or at least a leaching which affects with equal intensity different elements during weathering of bedrock.

INTRODUCTION

Geochemical exploration has been used increasingly during the last twenty years. However, in the same period, despite large improvements in applying techniques such as kriging to bedrock geochemistry, no real attempt has been made toward a better understanding of the relations between overburden and bedrock geochemistry. At the same time as the theory of regionalized statistics was developing, geochemical exploration surveys were still being interpreted by classical statistics, sometimes even with ambiguous definitions of the object of these statistics. The purpose of this paper is to pinpoint both the need for precision in statistical definition of geochemical objects and the necessity of examing geochemical surfaces with geostatistic methods.

THE THRESHOLD PROBLEM

Regional surveys

The universal definition of "anomalous threshold" or "first level of anomalies" or "level of possible anomaly" or even "level of 97.5% probability of an anomaly" is given by "mean (or log mean) + 2 standard deviations". This statement is not incorrect in itself, unlike its usual interpretation: "the probability of a value to be anomalous is 97.5% above $m + 2\sigma$". The only correct statistical meaning should be: "if we pick a random sample from the bulk population we are considering, the probability of obtaining a sample below $m + 2\sigma$ is 97.5% assuming a normal (log normal) population". But this has nothing to do with geochemistry. As far as geology is concerned, the correct meaning is: "if the population is normal (log normal), 97.5% of the samples fall below $m + 2\sigma$ and we choose the 2.5% remaining samples to be the anomalous ones". That is, instead of a certain metal content, we choose the ratio of:

$$\frac{\text{number of anomalous samples}}{\text{total number}} = \frac{\text{anomalous area}}{\text{total area}} = 2.5\%$$

as the arbitrary threshold level. But this has nothing to do with statistics.

Choosing a level for percentage of anomalous areas is no more rigorous than a metal content. Furthermore, this approach may be misleading. Thus, if there is no deposit we will still find 2.5% of the data to be anomalous. Alternatively, where the area is unusually rich in deposits, an anomaly of only 2.5% will underestimate the real situation. As long as sampling density remains constant there is only one situation in which we can choose an arbitrary ratio to define the threshold. This occurs in the unusual case when we already have the information from another survey completed under similar geological conditions.

Usually the ratio of mineral deposit/barren rocks in a geological environment is unknown although we may have an idea of the average metal content in ordinary rocks of the environment. Tables, such as those of Vinogradov (1959), giving average contents can be compared to regional background in the survey. There is no more arbitrariness in taking the mode or median value as a background than in choosing a 2.5% ratio for the anomalous surface. This median or modal value may be taken as a background in the first model of a "trial and error" procedure and anomalies are then defined through any kind of progression (arithmetic or geometric) which looks suited to description of the data. The geochemist is in the situation of a cartographer who wishes to describe a hilly or a mountainous area: he has no need of a statistic to know that he must take his level lines in a regular progression adapted to steepness of slope and to precision of his data. Cut-off values can be compared to the mean + 2 standard deviation values. For a normal population, 2.5% of the samples can occur randomly without indicating a geochemical anomaly. Clusters of anomalous samples are much less likely to exist. Consequently, isolated samples are usually rejected, whereas clusters are retained. According to the Hawkes' rule, 4 points are necessary to define an anomaly provided that sample density is sufficient. However, if sample densities are too low, dilution of anomalous metal values by background values can prevent identification of anomalies.

Should we wish to be more rigorous we need to use procedures which do not appear strictly necessary from a practical point of view. Let us assume (1) a preliminary threshold has been estimated, and (2) that geological background is uniform or can be split into uniform blocks. As a result samples from uniform geological formations can be tested statistically. What is the metal content of a uniform geological facies and which statistical law does it follow? We may postulate, and check in many circumstances, that background metal content is a constant affected by its own geological dispersion, and increased by the analytical and sampling dispersions, all of which can be shown to follow gaussian laws $[G\,(m,\sigma)]$ so that in a uniform environment:

$$\sigma^2 = \sigma_r^2 + \sigma_a^2 + \sigma_s^2$$

where σ^2 = observed deviation, σ_r^2 = metal content variance in rocks, σ_a^2 = analytical variance, and σ_s^2 = sampling variance

This is precisely what we observe when considering regional surveys with correct sampling density from which consistent anomalies have been separated. The geometrical interpretation of such conditions is very simple: a horizontal plane in X, Y, Z space where X, Y are the horizontal coordinates and Z the metal content (Fig.1). If the representative plane is not horizontal, which means that there is a linear geochemical trend within a uniform geological background, it can be shown that the data follow a new gaussian law with a higher variance and mean (Fig.2). If the trend is of more than the first order,

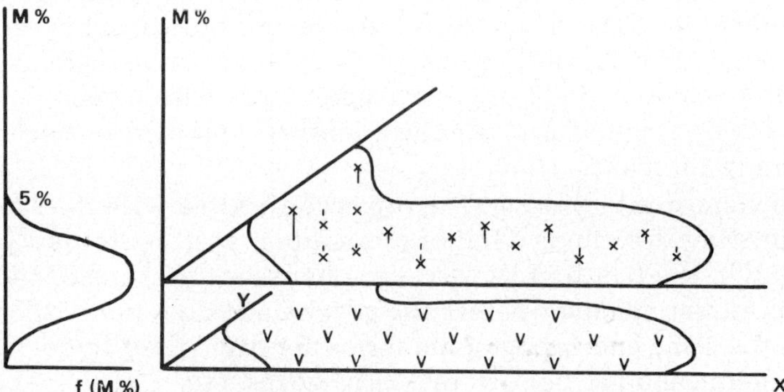

Fig.1. Geometric representation of background geochemical surface with anomalies taken apart: linear uniform model.

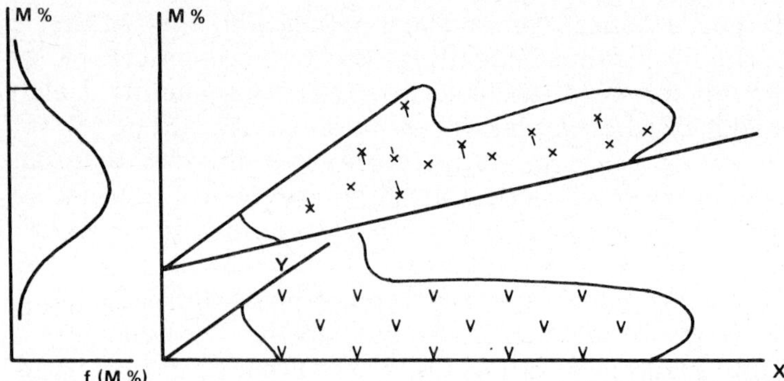

Fig.2. Geometric representation of background geochemical surface with anomalies taken apart: linear first-order model.

we come back to the case of an anomaly and geological background is far from uniform. We must find a smaller area where the gaussian model can be postulated (Fig.3).

We now have a background population described by a gaussian law and a total population described by an observed statistical law which represents background and anomalous samples. The deviation of the second law from the first expresses the influence of the anomalous values and we have a measure of the probability for value x to be anomalous; it is the following ratio:

$$\left(\frac{fa}{fa + fb} \right)_x$$

increasing from 0 to 1 with x increasing from the median to the highest values (Fig.4).

Detailed surveys

As already noted, the rule of mean (or median) + 2σ cannot be applied in detailed surveys of anomalous areas. Under these conditions our thoughts on threshold and background are different, and the only meaning of histograms or cumulative frequency curves is their strict correspondence with a real distribution, provided there is a uniform sampling density. This distribution can be used to determine anomalies.

Here again, the geochemist interpreting a survey is confronted with the same problem as the cartographer describing a hill or a mountain with contours. The only difference lies in the fact that the cartographer sees the shape of the landscape between his measurements whereas the geochemist does not. Otherwise the procedure is the same and topographic terms are often used to describe contoured geochemical data: hollow, plateaus, ridges, etc.

Spacing of contour lines must be compared to analytical + sampling variance. It is worthless and can be very misleading to have isolines separated by

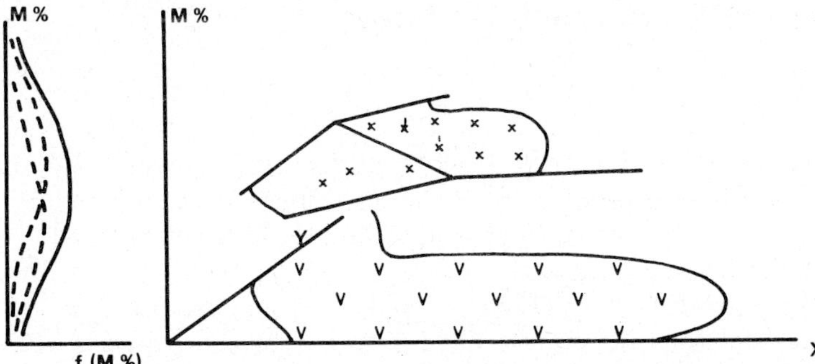

Fig.3. Geometric representation of background geochemical surface with anomalies taken apart: linear first-order model with more than one representative plane.

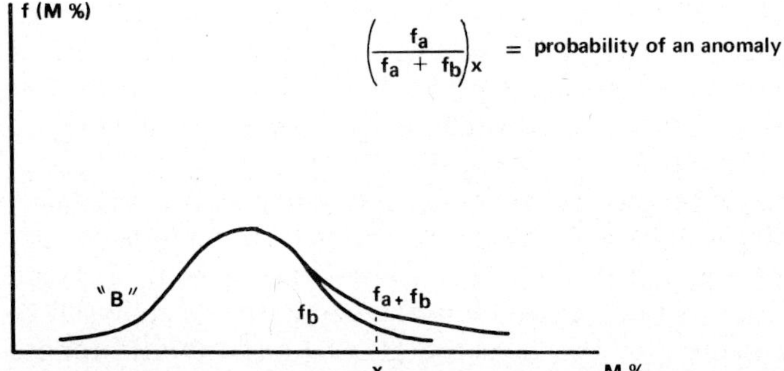

Fig.4. Statistical distribution of observed population, convolution of background "B" and anomaly "A".

less than the usual standard error, as determined by repeated analysis of a series of samples. The distance between the class-limits must be greater than the combined values of errors for analysis and sampling. Thus for a spacing of 2 standard deviation on either side of the class mean the probability of exceeding the margin of error will be less than 5%. If the spacing was decreased the probability of exceeding the margin of error would be greater.

QUANTITATIVE ESTIMATION IN DETAILED SURVEYS

Quantitative interpretation of lithological survey has been advocated by Solovov (1965). If solved, this important problem could transform practical utilization of geochemistry in mineral exploration. However, the problem has been neglected because of the apparent lack of correlation between the contents of metals in soils and in rocks. As anticipated, Solovov (1965) studying soil metal/bedrock metal ratios under various conditions found values of 1 in mainly clastic dispersion haloes; from 0.1 to 1 in heavily leached areas and between 1 and 10 in poorly drained soils. The K ratio:

$$K = \frac{(M)\ \text{soils}}{(M)\ \text{rock}}$$

can be estimated if we have data for both environments. This is usually the case for metal values in soils but seldom for the underlying rock. To provide a first estimation of the K ratio for residual soils in the Atlantic climate of France, two surveys have been conducted.

The first survey was undertaken by Costes (1973) in the Pyrenees. Soil data are for residual squelettic soils sampled at 0.2 m on the steep slopes of the western French Pyrenees, in an area subject to heavy rainfall. Bedrock results were obtained by regularly spaced grab samples in siltstones, sandstones, and limestones of Devonian age. Sphalerite occurs as fissure fillings within limestones. K ratios were 1 for lead and 2.2 for zinc.

The second survey was conducted in the Massif Central of France by a team of SNPA geologists (Colladon, 1972; Lasmartres, 1972). Soil information is for a residual soil developed on gentle slopes in agricultural and forest areas. Samples were taken at 0.2 m. Bedrock was sampled at 1—1.5 m by auger-drill holes in andesite and microgranite country rocks with siliceous veinlets. K ratios are summarized in Table I.

The most striking feature of these results is that K is always close to and slightly greater than 1. That is, a very slight concentration occurs in the soils and is of the same order of magnitude for all the elements analysed. This can be interpreted as a slight leaching of a major soluble element (not analysed for) and corresponding rearrangement of the trace metals without any differences in their leaching.

TABLE I

Mean and range of K ratios for the Massif Central, France (data after Colladon, 1972; Lasmartres, 1972)

Metal	K values	
	means	range (90%)
Pb	1.3	0.9—1.8
Zn	1.7—1.8	1.0—2.8
Cu	1.3—1.5	0.5—2.3
Bi	1.1—1.4	0.1—2.5
TiO_2	1.9	1.2—2.6
W*	1.5	1.3—1.7
W**		0.7—1.1

* Non-mineralized areas; **Mineralized areas.

CONCLUSIONS

The practical conclusion of the results for exploration in the climate of continental France is quantitative estimation of bedrock quantities of metals from soil survey data. At this stage the range of this ratio is still high and many other determinations will be necessary. Different surveys in other climates will also be necessary and will permit sound theoretical views on the leaching of elements according to temperature, rainfall, etc. Gathering all this data will probably need international cooperation in order to follow standard and adequate sampling procedures to obtain comparable results. In this way, the results of A.P. Solovov will be extended to non-Russian climates and provide geochemists with the quantitative key they have been seeking.

REFERENCES

Colladon, M., 1972. Rapport de stage E.N.S.G. (internal report)
Costes, B., 1973. Etude géologique et géochimique de l'indice zincifère du Val du Bitet (Pyrénées Atlantiques). Bull. Centre Rech. Pau, S.N.P.A., 7(2): 587—620
Lasmartres, B., 1972. Rapport de stage E.N.S.G. (internal report)
Solovov, A.P., 1965. Estimation des gites minéraux d'après les auréoles de dissemination secondaire des minerais. Izv. Acad. Nauk Kazakhskoi U.S.S.R., Serv. Geol., No.3 (translated into French)
Vinogradov, A.P., 1959. The Geochemistry of Rare and Dispersed Chemical Elements in Soils. Consultant Bureau, New York, N.Y., 209 pp. (translated from Russian)

A COMPUTER METHOD FOR DIVIDING A REGIONAL GEOCHEMICAL SURVEY AREA INTO HOMOGENEOUS SUBAREAS PRIOR TO STATISTICAL INTERPRETATION

RICHARD SINDING-LARSEN

Geological Survey of Norway, Trondheim (Norway)

ABSTRACT

This paper presents a computer procedure to divide a geochemically heterogeneous survey area into reasonably homogeneous subareas.

Humus (A_0) soils from an area in central Norway (70 km $\times$ 70 km) were sampled at the intersection of a 3-km square grid and 19 chemical elements were determined in each sample. Samples were considered as points in a 19-dimensional space with principal components as axes. In this space, groups were isolated by applying an algorithm which minimizes the mean square distance between samples within the same group. The samples were divided into an increasing number of groups and Wilks' criterion Λ (the ratio of the determinant of the within-group and total dispersion matrices), was determined for each partition. The number of groups, k, corresponding to the partition with the minimum value of the product Λk^2 was chosen as the number of geochemically homogeneous groups to be distinguished within the survey area.

Mapping these classes without considering geographical location produced maps with fragmented subareas and irregular boundaries. A regrouping of the samples was therefore performed according to a mixed distance measure, taking both geochemical and geographical proximity into account, resulting in more continuous and uniform subareas.

The difference in readily soluble K and P, pH, ash content and Co, Cr, Fe, Mo and Ni, as well as the difference in correlation between Pb, Cd, Cu, Zn, Ca and Mn between subareas showed that a better interpretation is achieved if threshold values and environmental effects are determined separately for each subarea.

INTRODUCTION

A regional geochemical survey area is often limited by arbitrary non-geological borders, such as the lines of a mapsheet. When considering a sample analysed for p elements from n localities within a survey area the geochemist would like to know if his n samples fall into some natural geochemical grouping and, if such groups exist, how these are distributed geographically. The use of regression technique, in estimation of environmental contribution to trace element content at each sampling site in order to identify samples with trace element content due to mineralization (Brundin and Nairis, 1972), can be misleading if the chemical elements show different geochemical behaviour within the survey area.

Numerical classification techniques can be used to overcome this problem. There are many different techniques that split a large number of multivariate observations into a smaller number of relatively homogeneous groups. The techniques can be either agglomerative (successive fusing of n samples into coarser partitions), or divisive (successive splitting of the totality of samples into finer partitions) or they can partition the n samples into a given number of groups, by optimizing a chosen criterion.

Much of the early work in clustering appeared in studies on numerical taxonomy (Sokal and Sneath, 1963) but have now spread to a number of other disciplines (Bolshev, 1969; Cole, 1969; Tryon and Bailey, 1970). In geology and geochemistry, maps have been constructed based on multivariate classification techniques, such as factor-analysis (Imbrie and Purdy, 1962) principal component analysis (McCammon, 1966; 1968; Garrett, 1973) and cluster analysis (Bonham-Carter, 1965; Parks, 1969; Hesp and Rigby, 1973; Obial and James, 1973). A comprehensive review of different classification techniques is given by Cormack (1971) and a review of pattern recognition techniques applicable to geochemistry by Howarth (1973).

A frequently used clustering procedure is to define a suitable distance function and compute the matrix of distances between all pairs of samples (Lance and Williams, 1967). For n samples this approach involves storage of $n(n-1)/2$ distances, which makes the method unsuitable for dealing with large numbers of samples, as is often the case in geochemical exploration.

In this study a classification technique closely related to the ISODATA (Ball, 1965) has been used, seeking groups with minimum within-group variance, by assigning the n samples to the nearest of a fixed number (k) of group centres. In this approach only nk distances are calculated which makes it a suitable technique for classifying large numbers of samples. To determine the number of groups (k) corresponding to a partition of the samples into natural clusters a criterion proposed by Mariott (1971), based on the determinant of the within-group scatter matrix, has been used.

The aim of this paper is to show how a survey area can be divided into subareas by use of multi-dimensional classification and how the geochemical characteristics of the resulting subareas can be evaluated.

The following aspects have been studied: (1) isolation of groups by means of a minimum variance algorithm in a space with chemical variables as axes, (2) geographical smoothing and mapping of samples, (3) plotting of variables and samples in the same diagram.

DESCRIPTION OF SURVEY AREA

The 70 km $\times$ 70 km area (with geographical coordinates 60°21', 61°00'N latitude and 9°30', 10°51'E longitude) straddles the border of the Caledonian overthrust in southern Norway, with youngest Precambrian and Cambro-Silurian rocks in the northeast and northwest, Permian rocks in the southeast, and Precambrian rocks in the central and southwestern parts (Fig.1 and 2).

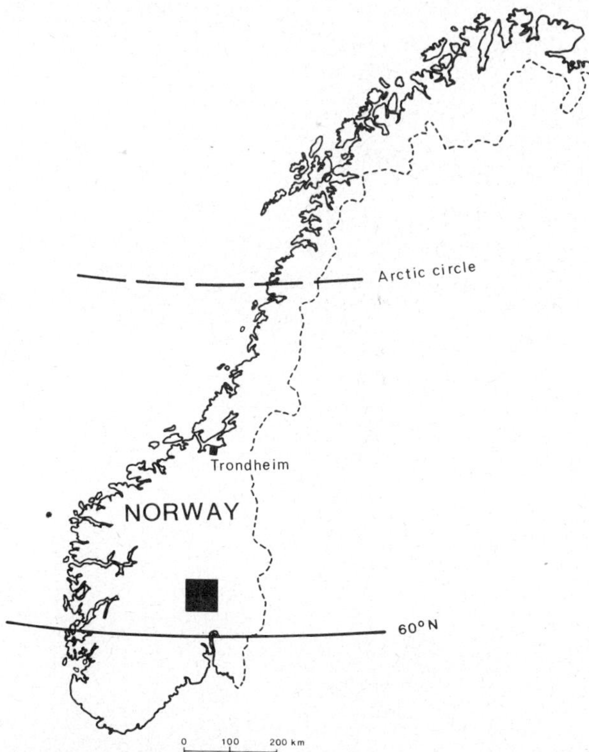

Fig.1. Location of the investigated area.

Topography varies from rather smooth in the north and east with altitudes ranging from 150 to 600 m a.s.l. to steep and hilly in the southwest with altitudes ranging from 150 to 1200 m a.s.l. Drainage is generally good, except in some elevated parts with large bog areas. Overburden consists mainly of till, generally thicker in the northeast than in the central and southwestern parts (Fig.3). Soils are mostly of the podzol type; both humus and bleached horizon thickness is normally less than 10 cm. Spruce and pine are common in the southwest whereas spruce is dominant in the northeast. The lowest part of the area is agricultural land. Climate varies with altitude, the mean temperature being 3—4°C and precipitation ranging from 300 to 700 mm per year.

METHODS

Composite humus (A_0) soils were collected in 1962 and 1963, by the National Forest Survey of Norway, from circular sample plots of 100 m^2 each at the intersection of a 3-km square grid. Within each sample plot information was gathered on soil conditions, such as depth and origin of soil material, profile type and humus thickness. At the Geological Survey of Norway samples were ashed overnight at 450°C and digested in hot HNO_3 1/1, for 3 hours,

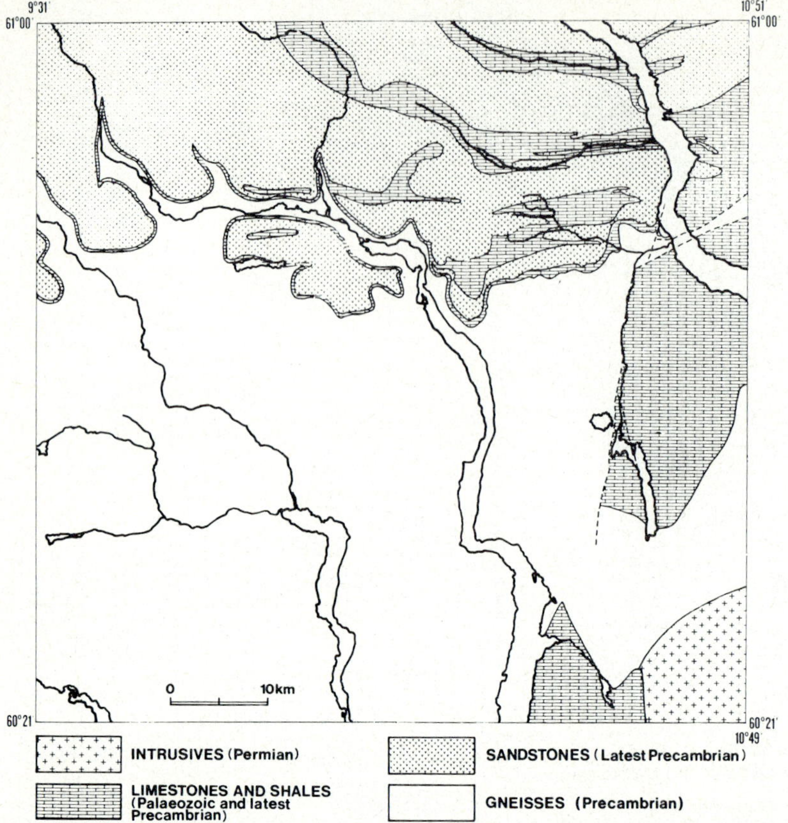

Fig.2. Geological map of the survey area. (After Holtedal and Dons, 1960).

followed by dilution to 20 ml. Ag, Ca, Cd, Co, Cu, Cr, Fe, Mn, Mo, Ni, Pb, V and Zn were determined by atomic absorption spectrometry on the diluted solution. Element content of ash was recalculated to content by dry weight. At the Agricultural University of Norway, loss on ignition, pH, the amount in milligrammes of readily soluble phosphorus (P-Al) and potassium (K-Al), and acid soluble phosphorus (P-HCl) and potassium (K-HCl) per 100 g of air-dried sample, were determined (Lag, 1968).

NUMERICAL CLASSIFICATION

Multi-dimensional representation of the data

In a common two-dimensional diagram with chemical variables as axes (R^2 space) n samples analysed for two variables may be represented as n points. In the same manner n samples analysed for p variables may be represented as n points in a p-dimensional space (R^p) with the p chemical variables as axes.

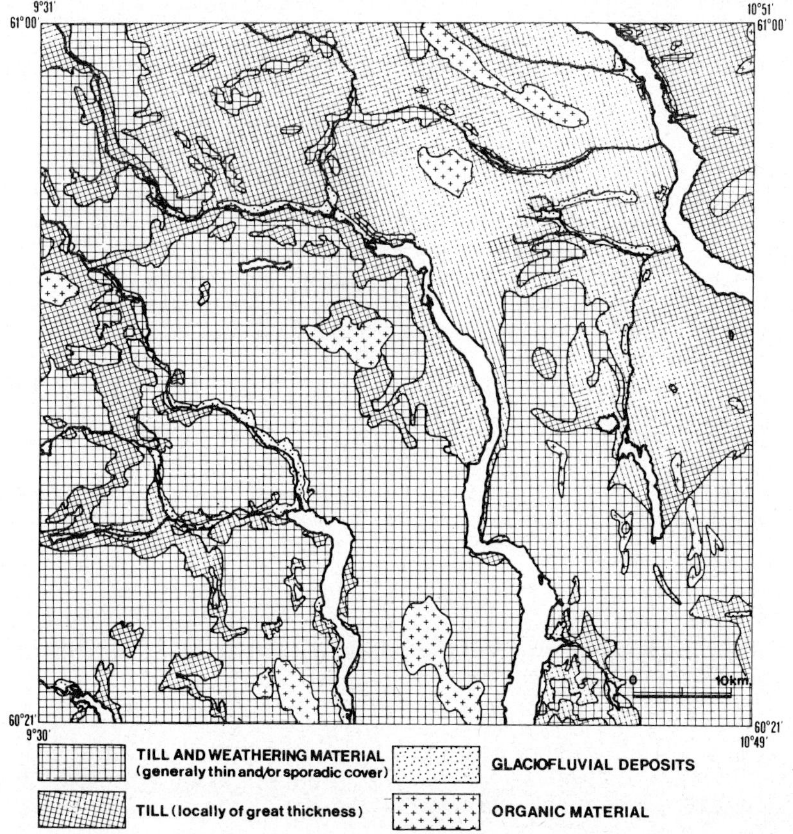

Fig.3. Simplified versions of 1:250 000 Quaternary geological map sheets Oppland and Hallingdal. After Holmsen (1954, 1955).

Relative position and interdistances between the n points in space R^p will depend on the scale along the axes. Chemical variables like Mn and Fe, which show a large range of variation, will produce an elongated cloud of sample points along the Mn and Fe axes. With no a-priori information about the relative importance of the chemical variables in characterizing different subareas it was decided to give equal weight to all variables. Samples were standardized accordingly to zero mean and unit variance.

Raw data for Ag, Ca, Cd, Co, Cu, Cr, Fe, Mn, Mo, Ni, Pb, V and Zn are approximately lognormal, the cumulative frequency distributions forming more or less straight lines when plotted on logarithmic probability paper. Data were, therefore, log-transformed before standardization. Data for ash-content, pH, P-Al, K-Al, P-HCl, K-HCl form more or less straight lines when plotted on normal probability paper and were left untransformed.

If one calculates the distances between sample points by the use of the Pythagorean theorem the chemical variables must be regarded as placed along orthogonal axes, i.e., the variables must be uncorrelated. From Table I we see

TABLE I

Pearson product-moment correlation coefficients between 19 chemical variables in 482 samples

	Cu	Pb	Zn	Cd	Ag	Ni	Co	Fe	Mn
	0.23	n.s.	0.17	n.s.	0.26	0.65	0.70	0.78	0.43
V	1.00	0.49	0.59	0.56	0.50	0.60	0.56	0.49	0.38
Cr	0.92	1.00	0.48	0.58	0.31	n.s.	n.s.	n.s.	n.s.
Mo	0.60	0.58	1.00	0.50	0.42	0.39	0.31	0.26	0.58
Ca	0.20	0.21	0.18	1.00	0.29	0.29	0.24	0.17	0.14
pH*	0.54	0.59	0.41	0.39	1.00	0.39	0.40	0.33	0.45
P-Al*	—0.41	—0.45	—0.34	n.s.	—0.40	1.00	0.80	0.79	0.45
K-Al*	—0.34	—0.36	—0.21	n.s.	—0.27	0.67	1.00	0.91	0.57
P-HCl*	0.12	n.s.	n.s.	n.s.	0.12	0.45	0.33	1.00	0.47
K-HCl*	n.s.	0.12	n.s.	n.s.	0.14	0.39	0.49	0.41	1.00
	V	Cr	Mo	Ca	pH*	P-Al*	K-Al*	P-HCl*	K-HCl*

Correlations not significant at the 5% level (r = 0.11) are indicated by n.s. All the variables except those marked by * have been log-transformed.

TABLE II

Eigenvalues and unit eigenvectors from principal component analysis on 482 standardized samples in variable space R^{19}

Eigenvector	u_1	u_2	u_3	u_4	u_5	u_6	u_7	u_8	u_9
Eigenvalue	7.65	3.37	1.89	1.13	0.72	0.68	0.65	0.50	0.43
Percent	40.3	17.7	9.9	5.0	3.8	3.6	3.4	2.7	2.3
Cumulative percent	40.3	58.0	67.9	73.9	77.7	81.3	84.7	87.4	89.7
Ash*	0.28	—0.20	—0.11	—0.15	—0.29	—0.02	0.23	0.12	0.02
Cu	0.24	0.22	—0.28	0.06	0.25	0.02	—0.21	—0.11	0.06
Pb	0.07	0.30	—0.42	—0.31	—0.29	—0.08	0.03	0.11	0.04
Zn	0.18	0.35	0.14	0.19	—0.29	—0.10	0.14	0.21	0.09
Cd	0.11	0.23	—0.48	—0.07	0.10	—0.29	—0.14	—0.04	—0.29
Ag	0.19	0.24	—0.00	—0.21	0.19	0.50	0.41	0.55	—0.24
Ni	0.32	—0.03	—0.02	0.03	0.11	—0.20	—0.07	—0.20	0.14
Co	0.33	—0.06	0.04	—0.03	0.04	—0.08	—0.15	—0.05	—0.11
Fe	0.33	—0.13	0.04	—0.08	0.00	—0.06	—0.15	—0.14	0.06
Mn	0.23	0.23	0.22	0.13	—0.45	0.09	0.10	0.18	—0.42
V	0.32	—0.06	0.10	—0.19	—0.04	0.11	0.10	0.01	0.38
Cr	0.33	—0.06	0.09	—0.14	—0.04	—0.05	0.07	—0.10	0.30
Mo	0.24	—0.04	0.01	—0.14	0.52	0.19	0.18	0.70	—0.15
Ca	0.13	0.21	—0.08	0.70	0.12	0.14	0.16	0.03	0.39
pH*	0.26	—0.06	0.21	0.39	0.01	0.01	—0.23	—0.09	—0.42
P-Al*	—0.17	0.40	0.19	—0.07	—0.12	0.18	0.09	0.09	0.16
K-Al*	0.14	0.35	0.23	—0.07	0.34	—0.26	0.15	0.04	—0.03
P-HCl*	0.05	0.33	0.28	—0.20	—0.01	0.33	—0.69	0.07	0.15
K-HCl*	0.04	0.28	0.45	—0.07	0.11	—0.57	0.11	—0.12	—0.02

All variables except those marked with * were log-transformed before standardization.

V	Cr	Mo	Ca	pH*	P-Al*	K-Al*	P-HCl*	H-HCl*	
0.79	0.80	0.52	n.s.	0.55	−0.52	−0.49	n.s.	n.s.	Ash*
0.45	0.49	0.43	0.45	0.38	−0.16	n.s.	0.24	n.s.	Cu
n.s.	n.s.	n.s.	n.s.	−0.19	0.18	n.s.	0.17	n.s.	Pb
0.30	0.32	0.22	0.53	0.25	0.17	n.s.	0.29	0.25	Zn
0.13	0.15	0.18	0.24	n.s.	n.s.	n.s.	n.s.	n.s.	Cd
0.43	0.44	0.36	0.25	0.24	n.s.	n.s.	0.29	0.20	Ag
0.73	0.82	0.54	0.30	0.62	−0.47	−0.34	n.s.	n.s.	Ni
0.80	0.83	0.60	0.26	0.67	−0.54	−0.36	n.s.	n.s.	Co
0.86	0.87	0.59	0.18	0.65	−0.60	−0.44	n.s.	n.s.	Fe
0.49	0.51	0.33	0.38	0.55	n.s.	n.s.	0.36	0.35	Mn

u_{10}	u_{11}	u_{12}	u_{13}	u_{14}	u_{15}	u_{16}	u_{17}	u_{18}	u_{19}
0.39	0.32	0.28	0.24	0.21	0.18	0.13	0.10	0.07	0.06
2.0	1.7	1.5	1.3	1.1	0.9	0.6	0.5	0.4	0.3
91.7	93.4	94.9	96.2	97.3	98.2	98.8	99.3	99.7	100.0
−0.05	−0.07	0.21	−0.36	−0.14	0.09	0.69	0.11	0.07	0.00
0.23	0.40	−0.23	0.00	−0.08	0.64	0.10	0.05	−0.00	0.01
0.43	0.22	0.34	0.30	0.03	−0.30	−0.03	−0.05	−0.06	−0.03
−0.30	0.20	0.48	0.23	−0.31	−0.26	−0.15	−0.14	0.10	0.02
−0.40	−0.54	−0.09	−0.10	−0.01	0.12	0.03	0.12	0.01	−0.00
−0.11	0.05	−0.03	0.05	−0.10	−0.16	0.04	0.00	−0.04	0.00
−0.10	0.13	0.26	−0.09	0.73	−0.28	0.06	0.01	−0.17	0.20
−0.20	−0.12	−0.18	0.35	0.04	−0.07	0.19	−0.50	0.57	0.11
0.10	−0.17	−0.01	0.24	−0.20	−0.04	0.06	−0.31	−0.69	−0.36
0.17	−0.08	−0.10	0.31	0.29	0.25	−0.08	0.31	−0.12	0.05
0.03	−0.24	−0.10	−0.07	−0.21	0.17	−0.36	0.04	−0.08	0.64
0.01	−0.11	0.03	−0.08	0.08	0.05	−0.36	0.26	0.36	−0.62
−0.14	0.13	−0.09	−0.05	0.08	−0.06	−0.06	−0.01	−0.01	−0.04
0.05	−0.16	−0.36	−0.03	0.00	−0.09	0.24	0.10	−0.02	−0.05
0.11	0.03	0.07	−0.58	−0.17	−0.12	−0.28	−0.14	0.00	0.02
−0.13	−0.19	0.07	−0.29	0.30	0.33	−0.05	−0.57	−0.02	−0.11
0.53	−0.35	0.37	0.02	−0.08	−0.16	0.07	0.14	0.01	0.02
−0.15	−0.03	−0.02	0.00	−0.07	−0.21	0.17	0.25	0.04	0.03
−0.26	0.34	−0.28	−0.02	−0.13	0.07	0.08	0.01	−0.05	−0.02

198

that many of the variables are highly intercorrelated which means that the variables form a space with oblique axes. To describe the n sample points in a euclidian space with orthogonal axes the standardized data were re-expressed in terms of principal components which mathematically are the eigenvectors of the correlation matrix. The normalized eigenvectors, u_k, are linear combinations of the standardized chemical variables (Table II). The first eigenvector, u_1, can be expressed by the standardized variables as:

$$u_1 = 0.28 \text{ ash} + 0.24 \text{ Cu} + 0.07 \text{ Pb} + \dots + 0.04 \text{ K-HCl}$$

The 482 sample points will by this transformation be expressed by 19 uncorrelated new variables (eigenvectors), each eigenvector accounting for a decreasing proportion of the total variance. The kth eigenvalue, λ_k, of the correlation matrix is equal to the variance of the 482 samples in the direction of corresponding eigenvector u_k.

As can be seen in Table II the first, second and third eigenvalues account for subsequently 40%, 18% and 10% of the total variance indicating that the cloud of 482 points form a hyperellipsoid elongated chiefly in the direction of the first three eigenvectors.

Classification of samples into k groups

Within the cloud of 482 sample points in the euclidian space spanned by the eigenvectors our problem is to find clusters corresponding to geochemically different subareas.

A geochemically homogeneous subarea is defined here as an area where samples from different sampling sites have similar values for all the chemical elements determined. The sample points from a geochemically homogeneous subarea will in space R^p (with the chemical variables as axes) form a cluster of near-spherical shape. The sample lying nearest to the centre of gravity of a cluster is termed the geochemical type sample of that cluster, and will hereafter be used as reference to the cluster.

To isolate clusters in space R^p an algorithm minimizing the sum of the squared lengths of the distances between samples and their group centre was chosen. This technique possesses variance constraints (Wishart, 1969) and is akin to the techniques described by Forgy (1965), Ball (1965), Jancey (1966) and MacQueen (1967), but is stripped of such features as "coarsening", "refinement", "hill-climbing", and can be regarded as a fast and simplified version of these techniques.

Let n_k be the number of sample vectors x_{ik} in group k and let g_k be the mean vector or group centre of the n_k points in space R^p:

$$g_k = \frac{1}{n_k} \sum_{i=1}^{n_k} x_{ik}$$

Sum S of the squared lengths between samples and their group centre for all groups is then defined by:

$$S = \sum_{k=1}^{k} \sum_{i=1}^{n_k} (x_{ik} - g_k)^2$$

which is equal to the trace of the total within-groups' sum of squares and cross-product matrix W, hereafter called the trace of the within-group scatter matrix (trW). The partitions obtained by minimizing S will thus have a minimum within-group variability according to the trW criterion. The iterative minimum variance partition has the advantage of producing no chain effect (Johnson, 1967) and of avoiding storage of $n(n-1)/2$ distances, but has the disadvantage of not taking within-group covariances into account. In this study however, within-group scatter matrices computed show low covariances which means that there are no elliptical clusters with axes not parallel to the coordinate axes.

The grouping procedure
When partitioning the samples into a given number, k, of groups, a random sample is chosen as first group centre and the sample with greatest distance from this group centre is chosen as the second group centre. In order to obtain evenly distributed group centres the remaining k centres are all successively chosen at a maximum distance from the existing centres. The n samples are then assigned to the nearest group centre. The centre of gravity of each group is then calculated and chosen as the new group centre. Calculation of distances from sample points to these new group centres as well as allocation of samples to the nearest group centre is then repeated. At each iteration the group centres will tend to move so as to minimize the sum of squared lengths, S, until a position is reached where they start to oscillate around a point from which any further displacement will increase the within-group variance of the partition. At each iteration (I) the group centre g_k (I) will move a distance $D_k(I)$ from $g_k(I-1)$ to $g_k(I)$. This displacement is compared with the total displacement $D_k = \sum_{I=1}^{I} D_k(I)$ and the iterations are terminated when the ratio $D_k(I)/D_k$ is less than 10^{-2}, or 10 iterations, which ever is reached first. When this criterion has been satisfied, (normally after 5 iterations), the programme prints out summary information about the final partition, as can be seen in Table III for the partition of the 482 geochemical samples into four groups.

In addition to the information given in Table III the program prints a list of samples in each group arranged according to increasing distance from the type sample. The pooled within-group scatter and total scatter matrices for the partition as well as the determinant of these two matrices are also printed.

TABLE III

Within-group variance, and within-group variance along the eigenvectors, for 482 samples partitioned into four groups

Geochemical type sample	207	1094	1175	931
Within-group variance along eigenvector				
u_1	2.54	0.75	2.44	1.10
u_2	2.81	12.10	2.55	2.34
u_3	1.41	1.48	2.41	1.38
u_4	1.37	0.57	0.64	0.95
u_5	0.85	0.22	0.83	0.62
u_6	0.83	0.88	0.48	0.61
u_7	0.87	0.35	0.32	0.61
u_8	0.52	0.01	0.69	0.43
u_9	0.62	1.50	0.29	0.35
u_{10}	0.50	0.95	0.26	0.35
u_{11}	0.41	0.55	0.33	0.25
u_{12}	0.37	0.30	0.21	0.25
u_{13}	0.31	0.00	0.17	0.23
u_{14}	0.23	0.36	0.26	0.18
u_{15}	0.23	0.14	0.16	0.16
u_{16}	0.17	0.31	0.12	0.09
u_{17}	0.10	0.42	0.08	0.10
u_{18}	0.08	0.08	0.06	0.06
u_{19}	0.06	0.01	0.09	0.04
Within-group variance	14.3	21.0	12.4	10.1
Number of samples	151	3	78	250

This final partition might represent only a local minimum for the within-group variance and can be improved upon by transferring samples from one group to another and searching for partitions with lower within-group variance. Rubin (1967) and Demirmen (1969) have developed elaborate methods for improving classifications by reallocating samples between groups but the resulting partition might still represent a local minimum and can depend on the positions of the initial group centres. For the classification problem described in this paper, however, it was not felt necessary to improve the classification by rearranging the samples to reduce within-group variance because the geochemical groups are smoothed geographically at a later stage in order to obtain more uniform subareas. This smoothing increases within-group variance and thus counteracts the effect of improving the classification. The classification procedure described gives us the best subdivision of the samples into (k) groups according to the minimum within-group variance criterion trW. It still remains to find the optimal value for k.

Optimal value for the number of groups (k)

To detect whether a given number k of groups correspond to regions of high local density of samples in space R^p (with the chemical variables as axes), Mariott (1971) has suggested partitioning R^p into an increasing number of groups ($k = 1-10$) each time calculating Wilks' Lambda (Λ) given by the determinantal ratio:

$$\Lambda = |W|/|T|$$

which is a relative measure of compactness of the grouping. For a geometrical interpretation of the determinant $|T|$ as the square of the scattering volume see Anderson (1958) and King (1967).

Wilks' Λ is not directly comparable for partitions into a different number of groups since its value for $k+1$ groups will be equal to, or less than, its value for k groups. With the successive introduction of new group centres the determinant of the within-group scatter matrix $|W|$ will tend to decrease rapidly when the partition corresponds to clusters, and only slightly when a cluster is split artificially. As the total scatter matrix $|T|$ is equal for all partitions, minimizing $|W|$ is equivalent to minimizing Λ. If the sample points are evenly distributed in space and if this space is divided into k equal-sized parts, Mariott (1971) has shown that Λ will decrease inversely as k^2, and that a natural grouping will appear by a sharp decrease well below a general trend of the criterion Λk^2.

To determine the best value for k the criterion Λk^2 was chosen and the samples were successively partitioned into 1, 2, 3, ... 10 groups.

In Fig.4 a plot of Λk^2 versus k is shown for the 10 partitions with approxi-

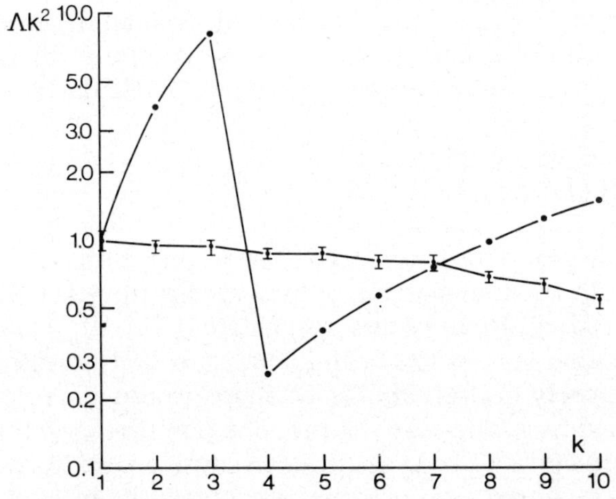

Fig.4. Plot of Λk^2 (where Λ is the ratio of the determinant of the within-group and total scatter matrices), versus the number of groups (k) for successive partitions into 1, 2, ... 10 groups of 482 humus (A_0) soil samples (in figure marked with dots) and 320 samples drawn from a uniform distribution (Mariott, 1971) (in figure marked with bars indicating approximate tolerance limits).

mate tolerance limits of Λk^2 for 320 samples drawn from a uniform distribution after Mariott (1971). As can be seen from the curve representing the uniformly distributed samples there is a bias in the criterion Λk^2 deviating from 1 for increasing values of (k). Mariott (1971) states that if the value of Λk^2 for different partitions lies in a band similar to that of the figure, or is greater than 1 for all possible subdivisions, there is no evidence of multi-modality. Comparing actual values for Λk^2 obtained from successive partition of the 482 geochemical samples into 2—10 groups with those from the uniform distribution, it can be seen that a partition of the geochemical samples into 3 groups appears to split clusters (Λk^2 is greater than for 2 groups), although the fairly sharp fall of Λk^2 for 4 groups well below the general trend, should indicate that 4 groups will represent a geochemically meaningful partition.

A partition of the samples into more than 4 groups gives only a slight reduction in the determinant of the within-group scatter matrix, W (Table IV). According to criterion Λk^2 it does not seem expedient to classify the samples into more than the 4 groups. Geochemical significance of the four-group partition will be deferred until the method of biplotting has been introduced, where a combined plot of both type samples and variables will be described.

Some structural characteristics of the four-group partition can, however, be seen in Table IV. Thus, the group containing 151 samples (207) consists of a stable core of about 80 samples as these are not lost when the 482 samples are split into more than six groups. Also, group (1175) retains its 80—90 samples up to the introduction of the 8th group when it starts losing samples to groups (931) and (685) with a subsequent change in type sample to samples (1074) and (1706). Marked deviation of a few samples is frequently observed in geochemistry and is caused by factors such as mineralization, contamination and environmental effects. Such samples will generally be trapped in small distinct groups. The two groups (1094) and (769) belong to this category, the former containing samples particularly low in metal content, the latter particularly high in metal content.

Stability of the four-group partition

To test the stability of the four-group partition 12 different partitions of this configuration were made, with random starting points, giving groups with geochemical type samples 931, 207, 1175 and 1094 respectively 11, 10, 5 and 5 out of 12 times. This indicates that groups (207) and (931) are fairly stable but that exchange of the more loosely bound samples of these groups provokes changes in the remaining group centres. The very distinct cluster corresponding to type sample 1094 does not appear in all the partitions because a partition that splits a large cluster may be favoured over one that maintains the integrity of a small distant cluster merely because the slight reduction in sum of squares achieved by splitting the large cluster is multiplied by many terms in the sum. This weakness of the minimum variance methods is stressed by Wishart (1969) who proposes an alternative procedure, mode analysis, which performs a search

TABLE IV

Type sample, group size, within-group variance, Λ and Λk^2 for each of 2 to 10 successive partitions of the 482 samples

Number of group centres	Type sample	No. of samples per group	Within-group variance	Λ	Λk^2
$k = 2$	548	477	17.1	0.9176	3.68
	1020	5	35.5		
$k = 3$	933	213	14.3	0.9009	8.10
	1094	3	21.0		
	685	266	11.5		
$k = 4$	207	151	14.3	0.0156	0.25
	1094	3	21.0		
	1175	78	12.4		
	931	250	10.1		
$k = 5$	207	124	13.7	0.0155	0.39
	1094	3	21.0		
	1175	70	11.4		
	685	182	9.3		
	667	103	8.8		
$k = 6$	207	78	12.8	0.0156	0.56
	1094	3	21.0		
	1175	94	10.2		
	649	118	8.9		
	667	100	8.3		
	933	89	11.1		
$k = 7$	207	77	12.3	0.0152	0.74
	1094	3	21.0		
	1175	93	10.6		
	737	124	8.9		
	667	94	8.1		
	933	87	1.0		
	769	4	13.0		

Number of group centres	Type sample	No. of samples per group	Within-group variance	Λ	Λk^2
$k = 8$	207	90	11.6	0.0153	0.98
	1094	3	21.0		
	926	37	11.0		
	931	158	7.6		
	1074	52	10.6		
	504	52	11.1		
	769	4	13.0		
	667	86	8.2		
$k = 9$	207	77	12.3	0.0150	1.22
	1094	3	21.0		
	532	53	7.3		
	685	116	7.5		
	1706	23	10.1		
	933	81	9.5		
	769	4	13.0		
	667	83	8.0		
	918	42	10.1		
$k = 10$	207	77	12.3	0.0147	1.47
	1841	2	10.9		
	532	53	7.3		
	685	116	7.5		
	1706	23	10.1		
	933	81	9.5		
	769	4	13.0		
	667	83	8.0		
	918	42	10.1		
	1020	1	0.0		

of dense points in hyperspace in terms of a "spherical" scanning device. For certain types of data this algorithm might prove superior and its relative merit for geochemical exploration data versus the method used in this paper should be evaluated.

Geographical smoothing

For the purpose of defining subareas the four-group partition, with geochemical type samples 931, 207, 1175 and 1094, was chosen. As can be seen in Fig.5 these groups gave three fragmented subareas with irregular boundaries, and 3 samples from group (1094) geographically scattered over the survey area. To obtain more continuous and uniform subareas a regrouping of samples was performed, taking into account both geochemical and geographical distance

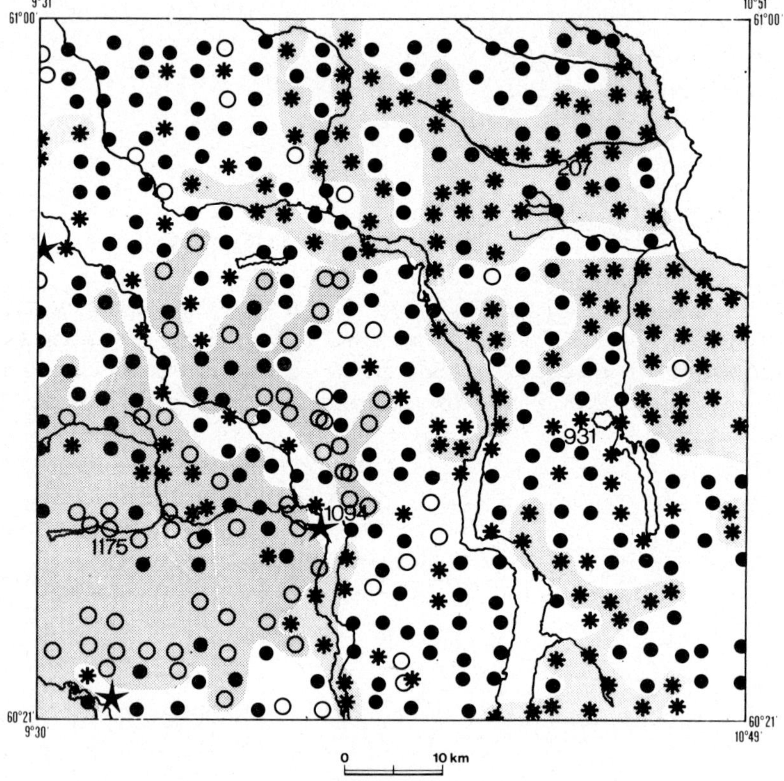

SUBAREAS WITH GEOCHEMICAL TYPE SAMPLE

Fig.5. The result of the minimum variance partition of 482 humus (A_0) samples into subareas. The number of samples in subareas (207), (931) and (1175) is 151, 250 and 78, respectively. For comparison the outlines of the geographically smoothed subareas (207) and (1175) are indicated by shading.

from the samples to the three type samples 207, 931 and 1175. A mixed distance measure (MD), similar to one used by Webster and Burrough (1972) in soil mapping, was used to smooth the subareas geographically. This mixed distance, MD, can be interpreted as an allocation rule assigning samples with equal geochemical distance from two type samples to the nearest group in terms of geographical distance. MD can be expressed by:

$$\mathrm{MD}_{ik} = D_{ik} \, [1 - \exp(-d_{ik} \cdot C/d_{min})]$$

where D_{ik} and d_{ik} are respectively the geochemical and geographical distance between sample (i) and type sample (k); d_{min} is the average distance between neighbours. If d_{ik} is less than d_{min} it is given the value d_{min}.

The effect of smoothing on the geochemical four-group partition (207), (931), (1175) and (1094) was studied by calculating Λ for several values of the geographical smoothing parameter, C.

A geographical smoothing will reallocate some of the samples classified on purely geochemical grounds and may thus increase the within-group variance. The smoothed map could however, show even lower values for Λ than the unsmoothed map indicating that a reallocation of samples according to the mixed-distance measure (MD) might help the partition to overcome a local minimum. A plot of Λ against C is shown in Fig.6. Comparing the value (Λ = 0.0164) of the smoothed partition (with smoothing parameter C, where C = 0.4) with that of the unsmoothed partition (Λ = 0.0156: Table IV), it can be seen that the smoothed four-group partition retains most of its geochemical significance when C is equal to, or greater than 0.4. For mapping purposes the value 0.4 for the geographical smoothing parameter, C, was chosen, reducing, as it can be seen in Table V, the geochemical distance by 65% and 20% for a sample being respectively neighbour and fourth nearest neighbour to a type sample. A sample belonging to one group must, therefore, be geochemically very distinct in order to plot near the centre of another group. The final smoothed map of the subareas with smoothing parameter C, where C = 0.4, can be seen in Fig.7. This map gives a clearer picture of the different subareas although there still are a number of single samples and smaller regions inside

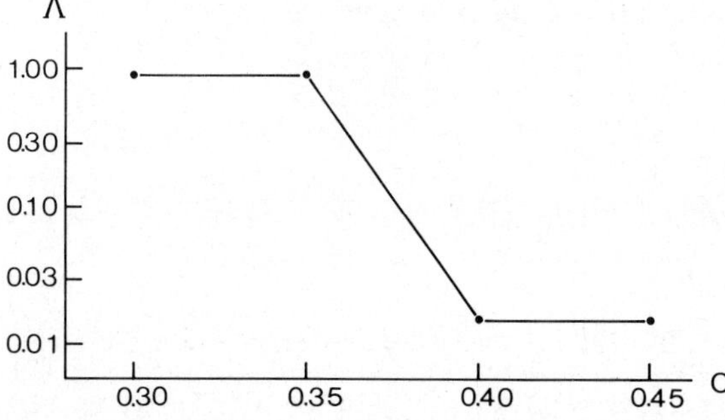

Fig.6. Plot of Wilks' Λ (the ratio of the determinant of the within-group and total scatter matrices), for different values of the geographical smoothing parameter C.

TABLE V

Reduction in geochemical distance as a function of the geographical distance, d_{ik}, between a given sample, i, and the geochemical type sample, k, for smoothing parameter C (= 0.4)

d_{ik}/d_{min}	1	2	3	4	5	6	8	10
$\exp(-d_{ik} \cdot C/d_{min})$	0.65	0.45	0.30	0.20	0.14	0.09	0.04	0.02

d_{min} = average geographical distance between neighbour samples.

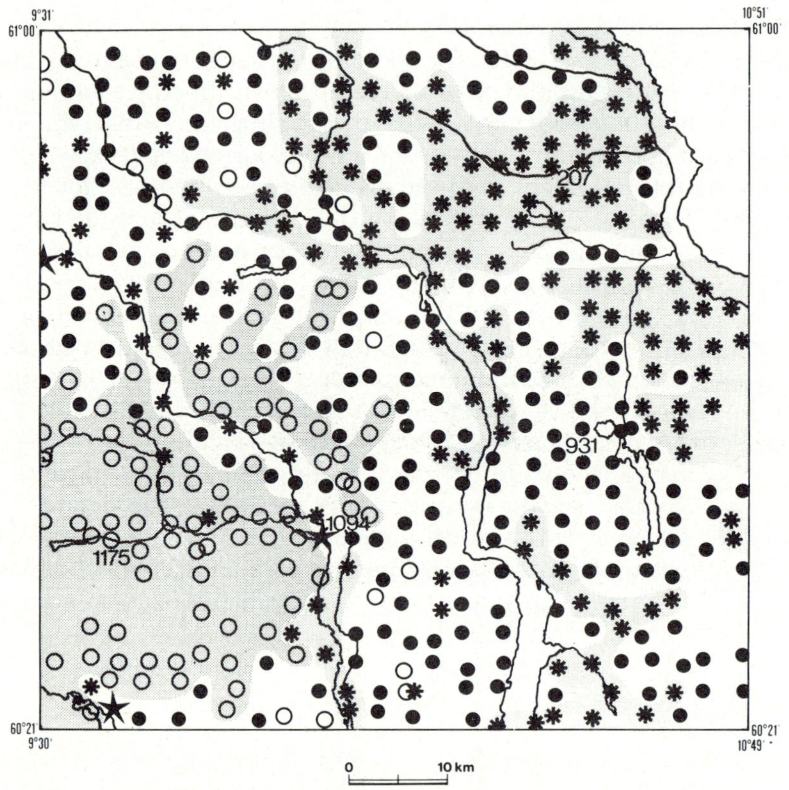

Fig.7. The result of the minimum variance partition of 482 humus (A_0) samples into subareas and subsequent geographical smoothing with smoothing parameter C, where $C = 0.4$. The number of samples in smoothed subareas (207), (931) and (1175) is 149, 223 and 107, respectively. The outlines of the smoothed subareas (207) and (1175) are indicated by the shading.

the subareas. Groups with less than five samples were ignored in the smoothing procedure.

EVALUATION OF CLUSTERS

Biplot of chemical variables and samples

As an aid to appreciation of geochemical differences between the four groups isolated by the grouping procedure, both type samples and chemical variables have been represented in the same diagram. This kind of plot (Figs.8 and 9) displays intersample distances as well as variances and correlations of the variables (Gabriel, 1971).

As n sample points in space R^p (having the chemical variables as axes) and p variable points in space R^n (having the samples as axes) are dual (Gower, 1966; Klovan and Imbrie, 1971; David et al., 1974) the eigenvector v_k in R^n

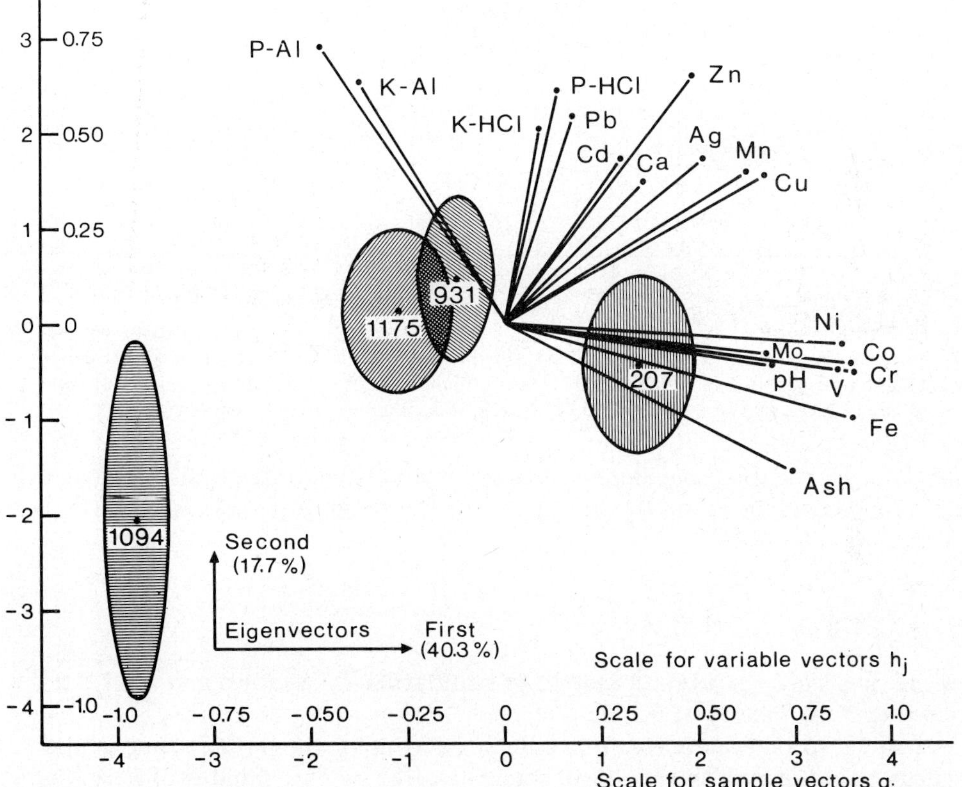

Fig.8. R-mode type of biplot in the plane of the first two eigenvectors v_1 and v_2 showing the type sample and the one standard deviation ellipse for each group together with the variances and correlations of the chemical variables. The number of samples in the groups (207), (931), (1175) and (1094) is 151, 250, 78 and 3, respectively.

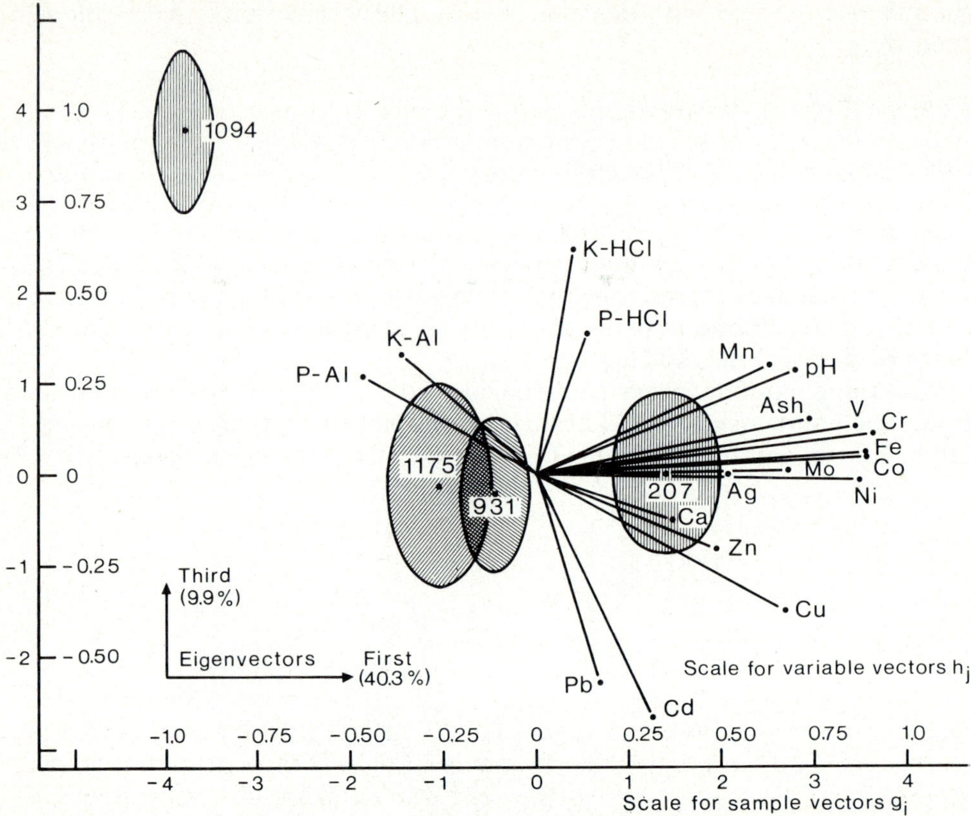

Fig.9. *R*-mode type of biplot in the plane of the first and third eigenvectors, v_1 and v_3, showing the type sample and the one standard deviation ellipse for each group together with the variances and correlations of the chemical variables. The number of samples in the groups (207), (931), (1175) and (1094) is 151, 250, 78 and 3, respectively.

may be expressed by the eigenvector u_k in R^p. If we impose the eigenvectors u_k and v_k corresponding to the eigenvalue λ_k to be unit vectors then (Gower, 1966):

$$u_k = \lambda_k^{-\frac{1}{2}} \, \mathbf{X}' v_k$$
$$v_k = \lambda_k^{-\frac{1}{2}} \, \mathbf{X} u_k$$

where $\mathbf{X}$ is the ($n \times p$) data matrix of standardized variables and $\mathbf{X}'$ the transposed ($p \times n$) matrix of $\mathbf{X}$.

If we let h_j be the vectors representing the chemical variables and g_i the vectors representing the samples, then the plotting coordinates of h_j and g_i can be expressed in terms of the eigenvectors u_k and v_k. Several different types of plots can be produced according to different scaling of the eigenvectors u_k and v_k. The choice:

$$h_j = (u_{1j}, u_{2j})$$
$$g_i = (\sqrt{\lambda_1}\, v_{1i}, \sqrt{\lambda_2}\, v_{2i})$$

gives a plot where the relation between the sample vectors, g_i, in the plane of the eigenvectors represents the relation between sample vectors x_i in space R^p. The euclidian distance between sample x_i and x_e in space R^p will be approximated by the distance between g_i and g_e in the plot. This will essentially be a Q-mode type of plot where both samples and chemical variables have been projected onto the eigenvectors, u, in space R^p. For plotting purposes in this paper variable vectors h_j and sample vectors g_i have been expressed in the alternative way:

$$h_j = (\sqrt{\lambda_1}\, u_{1j}, \sqrt{\lambda_2}\, u_{2j})$$
$$g_i = (v_{1i}, v_{2i})$$

This particular choice of plotting coordinates will give a R-mode type of plot where variables and samples are projected onto eigenvectors v in space R^n (having the samples as axes). In this plot the variances of the 19 chemical variables are represented by the length of the h_j vectors and their intercorrelation by the angle between them. As our chemical variables have been standardized to variance 1, the deviation of the length of the h_j vectors from 1 will indicate how much of the variance of jth chemical variable is contained in the plane of the first two eigenvectors, v_1 and v_2. From Fig.8 we see that the length of chromium is 0.9 and that the angle with the first axis is very small indicating that most of the variation in chromium is along the first eigenvector, v_1. If a chemical variable vector, h_j, is orthogonal to a sample vector, g_i, this means that the value of chemical variable j is 0 for sample i, which in the case of standardized variables, will be equal to the mean value.

The distance on the plot between sample points g_i and g_e will with the plotting coordinates chosen (Figs.8 and 9) be an approximation for the standardized distance in the plane of eigenvectors v_1, v_2. The squared distance $d^2_{g931,\ g207}$ in the plot between the geochemical type samples in the two largest groups will thus be an approximation for:

$$(x_{931} - x_{207})(X'X)^{-1}\,(x_{931} - x_{207})'$$

where x_{931} and x_{207} are the row vectors in the data matrix X, corresponding to the geochemical type samples 931 and 207, and where $(X'X)^{-1}$ is the matrix projecting the samples onto space R^n. By this projection, the spherical-shaped clusters isolated by the grouping procedure in space R^p are transformed into elliptical-shaped clusters in space R^n as can be seen from the plot in Fig.8.

In evaluating the interrelationship between variables and samples in the plots one must not forget that both samples and chemical variables are represented in the planes of the first, second and third eigenvectors accounting for 68% of the total variance (Table II).

As can be seen from the relation between the variable vectors, h_j (Figs.8

TABLE VI

Groups represented by type sample characterized by six sets of chemical variables

Type sample			Sets of chemical variables	
207	931	1175		
--	++	+	I	P-Al (0.7) K-Al
0	0	0	II	P-HCl (0.4) K-HCl
+	0	0	III	Pb (0.6) Cd
+	0	—	IV	Cu (0.6) Zn (0.5) Ca (0.3) Ag
++	—	--	V	Mo (0.6) Co (0.7) pH (0.7) Fe (0.9) Cr Cr (0.8) ash (0.8) V (0.7) Ni
+	0	—	VI	Mn

(0.0) = Pearson product-moment correlation coefficient (from Table I). ++ = highly positive influence, + = some positive influence, 0 = no influence, — = some negative influence, -- = highly negative influence.

and 9), some of the chemical variables are strongly correlated and have been arranged accordingly into six sets.

The relative influence of the sets of chemical variables in separating the three groups (207, 931, 1175) has been estimated and is shown in Table VI. Separation of the group (207) from the two other groups is mainly caused by the samples in group (207) having higher values for the chemical variables in set V. The two groups (1175) and (931) both have approximately mean values for the chemical variables in sets II and III as the vectors g_{931} and g_{1175} (Figs.8 and 9) are nearly orthogonal to the corresponding variable vectors, h_j. Separation of the two groups in the plane of the first, second and third eigenvectors is only about one and a half standard deviations and is chiefly due to the relatively greater number of samples in group (1175) with low values for the chemical variables in sets I and V.

Geochemical characteristics of the subareas

An evaluation of the differences in element behaviour within and between the three subareas will have to consider the complex interplay of conditions like soil moisture, Eh, pH, presence of antagonistic and stimulatory elements, difference in composition of the humus ash, climate and other edaphic factors, but this cannot be treated in this paper. Obvious differences between the subareas can be seen however, from Figs.8 and 9 and will be commented upon briefly.

The differences in mean value of the chemical elements between the three subareas fall into the following categories: those showing a marked difference in mean value between the subareas, and those showing minor to no difference. As it can be predicted from Figs.8 and 9 the chemical variables in sets I and V are of the former type (represented by iron and chromium in

Figs.10 and 11 and by ash content in Fig.12), whereas the elements in the other sets belong to the latter type, (represented by lead in Fig.13). Manganese in Fig.13 is an example of an intermediate type.

The practical consequences of the differences in element content between

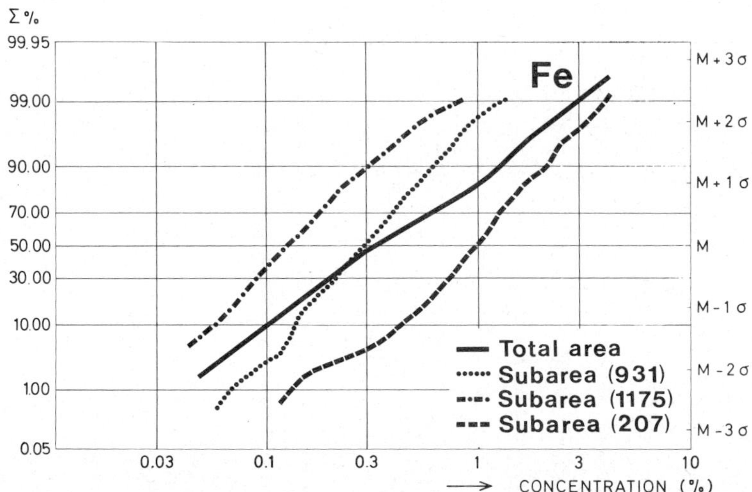

Fig.10. Cumulative frequency distribution curves for iron content in humus (A_0) samples from different subareas. The number of samples from subareas (207), (931) and (1175) is 149, 223 and 107, respectively. (Right-hand ordinate gives the scale for estimation of median, M, and standard deviation, σ.)

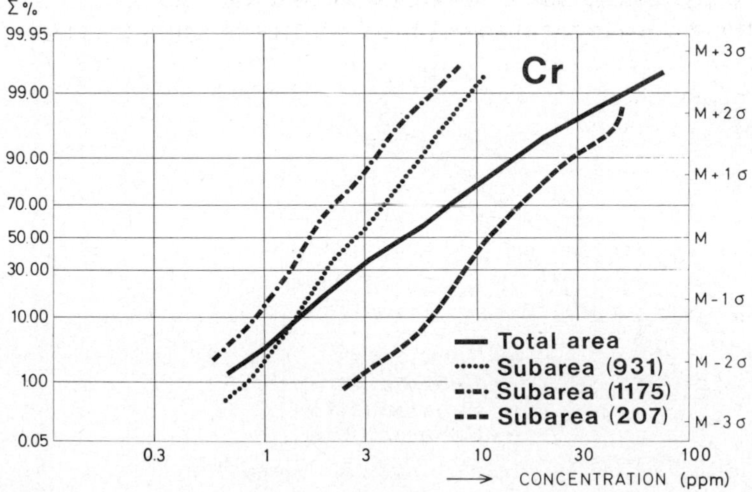

Fig.11. Cumulative frequency distribution curves for chromium content in humus (A_0) samples from different subareas. The number of samples from subareas (207), (931) and (1175) is 149, 223 and 107, respectively. (Right-hand ordinate gives the scale for estimation of median, M, and standard deviation, σ.)

212

the subareas are that threshold values should be set differently for each subarea.

Subareas differ both in mean value and correlation between chemical variables. Scattergrams of the chemical variables are of two types: those showing significant difference in correlation and those showing a minor to no difference in

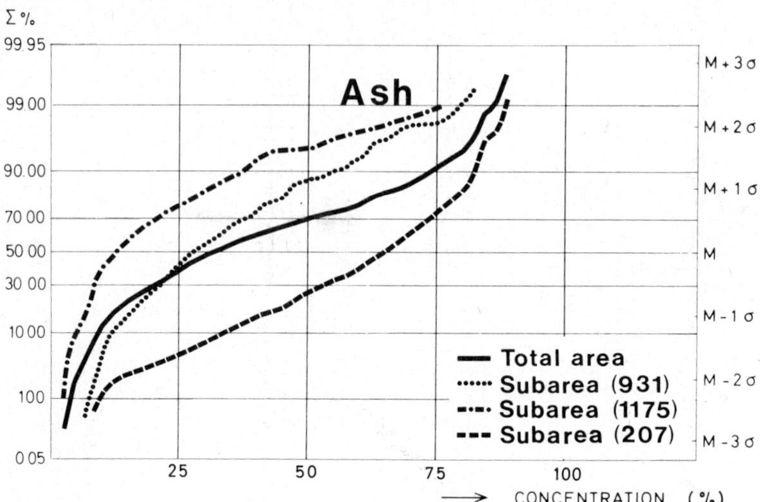

Fig. 12. Cumulative frequency distribution curves for ash content in humus (A_0) samples from different subareas. The number of samples from subareas (207), (931) and (1175) are 149, 223 and 107, respectively. (Right-hand ordinate gives the scale for estimation of median, M, and standard deviation, σ.)

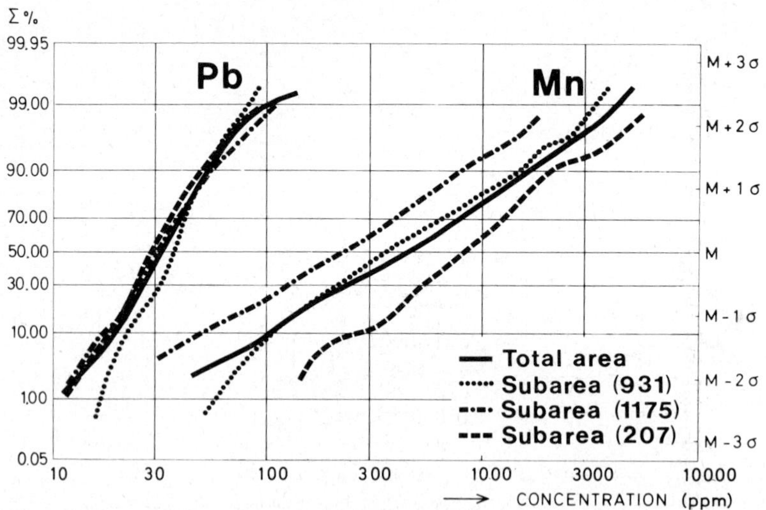

Fig. 13. Cumulative frequency distribution curves for lead and manganese content in humus (A_0) samples from different subareas. The number of samples in subareas (207), (931) and (1175) are 149, 223 and 107, respectively. (Right-hand ordinate gives the scale for estimation of median, M, and standard deviation, σ.)

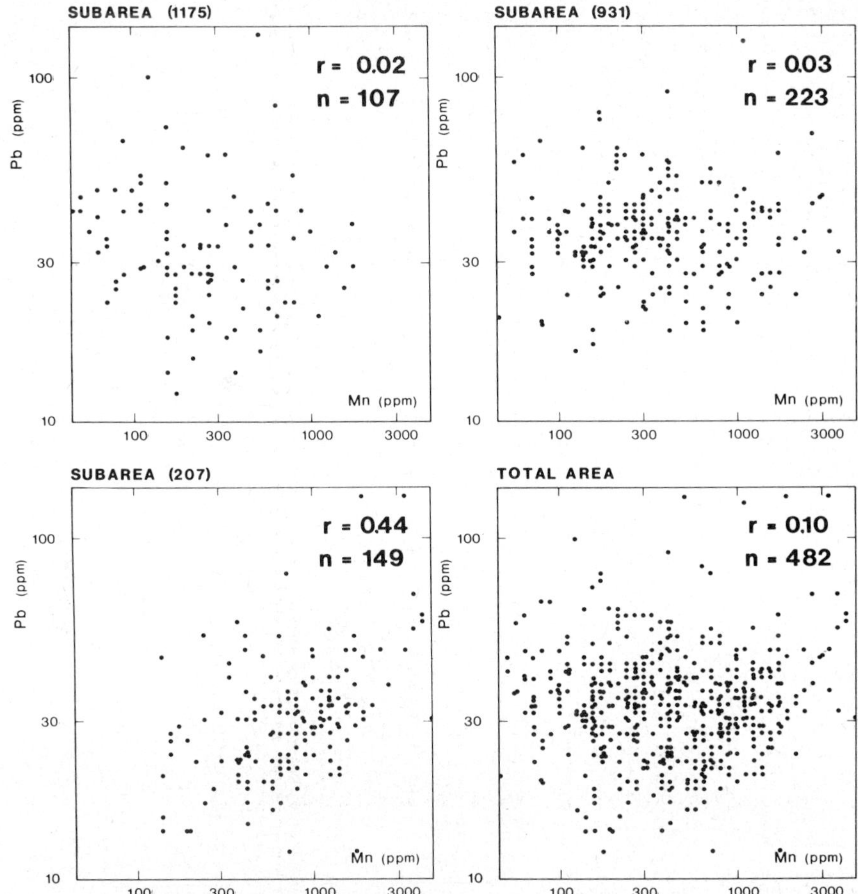

Fig.14. Scatter diagrams for Pb versus Mn for subareas and total area. The number of samples within each cell is indicated by dots; r = correlation coefficient; n = number of samples.

correlation between the subareas. As can be predicted from Figs.8 and 9, the correlation between the chemical variables in set V is nearly equal for all subareas, whereas the correlation between the chemical variables in sets III, IV and VI reveals differences. As an example of the latter the scattergrams for Pb versus Mn and Pb versus Fe for each subarea are shown in Figs.14 and 15. It can be seen that the Pb versus Mn relationship changes noticeably, showing no significant correlation at the 5% level in subareas (1175) and (931), a significant positive correlation in subarea (207) and no significant correlation in the total survey area.

This difference in correlation between the subareas for Pb versus Mn is not believed to be a result of the variation in mean value for manganese, as iron has both a larger difference in mean value than manganese (Figs.11 and 14) and lower correlation against lead (Fig.15).

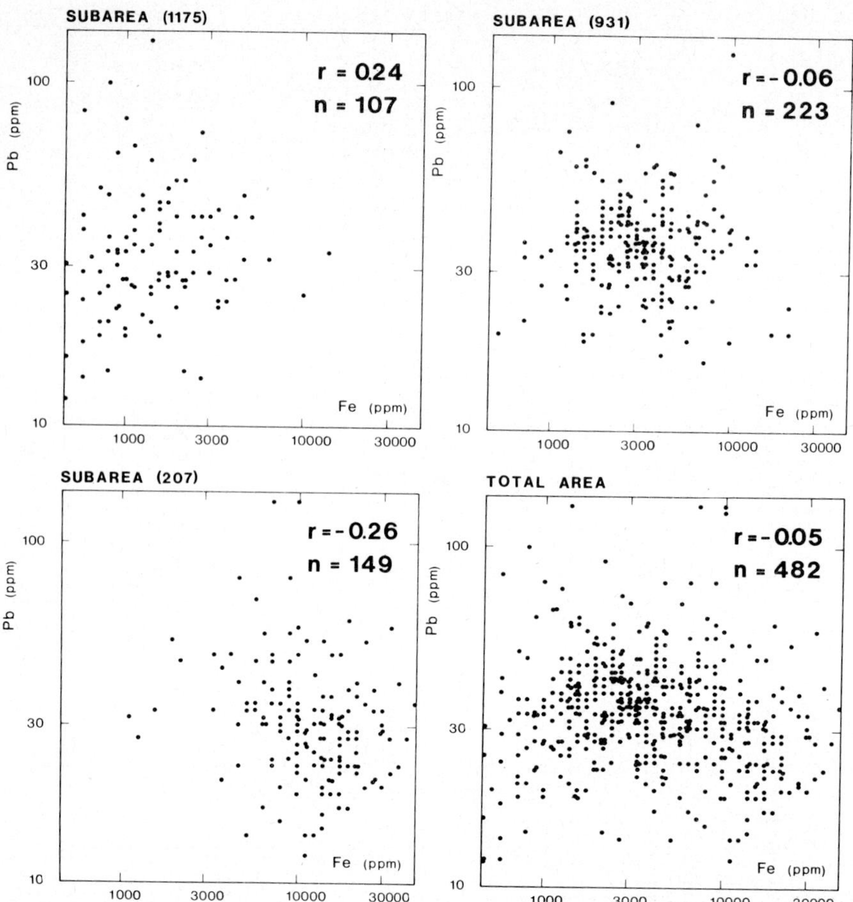

Fig.15. Scatter diagrams for Pb versus Fe for subareas and total area. The number of samples within each cell is indicated by dots; r = correlation coefficient; n = number of samples.

When regression analysis is used to estimate chemical element content in a sample due to environmental effects (Brundin and Nairis, 1972) in areas with varying interelement correlation, better results are obtained when each subarea is treated separately.

Soil conditions in the subareas such as depth and origin of soil material and profile type, are listed in Tables VII, VIII and IX. Comparing Figs.2 and 3 and Table VII most of the deeper overburden, which is dominant in subarea (207), seems to be connected with youngest Precambrian and Cambro-Silurian rocks, which, being less resistant to glacial erosion, have formed topographic lows. Till is dominant in all subareas, whereas soil material of glacio-fluvial origin is found chiefly in (207) (Table VIII).

A major part of the plots classified as brown earth belongs to subarea (207) compared to the other two areas (Table IX). This is in accordance with results

TABLE VII

Distribution in percent of the depth of soil material within the survey area

Subarea	0—20 cm	20—70 cm	>70 cm	Number of plots
207	0.7	11.4	87.9	149
931	4.0	28.7	67.3	223
1175	6.5	32.7	60.7	107

TABLE VIII

Distribution in percent of soil material according to origin

Subarea	Till	Glacifluvial deposits	Residual material (including talus)	Organic soil material	Number of plots
207	88.0	4.0	4.0	4.0	149
931	93.3	1.4	4.0	1.4	223
1175	90.7	0.9	4.7	3.7	107

TABLE IX

Distribution in percent of the main profile types within the different subareas

Subarea	Podzol	Brown earth	Transitions podzol-brown earth	Swamp soils	Number of plots
207	65.1	27.5	2.7	4.7	149
931	95.9	2.4	0.8	0.9	223
1175	96.3	0.9	0.0	2.8	107

obtained by Lag (1968) where brown earth samples from all Oppland county have markedly lower values for readily soluble potassium and phosphorus and higher values for pH and ash content than samples from podzol profiles, and should therefore according to Fig.8 be expected to fall within subarea (207).

CONCLUSIONS

By a combined use of an iterative grouping procedure giving a partition minimizing the trace of the within-group scatter matrix (trW) and of an evaluation of the resulting partition calculating the ratio of the determinant of the within-group and total scatter matrices $\Lambda = |W|/|T|$ it has been shown that 482 humus (A_0) soil samples from a regional geochemical survey form four geochemically distinct clusters in the 19-dimensional space of the standardized chemical variables.

An exchange of samples between the three largest groups according to a geographical smoothing parameter, C, has been shown to result in more uniform and homogenous subareas without significantly increasing the value of the overall classification as measured by Λ.

The R-mode type of plot used in this paper displaying chemical variables and type samples proved to be of considerable value in interpreting the results of the numerical classification and could be used beneficially in connection with any multi-dimensional interpretation of geochemical data.

Three subareas defined by the grouping procedure have been shown to be very different with respect to readily soluble phosphorus (P-Al) and potassium (K-Al), pH, ash content and Mo, Co, Fe, Cr and Ni. The strengths of inter-element correlation between Pb, Cd, Cu, Zn, Ca and Mn vary between subareas.

Differences between the subareas show that for geochemical exploration purposes a better interpretation is achieved if threshold values and environmental effects for the trace metals are determined separately within each subarea.

The computer procedure used in this paper represents a feasible but not a final and foolproof method and is not intended to replace more traditional methods for defining geochemically meaningful subareas on the basis of topography, geology and Quaternary geology. It does, however, permit the delineation of subareas within regional geochemical surveys when such primary information is lacking.

REFERENCES

Anderson, T.W., 1958. An Introduction to Multivariate Statistical Analysis. John Wiley and Sons, New York, N.Y.

Ball, G.H., 1965. Data analysis in the social sciences: What about the details? Proc. Fall Joint Computer Conference, pp.533—559

Bolshev, L.N., 1969. Cluster analysis. Bull. Int. Stat. Inst., 43: 411—425

Bonham-Carter, G.F., 1965. A numerical method of classification using qualitative and semi-quantitative data, as applied to the facies analysis of limestones. Can. Pet. Geol. Bull., 13: 482—502

Brundin, N.H. and Nairis, B., 1972. Alternative sample types in regional geochemical prospecting. J. Geochem. Explor., 1: 7—47

Cole, E.J., 1969. Numerical Taxonomy. Academic Press, London and New York, N.Y.

Cormack, R.M., 1971. A review of classification. J. R. Stat. Assoc., 134: 321—353

David, M., Campiglio, C. and Darling, R., 1974. Progresses in R- and Q-mode analyses: correspondence analysis and its application to the study of geological processes. Can. J. Earth. Sci., 11: 131—146

Demirmen, F., 1969. Multivariate procedures and Fortran IV program for evaluation and improvement of classifications. Kans. Geol. Survey Computer Contrib., 20: 1—51

Forgy, E., 1965. Cluster analysis of multivariate data: efficiency versus interpretability of classifications (abstract). Biometrics, 21: 768

Gabriel, K.R., 1971. The biplot graphic display of matrices with application to principal component analysis. Biometrika, 58: 453—467

Garret, R.G., 1973. Regional geochemical study of Cretaceous acidic rocks in the northern Canadian Cordillera as a tool for broad mineral exploration. In: M.J. Jones (Editor), Geochemical Exploration 1972. Institution of Mining and Metallurgy, London, pp.203—221

Gower, J.C., 1966. Some distance properties of latent root and vector methods used in multivariate analysis. Biometrika, 53: 325—338

Hesp, W.R. and Rigby, D., 1973. Cluster analysis of rocks in the New England igneous complex, New South Wales, Australia. In: M.J. Jones (Editor), Geochemical Exploration 1972. Institution of Mining and Metallurgy, London, pp.221—237

Holmsen, G., 1954. Description of the Quaternary geological map Oppland (scale 1:250 000). Norg. Geol. Unders., 187: 1—58

Holmsen, G., 1955. Description of the Quaternary geological map Hallingdal (scale 1:250 000). Norg. Geol. Unders., 190: 1—55

Holtedahl, O. and Dons, J.A., 1960. Geological map of Norway (bedrock). In: I. Holtedahl (Editor), Geology of Norway, Norg. Geol. Unders., 208: 1—540

Howarth, R.J., 1973. The pattern recognition problem in applied geochemistry. In: M.J. Jones (Editor), Geochemical Exploration 1972. Institution of Mining and Metallurgy, London, pp.259—275

Imbrie, J. and Purdy, E.G., 1962. Classification of modern Bahamian carbonate sediments. Am. Assoc, Pet. Geol., Mem., 1: 253—272

Jancey, R.C., 1966. Multidimensional group analysis. Aust. J. Bot., 14: 127—130

Johnson, S.C., 1967. Hierarchical clustering schemes. Psychometrica, 32: 241—245

King, B., 1967. Step-wise clustering procedures. J. Am. Stat. Assoc., 62: 86—101

Klovan, J.E. and Imbrie, J., 1971. An algorithm and Fortran IV program for large scale Q-mode analysis. Math. Geol., 3: 61—67

Lance, G.N. and Williams, W.T., 1967. A general theory of classificatory sorting strategies, 1, Hierarchical systems. Computer J., 9: 373—380 ·

Lag, J., 1968. Investigations on forest soils in Oppland county, Norway, in connection with the field work of the National Forest Survey. Medd. Norske Skogfors., 91: 332—393

MacQueen, J., 1967. Some methods for classification and analysis of multivariate observations. Proc. 5th Berkeley Symp. on Probability and Statistics, pp.281—297

Mariott, F.H.C., 1971. Practical problems in a method of cluster analysis. Biometrics, 27: 501—514

McCammon, R.B., 1966. Principal component analysis and its application in large scale correlation studies. J. Geol., 74: 721—733

McCammon, R.B., 1968. Multiple component analysis and its application in classification of environments. Am. Assoc. Pet. Geol. Bull., 52: 2178—2196

Obial, R.C. and James, C.H., 1973. Use of cluster analysis in geochemical prospecting, with particular reference to southern Derbyshire, England. In: M.J. Jones (Editor), Geochemical Exploration 1972. Institution of Mining and Metallurgy, London, pp.237—259

Parks, J.M., 1969. Multivariate facies maps. Kans. Geol. Survey Computer Contrib., 40: 6—12

Rubin, J.R., 1967. Optimal classification into groups: an approach for solving the taxonomy problem. J. Theoret. Biol., 15: 103—144

Sokal, R.R. and Sneath, P.H.A., 1963. Principles of Numerical Taxonomy. W.H. Freeman and Co., San Francisco, Calif.

Tryon, R.C. and Bailey, D.E., 1970. Cluster Analysis. McGraw-Hill, New York, N.Y.

Webster, R. and Burrough, P.A., 1972. Computer-based soil mapping of small areas from sample data, II. Classification smoothing. J. Soil Sci., 23: 222—234

Wishart, D., 1969. Mode analysis. In: A.J. Cole (Editor), Numerical Taxonomy. Academic Press, London and New York, N.Y., pp.282—308

APPLICATION OF DISCRIMINANT ANALYSIS TO THE GEOCHEMICAL EVALUATION OF GOSSANS

A.J. BULL and R.H. MAZZUCCHELLI

Western Mining Corporation Limited, Perth, W.A. (Australia)

ABSTRACT

Many of the nickel sulphide ore bodies discovered since 1965 in the Archean Shield of Western Australia are characterized by gossanous outcrops. Accordingly, successful identification of gossans played a prominent part in subsequent nickel exploration programmes.

One of the most widely employed techniques for evaluation of nickel sulphide gossans involves analysis and interpretation of various trace elements. In an attempt to define more clearly the trace element parameters a suite of 270 gossans and pseudo-gossans was analysed for five elements (Ni, Cu, Co, Cr, Zn), and subjected to discriminant analysis afforded by initial categorisation of the samples into six genetic groups. Application of the results to field data proved highly successful.

INTRODUCTION

The nickel sulphide deposits discovered between 1966 and 1972 in Western Australia introduced a new facet to exploration for ores of this type — that of gossan recognition. Most previously discovered Ni sulphide deposits occur in more temperate regions, where the weathering products of the ore tend to be removed progressively by erosion so that minimal gossan development takes place. The major Australian occurrences, on the other hand, have been found to be associated exclusively with Archean rocks occurring within an extensive inland plateau of low topographic relief, internal drainage and, having for the most part, a semi-arid climate. In fact, the area has been subjected to terrestrial weathering more or less continuously since the Permian, and the present land surface reflects not only the current arid cycle of weathering but also retains, almost intact, remnants of former climatic regimes, including lateritic and siliceous hardpans. As well as being favourable for the development and preservation of gossans over sulphide deposits, these factors also promote the development of numerous other types of ferruginous and jasperoidal outcrops which superficially resemble gossans. Consequently the search for and evaluation of gossanous outcrops has become an important aspect of exploration in the Archean of Western Australia.

NATURE OF THE PROBLEM

Properties that can be studied to provide a correct interpretation of gossanous material are: (1) field relationships — geological setting of occurrences and their physical characteristics, e.g. dimensions, structure, nature of contacts, etc.; (2) textural features — macroscopic and microscopic; (3) mineralogy; and (4) geochemical characteristics.

The importance of geochemical methods of gossan evaluation lies in the ease with which trace element analysis can be applied to the large numbers of samples generated in routine exploration programmes. For this reason geochemical evaluation is almost universally used as a screening technique, although very little information has been published to date.

Clema and Stevens-Hoare (1973) have developed a graphical method to discriminate sulphide-derived gossans from spurious occurrences, based on trace element abundances. G.A. Travis and D.E. Roberts reported (personal communication, 1973) that the study of textural features provides reliable discrimination in most instances. The present paper describes a system based on trace element analysis and multivariate statistical analysis, which was instigated by Western Mining Corporation Limited in 1969 and subsequently used and progressively refined as more data accrued from continued exploration.

The theoretical basis for the use of trace element criteria for discrimination of Ni—Cu sulphide gossans is that remnant traces of the constituent elements of the original sulphide ore are retained throughout oxidation and survive in the weathered derivative or gossan. A given element is leached from, or concentrated in, a gossan to an extent dependent upon the physico-chemical conditions which prevail during the weathering of the sulphides, and interactions with the other components present in the system. These processes are complex and little known, so that only very broad generalizations may be made about the behaviour of even the most common elements. Nevertheless, useful information may be derived from investigations of an empirical nature.

The principal base metals of nickel sulphide deposits, nickel and copper, are regarded as fairly mobile in the environment of oxidizing sulphide bodies, and thus appreciable depletion of these elements in gossans might be anticipated. On the other hand, elements such as tellurium, selenium and the platinoid group metals that are commonly concentrated in Ni—Cu sulphide bodies are regarded as relatively immobile elements under these conditions. These latter elements have, therefore, received considerable attention as possible indicator elements for nickel sulphide gossans.

Discrimination obviously necessitates careful consideration of the trace element characteristics of other gossan-like outcrops not related to nickel sulphides.

In the course of exploration between 1965 and 1969 the Exploration Division of Western Mining Corporation Limited sampled some thousands of gossans and pseudo-gossans. A proportion of these were sufficiently well documented by mapping, textural studies, geophysics and, in some cases, drilling, to enable conclusions to be drawn regarding their origins. It was recognized that this data could be used to derive quantitative statistical parameters, characteristic of the different geneses of the materials sampled, which could be of assistance in evaluating new samples of gossanous outcrops. A computer file was established to cater for the potential volume and complexity of the data and the addition of new data which would become available from time to time. The application of multivariate statistical techniques, particularly discriminant analysis, seemed especially pertinent to the problem in hand and this was an important factor influencing the decision to adopt a computer-based system.

COLLECTION AND ARRANGEMENT OF DATA

Trace element and other relevant data on gossanous samples were collated from numerous projects. Based on all the evidence available each sample was assigned to one or other of six groups, covering the main classes of gossans and pseudo-gossans relevant in nickel exploration. These are as follows:

Group 1: known to be related to nickel sulphide mineralization.

Group 2 : probably related to nickel sulphide mineralization.

Group 3 : probably related to sulphide-bearing tuff or sediment horizons within or adjacent to ultramafic rocks.

Group 4 : probably related to lateritic weathering of ultramafic rocks.

Group 5 : probably related to non- nickeliferous sulphide and/or oxide accumulations in ultramafic rocks.

Group 6 : significance unknown.

The essential difference between Groups 1 and 2 is one of confidence of classification. Group 1 includes only those gossans that have undergone sufficient testing by drilling to remove all doubt as to the validity of classification. Groups 2—5 have in general not been tested to the same extent but sufficient evidence of an independent nature exists to provide a reasonable probability for the correct classification of each sample. Individual samples within each group would have been tested by drilling. Group 6 is comprised of the unknowns, for which classification is sought by the method.

The first (1970) edition of the file consisted of some 270 gossans and pseudo-gossans, which had been analysed for up to 10 elements and assigned into Groups 1 to 6. A statistical report was initially produced in the form of cumulative frequency listings by groups for each element analysed. The results are summarized in Fig.1.

It is apparent from Fig.1 that the degree of overlap existing for each element makes discrimination of the different groups virtually impossible on the basis of single element data. Of the elements tested, tellurium shows the best separation between groups but even in this case it is impossible to select a cut-off value which would avoid the misclassification of a substantial proportion of the total samples. In any case tellurium analysis at the concentration levels sought is slow and expensive and it is hardly practicable to apply tellurium analysis on a routine basis. The same argument applies equally to selenium and elements of the platinoid group.

It was decided to investigate discriminant analysis using determinations for

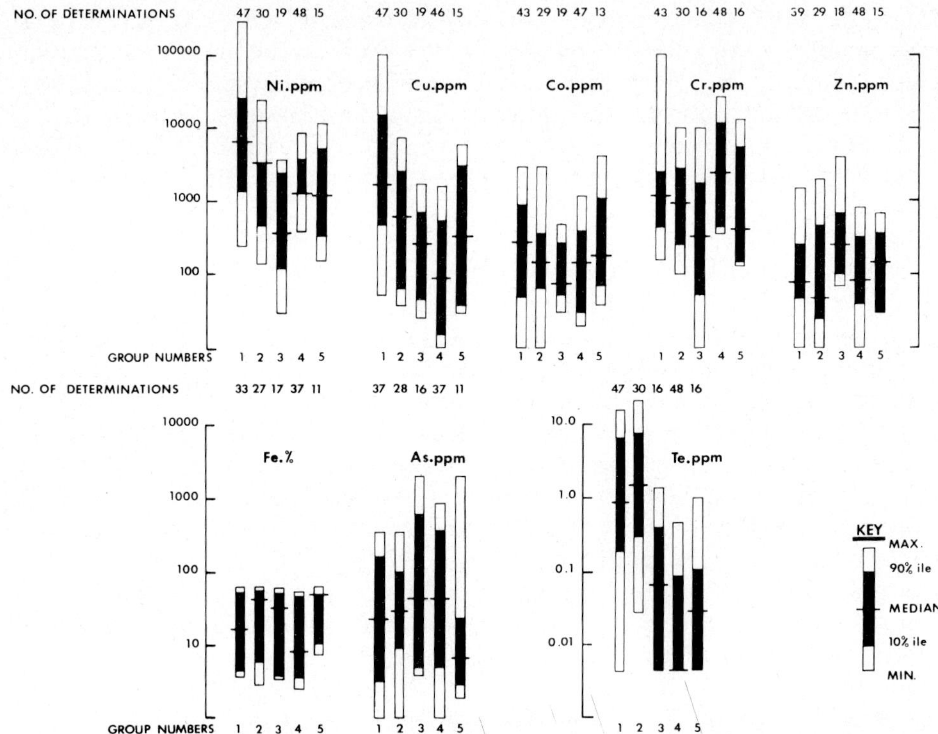

Fig.1. Element distribution in gossans and pseudo-gossans sampled from the Archean of Western Australia.

those elements which could be analysed quickly and cheaply by standard atomic absorption procedures.

DISCRIMINANT ANALYSIS

The decision to use multivariate discriminant analysis is a natural one. The technique is designed to answer the following question, "What is the major component of variation that distinguishes between groups comprising a population?". This differs importantly from factor analysis which is concerned only with major components of variation within a whole population and is in no way oriented toward characterization of known groupings within the population. Thus the analysis is directed specifically towards the particular properties upon which one is trying to discriminate.

There are two major approaches to discriminant analysis, both assume (1) the variables are normally distributed, (2) the groups have a common variance/covariance matrix, and (3) the groups of objects are in fact distinct.

We may suppose there are a set of G groups, and N_g, $g = 1, \ldots . G$ individual specimens in each group, each analysed for m geochemical variables.

The samples $x_1(g), \ldots x_{N_g}(g)$ are from independent normal populations π_g with means $\mu(g)$ and common variance/covariance matrix Σ.

The best linear estimate of $\mu(g)$ is $\bar{x}(g)$ the sample mean:

$$\bar{x}(g) = \frac{1}{N_g} \sum_\alpha x_\alpha(g)$$

and of Σ is S the joint sample variance/covariance matrix:

$$S = \sum_g \sum_\alpha [x_\alpha(g) - \bar{x}(g)] [x_\alpha(g) - \bar{x}_\alpha(g)]' / \sum_g (N_g - 1)$$

where summations in α are over all samples $1, \ldots N_g$ in group g.

Then W, the within group sums of squares and sums of products matrix (SSSP), is given by:

$$W = \sum_g (N_g - 1)S$$

and A, the between group SSSP matrix, is defined by:

$$A = \sum_g N_g [\bar{x}(g) - \bar{\bar{x}}][\bar{x}(g) - \bar{\bar{x}}]'$$

where $\bar{\bar{x}} = \sum_g N_g \bar{x}(g) / \sum_g N_g$ is the population mean.

Consider the discriminant factor $\gamma = v'x$ which is a function of the m variables, and the discriminant coefficients $v_i (i = 1, \ldots m)$.

The mean of γ for the gth group is:

$$\bar{\gamma}(g) = \frac{1}{N_g} \sum_\alpha \gamma_\alpha(g)$$

The sums of squares of γ within groups is $[\gamma(g) - \bar{\gamma}(g)][\gamma(g) - \bar{\gamma}(g)]' = v'Wv$.

For a given set of observed values on the m variables the problem resolves to that of determining the coefficients $v_1, \ldots v_m$, which maximize the ratio $\lambda = v'Av/v'Wv$.

By equating to zero the partial derivatives of λ with respect to v_i, we obtain:

$$\partial\lambda/\partial v_i = 0 \text{ implies } (A - \lambda W)v = 0$$

for which each solution v forms the eigenvector for the corresponding latent root $\lambda_j (j = 1, \ldots m)$ of $(W^{-1}A)$. These are the classical least-squares solution to the discriminant problem.

To test the significance of the discriminant factors we may use Rao's criteria:

$$z = \sum [N - 1 - \tfrac{1}{2}(G + m)] \ln(1 + \lambda_j) \sim \chi^2 \text{ on } G + m - 2k \text{ degrees of freedom}$$

where the summation is from $j = (k + 1)$ to m; and λ_j is the jth eigen-value associated with the jth discriminant factor.

The last $(m-k)$ linear discriminant factors (LDFs) are said to be due to chance if the value of z is less than the tabulated value of χ^2 for $G + m - 2k$ degrees of freedom.

Having determined the significant number of discriminant factors, new samples may be categorized by evaluation of the LDFs of the sample and plotting the point in the LDF space, assigning the sample to the group to which it lies nearest.

Alternatively, a log likelihood approach may be used. Let $P_i(x)$ be the density function of the ith population, π_i. Then:

$$P_i(x) = (2\pi)^{-m/2}|\Sigma|^{-\frac{1}{2}}\exp\ \{-\tfrac{1}{2}[x-\mu(i)]\ '\ \Sigma^{-1}[x-\mu(i)]\}$$

and we may define:

$$U_{jk}(x) = \ln\ |P_j(x)/P_k(x)|$$

Under the assumption that the penalties of misclassification are equal we may define the region R_j by those x satisfying $R_j : U_{jk}(x) \geqslant 0, k = 1, \ldots G$, and classify as belonging to population π_j all points falling in R_j.

The system can be adjusted to the case of unequal penalties of misclassification. Σ and μ_i, as shown are estimated by S and $\bar{x}_i$.

In the system just propounded we have $U_{jk} = U_{kj}$; however, there remains a total of $g(g-1)/2$ distinct discriminant factors. The following technique provides the same solution whilst reducing the number of generated factors.

For each group, g, define a function:

$$f_g = c_{0,g} + x'c_g$$

where $c_{0,g} = -\tfrac{1}{2}\mu(g)'\Sigma^{-1}\mu(g)$ and $c_g = \Sigma^{-1}\mu(g)$.

A new sample is classified as belonging to the group g for which f_g is the largest.

The following considerations were acknowledged in applying the analysis to the file of gossanous material.

(1) To produce distributions approximating normality, a log transform was found to be satisfactory.

(2) The constraint of equal variance/covariance matrices was not felt to be too restrictive for it has been shown (Dempster, 1964), that the analysis is reasonably robust against unequal variance/covariance matrices. However, in estimating the population parameters, little credence was placed upon a population with few samples in which one or two outliers could create a strong bias. Anderson and Bahadur (1962) give a modification for the case of unequal variance/covariance matrices, but this was not used.

(3) The technique of Cochran and Bliss (1948) was incorporated, where by consideration of the partial correlations between the elements assayed a subset of elements could be extracted with the criteria that the subset had a maximum power to discriminate between the groups, in the sense that the

addition of further elements would not give rise to any statistically significant increase in the discriminatory power.

However, one of the most important considerations was not the actual analysis, but rather, the presentation of the results in a form suitable for the field geologist. For this purpose the discriminant analysis with covariance proved ideal, for traditionally, to obtain a simple two dimensional picture of discrimination upon five elements, e.g. Ni, Cu, Co, Cr, Zn, one has available the first two factors of discrimination as defined by the least squares technique. However, to assess a sample by these factors requires some initial calculation on the part of the geologist. Alternatively by employing as coordinates the first two elements incorporated in the analysis with covariance, one can ascribe, for various levels of these two elements, a marginal distribution of the remaining elements. Consequently, utilising the set level of the given two elements, and the appropriate mean value of the marginal distribution of each other element upon the first two, one is able to use the full discriminant function on all elements to define a partitioning of the two element space by population boundaries as developed through the maximum likelihood approach.

The technique is readily adopted to higher dimensional analysis. Thus marginal distributions could be taken upon the first three or even four elements of the analysis. The final presentation of this particular analysis was a two dimen-

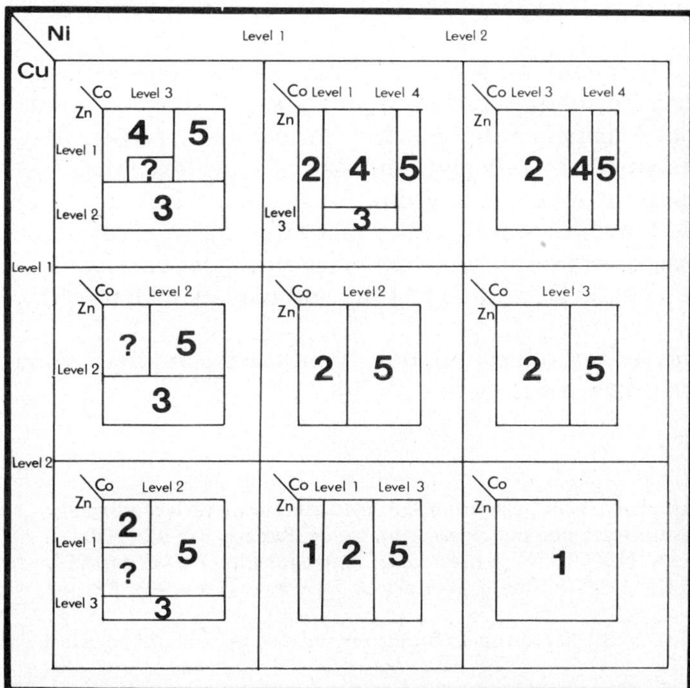

Fig.2. An idealization of the discriminant function on Ni, Cu, Co, Cr, and Zn for determining the genetic category of an ironstone sample.

sional presentation of the four dimensional sub-space using the marginal distributions for chromium. The actual chart with assay levels omitted, is presented as Fig.2.

Some portions of the space depended heavily on the exact proportion of the individual elements and so were left as "pictorially unclassified", although, of course a classification is available by resort to the original discriminant factors from which the chart was derived.

RESULTS

Assessment of the validity of the scheme may be based on a re-classification of the file using the discriminant function derived. Consideration of Groups 1 and 2 show that: 88% of these samples are recognized as Groups 1 or 2, 8% are indicated to belong to other groups, and 4% are unclassified by the function.

Consideration of Groups 1 and 2 combined, 3, 4 and 5 show that: 69% of all samples are classified correctly, 18.5% are incorrectly classified, and 12.5% are unclassified by this scheme.

In routine exploration, samples of previously unknown origin which are given a favourable classification (Groups 1 or 2) by the discriminant function are singled out for more detailed work. This often involves ground acquisition, geological mapping, geochemical or geophysical surveys and, if results are encouraging, drilling.

The outcome of any subsequent work may result in the need for re-classification of samples already on the file. For example, Group 6 samples which are accorded a Group 1 or 2 classification which is borne out by further exploration would be re-classified accordingly. Similarly new evidence may be gathered which suggest a more appropriate group for samples which appeared to be incorrectly classified in Groups 1—5 at the first attempt. In addition, new data arising from discoveries of nickel sulphide occurrences unrelated to the original file come to hand from time to time and may be added to the file.

In this way the file may be periodically updated and the discriminant function will thereby become progressively refined.

REFERENCES

Anderson, T.W. and Bahadur, R.R., 1962. Classification into two multivariate normal distributions with different covariance matrices. Ann. Math. Statist, 33: 420—431
Clema, J.M. and Stevens-Hoare, N.P., 1973. A method of distinguishing nickel gossans from other ironstones on the Yilgarn Shield, Western Australia. J. Geochem. Explor., 2: 393—402
Cochran, W.G. and Bliss, C.I., 1948. Discriminant functions with covariance. Ann. Math. Statist, 9: 151—176
Dempster, A.P., 1964. Tests for the equality of two covariance matrices in relation to a best linear discriminator analysis. Ann. Math. Statist, 35: 190—199

A TOPOLOGICALLY OPTIMUM PROSPECTING PLAN FOR STREAMS

W.E. SHARP and THOMAS L. JONES, Jr.

Department of Geology, University of South Carolina, Columbia, S.C. (U.S.A.)

ABSTRACT

Study of the branching network of streams suggests that an optimum sequential prospecting plan for a stream can be obtained by successive halving of the basin at the centroid. The modal number of sequential samples to locate a source is given by $1 + [\log_2 M_O]$ where M_O is the link magnitude of the stream at the outlet. This sample design can serve as a model against which sampling efficiency of any practical program may be judged.

The sequential plan is illustrated with a field example from the drainage basin containing the Brewer gold mine (S.C.). This mine has an alteration halo with an unusual chert-like topaz. Sampling of the basin using the sequential plan gave results in agreement with the theory.

The sequential plan serves as a good basis for designing a simultaneous sampling plan. Good reconnaissance coverage of a basin can be obtained when about 1/2 of a sequential plan is carried out as a simultaneous plan. The number of sites needed is given by the power of two having a value nearest to $\sqrt{M_O}$.

INTRODUCTION

It has been postulated that the topologically optimum sequential sampling procedure for a stream network is through centroid sampling obtained by successive halving of the network (Sharp, 1971). To test the sampling procedure with a field example, a moderate-sized drainage basin was chosen which drains the area surrounding the Brewer gold mine. The stream network of this basin carries a unique variety of topaz from the alteration halo around the mine.

SEQUENTIAL SAMPLING

Prospecting for the source of mineral float in a drainage basin is analogous to the mathematical problem of locating the proper position to place a new object among an ordered set of objects (Steinhaus, 1969). Such a problem is solved by sequential centroid sampling of the list. In a stream, a sample is first taken at the outlet of a basin and if mineral float is detected, the stream is prospected by taking a series of sequential samples in which each sample is analyzed before taking the next. The optimum location of these samples is at successive centroids of the stream network.

To order a drainage basin so that centroid sampling can be carried out, the

stream segments of a basin are numbered such that each interior stream link
is assigned a rank equal to the number of fingertip tributaries (exterior links)
upstream from that link (Scheidegger, 1965; Shreve, 1967). Each exterior
link (source) is assigned a magnitude of 1, and extends from its point of origin
or source to the first fork or junction downstream. Interior links are bounded
by forks on each of the two ends. The magnitude of the outlet or mouth of
the drainage basin numbered by this method will be equal to the number of
exterior links in the basin (Fig.1).

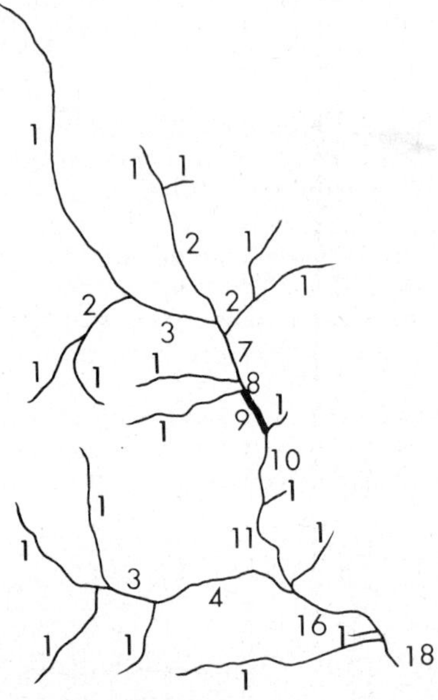

Fig.1. An example of stream branching showing the assignment of link magnitude and the
position of a centroid link (9) on Fork Creek, Jefferson Quadrangle, South Carolina.

The centroid of a stream network is that link which most nearly divides
the network into two equal halves (Sharp, 1971); in other words, if the
centroid link is removed, both of the newly formed stream networks ideally
should have the same number of exterior links. The ideal centroid is given by:

$$M_c = [(M_o + 1)/2]$$

where M_c is the centroid magnitude, M_o is the outlet magnitude, and [ ] is
the integral value (the closest integer less than the bracketed value). However,
in a real stream network, only under fortunate circumstances will it divide into exactly equal parts (Fig.1). Usually the centroid will vary randomly about
$M_o/2$ so that the modal number of tests needed to locate a source of mineral
float should be given by (Sharp, 1971):

$$S_{mode} = 1 + [\log_2 M_o]$$

This relation assumes that the source of float is on an exterior link; if interior links are included, the modal number of tests is simply increased by one. This increase covers the possibility of having to choose between exterior and interior links when the last exterior link has been found through rejection of any alternate choices.

Due to varying geometries of stream networks which can increase or decrease the deviation of the actual centroid from the ideal centroid the number of actual tests required to search a network can vary between a minimum of $[\log_3 M_o]$ to a maximum of $2\{1 + [\log_3 M_o]\}$ (Sharp, 1971).

FIELD TEST

For a field trial of the centroid sampling method the drainage basin surrounding the Brewer gold mine (Pardee and Park, 1948) in Chesterfield County, South Carolina, was chosen. This deposit has a rather unique occurrence of abundant chert-like, massive topaz (Pardee et al., 1937) in the alteration halo. The topaz pebbles survive transport relatively well and are visually distinctive because of their blue-grey color.

The basin containing the Brewer mine is approximately 109 km^2 in areal extent and is generally underlain by an intrusive quartz monzonite. The southern third of the basin, however, is underlain by laminated slates, sericite phyllite, and mafic and felsic volcanics of the South Carolina slate belt (Nystrom, 1972); the stream interfluves are capped by relatively flat lying coastal plain sands and pebble conglomerates (Fig.2).

Using the Jefferson and Jefferson NE quadrangle maps (scale 1:24,000) as a base, the basin had 145 sources or exterior links divided mostly into two dominant streams: Fork Creek to the east and Little Fork Creek to the west. These southward flowing streams converge and discharge into the Lynches River.

Sampling for the topaz was accomplished using a simple hand sieve. It had a volume of approximately 10 liters and was covered on the bottom with 6 mm screen (¼ inch hardware cloth). This sieve was filled once at each site using the coarsest bar gravel available. Winnowing was accomplished through immersion and shaking of the sample in the stream water. The recovered plus 6-mm fraction varied from 3 to 10 kg but averaged about 6 kg. The high variance resulted from the variability of the available gravel. The plus 6-mm fraction was visually inspected for topaz. Questionable pebbles were given a float test in a heavy liquid (tetrabromoethane) to confirm that they had a high specific gravity (~ 3.5) of topaz. Extra care was exercised in testing those sites yielding no topaz.

Sample sites

Initially the basin was tested near its mouth, 100 m upstream from the Lynches River (Fig.3). The sieve test here yielded 11 pieces of topaz weighing

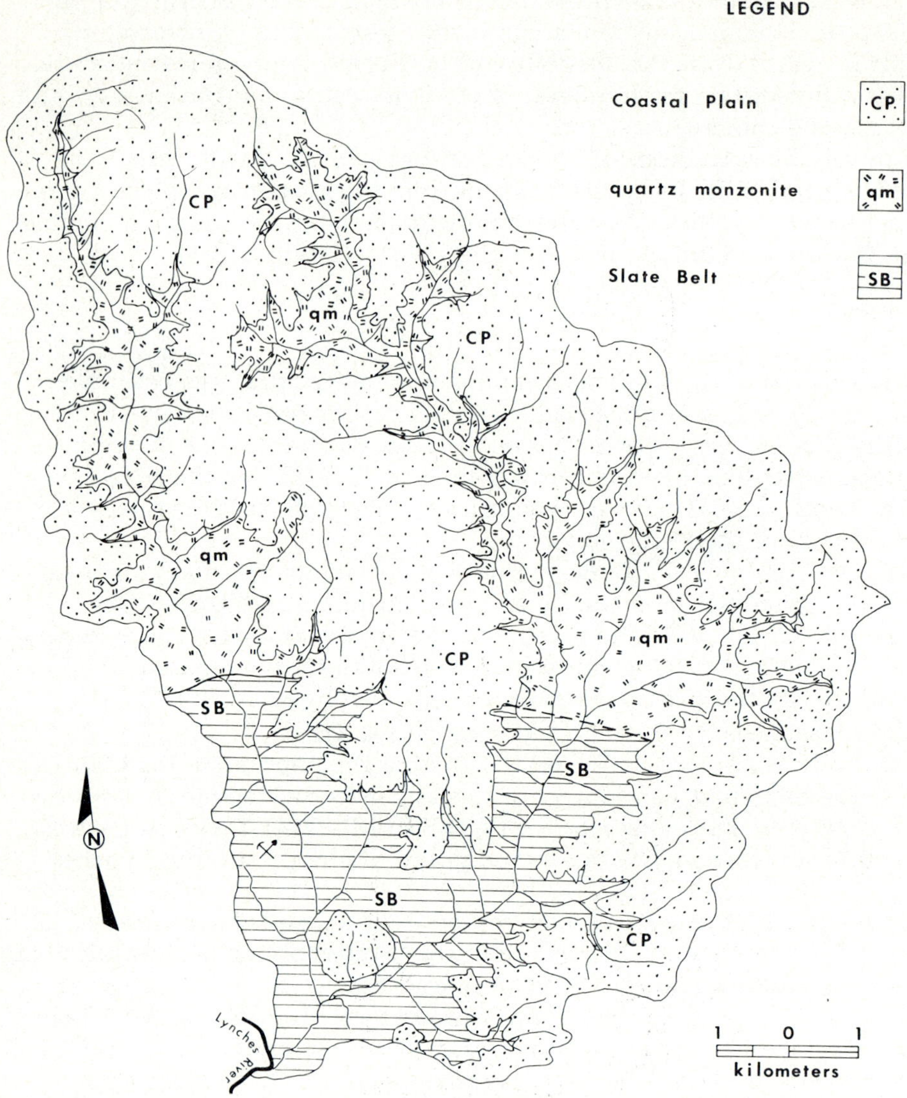

Fig.2. Geology of the stream basins around the Brewer mine, South Carolina.

20 g, thereby verifying that there was a topaz source within the basin. The basin was then sampled at the first centroid (Fig.3) which had a calculated value of 73 obtained from the relationship:

$$M_c = [(M_o + 1)/2] = [(145 + 1)/2] = 73$$

The closest link on the map to 73 had a magnitude of 72 (Fig.3); this was used as the first sample site. No topaz was found here so all links above (up-

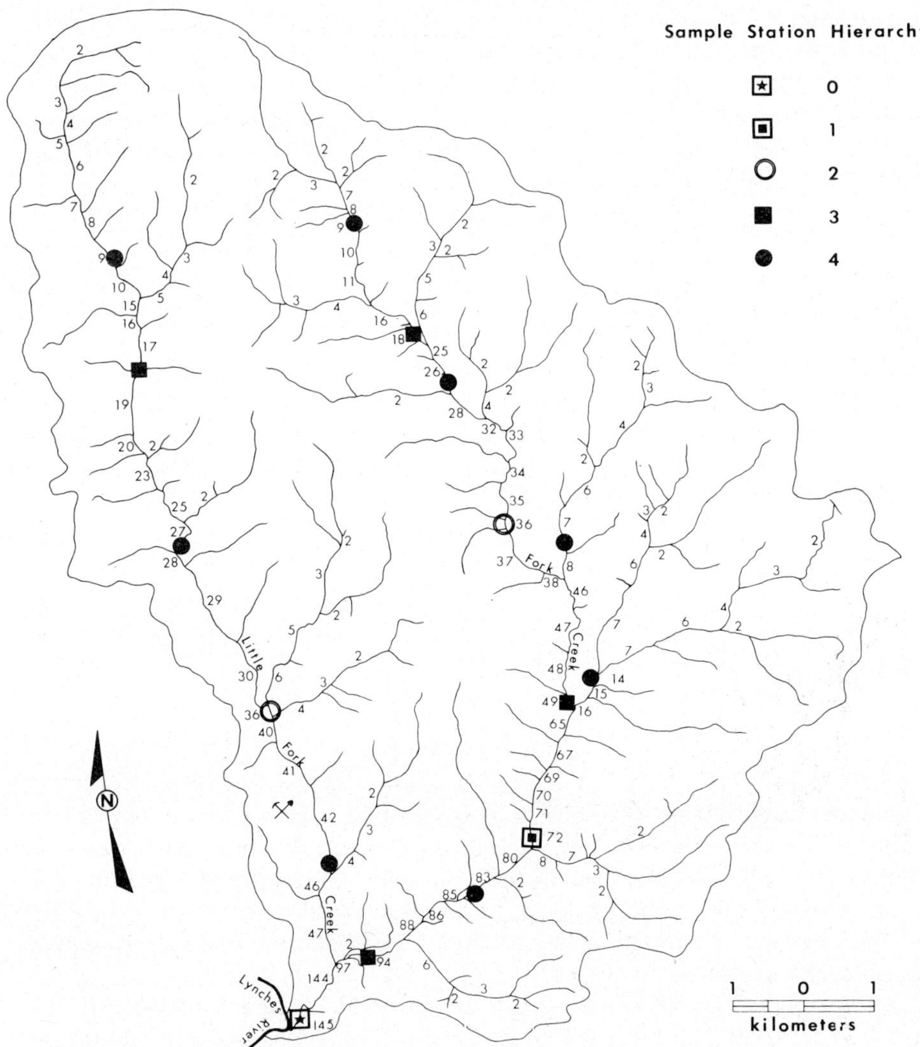

Fig.3. A topologically optimum sampling plan for Fork and Little Fork Creeks, showing sample sites for the first 4 stations of a sequential plan or a 16-station simultaneous plan.

stream) this point were eliminated as possible sources. When the basin has been divided on the upstream side as just described, then the next centroid for the retained portion of the network is given by the link magnitude closest to either (Sharp, 1971):

$$M_c = [(M_o - M_u + 1)/2] = [(145 - 72 + 1)/2] = 37$$

or:

$$M'_c = M_u + M_c = 109$$

In this instance the closest match is a link of magnitude 36 on Little Fork Creek and this was used as the second sample site. A negative result here again eliminated all sources upstream. The stream basin has now been divided at two different places on the upstream side. The centroid for the retained portion of the network is now found by locating that link closest to any of the following values (Sharp, 1971):

$$M_c = [\,(M_o - M'_u - M_u + 1)/2] = [\,(145 - 36 - 72 + 1)/2] = 19$$

$$M'_c = M'_u + M_c = 36 + 19 = 55$$

$$M''_c = M_u + M_c = 72 + 19 = 91$$

The best match from the basin is found to be either link 94 or 88 on Fork Creek (Fig.3). Since either choice could be made the link of magnitude 94 was chosen because of its accessibility (closeness to the road). No topaz was detected at this third site.

For the next centroid, there is again division at two places on the upstream side of the basin so that the possible centroids are:

$$M_c = [\,(145 - 94 - 36 + 1)/2] = 8$$

or:

$$M'_c = 36 + 8 = 44$$

or:

$$M''_c = 94 + 8 = 102$$

The closest possibilities were either link 42 or link 46 on Little Fork Creek. Link 42 was chosen for accessibility reasons again. In this case both were equally close to the road but the route to 42 had less undergrowth. Sieving at this site yielded 106 pebbles of topaz weighing 420 g. This indicated the source was upstream from this site and the source has now been narrowed to a small segment of the basin (Fig.4).

The remaining centroids were readily determined by visual inspection of the map and samples 5 and 6 yielded no topaz. Test 6 has isolated the last exterior link so that the source of topaz is either on the remaining exterior link or the interior link immediately downstream. To determine which of these was the source necessitated an additional sample. A negative result here showed the source to be the interior link adjacent to the Brewer gold mine and its topaz-rich spoil piles.

Results

The expected modal number of samples needed to search a basin with 145 sources is found to be:

$$S_{\text{mode}} = 1 + [\,\log_2 M_o\,] = 1 + [\,\log_2 145] = 1 + 7 = 8$$

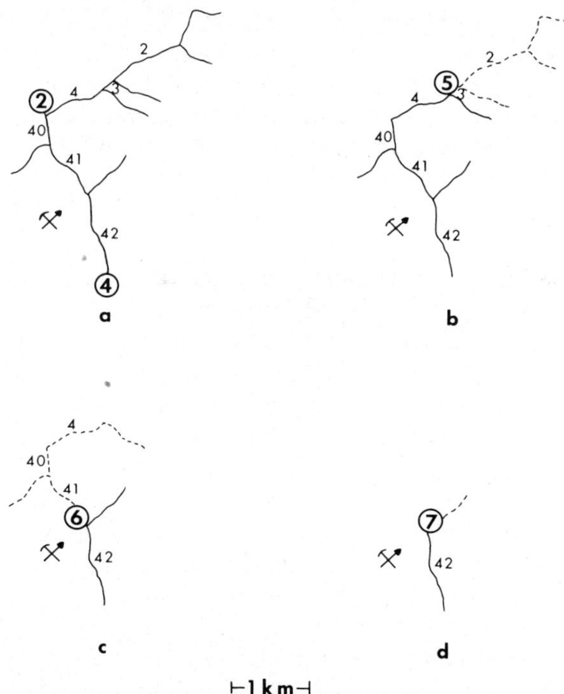

Fig.4. A portion of Little Fork Creek showing the locations for sampling sites 5, 6, and 7 used to complete the search for the topaz source.

when the source is on an exterior link or 9 if on an interior link. The actual number of samples was 7, the difference between the actual and expected numbers is due to fortunate choices of sample sites at centroids 3, 4, and 6. If unfortunate choices had been made at these sites, the total number of samples would have been 9. These values are well within the theoretical bounds of a minimum of 5 and maximum of 11.

This example worked out quite well because there was only one unique source. Multiple sources will introduce added difficulties but do not prevent location of at least one source. Dilution of the substance for which testing is made is important and determines amount and size of sediment to be examined at each site. Variable energy conditions affect the grain-size distributions of the stream and on one site might yield 10 kg of gravel while on another only 3. In the example used here, the topaz was being transported by the stream and the ease with which the tests were made suggests the bed load was artificially enriched by runoff from the tailings of the Brewer mine so that only a small amount of gravel (6 kg) was examined at each sample site. The presence of topaz pebbles in ancient terrace gravel along Little Fork Creek do attest that not all of the topaz was due to mine tailings.

The time required to conduct this search included 4 hours of preliminary office work which included securing the appropriate maps, tracing the drainage network, and ordering of the network. Actual sampling required 16 hours of field work, this included 6 hours spent in travel to and from the area on three different occasions. This is about 1 hour of actual field time required per sample site.

SIMULTANEOUS SAMPLING

In many practical instances, dilution of the mineral float prevents its detection at the mouth of a stream. As a result, the common practice in geochemical prospecting is to carry out a simultaneous search in which a predetermined number of samples are collected and brought into the laboratory for analysis.

A simultaneous plan for a stream network can be obtained simply by dividing a stream basin into a series of segments of equal magnitude. The number of intervals is determined by the completeness of the coverage needed and the time. A convenient way to organise a simultaneous plan is to use the sequential plan to subdivide all portions of the basin into as many segments as desired (Fig.3). This procedure produces uniform coverage of the network, but in such an arrangement the number of sample sites increases as a power of two for each step in the plan.

To illustrate how a simultaneous plan can be used in conjunction with a sequential plan, a 16-station simultaneous sampling plan using the sequential method of site selection was obtained (Fig.3).

The results yield topaz at the mouth and at only one of the 16 sites (site 4 of the previous sequential plan). The final search for the source could then be completed just as in the sequential plan. These results confirm that no unexpected topaz sources occur in the basin. In this instance a total of 19 samples were required to accomplish the same result that was accomplished in 7 by sequential sampling. Although efficiency of search has been lost, confidence as to overlooking a source has been gained and it seems that in reconnaissance work a good compromise consists of using a simultaneous plan that is equivalent to half of a sequential search. If the sequential search would require an expected mode of n samples, the number of simultaneous samples for 50% coverage is given by $2^{\lfloor n/2 \rfloor}$. If M_O is the magnitude at the starting location, then substitution of $n = \log_2 M_O$ yields $\sqrt{M_O}$ as a first estimate of the number of simultaneous samples needed. A more precise estimate of the number of sites is given by finding the power of two having its value nearest to $\sqrt{M_O}$.

ACKNOWLEDGEMENTS

Thanks are owing to Dr. L.R. Gardner of the University of South Carolina for his helpful criticism of the manuscript.

REFERENCES

Nystrom, P.G., 1972. Geology of the Catarrh N.W. Quadrangle, University of South
    Carolina, Columbia, S.C., Master's Thesis, 49 pp. (unpublished)
Pardee, J.T. and Park, Jr., C.F., 1948. Gold deposits of the southern Piedmont. U.S. Geol.
    Survey Prof. Paper 213, pp. 106—111
Pardee, J.T., Glass, J.J. and Stevens, R.E., 1937, Massive low-fluorine topaz from the
    Brewer Mine, South Carolina. Am. Mineral., 22: 1058—1064
Scheidegger, A.E., 1965. The algebra of stream-order numbers. U.S. Geol. Survey Prof.
    Paper 525B, pp. B187—B189
Sharp, W.E., 1971. A topologically optimum water-sampling plan for rivers and streams.
    Water Resour. Res., 7: 1641—1646
Shreve, R.L., 1967. Statistical law of stream numbers. J. Geol., 75: 179
Steinhaus, H., 1969. Mathematical Snapshots. Oxford University Press, New York, N.Y.,
    311 pp.

Chairmen:

R.W. BOYLE
Geological Survey of Canada, Ottawa, Ont., Canada
F. CACHAU-HERREILLAT
Société Nationale des Pétroles d'Aquitaine, Pau, France

INTEGRATED GEOLOGIC AND GEOCHEMICAL STUDIES, EDNA MOUNTAIN, NEVADA

S.P. MARSH and R.L. ERICKSON

U.S. Geological Survey, Denver, Colo. (U.S.A.)

ABSTRACT

Detailed geologic and geochemical studies of the Edna Mountain quadrangle, Humboldt County, Nevada, were undertaken to determine: (1) the stratigraphic and structural history of this complex area; (2) the regional distribution and abundance of metals in rocks of the area; and (3) the factors that control the distribution and abundance of the metals.

Host rocks in the area range in age from Cambrian to Permian and include near-shore, transitional, and deep-basin clastic rocks telescoped by thrust faults related to at least four different orogenic episodes. The sedimentary section has been intruded by small plutons, dikes, and sills of Mesozoic age, and is locally concealed by Tertiary volcanic rocks. The entire rock complex is repeatedly offset by later normal faulting.

There are no active mines of economic significance in the area, but numerous occurrences of mineralized ground, combined with structural complexity, various types of mineralization, abundance of altered rock, and the presence of intermediate to silicic igneous intrusive rocks, suggest that concealed or heretofore unrecognized mineral deposits may exist.

INTRODUCTION

Detailed geological studies of the Edna Mountain quadrangle, Humboldt County, Nevada, were undertaken to determine the stratigraphic and structural history of this complex area, the regional distribution and abundance of metals in rocks of the area, and the factors that control the distribution and abundance of the metals. The ultimate objective is the identification of broad target areas and guidelines for mineral exploration in this region.

There are no active mines of economic significance in the Edna Mountain quadrangle, but numerous occurrences of mineralized ground suggest that concealed mineral deposits may exist.

To achieve our objectives we utilized a multi-disciplinary approach employing a combination of geologic mapping, geochemistry, and geophysics. Two geologists and two chemists were assigned to the project, with field and laboratory support from scientists and technicians from other disciplines.

Geologic and geochemical mapping, at a scale of 1:24,000, of the four 7½-minute quadrangles that make up the Edna Mountain 15-minute quadrangle were completed during the 1970—1973 field seasons. Composite geochemical, aeromagnetic, and generalized geologic maps showing the distribution of Cu, Pb, Zn, Mo, Ag, Au, Hg, As, Sb, and W have been published for all four quadrangles (Erickson and Marsh, 1971, 1972, 1973, 1974a). Detailed geologic map compilation, petrographic and mineralogic studies, and geochemical data interpretation are continuing.

GEOLOGIC SETTING

The pre-Tertiary sedimentary rocks in the Edna Mountain quadrangle may be divided into four major structural and stratigraphic blocks or sequences separated by thrust faults (Fig.1). These rocks range in age from Cambrian to Permian and include near-shore, transitional, and deep-basin clastic rocks telescoped by thrust faults related to at least four different orogenic episodes.

The oldest major rock sequence is Cambrian(?) and Cambrian in age and consists of two formations: Osgood Mountain Quartzite, a relatively pure, light-brown-weathering, white to light-gray, thin to massive crossbedded, medium-grained quartzite, and the overlying Preble Formation, a greenish-

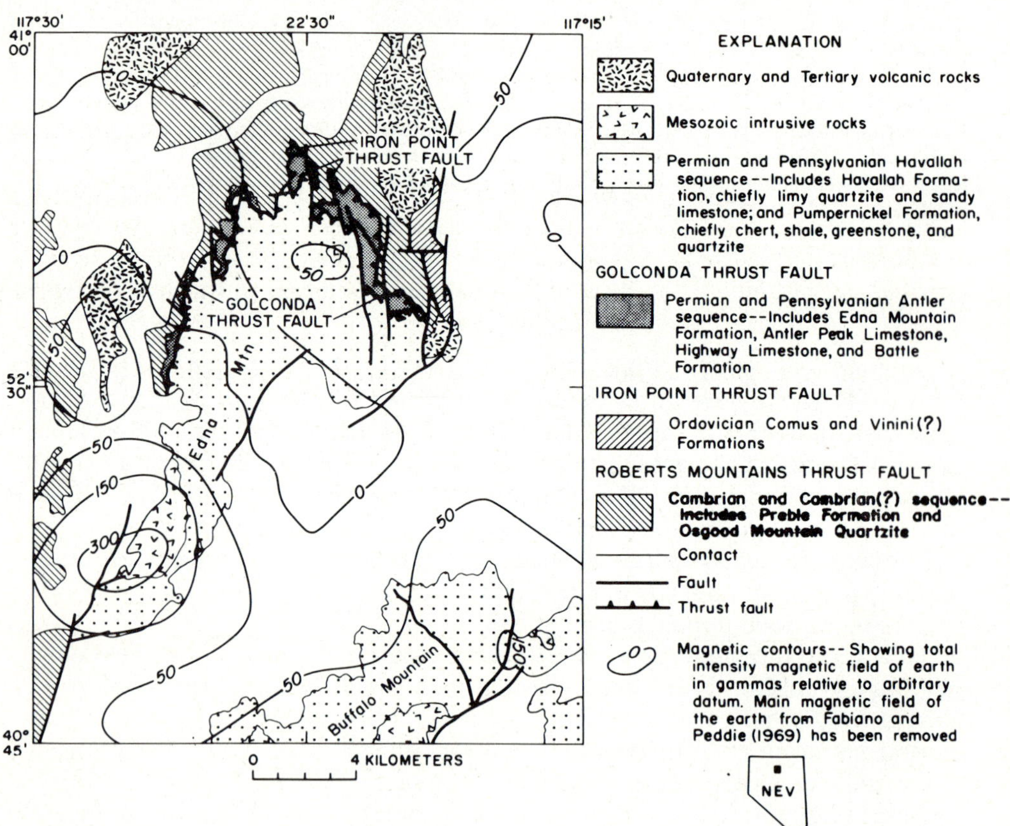

Fig.1. Aeromagnetic and generalized geologic map of Edna Mountain quadrangle, Nevada.

and yellowish- to dark-gray phyllitic shale, containing minor blue-gray, fine-grained, coarsely recrystallized dolomitic limestone. The Cambrian rocks are regionally metamorphosed and folded into south-plunging asymmetric folds overturned to the west. This period of deformation clearly predates deposition of Lower Pennsylvanian rocks in the area and may be as old as Late Cambrian (Erickson and Marsh, 1974b).

The second major rock sequence is an isolated fault-bounded block of Ordovician rock at the northeast end of Edna Mountain (Fig.1). Two distinctly different lithologic and structural units are present in this block. One unit, which we have tentatively referred to as the Vinini(?) Formation, is composed of isoclinally and recumbently folded beds of chert, siliceous shale, and quartzite. Overriding of beds was from west to east. The second lithologic unit in the Ordovician block, the Comus Formation, is about 610 m thick and is composed of finely laminated dolomitic siltstone and silty dolomite in the upper part, giving way to interbedded chert and dolomite, and graptolite-bearing tuffaceous siltstone in the lower part. From east to west, the beds dip from moderately to steeply west, to slightly overturned; they are not tightly folded. Autobreccias and quartz-rich shear zones parallel or subparallel to bedding are common, particularly in the lower part, and probably have tectonic significance. The contrast in lithology and structure between the Vinini(?) and Comus suggests that the Vinini(?) Formation belongs to a eugeosynclinal facies that has moved into the area as a thrust sheet from the west. The lithology and structural style of the Comus Formation suggest that it belongs to a transitional facies and has not moved far from its site of deposition. The Comus exposure is interpreted as a window in the thrust sheet that carried the Vinini(?) into this area. Deformation and sliding within the Comus, and the eastward-directed overriding thrust sheet of Vinini(?) are assigned to the Antler orogeny of Late Devonian or Early Mississippian age.

The third major rock sequence includes the Battle Formation, Highway Limestone, and Antler Peak Limestone ranging in age from Early Pennsylvanian to Early Permian. These rock units and, in addition, the Edna Mountain Formation of Late Permian age are part of the overlap assemblage that was deposited on the folded and faulted strata involved in the Antler orogeny, and are named the Antler sequence (Roberts et al., 1958). The Battle Formation is a conglomerate composed chiefly of boulders and cobbles of Cambrian(?) Osgood Mountain Quartzite. The Battle Formation is overlain by the Highway Limestone, which is predominantly a yellowish-gray to gray, fine-grained limestone, with quartz pebble conglomerates and sandy beds. Scattered throughout the Highway are irregular lenses of poorly sorted intraformational conglomerate composed mostly of subangular pieces of Cambrian Preble Formation. Above the Highway Limestone lies the Antler Peak Limestone, a gray to dark-gray, medium- to coarse-grained, bioclastic limestone, with local fine-grained sandy beds, which is Early Permian and Late Pennsylvanian in age. These three formations of the Antler sequence are asymmetrically folded, in places overturned, and have been moved into the area on the Iron Point thrust fault of probable

Early Permian age (Erickson and Marsh, 1974b). The Edna Mountain Formation, the uppermost unit of the Antler sequence, is a limy, brown coarse grit to fine-grained quartzite composed of quartz and black chert grains. This unit has not been folded — only tilted — and occurs in depositional overlap contact on an erosion surface cut across the older folded rocks in the Iron Point plate; locally the plate was completely eroded, and the Edna Mountain Formation rests with angular unconformity on the Preble Formation of Cambrian age. The Edna Mountain Formation is probably the only pre-Tertiary stratigraphic unit that is truly autochthonous.

The uppermost sequence of rocks in the Edna Mountain quadrangle is the Havallah sequence of Early Pennsylvanian to Early Permian age. This sequence, consisting of the Pumpernickel and Havallah Formations, was brought into the area from the southwest on a thrust fault of great magnitude, the Golconda thrust. Rocks in the plate are intensely deformed, with large tight northwest-plunging folds overturned to the northeast. The Pumpernickel Formation is chiefly pale olive, siliceous shale and thin-bedded chert, with greenstone and minor quartzite and lime units. The Havallah Formation overlies the Pumpernickel Formation and consists mainly of interbedded, fine-grained quartzite and limestone, and interbedded dark-gray chert and limestone.

Igneous rocks are widespread in the Edna Mountain area (Fig.1). Mesozoic granodiorite and quartz monzonite plutons and related dikes (not shown in Fig.1) intrude the section; volcanic rocks of Tertiary age locally conceal the section.

The entire complex in the Edna Mountain quadrangle is repeatedly offset by later normal faults.

MINERALIZATION

There are no active mines of economic significance in the Edna Mountain area, but numerous occurrences of mineralized ground, with a broad spectrum of types of mineralization, suggest that heretofore unrecognized mineral deposits may exist.

Some of the observed types of mineralization are tungsten-bearing hot spring tufas, metalliferous black shale, low-grade disseminated gold, copper associated with greenstones, base metal and barite deposits associated with Paleozoic sedimentary rock, and base metals associated with Mesozoic igneous rocks. Geochemical studies have sought to identify the genetic relationships of these varied types of mineralization and to identify broad target areas for mineral exploration.

Geochemical survey

Geochemical studies were undertaken with two broad objectives in mind: (1) the establishment of the trace element signature or fingerprint of the various units in the Edna Mountain quadrangle, and (2) the identification of possible target areas for concealed or unrecognized ore deposits.

Over 4000 rock samples were collected in the course of the project. All samples were prepared and analyzed in truck-mounted mobile laboratories stationed at the project field headquarters in Winnemucca, Nevada. Cu, Pb, Zn, Ag, Au, and Hg were determined by atomic absorption methods and results were available to the geologists daily. In the latter part of each field season, the area was visited by a truck-mounted spectrograph and all samples were analyzed for 30 elements by semiquantitative spectrographic methods.

Two types of rock samples were collected: (1) barren rocks selected to determine trace element signature and background values, and (2) mineralized or altered rocks selected to enhance the probability of detection of leakage halos and zoning patterns, which help outline concealed target areas. Most of these samples were from shear or fault zones, fractures, jasperoid, breccia reefs, veins, and altered zones. Stream sediment sampling was not used in the program, because the district had been thoroughly prospected on several occasions, and it was felt that it would only lead back into areas of known surficial mineralization. Analytical results for barren rocks are not reported in this paper.

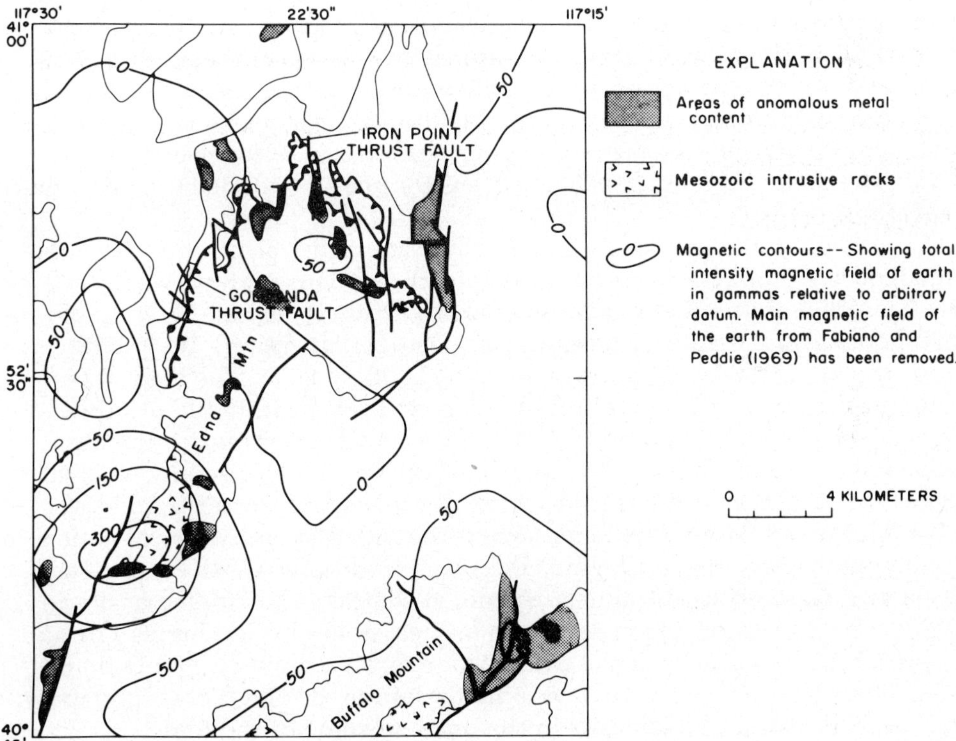

Fig.2. Aeromagnetic and generalized geologic map of Edna Mountain quadrangle, Nevada, showing areas of anomalous metal content.

244

RESULTS

Geochemical studies in the Edna Mountain quadrangle have revealed numerous areas that contain anomalous amounts of various metals (Fig.2). Three episodes of mineralization have been recognized, each characterized by a distinctive metal suite. In some places the same block of structurally prepared ground has been invaded by all three episodes. The earliest period of mineralization resulted in a stratabound vanadium deposit in black, finely laminated, carbonaceous siltstone in the Vinini(?) Formation of the Ordovician sequence. These vanadiferous beds also contain anomalous amounts of beryllium, chromium, mercury, nickel, and tungsten (Table I). High vanadium concentrations are believed to be the result of leaching of chert and shale beds of the Vinini(?) Formation, causing residual enrichment of vanadium, chromium, and nickel. Unleached beds contain 300—1000 ppm V.

A second period of mineralization resulted in deposits of a base-metal suite of copper, molybdenum, and bismuth, with satellite deposits of lead, zinc, and silver (Fig.3). These deposits are associated with Mesozoic quartz monzonite and granodiorite intrusives and are crudely zoned about them. Copper, molybdenum, gold, and silver anomalies occur in and along the contacts of a small granodiorite mass and associated dikes cutting altered greenstone of the Pumpernickel Formation (upper plate of the Golconda thrust) on the southeast flank of Edna Mountain (Fig.3). The altered greenstone contains abundant euhedral pyroxene metacrysts and small amounts of disseminated pyrrhotite. Oxidation of pyrrhotite stains the rock red brown over a roughly square area about 300 m on a side. Basinward extent of alteration is concealed by fan gravel. The rock types and geochemical suite suggest concealed Cu—Mo mineralization.

Copper, molybdenum, bismuth, and silver anomalies occur in and around a small quartz monzonite pluton on the northeast flank of Buffalo Mountain (Fig.3). Although the rocks in the area are hydrothermally altered (chiefly silica flooding and argillic alteration with some sericitization), the quartz monzonite itself rarely contains as much as 1000 ppm Cu. The highest copper values (greater than 1%) are found in shears and veins within the pluton and in adjacent interbedded lime and quartzite. A lead, silver, bismuth anomaly is peripheral to the central Cu—Mo zone and is particularly pronounced in an area of silicified Havallah sequence rocks to the west of the intrusive.

Another broad base-metal target area is the belt of occurrences of anomalous amounts of copper, lead, zinc, and silver around the north flank of the Mesozoic granodiorite pluton near the north-central part of the quadrangle (Fig.3). Pb—Ag also occurs in veins and in shear zones in the Comus Formation of the Ordovician sequence and in some of the Cambrian Preble lime units, where they are overturned and highly fractured. The Preble lime units also contain small barite deposits in the noses of some of the folds.

Abundant shows of secondary copper minerals occur in the Pumpernickel Formation along shallow fractures and shear zones at and near the contact

Analytical data for vanadiferous shale, Ordovician Vinini(?) Formation, Edna Mountain, Nevada

Chemical analysis (wt.%) (Analysis by P. Elmore, J. Glenn, J. Kelsey, and H. Smith)		Semiquantitative spectrographic analysis (ppm)[1] (Analysis by J.L. Harris)		Atomic absorption analysis (ppm) (Analysis by R.M. O'Leary and M.S. Erickson)	
element	sample number IP-0428	element	sample number IP-0428	element	sample number IP-0428
SiO_2	52.7	Ag	10	Au	$N(0.02)^2$
Al_2O_3	14.3	As	$N(1000)^2$	Hg	1.4
Fe_2O_3	4.7*	B	30		
FeO	*	Ba	10,000		
MgO	6.3	Be	5		
CaO	0.42	Bi	$N(10)^2$		
Na_2O	0.31	Co	$N(5)^2$		
K_2O	3.4	Cr	150		
H_2O^+	4.6	Cu	200		
H_2O^-	1.3	La	$N(50)^2$		
TiO_2	0.92	Mo	$N(3)^2$		
P_2O_5	0.92	Nb	$N(10)^2$		
MnO	0.07	Ni	300		
CO_2	0.11	Pb	50		
Other volatiles	8.7	Sb	$N(200)^2$		
		Sc	20		
Sum	99	Sn	$N(10)^2$		
		Sr	7000		
		V	7000		
		W	100		
		Y	200		
		Yb	20		
		Zn	$N(300)^2$		
		Zr	300		

* Total iron as Fe_2O_3.
[1] Elements were determined by semiquantitative spectrographic methods described by
Myers et al. (1961). Results of the spectrographic analyses are to be identified with geometric
intervals having the boundaries 1200, 830, 560, 380, 260, 180, 120, etc., in ppm, but are
reported in the table by approximate geometric midpoints such as 1000, 700, 500, 300, 200,
150, 100, etc. Precision of a reported value is approximately plus or minus one interval at 68%
confidence, or plus or minus two intervals at 95% confidence. N = not detected at limit of
detection.
[2] N( ) = not detected at value shown.

245

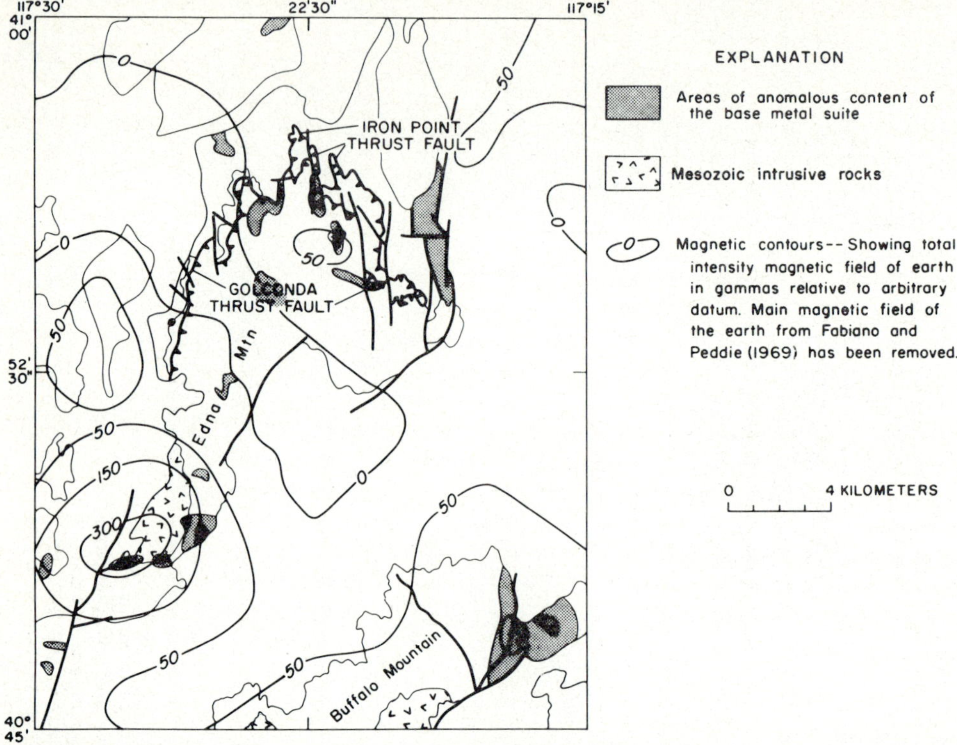

Fig.3. Aeromagnetic and generalized geologic map of Edna Mountain quadrangle, Nevada, showing areas of anomalous content of the base-metal suite.

between greenstone and chert shale beds. These shows may be leakage halos from concealed copper deposits associated with adjacent intrusive rocks or they may be concentrations of copper sweated out of greenstone units and redeposited in fractures and shears. If the metal suite in rocks from these areas contains anomalous amounts of other metals such as zinc, lead, silver, and mercury, it is suspected that the observed secondary concentrations derived from intrusive rocks. If the anomalous metal suite contains copper, nickel, and chromium, with only minor amounts of the other base metals, the anomaly may well be derived from greenstone units (Fig.3).

In the southwest part of the Edna Mountain quadrangle, a large granodiorite pluton has metamorphosed Paleozoic sediments forming several small weakly mineralized tactite zones (Fig.3). Chalcopyrite, related secondary copper minerals, and sparse molybdenite occur with wollastonite, diopside, and garnet in metamorphosed lime units.

The third episode of mineralization, probably late Tertiary, is characterized by the same metal suite (Hg, As, Sb, and W) that is associated with the Carlin-type gold deposit and that probably was emplaced by thermal-spring activity.

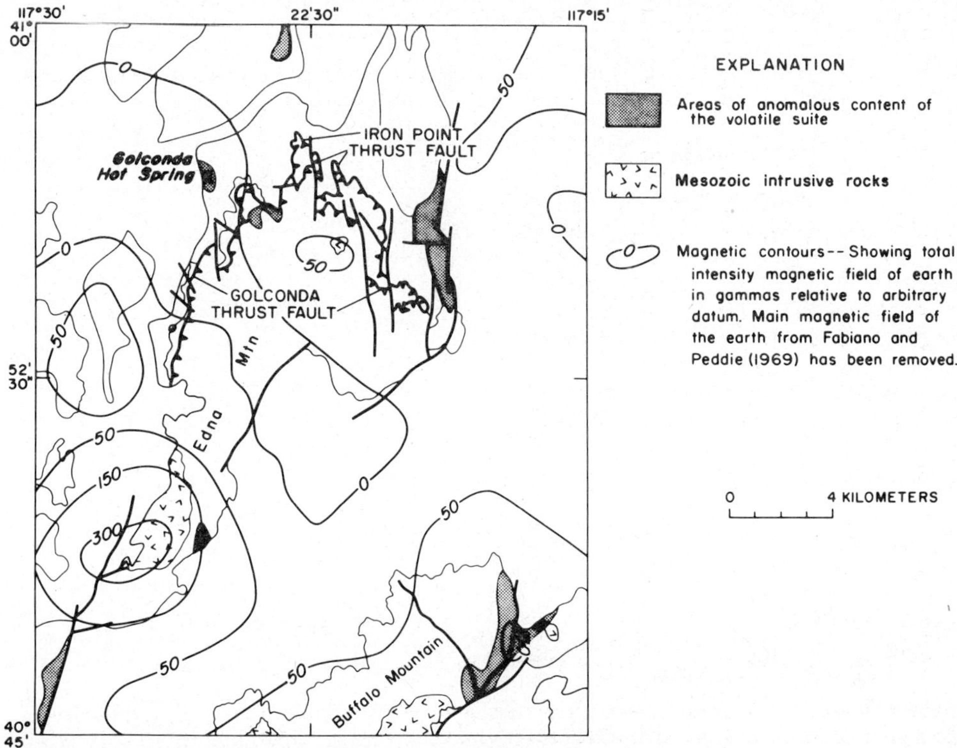

Fig.4. Aeromagnetic and generalized geologic map of Edna Mountain quadrangle, Nevada, showing areas of anomalous content of volatile suite.

This metal suite is particularly well developed in the Ordovician block where anomalous amounts of metals occur throughout both the Comus and Vinini(?) Formations (Fig.4). No active hot springs are present but evidence of past thermal activity is abundant. These metal anomalies are believed to be leakage halos and more favorable host rocks may be present below the thrust (Roberts Mountain?) plane. The gravel-veneered, downfaulted block immediately to the east of the Ordovician exposure should contain similar mineralized rocks. Anomalies also occur along and west of the northeast-trending fault marked by active hot springs in the southwest corner of the quadrangle, where metal deposition is going on at present (Fig.4). In the west-central part of the Edna Mountain quadrangle, the Golconda Hot Spring area was mined for tungsten during World War II (Fig.4). Tungsten, in amounts as much as 7%, occurs in black and orange-brown irregular lenses and pods of manganese and iron oxides in tufa. Although tungsten is the most abundant metal, anomalously high amounts of arsenic, cobalt, germanium, thallium, and beryllium are also present. The metals are strongly partitioned, such that most of the tungsten, cobalt, and strontium occur in the black manganese lenses and

TABLE II

Semiquantitative spectrographic analyses showing partitioning of elements in two samples from Golconda tungsten mine[1] (Analyses by Nancy M. Conklin)

Element	Sample number	
	G1 (soft, black manganese,oxide)	G2 (soft, orange-brown iron oxide)
Fe (%)	1.5	M($>$ 10)
Mn (%)	M($>$ 10)	3.0
As (ppm)	1500	15,000
B (ppm)	150	300
Ba (ppm)	70,000	1500
Be (ppm)	30	30
Co (ppm)	3000	30
Cu (ppm)	70	7
Ge (ppm)	70	150
Nb (ppm)	150	30
Ni (ppm)	300	30
Sr (ppm)	7000	1500
Tl (ppm)	70	0
V (ppm)	7	150
W (ppm)	70,000	7000

[1] Elements were determined by semiquantitative spectrographic methods described by Myers et al. (1961). Results of the spectrographic analyses are to be identified with geometric intervals having the boundaries 1200, 830, 560, 380, 260, 180, 120, etc., in ppm, but are reported in the table by approximate geometric midpoints such as 1000, 700, 500, 300, 200, 150, 100, etc. Precision of a reported value is approximately plus or minus one interval at 68% confidence, or plus or minus two intervals at 95% confidence. The following elements were looked for but not detected at value shown: K (0.7%), P (0.2%), Ag (0.5), Au (20), Bi (10), Cd (50), La (50), Mo (3), Pb (10), Sb (200), Sn (10) Zn (300).

most of the arsenic occurs in the orange-brown iron oxide lenses (Table II). The mineralogy of the trace elements in these materials is complex and virtually unknown.

The Cambrian Preble Formation in the north-central part of the quadrangle contains anomalous amounts of the "volatile" suite; also, a low-grade gold deposit has been discovered recently in the area. A gold, mercury, arsenic, antimony anomaly occurs on the southwest flank of Buffalo Mountain, but the gold apparently occurs only in narrow northwest-striking silicified shear zones and quartz veins (Fig.4). Some highly faulted and fracture areas adjacent to the Mesozoic intrusives also contain anomalous amounts of the volatile suite and may represent a late-stage thermal process.

In several areas in the Edna Mountain quadrangle, metal suites are superimposed on each other or overprinted, particularly where structural complexity is associated with thermal igneous activity (Fig.5). All three periods of mineralization have soaked the structurally broken Ordovician block.

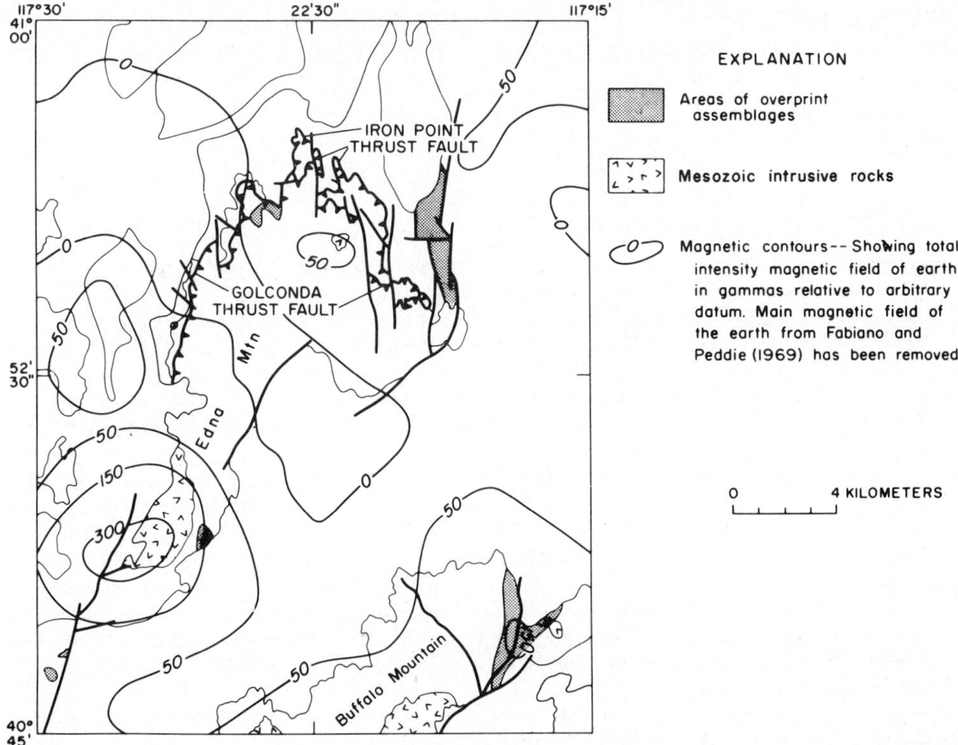

Fig.5. Aeromagnetic and generalized geologic map of Edna Mountain quadrangle, Nevada, showing areas of overprint assemblages.

CONCLUSIONS

The geologic and geochemical studies to date have identified four structural blocks of Paleozoic sedimentary rocks separated by thrust faults and intruded by Mesozoic granodiorite and quartz monzonite plutons and associated dikes and sills. The regional distribution and abundance of several metals in rocks in the area, and broad target areas and guidelines for mineral exploration have been identified and related to geologic and aeromagnetic features.

The distribution and abundance of metals are structurally controlled and crudely zoned. Three episodes of mineralization can be recognized, each characterized by a distinctive metal suite. In some places the same block of structurally prepared ground has been invaded by all three episodes. The earliest period of mineralization resulted in a stratabound vanadium deposit in black carbonaceous siltstone of Ordovician age. The second episode resulted in deposits of copper, molybdenum, and bismuth, with satellite deposits of lead, zinc, and silver associated with Mesozoic granodiorite and quartz monzonite plutons. The third episode of mineralization, probably late Tertiary, is characterized by mercury, arsenic, antimony, and tungsten — the same metal suite that is associated with the Carlin-type gold deposit. This third group of deposits has been emplaced by thermal-spring activity.

REFERENCES

Erickson, R.L. and Marsh, S.P., 1971. Geochemical, aeromagnetic, and generalized geologic maps showing distribution and abundance of mercury, arsenic, antimony, tungsten, gold, copper, lead, and silver, Golconda and Iron Point quadrangles, Humboldt County, Nevada. U.S. Geol. Survey Misc. Field Studies Maps MF-312, MF-313, MF-314, and MF-315

Erickson, R.L. and Marsh, S.P., 1972. Geochemical, aeromagnetic, and generalized geologic maps showing distribution and abundance of molybdenum and zinc, Golconda and Iron Point quadrangles, Humboldt County, Nevada. U.S. Geol. Survey Misc. Field Studies Map MF-345

Erickson, R.L. and Marsh, S.P., 1973. Geochemical, aeromagnetic, and generalized geologic maps showing distribution and abundance of lead, silver, gold, copper, mercury, arsenic, antimony, tungsten, molybdenum, and zinc, Goldrun Creek quadrangle, Humboldt County, Nevada. U.S. Geol. Survey Misc. Field Studies Maps MF-506, MF-507, and MF-508

Erickson, R.L. and Marsh, S.P., 1974a. Geochemical, aeromagnetic, and generalized geologic maps showing distribution and abundance of mercury, arsenic, antimony, tungsten, gold, copper, molybdenum, lead, bismuth, and silver, Brooks Spring quadrangle, Humboldt County, Nevada. U.S. Geol. Survey Misc. Field Studies Maps MF-563, MF-564, MF-565, MF-566, and MF-567

Erickson, R.L. and Marsh, S.P., 1974b. Paleozoic tectonics in the Edna Mountain quadrangle, Nevada. U.S. Geol. Survey. J. Res., 2(3): 331—337

Fabiano, E.B. and Peddie, N.W., 1969. Grid values of total magnetic intensity IGRF-1965. E.S.S.A. Tech. Rept. C and GS 38

Myers, A.T., Havens, R.G. and Dunton, P.J., 1961. Spectrochemical method for the semi-quantitative analysis of rocks, minerals and ore. U.S. Geol. Survey Bull., 1084-I: 207—229

Roberts, R.J., Hotz, P.E., Gilluly, J. and Ferguson, H.G., 1958. Paleozoic rocks of north-central Nevada. Am. Assoc. Pet. Geol. Bull., 42(12): 2813—2857

THE CORRAL DE PIEDRA MOLYBDENITE STOCKWORK, DURANGO, MEXICO

J.A. RANDALL

Department of Geological Sciences, University of Saskatchewan, Saskatoon, Sask. (Canada)

ABSTRACT

The Corral de Piedra stockwork is located on the Piaxtla River on the highly dissect⌐ ⌐,
western edge of the Sierra Madre Occidental. It is some 900 km northwest of Mexico City
and about 850 km south of El Paso, Texas, and near the Ag—Au mine at Tayoltita, Durango.
As no airphoto, nor map coverage was available, geochemical sampling was carried out con-
currently with the surveys for geologic mapping. The property lies in a doubly-plunging
anticline, delineated by Miocene ignimbrites, whose long axis trends northeast. The mineralization occurs in early Tertiary(?) rhyolite and granodiorite in the core of the dome.
Roughly concentric alteration zones of K-feldspar, quartz, sericite, and propylite overprint
the host rocks and enclose a sequence of northeast, north, and northwest-striking molybdenite-pyrite veinlets. Chalcopyrite occurs sparingly with molybdenite in the early stages,
but becomes more common in the later northwest veinlets, and is there partly associated
with cassiterite and tungsten. Silver is present in small amounts, but gold content is negligible. Two sets of dykes intrude the zone: a partly pre-mineral northeast-striking rhyodacite
group, and an essentially post-mineral northwest-striking andesite set.

Background for molybdenum is less than 5 ppm, and the resulting anomalies show a
steep gradient from 50 to over 250 ppm. The 100 ppm molybdenum threshold delineates
an anomaly of about 2 km by 400 m, oriented north-northeast. Tin is unusually common,
regionally; in zones surrounding the molybdenum anomaly, rock samples often carry 100
ppm tin. Generally, molybdenum and tin anomalies are mutually exclusive. The integration
of geochemistry and geology permitted a basic understanding of this previously unknown
property in unmapped terrain.

INTRODUCTION

Exploration of the Corral de Piedra molybdenite stockwork began on a
virgin property on the highly dissected western margin of the Sierra Madre
Occidental in the State of Durango, Mexico. It is located some 900 km (550
miles) northwest of Mexico City and about 850 km (500 miles) south of El
Paso, Texas, U.S.A. The area is crossed by the trail between the long-estab-
lished silver—gold camp of Tayoltita and the State capital, Durango; a pack
route used since colonial times. 14 km from Tayoltita, the region has been
mainly prospected for precious metals, and molybdenite was not previously
identified as such. Occasional copper stains are not unusual for the region,
and thus there was little to draw prospectors to the locality.

This report is a case history of the geochemical exploration of the property
and a compilation of the primary metal dispersion patterns as known at this time.

GENERAL GEOLOGY

The regional stratigraphy includes rocks from middle Cretaceous to Miocene in age (Randall, 1972; McDowell and Clabough, 1972). The former are meta-volcanics and fossiliferous carbonates, but the entire 4000 m thick Tertiary sequence consists of continental intermediate to acid volcanics and related intrusives. The stratified rocks have been mildly folded in northwest-trending structures with occasional north-striking cross-folds. The intersections form several domes; in one of these is the Corral de Piedra stockwork (Fig.1). The mineralization occurs in Tertiary(?) rhyolite and the intruding Piaxtla grano-diorite batholith. Northeast-trending rhyodacite dykes, which served as feeders for the upper Tertiary ignimbrites, cut the previously mentioned rock units as shown in Fig.2. The youngest magmatic rocks are northwest-striking andesite dykes; these apparently cut the bulk of the molybdenite mineralization.

K—Ar dates on the granodiorite batholith range from about 36 million years at Tayoltita, to 50—90 million years in the southern part of Sinaloa

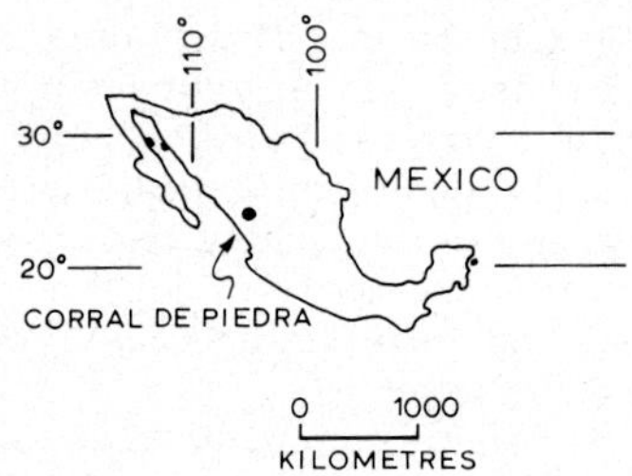

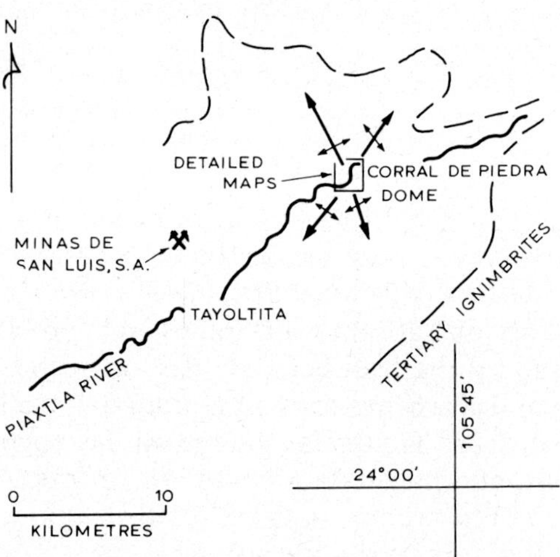

Fig.1. Location map, Corral de Piedra, Durango, Mexico.

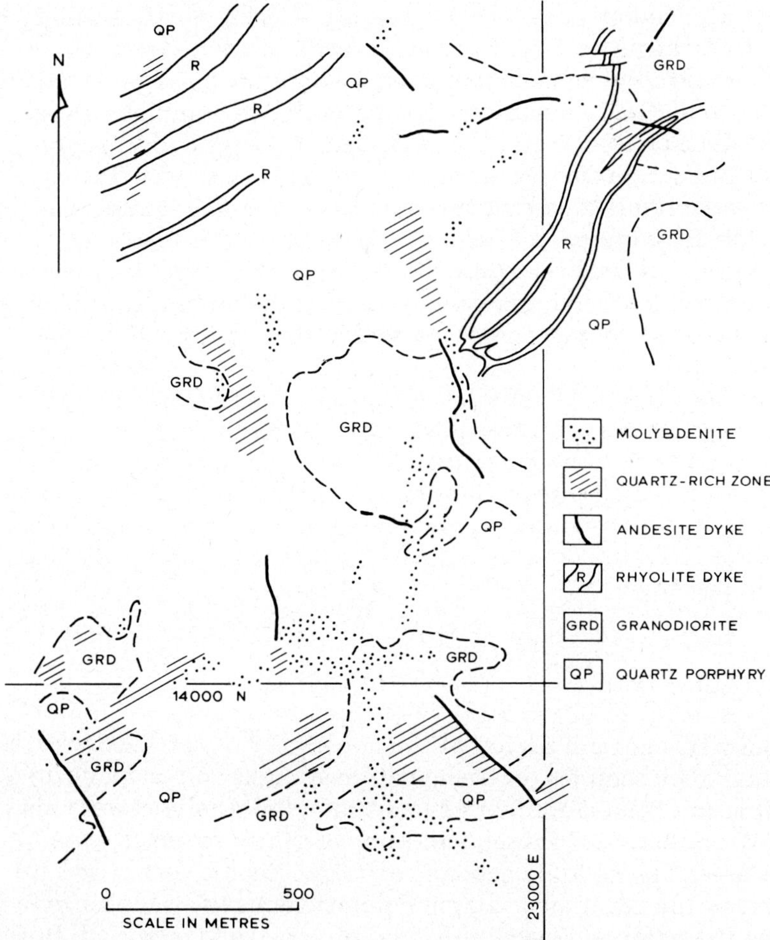

Fig. 2. Geologic map.

(Henry and Fredrikson, 1972). Therefore, the intrusive activity may have started in Laramide time. The younger rhyolite ignimbrites are bracketed between 21 and 34 million years (Oligocene/Miocene) by K—Ar dates (McDowell and Clabaugh, 1972).

MINERALIZATION

Early alteration included both massive silicification and K-feldspar metasomatism (K/Na ratios of 10/1 exist in highly altered rocks). The silicification, forming bodies of "quartz-rock", and extensive sericitization are found bordering the strongest metallized zones. Propylitic alteration is often present on the margins. The zonal pattern of alteration is roughly similar to that described by Lowell and Guilbert (1970).

The first wave of mineralization consisted of fine-grained pyrite, molybdenite, and traces of chalcopyrite in quartz veinlets striking northeast and dipping steeply to the northwest. Occasionally, medium-grained molybdenite in quartz veins is found toward the margins of the deposit. Later molybdenite followed northerly and finally northwesterly structures to complete the stockwork form of the deposit. Almost all the mineralized stringers are steeply-dipping. Specularite-quartz veins are mostly post-molybdenite and have northwest strikes with moderately steep southwest dips. These commonly carry cassiterite, occasionally copper, and (probably) tungsten. The few, vuggy, black sphalerite veinlets found, seem also to belong to this stage. It appears that the metals formed in several pulses with most of the molybdenum being early, tin mostly late, and copper erratically distributed during most of the time of mineralization.

Surface alteration of the stockwork has produced much yellow limonite and some jarosite; certain areas on the north edges of the zone show extensive, red hematite. Ferrimolybdite can be recognized at numerous mineralized outcrops. Malachite and lesser azurite staining is widespread; in certain zones red cuprite is also present. This secondary enrichment gives assays of 0.15% Cu or more. Copper seems to be the only metallic element that shows a notable secondary dispersion pattern.

GEOCHEMICAL PROGRAMME

In conjunction with a general survey of the apparently mineralized area, it was decided to take rock samples on triangulation stations and claim corners to amplify information from a number of scout samples. Analyses were made for Mo, Sn, and W on these and subsequent samples. Since no maps were available for this area, a plane table geologic and topographic survey was utilized to spot most of the additional sampling points. Thus the samples were taken at the same time that geological information was being recorded. Both mineralized material and rock chips were sent for assay; however, only the latter were used for the accompanying geochemical maps. A combination of points using the various surveys for location control allowed the preparation of accurate contour maps. On the other hand, the lack of base maps and the extreme local relief made an initial grid layout nearly impossible.

Channel samples were cut in selected areas of obvious mineralization to provide further information for the contour maps. Sample spacing was closer for molybdenum than for tin as not all samples were analysed for the latter. Most of the geochemical analyses were done by atomic absorption spectrometry. In total, nearly 1200 samples were used.

Molybdenum

Although molybdenum is present in trace amounts in precious and base metal ores of the region (and rare molybdenite veins are known 15—20 km to the north), the element does not seem to be unusually concentrated in the rock sequence here. This geologic setting should give a background content of 0.7—1.5 ppm Mo (Taylor, 1969, p.61), and the background at Corral is probably no more than 5 ppm at the most. The threshold value of 100 ppm (Fig.3) is conservative; however, lowering the threshold to 50 ppm does not materially add to the area covered by the higher value. Indeed, the geochemical gradient is very steep from less than 50 ppm to over 250 ppm. The gradient continues to be relatively steep to levels of 350 ppm.

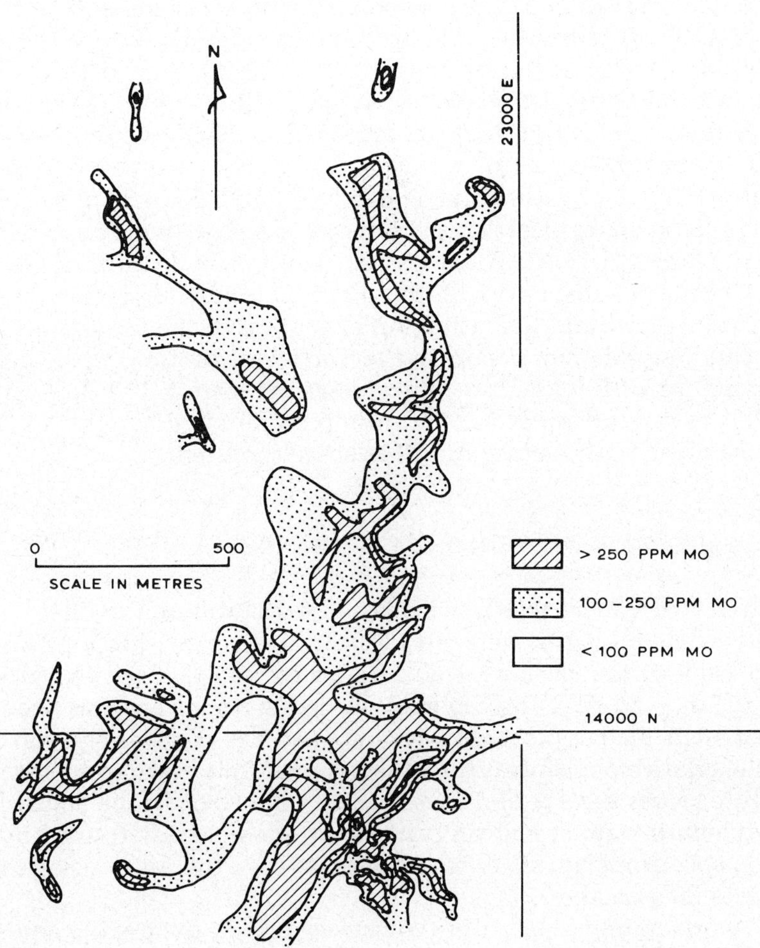

Fig.3. Molybdenum content in rock.

The steep geochemical gradient in a region of low background produces very marked anomalies. The main anomaly forms a north-northeast trending zone, nearly 2 km long and about 400 m wide. This strong anomaly lies within a more generalized one marked by the plus 5 ppm level, comprising a zone some 10 km × 3 km oriented about northeast. This local background "anomaly" is truncated to the southwest by a major normal fault.

Tin

Cassiterite mineralization is widespread in the Sierra Madre Occidental of Durango although workable deposits are rare. Nevertheless, the region is enriched in tin; this is certainly most notable in the more felsic rocks. Tin contents of 100 ppm are common in the ignimbrites; elsewhere, where tin deposits are worked, they consist of cassiterite associated with iron oxides, and occur in rhyolite tuffs. Whereas granites in other tin districts carry 20—60 ppm Sn (Sainsbury, 1968, p.1569, or Cotelo Neiva, 1972, p.285), the Corral de Piedra perimeter probably averages well over 50 ppm with many rock tin contents exceeding 100 ppm. This is, of course, far above normal granite averages of 2—3 ppm Sn (Taylor, 1969, p.61).

The molybdenite zone shows lower tin averages (note Sn low in Fig.4); values of 25—40 ppm Sn are typical. This is somewhat higher than the 10—20 ppm Sn content at Climax, U.S.A. (Wallace et al., 1968). The pattern of the tin anomalies at Corral is almost antipathetic to the molybdenum concentrations and thus tends to surround the latter. An exception to the usual mutual exclusion of tin and molybdenum occurs in the northwest portion of the detail maps where a minor >250 ppm Mo zone crosses the main >200 ppm Sn anomaly. The molybdenite—cassiterite relationship is amplified here by a few specularite veins carrying both metals as well as some tungsten.

PRIMARY METAL DISPERSION WITH SECONDARY CHANGES

Fig.5 was plotted to compare relative changes in metal values. Channel samples, generally 10 m in length, were cut at intervals from a 1 km long line on the northwest bank of the Piaxtla River. These were analysed for five elements: Mo, Cu, Sn, Ag, and W. Tungsten gave negligible results and has been omitted. The four elements are plotted in separate profiles (Fig.5). The vertical scale shows the relative concentration of each metal, but is different for each as the absolute values are not considered to be important to the following discussion. Molybdenum, copper and tin concentrations were determined by atomic absorption spectrometry; silver was determined by fire assay using the average of duplicate 20-g charges.

Molybdenum is moderately high on the southwest end of the profile and appears as a very small anomaly on the northeast end. None of the other three

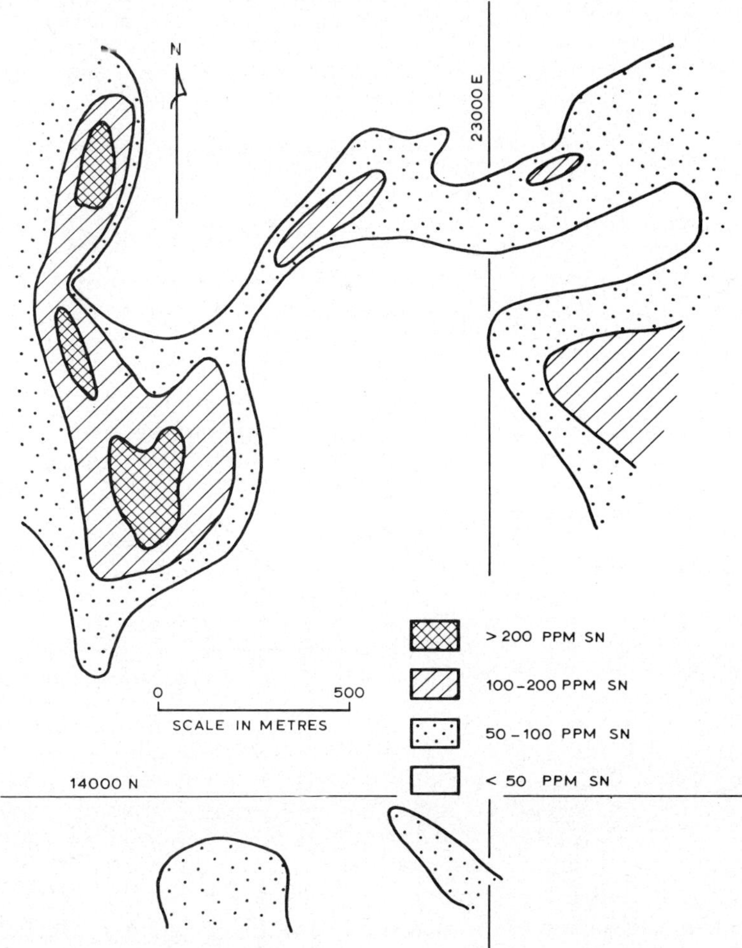

Fig.4. Tin content in rock.

metals reflects these anomalies. The copper plot shows one anomaly toward the
centre of the profile: this is accompanied by tin. The reason for the coincidence
of Cu and Sn here seems to be a combination of considerable fracturing in the
zone, formation of various specularite veinlets, and permeable ground allowing
easy movement of groundwater carrying oxide copper minerals. Thus, copper
has been concentrated by secondary enrichment in the broken ground which
carried the Sn-bearing specularite veins. The copper low, occurring where
there is only primary chalcopyrite on the southwestern part of the profile,
contrasts with the high molybdenum values in that area. This relationship of
at least partial mutual exclusion of copper and molybdenum is common in

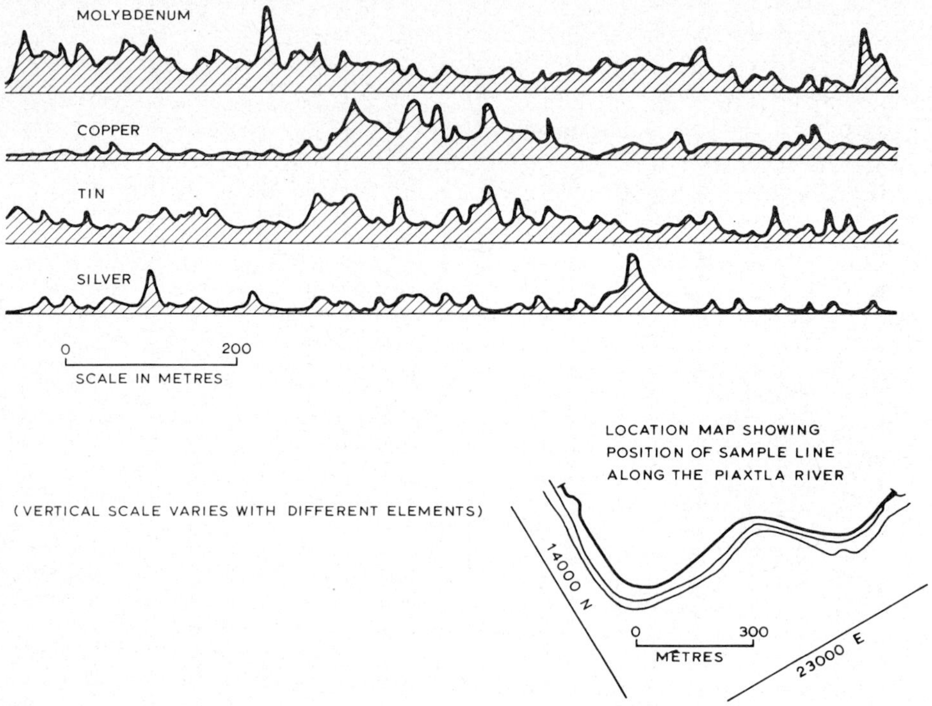

Fig.5. Four profiles of metal values.

this stockwork. On the other hand, it appears that tin and copper do have a
primary coincidence as well as the supergene one previously described. The
two silver peaks on the profile show no known relationship to geology; it is
possible that the more northerly one is supergene silver related to local faulting.

CONCLUSION

The Corral de Piedra property has many similarities to the Climax, Colorado,
U.S.A., stockwork, especially in the nature of the molybdenite-pyrite veinlets
and in alteration zoning. The metallic elements molybdenum and tin coincide
in time and space in the two localities, but the Climax-type of tungsten zone
does not appear to have been developed. The tungsten at Corral is thought to
be a wolframite as no scheelite or powellite has been found locally; although,
there is considerable powellite at another prospect a few kilometres distant.
Copper concentrations at Corral, though minor, are distinctly higher than
that at Climax. Nevertheless, the multi-stage intrusion and mineralization at
Climax is certainly reflected at Corral de Piedra.
The deeply dissected topography fosters rapid erosion and ensures excellent

rock exposures. This allowed an effective rock sampling programme, the results of which appear to coincide very well with observed geology, and to have delineated the metal concentrations. During much of the exploration; surveys, geochemistry, and geology were done at the same time with reasonable success. Hindsight might indicate that sampling on an airphoto mosaic would have been more efficient, but this was beyond our capabilities to do with any accuracy. Rather, the conjunction of geology and geochemistry was valuable in giving a preliminary understanding of an apparently large stockwork deposit in previously unmapped terrain.

ACKNOWLEDGEMENTS

It gives me great pleasure to thank those who helped me in the field on this property, especially my assistants: T. Meza S., T. Lares L., M. Pérez C., M.J. Hernandez, J.M. Bueno R.; my wife, Shirley J. Randall, and Ing. C. Bovio who did the surveys. Mr. P.W. Richardson, then Resident Manager at Tayoltita, gave much encouragement and advice throughout the project.

REFERENCES

Cotelo Neiva, J.M., 1972. Tin—Tungsten deposits and granites from Northern Portugal. 24th Int. Geol. Congr., Montreal, 1972, Sec. 4, pp.282—288

Henry, C.D. and Fredrikson, G., 1972. Edades de intrusiones, Sinaloa, México (abstract). Mem. IIa. Conv. Nac. Soc. Geol. Mexicana, Mazatlán, p.155

Lowell, J.D. and Guilbert, J.M., 1970. Lateral and vertical alteration-mineralization zoning in porphyry ore deposits. Econ. Geol., 65(4): 373—408

McDowell, F.W. and Clabaugh, S.E., 1972. Edades potasio—argon de rocas volcanicas en la Sierra Madre Occidental, al noreste de Mazatlán (abstract). Mem. IIa. Conv. Nac. Soc. Geol. Mexicana, Mazatlán, pp.183—184

Randall, J.A., 1972. Metallization sequence in the Tayoltita region, San Dimas, Durango, México. 24th Int. Geol. Congr., Montreal, 1972, Sec. 4, pp.309—317

Sainsbury, C.L., 1968. Tin and beryllium deposits of the Central York Mountains, Alaska. In: J.D. Ridge (Editor), Ore Deposits in the United States. Graton-Sales, Vol. II, p.1569

Taylor, S.R., 1969. Trace element geochemistry of andesites and associated calc-alkaline rocks. In: A.R. McBirney (Editor), Proc. of the Andesite Conference. State of Oregon, Dept. of Geol. and Min. Ing. Bull., 65: 61

Wallace, S.R. et al., 1968. Multiple intrusion and mineralization at Climax, Colorado. In: J.D. Ridge (Editor), Ore Deposits in the United States. Graton-Sales, Vol. I, pp.605—640

ZINC ABUNDANCE IN EARLY PRECAMBRIAN VOLCANIC ROCKS: ITS RELATIONSHIP TO EXPLOITABLE LEVELS OF ZINC IN SULPHIDE DEPOSITS OF VOLCANIC-EXHALATIVE ORIGIN[*]

W.J. WOLFE[1]

Ontario Division of Mines, Ministry of Natural Resources, Toronto, Ont. (Canada)

ABSTRACT

A survey of 1041 complete and 1472 partial chemical analyses of Early Precambrian tholeiitic and calc-alkaline volcanic rocks from the Abitibi, Wawa, Uchi,and Wabigoon belts of the Canadian Shield has demonstrated certain regularities in the abundance and distribution of zinc. Zinc abundance ranges from 3 to 200 ppm in background volcanic rocks remote from mineralization and much of this variation is accounted for by fractional crystallization processes represented by the variables SiO_2 and total iron. Within this range, zinc values of 120—200 ppm are related to Fe-rich tholeiitic basalts identified by levels of total FeO (FeO + 0.9 Fe_2O_3) in excess of 10%.

The average zinc contents of sampled areas in the volcanic belts vary from 80 to 95 ppm and zinc frequency distributions closely approximate a normal form. Small differences in the statistical characteristics of zinc populations may be explained by systematic interlaboratory analytical error and by the use of unweighted averages in localized areas containing unrepresentative proportions of tholeiitic and calc-alkaline rocks.

The chemical data have been compiled and plotted to develop diagnostic charts showing zinc variability in unmineralized differentiated sequences of Early Precambrian volcanic rocks. By relating zinc abundances to SiO_2 and total iron (as FeO), components of regional background zinc variability due to the silicate fractional crystallization process can be isolated from local anomalous zinc variation attributable to mineralizing processes connected with the localization of sulphide deposits of volcanogenic origin.

Zinc variation in Precambrian volcanic rocks has been mapped in detail in a 90-km² test area situated in the central Abitibi belt, 48 km west of Noranda and 34 km northeast of Kirkland Lake. Data collected from sites spaced with a mean density of 11 per km², are used to illustrate regional and residual patterns of zinc variability related to normal differentiation processes and mineralizing processes respectively.

INTRODUCTION

The volcanic-exhalative concept is now a widely accepted model for the origin of Precambrian stratabound base-metal deposits (Derry, 1973) and the recently published papers of Sangster (1972) and Hutchinson (1973) give

[*]Published by permission of the Director, Geological Branch, Ontario Division of Mines, Ministry of Natural Resources.
[1]Present address: Cominco Limited, Exploration Division, 120 Adelaide St. West, Toronto, Canada.

comprehensive reviews of the distribution, internal features, mineralogy, bulk chemistry, alteration, host rock depositional environment and genesis of volcanogenic massive Zn—Cu—Pb—Ag sulphide deposits — all explained in terms of metal-exhalative processes operating at or near the sea-floor surface.

Because submarine metal-exhalative processes profoundly influenced the localization of base-metal sulphide ores, and controlled the nature and extent of metal dispersion at the time of ore formation, an examination of these processes and their effects is basic to the understanding of bedrock geochemical dispersion halos in Archean volcanic rocks of the Canadian Precambrian Shield. Geochemical abnormalities directly associated with these deposits are likely to be confined to the walls of the pipes through which the mineralizing solutions passed and to the stratigraphic zones in the volcanics and associated sediments where the metals were precipitated. The metal-exhalative process, therefore, superimposes locally anomalous geochemical patterns on regional trends of geochemical variation that are explained by normal magma differentiation.

In 1972, the Geological Branch of the Ontario Division of Mines initiated a geochemical investigation of Early Precambrian (Archean) volcanic rocks with special emphasis on a study of metal distribution in these rocks and its relation to the existence of stratiform volcanogenic sulphide deposits. In the early stages of this work, it became apparent that from a mineral exploration standpoint, the value of the information extracted from a primary geochemical dispersion study was closely related to the type and especially the scale of geological, tectonic, and ore-forming processes operating to produce a mineral deposit, and had very little to do with overall metal abundance at the regional level. Indeed, the data gathered to date, suggest that unless geochemical rock sampling of Archean greenstone belts is carried out in detail consistent with the scale of mineralized pipes and stratigraphic zones, the results will seldom be of value in making mineral exploration decisions.

THE CHEMISTRY OF EARLY PRECAMBRIAN VOLCANIC ROCKS

In the Canadian Shield, Early Precambrian (Archean) volcanic assemblages of approximate age 2750 million years old show well-defined petrochemical progressions from lower primitive peridotitic lavas (Pyke et al., 1973) and Ca—Mg-rich basaltic komatiite lavas (Brooks and Hart, 1972, 1974) through tholeiitic lavas of moderate iron enrichment trends, to upper high-alumina calc-alkaline lavas (Goodwin, 1973) (Fig.1). The peridotitic and basaltic komatiites are not volumetrically important in most Archean volcanic successions. Their existence is mainly significant in terms of explaining tectonic environments (e.g. steep geothermal gradient, high convection rate, thin lithospheric plates) peculiar to Early Precambrian time. Sub-alkaline mafic to felsic flows and pyroclastics of tholeiitic to calc-alkaline chemical affinity predominate in the volcanic belts but some alkalic volcanic rocks are present in the Kirkland Lake region of the Abitibi belt (Cooke and Moorhouse, 1969).

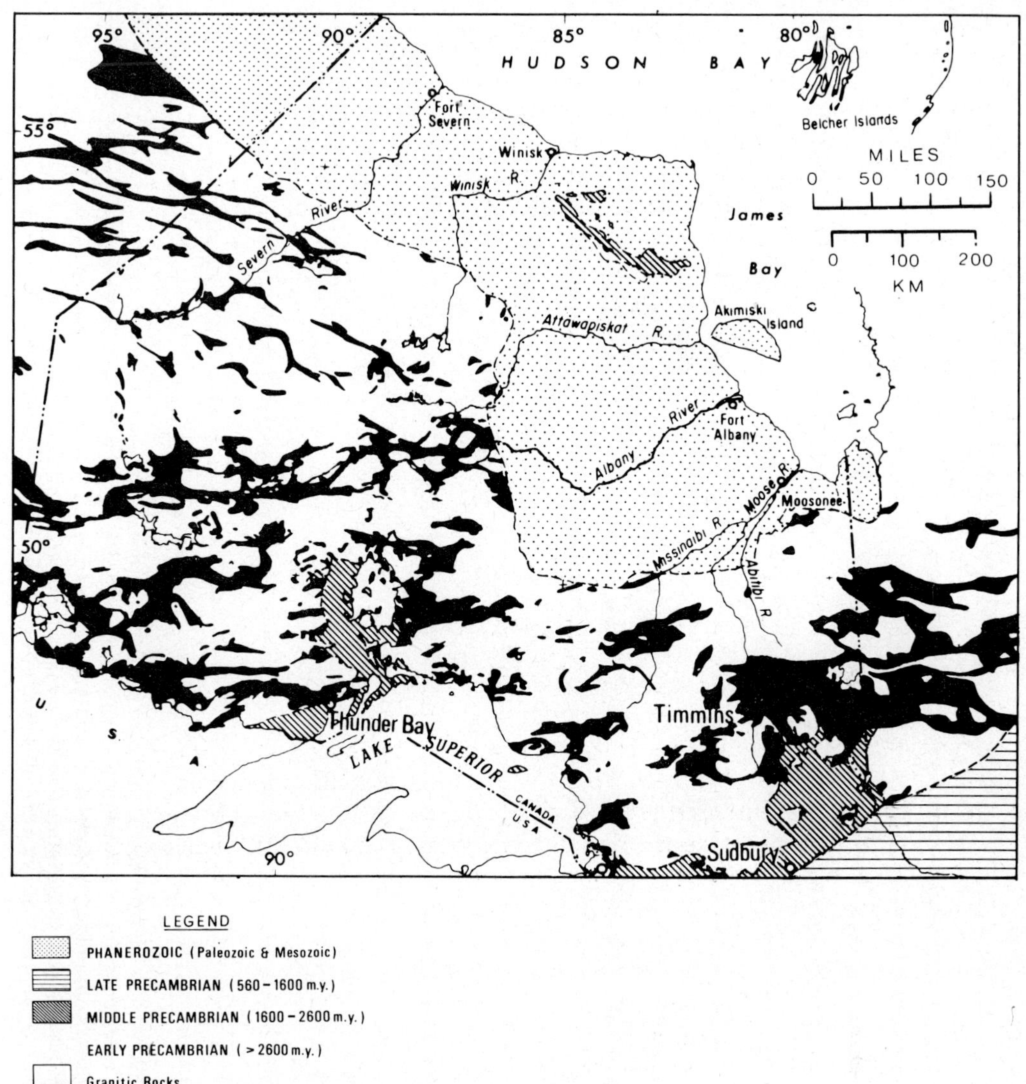

Fig.1. Distribution of Archean metavolcanic-metasedimentary belts in the Superior province of the Canadian Shield.

Large numbers of chemical analyses of Archean volcanic rocks from widespread areas of the Precambrian Shield have consistently shown the same petrochemical characteristics (Wilson et al., 1965; Goodwin, 1967, 1968, 1970; Baragar, 1966, 1968; Descarreaux, 1972; Barlow, 1973; Wolfe, 1973). Chemical trends originate well within the tholeiitic field when projected onto the

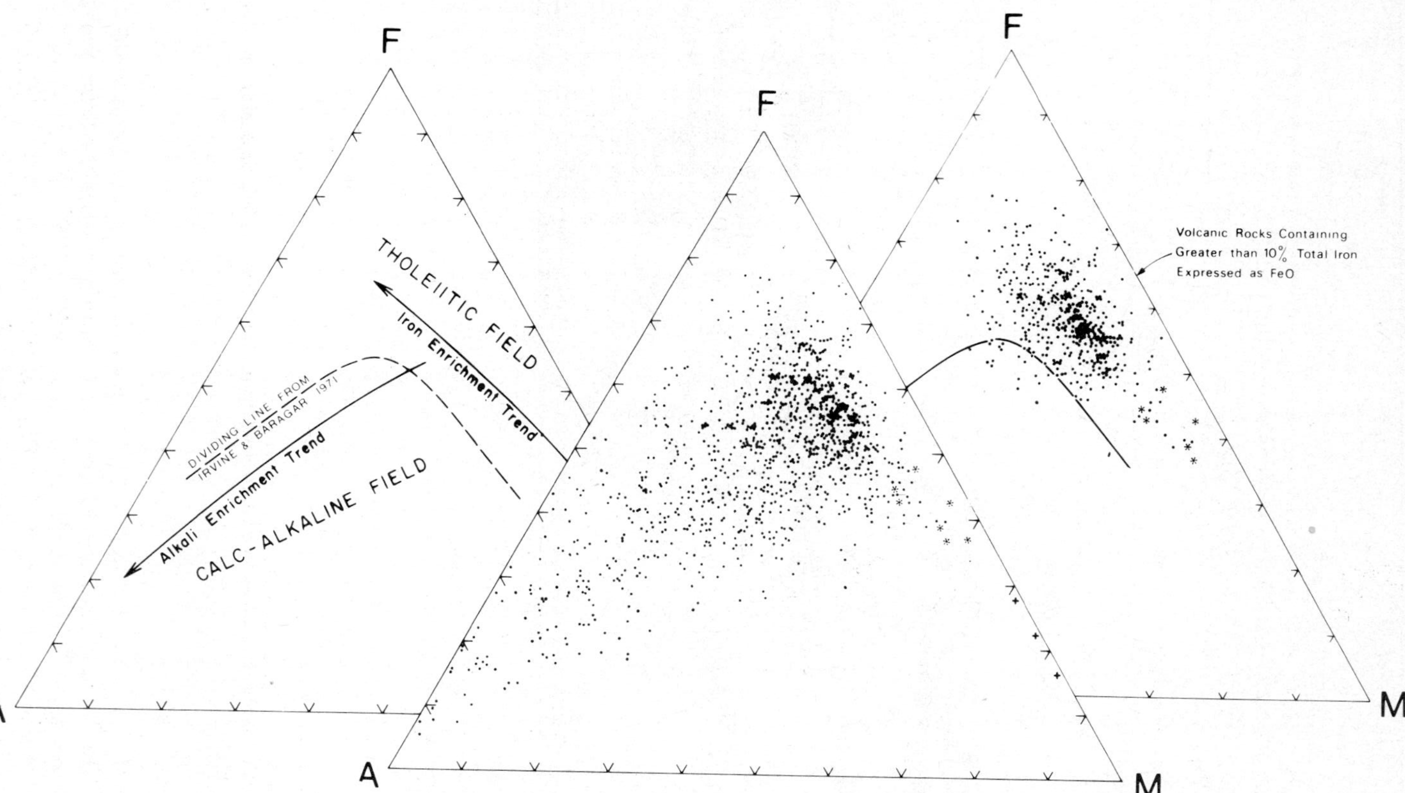

Fig.2. AFM plots showing patterns of variation in 1041 samples of Archean tholeiitic and calc-alkaline volcanic rocks. $A = Na_2O + K_2O$; $F = FeO + 0.9\ Fe_2O_3$; $M = MgO$, all in weight percent. Samples containing F greater than 10.0% are plotted separately in the diagram at the right. ★ = basaltic komatiite defined by MgO greater than 9.0% and CaO/Al_2O_3 ratio greater than 1. + = analyses of extrusive ultramafic flows from Pyke et al. (1973).

base of the diopside—olivine—nepheline—quartz tetrahedron of Yoder and Tilley (1962) and the major trend is sharply directed towards quartz enrichment (Baragar, 1972). Archean volcanic rocks compare in part, with continental tholeiites and in part with the calc-alkaline basalt—andesite—rhyolite association common in the orogenically active regions of the continental margins including the island-arc systems (Wilson et al., 1965). When chemical analyses of Archean volcanics are plotted on the standard AFM diagram (Fig.2), elements of both tholeiitic and calc-alkaline trends are recognized. Tholeiitic trends represented by iron enrichment relative to magnesium can usually be easily identified in the thick sequences of sub-aqueous, Fe-rich, pillowed basalts and andesites in the lower part of Archean volcanic cycles. Calc-alkaline trends represented by alkali enrichment without significant iron enrichment are recognizable in the upper explosive felsic pyroclastic parts of the cycles. Although both major trends are evident in Fig.2, there is a wide scatter of points and general lack of coherent differentiation trends that characterize younger volcanic suites. These features lead to uncertainty in the classification of analyses that plot near the tholeiite—calc-alkaline boundary as defined by Irvine and Baragar (1971). Recognition of these two differentiation trends at the regional level is of some importance because most of the important Precambrian volcanogenic massive sulphide deposits of the Canadian Shield are associated with rocks of calc-alkaline affinity.

In Fig.2, analyses of 1041 Archean volcanic rocks obtained from reports by Goodwin (1967, 1970) and Wolfe (1973) and from unpublished thesis studies by Barlow (1973) and Descarreaux (1972) are plotted in the standard AFM ternary diagram. From this group of chemical analyses, those containing (FeO + 0.9 Fe$_2$O$_3$) in excess of 10.0% by weight were extracted and re-plotted separately in the diagram shown at the right of Fig.2. All but 16 of these high-iron Archean volcanic rocks plotted in the tholeiitic field of the AFM diagram as defined by the dividing line suggested by Irvine and Baragar (1971). For Archean volcanic rocks of low to moderate Mg/Fe ratio, the total iron content expressed as FeO appears to discriminate well between rocks of tholeiitic and calc-alkaline affinity.

ZINC ABUNDANCE IN ARCHEAN VOLCANIC ROCKS

A survey of 1041 complete and 1472 partial chemical analyses of Early Precambrian tholeiitic and calc-alkaline volcanic rocks from the Abitibi, Wawa, Wabigoon and Uchi belts of the Superior province of the Canadian Shield has demonstrated certain regularities in the abundance and distribution of zinc. Statistical distributions of zinc in groups of volcanic rocks from six areas within these belts are shown in Fig.3. Zinc abundances range from 3 to about 200 ppm in background volcanic rocks remote from mineralization and much of this variation is explained by processes of magma differentiation represented by the variables SiO$_2$ and total iron (expressed as FeO).

The average zinc contents of sampled areas in the volcanic belts vary from

80 to 95 ppm and zinc frequency distributions closely approximate a normal form. In the upper tails of the histograms (Fig.3), zinc values in the range of 120—200 ppm are invariably identified with high-iron tholeiitic basalts containing total FeO (FeO + 0.9 Fe_2O_3) in excess of 10.0%. In the lower tails, zinc values in the 3—50 ppm range represent siliceous dacites and rhyolites of low iron content. The small differences in the statistical characteristics of the six zinc populations may in part be due to systematic inter-laboratory error. More generally, however, these differences are caused by unweighted comparisons between different areas containing unrepresentative proportions of high Fe—Zn tholeiitic basalts and low Fe—Zn calc-alkaline volcanics. The Wabigoon, Pukaskwa and Mattagami-Rouyn populations include a significant number of Fe-rich tholeiites and have high average zinc abundances of 85, 95 and 85 ppm, respectively. Samples in these 3 groups were collected over large areas of 1000 km^2 or more,and, therefore, accurately represent the overall composi-

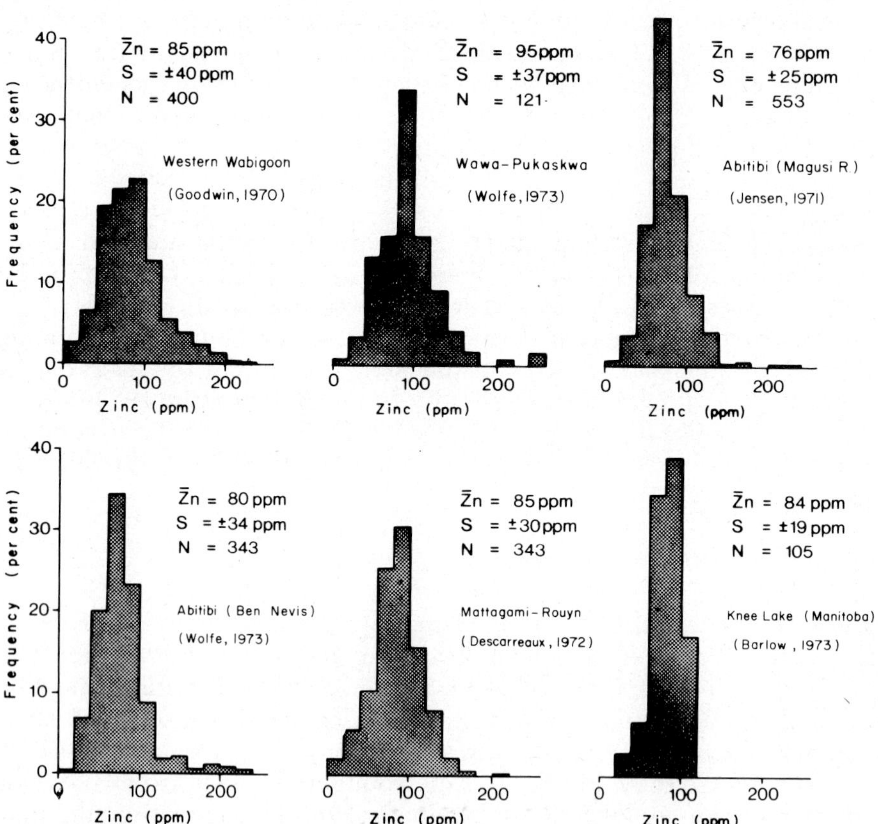

Fig.3. Histograms of zinc abundance in volcanic rocks from six areas within Archean belts of the Superior province.

tion of Archean volcanic belts. In comparison, the Magusi River and Ben Nevis populations represent localized areas within the Abitibi belt that are dominated by calc-alkaline volcanic rocks of relatively lower average Fe—Zn content.

Statistical correlations of the type shown in Table I were used to test the degree to which metal concentrations were dependent on chemical parameters which could be tied to the fractional crystallization process. As expected, total iron as FeO, MgO, normative Color Index (C.I.), zinc and log Zn all show strong negative linear correlation coefficients with respect to SiO_2. The degree of correlation between zinc and SiO_2 is a good deal better than that between copper and SiO_2 because in background volcanic rocks, sphalerite is a relatively rare mineral and zinc occurs primarily in mafic silicates and oxides as a substitute for iron. In comparison, our data obtained by sulphide-selective digestion of rock samples, indicates that copper occurs almost exclusively in the form of chalcopyrite. On the basis of the above information, SiO_2 was selected as the best single indicator of the differentiation process and plots of SiO_2 versus Zn were used to express zinc variability in unmineralized differentiated sequences of Archean volcanic rocks.

TABLE I

Chemical correlations in 400 Archean volcanic rocks — Western Ontario: table of correlation coefficients

	SiO_2	FeO	MgO	FeO + MgO	C.I.	Zn
FeO	—0.87					
MgO	—0.74	0.56				
C.I.	—0.92	0.83	0.88	0.96		
Zn	—0.62	0.66	0.38	0.61	0.55	
log Zn	—0.67	0.66				
Cu	—0.40	0.37	0.27	0.37	0.38	0.32

In Fig.4 zinc variation is plotted as a function of the SiO_2 content in 1575 samples of Archean volcanic rocks. The chemical data used to construct this plot were obtained from various published and unpublished sources including Goodwin (1970); Descarreaux (1972); Wolfe (1973); Barlow (1973) and Ontario Division of Mines geochemical files. The zinc content of basaltic rocks (45—52% SiO_2) varies widely between about 50 and 200 ppm and in a general way shows a positive relationship to the iron content of these rocks. The average zinc content of andesites remains essentially constant at about 85 ppm over the silica range of 52—60%. At silica levels above 60%, zinc decreases linearly to a mean value of about 35 ppm in rhyolitic volcanics containing 75% SiO_2. The trend shown by the heavy line in Fig.4 represents zinc variation attributable to differentiation processes in unmineralized suites of Archean tholeiitic and calc-alkaline volcanics. A field drawn to enclose the plotted

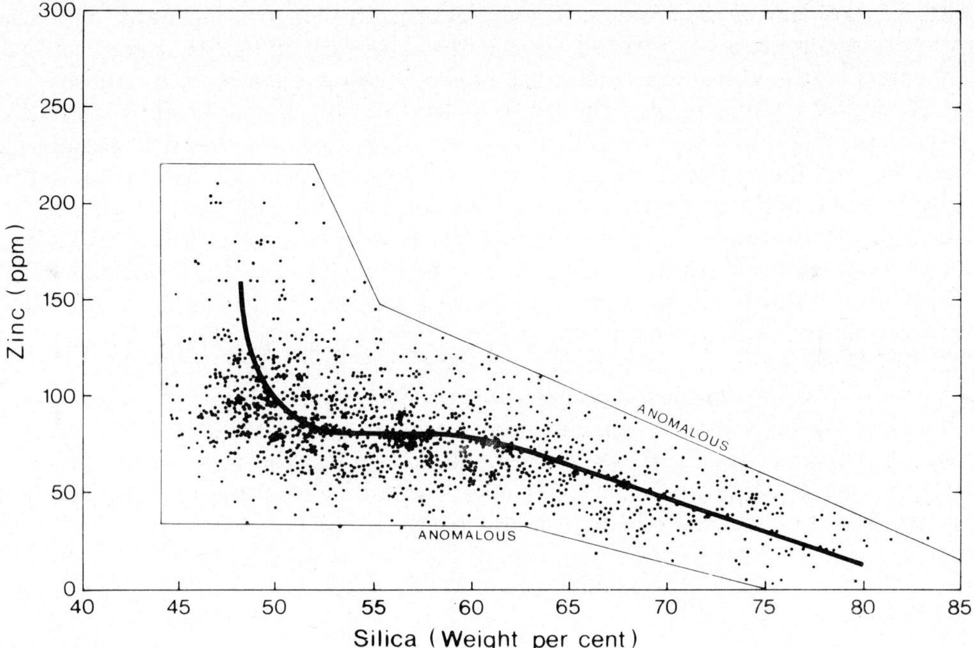

Fig.4. Plot of Zn versus SiO$_2$ in 1575 samples of Archean volcanic rocks.

points defines the range of expectable Zn—SiO$_2$ variation in normal Archean volcanic rocks. Analyses falling outside the upper boundary of this field therefore are presumed to reflect abnormal zinc concentrations connected with a mineralizing process.

Because Precambrian volcanogenic massive sulphide deposits are typically associated with calc-alkaline volcanic sequences and are notably absent in the Fe-rich basaltic lavas in the lower parts of Archean volcanic successions, Fig.4 can be further refined by discarding all analyses containing more than 10% total iron as FeO. When this is done, as in Fig.5, the average zinc content of the remaining low-iron mafic volcanic rocks remains essentially constant at 85 ppm throughout the silica range of 45—60%. The upper limit of normal zinc variation in a calc-alkaline volcanic sequence is, therefore, not likely to exceed 125 ppm.

To test the ability of plots such as that shown in Fig.5 to discriminate between barren and mineralized environments, a number of samples collected by A.M. Goodwin from acid volcanic cycles in the Uchi-Confederation Lake area of northwestern Ontario were analysed for zinc. Silica values for these samples were previously reported by Goodwin (1967). Both the upper and lower acid volcanic cycles contain massive Zn—Cu—Ag sulphide deposits but the only producing mine, the Selco South Bay deposit, is located within the upper acid volcanic cycle. As shown in Fig.6, all of the analysed samples from

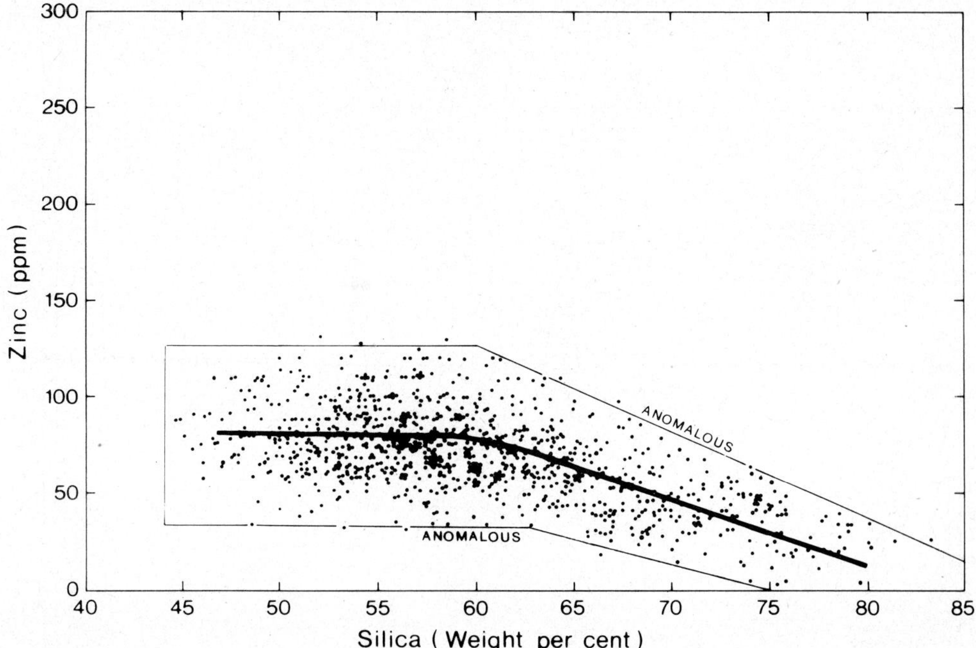

Fig.5. Plot of Zn versus SiO_2 in Archean volcanic rocks containing less than 10.0% FeO + 0.9 Fe_2O_3.

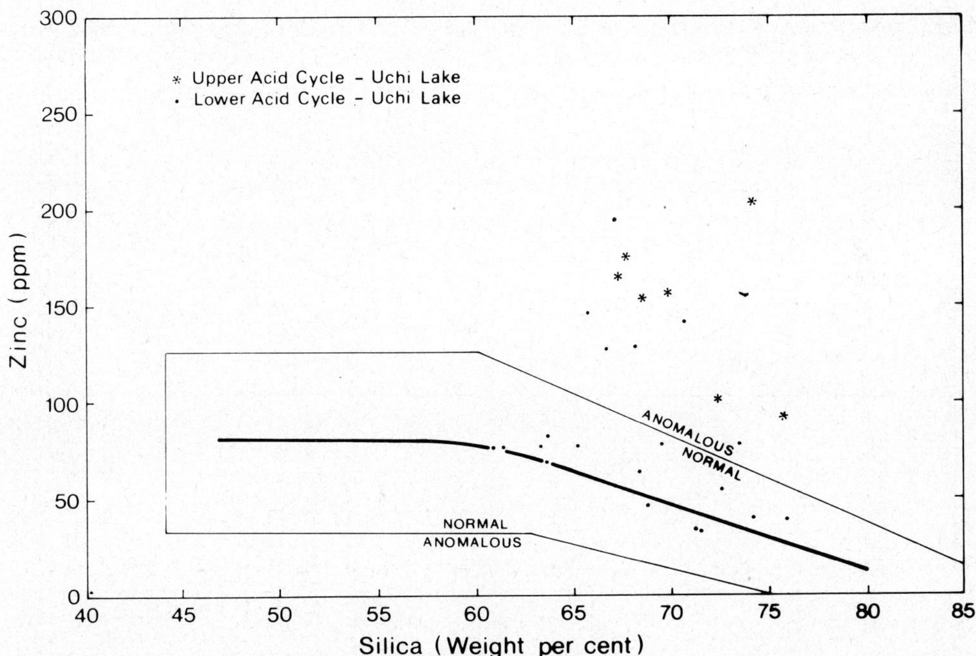

Fig.6. Plot of Zn versus SiO_2 in samples from the upper and lower acid cycles of the Uchi Lake—Confederation Lake metavolcanic belt.

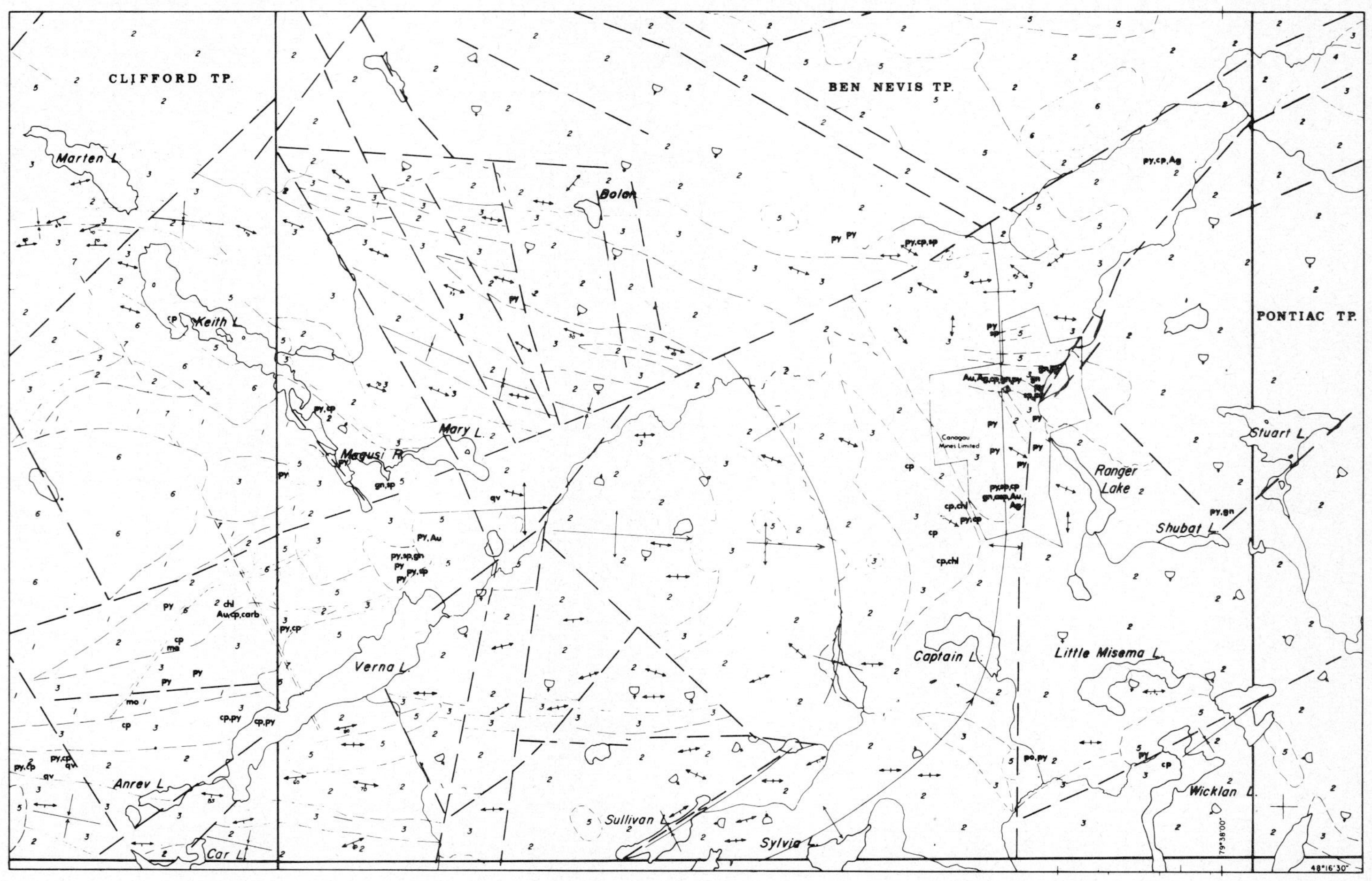
CLIFFORD TP.
BEN NEVIS TP.
PONTIAC TP.
Marten L.
Keith L.
Magusi R.
Mary L.
Verna L.
Anrev L.
Car L.
Bolon
Sullivan L.
Sylvia
Captain L.
Little Misema L.
Wicklan L.
Shubat L.
Ranger Lake
Stuart L.
Canagau Mines Limited
py,cp,Ag
py,cp,sp
py,gn
48°16'30"
79°38'00"

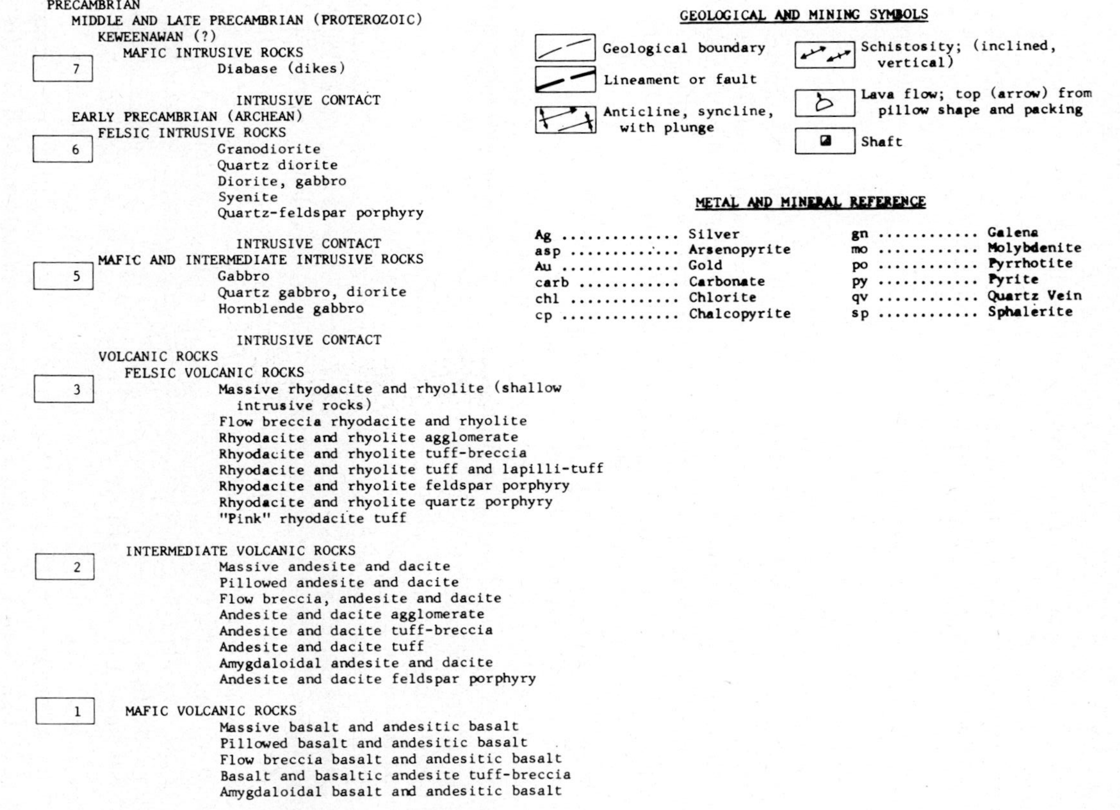

Fig.7. Geological map of the Ben Nevis area (after Jensen, 1971c,d).

the upper cycle contained anomalously high levels of zinc. In addition, 7 of 21 samples from the lower cycle were identified as being anomalously high in zinc. In a more extensive study of the geochemistry of volcanic rocks in the Birch-Uchi Lakes area, Davenport and Nichol (1973) reached similar conclu sions regarding the anomalous nature of the upper volcanic cycle.

ZINC VARIATION IN THE BEN NEVIS STUDY AREA

Geology and mineral deposits

The Ben Nevis area is located in the central part of the Abitibi Archean "greenstone" belt, 34 km northeast of Kirkland Lake, Ontario, and 48 km west of Noranda, Quebec. The geology of Ben Nevis Township and of Clifford Township, immediately west of Ben Nevis, has been mapped and fully described by Jensen (1971b). The following geological descriptions are largely summarized from the work of Jensen, and the geological map shown in Fig.7 is derived more or less directly from his maps (Jensen, 1971c,d).

With the exception of a single northwest-striking diabase dike of possible Proterozoic age, all of the bed-rock in the map-area is Archean. The geological mapping of Jensen (Fig.7) outlined a sequence of mafic to felsic volcanic flow and pyroclastic units flanking a massive to porphyritic, sub-volcanic granodiorite-quartz diorite stock located in the western part of the map-area. Intermediate volcanic rocks consist of massive, pillowed, and flow breccia flows, and pyroclastic units of agglomerate, tuff-breccia, lapilli tuff, and tuff. Many of the intermediate pyroclastic units contain felsic volcanic rock fragments in a predominantly intermediate matrix. The felsic volcanic rocks include massive dacite-rhyolite, rhyolite feldspar porphyry and rhyolite quartz porphyry plugs, and pyroclastic units consisting of flow-breccia, tuff-breccia, lapilli tuff, and crystal tuff. Felsic pyroclastic rocks in central Ben Nevis Township become finer-grained eastward with increasing distance from the granitic stock in Clifford Township. Massive and pillowed mafic volcanic flows occur south of Keith Lake and west of Verna Lake in eastern Clifford Township. The Ben Nevis volcanic rocks in general display a calc-alkaline chemical affinity consistent with significantly lower than normal measurements of iron content in the basalts and andesites. With the exception of 6 samples of highly chloritized, metasomatically altered mafic volcanics, all of the analysed extrusive igneous rocks in the map-area contained less than 10% total iron expressed as FeO. In a group of 343 randomly distributed samples of volcanic rocks from the Ben Nevis map-area the average SiO_2 content was 58.0% and the average iron content (total as FeO) was 6.56%. The partial analyses could be grouped as shown in Table II.

The layered volcanic assemblage is intruded by sills, dikes and stocks of gabbro, quartz gabbro, hornblende gabbro and diorite. The granodiorite-quartz diorite stock at the western edge of the area is approximately 1800 m in diameter and cuts all rock types of the map-area excepting a younger north-

TABLE II

Results of partial analysis of volcanic rocks from the Ben Nevis area

	SiO_2 (%)	Frequency	Total Fe as FeO (%)	Frequency
Basalt	< 54.0	121	8.0—10.0	79
Andesite	54.0—62.0	142	6.0—8.0	152
Dacite	62.0—70.0	44	3.0—6.0	79
Rhyolite	> 70.0	36	< 3.0	33

west trending olivine diabase dike. The contact between the granodiorite intrusive and surrounding volcanic strata is sharp and dips outward from the centre at 40—60 degrees. Regional zeolite facies metamorphism has been upgraded to albite-epidote hornfels rank near intrusive masses.

In the west central part of Ben Nevis Township, the volcanic units are folded concentrically in an open, east trending anticlinal dome that is flanked to the north, east and south by a complimentary syncline. A second anticline with an arcuate, north to northeast trending, double plunging axis is situated to the east of the syncline and it is along the hinge line of this structure that much of the base metal exploration has so far occurred. The volcanic strata are displaced by a regional northeast trending fault extending across Ben Nevis and Clifford Townships. The radial faults in the western half of the map area are intruded by feldspar porphyry related to the Clifford granodiorite stock. Short, parallel block faults that are subparallel to the boundaries of the Clifford stock may be volcanic explosive structures related to a central eruptive vent located in east-central Clifford Township.

The base metal sulphide deposits in east-central Ben Nevis Township are closely associated with rhyolite tuffs, tuff-breccias and lapilli tuffs that occupy the core of a south plunging anticlinal structure. Sphalerite, galena, chalcopyrite, pyrrhotite, silver and gold occur as massive replacement deposits in shear zones in altered rhyolite tuffs and as disseminated minerals in adjacent tuffs. Massive sulphide deposits from 15 to 45 cm wide occur on the Canagau Mines Limited property and the largest of these has been traced to a depth of 100 m over a length of 150 m by underground exploration (Jensen, 1971b). Extensive sericite, chlorite, talc and carbonate alteration is developed in the mineralized rhyolite tuffs near the Canagau shafts. South of the shafts, the sheared tuffaceous metavolcanics along the axial trace of the major south-trending fold structure contain widespread disseminated sulphide minerals. The metavolcanic units flanking the central rhyolitic core rocks to the southwest are highly chloritized and contain vesicle fillings of chalcopyrite and chlorite.

Geochemical sampling and analysis

During 1972, 335 samples of bedrock were collected in southern Ben Nevis Township for the purpose of determining patterns of metal variation in relation to mineralized volcanic strata. Representative volcanic bedrock was sampled at more or less evenly spaced, but randomly distributed sites. Sample density was increased near known occurrences of base-metal sulphide minerals north of Verna Lake and west of Ranger Lake. This collection of 335 samples was augmented by 653 specimens collected by L.S. Jensen and D.S. Hunt during a 1968 programme of geological mapping in Clifford and Ben Nevis Townships. The combined groups of samples, therefore, represented 988 outcrop locations in a 90-km^2 area covering the southern two-thirds of Ben Nevis Township and the extreme southeast part of Clifford Township. All samples were initially analysed for Zn, Cu, Ni, Pb, Co and Mn.

The rock specimens were successively passed through a jaw crusher and a cone crusher set to deliver material of about 10 mesh size. After mixing by riffling on paper sheets, a sample split was taken for grinding to minus 200 mesh size in a ceramic ball pulverizer. Samples of rock powder weighing 300 mg were treated with hot HNO_3/HCl leach, and the resulting solutions were diluted to 15 ml and aspirated for the determination of zinc by atomic absorption spectrophotometry. The detection limit of the routine analytical method was 3 ppm for zinc and the precision at zinc concentrations above 10 ppm was ± 15% at the 95% confidence level.

Sample identifications and chemical data for each of the 988 samples in the Ben Nevis area were transferred to 80 column computer cards for electronic data processing. The positions and sample identifications were digitized from 1 inch to ¼ mile base maps to assign geographic X and Y co-ordinates to the samples. These results were later merged on magnetic tape to produce a complete record of locational information and chemical data.

Results

The aqua regia soluble zinc content of sampled volcanic rocks in the Ben Nevis area ranges from 8 to 11,550 ppm. To obtain an unbiased estimate of the average zinc abundance in unmineralized background volcanic rocks, anomalously high zinc values attributable to the existence of sphalerite mineralization were discarded prior to the calculation of statistical parameters. Within a group of 920 samples of unmineralized volcanic rocks, the average HNO_3/HCl-soluble zinc content is 70 ppm — compared with an average value of 80 ppm obtained by analysing a selection of 343 samples for total zinc using a three-acid (HF/HNO_3/HCl) decomposition technique.

In background volcanic rocks remote from mineralization, the aqua regia soluble zinc abundance ranges from 8 to 125 ppm and much of this variation is accounted for by magmatic differentiation processes which produce a trend of zinc depletion with increasing silica and decreasing total iron. The frequency distribution of the background population closely approximates a normal, bell-shaped form. The upper tail of the distribution curve is truncated at about 125—130 ppm and zinc concentrations above this limit presumably reflect a secondary mineralizing process superimposed on a primary background pattern.

Zinc variations in the Ben Nevis volcanic rocks are contoured in Fig.8 at intervals of 10 ppm between 50 and 100 ppm and at higher levels of 125, 150, 200 and 250 ppm. The contours display regional zinc variation within the volcanic rocks, and are generated from an orthogonal grid of regularly spaced data points derived from the original irregularly spaced data by a moving average technique. Extremely high zinc values were truncated by the contouring programme at 500 ppm to avoid undue bias in averaging caused by single spikes in the data.

Fig.8 outlines a broad zone of anomalous zinc in volcanic rocks situated along the arcuate, north trending axis of the doubly plunging anticlinorium in east-central Ben Nevis Township. The zinc anomalies occur within a zone which cross-cuts stratigraphic boundaries and extends from the Canagau shaft southward for a distance of 2.0 km. A large number of abnormally high zinc values cluster within the southern part of the unit of rhyolite tuff, tuff-breccia and lapilli tuff that underlies much of the Canagau mines property. In comparison, the part of this felsic volcanic unit situated north of the Canagau shaft, contains low concentrations of zinc that are typical of unmineralized rhyolitic rocks.

A second cluster of zinc anomalies is related to rhyolite quartz-feldspar porphyry and associated rhyolite breccia in the area between the northeast end of Verna Lake and the southeast end of Keith Lake. In the northwest part of the map-area, discontinuities in the zinc contour pattern reflect the existence of north- to northwest-trending transverse faults that cross-cut the volcanic stratigraphy.

SUMMARY

Analyses of Archean volcanic rocks for zinc, SiO_2, and total iron can be used to isolate components of normal zinc variability due to the silicate fractional crystallization process from anomalous zinc variation attributable to metal-exhalative mineralizing processes connected with the formation of volcanogenic massive sulphide deposits. The calc-alkaline volcanics which host these deposits normally contain less than 10.0% total FeO (FeO + 0.9 Fe_2O_3). The upper limit of normal zinc variation in mafic to intermediate Archean volcanics containing 45—60% SiO_2 and less than 10.0% total FeO, is 125 ppm. In felsic Archean volcanics containing 60—80% SiO_2 the upper limit of normal zinc variation decreases linearly with increasing SiO_2. In rocks of dacite-rhyolite composition, the recognition of zinc anomalies within the interval of 50—125 ppm is therefore dependent on a knowledge of the silica content of the rock.

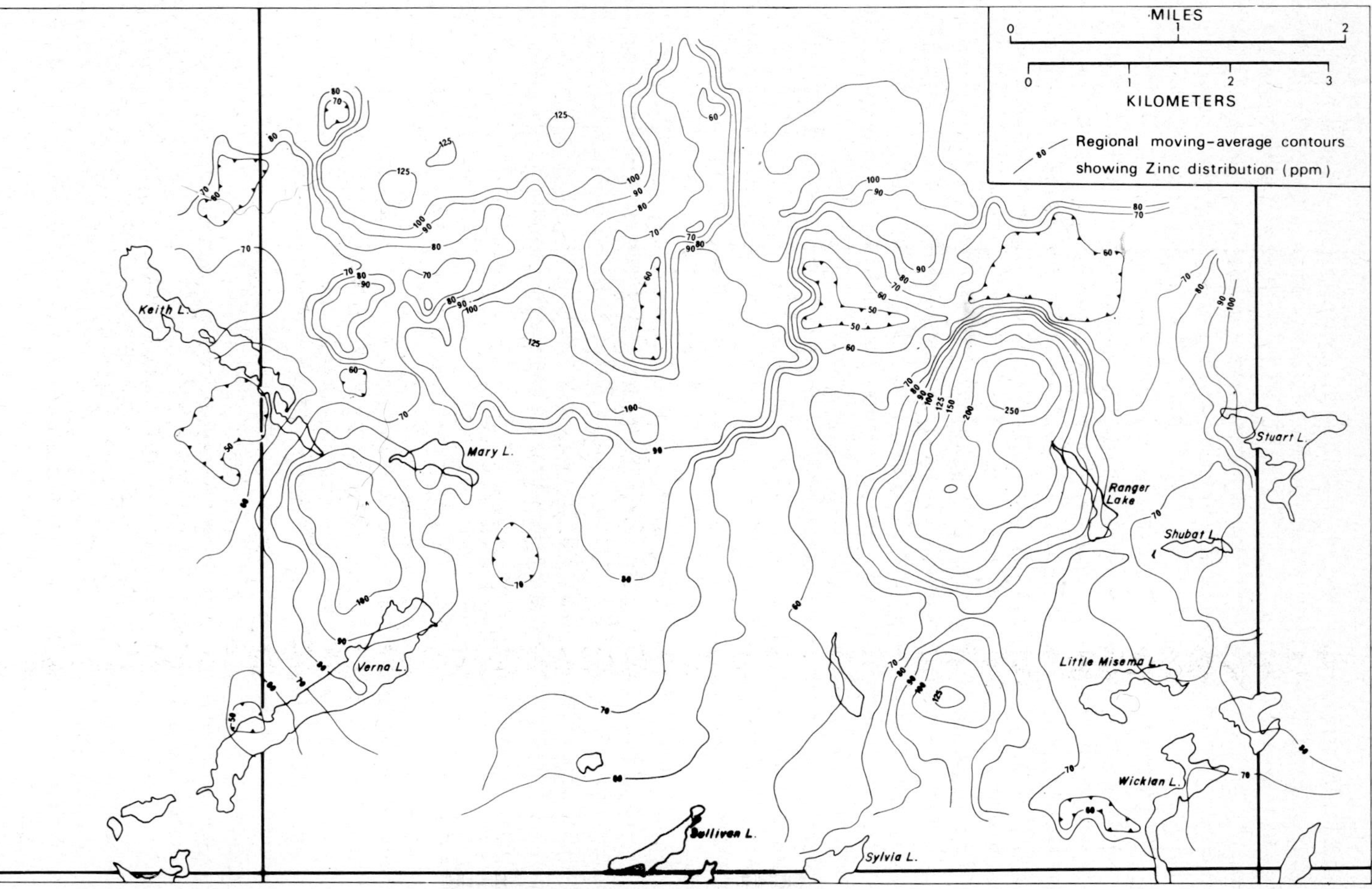

Fig.8. Regional moving-average trends of zinc distribution in Archean volcanic rocks of the Ben Nevis area (after Wolfe, 1974).

ACKNOWLEDGEMENTS

The geochemical study of Archean volcanic rocks in the Ben Nevis area was only possible because of the interest and co-operation of L.S. Jensen whose geological mapping of the area formed the groundwork for the geochemical sampling and interpretation. I also wish to extend acknowledgement to L.S. Jensen for significantly enhancing the completeness of the Ben Nevis study by making his samples available for chemical analysis. R.B. Barlow kindly permitted the writer to use certain chemical data from his unpublished M.Sc. thesis.

Chemical analyses connected with the Ben Nevis study were performed by the staff of the Mineral Research Branch of the Ontario Division of Mines, Ministry of Natural Resources. S.R. Parmar made a substantial contribution in the compilation and plotting of chemical data and the preparation of the figures.

REFERENCES

Baragar, W.R.A., 1966. Geochemistry of the Yellowknife volcanic rocks. Can. J. Earth Sci., 3: 9—30

Baragar, W.R.A., 1968. Major element geochemistry of the Noranda volcanic belt, Quebec—Ontario. Can. J. Earth Sci., 5: 773—790

Baragar, W.R.A., 1972. Some physical and chemical aspects of Precambrian volcanic belts of the Canadian Shields. In: E. Irving (Editor), The Ancient Oceanic Lithosphere. Publ. Earth Phys. Branch, Dept. Energy, Mines, Res., 42(3): 129—140

Barlow, R.B., 1973. Major and minor element geochemistry of the Archean volcanic rocks of the Southern Knee Lake area, Manitoba. M.S. Thesis, Michigan Technological Univ., Houghton, Mich., 124 pp. (unpublished)

Brooks, C. and Hart, S.R., 1972. An extrusive basaltic komatiite from a Canadian meta-volcanic belt. Can. J. Earth Sci., 9: 1250—1253

Brooks, C. and Hart, S.R., 1974. On the significance of komatiite. Geology, 2(2): 107—110.

Cooke, D.L. and Moorhouse, W.W., 1969. Timiskaming volcanism in the Kirkland Lake area, Ontario, Canada. Can. J. Earth Sci., 6: 117—132

Davenport, P.H. and Nichol, I., 1973. Bedrock geochemistry as a guide to areas of base-metal potential in volcano-sedimentary belts of the Canadian Shield. In: M.J. Jones (Editor), Geochemical Exploration 1972. Institution of Mining and Metallurgy, London, pp.45—57

Derry, D.R., 1973. Ore deposition and contemporaneous surfaces. Econ. Geol., 68: 1374—1380

Descarreaux, J., 1972. Geochimie des roches volcaniques de l'Abitibi. Doctorate Thesis, Univ. Laval, Quebec, Que., 283 pp. (unpublished)

Descarreaux, J., 1973. A petrochemical study of the Abitibi volcanic belt and its bearing on the occurrences of massive sulphide ores. Bull. Can. Inst. Min. Metall., 66(730): 61—69

Goodwin, A.M., 1967. Volcanic studies in the Birch—Uchi Lakes area of Ontario. Ont. Dept. Mines Misc. Paper 6, 96 pp.

Goodwin, A.M., 1968. Evolution of the Canadian Shield. Proc. Geol. Assoc. Can., 19: 1—14

Goodwin, A.M., 1970. Archean volcanic studies in the Lake of the Woods-Manitou Lakes—Wabigoon region of western Ontario. Ont. Dept. Mines, Open File Rept. 5042, 47 pp.

Goodwin, A.M., 1973. Petrochemical trends in Archean volcanic assemblages, Abitibi Belt, Ontario and Quebec, Canada (abstract Geol. Soc. Am. Ann. Meet., Dallas, November 12—14, 1973, pp.641—642

Hutchinson, R.W., 1973. Volcanogenic sulphide deposits and their metallogenic significance. Econ. Geol., 68: 1223—1246

Irvine, T.N. and Baragar, W.R.A., 1971. A guide to the chemical classification of the common volcanic rocks. Can. J. Earth Sci., 8: 523—548

Jensen, L.S., 1971a. Geology and geochemistry of Melba and Bisley Townships, District of Timiskaming, Ontario. M. Sc. Thesis, Univ. of Saskatchewan, Saskatoon, Sask., 80 pp. (unpublished)

Jensen, L.S., 1971b. Geology of Clifford and Ben Nevis Townships, District of Timiskaming. Ont. Dept. Mines and Northern Affairs, Open File Rept. 5061, 91 pp.

Jensen, L.S., 1971c. Clifford Township, District of Timiskaming. Ontario Dept. Mines and Northern Affairs Prelim. Map P.692

Jensen, L.S., 1971d. Ben Nevis Township, District of Timiskaming. Ont. Dept. Mines and Northern Affairs Prelim. Map P.693

Pyke, D.R., Naldrett, A.J. and Eckstrand, O.R., 1973. Archean ultramafic flows in Munro Township, Ontario. Geol. Soc. Am. Bull., 84: 955—978

Sangster, D.F., 1972. Precambrian volcanogenic massive sulphide deposits in Canada: a review. Geol. Surv. Can. Paper 72-22, 44 pp.

Wilson, H.D.B., Andrews, P., Moxham, R.L. and Ramlal, K., 1965. Archean volcanism of the Canadian Shield. Can. J. Earth Sci., 2: 161—175.

Wolfe, W.J., 1973. Regional geochemical reconnaissance of Archean metavolcanic-metasedimentary belts in the Pukaskwa region. Ont. Div. Mines, Open File Rept. 5091, 62 pp.

Wolfe, W.J., 1974. Geochemical distribution of zinc, nickel and copper in volcanic rocks, Ben Nevis Township and parts of Clifford and Pontiac, District of Timiskaming. Ont. Div. Mines Prelim. Geochem. Maps, P.915, P.916, and P.917

Yoder, H.S. and Tilley, C.E., 1962. Origin of basalt magmas: an experimental study of natural and synthetic rock systems. J. Petrol., 3: 342—532

MAJOR AND MINOR ELEMENT HALOS IN VOLCANIC ROCKS AT BRUNSWICK NO.12 SULPHIDE DEPOSIT, N.B., CANADA

W.D. GOODFELLOW

Department of Geology, University of New Brunswick, Fredericton, N.B. (Canada)

ABSTRACT

Preliminary data from the Brunswick No.12 deposit show that there are significant variations in Ca, Mg, Na, K, Mn, and Fe which can be spatially related to the sulphide zone.

The Brunswick No.12 deposit, located in northern New Brunswick, contains at least 82×10^6 tons of massive sulphides containing 9.69% Zn, 3.77% Pb, 0.29% Cu, and 2.46 oz. Ag per ton; the main sulphide minerals are pyrite, sphalerite, galena, pyrrhotite, chalcopyrite, tetrahedrite, and bornite. The ore zone lies within a Cambro-Ordovician sequence of pyroclastic volcanic rocks, sedimentary rocks with associated sulphides and iron formation, and intermediate to basic volcanic rocks, all of which have been subjected to lower greenschist regional metamorphism. The only alteration described in the literature consists of chloritization, sericitization, and silicification of the sedimentary rock units immediately associated with the ore deposit and extending up to 200 ft into the footwall volcanic rocks. Mineralogical alteration extending laterally along the ore horizon is also localized in the immediate area of sulphide mineralization; this is referred to as the "Intense Alteration Zone".

The acid volcanic rocks below the deposit show the following distributions of major and minor elements in a halo zone up to 1500 ft below the deposit and 1500 ft laterally north and south of the deposit.

(1) Manganese, magnesium and iron markedly increase in concentration in the Intensely Altered Zone, and magnesium and iron increase in concentration in the geochemical Halo Zone.

(2) Calcium and sodium show a marked concentration decrease in the Intensely Altered Zone, and calcium decreases in concentration in the geochemical Halo Zone.

(3) The concentration of potassium generally decreases towards the deposit. The Na/K ratio, however, consistently decreases toward the deposit.

(4) Aluminium shows no apparent variation with proximity to the sulphide deposit; fluctuations in aluminium concentrations are attributed to sample variation.

Acid volcanic rocks above the deposit reveal no anomalous trends in major element distribution. They represent apparent background conditions.

The recognition of a major and minor element halo in the footwall volcanic rocks of Brunswick No.12 deposit suggests several applications useful to exploration in the Bathurst area. These are: (1) to determine the stratigraphic "up" direction from the volcanic rocks associated with massive sulphide deposits; (2) to outline significantly larger targets in geochemical exploration for massive sulphide deposits; (3) to establish a close genetic and spatial relation between the source of metals from late-stage volcanic activity and the formation of massive sulphide deposits; and (4) to locate favourable stratigraphic horizons in the exploration for unknown sulphide deposits and to locate extensions of the zones of mineralization around known ore deposits.

INTRODUCTION

Chloritic and sericitic alteration of the sedimentary and volcanic rocks associated with sulphide deposits of the Bathurst region of New Brunswick has been recognized since 1958 by Lea and Rancourt (1958), although McAllister (1960) noted that around most deposits no significant hydrothermal alteration had been observed. Later work (Boyle and Davies, 1964; Boyle, 1965, 1969; Davies and Smith, 1966) has substantiated the presence of mineralogical alteration patterns in the rock units immediately associated with some sulphide deposits of the Bathurst district; Boyle (1965) noted that there is recognizable hydrothermal alteration (mainly chlorization, sericitization, and pyritization) at many of the Bathurst deposits. Boyle (1969) also recognized chloritic and sericitic wall-rock alteration patterns flanking and surrounding the sulphide deposits of the area. This type of mineralogical alteration has been recognized by the writer at Brunswick No.12 deposit, but it appeared to be of very local distribution (extending up to 200 ft into the footwall volcanic sequence). While areas of intense hydrothermal alteration can be recognized through characteristic mineralogical assemblages, it is suggested, in analogy with the work of Govett in Cyprus (Govett and Pantazis, 1971; Govett, 1972; Constantinou and Govett, 1972, 1973), that areas of less intense alteration, further from the deposit, may be recognized only through chemical variations in the volcanic rocks. This paper shows that a major and minor element alteration halo (1500 ft X 3000 ft) can be recognized in the volcanic rocks below the Brunswick No.12 deposit; this geochemical halo is far greater in extent than the mineralogical alteration halo and therefore has considerable significance.

GEOLOGY

Brunswick No.12 sulphide deposit is situated approximately 120 miles north of Fredericton, New Brunswick (Fig.1) in the Tetagouche Group of rock formations which are believed to be of Cambro-Ordovician age (Alcock, 1941; Smith and Skinner, 1958). The Tetagouche Group (of rocks) has been generally subjected to three periods of deformation (Helmstaedt, 1970; Davies, 1972; Luff, 1973). The major periods of folding were associated with the earlier periods of deformation; later periods of deformation produced only minor overprinting structures. Metamorphism of the Tetagouche Group is characterized by lower greenschist metamorphic assemblages consisting of chlorite, sericite, and minor biotite. The stratigraphic sequence of rocks, most recently described by Rutledge (1972) and Luff (1973) in the area of Brunswick No.12 deposit, consists of intermediate to basic volcanic rocks: siliceous sericitic chloritic schist with associated sulphides and overlying iron formation; augen schist; and metasedimentary rocks (Table I, Fig.2). The individual rock units associated with the massive sulphide deposit are described below.

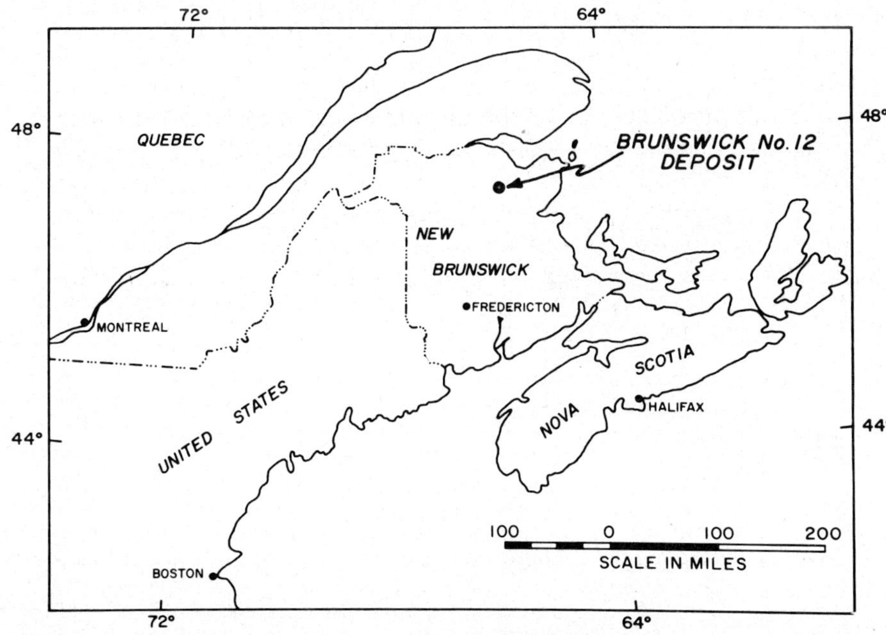

Fig.1. Location of Brunswick No.12 sulphide deposit.

TABLE I

Stratigraphic sequence of rock formations near Brunswick No.12 deposit

Epoch	Group	Lithology
Ordovician		quartz-feldspar porphyry dyke
Cambro-Ordovician	Tetagouche	(a) intermediate to basic volcanic rocks (b) siliceous chloritic sericitic schist and iron formation (c) augen schist (d) metasedimentary rocks

Metasedimentary rocks

This unit consists of finely bedded, highly contorted grey to green phyllitic schist, fine- to medium-grained greywacke, graphitic schist, and siliceous argillite; it is generally barren of economic mineralization.

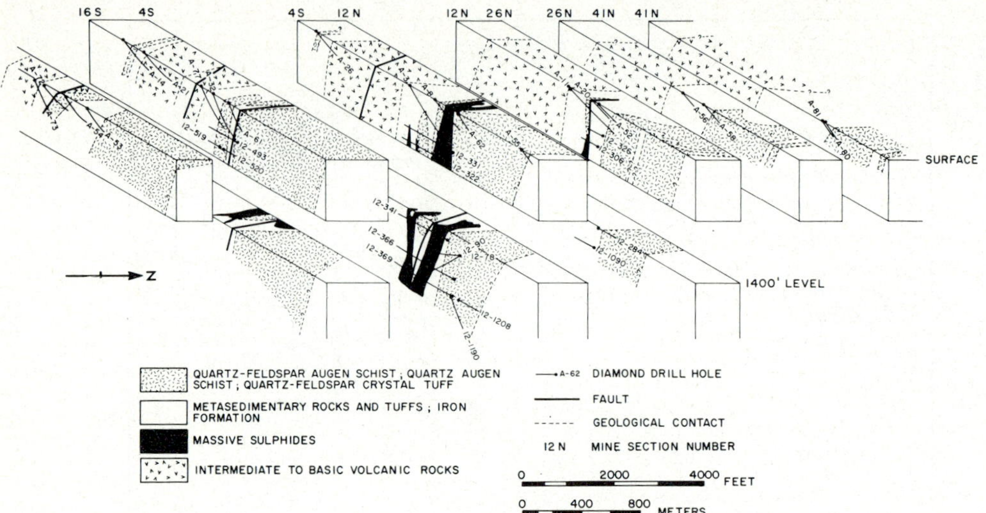

Fig.2. Block diagram illustrating the geology and diamond drill hole locations at Brunswick No.12 deposit.

Augen schist

The augen schist unit, which consists of quartz augen schist, quartz-feldspar augen schist, and crystal tuff, is believed to be of volcanic origin. The quartz-feldspar augen schist is composed of euhedral to subhedral crystals of plagioclase and quartz, set in a fine ground mass of quartz, chlorite, and sericite. Carbonate assemblages are common in hydrothermally altered areas associated with the sulphide deposit. Plagioclase feldspars, which have a mineralogical composition of $Ab_{95}An_5$ (Stewart, 1954) and exhibit albite twinning, range in size from microscopic to 2 cm in length and show no preferred orientation. Quartz appears much more uniform in size, ranging up to 1 cm in length.

The crystal tuff is composed of essentially the same mineralogical assemblages as the quartz-feldspar augen schist, and it is believed to be genetically related. However, the quartz and plagioclase crystals are not subhedral; they occur as irregular and shattered crystal fragments of generally finer grain size. This rock unit, which is located stratigraphically above the quartz-feldspar augen schist, is characteristic of the late steam explosive stage of volcanic activity similar to that postulated for the Kosaka district of Japan (Horikoshi, 1969).

Chloritic sericitic schist and iron formation

Chloritic sericitic schist is abundant in the immediate area of the ore deposit. On the footwall, the unit becomes much more chloritic and is impregnated with pyrite disseminations and stringers with minor chalcopyrite, sphalerite, and galena. On the hanging wall, it is generally less chloritic and more sericitic and contains only minor sulphide mineralization. Microscopically, it appears to be of sedimentary or tuffaceous origin and is composed of fine-grained aggregates of quartz, chlorite, and sericite. Chlorite and sericite may be concentrated either along cleavage planes, giving the rock a striped appearance, or uniformly distributed throughout the rock, producing a more uniform schistose fabric.

Iron formation, which overlies the massive sulphide and extends laterally north and south of the deposit, is rhythmically banded with siderite, magnetite, quartz, and chlorite. Directly over the sulphide deposit, significant concentrations of ore metals are present in the iron formation. Microscopically, the magnetite appears to replace siderite and probably originated from the decomposition of siderite during regional metamorphism.

GEOLOGY OF THE ORE

Brunswick No.12 deposit contains at least 82×10^6 tons of massive sulphides containing 9.69% Zn, 3.77% Pb, 0.29% Cu, and 2.46 oz. Ag per ton (Davies and Smith, 1966); the main sulphide minerals are pyrite, sphalerite, galena, pyrrhotite, chalcopyrite, tetrahedrite, and bornite. The No.12 main zone is a generally lenticular mass, dipping at 75° to the west and plunging steeply to the south. With depth, the size of the pyrite body increases with a corresponding decrease in sphalerite and galena. The west ore zone is also of lenticular mass outline; it dips steeply to the west to a depth of approximately 1900 ft, where a flexure occurs and the dip becomes vertical. Consequently, the main and the west ore zones converge at depth.

Both the main and the west ore zones exhibit similar trends in metal zoning, with copper having its highest tenor in the footwall and south end of the deposit, and lead and zinc on the hanging wall and north end of the deposit. Thus, both a vertical and a lateral zoning of copper, lead, and zinc is present in the deposit.

ALTERATION

Mineralogical alteration consists of chloritization, sericitization, and silicification of the sedimentary and volcanic rocks immediately associated with the ore deposit and extending up to 200 ft into the footwall volcanic rocks. Mineralogical alteration extending laterally along the ore horizon is also localized to the immediate area of sulphide mineralization; this is referred to as the "Intense Alteration Zone". In this zone the euhedral to subhedral plagioclase feldspars characteristic of the less intensely altered quartz-feldspar augen schist are ragged and in certain areas almost completely obliterated, except for ghost structures of fine-grained platy sericite.

In the stockwork area, which is characterized by pyrite and pyrrhotite stringers and by intense metasomatic hydrothermal alteration, even these ghost structures are absent. At the Boliden sulphide deposit in Sweden, Nilsson (1968) noted that sericite directly replaces all the primary silicate minerals, and the plagioclase of the fresh acid volcanic rocks is destroyed towards the centre of alteration. Alteration, consisting of chloritization, sericitization, and silicification of the host rocks associated with massive sulphide deposits, has also been described at the Lake Dufault deposit (Sakrison, 1966), the Matagami Lake mine (Descarreaux, 1973), the Millenbach deposit (Simmons, 1973), and at Japanese deposits (Tatsumi and Clark, 1972).

A silicified zone, consisting of highly siliceous sedimentary rocks and tuffs and associated sulphide mineralization along stringers and in disseminations, is located directly below the Brunswick No.12 deposit. This zone has its greatest thickness at the south end of the deposit, where it appears spatially associated with the Cu-rich pyrite-pyrrhotite zone of the ore deposit. This zone of intense silicification is localized in the contact areas between the sulphides and footwall sedimentary rocks and tuffs, and extends up to 100 ft into the footwall sedimentary rocks and tuffs. Intense silicification in the underlying pyroclastic acid volcanic rocks is not megascopically apparent; geochemical variations of silica in these rocks have not been considered in this paper.

CHEMISTRY OF THE ACID VOLCANIC ROCKS

Analytical procedures

Rock samples of quartz-feldspar augen schist, quartz augen schist, and crystal tuff were taken from diamond drill core through acid volcanic rocks associated with Brunswick No.12 deposit and from volcanic rock units located as far as 2 miles distant from the deposit along the same stratigraphic horizon and as far as 2000 ft stratigraphically below and above the deposit. The samples, each consisting of approximately 6 inches of core, were crushed and ground to minus 80-mesh size, digested using a hydrofluoric-perchloric digestion technique (Langmyhr and Paus, 1968a, b), and analysed with a Perkin Elmer 306 atomic absorption spectrophotometer for Al, Ca, Mg, Na, K, Mn, and Fe. A solution containing 0.1% lanthanum and potassium chloride (S. Pajari, University of New Brunswick Exploration Geochemistry Group, personal communication, 1973) was added to each sample analysed for calcium, magnesium, and aluminium to suppress the ionization of aluminium in a nitrous oxide flame and to reduce the interference of aluminium on calcium and magnesium. The accuracy and precision for triplicate analysis of standard rocks G-2, AGV-1, and BCR-1 are presented in Table II with ranges and average values collected and recorded by Flanagan (1969). All element concentrations for the different standard rocks fell within the reported ranges and, except for aluminium, which was consistently low, the element concentrations approached the reported average values.

Distribution of major and minor elements

Table III, which gives the means and coefficients of variation for element concentrations in the different alteration zones, illustrates the distribution of

TABLE II

Results of analyses of U.S.G.S. standard rocks with reported values

Standard rock	Oxide	Reported values (%)			Replicate analysis (%) (this paper)		
		recommended	minimum	maximum			
G-2	Al_2O_3	15.34	14.64	15.98	14.63	14.72	14.63
	CaO	1.98	1.80	2.30	1.92	1.90	1.92
	MgO	0.78	0.35	1.08	0.76	0.76	0.78
	MnO	0.03	0.03	0.08	0.025	0.025	0.025
	Fe_2O_3	2.76	2.61	2.90	2.62	2.62	2.62
	Na_2O_3	4.15	3.80	4.75	4.21	4.21	4.27
	K_2O	4.51	4.16	5.10	4.66	4.60	4.64
AVG-1	Al_2O_3	17.01	15.78	17.65	16.33	16.72	16.91
	CaO	4.98	4.44	5.98	4.76	4.76	4.76
	MgO	1.49	0.92	2.01	1.49	1.47	1.47
	MnO	0.09	0.05	0.12	0.089	0.090	0.085
	Fe_2O_3	6.80	6.62	7.05	6.77	6.81	6.81
	Na_2O	4.33	4.04	4.84	4.48	4.35	4.37
	K_2O	2.89	2.78	3.28	3.01	3.02	3.01
BCR-1	Al_2O_3	13.65	12.32	14.15	13.49	13.30	12.83
	CaO	6.95	6.14	8.33	6.59	6.54	6.74
	MgO	3.49	1.93	3.81	3.47	3.42	3.45
	MnO	0.17	0.11	0.23	0.178	0.171	0.174
	Fe_2O_3	13.50	13.20	14.40	13.26	13.15	13.30
	Na_2O	3.31	3.10	3.84	3.57	3.44	3.42
	K_2O	1.68	1.49	1.82	1.72	1.72	1.74

All reported analyses from Flanagan (1969) except recommended value for MgO in BCR-1 which is from Abbey (1973)

elements in the acid pyroclastic volcanic rocks located stratigraphically below and above Brunswick No.12 massive sulphide deposit. The three major zones outlined are the Intensely Altered Zone, characterized by a stockwork of pyrite stringers and dissemination, and consequently very high iron; an intermediate geochemical Halo Zone (less intensely altered zone); and an outer zone representing background or semi-background concentrations for the different elements. The spatial distribution of the zones of alteration in relation to the ore deposit are schematically illustrated in Fig.3.

A plot of the Mg/Ca ratio in footwall volcanic rocks (Fig.4) illustrates a broad dish-shaped geochemical Halo Zone extending up to 1500 ft stratigraphically below the deposit and 1500 ft laterally both north and south of the deposit; a mineralogical-geochemical Intensely Altered Zone is located up to 200 ft stratigraphically below the deposit and within the lateral range of the deposit.

TABLE III

Means and coefficients of variation of element concentrations in the Intensely Altered Zone, geochemical Halo Zone and background areas

Zones and areas	Characteristic mineral assemblage	Number of samples	Al%		Ca%		Mg%		Mn%		Fe%		Na%		K%	
			m	cv	m	cv	m	cv	m	cv	m	cv	m	cv	m	cv
Intensely Altered Zone and associated stockwork zone	chlorite, sericite, sulphides, quartz, minor plagioclase feldspar	54	8.010	0.150	0.373	0.484	1.732	0.533	0.055	1.011	5.416	0.681	0.478	0.959	2.650	0.454
Geochemical Halo Zone (less intensely altered)	chlorite, sericite, quartz, plagioclase feldspars, minor sulphides	53	8.193	0.149	0.378	0.722	2.079	0.560	0.043	0.711	3.011	0.563	1.147	0.356	3.228	0.309
Background	chlorite, sericite, plagioclase feldspar, quartz	29	8.190	0.129	1.068	0.789	1.221	0.413	0.051	0.727	2.765	0.692	1.062	0.692	3.712	0.208

m = mean, cv = coefficient of variation.

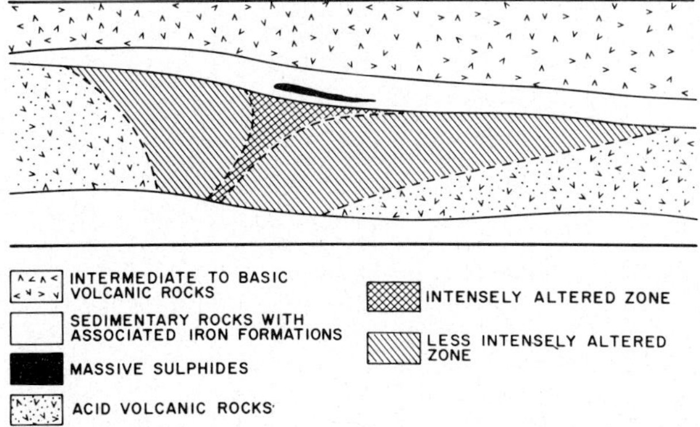

Fig.3. Schematic illustration of geology and spatial distribution of zones of alteration below Brunswick No.12 deposit.

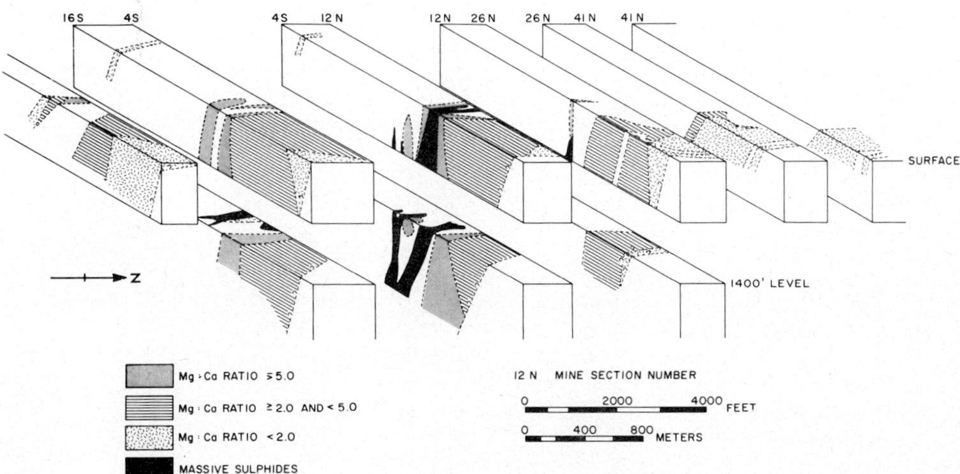

Fig.4. Spatial distribution of the Mg/Ca ratio in the acid volcanic rocks below Brunswick No.12 deposit.

The variation of major and minor elements with proximity to the ore deposit is illustrated in Fig.5. The geochemical relations among the elements Na, Ca, K, Mg, and Fe in the altered and unaltered volcanic rocks are further illustrated on a ternary plot (Fig.6). Although arbitrary boundaries were drawn between the different zones, the transition from intensely altered volcanic rocks to unaltered volcanic rocks is gradational.

Aside from aluminium, which shows no apparent variation with proximity to the sulphide deposit (fluctuations are attributed to sample variation), the other elements investigated show varying patterns of distribution as the ore body is approached:

(1) Magnesium increases in concentration in the broad geochemical Halo

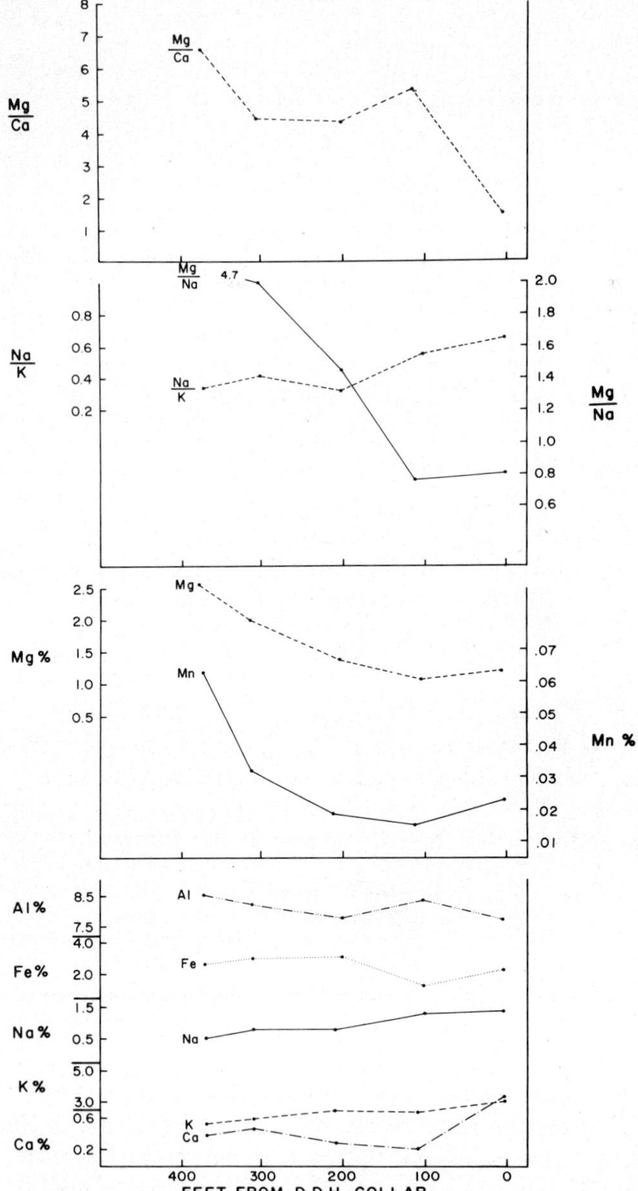

Fig.5. Variation of major and minor elements in the Intensely Altered Zone with proximity to the ore deposit in DDH 12-78 (see Fig.2 for location of drill holes).

Zone; it has a lower mean concentration in the Intensely Altered Zone, but generally increases in concentration as the ore deposit is approached.

(2) Iron increases in concentration in the Halo Zone, and there is an even more marked increase in concentration in the Intensely Altered Zone.

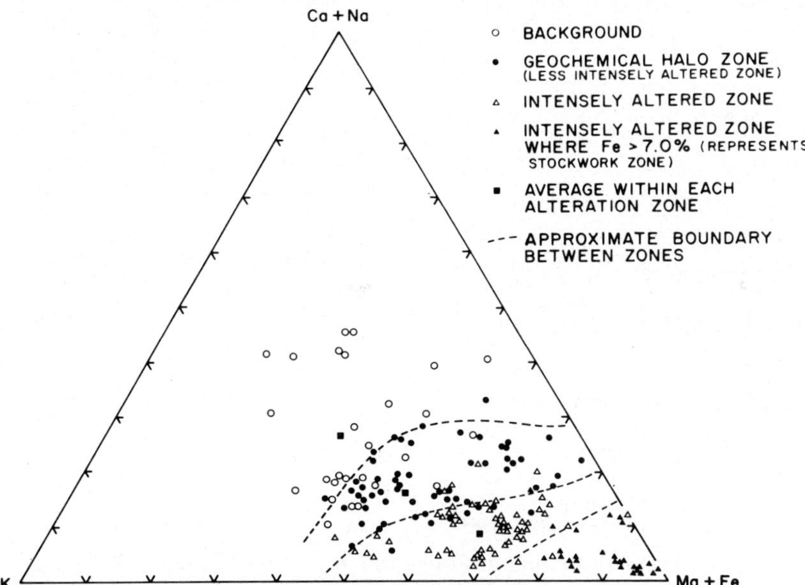

Fig.6. Ternary plot of Na + Ca, K, and Mg + Fe illustrating the chemical variation between the Intensely Altered Zone, geochemical Halo Zone, and Background Area.

(3) The concentration of manganese decreases slightly, relative to background levels, in the Halo Zone, but it shows an overall increase in concentration in the Intensely Altered Zone where concentrations increase as the ore deposit is approached.

(4) Calcium decreases markedly in the Halo Zone and in the Intensely Altered Zone relative to background concentrations. Within the Intensely Altered Zone, calcium decreases towards the ore deposit.

(5) Sodium increases slightly in concentration in the Halo Zone but decreases markedly in the Intensely Altered Zone as the ore deposit is approached.

(6) The concentration of potassium generally decreases in the Halo Zone and in the Intensely Altered Zone compared with background concentrations. In the Intensely Altered Zone, which also has a lower potassium mean concentration than the Halo Zone, the potassium concentration may increase or decrease towards the deposit.

The Na/K ratio consistently decreases towards the deposit. The absolute decrease of both potassium and sodium in the Intensely Altered Zone may have resulted from an extensive leaching by hydrothermal solutions. The distribution of potassium and sodium, illustrated on a binary diagram (Fig.7), outlines the three major zones noted above. The volcanic rocks stratigraphically above the deposit belong to the unaltered class.

Major and minor element distribution patterns in altered rocks below massive sulphide deposits, similar to those present at Brunswick No.12 deposit, have been described by Goodwin (1964), Sakrison (1966), Nilsson (1968),

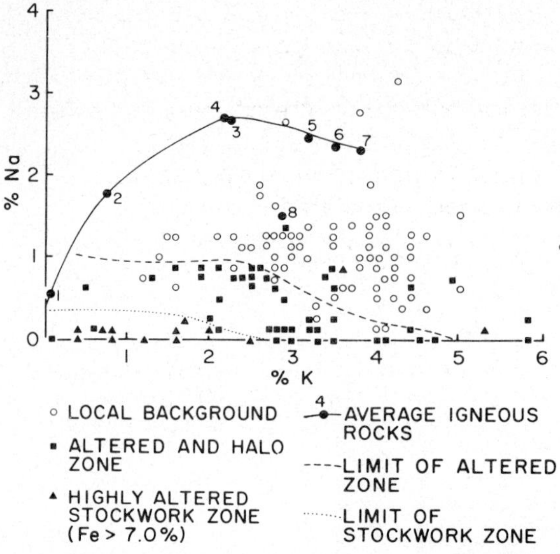

Fig.7. Binary plot of potassium and sodium outlining the Intensely Altered Zone, geochemical Halo Zone, and Background Area: 1 = ultramafic rocks average (Krauskopf, 1967); 2 = mafic rocks average (Krauskopf, 1967); 3 = dacite average (Daly, 1933); 4 = intermediate rocks (Krauskopf, 1967); 5 = felsic rocks (Krauskopf, 1967); 6 = granite average (Daly, 1933); rhyolite average (Daly, 1933).

TABLE IV

Mineralogy and geochemistry of rock units associated with some ore deposits

Deposit	Mineralogical assemblage	Elements added	Elements removed	Elements unchanged
Millenbach (Simmons, 1973)	chlorite, sericite, or anthophyllite	Mg, Fe	Na, Ca, Si	
Abitibi Volcanic Belt (Descarreaux, 1973)	chlorite, sericite, quartz	Mg, K	Na, Ca	Si
Killingdal (Rui, 1973)	chlorite, biotite, quartz	Mg, K, Mn	Na, Ca, Si	Al, Ti, Fe total
Kuroko (Tatsumi and Clark, 1972)	sericite, quartz, or calcite	Mg, K	Na, Ca, Fe	Al
Boliden (Nilsson, 1968)	chlorite, sericite, quartz, andalusite	Mg, K, Al, Si, Ti	Na, Ca	
Lake Dufault (Sakrison, 1966)	chlorite, sericite	Mg, Fe, Mn	Na, Ca	Al, Ti, K, Si
Helen Iron Range (Goodwin, 1964)	sericite, chlorite, carbonate, quartz, zoisite, chloritoid	Mg, Fe, Mn	Na, K, Si	
Hitachi, Japan (Kuroda, 1961)	anthophyllite and/or cordierite	Mg, Fe, Ba	Na, Ca, Sr	
Brunswick No.12	chlorite, sericite quartz	Mg, Fe, Mn, K	Na, Ca	Al

Gale (1970), Descarreaux (1972), Tatsumi and Clark (1972), and Simmons (1973). The generalized trends in element distribution are summarized in Table IV and below: (1) magnesium, manganese and iron are markedly concentrated in the zone of alteration compared with unaltered areas; (2) sodium and calcium show a marked concentration decrease in the alteration zones; and (3) potassium is generally more concentrated in the altered zones.

Aluminium concentrations in altered zones associated with massive sulphide deposits appear to vary from one deposit to another. Nilsson (1968) noted that aluminium markedly increases in concentration in the zone of alteration present at the Boliden deposits. However, in a study of the Lake Dufault mines, Sakrison (1966) noted that aluminium shows only minor variations in the altered zone and is considered to be chemically stable. Variations of aluminium trends may reflect inherent differences in the development of zones of alteration associated with massive sulphide deposits.

DISCUSSION

The distribution of major and minor elements in the footwall volcanic rocks at Brunswick No.12 sulphide deposit suggests that the sequence of pyroclastic volcanic rocks, consisting of readily leachable glass shards and crystal fragments, has been subjected to metasomatic hydrothermal alteration, probably from late-stage metal-bearing fumarolic fluids, similar to those described for the Kuroko-type deposits of Japan (Horikoshi, 1969). The late period of pyroclastic volcanism is characterized by an explosive stage, caused by the rapid escape of gases, where the euhedral to subhedral crystals, characteristic of the earlier quartz-feldspar porphyry, have been shattered and brecciated and deposited as fragmental crystal tuff. This explosive stage could supply the necessary channelways for the escape of S-rich metal-bearing hydrothermal solutions. Because of the primary porosity and permeability of the pyroclastic volcanic sequence, these fluids could diffuse laterally, (chemically) leaching available Na and Ca from glass and plagioclase feldspars and simultaneously (metasomatically) depositing magnesium and potassium in chlorite and sericite mineral assemblages. The concentration of metasomatic hydrothermal solutions will vary with: (1) proximity to the escape channelways of hydrothermal solutions; (2) changes in the source area (whether it be a volcanic magma chamber with a distinct fluid phase or merely a heat source to recirculate connate waters down thermal gradients); and (3) permeability which will be significantly affected by the formation of new metasomatic mineral assemblages and the chemical precipitation of minerals, such as metal sulphides, in solution conduits. In the zone of intense alteration the occurrence of abundant stringers of pyrite and pyrrhotite with generally minor chalcopyrite, galena, and sphalerite is similar to stockwork zones below sulphide deposits most recently described by Constantinou and Govett (1973) and Simmons (1973).

Marimoto and Osaka (1955), in their observation of present-day submarine

292

eruptions, noted that the activity of S-rich hydrothermal fluids is most intense immediately before or after the breakdown of the lava dome, suggesting a genetic time relation between the source of the hydrothermal fluids and volcanic activity. Whitehead (1973), in a study of the distribution of manganese and iron in a sedimentary basin during sulphide deposition, suggested that a decrease in the activity of sulphur in the waning stages of fumarolic activity caused a rapid change from iron sulphide deposition to iron carbonate deposition. Sulphur isotope ratios of the Bathurst district (Tupper, 1960) further support the hydrothermal source of sulphur.

Stratigraphically below and laterally further from the zone of intense mineralogical alteration, any mineralogical assemblages resulting from metasomatic alteration by hydrothermal fluids may have been masked or indeed destroyed by later greenschist metamorphism exhibiting similar mineralogical assemblages (Schermerhorn, 1970; Roscoe, 1971). Nevertheless, the chemical differences in element distribution are no less significant in outlining a broad geochemical alteration aureole.

While the problem of outlining the lateral extent of a geochemical alteration halo in the footwall acid volcanic sequence is complicated by poor diamond drill hole control in an area south of the Brunswick No.12 deposit, good diamond drill hole control north of the deposit makes it possible to extrapolate the distribution of the halo zone south of the deposit. Further complications resulting from structural deformation by low angle faulting causes repetition of the stratigraphic sequence in an area of the mine and a distortion as well as a displacement of the alteration and associated stockwork zones. Since these complications are most significant on a local scale only in the immediate area of the mine, the broad geochemical halo in the zone of less intense alteration is little affected. Because of the spatial relation between sulphide mineralization and geochemical halos, any repetition of the ore horizon will also cause a repetition of the alteration zones (Fig.2) and consequently increase the size of an exploration target.

In this study a sufficient number of samples were available to establish apparent background concentrations for the different elements. It was, therefore, possible to outline a geochemical halo zone below the Brunswick No.12 deposit, where the differences in major and minor element concentrations are large enough to distinguish it from relatively background concentrations.

The distribution of geochemical alteration zones stratigraphically below Brunswick No.12 deposit suggests that the sulphide mineralization was deposited from late-stage metal-rich solutions shortly after a period of volcanic activity in which great thicknesses of pyroclastic volcanic rocks were deposited. It is the opinion of the writer that this distribution is totally consistent with the syngenetic theory of deposition of massive sulphides, and that it should not be construed to support an epigenetic theory of sulphide deposition.

CONCLUSIONS

The recognition of major and minor element halos in the footwall volcanic rocks of the Brunswick No.12 deposit has several applications useful to exploration in the Bathurst area. These are:

(1) To determine the stratigraphic top of the volcanic rocks associated with massive sulphide deposits. The alteration zone is confined to the footwall side and increases in intensity as the ore deposit is approached.

(2) To outline significantly larger targets in geochemical exploration for massive sulphide deposits. A major and minor element halo of 1500 ft × 3000 ft in size is much larger than that outlined by mineralogical alteration assemblages in the immediate area of the deposit.

(3) To establish a close genetic and spatial relationship between the source of metals from late stage volcanic activity and the formation of massive sulphide deposits. The confinement of the alteration halo to the footwall volcanic sequence suggests that the sulphides were deposited after the deposition of the footwall volcanic sequence, and before the deposition of hanging wall acid volcanic rocks.

(4) To locate favourable stratigraphic horizons in the exploration for unknown sulphide deposits and to locate extensions of the zones of mineralization around known ore deposits. This is a corollary of (2) and might be most useful in the regional evaluation of a volcanic belt.

ACKNOWLEDGEMENTS

This study was supported by Brunswick Mining and Smelting Ltd., by an International Nickel Company of Canada Graduate Fellowship, and by the National Research Council of Canada Grant No.A-8858 awarded to G.J.S. Govett to develop geochemical methods, useful in the exploration for deeply buried massive sulphide deposits. This paper is part of a thesis being prepared as a partial requirement for a Ph.D. degree at the University of New Brunswick. Special thanks are extended to Professor G.J.S. Govett, research supervisor, for many critical discussions and assistance in preparing the manuscript. Mr. R.N. Smith, Chief Geologist, and Mr. D. Rutledge, Research Geologist, both of Brunswick Mining and Smelting Ltd. are thanked for their informative discussions and assistance. Thanks are also extended to Dr. R.E.S. Whitehead, Assistant Professor, Laurentian University for stimulating discussions during his tenure as Research Associate at the University of New Brunswick. Grateful acknowledgement is made to Mr. R. Phillips of the University of New Brunswick for preparing the diagrams and to Mrs. M.H. Govett for editorial assistance.

REFERENCES

Abbey, S., 1973. Studies in "standard samples" of silicate rocks and minerals, 3. 1973 extension and revision of "usable values". Geol. Survey Can. Paper 73-36, 19 pp.

294

Alcock, F.J., 1941. Jacquet River and Tetagouche River map-areas, New Brunswick. Geol. Surv. Can. Mem. 227'

Boyle, R.W., 1965. Origin of the Bathurst-Newcastle sulfide deposits, New Brunswick. Econ. Geol., 60: 1529—1532

Boyle, R.W., 1969. Further remarks on the origin of massive sulfide deposits. Econ. Geol., 64: 829—830

Boyle, R.W. and Davies, J.L., 1964. Geology of Austin Brook and Brunswick No.6 sulphide deposits, Gloucester County, New Brunswick. Geol. Surv. Can. Paper 63-24, 23 pp.

Constantinou, G. and Govett, G.J.S., 1972. Genesis of sulphide deposits, ochre and umber of Cyprus. Trans. Inst. Min. Metall. (London), 81: B34—B46

Constantinou, G. and Govett, G.J.S., 1973. Geology, geochemistry, and genesis of Cyprus sulfide deposits. Econ. Geol., 68: 843—858

Daly, R.A., 1933. Igneous Rocks and the Depth of the Earth. McGraw-Hill, New York, N.Y.

Davies, J.L. and Smith, R.N., 1966. Geology of the Brunswick No.6 and No.12 sulphide deposits. Geol. Assoc, Can. Guideb., pp.45—47

Davis, G.H., 1972. Deformational history of the Caribou stratabound sulfide deposit, Bathurst, New Brunswick, Canada. Econ. Geol., 67: 634—655

Descarreaux, J., 1973. A petrochemical study of the Abitibi volcanic belt and its bearing on the occurrences of massive sulfide ore. Can. Inst. Min. Metall. Bull., 66: 61—69

Flanagan, F.J., 1969. U.S. Geological Survey standards, II. First compilation of data for the new U.S.G.S. rocks. Geochem. Cosmochim. Acta, 33: 81—120

Gale, G.J., 1970. The primary dispersion of Cu, Zn, Ni, Co, Mn, and Na adjacent to sulphide deposits, Springdale Peninsula, Newfoundland. M.Sc. Thesis, Memorial University of Newfoundland.

Goodwin, A.M., 1964. Geochemical studies at the Helen Iron Range. Econ. Geol., 59: 684—718

Govett, G.J.S., 1972. Interpretation of a rock geochemical exploration survey in Cyprus — statistical and graphical techniques. J. Geochem. Explor., 1: 77—102

Govett, G.J.S. and Pantazis, Th.M., 1971. Distribution of Cu, Zn, Ni and Co in the Troodos Pillow Lava Series, Cyprus, Trans. Inst. Min. Metall. (London), 80: B27—B46

Helmstaedt, H., 1970. Structural geology of Portage Lakes area, Bathurst-Newcastle district, New Brunswick. Geol. Surv. Can. Paper 70-28, 42 pp.

Horikoshi, E.I., 1969. Volcanic activity related to the formation of the Kuroko-type deposits in the Kosaka district, Japan. Mineral. Deposita (Berl.), 4: 321—345

Krauskopf, K.K., 1967. Introduction to Geochemistry. McGraw-Hill, New York, N.Y. 721 pp.

Kuroda, Y., 1961. Minor elements in a metasomatic zone related to a copper-bearing pyrite deposit. Econ. Geol., 56: 847—854

Langmyhr, F.J. and Paus, F.E., 1968a. The analysis of the inorganic siliceous materials by atomic absorption spectrophotometry and the hydrofluoric acid decomposition technique. Anal. Chem. Acta, 43: 397—408

Langmyhr, F.J. and Paus, F.E., 1968b. Hydrofluoric acid decomposition — atomic absorption analysis of inorganic siliceous materials. At. Absorption Newsl., 7: 103—106

Lea, E.R. and Rancourt, C., 1958. Geology of Brunswick Mining and Smelting ore bodies, Gloucester County. Can. Inst. Min. Metall. Bull., 51: 167—177

Luff, W., 1973. Structural geology of Brunswick No.12 open pit. M.Sc. Thesis, University of New Brunswick, Fredericton, N.B.

Marimoto, R. and Osaka, J., 1955. The 1952—1953 submarine eruption of the Myojin reef near the Bayonaise rocks, Japan. Earthquake Res. Inst. Bull., 33: 221—250

McAllister, A.L., 1960. Massive sulphide deposition in New Brunswick. Can. Inst. Min. Metall. Bull., 53: 88—98

Nilsson, C.A., 1968. Wall rock alteration at the Boliden deposit, Sweden. Econ. Geol., 63: 472—494

Roscoe, W.E., 1971. Geology of the Caribou deposit, Bathurst, New Brunswick. Can. J. Earth Sci., 8: 1125—1136

Rui, I.J., 1973. Structural control and wall rock alteration at Killingdal Mine, Central Norwegian Caledonides. Econ. Geol., 68: 859—883

Rutledge, D.W., 1972. Brunswick Mining and Smelting Corporation, No.6 and No.12 mines. In: A.L. McAllister and R.Y. Lamarche (Editors), Mineral Deposits of Southern Quebec and Northern New Brunswick. 24th Int. Geol. Congr., Field Excursion A58-C58 Guidebook.

Sakrison, H.C., 1966. Chemical studies of the host rocks of the Lake Dufault Mines, Quebec. Ph.D. Thesis, McGill University, Montreal, Ont.

Schermerhorn, L.J.G., 1970. The deposition of volcanics and pyritite in the Iberian pyrite belt. Mineral. Deposita (Berl.), 5: 273—279

Simmons, B.D., 1973. Geology of the Millenbach massive sulphide deposit, Noranda, Quebec. Can. Inst. Min. Metall. Bull., 66: 67—78

Smith, C.H. and Skinner, R., 1958. Geology of Bathurst-Newcastle mineral district, N.B. Can. Inst. Min. Metall. Bull., 51: 150—155

Stewart, K.J., 1954. Geology of Brunswick Mine, Gloucester County, New Brunswick. M.Sc. Thesis, University of New Brunswick, Fredericton, N.B.

Tatsumi, T. and Clark, L.A., 1972. Chemical composition of acid volcanic rocks genetically related to formation of the Kuroko deposits. J. Geol. Soc. Japan, 78: 191—201

Tupper, W.M., 1960. Sulfur isotopes and the origin of sulfide deposits of the Bathurst-Newcastle area of northern New Brunswick. Econ. Geol., 55: 1676—1707

Whitehead, R.E.S., 1973. Environment of stratiform sulphide deposition; variation in Mn/Fe ratio in host rocks at Heath Steele Mine, New Brunswick, Canada. Mineral. Deposita (Berl.), 8: 148—160

CHEMICAL ZONING ASSOCIATED WITH THE INGERBELLE—COPPER MOUNTAIN MINERALIZATION, PRINCETON, BRITISH COLUMBIA

J.E. GUNTON and IAN NICHOL

Department of Geological Sciences, Queen's University, Kingston, Ont. (Canada)

ABSTRACT

Zoning of mineral assemblages reflecting alteration of the cooling intrusive phase and adjacent wall rocks has been shown to be associated with mineralization in many porphyritic intrusions. These alteration zones are normally broader than the mineralization and thus constitute a larger exploration target for porphyry-type mineralization. The delineation of these zones normally has been defined on a mineralogical basis, but rarely on a chemical basis. The chemical zoning of major, minor and trace elements associated with the alteration zones adjacent to the Ingerbelle—Copper Mountain ore bodies is described. An evaluation is made of the extent to which these chemical alteration zones constitute a broader exploration target that may be useful in mineral exploration. The mineral deposits in Nicola volcano-sedimentary stratigraphy associated with the Copper Mountain Intrusions most closely resemble porphyry-type copper deposits. Deposits of a similar type occur in the Intermontane Belt of British Columbia and thus there is a potential for locating further deposits. The Nicola formations adjacent to the Copper Mountain Intrusions were sampled together with a background area of similar original lithology. The samples were analyzed by X-ray fluorescence and atomic absorption for eighteen major, minor and trace elements. The most notable feature from a reconnaissance exploration viewpoint is the association of increased contents of phosphorus, rubidium and strontium and to a lesser extent soda and potash in the Nicola formations intruded by the Copper Mountain Stock. These features constitute parameters for the recognition of potentially mineralized areas of altered Nicola rocks associated with alkalic igneous centres.Considering more local chemical zoning associated with mineralization, the Nicola formations north of the stock, containing the Ingerbelle and Copper Mountain deposits, have higher strontium and copper contents than less intensely mineralized formations south of the stock. Variation in the nature of alteration associated with economic mineralization is indicated by a broad zone of soda enrichment related to the Ingerbelle deposit as opposed to localized zones of potash enrichment associated with the Copper Mountain deposits. This chemical zoning is considered to reflect varying degrees and types of metasomatism and mineralization. The distribution of the chemical zoning with respect to areas of discrete mineralization in some cases constitutes larger and thus more attractive detailed scale exploration targets than the more localized areas of mineralization.

INTRODUCTION

The usefulness of primary geochemical dispersion patterns has been demonstrated in the search for a variety of types of mineral deposits. These dispersion patterns have been found to be particularly well developed in rocks associated with the formation of mineral deposits related to igneous processes. Of particular interest in the exploration for porphyry-type deposits in British Columbia are the dispersion patterns related to high-level intrusive complexes. These complexes may be regarded as remnants of subvolcanic parent magma to volcanic centres from which sequences of lava were extruded and pyroclastic and sedimentary units derived. Mineralization in a geological setting of this type is often intimately associated with broader alteration patterns of the intrusive and surrounding host rock.

This alteration zoning frequently constitutes a larger exploration target than the discrete mineralization and thus offers a more easily detectable target in reconnaissance level exploration. In many cases alteration zoning can be readily distinguished by petrographic investigations or by geological mapping (Hausen and Kerr, 1971). In the case of the Ingerbelle—Copper Mountain deposits the complexity of the alteration is such that it is almost impossible to detect any meaningful zoning in the alteration that might serve to indicate mineralization. The present investigation was undertaken to establish whether it was possible to establish alteration zoning associated with the Ingerbelle—Copper Mountain deposits on the basis of bedrock geochemistry and to evaluate the extent to which this approach might contribute to the future search for similar types of deposits throughout the Intermontane Belt (Gunton, 1974).

DESCRIPTION OF AREA

The Copper Mountain area is located 9 miles south of Princeton in southern British Columbia (Fig.1). The area occurs at the southern extremity of

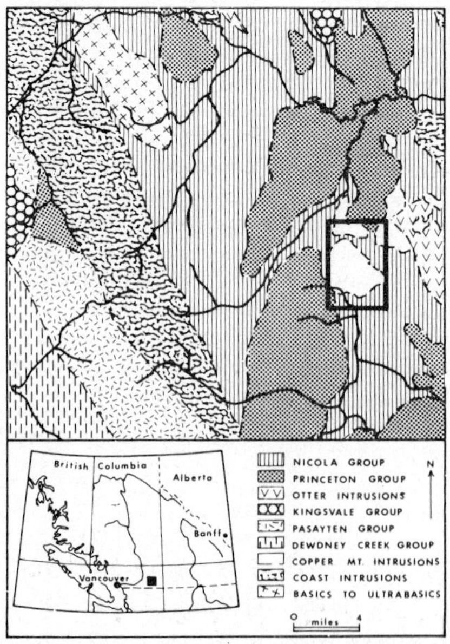

Fig.1. Area location and regional geology (after Rice, 1947).

the Intermontane physiographic subdivision of the Cordillera which extends
north-northwest to the east of the Alaska Panhandlc (Sutherland-Brown et
al., 1971). The Intermontane Belt is composed principally of Late Paleozoic,
Triassic and Jurassic eugeosynclinal volcanic and clastic rocks which have
been intruded by both Si-poor (alkalic-syenite clan) and Si-rich (calc-alkalic
clan) intrusive bodies. In places Tertiary continental volcanic rocks unconformably overlie the earlier formations. Low-grade disseminated and fracture-
filling copper mineralizations are often associated with these intrusive complexes and include deposits classified as porphyry, skarn and pyrometasomatic
types.

The Copper Mountain area is underlain by Triassic plagioclase and augite
andesite porphyry flows with coarse pyroclastics, tuffs and sediments of variable thickness (Preto, 1972). These lithologies comprise the Nicola Group
which has been intruded by Si-poor bodies, collectively called the Copper
Mountain complex. The two most important members of the intrusive complex are the Copper Mountain Stock and the Lost Horse Intrusions (Fig.2).
These have been dated at 193 ± 8 million years which indicates they only
slightly post-date the formation of the Nicola host rocks (Sinclair and White,
1968; Preto et al., 1971).

The Mine Series of the Nicola, bounded to the south by the Copper Mountain Stock and to the north by the Lost Horse Intrusions (Fig.2) (Macauley,
1973) contains all the known major copper deposits totalling about 76 million tons of 0.53% Cu in a 14,000 ft × 3500 ft belt. In the east of the belt
the Copper Mountain Pit 1 and Subsidence Zone deposits occur adjacent to
the stock while the Copper Mountain Pit 2 deposit occurs in contact with the
Lost Horse Intrusions (Fig.3). In the west of the belt, the Ingerbelle deposit
lies in contact with the Lost Horse Intrusions. The mineralization occurs as
dominantly disseminated and fracture filling chalcopyrite-pyrite with bornite
becoming important in rocks adjacent to the Copper Mountain Stock, Extensive pyrite haloes have only been identified south of the Ingerbelle deposit.

Two types of metasomatic mineral assemblages have been recognized with
K-rich assemblages occurring in the east of the belt and Na-rich assemblages
in the west (Preto, 1972; Macauley, 1973). Clinozoisite-epidote and apatite
occur throughout the Mine Series in above average concentrations compared
with Nicola rocks remote from the area. Alteration zones are difficult to
identify by mapping as a result of extreme local heterogeneity in alteration
type. This is in part due to the superposition of the effects of several phases
of alteration. Original lithological variability over small thicknesses and short
strike lengths may have contributed to the alteration heterogeneity by virtue
of the different permeabilities of the lithologies and their varying responses
to strain, giving rise to complex fracture patterns. Thus, the channelways
through which the hydrothermal fluids permeated would be irregular resulting
in local variation in alteration intensity and type. Nicola rocks of the Kennedy
Mountain Series lying east of the Boundary Fault (Fig.3) are believed to be
younger and hence less altered than the Mine Series.

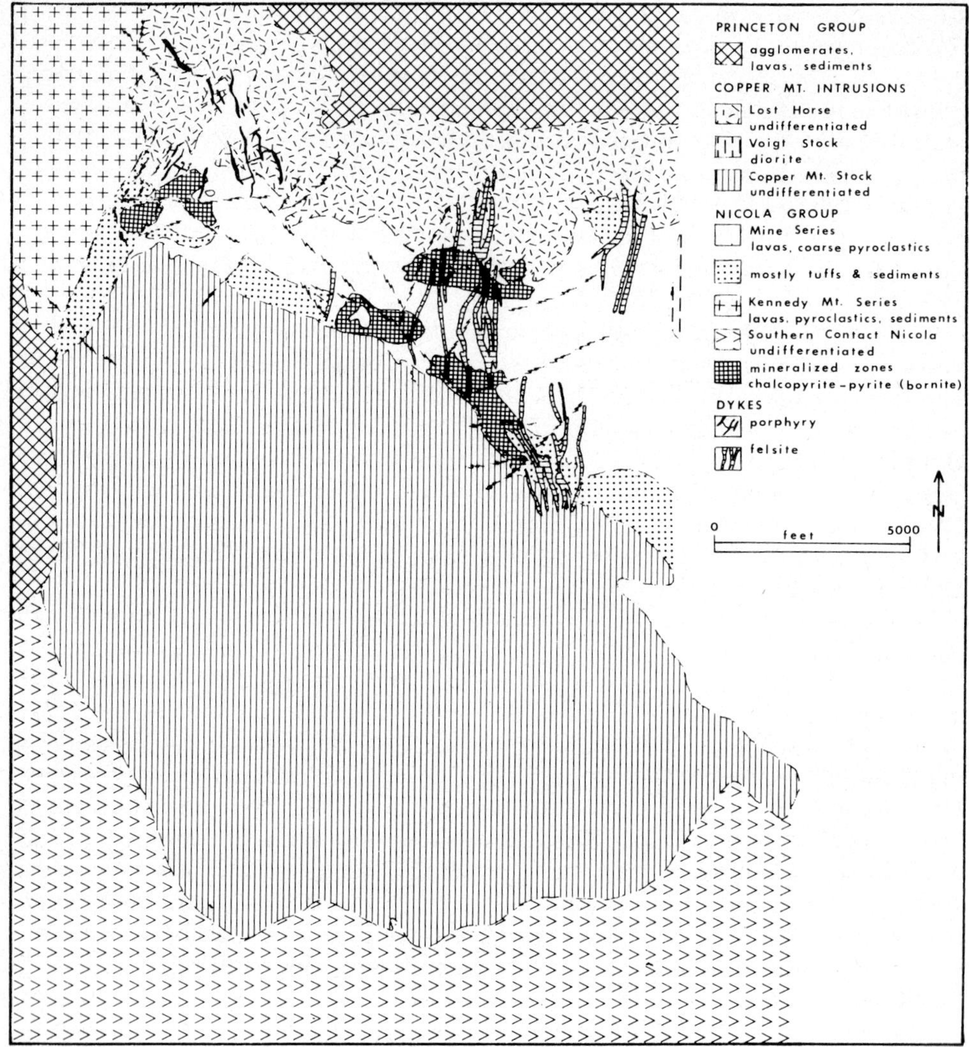

Fig.2. Geology of Copper Mountain area (after Preto, 1972).

APPROACH TO THE PROBLEM

An orientation study was carried out to establish the nature of metasomatic mineral assemblages and chemical variability as a basis for establishing appropriate sampling and analytical procedures for the determination of chemical zoning associated with the different deposits. On the basis of the results of this orientation survey, bedrock samples were collected over a 14,000 × 3500 ft^2 area of the Mine Series lying north of the Copper Mountain Stock and from Nicola rocks lying south of the stock (Fig.3). Background material was

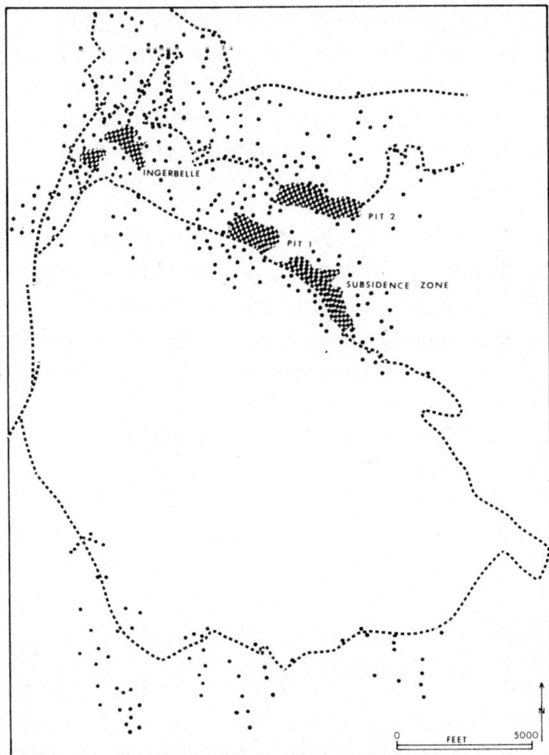

Fig.3. Sample location map.

sampled from Nicola Group rocks up to 15 miles from the intrusive complex and from the Kennedy Mountain Series. These latter two groups comprise background or "normal" Nicola lithologies devoid of any metasomatic mineral assemblages related to intrusive activity. These background samples formed the basis for a comparison between Nicola rocks south of the Copper Mountain Stock and mineralized and heavily altered Nicola rocks lying north of the stock. The gabbro-diorite differentiate of the Copper Mountain Stock was sampled along the northern and southern contact and samples were also taken from the Lost Horse Intrusions. Samples were collected from 500 ft × 500 ft grid cells. Insofar as was logistically possible, all the outcrops within each grid cell were chip sampled over the exposed surface. Every attempt was made to exclude weathered material; chips were taken randomly over the surface of the outcrop. A total of 443 samples were collected with an average sample weight of approximately 5 lbs.; each sample comprised an average of 20 chips. This approach was adopted in order to achieve representative material as a basis for identifying chemical zoning patterns related to the ore deposits.

All rock samples were first crushed to ½-inch chips using a "Braun-Chip-

munk" jaw crusher with steel plates followed by further crushing of each
sample to ¼-inch chips using a "Lemaire" jaw crusher with alumina plates.
In both stages the entire sample was crushed in order to attain maximum
representativity. A 5-g portion of the sample was obtained using a riffle
splitter prior to final grinding. The final grinding process was carried out
using a set of seven "Spex Industries Shatterbox" hardened steel grinding
vials.

The samples were analyzed for Si, Al, Fe (Fe_2O_3), Ca, Mg, K, Na, Ti,
P (P_2O_5), Mn (MnO), S, Sr and Rb by X-ray fluorescence. Total copper and
cobalt were determined following hot digestion with $4HClO_4/1HNO_3$ using
an atomic absorption unit. The copper, cobalt and iron contents of the sulphide mineral phases were determined following extraction with ascorbic
acid/hydrogen peroxide (Cameron et al., 1971). The precision of the analyses
for the major and minor elements was better than ± 10%, with the exception
of manganese and sulphur which were ± 15% at the 95% confidence level.
The analyses of elements determined by atomic absorption were within ±
25% precision at the 95% confidence level. For the most part the accuracy
of the analysis was within the precision limits.

MINERAL ASSOCIATIONS AND ZONING

Petrographic studies supplemented by feldspar-staining of slabbed specimens
and field observations confirmed the main alteration types and the variation
in intensity of this alteration in the Nicola lithologies associated with the
ore deposits. This work endorsed the findings of previous workers (Hausen,
1968; Macauley, 1970; Preto, 1972). Relative to Nicola rocks remote from
the Copper Mountain Intrusions, the Mine Series rocks contain large amounts
of albite, scapolite, apatite, zeolite and clinozoisite-epidote in the area of
the Ingerbelle deposit. Potash feldspar, biotite, apatite and clinozoisite-
epidote are found predominantly in the rocks associated with the Copper
Mountain deposits. Disseminated and fracture filling sulphides exist in above
average concentrations in the rocks immediately adjacent to the ore bodies
with a pronounced pyrite-pyrrhotite zone displayed south of the Ingerbelle
deposit. An area of bleached rocks extends north and east from the Ingerbelle
deposit bearing no relation to the Mine Series/Lost Horse Intrusion contact;
these rocks are lacking in ferromagnesian and sulphide minerals.

If these minerals are a result of metasomatism then enrichments in Na,
K, Ca, Sr, Rb and P and depletions of Fe, Mg, Co and Mn might be expected
relative to Nicola rocks remote from the Copper Mountain Intrusions. Variations in the distributions of these elements might, therefore, be spatially
related to the ore deposits in as much as the alteration process was associated
with mineralization and the element abundances reflect the proportion of
metasomatic minerals present in the rocks. Hence it would be expected
that the mineralogical changes, resulting from metasomatism, which have
been recognized by petrographic work and in field observations, would be

recognizable and to a certain extent quantified by the chemistry of the bedrock samples.

DISTRIBUTION OF ELEMENTS

Silica

The mean silica content of all areas underlain by Nicola rocks investigated is about 54% indicating no significant variation in silica content between areas (Table I). Within the northern contact zone an area of high silica (> 56%) occurs between the Ingerbelle and Copper Mountain deposits (Fig.4). The deposits occur in areas of slight silica depletion with respect to the overall mean content of 54% SiO_2.

Sodium

The mean soda contents of Nicola rocks adjacent to the north and south of the Copper Mountain Stock are higher (> 3%) than equivalent rocks remote from the stock (< 3%) (Table I). With reference to the Nicola for-

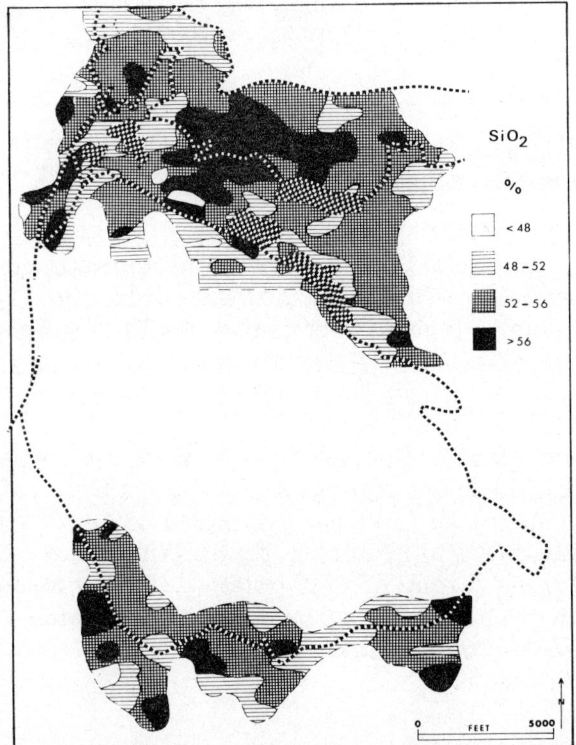

Fig.4. Silica distribution in the Ingerbelle—Copper Mountain area.

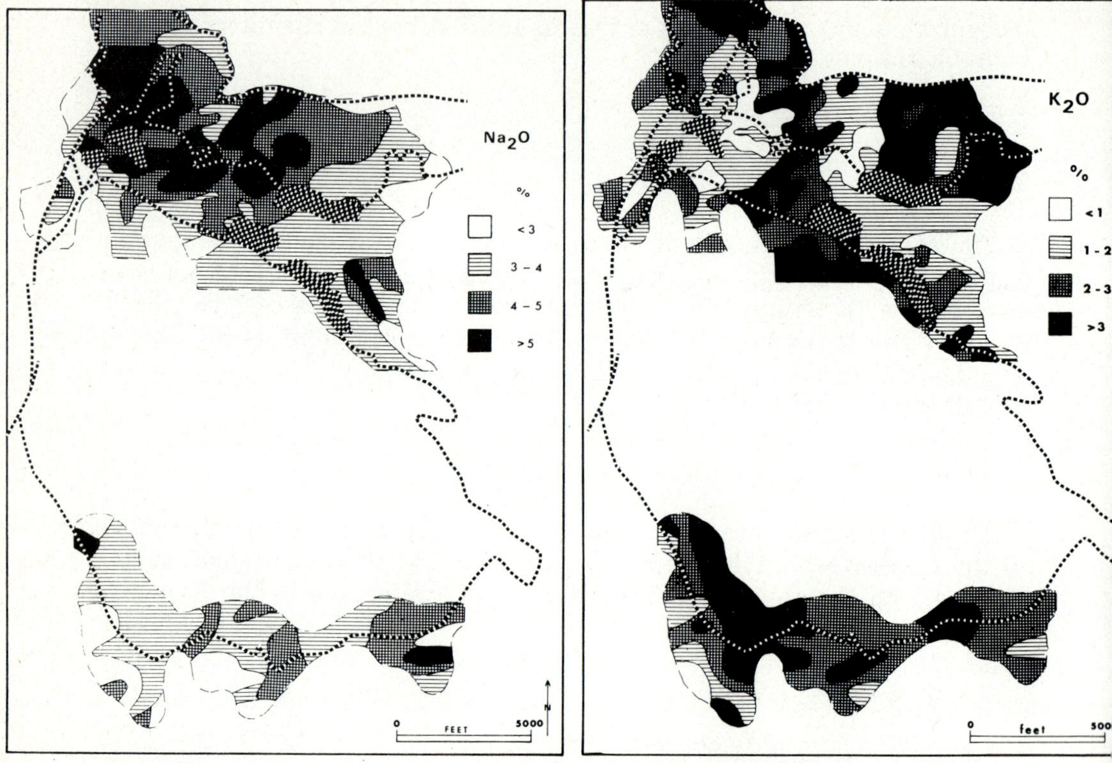

Fig.5. Sodium distribution in the Ingerbelle—Copper Mountain area.
Fig.6. Potassium distribution in the Ingerbelle—Copper Mountain area.

mations adjacent to the Copper Mountain Stock, a broad zone approximately 10,000 ft wide near the west central portion of the northern contact including the Ingerbelle deposit contains high soda contents (> 4%) (Fig.5).

Potassium

Slightly higher potash contents occur in the Nicola formations surrounding the Copper Mountain Stock than in background areas. The Nicola south of the stock has an overall higher content (2.7%) than that associated with the mineral deposits north of the stock (2.1%) (Table I) (Fig.6). Within the Nicola formations north of the stock a zone of high potash (> 2%) is associated with the Copper Mountain ore bodies and a zone some 2000 ft wide trends northeast in the central portion of the Mine Series belt.

Phosphorus

The mean phosphorus contents of Nicola rocks north and south of the Copper Mountain Stock are significantly higher (0.55 and 0.59%) than

TABLE I

Arithmetic means ($\bar{x}$) and standard deviation ranges (R) of elements in bedrock from the Ingerbelle—Copper Mountain area.

			Northern contact Nicola Mine Series ($n = 173$)	Southern contact Nicola ($n = 51$)	Background Nicola ($n = 40$)	Nicola (Kennedy Mt. Series) ($n = 14$)	Fault zone Nicola ($n = 13$)	Lost Horse Nicola ($n = 96$)	Copper Mountain Stock diorite (north) ($n = 38$)	Copper Mountain Stock diorite (south) ($n = 18$)
SiO_2	(%)	R	50.0 —56.0	50.7 —59.3	43.5 —59.3	45.3 —61.7	44.7 —66.3	51.8 —58.2	48.7 —55.2	49.3 —54.7
		$\bar{x}$	53.0	55.0	54.4	54.0	55.5	55.0	52.0	52.0
Al_2O_3	(%)	R	15.8 —18.2	13.4 —18.2	13.8 —17.0	13.5 —17.1	11.4 —15.2	16.5 —18.5	16.8 —19.2	13.0 —19.4
		$\bar{x}$	17.0	15.8	15.4	15.3	13.3	17.5	18.0	16.2
Fe_2O_3	(%)	R	4.0 —10.0	6.9 —10.3	7.6 —11.8	6.3 —11.3	6.2 —12.8	3.0 —10.0	5.3 — 9.7	7.0 —10.7
		$\bar{x}$	7.0	8.6	9.7	8.8	9.5	6.5	7.5	8.9
CaO	(%)	R	8.4 —12.0	6.4 —11.4	5.7 —12.9	6.6 —17.0	2.2 —20.0	6.9 —10.3	8.3 —11.5	2.5 — 6.7
		$\bar{x}$	10.2	8.9	9.3	11.8	11.1	8.6	9.9	10.3
MgO	(%)	R	3.1 — 5.1	2.4 — 4.4	3.0 — 7.0	2.2 — 5.6	2.4 — 6.8	2.4 — 4.2	2.6 — 5.0	2.5 — 6.7
		$\bar{x}$	4.1	3.4	5.0	3.9	4.6	3.3	3.8	4.6
K_2O	(%)	R	0.8 — 3.4	1.9 — 3.5	0.9 — 2.3	1.5 — 2.5	0.7 — 2.9	1.3 — 4.7	1.7 — 3.7	1.9 — 4.1
		$\bar{x}$	2.1	2.7	1.6	2.0	1.8	3.0	2.7	3.0
Na_2O	(%)	R	3.2 — 5.6	2.7 — 4.5	1.9 — 3.7	1.5 — 3.1	1.4 — 3.4	3.4 — 5.8	2.8 — 4.3	2.2 — 4.0
		$\bar{x}$	4.4	3.6	2.8	2.3	2.4	4.6	3.6	3.1
TiO_2	(%)	R	0.60— 0.82	0.64— 1.06	0.72— 1.06	0.70— 1.30	0.60— 0.92	0.53— 0.75	0.59— 0.81	0.64— 0.92
		$\bar{x}$	0.76	0.85	0.89	1.0	0.76	0.64	0.70	0.78
P_2O_5	(%)	R	0.51— 0.67	0.44— 0.66	0.36— 0.58	0.39— 0.53	0.32— 0.88	0.48— 0.68	0.48— 0.74	0.52— 0.80
		$\bar{x}$	0.59	0.55	0.47	0.46	0.60	0.58	0.61	0.66
MnO	(%)	R	0.07— 1.1	0.08— 0.14	0.08— 0.14	0.07— 0.11	0.05— 0.19	0.05— 0.09	0.11— 0.13	0.1 — 0.18
		$\bar{x}$	0.09	0.11	0.11	0.09	0.12	0.07	0.12	0.14
S	(%)	R	N/A	0.02— 0.76	N/A	N/A	N/A	N/A	N/A	N/A
		$\bar{x}$	0.48	0.39	0.32	0.13	0.30	0.16	0.14	0.16
Sr	(ppm)	R	547 —1291	401 — 897	214 — 584	N/A	180 —449	556 —1700	722 —2206	542 —2308
		$\bar{x}$	919	649	399	567	316	1128	1464	1425
Rb	(ppm)	R	3.7— 33.1	9.7—22.5	2.2—14.4	3.4—43.5	13.2—32.2	8.0—32.4	4.5—31.2	8.4—24.6
		$\bar{x}$	18.4	16.1	8.3	23.4	22.7	20.2	17.8	16.5
Cu tot	(ppm)	R	N/A	N/A	N/A	33 —63	N/A	N/A	N/A	N/A
		$\bar{x}$	1184	140	242	48	161	429	267	299
Cu ex	(ppm)	R	N/A	N/A	N/A	N/A	N/A	N/A	N/A	N/A
		$\bar{x}$	779	111	55	14	87	273	192	264
Co tot	(ppm)	R	N/A	7.2—20.8	11.2—23.2	8 —24	7.4—30.6	N/A	68 —20.2	12 —21.4
		$\bar{x}$	15.4	14.0	17.2	15.7	19.0	7.2	13.5	16.7
Co ex	(ppm)	R	N/A	4.2—14.8	2.3— 8.9	N/A	N/A	0.9— 8.0	1.1— 8.5	1.7— 7.5
		$\bar{x}$	9.2	9.5	5.6	4.7	8.5	4.5	4.8	4.6
Fe ex	(ppm)	R	831 —11797	5249 —14739	870 —8850	N/A	874 —11450	892 —5852	1141 —4883	1379 —8009
		$\bar{x}$	6314	9994	4860	2600	6162	3372	3012	4697

n = number of samples; N/A = not applicable; tot = nitric perchloric extractable; ex = ascorbic acid/ hydrogen peroxide extractable.

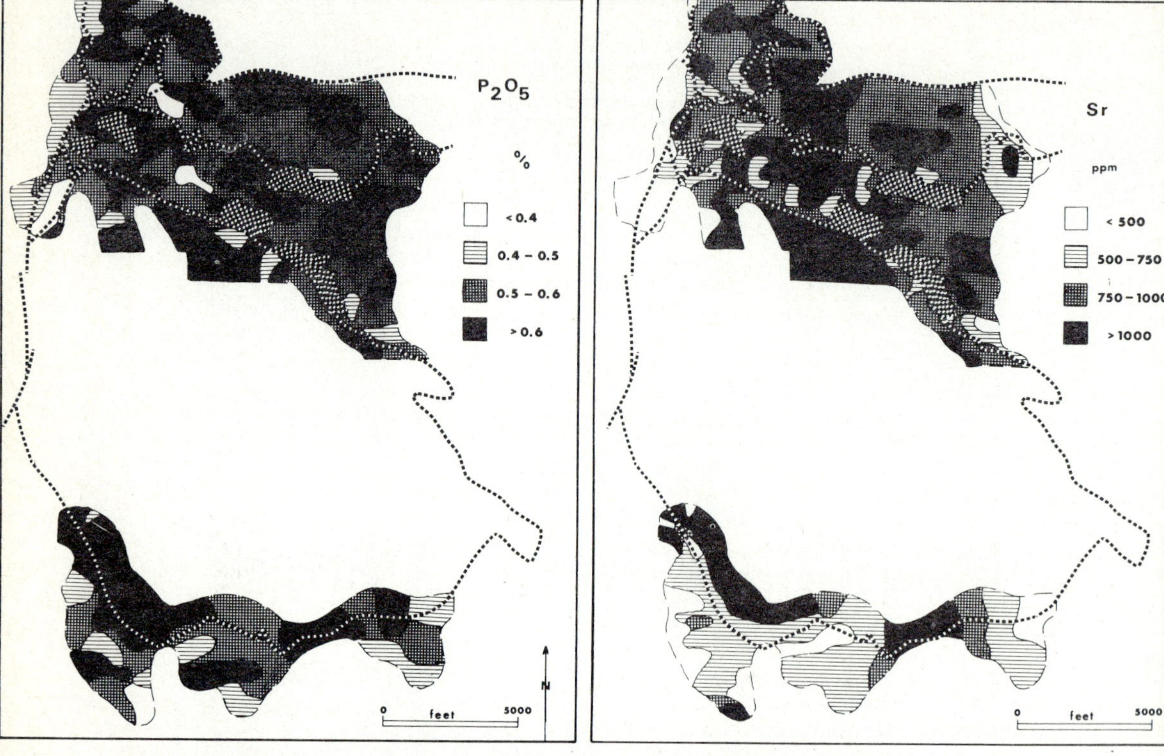

Fig.7. Phosphorus distribution in the Ingerbelle—Copper Mountain area.
Fig.8. Strontium distribution in the Ingerbelle—Copper Mountain area.

Nicola formations remote from the stock (0.46%) (Table I). The Nicola associated with the main mineralization north of the stock has slightly higher phosphorus contents (0.59%) than equivalent formations south of the stock (0.55%). Within the rocks along the northern contact locally high phosphorus zones are associated with the Ingerbelle and part of the Copper Mountain Pit 2 deposit (Fig.7).

Strontium

The average strontium contents of Nicola formations surrounding the stock are markedly higher (650—900 ppm) than formations remote from the stock (400 ppm) (Table I). In addition, the Nicola associated with the main deposits north of the stock has a significantly higher mean strontium content (900 ppm) than that of south of the stock (650 ppm) (Fig.8, Table I).

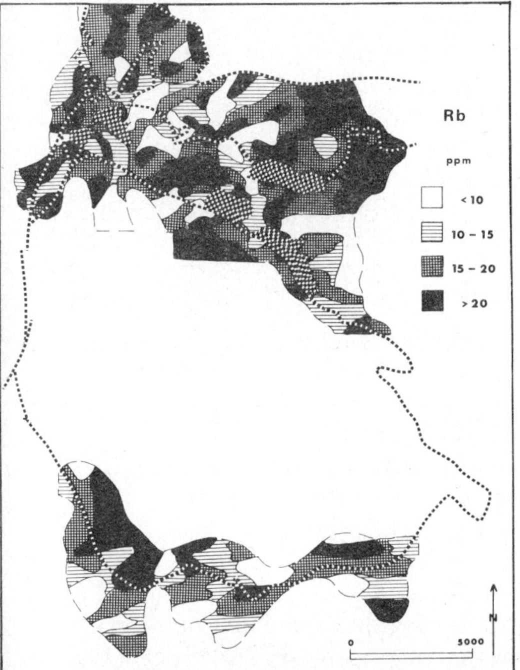

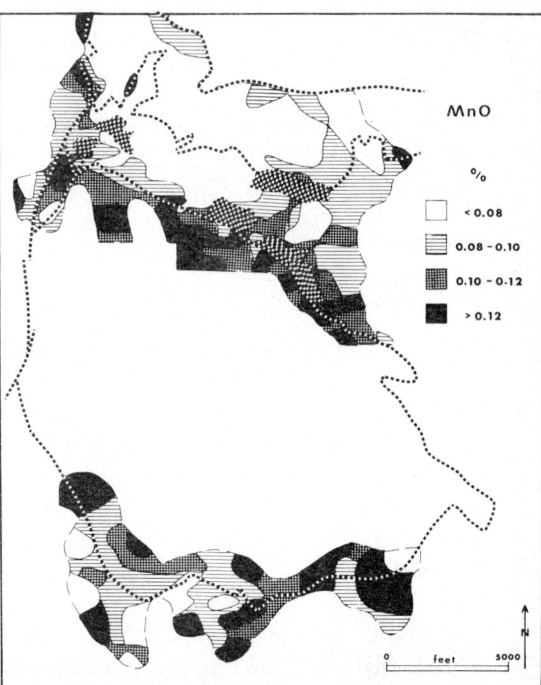

Fig.9 Rubidium distribution in the Ingerbelle—Copper Mountain area.
Fig.10. Manganese distribution in the Ingerbelle—Copper Mountain area.

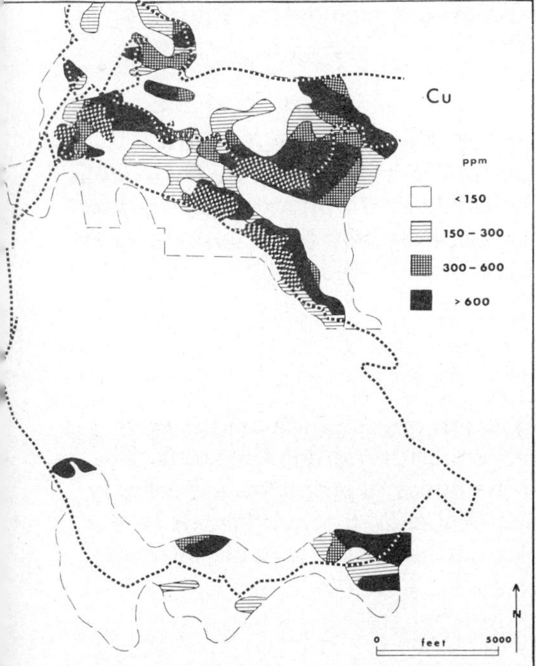

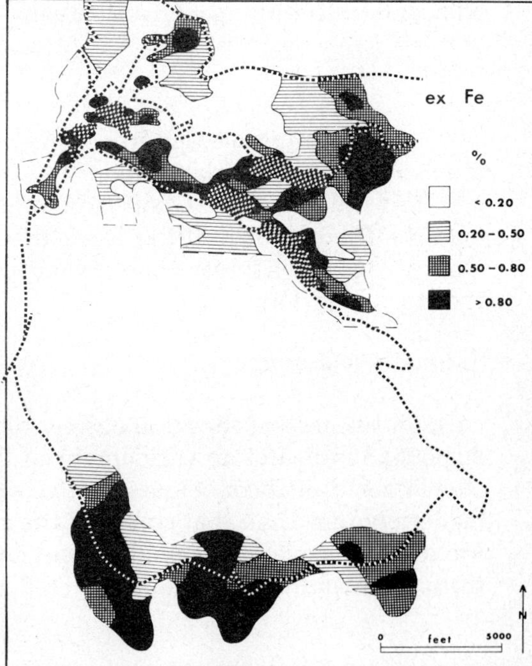

Fig.11. Copper distribution in the Ingerbelle—Copper Mountain area.
Fig.12. Sulphide-held iron distribution in the Ingerbelle—Copper Mountain area.

Rubidium

The mean rubidium content of the Nicola surrounding the stock is higher
(16.1 and 18.4 ppm) than background Nicola (8.3 ppm) although the Kennedy Mountain Series of the Nicola have a higher mean content of rubidium
(23 ppm) (Table I). There does not appear to be any significant variation
in the rubidium distribution within the Nicola surrounding the stock (Fig.9).

Manganese

There are no significant differences in mean manganese content between
Nicola formations adjacent and remote from the stock (Table I). However,
the Nicola adjacent to the northern contact has a lower mean manganese
content (900 ppm) than that south of the stock (1100 ppm). In addition,
within the area of the northern contact low manganese contents coincide
with the Ingerbelle deposits and also occur over a broad central zone (Fig.10).

Copper

As is to be expected the Nicola formations surrounding the stock have
significantly higher copper contents than areas remote from the stock (Fig.11).
In addition the Nicola north of the stock has a markedly higher mean copper
content (1200 ppm) than that south of the stock (140 ppm) (Table I). Linear zones of high copper contents (> 600 ppm) up to 1500 ft wide occur
adjacent to the Mines Series/Intrusive contact and include the four deposits.

Sulphide-held iron

The mean sulphide-held iron contents of the Nicola adjacent to the
stock are somewhat higher than the background Nicola (Table I). The Nicola
south of the Stock has a uniformly high sulphide-held iron content, whereas
over the Nicola north of the stock the high sulphide-held iron contents (0.5%)
are restricted to a linear zone extending along the entire northern contact of
the stock (Fig.12).

Sulphide-held copper

As in the case of the total copper distribution the sulphide-held copper
dispersion indicates an enrichment in the Nicola surrounding the stock. The
Nicola north of the stock associated with the major deposits has a markedly
higher content than that south of the stock (Table I). Rocks adjacent to the
ore deposits have high sulphide-held copper contents over variable distances
from the margins of the deposits (Fig.13).

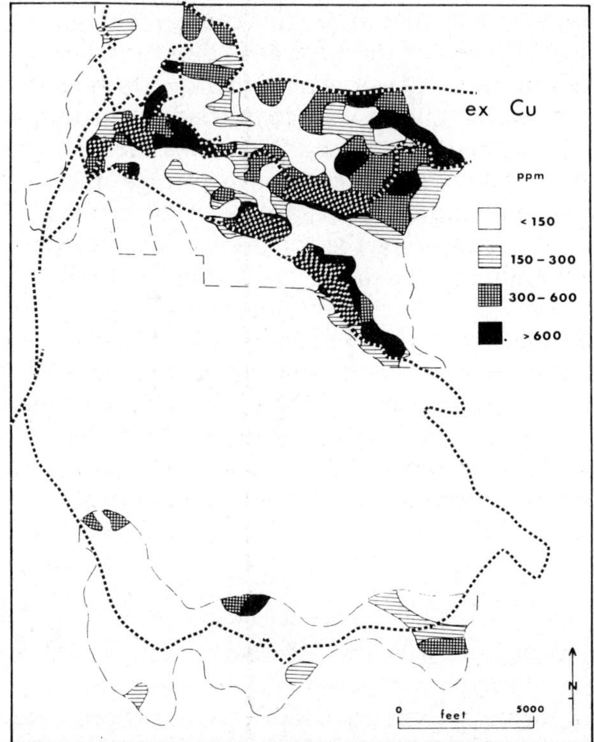

Fig.13. Sulphide-held copper distribution in the Ingerbelle—Copper Mountain area.

RELATIONSHIP OF GEOCHEMISTRY TO ALTERATION

From the foregoing description of some of the single element distributions it is apparent that there are similarities in the distributions of certain elements while at the same time there are marked variations in detail.

The principal element associations are as follows:

(1) The Na, K, P, Rb, Sr, Cu and sulphide-held Cu, and Fe distributions are variously enriched in the Nicola formations surrounding the Copper Mountain Stock relative to background areas.

(2) P, Sr, Cu and sulphide-held Cu are enriched in the Nicola on the north side of the stock including the area of the four major deposits relative to the less intensely mineralized area of the Nicola south of the stock. In contrast, however, higher contents of potash and manganese occur in the Nicola on the south side of the stock than on the northern side.

(3) Marked zoning in the distribution of Na, K, P, Si, Mn, Cu and sulphide-held Cu, and Fe is noted within certain areas of the Nicola formations north and south of the stock. This zoning appears in part at least to be related to the location of the mineral deposits.

The individual elemental distributions reflect the response of elements to

310

a number of independent processes as reflected in the mineralogy. For example soda enrichment occurs in the rocks of the Ingerbelle deposit where
Na-rich phases of scapolite and albite were identified as being the dominant
metasomatic minerals. Conversely, K-rich phases were identified in the immediate vicinity of the Copper Mountain deposits and these rocks are enriched in potash relative to background Nicola rocks. Above average silica
and below average manganese concentrations occur in rocks of the central
area coinciding with the bleached appearance of the rocks in which there is
an absence of ferromagnesian minerals and the occurrence of insterstitial
amounts of quartz.

As the dispersion of certain elements appears to coincide with the distribution of metasomatic minerals it appears that the distribution of these minerals can be estimated on the basis of the chemistry of the rocks. The usefulness
of this relationship is based on the more extensive distribution of alteration
types, and hence the anomalous chemical zones, than on the ore bodies themselves. In this way broader exploration targets can be identified relatively
rapidly and cheaply.

SUMMARY AND CONCLUSIONS

The complexity of alteration associated with the Ingerbelle—Copper Mountain deposits has precluded the identification of the extent of alteration
zoning normally associated with porphyry-type mineralization. Previous investigations have revealed the presence of distinctive and normal alteration assemblages resulting from metasomatism probably associated with the emplacement of the main deposits and the intrusive bodies. The present investigation
was undertaken to establish whether it was possible to identify alteration
zoning on the basis of the chemical composition of the rocks. Bedrock sampling was carried out over Nicola and adjacent formations surrounding the
Copper Mountain Stock and the samples analyzed for eighteen elements. The
distributions of the elements indicated variously the presence of high concentrations of Na, K, P, Rb, Sr, Cu and sulphide-held Cu in the Nicola rocks surrounding the Copper Mountain Stock relative to background areas. The presence
of these features on a broader scale than the mineral deposits provides a larger
target for reconnaissance-level exploration. The recognition of the metasomatized
lithologies need not necessarily imply an association with sulphide mineralization
although in the Copper Mountain area the alteration is believed to be genetically
related to the formation of the sulphide bodies. The basis for the use of bedrock
geochemistry at the reconnaissance level of exploration for Ingerbelle—Copper
Mountain-type deposits is dependent on the identification of igneous alkalic
centres and related metasomatism of the host rocks. As such three fundamental requirements are necessary:

(1) The hydrothermal solutions should be able to escape from the intrusions
and permeate the adjacent Nicola rocks.

(2) The solutions should be copper-bearing.

(3) The properties of the Nicola rocks in contact with the intrusions should

be such that copper will be precipitated from solutions as soon as the Nicola rock is permeated by the solution.

If these conditions are not met it is suggested that mineralization of the Ingerbelle—Copper Mountain type would not be formed though it is possible that some other type of deposit might be formed if at least the second condition were fulfilled.

Hence the extent to which rocks adjacent to the intrusion are altered is an important but by no means definitive criterion for the occurrence of mineralization. The importance of the alkalic character of the alteration patterns is not known and the possibility remains for the formation of a similar-sized alteration halo from calc alkalic intrusions though such alteration would undoubtedly be of different mineralogical and chemical type.

On a more detailed scale certain conclusions may be drawn where the requirement is to identify areas within Nicola rocks in contact with the stock with ore-bearing potential. In discriminating between the mineralized Nicola rocks north of the stock and only weakly mineralized Nicola rocks adjacent to the southern contact, the strontium content appears to be most useful. In addition the distribution of copper provides a means of delineating those areas where the rocks are significantly mineralized; in these areas the broad distribution of high soda concentrations would indicate Ingerbelle-type mineralization, whereas an association with high potash contents would suggest mineralization of the Copper Mountain type. On this basis the determination of bedrock chemistry in the Copper Mountain area has indicated the presence of marked chemical zones variously related to the mineral deposits. The application of this procedure offers a method for determining the ore-bearing potential of remaining areas of the Nicola rocks.

The following conclusions may be drawn regarding procedures for the search for mineralization similar to the Ingerbelle—Copper Mountain type. At the reconnaissance level, areas of Nicola formation in contact with known alkalic intrusive complexes should be sampled from 500×500-ft^2 cells at a density of 10 samples per square mile with 20 chip samples collected from each cell. The dominant alteration types, degree of fracturing and sulphide content should be documented for each sample. The samples should be analyzed for P, Sr, Na, K and Cu.

At the detailed level of exploration, areas identified as being anomalous on the basis of the regional programme should be sampled in a similar manner at a density of approximately 80 samples per square mile. The lack of bedrock exposure may preclude sampling at this density in which case samples of the suboutcrop may be sampled by rotary drilling. Samples should be analyzed only for copper.

It is believed that adoption of a procedure such as this will provide useful information on areas underlain by alkalic intrusions and associated Nicola rocks.

ACKNOWLEDGEMENTS

We would like to express our appreciation to the management of Newmont
Mining Corporation of Canada for technical and financial support towards the
investigations which formed part of the doctoral research program of one of
us (John E. Gunton) at the Department of Geological Sciences, Queen's University, Kingston, Ontario.

The active interest and cooperation of company representatives has been
of great assistance to us and in particular would express our gratitude to Dr.
J.A. Coope who initially drew the authors' attention to the problem and to
T.N. Macauley and R.F. Sheldon of Newmont Mining Corporation of Canada
and J. McCue of Similkameen Mining Company Ltd.

REFERENCES

Cameron, E.M., Siddeley, G. and Durham, C.C., 1971. Distribution of ore elements in
rocks for evaluating ore potential: nickel, copper, cobalt and sulphur in ultramafic rocks
of the Canadian Shield. Can. Inst. Min. Metall., Spec. Vol., 11: 298—313

Gunton, J.E., 1974. Geochemical dispersion associated with porphyry-type mineralization
in the Canadian Cordillera. Ph.D. Thesis, Queen's University, Kingston, Ont. (unpublished)

Hausen, D.M., 1968. Classification of rock types from the Similkameen Mining Company,
Princeton, B.C. Newmont Exploration Ltd. (private report)

Hausen, D.M. and Kerr, P.F., 1971. X-ray diffraction methods of evaluating potassium
silicate alteration in porphyry mineralization. Can. Inst. Min. Metall., Spec. Vol., 11:
334—340

Macauley, T.N., 1970. Geology of the Ingerbelle-Similkameen deposits at Copper Mountain, B.C. Similkameen Mining Co. Ltd., (private report)

Macauley, T.N., 1973. Geology of the Ingerbelle and Copper Mountain deposits at Princeton,
B.C. Bull. Can. Inst. Min. Metall., 66(732): 105—112

Preto, V.A.G., 1972. Geology of Copper Mountain. Bull. B.C. Dep. Min. Pet. Res., 48: 87

Preto, V.A.G., White, W.H. and Harakal, J.E., 1971. Further potassium—argon age dating
at Copper Mountain, B.C. Bull. Can. Inst. Min. Metall., 64(708): 58—61

Rice, H.M.A., 1947. Geology and mineral deposits of the Princeton map-area British
Columbia. Geol. Survey Can. Mem., 253: 136 pp.

Sinclair, A.J. and White, W.H., 1968. Age of mineralization and post-ore hydrothermal
alteration at Copper Mountain, B.C. Bull. Can. Inst. Min. Metall., 61(673): 633—636

Sutherland-Brown, A., Cathro, R.J., Panteleyev, A. and Ney, C.S., 1971. Metallogeny
of the Canadian Cordillera. Can. Inst. Min. Metall., Trans., LXXIV: 121—145

VARIATIONS IN THE MERCURY CONTENT OF SPHALERITE FROM SOME CANADIAN SULPHIDE DEPOSITS

I.R. JONASSON and D.F. SANGSTER

Geological Survey of Canada, Ottawa, Ont. (Canada)

ABSTRACT

Mercury levels have been determined in sphalerites and zinc concentrates from some Canadian sulphide deposits primarily of the volcanogenic and "Mississippi Valley" (carbonate-hosted) types but also including other types. Samples were collected from deposits in the Canadian Shield, Cordillera, Arctic Islands, and the Appalachian region. Host rocks range in age from Archean to Mesozoic and the ores are of different compositions, e.g., Cu—Zn, Cu, Pb—Zn—Cu and Pb—Zn.

Based on preliminary data, mercury content appears to vary according to genetic type, regions in Canada, age of host rocks, and to a lesser extent, bulk ore compositions. In general, sphalerite in volcanogenic deposits contains more mercury than that in carbonate-hosted stratabound deposits. The Flin Flon—Snow Lake—Lynn Lake volcanogenic deposits (considered here to be of Proterozoic age) contain sphalerite of higher mercury content (up to 450 ppm) than similar deposits elsewhere in the Shield. The mercury content of sphalerite in stratabound deposits of the Grenville province, irrespective of their genetic type, is high (up to 160 ppm) relative to other tectonic provinces. With respect to age of host rocks, mercury in sphalerite of volcanogenic-type deposits varies in relative abundance as follows: Proterozoic > Phanerozoic > Archean.

In carbonate-hosted deposits, the order of abundance is similar: Proterozoic > Phanerozoic. Carbonate-hosted deposits have not been recognized in the Archean.

Sphalerite in deposits which contain sulphosalt minerals (generally the Pb—Zn—Ag ores) were found to contain high mercury levels. The mercury content of sphalerite in Zn-rich volcanogenic deposits is generally higher than that in the Cu-rich ones.

Although these data are preliminary and further analyses may ultimately result in a revision of the trends noted above, some interpretation is proposed and the application of these data to the use of mercury as a geochemical prospecting aid is discussed.

INTRODUCTION

General remarks

Mercury levels have been determined in sphalerites and zinc concentrates from some Canadian massive sulphide deposits primarily of the volcanogenic and Mississippi Valley types. Samples were collected from deposits in the Canadian Shield, Cordillera, Arctic Islands and the Appalachian region. Host rocks range in age from Archean to Mesozoic and the ores are of different bulk composition, e.g., Cu—Zn; Cu; Pb—Zn, Ag. Many of the geological and chemical features of volcanogenic deposits have been described by Sangster (1972a); and for Mississippi Valley types by Sangster (1970) and Sangster and Liberty (1971). In this paper the term "deposit" is used quite loosely and in no way implies economic concentrations of ore.

The ultimate objective of the project is to determine the abundance of accessory elements in Canadian ore assemblages of *all* types and in sulphide mineral separates from those ores. This preliminary assessment of mercury values represents the first efforts to interpret some of the limited information gathered to date.

The collection of data is timely. Not only is mercury of more than passing interest to exploration geologists as a prospecting aid in geochemical survey programmes, but it is also currently under the scrutiny of biologists and related environmental scientists. The implications of the present work are discussed only cursorily with regard to this latter aspect of geochemistry, but which nonetheless do remain of considerable relevance and significance to environmentalists.

The approach of geologists towards the use of mercury in geoexploration surveys has not always appeared to be logical. Little consideration has been accorded to such factors, as the abundance of mercury in different geological materials including ores, which would provide a more rational basis for optimum exploitation of the unique physical and chemical properties of mercury. Some of the disrepute into which the geochemical use of mercury has subsided is undoubtedly due to its use, or rather its mis-use, in geological terranes wherein it should never have been considered as a pathfinder element in the first place.

The major aspect of this study is, therefore, an attempt to rationalize such use of mercury both as a prospecting aid and as an indicator for certain ore types in selected parts of Canada in terms of geology, mining camps and mineralogy therein. Neither the restrictions on its geochemical uses due to terrain and landscape nor its geochemical methodology are discussed. These must, of course, be considered carefully prior to undertaking any surficial geochemical survey.

Some aspects of previous work

Previous work on mercury abundance in stratabound sulphide deposits of Canada has been limited when the Canadian scene as a whole is considered. Some work on ores from Quebec has been published (Sears, 1971) in which an attempt was made to equate measured mercury levels with the zinc contents of the ores. In most cases a significant positive correlation was recorded for each mine considered. Little other detailed work on a regional basis is available. R.W. Boyle (G.S.C., personal communication, 1974; and 1971) has investigated mercury contents of numerous gold and silver ores in Canada, including the Yukon and also the Cobalt area of Ontario where occurrences of mercury tellurides are known. In general, high amounts of mercury were found in sphalerites and other sulphides (~50 ppm). At this stage of the present work it is not practicable to make more than casual comparisons of data from the vein types he dealt with and the massive sulphides under discussion here. Such vein types are under consideration within the scope of the full project, but will not be described here in detail.

A few isolated reports exist on the mercury levels measured in some massive sulphides from specific deposits, e.g., Manitou-Barvue, Quebec (Ripley, 1964), but in general it has proved easier to repeat the work rather than track down elusive references which lie largely in company or Federal and Provincial government reports. This article, therefore, does not purport to contain a section reviewing all previous work on mercury in sulphides.

The diversity and often, the unreliability, of mercury analytical methods (particularly those used prior to the advent of atomic absorption spectroscopy) may preclude direct comparison of results of this work with those from some previous studies. The comparisons made herein between and within mining camps and districts are, therefore, more meaningful than they might otherwise be because all samples have been analysed in the same laboratories by the same analytical technique. Details of the procedure are described by Jonasson et al. (1973) and need not be repeated here. Suffice to say that the technique falls into the group loosely described as cold vapour, atomic absorption methods.

The environmental impact of some of the data presented here has already been discussed in an article by MacLatchy and Jonasson (1974). Samples of zinc concentrates, copper concentrates, ore heads and tailings were analysed for mercury with a view to ascertaining the significance of the contribution mining activities in northwestern Quebec, i.e., Matagami and Chibougamau areas, make to the mercury levels observed in drainage basins of the Bell, Rupert and Nottaway Rivers. Summaries in tabular form of some aspects of this study which are relevant to geochemical prospecting are discussed subsequently (Table I).

Ore geochemistry

Results of the above study further confirmed the proposition, long held, that mercury is preferentially contained (or trapped) in zinc sulphide lattices rather than in co-existing pyrite, galena or chalcopyrite. Summary information on this subject is widely available; for example, in a recent review of the geochemistry of mercury by Jonasson and Boyle (1972).

Further examples from Canada are presented in Table I. A full study of the ore mineralogy of each deposit would be necessary to confirm or deny whether all of the mercury reported for sphalerite is in fact due only to sphalerite. Sphalerite was separated from the host ore assemblage by a combination of hand-picking, super-panning and magnetic separation methods. The minerals analysed were obtained as pure as could be determined visually. All of the data presented are for specimens at least 95% pure as determined

TABLE I

Mercury contents (ppm) of some co-existing sulphides

Deposit	Area location	Orogen, platform or province	Chalcopyrite	Sphalerite	Galena	Pyrite
Orchan	Matagami, Que.	Superior	0.92	3.25	—	—
Mattagami L.	Matagami, Que.	Superior	0.44	2.56	—	1.20
Poirier	Val d'Or, Que.	Superior	0.33	0.80	—	0.90
Coniagas	Val d'Or, Que.	Superior	—	8.6	2.71	3.90
Manitou-Barvue	Val d'Or, Que.	Superior	1.40	154	60	1.27
Lac Dufault	Noranda, Que.	Superior	8.20	10.6	—	—
Normetal	Noranda, Que.	Superior	3.31	21.5	—	0.50
Kidd Creek	Timmins, Ont.	Superior	1.42	2.90	—	3.60
Geco	Manitouwadge, Ont.	Superior	1.23	3.9	—	0.90
Mattabi	Sturgeon L., Ont.	Superior	3.35	53	1.10	5.9
Fox Lake	Lynn Lake, Man.	Churchill	35.4	238	—	—
Chisel Lake	Snow Lake, Man.	Churchill	—	106	3.0	3.3
High Lake	High Lake, N.W.T.	Slave	0.24	5.1	—	0.34
Balmat	N.Y., U.S.A.	Grenville	—	1200	16.5	—
Long Lake	Kingston, Ont.	Grenville	—	31	2.2	1.10
New Calumet	Calumet Is., Que.	Grenville	—	53	4.7	1.90
Gays River	Gays River, N.S.	Appalachian	—	0.31	0.28	—
Walton	Walton, N.S.	Appalachian	0.35	0.55	0.30	0.10
Bankeno	Cornwallis Is. N.W.T.	Innuitian	—	0.40	0.40	0.20
Prairie Ck.	N.W.T.	Cordilleran	—	933	9.0	—
Anvil	Faro, Y.T.	Cordilleran	—	114	2.2	3.3
Monarch	Field, B.C.	Cordilleran	—	12.8	1.32	—
Pine Point	Pine Point, N.W.T.	Interior	—	0.33	0.06	—

by analysis of gross contaminants such as Cu, Pb, As, Sb and Fe, as well as percentage Zn.

The most common mineral impurity was usually pyrite and less commonly, pyrrhotite.

It is possible that some of the zinc sulphide minerals contain small inclusions of other sulphides which may be mercuriferous, e.g. tetrahedrite, etc., but the amounts of these inclusions in most of the sulphide minerals studied are minimal and would not seem to have a major influence on the data presented.

RESULTS AND DISCUSSION

Archean volcanic types

The results of mercury analyses for sphalerites from volcanogenic sulphide deposits and occurrences in the Archean Slave, Superior and Churchill structural provinces of the Canadian Shield are presented in Table II and Fig.1.

It can be seen that values range from a low of 0.039 ppm (Zenmac, Ont.) to a high of 229 ppm (Indian Mountain Lake, N.W.T.) with a mean of 33 ppm for all deposits.

If the deposits are further divided into their different geographic (and probably volcanic belt) units, the following means obtain:

Slave province:	65 ppm	(4 deposits)
Manitouwadge:	2.7 ppm	(3 deposits)
Northwestern Ontario:	53 ppm	(3 deposits)
Timmins– Val d'Or region:	37 ppm	(8 deposits)
Matagami region:	2.2 ppm	(3 deposits)
Spi Lake, Keewatin:	1.5 ppm	(1 deposit)

The average for the Slave province is strongly influenced by some very high values from Indian Mountain Lake. In the northern and eastern parts of the Slave mercury values are much lower and more comparable with those of the Superior province. It may be that the proximity of a deep rift, the post-Proterozoic (?) Macdonald fault, has had some positive influence on the mercury content of ore minerals emplaced before faulting or fault reactivation. It has been postulated that deep rifting causes mercury to be released from its mantle sources (Jonasson and Boyle, 1972; Ivankin and Turkin, 1972) and which may be absorbed or incorporated into favourable host materials such as sphalerite or tetrahedrite during the course of the migration along those structures. The latter mineral which has been shown to be present and to contain about 0.2% Hg (A.G. Plant, G.S.C., personal communication, 1974) in the Indian Mountain Lake ores, tends to be collected along with sphalerite upon separation from ores by the super-pan and electromagnetic methods employed in this study. Sampling of other occurrences in the immediate vicinity of the fault may yield worthwhile information to help resolve this point. Some preliminary data on vein occurrences (Table IV) suggests that mercury enrichment may be a general feature of the area.

318

TABLE II

Mercury content of sphalerites from volcanogenic-type ores

Name of occurrence	Location	Host rock	Ore type	No. of samples	Range (ppm)	Mean (ppm)
Phanerozoic						
Faro[1]	Y.T.	meta-seds.	Pb—Zn, Ag	2	80—157	114
Weedon	S. Que.	meta-tuffs	Cu—Zn	2	26—96	61
Key Anacon	N.B.	meta-volcs.	Pb—Zn	2	70—161	116
Wedge	N.B.	meta-volcs.	Pb—Zn	1	59—	59
Heath Steele	N.B.	meta-volcs.	Pb—Zn, Cu	1	11—	11
Que. Sturgeon River	N.B.	meta-volcs.	Pb—Zn, Cu	1	3.2—	3.2
Brunswick No. 12	N.B.	meta-tuffs	Pb—Zn, Cu	1	22—	22
Brunswick No. 6	N.B.	meta-tuffs	Pb—Zn, Cu	1	26—	26
Seneca	B.C.	andesite	Pb—Zn, Cu	2	15—42	28
Western	B.C.	volcs.	Pb—Zn, Cu	1	19—	19
Big Bull	B.C.	volcs.	Cu—Zn, Pb	1	23—	23
Buchans	Nfld.	volcs.	Pb—Zn, Cu, Ag, Ba	1	12—	12
Proterozoic[2]						
Fox Lake[3]	N. Man.	meta-seds.	Cu—Zn	2	110—238	174
Sherridon	N. Man.	meta-seds.	Cu—Zn	5	2.9—40	18
Ruttan Lake	N. Man.	meta-seds.	Cu—Zn, Pb	2	130—140	135
Osborne Lake	N. Man.	meta-seds.	Cu, Zn	2	121—452	238
Chisel Lake	N. Man.	meta-seds.	Pb—Zn, Cu	3	40—193	106
Schist Lake	N. Man.	meta-volcs.	Cu—Zn	3	11—86	41
Western Nuclear	N. Sask.	meta-volcs.	Pb—Zn, Cu	1	73—	73
N. Contact Lake	N. Sask.	meta-seds.	Cu—Zn	1	200—	200
Flin Flon	N. Man. and Sask.	meta-volcs.	Cu—Zn	2	32—54	43
Sulphide Lake	N. Sask.	meta-seds.	Cu—Zn	1	3.6—	3.6
Errington	N. Ont.	tuffs	Cu—Zn, Pb	3	140—194	161
New Calumet	W. Que.	meta-seds.	Pb—Zn, Ag	5	31—125	53
Tetrault[1]	W. Que.	para-gneiss	Pb—Zn, Ag	2	6.1—6.3	6.2
Archean						
Indian Mt. Lake	Mack.	meta-rhyolite	Pb—Zn	3	90—312	229
N. Slave province	Mack.	meta-rhyolite	Pb—Zn	4	1.4—13	5.1
Hackett River	Mack.	meta-rhyolite	Pb—Zn, Ag	3	14—28	23
High Lake	Mack.	meta-rhyolite	Cu—Zn	2	4.2—5.9	5.1
Spi Lake	Keew.	meta-volcs.	Cu—Zn, Pb	3	1.0—2.0	1.5
Mattabi	N.W. Ont.	xl. tuffs	Cu—Zn	6	27—100	53
South Bay	N.W. Ont.	volcs.	Cu—Zn	4	12—38	19
Sturgeon Lake	N.W. Ont.	volcs.	Cu—Zn	1	83—	83
Zenmac	W. Ont.	meta-gabbro	Zn	2	0.030—0.048	0.03
Geco	W. Ont.	meta-seds.	Cu—Zn, Pb	2	3.6—4.2	3.9
Willecho	W. Ont.	meta-seds.	Cu—Zn	3	1.3—8.9	4.2
Kam-Kotia	N. Ont.	andesite	Cu—Zn	1	32—	32
Cdn.-Jamieson	N. Ont.	rhyolite	Cu—Zn	3	40—70	60
Kidd Creek	N. Ont.	meta-volcs.	Cu—Zn, Ag	5	1.1—6.3	2.9
Normetal[3]	Que.	volcs.	Cu—Zn	8	19—31	23

TABLE II (continued)

Name of occurrence	Location	Host rock	Ore type	No. of samples	Range (ppm)	Mean (ppm)
Lac Dufault	Que.	rhyolite-andesite	Cu—Zn, Au	5	4—17	9.0
Delbridge	Que.	rhyolite	Zn	2	1.60—1.70	1.65
Coniagas[3]	Que.	tuffs	Pb—Zn, Ag	1	8.6—	8.6
Manitou-Barvue[3]	Que.	tuffs	Cu—Pb- Zn	5	152—166	156
Mattagami Lake[3]	Que.	rhyolite	Cu—Zn	10	2.36—4.30	2.93
Orchan[3]	Que.	rhyolite	Cu—Zn	5	1.60—3.78	2.88
Poirier	Que.	rhyolite	Cu—Zn	1	0.80—	0.80

[1] Genetic type in doubt.
[2] Proterozoic age of many listed deposits remains in contention.
[3] Monthly composite zinc sulphide concentrates.

Little information is available from the deposits in the interior of the Churchill province. Only one, Spi Lake, has been sampled. The very low figure found from two small samples may not be representative.

Most Archean volcanogenic sphalerites occur in Cu—Zn deposits within the Superior province. The vast greenstone belts of the Abitibi area of northern Ontario and of western Ontario show similarly high mercury values for sphalerites, but it is interesting to note that there seems to be much less mercury in sphalerites from the Noranda and Matagami areas. Rather, the highest mercury values are found in the Val d'Or area and in the northern portion of the Noranda camp where zinc is the more dominant metal in the ores; and where lead and silver values also increase markedly, for example, in the Manitou-Barvue deposit. Preliminary data indicate that the same pattern is followed for mercury contents in chalcopyrites ($\geqslant 98\%$ pure) from the same deposits. The Noranda and Matagami camps average about 0.8 ppm, the Val d'Or camp averages about 3.2 ppm, as do samples from Normetal mine, and the Timmins area averages about 1.5 ppm. Some further comparisons between sulphides for individual deposits are presented in Table I. Thus, within the mining camps of Timmins, Noranda, and Val d'Or, but not of Matagami, which lie within connecting belts of volcanic rocks, the mercury content seems to depend to some extent directly on the bulk zinc content of the ore body and on the presence or absence of Ag—Pb-bearing minerals such as sulphosalts. These are known to be important accumulators of mercury, in addition to other heavy metals. A result of these observations is to further emphasize the necessity for making detailed ore mineralogical investigations in each ore deposit studied for its mercury content.

At this point in the discussion it is worth noting that the ore mineralogy of the Manitou-Barvue deposit more closely resembles those more commonly found in Proterozoic deposits, viz; Pb—Zn—Cu-rich types, which are somewhat higher in mercury content than most Archean Cu—Zn types.

In western Ontario, the two deposits near Manitouwadge showed similar,

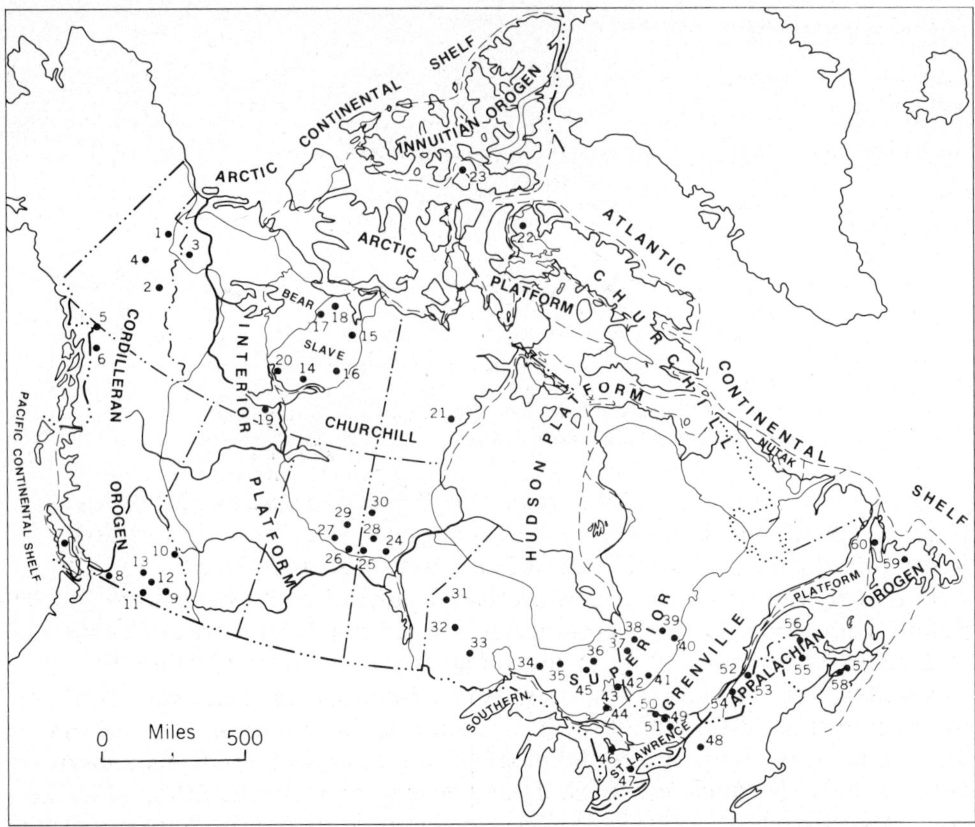

Fig.1. Location of some Canadian sulphide deposits.

1 = Hart River, Y.T.
2 = Faro (Anvil), Vangorda, Y.T.[3]
3 = Prairie Creek, N.W.T.[1,3]
4 = Keno Hill (United Keno, Galkeno, Tom, Elsa) Y.T.[2]
5 = Arctic Silver, Y.T.[2]
6 = Big Bull, Tulsequah, B.C.
7 = Western, B.C.
8 = Seneca, B.C.
9 = Sullivan, B.C.
10 = Monarch, Kicking Horse, B.C.[1]
11 = Reeves Macdonald, H.B.[1], Jersey, Blue Bell[2], B.C.
12 = Duncan, B.C.
13 = Big Ledge, B.C.[1]
14 = Indian Mt. Lake, Homer Lake[2], Turnback Lake[2], N.W.T.
15 = Hackett River (Bathurst-Norseman), N.W.T.
16 = Clinton-Colden Lake, N.W.T.

17 = Takiyuak Lake area (massive and vein), N.W.T.
18 = High Lake, N.W.T.
19 = Pine Point, N.W.T.[1]
20 = Yellowknife area, N.W.T.
21 = Spi Lake, Keew.
22 = Baffin (Strathcona), N.W.T.[1]
23 = Bankeno, N.W.T.[1]
24 = Snow Lake (Osborne L., Stall L., Anderson L., Chisel L.), Man.
25 = Flin Flon, Flexar, Coronation, Schist L., Birch L., Man.; Sask.
26 = Hanson Lake (Western Nuclear), Sask.
27 = La Ronge Lake (Sulphide L., N. Contact L., Anglo-Rouyn), Sask.
28 = Sherridon, Sask.
29 = Brabant, Sask.
30 = Fox Lake, Ruttan Lake, Lynn Lake, Man.
31 = Berens River, Severn River, N. Spirit Lake, Ont.[2]

low mercury contents, but Zenmac, which is farther west, showed the lowest mercury values found so far in Archean sphalerites. The only unusual aspect of this deposit compared with its neighbours is that the host rock is meta-gabbro rather than the more common meta-volcanics or meta-sediments. The reason for the low value, however, is not yet clear; perhaps the deposit has been mis-classified as volcanogenic. Its unusual geological nature appears to be reflected in the mercury content of its sphalerite. A value of 0.039 ppm is far more typical of a gabbro host rock than of a sulphide so it is reasonable to presume that there has been little or no input of mercury into this ore body at all.

Proterozoic volcanic types

Table II and Fig.1 also include data for sphalerites from Proterozoic volcanogenic ores. Samples were collected from the Cordillera, the Grenville structural province, and from presumed Proterozoic (Aphebian) units *within* the Churchill and Southern provinces. Values range from a low of 3.6 ppm (Sulphide Lake, Sask.), to a high of 238 ppm (Osborne Lake, Man.) both of which are highly metamorphosed. The mean for all deposits is 99 ppm. No samples are presently available from the Bear structural province wherein massive sulphide deposits appear to be quite rare and where vein-type ores predominate. It is interesting to compare means for sphalerites from the three regions represented.

---

32 = Uchi Lake (South Bay), Ont.

33 = Sturgeon Lake (Mattabi, Sturgeon L.), Ont.

34 = Zenmac, Ont.

35 = Manitouwadge (Geco, Willecho, Willroy), Ont.

36 = Kidd Creek, Ont.

37 = Normetal, Que.

38 = Joutel, Poirier, Que.

39 = Mattagami Lake, Orchan, Que.

40 = Coniagas, Que.

41 = Manitou-Barvue, Louvem, Que.

42 = Noranda (Vauze, W. Macdonald, Horne, Quemont, Delbridge, Lac Dufault), Que.

43 = Cobalt area, Ont.

44 = Sudbury basin (Errington, Vermilion), Ont.

45 = Jameland, Kam-Kotia, Canadian Jamieson, Genex, Ont.

46 = Bruce Peninsula, Ont.[1]

47 = Niagara Peninsula, Ont.[1]

48 = Balmat, N.Y., U.S.A.[1]

49 = Long Lake, Thirty Island Lake, Ont.[1]

50 = New Calumet, Que.

51 = Kingdon, Ont.[2]

52 = Montauban (Tetrault), Que.[3]

53 = Weedon, Cupra, Que.

54 = Moulton Hill (Aldermac), Que.

55 = Bathurst (Brunswick No. 6 and No. 12, Heath Steele Wedge, Key Anacon, Que., Sturgeon River), N.B.

56 = Bathurst (Nigadoo, Keymet), N.B.[2]

57 = Gay's River (Cuvier), N.S.[1]

58 = Walton (Magnet Cove), N.S.[1]

59 = Buchans, Nfld.

60 = Daniels Harbour (Newfoundland Zn), Nfld.[1]

---

[1] Carbonate-hosted types, [2] vein types, [3] classification in doubt.

All other deposits (occurrences) are classed as volcanogenic massive sulphides.

Grenville:	30 ppm	(2 deposits)
Sudbury basin:	150 ppm	(2 deposits)
Northern Manitoba and		
Saskatchewan:	103 ppm	(10 deposits)

Inspection of the mean values indicates that sphalerites of Proterozoic age (Aphebian and Helikian) are considerably enriched in mercury compared with those of Archean age. This point is quite apparent even allowing for the somewhat limited sampling which is confined at present to the margins of the Grenville province and to a few inliers in Archean rocks.

It may be possible with further information to subdivide the samples into Aphebian and Helikian and postulate that the former (older) are the more mercury rich.

The sphalerites of the Flin Flon—Snow Lake—Lynn Lake areas of northern Saskatchewan and Manitoba are quite well sampled. In this paper it is presumed, on the basis of recent lead isotope work by Sangster (1972b), that the deposits are of Aphebian age. The high mercury values tend to support this proposed Aphebian age of these deposits. If this is the case, then it would seem likely that stratabound deposits of Proterozoic age contain, on the average, more mercury than do their Archean analogues. The presence of Pb-bearing minerals and Sb-rich sulphosalt minerals is also more common. The latter are very favourable hosts for mercury as noted earlier.

Similar observations are valid for the confirmed Aphebian age volcanics of the Sudbury basin and once again the concept of enrichment by mercury in regions close to rifting or deep faulting seems to be confirmed. The proximity of the Superior rift, Ottawa Valley rift, Murray and Temiscaming faults lends good support to this idea (Card and Hutchinson, 1972). Certainly, the mercury contents of sphalerites from the Errington and Vermillion mines are relatively high.

In the Grenville province, both deposits sampled, New Calumet and Tetrault, lie close to the Ottawa Valley—St. Lawrence Valley rift system (Kumarapeli and Saull, 1966). Clearly, sampling of other occurrences is required to substantiate whether this spatial relationship has a real and positive influence on the mercury levels in sphalerites, and for that matter, other sulphides.

Phanerozoic volcanic types

Table II and Fig.1 display the distribution and values for samples collected in the Maritimes, Quebec and in the Yukon. The mercury contents of sphalerites range from 3 ppm (Sturgeon River, N.B.) to 116 ppm (Key Anacon, N.B.) with a mean of 50 ppm.

In terms of geographic locations the following subdivision is relevant:

British Columbia:	23 ppm	(3 deposits)
Yukon:	114 ppm	(1 deposit)
Southern Quebec:	61 ppm	(1 deposit)
New Brunswick:	40 ppm	(6 deposits)

Data on the Phanerozoic deposits are presently limited; therefore, it is not particularly useful, at this stage of the project, to consider further subdivision into specific geological epochs. Influences of tectonic disturbances would seem to be more significant factors than ages of deposits. For example, the only major occurrences of mercury sulphides in Canada are in the Cordillera where they are clearly related to deep faulting (Pinchi fault, Tintina trench). The Anvil (Faro) deposit in the Yukon, situated close to the Tintina trench, contains above-average quantities of mercury in its sphalerites.

Inspection of data for carbonate-hosted type sphalerites (Table III), which will be discussed in the next section, indicates a similar trend in the Cordilleran Orogen. A paucity of data from deposits in New Brunswick is likely to lead to a revision of the values quoted for mercury in those deposits although the orders of magnitude are not expected to vary. In general, however, it can be seen that there is more mercury associated with Phanerozoic age sphalerites than with those of Archean age, with a few notable exceptions, and less than in Proterozoic age sphalerites.

Proterozoic carbonate-hosted types

Table III and Fig.1 show the sample locations and mean values for mercury in sphalerites from carbonate-hosted type ores, collected within the Shield.

The values are quite variable with little evidence of a characteristic pattern. To quote a mean value would be somewhat pointless, so each of the districts in which the occurrences lie are averaged instead.

Southern Ontario:	19	ppm	(2 deposits)
Baffin Island:	3.5	ppm	(1 deposit)
New York State:	1198	ppm	(1 deposit)

Although the last deposit, at Balmat, N.Y., does not lie in Canada it is interesting to note the unusually high value for mercury which reaches 0.1% in zinc concentrates, an economically significant quantity. However, mercury may have been added subsequent to ore emplacement and possibly even subsequent to metamorphism. Similar comments may apply also to Long Lake where the mercury content is quite high, although much less than at Balmat.

The first important observation to be made is that Mississippi Valley-type sphalerites generally contain much less mercury than volcanogenic deposits with the one extreme exception at Balmat. Samples have been collected to date only from the Grenville province and from Baffin Island. The Grenville samples, which are largely from the Frontenac axis, are much higher than those measured from Baffin Island.

Phanerozoic carbonate-hosted types

Table III and Fig.1 summarize the data obtained for Phanerozoic Mississippi Valley-type ores. With one exception (Prairie Creek, N.W.T.) all mean values lie between 0.075 ppm (Little Pike Bay) and 26 ppm (H.B., B.C.). The mean

TABLE III

Mercury content of sphalerites from Mississippi Valley-type ores

Name of occurrence	Location	Host rock	Ore type	No. of samples	Range (ppm)	Mean (ppm)
Phanerozoic						
Walton	N.S.	limestone	Cu—Pb—Zn, Ag, Ba	2	0.180—0.550	0.365
Prairie Creek[1]	Mack.	carbonates	Pb—Zn	2	800—1067	933
H.B.	B.C.	dolomite	Pb—Zn	1	20—	26
Kicking Horse	B.C.	dolomite	Pb—Zn	5	6.49—21.5	11.6
Monarch	B.C.	dolomite	Pb—Zn	4	12—15.1	12.8
Pine Point[2]	Mack.	dolomite	Pb—Zn, Cd	26	0.010—2.59	0.326
Bankeno	Frnk.	dolomite	Pb—Zn	1	0.40	0.40
Wiarton	S. Ont.	dolomite	Zn	6	0.108—0.158	0.120
Tobermory	S. Ont.	dolomite	Zn	5	0.138—0.450	0.311
Little Pike Bay	S. Ont.	dolomite	Zn	1	0.075—	0.075
Ferndale	S. Ont.	dolomite	Zn	1	1.35—	1.35
Kaladar Rd.	S. Ont.	dolomite	Zn	2	4.60—6.60	5.60
Schoolhouse	S. Ont.	dolomite	Zn	17	1.64—10.4	5.22
Gays River	N.S.	limestone and dolomite	Zn, Pb	5	0.020—0.875	0.305
Newfoundland Zinc	Nfld.	dolomite	Zn	7	0.178—0.964	0.520
Proterozoic						
Strathcona	Frnk.	dolomite	Pb—Zn	3	2.2—4.7	3.5
Long Lake	S. Ont.	limestone	Pb—Zn	5	21—49	31
Thirty Is. Lake·	S. Ont.	limestone	Pb—Zn	1	6.0—	6.0
Balmat[2]	N.Y., U.S.A.	limestone	Pb—Zn, Hg	2	1118—1278	1198

[1] Genetic type classification in doubt.
[2] Includes some monthly composite zinc sulphide concentrates.

for all sphalerites, excepting Prairie Creek, which may or may not be of the Mississippi Valley-type, is 3.7 ppm. But once again it makes more sense if averages are quoted for specific areas rather than for the group as a whole.

Maritimes:	0.42 ppm	(3 deposits)
Bruce Peninsula:	0.46 ppm	(4 deposits)
Niagara Peninsula:	5.4 ppm	(2 deposits)
British Columbia:	16.8 ppm	(3 deposits)
Pine Point:	0.39 ppm	(5 deposits)
Arctic Islands:	0.40 ppm	(1 deposit)
MacKenzie Mountains:	933 ppm	(1 deposit)

It is immediately apparent that almost all Phanerozoic age Mississippi Valley sphalerites contain much less mercury than do those in the Proterozoic. The highest contents are to be found in those samples collected in the Cordillera where there is evidence of fault control of some of the ore bodies studied; an observation which is consistent with the pattern already noted for volcanogenic types. However, Prairie Creek is decidedly unusual, even for the Cordillera. When values around 0.1% Hg are recorded in the Cordillera, it may be advisable to look for mercury minerals or highly enriched sulphosalts. The recent discovery of new deposits of Pb—Zn in nearby areas of the MacKenzie and Selwyn Mountains may provide further insight into the problem of whether the high mercury content is locally or regionally distributed. Samples from the Maritimes, which include some sphalerites from the Gay's River area are all typically low and comparable to Pine Point, Bruce Peninsula and the Arctic Islands. The presence of associated hydrocarbons in the zinc deposits of the Bruce Peninsula (Sangster and Liberty, 1971) or at Pine Point does not appear to have influenced the mercury values in sphalerites whatever, although there are a few local highs to be found in the Pine Point ores. Data in Table I suggest that mercury distribution is uniform for all types of Mississippi Valley sulphides if they occur in areas removed from post-depositional tectonic disturbances; e.g., in the Maritimes. From this viewpoint, it is interesting to contemplate the reasons for the relatively elevated values reported for the Niagara Peninsula area sphalerites which approach the levels found for some of their volcanogenic neighbours in the nearby Grenville structural province.

APPLICATIONS TO GEOEXPLORATION

Deposits in the Canadian Shield

The most important inference which can be drawn from the data of Tables II and III is that certain parts of the Canadian Shield emerge as highly favourable for the use of mercury as a prospecting aid, because of its enrichment in zinc sulphides within showings and deposits of those areas; whilst others can be eliminated from consideration by converse reasoning.

All areas of Proterozoic age volcanic and sedimentary terranes must be deemed favourable. If sulphide mineralization does exist in those areas it is likely mercury enriched. Caution should however be observed in interpretation procedures when using mercury because there is some evidence to suggest that certain host rocks such as carbonaceous black shales of Proterozoic (Aphebian) age are considerably enriched in mercury compared with both younger Phanerozoic and older Archean equivalents (Cameron and Jonasson, 1972).

The same observations have not been made in detail for other rock types. Mercury is not strongly held in silicate lattices, in fact no silicate minerals of mercury are presently known. One of the reasons why black shales gather and retain mercury is that these are often pyritic (or at least sulphur-containing). Mercury is incorporated very efficiently into sulphides either during or after their deposition; and especially in the pyrite lattice, is retained during heating and other metamorphic effects. Thus in sulphide-poor rocks mercury is likely to be lost almost completely. Another strong binding influence on mercury is carbonaceous matter itself. Mercury has been shown to survive in measurable quantities through to the graphitic end member of hydrocarbon maturation (Cameron and Jonasson, 1972). Thus sulphidic carbonaceous shales are considered to be highly stable repositories for mercury should it become incorporated into the sediments at some stage during their deposition and diagenesis. The Proterozoic eon (and the Aphebian epoch in particular) seems to have been a time in which the mercury abundance in the Shield increased sharply. This increase is reflected in certain groups of sedimentary rocks (Cameron and Jonasson, 1972) and zinc sulphide ores (this paper).

Fig.2 presents the areas of Canada where it is considered that mercury can be employed in exploration, for genetic types of deposits considered in this paper, with a reasonable chance of success. The Hanson Lake—Flin Flon—Snow Lake—Lynn Lake volcanic belts seem to be particularly favourable areas. Other areas where Proterozoic age sulphides, both volcanogenic and carbonate-hosted, are known, viz., the Frontenac axis—Adirondack dome region of the Grenville province, the Ottawa—St. Lawrence Valley rift zones within the Grenville province, and the Sudbury basin, are also strongly favoured.

Although no data are presented here in support, preliminary sampling of the sulphide vein materials of the Ag-rich areas (Proterozoic) of the Great Bear Lake shows a very high mercury content (up to 0.5%) so this region can also be added to Fig.2. In fact all of the major silver-producing areas of Canada sampled to date, e.g., Cobalt and Silver Islet contain native silver and sulphides with percentage contents of mercury.

When terranes which host Archean age volcanogenic deposits are considered for possible mercury surveys of rocks, soils and sediments, only parts of the mineralized belts emerge as favourable. The most interesting of these is the Noranda—Val d'Or region and the whole belt of volcanics west to Timmins and Cobalt. Although there is considerable variation within orebodies across this zone, it is nevertheless the best area of the Archean found to date in which to use mercury as a pathfinder. The Cu—Au belt near Chibougamau is not considered a favourable area for mercury prospecting; copper sulphides

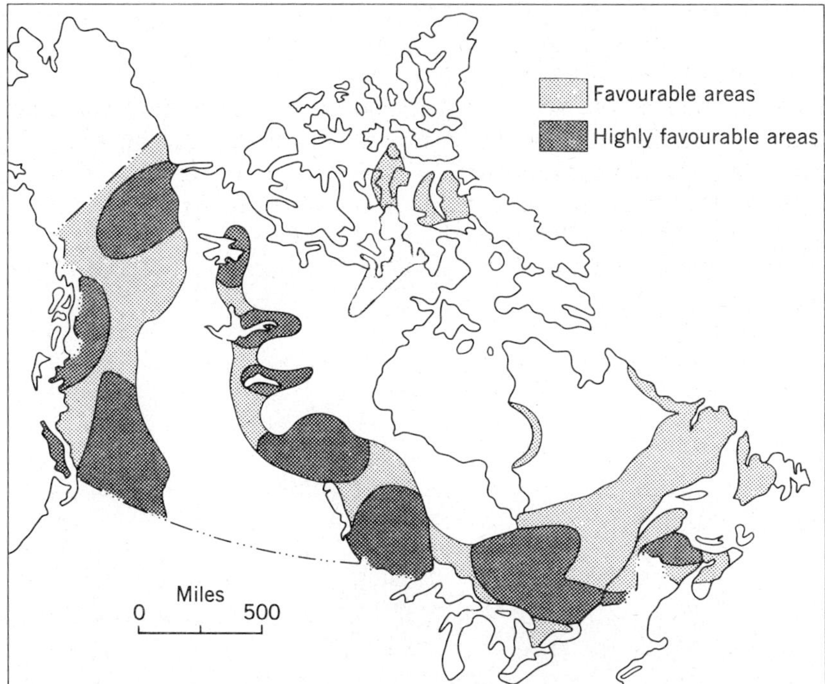

Fig.2. Areas of Canada amenable to geochemical usage of mercury.

from this region are very low in mercury (MacLatchy and Jonasson, 1974).
The low values encountered in sphalerites from Kidd Creek compared with
nearby mines are striking but as yet unexplainable; particularly in view of the
high levels of elemental silver present in the Kidd Creek sulphides. The
Manitouwadge region is not considered very suitable for mercury usage, but
to the west, near Thunder Bay where Cobalt-type silver deposits occur,
preliminary data have shown considerable enrichment of the vein sulphides
in mercury (Table IV). This area is, therefore, also designated a favourable
one (Fig.2). On the other hand, data for the Proterozoic age Mississippi Valley
deposits of Baffin Island suggest that similar terranes in that region of the
Shield are not so favourable for mercury prospecting methods.

Significantly high levels of mercury are observed in the Sturgeon Lake area
of northwestern Ontario to make it good one for mercury usage. Some work
by E.M. Cameron (G.S.C., personal communication, 1974), has shown that
the volcanic (greenstone) rocks of northwestern Ontario are higher in mercury
than their analogues in the Noranda area, an interesting observation in view
of the suitability of mercury in the latter area. It is probable, therefore, that
both regions are favourable for mercury usage. Some of the environmental
problems encountered with mercury in northwestern Ontario may, in part,
be due to naturally high levels in all rocks and sulphides of the area. Although
no detailed information is presently available on the subject, it is known that

TABLE IV

Mercury contents of some sphalerites from vein-type deposits

Deposit	Location	Ore type			No. of samples	Mean (ppm)
Paleozoic[1]						
United Keno	Keno Hill, Y.T.	Pb—Zn, Ag,	Cd		2	20.0
Galkeno	Keno Hill, Y.T.	Pb—Zn, Ag			1	7.8
Arctic Silver	Y.T.	Pb—Zn, Ag,	Au		1	14.5
Blue Bell	B.C.	Pb—Zn, Ag			1	27.5
Smithers area	B.C.	Pb—Zn, Cu			3	11.0
Keymet	Bathurst, N.B.	Cu—Zn			2	0.30
Dorchester	N.B.	Pb—Zn			8	2.80
Proterozoic						
Ramah shales	Nth Labrador	Pb—Zn, Cu			1	28.7
Cobalt area	Nth Ontario	Pb—Zn, Ag			5	200
Frontenac Lead	Kingston, Ont.	Pb, Zn			1	1.08
Kingdon Lead	Arnprior, Ont.	Pb, Zn			1	11.3
East Silver Mtn.	Thunder Bay, Ont.	Pb—Ag, Zn			2	39.0
Silver Lake	Thunder Bay, Ont.	Pb—Ag, Zn			2	9.1
Porcupine	Thunder Bay, Ont.	Pb—Ag, Zn			2	32.5
Beaver	Thunder Bay, Ont.	Pb—Ag, Zn			2	33.0
Archean						
Thubin Lake	N.W.T.	Zn			1	50.0
Homer Lake	N.W.T.	Zn, Pb			1	19
Turnback Lake	N.W.T.	Zn			2	22
Cobalt area	N. Ont.	Pb—Zn, Ag			1	55
Box (Uranium City)	N. Sask.[2]	Pb—Zn, Au			7	0.04
Severn River	N.W. Ont.	Cu—Zn			1	20
Berens River	N.W. Ont.	Cu—Zn			2	70

[1] The ages of some of the deposits listed as Paleozoic remain in contention.
[2] Data from R.W. Boyle, Geological Survey of Canada.

certain gold—quartz deposits enriched in sulphosalts and stibnite contain significant amounts of mercury in both western Ontario and in the Yellow-knife area (Archean) of the Slave province (R.W. Boyle, G.S.C., personal communication, 1974; and 1961). Information on the volcanogenic massive sulphides of the Slave province is somewhat sparse. To date only the area close to the Macdonald fault and its subsidiaries looks a good prospect for mercury geochemical surveys. Mercury contents of sphalerites from the northern parts of the Slave are low in comparison.

The possible relationship between the mercury enrichment of the sphalerites and tetrahedrites of Indian Mountain Lake and the Macdonald fault system raises an interesting possibility. It does seem that many of the elevated mercury levels in both Proterozoic and Archean sphalerites are found in deposits close to major lineaments or faults, and this finding parallels a worldwide trend noted in mercury deposits of all ages (Jonasson and Boyle, 1972). In the Canadian Shield, at least, there is little evidence of high mercury levels in sulphides in Archean time, except in deposits in the vicinity of extensive fault structures. However, Proterozoic volcanism may have subsequently led to considerable mercury enrichment in ore zone materials (sulphides and carbonaceous shales) which can retain it. Could it be that the high mercury contents of some sphalerites in Archean rocks are related to mercury input from faults that were either formed in or re-activated in, later geological epochs.

The Cretaceous through Tertiary epochs also seem to be a time of mercury mobilization and incorporation into geological materials. These influences are probably manifest in deposits of all ages in the Cordillera in particular.

The only known occurrence of cinnabar in the Canadian Shield occurs associated with vein-type mercurian tetrahedrite (4% Hg) and a number of other Hg-rich sulphides (chalcopyrite, pyrite, stibnite) in the Clyde Forks area of southeastern Ontario (Grenville province). It and a number of other Hg-rich tetrahedrite showings lie along a zone, some 30 miles in length, that may well have been subjected to effects from deep faulting related to the formation of the Ottawa Valley graben to the north. In particular, the Clyde Forks occurrence lies in an ancient Precambrian fault which is cut nearby by the Long Lake fault of possible Cretaceous age (Nikols, 1972).

One final point of interest is that many of the major zones of interest with respect to mercury occurrence in the Shield lie close to the southern margin of the exposed Shield. Perhaps this is not unusual since many other base metals are more plentiful near the southern perimeter of the Archean than in the centres of the Precambrian masses. But where sulphides do occur in the interior of the Shield, the mercury content is usually low.

Consequently, *all* areas of the Shield where late, regional tectonic disturbances are known or suspected, should be considered favourable for the use of mercury geochemistry. Some of these have been marked in Fig.2.

Phanerozoic age deposits

As a general rule, Mississippi Valley (carbonate-hosted)-type sphalerites have very low mercury contents. In fact, the mercury content of the sulphides is usually the same as that of the carbonate or evaporite host-rocks. Bearing this in mind, it is interesting to note the striking differences between average mercury contents of sphalerites from volcanogenic deposits and Mississippi Valley deposits in the Appalachian regions of Canada. These differences represent one to two orders of magnitude in favour of volcanogenic types. Therefore, the areas of Appalachia, viz., New Brunswick, wherein volcanogenic ores occur which are enriched in mercury, are considered quite favourable for mercury usage in prospecting for deposits. (Similar considerations apply to the Cordillera.) However, Mississippi Valley types in the same areas are unlikely to be detected by these methods, simply because mercury is of such low abundance in these ore types.

Some comment ought to be made on the relative levels of mercury in sphalerites from volcanogenic and Mississippi Valley-type deposits in Appalachia. Two inferences are possible: that mercury was not available in the source rocks for its eventual incorporation into the sulphides of Mississippi Valley-type deposits; or that mercury was available but did not participate in the geochemical transportation processes by which zinc, and possibly lead, were carried to their depositional sites. It is significant that other highly chalcophilic metals such as silver, arsenic and copper are also generally absent from Mississippi Valley-type deposits. Consequently, most Phanerozoic Mississippi Valley types should not be sought with mercury geochemistry. The sole exceptions seem to be the deposits of the Cordilleran Orogen, e.g., Kicking Horse, where some fault control of the deposits may be in evidence. It is considered then, that mercury could be usefully employed in geochemical surveys in most places in the Cordilleran regions of western Canada (including the Keno Hill area where mercury ranges from 8 to 25 ppm in vein-type sphalerites; Table IV).

The moderate levels observed in occurrences of sphalerites in the zinc sulphides of the Niagara Peninsula remain unexplained but the values recorded herein may represent threshold levels when mercury is to be considered as a prospecting aid in such terranes.

CONCLUSIONS

Assessment of the mercury contents of sphalerites and zinc concentrates from deposits and occurrences of stratabound sulphides, both volcanic types and carbonate hosted types, indicates a wide variance. The significant factors which influence the magnitude of mercury levels include age and type of deposit, bulk composition of the ore, i.e., proportion of zinc and perhaps lead, and proximity to post-ore activation or re-activation of regional fault structures.

With respect to the age of host rocks, mercury in sphalerite of volcanogenic type deposits varies in its mean relative abundance as follows: Proterozoic (99 ppm) > Phanerozoic (50 ppm) > Archean (33 ppm).

In carbonate-hosted deposits, the order of abundance is similar: Proterozoic (~10 ppm) > Phanerozoic (3.7 ppm); similar deposits in the Archean have not been recognized.

The mercury contents of sphalerite in Zn- and Pb-rich volcanogenic deposits are generally higher than those in the Cu-rich ones. Sphalerites in deposits which contain sulphosalt minerals (generally Pb—Zn, Ag ores) e.g., tetrahedrite, were found to be high in mercury. The highest mercury contents in sphalerites from Mississippi Valley-type deposits generally came from those enriched in lead; but although this may not be a general rule, it may offer some clues to ore genesis mechanisms.

The influences of regional faulting and related tectonic disturbances may well be important, particularly if the disturbances are post-ore. Two periods of mercury mobility and deposition are considered possible — the Aphebian epoch and the Cretaceous epoch. Many of the very high values found in sphalerites of all ages and types may possibly be related to tectonic events within these time periods.

From the viewpoint of the geochemical usage of mercury in prospecting surveys it is recommended that all areas where ores of Proterozoic age occur, whether vein-type, volcanogenic or carbonate-hosted, are suitable for such mercury usage. Certain areas within Archean terranes are also recommended as suitable, as are also parts of the Appalachian Orogen and the whole of the Cordilleran Orogen. In general, the interior areas of Precambrian masses, in particular the Archean, are not considered amenable to the exploitation of mercury as a prospecting aid.

ACKNOWLEDGEMENTS

The authors gratefully acknowledge the invaluable contribution of P.J. Lavergne who devised and implemented the sphalerite separation procedures employed in this study. We also thank our colleagues, R.W. Boyle, R.V. Kirkham, A.G. Plant and R.I. Thorpe who contributed both of samples to the project and of their expertize to aspects of the discussion.

REFERENCES

Boyle, R.W., 1961. The geology, geochemistry and origin of the gold deposits of the Yellowknife District. Geol. Survey Can., Mem., 310: 193 pp.

Boyle, R.W. and Dass, A.S., 1971. Origin of the native silver veins at Cobalt, Ontario. Can. Mineral., 11(1): 414—417

Cameron, E.M. and Jonasson, I.R., 1972. Mercury in Precambrian shales of the Canadian Shield. Geochim. Cosmochim. Acta, 36: 985—1005

Card, K.D. and Hutchinson, R.W., 1972. The Sudbury structure: its regional geological setting. Geol. Assoc. Can., Spec. Paper 10: 67—78

Ivankin, P.F. and Turkin, I.S., 1972. Mercury mineralization genesis. Sov. Geol., 15(1): 53—61 (in Russian)

Jonasson, I.R. and Boyle, R.W., 1972. Geochemistry of mercury and origins of natural contamination of the environment. Can. Inst. Min. Metall., Bull., 65: 32—39

Jonasson, I.R., Lynch, J.J. and Trip, L.J., 1973. Field and laboratory methods used by the Geological Survey of Canada in geochemical surveys, No. 12. Mercury in ores, rocks, soils, sediments and water. Geol. Survey Can., Paper 73-21: 22 pp.

Kumarapeli, P.S. and Saull, V.A., 1966. The St. Lawrence Valley system; a North American equivalent of the East African rift valley system. Can. J. Earth Sci., 3: 639—658

MacLatchy, J.E. and Jonasson, I.R., 1974. The relationship between mercury occurrence and mining activity in the Nottaway and Rupert River basins of northwestern Quebec. Geol. Survey Can. Paper Ser. (in press)

Nikols, C.A., 1972. Geology of the Clyde Forks mercury, antimony, copper deposit and surrounding area, Lanark County, Ontario. M.Sc. Thesis, Queen's University, Kingston, Ont., 116 pp.

Ripley, L.G., 1964. The spectrophotometric determination of mercury in a zinc concentrate from Manitou-Barvue Mines, Ltd., Val d'Or, Quebec. Min. Bur. Invest. Rept., No.IR 64-83 (Ottawa), 10 pp.

Sangster, D.F., 1970. Metallogenesis of some Canadian lead—zinc deposits in carbonate rocks. Geol. Assoc. Can., Proc., 22: 27—36

Sangster, D.F., 1972a. Precambrian volcanogenic massive sulphide deposits in Canada: a review. Geol. Survey Can., Paper 72-22: 44 pp.

Sangster, D.F., 1972b. Isotopic studies of ore-leads in the Hanson Lake—Flin Flon—Snow Lake mineral belt, Saskatchewan and Manitoba. Can. J. Earth Sci., 9: 500—513

Sangster, D.F. and Liberty, B.A., 1971. Sphalerite concretions from Bruce Peninsula, southern Ontario, Canada. Econ. Geol., 66: 1145—1152

Sears, W.P., 1971. Mercury in base metal and gold ores of the Province of Quebec. In: R.W. Boyle and J.I. McGerrigle (Editors), Geochemical Exploration. Canadian Institute of Mining and Metallurgy, Montreal, Ont., pp.384—390

RUBIDIUM AND STRONTIUM AS GUIDES TO COPPER MINERALIZATION EMPLACED IN SOME CHILEAN ANDESITIC ROCKS

JORGE OYARZUN M.

IRITUN, Universidad del Norte, Antofagasta (Chile)

ABSTRACT

It has proved possible to use rubidium and strontium as guides to copper mineralization of both porphyry (Braden, Rio Blanco) and stratiform (Buena Esperanza) types emplaced in andesitic and basaltic-andesitic types.

Both elements follow variations of major elements (Rb—K and Sr—Ca) which were themselves added or removed during hydrothermal alteration associated with mineralization. However, their high sensitivity to hydrothermal processes, and the facility with which they may be determined by X-ray fluorescence, renders rubidium and strontium preferential indicators.

Where porphyry copper deposits are emplaced in andesitic rocks rubidium is a sensitive and semi-quantitative indicator of alteration (G. Armbrust et al., 1971; J. Oyarzun M., 1971). The absolute rubidium content and the Rb/K ratio both increase passing from unaltered rock through to propylitic zone to the potassic zone — the latter being usually associated with primary mineralization. However, this effect apparently does not apply where porphyry copper deposits are emplaced in granitic rocks, e.g. El Abra (B.G.N. Page and H. Conn, 1973).

Where stratiform copper deposits occur within Jurassic andesitic and basaltic volcanics, e.g. Buena Esperanza (J. Losert, 1972), strontium is a good indicator of local alteration associated with mineralization in that it decreases to 1/10 of the "normal" content of 200—300 ppm.

INTRODUCTION

This paper summarizes an investigation into the use of rubidium and strontium distributions in altered volcanic rocks as a guide to copper mineralization.

The general geochemical behaviour of rubidium and strontium is well known. Rubidium replaces potassium in silicates, chiefly feldspars and micas, and due to its higher ionic radius, with consequent lower bonding energy, concentrates in the residual Si-rich magmatic differentiates (Volkov and Savinova, 1961; Voskresenkaya et al., 1962) (Table I). For example, a 10-fold rubidium enrichment and 4-fold decrease in K/Rb ratio with respect to the first low silica differentiates, has been established for some New Zealand rhyolites (Taylor et al., 1956).

Although strontium and calcium share similar crystallochemical character-

TABLE I

K and Rb concentrations of rocks of the circum-Pacific belt (Taylor, 1968)

	K (%)	Rb (ppm)	K/Rb
Basic andesite	0.91	14	650
Andesite	1.33	31	430
Dacite	1.70	44	386
Rhyolite	2.81	108	250

TABLE II

Rb/Sr ratios in fresh and altered andesites

Andesites	Rb/Sr
Quaternary (Chile)	0.06
Circum-Pacific belt	0.08
Tertiary (Chile)	0.16
Cretaceous (Chile)	0.18
Rio Blanco	
potassic alteration	2.55
El Teniente	
propylitic alteration	0.31
potassic alteration	0.51

istics, they do not correlate as well as rubidium and potassium and may even present an inverse relationship (Turekian and Kulp, 1956).

Hydrothermal processes generally cause silicate rocks to become depleted in calcium (except in the case of epidote alteration). This is followed by precipitation of calcium as carbonate or sulfate in veinlets or vesicles. Consequently, strontium should also be depleted in most altered rocks. Potassium behaves differently since it is enriched in an important alteration type associated with porphyry copper mineralization, and is generally less depleted than calcium due to its bonding to secondary micas and argillic minerals. Similarly, rubidium may also be enriched or preserved but strontium is generally removed from most hydrothermally altered rocks. The Rb/Sr ratio, characteristic for each magmatic series, should be considerably higher in hydrothermally altered rocks or in those affected by low grade metamorphism. For example, Chilean Quaternary andesites have an average Rb/Sr ratio of 0.06, whilst the figure for circumpacific andesites (Taylor, 1968) is 0.08. However Tertiary and Cretaceous rocks of Chile, that are either slightly altered or display burial metamorphism, have Rb/Sr ratios of 0.16 and 0.18, respectively.

The Rb/Sr ratio is substantially greater for hydrothermally altered rocks in the vicinity of some Chilean porphyry copper deposits (Table II).

Reasons for studying rubidium and strontium in preference to potassium and calcium, their geochemically associated major elements, include:

(1) greater sensitivity to hydrothermal processes expressed as variations of the K/Rb and Ca/Sr ratios; (2) wider ranges of variation than major elements; (3) more readily amenable to rapid and precise analytical determination.

RUBIDIUM AND STRONTIUM IN PORPHYRY COPPER DEPOSITS

Porphyry copper mineralization is frequently closely related to potassic alteration, and "productive" porphyries may have a well-defined potassium enrichment when compared with barren ones (Davis and Guilbert, 1973).

If mineralization and alteration associated with porphyries is superimposed on andesitic or basaltic rocks that are originally low in rubidium and high in strontium, potassic alteration may produce a great change in both elements with respect to background values. This is the case for many porphyries emplaced in andesitic formations of South America and the South East Asian islands.

As part of a larger lithogeochemical study, samples of altered and unaltered andesites from the Rio Blanco porphyry copper deposit (33°9'S; 70°16'W) were analysed for major and trace elements, including rubidium and strontium (Oyarzun, 1971). The Rio Blanco deposit includes a granodiorite-dacite complex, intruded into andesites of the Farellones Formation (Upper Cretaceous to Lower Tertiary). The mineralization (Cu and Mo) is mainly in the potassic altered andesites and has an age of about 4 million years (K—Ar in magmatic biotite; Quirt et al., 1971).

A 3 times enrichment in rubidium and an 8 times depletion in strontium was found. However, the K/Rb ratio was in the same order and the Ca/Sr ratio was about half that of the fresh andesite (Table III).

To test possible systematic variation in rocks altered by porphyry copper mineralization, El Teniente (Braden) deposit (34°6'S, 70°20'W) was studied by Armbrust et al. (1971). This deposit is emplaced in the lower member of the Farellones Formation. Economic mineralization comprises copper and molybdenum sulphides peripheral to a breccia pipe structure whose shape is that of an inverted cone 1200 m in diameter and 1560 m deep (Fig.1).

TABLE III

Rb/Sr ratios in altered rocks adjacent to porphyry copper mineralization

	Rb (ppm)	K/Rb	Sr (ppm)	Ca/Sr
Fresh andesite	50	330	480	77
Rio Blanco				
potassic	155	335	60	35
El Teniente				
propylitic	105	167	338	
potassic	195	160	380	
leached cap	194		27	

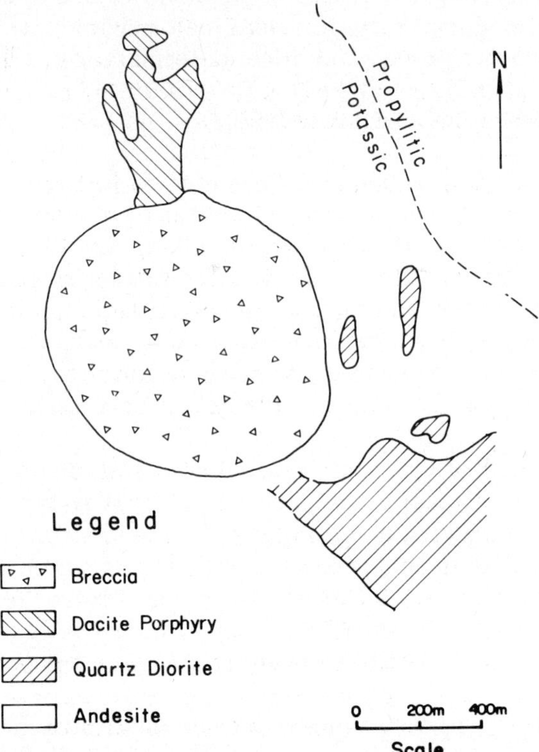

Fig.1. Generalized geological map, El Teniente deposit.

Fragments in the breccia are andesites from the Farellones Formation and tonalites and porphyritic rocks from the intrusive complex emplaced around the pipe. Age of El Teniente (K—Ar, whole rock; Quirt et al., 1971), is about 4—5 million years. Mineralization is closely related to a biotite-anhydrite alteration assemblage. In andesites of the Farellones Formation peripheral to the ore, this potassic alteration grades outward into a **propylitic assemblage.**

Sixty-two samples from mine workings and drill cores were analysed for potassium, rubidium,and strontium (Table III). Fresh andesites of the Farellones Formation average 50 ppm Rb and have a K/Rb ratio of 330 (Oyarzun, 1971). Andesites of the biotitic zone at El Teniente average 195 ppm Rb, and those of the propylitic zone 105 ppm. The K/Rb ratio in the two zones are 160 and 167, respectively. Therefore, a well-defined increase in rubidium was **observed from the** unaltered andesites to those of the potassic zone, while propylitized andesites have an intermediate value. In contrast to the Rio Blanco deposit no significant primary depletion in strontium was found, probably due to the abundant anhydrite associated with the potassic zone at El Teniente.

Strongly leached samples from the supergene zone have preserved an original high rubidium content but are markedly depleted in strontium. The latter was presumably removed by acidic meteoric water whilst rubidium was retained in the clay minerals. Thus, rubidium analysis may be useful even where weathering has masked the mineralogical effects of hydrothermal alteration.

In porphyry copper deposits that are intruded by intermediate or high rubidium content granitoids, different results have been obtained. At El Abra (21°56'S, 68°51'W) Page and Conn (1973) found no significant rubidium anomalies probably due to the lack of chemical contrast between intrusive and the intruded rocks. J. Arias (personal communication, 1974), however, detected highly anomalous rubidium concentrations in granitoids from the Chuquicamata deposit (22°18'S, 68°55'W). Sixteen out of 38 samples had rubidium values of around 1% which are very unusual rubidium values for any rock type. The potential use of rubidium and strontium as geochemical tools seems most suited to exploration in andesitic-basaltic terrains, a condition that prevails over large areas of the world, particularly the circum-Pacific region.

RUBIDIUM AND STRONTIUM IN STRATIFORM COPPER DEPOSITS

Losert (1972) carried out a study of the Buena Esperanza copper deposit (22°11'S, 70°14'W) which typifies manto-type mineralization. The deposit is emplaced in 28 andesitic flows of the mid to upper Jurassic La Negra formation. This 5—10 km thick formation has a fairly homogeneous andesitic lithology with minor ignimbrite intercalations. Sulphides, mainly bornite and chalcocite, occur as vesicles in the upper part of the flows and as disseminations in the ground mass. The rocks are affected by greenschist facies regional metamorphism. A gabbroic intrusion occurs in the mine whilst granodioritic intrusions occur nearby.

The mineralization, considered by Losert to be epigenetic, lies stratigraphically above a 100 m thick epidotized horizon which has been strongly depleted in copper and sulphur. The mineralized flows have a mean of 65 ppm Rb which is 1.5 times the content of "normal" andesites and 7 times that found in epidotized rocks. The strontium content, 25 ppm, is very low (Table IV).

TABLE IV

Rb/Sr ratios in altered rocks adjacent to stratiform copper mineralization

	Rb (ppm)	K/Rb	Sr (ppm)	Ca/Sr	Rb/Sr
La Negra					
Formation	45	270	230	260	0.20
Epidotized	10	300	1000	180	0.01
Mineralized					
(potassic)	65	280	25	660	2.60

In the barren epidotized zone, strontium is about 1000 ppm and the Ca/Sr value 180. Thus a 4-fold enrichment in strontium and a decrease in the Ca/Sr ratio is evident with respect to normal andesites from the La Negra Formation, i.e. 230 ppm Sr and Ca/Sr 260. In this zone the behaviour of rubidium is anti-pathetic to that of strontium and many samples contain less than the 10-ppm detection limit.

Thus normal andesites have an Rb/Sr ratio of 0.2; this falls to 0.01 in the epidotized zone but rises to 2.6 in the mineralized zone. Rubidium and strontium analysis is considered by Losert to be a valuable exploration guide for manto-type mineralization in volcanic rocks since Rb/Sr anomalies may indicate a mineralized horizon even though the deposit itself does not outcrop. Alfaro (1973) has used rubidium and strontium analysis for exploration in the La Negra Formation with positive results.

REFERENCES

Alfaro, G., 1973. Geologia y Prospeccion de yacimientos de cobre en la zona de Tocopilla. Instituto de Investigaciones Geologicas, Santiago, Intern. Rept., 26 pp.

Armbrust, G., Oyarzun, J. and Arias, J., 1971. Rubidium as a guide to ore at El Teniente (Braden), Chile. Geol. Soc. Am. 1971 Ann. Meet., Abstracts, p.494

Davis, J.D. and Guilbert, J.M., 1973. Distribution of the radioelements potassium, uranium, and thorium in selected porphyry copper deposits. Econ. Geol., 68(2): 145—160

Losert, J., 1972. Genesis of copper mineralisations and associated alterations in the Jurassic volcanic — rocks of the Buena Esperanza mining area (Antofagasta Province, Northern Chile). Dept. of Geology, Univ. of Chile, 104 pp.

Oyarzun, J., 1971. Contribution à l'étude des roches volcaniques et plutoniques du Chili. Thèse, Univ. of Paris, 195 pp.

Page, B.G.N. and Conn, H., 1973. Investigacion sobre metodos de prospeccion geoquimica en el yacimiento tipo cobre porfidico, El Abra, provincia de Antofagasta—Chile. Instituto Investigaciones Geologicas, Santiago, 40 pp.

Quirt, S., Clark, A.H. and Farrar, E., 1971. Potassium—argon ages of porphyry copper deposits in Northern and Central Chile. Geol. Soc. Am.,1971 Ann. Meet., Abstracts, p.676

Taylor, S.R., 1968. Trace element chemistry of andesites and associated calc-alkaline rocks. In: Upper Mantle Project, Proc. Andesite Conf., Eugene, Oreg., pp.43—63

Taylor, S.R., Emeleus, C.H. and Exley, C.S., 1956. Some anomalous K/Rb ratios in igneous rocks and their petrological significance. Geochim. Cosmochim. Acta, 10: 224—229

Turekian, K.L. and Kulp, J.L., 1956. The geochemistry of strontium. Geochim. Cosmochim. Acta, 10: 245—296

Volkov, V.P. and Savinova, E.N., 1961. Variation in the potassium—rubidium ratio during the evolution of calc-alkalic and alkalic magmas. Geochem. Int., 12: 1227—1236

Voskresenskaya, N.T., Titkova, N.F. and Shulyakovskaya, N.S., 1962. Geochemistry of thallium, rubidium and lithium in the magmatic process. Geochem. Int., 3: 282—292

Part 4
PRIMARY DISPERSION, II

Chairmen:

A.J. SINCLAIR
University of British Columbia, Vancouver, B.C., Canada
B. BÖLVIKEN
Geological Survey of Norway, Trondheim, Norway

RUBIDIUM: A PRIMARY DISPERSION PATHFINDER AT OGOFAU GOLD MINE, SOUTHERN WALES

M.J. AL-ATIA and J.W. BARNES

Department of Geology, University College of Swansea, Swansea (Great Britain)

ABSTRACT

Elements which are suitable as pathfinders to ore share certain characteristics, namely low melting points and high vapour pressures. Rubidium falls within these parameters and work on the primary dispersion aureole surrounding the ancient Ogofau gold mine in southern Wales suggests it has promising applications. At Ogofau the rubidium aureole in the shaley country rocks is better developed than the aureoles for lead, zinc, copper and mercury. The gold-pyrite mineralization is surrounded by a zone in which 90% of the 51 samples taken show rubidium enhancement greater than two standard deviations (20 ppm) above the mean background value for the area of 125 ppm, and 84% show enhancements greater than three standard deviations (30 ppm). There is no accompanying change in potassium content which might account for this enrichment. It is also significant that no rubidium aureole surrounds the low-temperature pyrite-free lead and zinc ores at Rhydymwyn, which occur only 7½ miles away in similar rocks, but there is a well-developed primary dispersion aureole of lead and zinc.

Rubidium has also been noted in the rocks surrounding the Braden porphyry copper deposit in Chili (G.A. Armbrust et al., 1971). The element may well be a selective pathfinder to certain types of mineralization but still requires further investigation. It has the advantage of ease of chemical determination in rock samples by simple XRF procedures.

INTRODUCTION

Primary dispersion aureoles which surround areas of mineralized rock, or which form a front to the paths of mineralizing fluids, may be used as prospecting targets which can lead to ore, and show especial promise in the search for blind ore bodies. For this reason there has been an increasing interest in the nature and patterns of primary dispersion and, in particular, in the distribution of "pathfinder" elements, that is, those which are related to mineralization without being of economic significance in themselves. Such studies have already shown promise (Broderick, 1929; Tauson and Petrovskaya, 1971; Chaffee, 1972). The elements most studied as pathfinders to date are As, Hg, Se, Bi, Zn and Pb. Rubidium, the subject of this paper, has not so far been considered in such a role, with the possible exception of a brief mention of anomalous contents surrounding porphyry copper deposits in Chili (Armbrust et al., 1971). The present study shows that rubidium has a theoretical potential as a pathfinder and that it is relatively easy to detect at the concentrations required. Practical work on two localities in southern Wales, the Ogofau gold mine and the Rhydymwyn Pb—Zn area, shows that rubidium is discriminatory, and this discrimination may itself have uses.

RUBIDIUM AND THE ROLE OF PATHFINDERS

The most significant elements in an aureole surrounding an ore body are obviously those metals most abundant in the ore, for instance lead and zinc in Pb—Zn deposits or copper in the wall rocks of copper deposits. However, associated elements present in only minor amounts in the ore may often define the aureole of mineralization better than the ore metals themselves, because they are more widely dispersed, or even more abundant in the aureole, than the ore metals. Many pathfinders are known and some which were listed by Hawkes and Webb (1962) are given in Table I.

TABLE I

Pathfinders for ore deposition (Hawkes and Webb, 1962)

Pathfinder elements	Material samples	Ore type
As	wallrock, residual soil, stream sediments	vein-type Au ore
Hg, Zn, Pb	wallrock, soil	complex Pb—Zn—Ag ores
Se	gossan, residual soil	epigenetic sulphides
Ag	residual soil	Ag-bearing Au ore
Mo	water, stream sediments, soil	porphyry Cu deposits

The presence of a pathfinder element in an aureole depends on many factors, such as: (1) its abundance in the mineralizing fluid; (2) its chemical state, whether elemental or as a compound, since this governs its chemical behaviour; (3) its melting point, boiling point and solubility, for these control its mobility in the fluid state; and (4) its vapour pressure, because this is a dynamic factor controlling its dispersion.

On the basis of these factors, rubidium would appear to be a promising pathfinder for it is volatile, it has a low melting point (38.5°C), and it also has a relatively high vapour pressure (1.56×10^{-4} N/m^2 at the melting point*).

---

*N/m^2 = 0.75×10^{-2} mm of Hg.

The chemical state in which it occurs in an ore fluid is, however, unknown
but many of its salts, for instance the chlorides, the nitrates and the carbonates,
are highly soluble in aqueous solutions, far more so than other pathfinders,
such as arsenic and mercury. It was because of its theoretical promise as a
pathfinder that it was decided to include rubidium as one of the elements to
be studied in the area investigated.

Little appears to be known of rubidium in its relation to mineralization
although it has been reported as a characteristic element in quartz-cassiterite
and quartz-topaz veins (Vlasov, 1966). It is not known as an element associated
with sulphide ores although there is a brief report of its presence in the aureole
surrounding the Braden porphyry copper deposit in Chili (Armbrust et al.,
1971). However, rubidium has a well known association with potassium in
nature which allows it to substitute for potassium owing to its similarity of
ionic radii (K = 1.35 Å; Rb = 1.49 Å). Vlasov (1966) reports enrichments of
up to 2.88% Rb in microcline and 1.6% in muscovite. Therefore, if rubidium
is to be used as a pathfinder, it is also necessary to determine potassium to
ensure that any enrichment is not merely a result of a corresponding change
in potassium content. That certain types of mineralization appear to reflect
an increase in rubidium in the surrounding rocks, independent of potassium, is
shown by the results given below.

GEOLOGY OF THE STUDY AREA

The study area lies between the town of Llandovery and the village of
Pumpsaint (Fig.1) on the western limb of the "Towy Anticline", a broad north-
easterly-trending feature striking northeasterly. Although the structure would
appear to be relatively simple, there are local zones of intense folding and
faulting. The only rocks seen in the study area are medium to dark shales
with sandstones, grits and conglomerates, of Bala (Upper Ordovician) and
Llandovery (Lower Silurian) age (Fig.2). The boundary between the rocks of
the two series is difficult to distinguish owing to similarities in lithology and
there appears to have been no distinct break in sedimentation between them
during their deposition in this area. No igneous rocks are exposed, nor is there
any indirect evidence to suggest any occur in the vicinity. As a study area it
is ideal, with a limited number of rock types, and free from any influences
which may disturb trace element distribution other than those due to mineralization.

ORE BODIES

There are two general areas of mineralization in the region considered, one
of gold, the other of lead and zinc. They are separated by about 12 km and
although there are differences in the characters of their mineralization, both
occur in rocks of Bala age. The gold ores at Ogofau were first exploited as far
as we know, during the Roman occupation of Britain between the 1st and

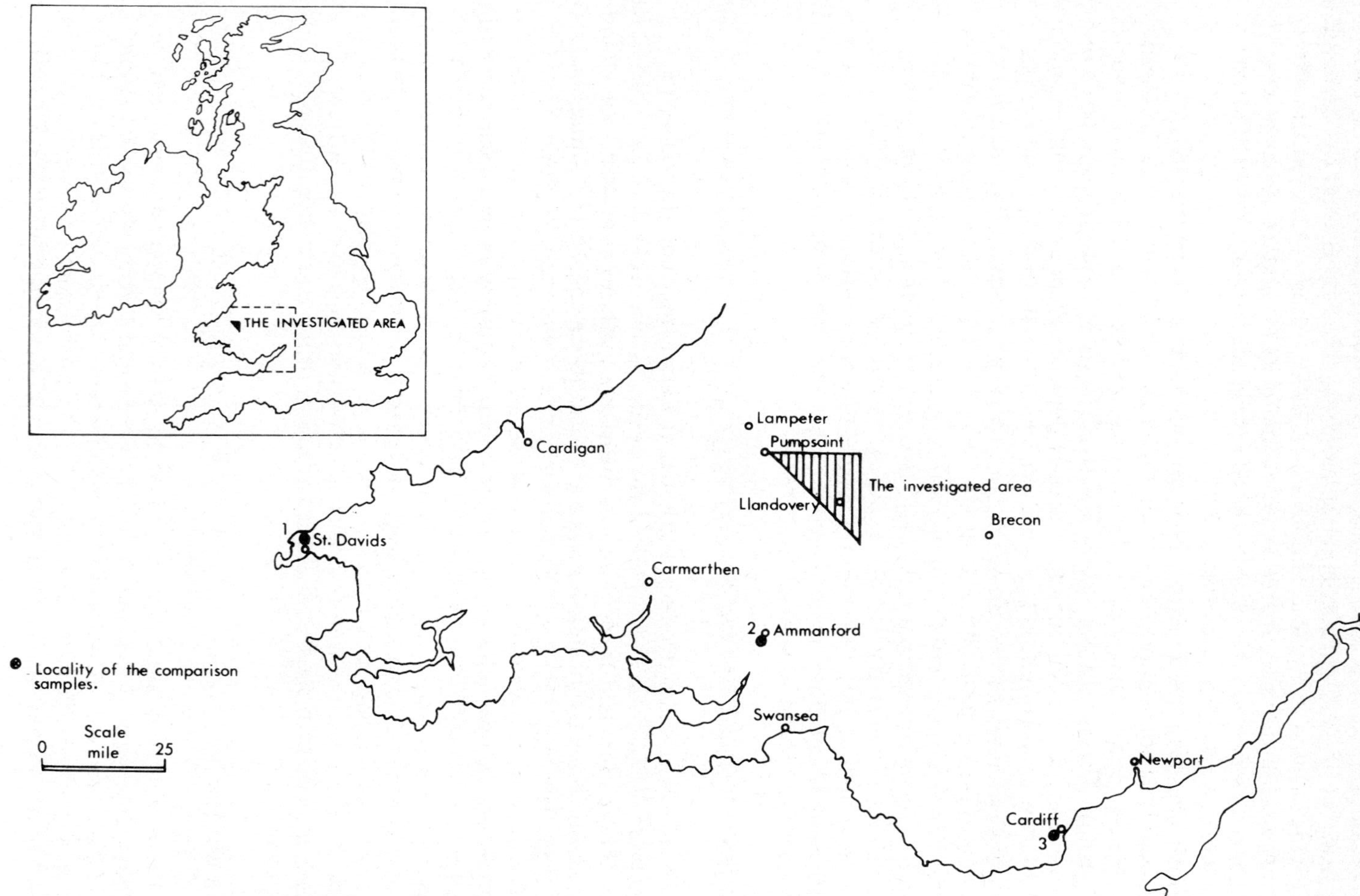

Fig.1. Locality map of the area investigated.

5th centuries A.D. (Jones and Lewis, 1971), and they have been worked
sporadically ever since the last phase ending in 1939. The gold occurs in
quartz-pyrite veins in Upper Bala shales and quite possibly in Lower Llandovery
shales too, for here the two series are difficult to distinguish in the intensely
folded rocks (Davies, 1933). A small amount of sphalerite and galena also
occur. The veins lie along faults and dilations in the shales and are often
irregular, with pinches and swells. Gold occurs free, and also in pyrite. Pyrite,
galena and sphalerite also form minor local enrichments without quartz in the
shales.

The Pb—Zn ores occur in the district surrounding the village of Rhydymwyn.
The main group of veins was mined as the Nantymwyn mine, ceasing work in
1939 after some 100,000 tons of ore had been won (Hall, 1971). There are
also some smaller showings in the area and one of these, Cynnant, enjoyed a
short life as a one-man mine until four years ago. Otherwise there is now no
mining activity in the area. The Nantymwyn ores occur as mineralized breccias
along faults which cut grit bands in the Bala shales, mineralization becoming
uneconomic as the faults pass into the shales. The veins carry galena and pale
to medium brown sphalerite in a quartz gangue and this vein material cements
angular breccia fragments of grit and sometimes shale. The quartz shows
crustification around breccia fragments, at times leaving small vugs lined with
tiny quartz crystals, and there is no sign of any replacement even where fine
shaley flakes occur. All the indications are that the ores were deposited at
very low temperature and there appears to be some similarity to the "tele-
thermal" Mississippi-type ores in that there is almost a complete absence of
pyrite, and that they are not obviously related to any igneous source. The
gangue is admittedly quartz, but no carbonate rocks are known in the vicinity
which could have supplied carbonate. Cynnant is merely a small-scale example
of Nantymwyn.

The gold and the Pb—Zn ores are thus very different in character. One is
rich in sulphide and gives every appearance of a mesothermal iron sulphide-
rich ore deposited with at least some replacement along fractures and dilations
in shale; the other, a typically epithermal fracture filling in grits with no
replacement and a very low iron content. Neither however, can be related to
any igneous rocks.

SAMPLING AND ANALYTICAL TECHNIQUES

The sampling programme covered both Bala and Llandovery rocks at
22 localities throughout the study area and some 181 samples were collected,
all of shale.

Localities were selected so that a trace element background for the area
could be established (100 samples), and so that disturbances to the regional
pattern caused by mineralization could be determined. (51 samples at Ogofau,
30 at Rhydymwyn: see Fig.2.)

Samples were prepared for analysis by crushing to 200 mesh in a Tema

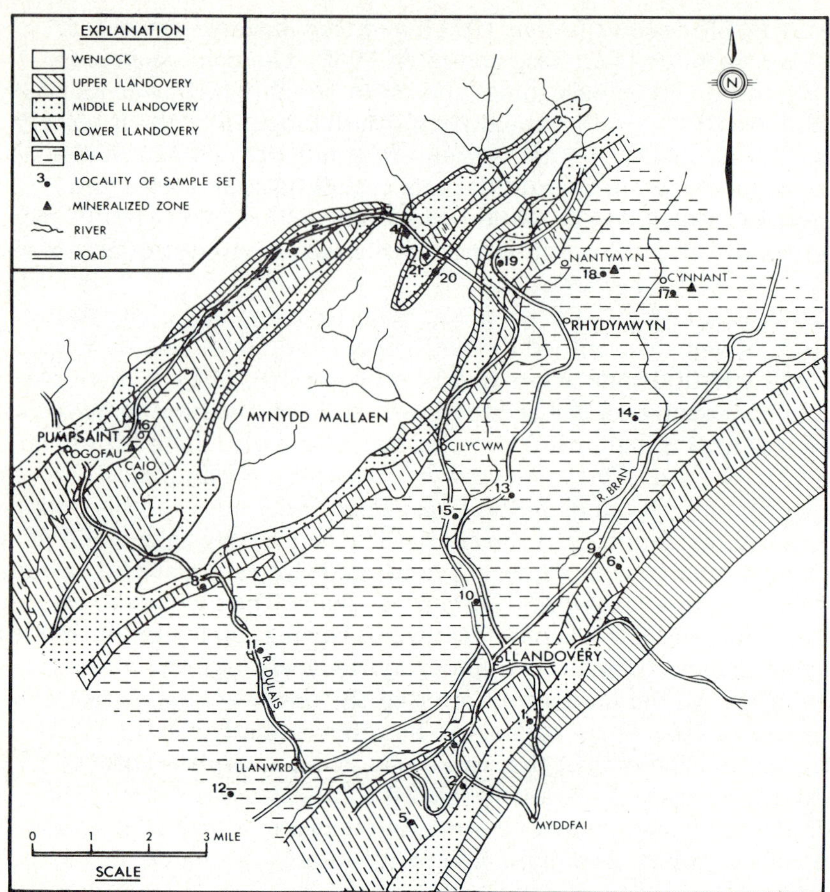

Fig.2. Geological map of Pumpsaint-Llandovery area (after Davies, 1933).

rotary mill with tungsten carbide liners. One gram of each sample was pelletized with stearic acid and analysed for Rb, Pb, Zn, Cu and K on a Phillips PW 1540 XRF spectrometer. A tungsten tube at 40 keV, a (220) crystal LiF, and a scintillation detector were used except for potassium determinations, when a chromium tube, a PE crystal, and a flow counter were substituted. Mercury was measured on a "Southern Analytical" atomic absorption spectrometer by the method described by Cameron and Jonasson (1972). Sensitivity, precision and accuracy were within the limits considered desirable for a study of this type and are reported elsewhere (Al-Atia, 1973).

PRIMARY DISPERSION AT OGOFAU

There is abundant evidence of mineralization at Ogofau still to be seen but there is little information to be found in literature concerning the nature of the ores, or indeed, about their geology at all. Field observations show that

the quartz veins occur mainly in Bala shales, but possibly also in shales of
Llandovery age too. Fortunately, it was found that the trace element background in both these groups of rocks was very similar, thus simplifying
evaluation of anomalous results. Fifty one samples were collected at Ogofau
and analysed for Rb, Pb, Zn, Cu, K and Hg. Gold was not determined owing
to the lack of a reliable method for such low concentrations except for
neutron activation, which was not available. The results are given below and
for the purposes of this study the threshold for anomalous results was taken
arbitrarily at two standard deviations above mean background.

Base metals

As the Ogofau gold-bearing sulphide ore contains both lead and zinc these
metals also occur in the aureoles surrounding the veins. Results show that
18% of the samples show anomalous zinc and 31% anomalous lead (Fig.3). It
will be seen in Fig.4 that lead tends to lie closer to the veins whilst zinc
gathers strength away from the vein to reach an average maximum of
250—300 ppm.

No copper enrichment above the background level of 75 ppm was found
in the rocks in this vicinity and no copper minerals occur in any ore which

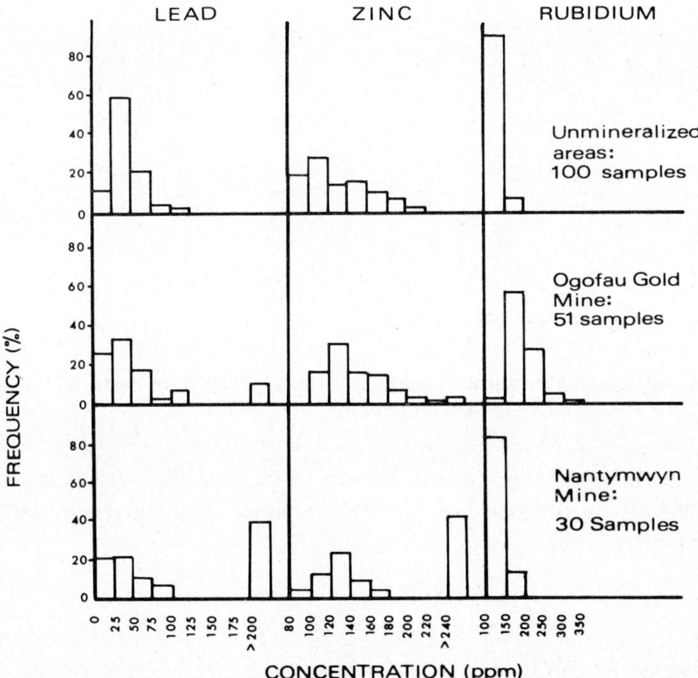

Fig.3. Frequency distribution of lead, zinc, and rubidium in samples representing Ogofau,
Nantymwyn, and unmineralized areas.

348

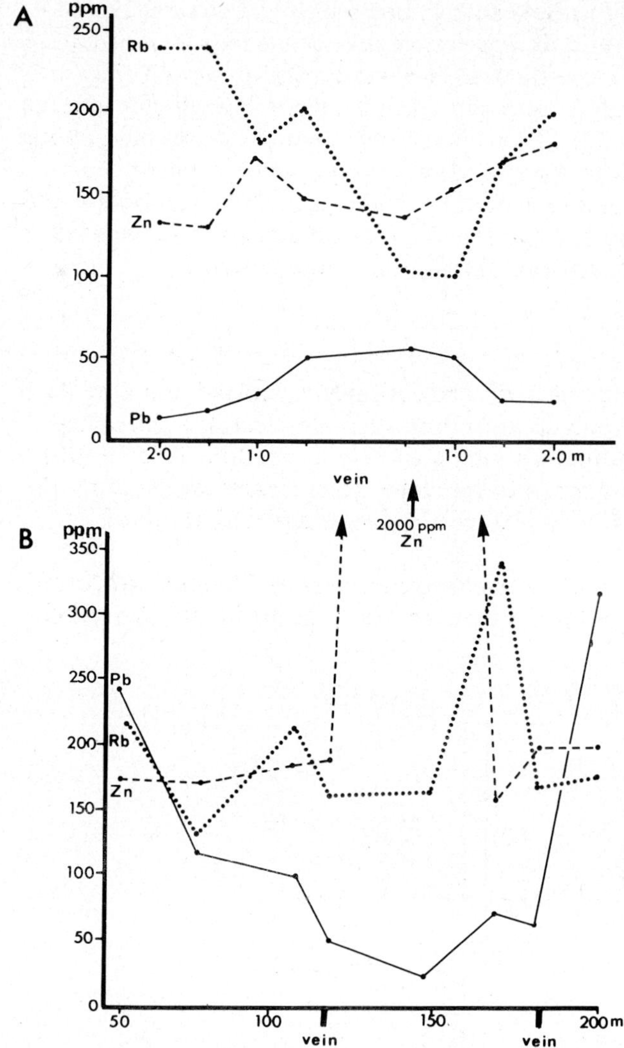

Fig.4. Distribution of lead, zinc, and rubidium in samples from Ogofau gold mine: (a) across a small isolated (20 cm) vein; and (b) across two large veins.

can be seen today. Because there is a wide variation in value (20—180 ppb), mercury is also of little help in delineating an aureole here, even in rocks well outside the mineralized zone.

Rubidium

Of the 51 samples taken from Ogofau 90% of them are anomalous at over two standard deviations (20 ppm) above the regional background of 125 ppm, and 84% are over three standard deviations (30 ppm) higher. Potassium, on

TABLE II

Mean concentrations of Pb, Zn, Cu, Hg, Rb and K and their standard deviations from the mean, for sample sets from the area investigated

Set No.	No. of samples in set	Formation (and Period)	Pb (ppm)		Zn (ppm)		Cu (ppm)		Hg (ppb)		Rb (ppm)		K (%)		Remarks
			M	S	M	S	M	S	M	S	M	S	M	S	
1	11	U./M. Llandovery (L. Silurian)	40	10	100	10	70	7	130	20	110	10	3.6	0.3	unmineralized area
2	5	U./M. Llandovery (L. Silurian)	55	10	110	7	60	15	125	40	125	7	3.8	0.1	unmineralized area
3	4	U./M. Llandovery (L. Silurian)	50	6	95	15	75	8	135	110	110	40	3.7	0.3	unmineralized area
19	3	U./M. Llandovery (L. Silurian)	30	15	160	25	50	15	88	5	130	8	3.8	0.05	unmineralized area
4	3	M. Llandovery (L. Silurian)	35	10	185	7	65	8	90	10	100	6	3.6	0.1	unmineralized area
20	11	M. Llandovery (L. Silurian)	65	60	330	525	65	10	100	8	115	15	3.6	0.25	1 mile west of Nantymwyn
5	4	L. Llandovery (L. Silurian)	40	12	110	15	70	6	135	15	135	2	3.7	0.4	unmineralized area
6	4	L. Llandovery (L. Silurian)	50	12	145	7	60	9	70	10	120	7	3.3	0.1	unmineralized area
7	3	L. Llandovery (L. Silurian)	35	12	145	20	75	12	35	5	120	5	4.3	0.1	unmineralized area
8	4	L. Llandovery (L. Silurian)	70	12	145	30	90	15	80	35	135	20	5.0	0.3	unmineralized area
21	3	L. Llandovery (L. Silurian)	38	4	185	5	50	7	100	25	105	7	3.5	0.25	unmineralized area
9	12	Bala (Ordovician)	36	13	135	45	70	7	85	25	125	20	3.5	0.2	unmineralized area
10	10	Bala (Ordovician)	45	8	95	15	60	20	60	20	120	8	3.7	0.1	unmineralized area
11	8	Bala (Ordovician)	55	25	110	25	70	8	80	45	125	10	3.2	0.3	unmineralized area
12	5	Bala (Ordovician)	40	20	160	20	90	3	55	10	140	18	4.4	0.25	unmineralized area
13	4	Bala (Ordovician)	55	7	135	6	100	7	50	35	125	9	4.6	0.05	unmineralized area
14	6	Bala (Ordovician)	55	8	125	15	90	15	45	20	130	4	4.5	0.05	unmineralized area
15	5	Bala (Ordovician)	65	20	170	77	65	5	35	7	130	9	4.7	0.1	unmineralized area
16	51	Bala/L. Llandovery	130	456	180	250	75	15	155	100	190	40	3.9	0.4	Ogofau gold mine
17	10	Bala	2000	1700	675	700	75	15	75	12	155	14	4.3	0.2	Cynnant mine
18	9	Bala	330	730	250	120	70	8	70	15	130	20	3.9	0.25	Nantymwyn mine
22	6	Bala	120	110	150	15	75	15	125	30	125	15	3.9	0.5	2 miles south of Nantymwyn

M = mean values of sets, S = standard deviations from those means.

the other hand remains relatively constant at about 4%, and the Rb/K ratio doubles in value. Thus there is clear evidence that the rocks surrounding the Ogofau gold mineralization have been enriched in rubidium and that this enrichment is quite independent of potassium, an element which shows no enhancement here. Rubidium enrichment appears to follow a trend similar to that of zinc, tending to increase to a maximum away from the vein margins. This is well shown in Fig.4a, but is less clear in Fig.4b owing to the interference of several closely spaced veins. Thus there is a well-marked primary dispersion aureole surrounding this gold deposit and, in this particular instance rubidium shows by far (up to 40 cm from the veins) the most distinctive enhancement of the six elements measured (see also Table II).

PRIMARY DISPERSION AT RHYDYMWYN

As with Ogofau, the geology of the ores is poorly reported and as previously mentioned mineralization occurs entirely in Bala grits and shales. There are slight differences in mineralization between Nantymwyn and Cynnant; at Cynnant galena is almost the only ore mineral, sphalerite occurring only sparsely, and mainly where the vein passes from grit into shale before pinching out. At Nantymwyn, on the other hand, sphalerite was obviously present in the ore in some abundance, judging by the amount seen on the dumps, and was of at least two generations, a very pale brown, and a darker variety. Unfortunately, in the vicinity of Nantymwyn mine reliable systematic sampling is difficult owing to soil cover, and to extensive dumps which cover the surface in the immediate mine area: there could thus be some secondary interference. Despite this some 30 samples were collected in the district and analysed for the same elements as before.

Base metals

At Cynnant, where galena was the main ore mineral, the average lead concentration in the surrounding rocks was 2000 ppm, some fifty times greater than the normal background figure of 45 ppm. This lead enhancement persisted for up to 35 ft from the vein. Zinc on the other hand was enriched merely by a factor of five, increasing to 675 ppm against a background of 135 ppm. Somewhat lower values were found at Nantymwyn for both lead and zinc, but this was because reliable samples could not be collected closer than 80 m from the veins. Rocks elsewhere in the area were also found to carry anomalous zinc (sample set 20 averaged 330 ppm Zn), suggesting that this was part of the Nantymwyn aureole itself, or perhaps indicating a blind ore body: no nearby vein could be found.

As at Ogofau, no copper enhancement was noted above the background level of 75 ppm, nor have any copper minerals been reported in the ores of the areas examined. Mercury was, however, more consistent than at Ogofau at about 80 ppb. Presumably this ore was free from mercury, and this may be another pointer to the nature of the ore fluid.

Rubidium

Rubidium shows very little evidence of concentration at Rhydymwyn except in a few rocks containing very high lead and zinc (set 17): these samples averaged 155 ppm Rb. Generally the rubidium concentration was within the background range for the region (Fig.3).

DISCUSSION

Pathfinder elements in general share some of the common characteristics of low melting points, high solubility and high vapour pressures. Rubidium was selected for trial on the basis of such properties and the work reported here suggests that there is some justification for believing that this element may have uses in defining exploration targets for certain types of mineralization. At Ogofau rubidium appears to show a better defined aureole than do lead, zinc, copper or mercury and there is no possibility that this aureole is unrelated to ore, nor is it related to potassium metasomatism or any other enrichment in potassium from any other cause. No such aureole was detected at Rhydymwyn. The difference in behaviour of rubidium in these two areas presumably reflects a difference in the genesis of the ores. At Rhydymwyn the fluids would appear to have been rich in lead and zinc, poor in iron, and of very low temperature: at Ogofau the ore appears to be a typically meso-thermal epigenetic sulphide body, rich in iron, with a little lead and zinc, some gold, and although surprisingly low in copper, the type of body often found close to igneous rocks. However, in neither case are any igneous rocks known or even indicated.

Further work is now being directed to discover whether rubidium normally forms aureoles around other gold deposits in the United Kingdom, few and small though these may be, and also to determine whether such aureoles occur around other types of epigenetic mineralization, such as Pb—Zn mineralization formed at higher temperatures, and tin veins. Another application of rubidium may be in distinguishing between ores of different genetic types. The authors speculate that perhaps rubidium is an element which may distinguish ores related to igneous sources from those where the ore fluids are derived from low temperature brines of meteoritic origin.

ACKNOWLEDGEMENTS

The authors would like to thank the Ministry of Higher Education and Scientific Research of Iraq who sponsored this research by a scholarship to one of its authors (M.J.A.).

REFERENCES

Al-Atia, M.J., 1973. A preliminary study of the variations of trace heavy metals and some other elements in certain Paleozoic rocks of South Wales. M.Sc. Thesis, Univ. of Wales Cardiff

Armbrust, G.A., Munoz, J.O. and Farias, J.A., 1971. Rubidium as a guide to ore at El Teniente (Braden), Chile (abstract). Econ. Geol., 66: 977

Broderick, T.M., 1929. Zoning in Michigan copper deposits and its significance. Econ. Geol., 24: 149—162

Cameron, E.M. and Jonasson, I.R., 1972. Mercury in Precambrian shales of the Canadian Shield. Geochim. Cosmochim. Acta, 36(9): 985—1005

Chaffee, M.A., 1972. Distribution and abundance of gold and other selected elements in altered bedrock, Empire mining district, Clear Creek County, Colorado. U.S. Geol. Survey Bull., 1278c: 23 pp.

Davies, K.A., 1933. The geology of the country between Abergwesyn (Breconshire) and Pumpsaint (Carmarthenshire). Q. J. Geol. Soc. Lond., 89: 172—201

Hall, G.W., 1971. Metal Mines of Southern Wales. J. Fennings, Gloucester, pp.39—59

Hawkes, H.E. and Webb, J.S., 1962. Geochemistry in Mineral Exploration. Harper and Row, New York, N.Y., pp.45—72

Jones, G.D.E. and Lewis, P.R., 1971. The Roman Gold Mines of Dolaucothi, 10—11. Carmarthen County Museum

Tauson, L.V. and Petrovskaya, S.G., 1971. Endogenic halo type of hydrothermal molybdenum deposits. In: R.W. Boyle and J.I. McGerrigle (Editors), Geochemical Exploration. Can. Inst. Min. Metall., Spec. Vol., 11: 394—396

Vlasov, K.A., 1966. Geochemistry of Rare Elements, 1. Israel Programme for Scientific Translations, Jerusalem, pp.128—152

THE USE OF RUBIDIUM/STRONTIUM RATIOS AS A GUIDE TO MINER-ALIZATION IN THE GALWAY GRANITE, IRELAND

G. LAWRENCE

Hunting Geology and Geophysics Limited, Boreham Wood, Herts. (Great Britain)

ABSTRACT

Minor molybdenite mineralization occurs in the Lower Devonian Galway Granite of western Ireland. Detailed geology and petrography have been studied in 11 km² of the southwest corner of the pluton and a 1:10,560 scale geological map prepared. A central even-grained adamellite and a porphyritic adamellite separated by a 200 m wide transition zone, make up the bulk of the pluton. A leucogranite — the Murvey Granite — occurs marginally, and in places is garnetiferous; aplite veins which cut the leucogranite are very locally garnetiferous. Sulphide mineralization is confined to the garnetiferous rocks and to quartz veins, and together they represent the last magmatic fraction to crystallize.

The variation of 10 major elements and 17 trace elements in 70 samples of the Galway Granite, determined by automatic X-ray fluorescence analysis, have been examined in a series of profiles across the pluton and in element plots.

Alkali element variation and especially Rb/Sr ratio accurately characterize the fractionation sequence of the Galway pluton; the mean Rb/Sr × 100 ratio is 44 for the central adamellite, 60—63 for the porphyritic adamellite and 350 for the leucogranite. The mean Rb/Sr × 100 ratio for the aplite veins is 1780 and isolated values reach almost 4300. Several other elements, notably V, Cr, Cu, Mo and Ce, show a relative enrichment in the aplite veins and garnetiferous Murvey Granite, and a very small residual volatile fraction is inferred. The correlation of high Rb/Sr ratio with the volatile fraction and sulphide mineralization is confirmed. The localized occurrence of the high ratios, however, coupled with the paucity of volatiles and overall dryness of the parental magma (indicated by low S, Cl, La, Ce and Zr), suggest that sulphide mineralization in the pluton is predictably of low volume. Consideration of published data for the mineralized granites of southwest England demonstrates that wide areas of high Rb/Sr ratio can indicate zones of K_2O enrichment due to fractionation or alteration. The use of Rb/Sr ratios in regional geochemical programmes is proposed as a more consistent guide to the ore potential of acid igneous rocks than the variation of base or precious metals.

INTRODUCTION

The Galway Granite pluton occupies over 1000 km² of Connemara, western Ireland. (Fig.1) The post-tectonic batholith was emplaced between 380 and 390 million years ago (Leggo et al., 1966) and shows broad similarities with several other batholiths emplaced in the Dalradian and Lower Palaeozoics of the British Isles. Mineralization is unremarkable in the pluton; molybdenite in quartz veins was reported near Roundstone as early as 1917 (Cole, 1922,

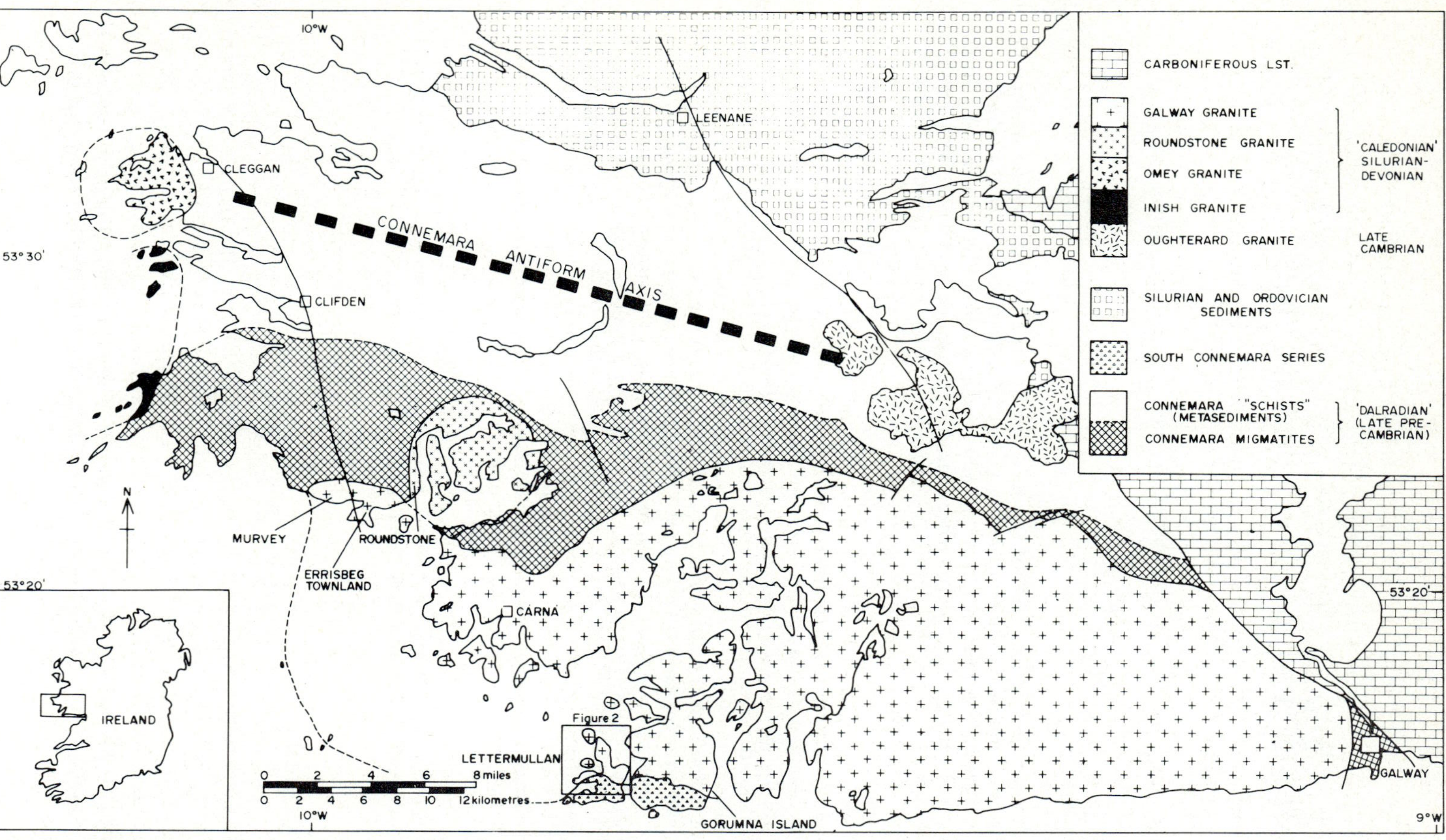

Fig.1. Geological sketch map of Connemara, western Ireland.

p.132), and 250,000 tons averaging 0.13% MoS_2 have been outlined in the Carna—Mace area (O'Brien, 1959, p.16). The latter occurrence is the most important of several mineral localities where molybdenite and pyrite, with very rare pyrrhotite, chalcopyrite, galena and fluorite, occur in quartz veins or as weak disseminations in granite. Rare sulphides are also found in microgranite porphyry dykes and along fracture zones. The mineralization near Carna is confined to the Murvey Granite, a marginal leucogranite facies of the Galway pluton. The bulk of the pluton comprises a central even-grained adamellite of granodioritic aspect — the Carna Granite — succeeded by a porphyritic adamellite — the Errisberg Townland Granite — both showing an overall easterly strike (Wright, 1964; Aucott, 1968; Claxton, 1968; Coats and Wilson, 1971; Leake, 1974).

The aim of this study has been to establish geochemical criteria, complementary to the geological features, which characterize the mineralization of the Galway pluton.

Alkali elements are rarely used to define ore potential of acid igneous rocks; most regional geochemical programmes involve base or precious metals only. Many research workers have shown that the behaviour of alkali elements can indicate both the degree of fractionation (Mursky, 1973) and the presence of volatiles in the parent magma. As the volatile fraction plays an essential role in ore genesis, element patterns that disclose its presence could be of considerable use in the assessment of ore potential of acid igneous rocks.

THE GALWAY GRANITE OF LETTERMULLAN

Rock exposure on the islands of Lettermullan is extremely good; low ice-smoothed crags are interrupted only by shallow peat occupying depressions along eroded fracture zones. Detailed geology has been mapped at 1:10,560 scale (6 inches to 1 mile) over 11 km^2 and is presented in Fig.2..

The modal compositional variation between the central Carna Granite and the marginal Murvey Granite is summarized in Table I. The marginal Murvey Granite occupies a 2 km wide embayment in the amphibolite wall rock. It meets the porphyritic adamellite of the Errisbeg Townland Granite in a steep, sharp undisturbed junction along part of which the adamellite is non-porphyritic, probably representing a chilled-margin. The Murvey Granite is a pink coarse even-grained albite—quartz—K-feldspar leucogranite, containing rare ragged chloritized biotite but no other ferromagnesian material. A finer-grained, white, highly siliceous granite intervenes between the main Murvey Granite and the wall rock, and contains rare spessartine—almandine garnets. Aplite veins cut the marginal leucogranite and also contain garnets.

Molybdenite, chalcopyrite and pyrite occur as rare isolated grains and clusters within the garnetiferous Murvey Granite and aplite veins, but are more common in quartz veins. The quartz veins commonly carry tourmaline and fluorite and are the last fraction of the parental magma to crystallize. Hydrothermal alteration is restricted to very minor sericitization of K-feldspar and chloritization of biotite.

TABLE I

Average modal compositions of the granites of Lettermullan

Modal volume %		Carna	Transitional	Errisbeg Townland	Marginal E.B.T.	Murvey	Aplites
Quartz	range			20—32		30—36	40—45
	mean	25	n.d.	27	n.d.	34	42
K-feldspar	range	25—30			20—25	29—48	35—40
	mean		n.d.	31		36	37
Plagioclase	range	35—40		27—37		20—32	15—25
	mean	38	n.d.	34	n.d.	28	ȯ0
Anorthite	mean	27—33	25—29	21—27	22—26	9—14	5—11
Biotite/chlorite	range	5— 7		4— 7		0— 2	tr
	mean	6	n.d.	5	10		
Hornblende	range	2		<1	1— 2	none	none
	mean		n.d.	often <0.5			
Accessories							
magnetite	range	up to 1		<0.5	<0.7	tr	none
	mean		n.d.				
apatite	mean	up to 0.5	n.d.	tr	0.2	tr	tr
sphene/rutile	mean	0.02	n.d.	0.05	0.07	tr	rare tr
allanite/clinozoisite/ zoisite	mean	0.02	n.d.	0.05	n.d.	tr	none
prehnite/epidote	mean	0.05	n.d.	tr	n.d.	none	none
zircon	mean	0.02	n.d.	0.02	n.d.	tr	tr
pyrite	mean	none	none	none	none	tr	tr
garnet	mean	none	none	none	none	tr	tr

On geological and petrographic evidence the granites on Lettermullan exhibit a familiar fractionation sequence, in which there is increase of K-feldspar, quartz and sodic plagioclase and simultaneous decrease of hornblende and biotite. The dryness of the parental magma is inferred from the absence of major pegmatites or alteration zones, although volatiles may have concentrated in the ultimate residual fraction from which the aplite and quartz veins crystallized. Molybdenum mineralization in the Galway Granite pluton is entirely restricted to this ultimate magmatic fraction, the meagreness of which might suggest that sulphide mineralization would be of low volume in the pluton.

GEOCHEMICAL VARIATION OF THE GRANITES ON LETTERMULLAN

Rocks offer the only useful sampling medium on Lettermullan. Samples weighing approximately 2 kg were collected and broken up in a jaw-crusher after removal of any weathered material. The pieces were reduced to 1 mm size in a hardened steel rolling mill and ground to 60 microns (250 B.S.S. mesh) in a Tema agate swing mill. Analysis for 10 major and 17 trace elements, in pellets of the rock powders, was carried out using automatic X-ray fluorescence techniques, described by Leake et al. (1969); ferrous iron was analysed by standard potassium dichromate titration, water by a modified Penfield method and molybdenum was determined colorimetrically, in a few representative samples, using dithiol. Analytical reproducibility and accuracy of the X-ray fluorescence analysis are fully discussed by Leake et al. (1969).

Arithmetic means and ranges of element concentration in 70 samples of granite, aplite and microgranodioritic xenoliths are presented in Table II. Geochemical variation across the pluton is expressed in the north—south profiles of Fig.3.

The Rb/Sr × 100 ratio variation (Fig.4) shows a steady increase southward away from the central Carna Granite with a 6-fold increase in the Murvey Granite and a 20-fold increase in the garnetiferous Murvey Granite. The aplite veins exhibit a particularly anomalous Rb enrichment giving Rb/Sr × 100 ratios as high as 4271 (Fig.5).

The consanguinity of the granites of Lettermullan, their aplites and the associated dykes (defined by 30 samples) is inferred from the Rb—Sr plot (Fig.5). The usefulness of the Rb/Sr ratio is that it is an indicator of the whole fractionation sequence of a magma, as both rubidium and strontium participate continuously in the fractionation process (Taylor, 1965).

As molybdenum mineralization, although sparse, is restricted to garnetiferous Murvey Granite, aplite veins and to quartz veins, the anomalously high Rb/Sr ratio seems to be a diagnostic indicator of potential sulphide mineralization, as well as of extreme magmatic fractionation.

An examination of other trace element distributions discloses further anomalous concentration increases in the aplite veins. The parallel variation of Cr, Ni, Zn and Cu correlates closely with ferromagnesian mineral variation (Fig.6) although chromium and nickel show steeper rates of decrease than copper or

TABLE II

Mean and range of major and trace elements in the granites of Lettermullan

	Microgranodiorite xenoliths			Carna Granite			Transitional Granite			Errisbeg Townland Granite		
	mean	range		mean	range		mean	range		mean	range	
SiO_2 (%)	61.19	57.40—	66.26	67.59	66.66—	68.84	67.40	65.20—	68.43	69.10	68.02—	70.06
Al_2O_3	16.16	15.12—	16.79	15.67	15.39—	16.02	15.45	14.92—	16.50	15.19	14.68—	15.89
TiO_2	0.83	0.46—	1.61	0.42	0.41—	0.44	0.41	0.36—	0.50	0.35	0.32—	0.37
Fe_2O_3	1.89	1.21—	3.04	1.17	1.02—	1.28	1.22	0.96—	1.50	1.04	0.77—	1.30
FeO	3.11	1.92—	5.39	1.81	1.72—	1.88	1.72	1.37—	2.04	1.44	1.24—	1.63
MgO	3.02	1.40—	5.50	1.58	1.42—	1.86	1.63	1.50—	1.87	1.38	1.12—	2.35
CaO	4.00	2.75—	6.59	3.30	3.12—	3.45	3.08	2.77—	3.52	2.43	1.24—	2.87
Na_2O	3.89	3.21—	4.57	3.84	3.62—	4.02	3.85	3.81—	3.91	4.07	3.71—	4.54
K_2O	4.05	1.66—	5.34	3.91	3.81—	4.05	3.98	3.79—	4.15	4.10	3.51—	4.36
MnO	0.16	0.11—	0.25	0.08	all 0.08		0.09	0.08—	0.11	0.08	0.05—	0.10
P_2O_5	0.52	0.24—	1.19	0.15	0.14—	0.17	0.17	0.14—	0.23	0.13	0.11—	0.15
H_2O	1.37	0.74—	2.52	0.86	0.70—	0.96	0.84	0.72—	0.99	0.86	0.66—	1.13
S (ppm)	164	101 —	256	91	54 —	119	93	64 —	135	98	41 —	166
Cl	205	124 —	389	185	170 —	202	191	172 —	240	181	151 —	277
Sc	7	3.5 —	16	3.5	3 —	5	5	4 —	6	4	3 —	5
V	105	58 —	170	59	36 —	77	42	42 —	69	66	34 —	108
Cr	66	39 —	110	34	33 —	35	33	19 —	44	32	17 —	42
Ni	61	35 —	101	36	29 —	39	28	19 —	35	28	21 —	36
Cu	38	9 —	106	25	20 —	31	19	16 —	23	21	8 —	82
Zn	90	60 —	150	60	59 —	61	60	57 —	64	61	51 —	91
Rb	149	63 —	193	135	129 —	140	138	134 —	141	146	126 —	182
Sr	249	166 —	366	304	298 —	314	308	298 —	316	242	223 —	303
Zr	207	154 —	346	154	141 —	158	156	132 —	187	142	123 —	163
Cs	5.5	4 —	9.5	5	4 —	7	4.5	4 —	5	5	3.5 —	6
Ba	942	456 —	1449	971	873 —	1166	1035	804 —	1262	820	630 —	1280
La	34	16 —	75	17	13 —	27	22	14 —	35	18	9 —	34
Ce	93	59 —	192	55	48 —	68	58	43 —	74	54	38 —	75
Th	30	21 —	37	31	27 —	35	36	27 —	42	37	30 —	45
U	2	b.d.l. —	7	2	1 —	2	2	b.d.l. —	4	3	b.d.l. —	8
Rb/Sr × 100	60	28 —	116	44	42 —	46	45	43 —	46	60	47 —	82
No. of samples	13			5			5			19		

TABLE II (*continued*)

	Non-porhyritic E.B.T. Granite		Murvey Granite		Garnetiferous Murvey Granite		Aplites	
	mean	range	mean	range	mean	range	mean	range
SiO_2 (%)	68.57	66.57— 70.37	75.68	74.49— 76.66	75.89	75.64— 76.29	76.77	75.42— 77.66
Al_2O_3	15.31	15.02— 15.56	13.35	12.68— 13.76	13.43	13.40— 13.45	12.60	11.94— 13.34
TiO_2	0.38	0.31— 0.45	0.13	0.10— 0.20	0.09	0.03— 0.14	0.07	0.03— 0.15
Fe_2O_3	1.12	1.00— 1.34	0.43	0.28— 0.68	0.22	0 — 0.44	0.33	0.22— 0.49
FeO	1.61	1.19— 2.15	0.57	0.41— 0.67	0.55	0.45— 0.64	0.28	0.19— 0.56
MgO	1.44	1.15— 1.83	0.16	0.12— 0.24	0.07	b.d.1. — 0.15	0.05	b.d.1. — 0.17
CaO	2.43	1.93— 3.09	0.58	0.23— 0.93	0.46	0.30— 0.61	0.42	0.25— 0.83
Na_2O	4.84	4.03— 4.50	4.39	3.18— 5.06	4.71	4.22— 5.20	4.32	3.90— 4.74
K_2O	3.68	3.14— 4.21	4.84	4.44— 6.73	4.53	4.34— 4.72	5.12	4.38— 5.55
MnO	0.12	0.08— 0.23	0.04	0.03— 0.04	0.07	0.06— 0.07	0.05	0.02— 0.17
P_2O_5	0.12	0.10— 0.15	0.04	0.03— 0.06	0.04	0.03— 0.04	0.02	0.01— 0.04
H_2O	0.92	0.60— 1.28	0.45	0.36— 0.56	0.36	0.24— 0.47	0.37	0.21— 0.51
S (ppm)	101	51 —217	79	27 —194	74	50 — 98	53	25 — 91
Cl	173	132 —222	172	137 —213	142	126 — 157	140	104 — 324
Sc	6	3 — 10	4.5	2 — 6	7	6 — 8	4	2 — 6
V	68	45 — 90	32	19 — 47	25	19 — 31	24	13 — 36
Cr	47	35 — 65	13	7 — 32	16	8 — 24	16	5 — 31
Ni	38	32 — 47	9	2 — 16	10	8 — 12	11	5 — 16
Cu	18	7 — 27	6	2 — 10	19	5 — 42	9	3 — 24
Zn	68	60 — 78	39	33 — 46	40	20 — 50	27	16 — 38
Rb	149	136 —173	212	146 —263	235	225 — 244	262	217 — 363
Sr	236	202 —257	60	46 —123	33	12 — 54	26	9 — 57
Zr	144	129 —163	99	73 —125	80	64 — 96	62	41 — 108
Cs	5	4.5 — 5	4	3 — 4.5	4	4 — 4.5	4.5	3 — 10
Ba	818	770 —871	329	197 —594	194	95 — 293	107	48 — 311
La	17	12 — 24	12	5 — 26	8	4 — 12	2	b.d.1. — 5
Ce	53	46 — 64	45	39 — 63	40	30 — 49	27	21 — 39
Th	30	18 — 39	62	56 — 95	50	41 — 60	44	32 — 59
U	2	b.d.1. — 5	3	b.d.1. — 8	1	b.d.1. — 2	5	b.d.1. — 8
Rb/Sr × 100	63	55 — 85	353	172 —571	1225	417 —2033	1781	393 —4271
No. of samples	7		11		2		8	

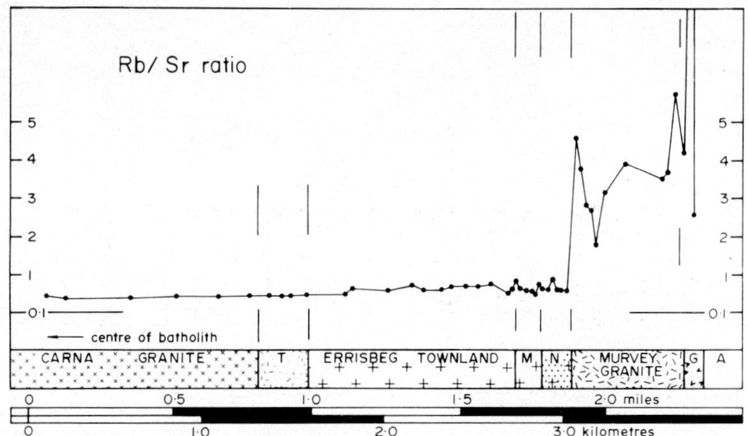

Fig.4. Rb/Sr ratio across the Galway Granite of Lettermullan. Abbreviations as in Fig.3, M = marginal E.B.T. Granite.

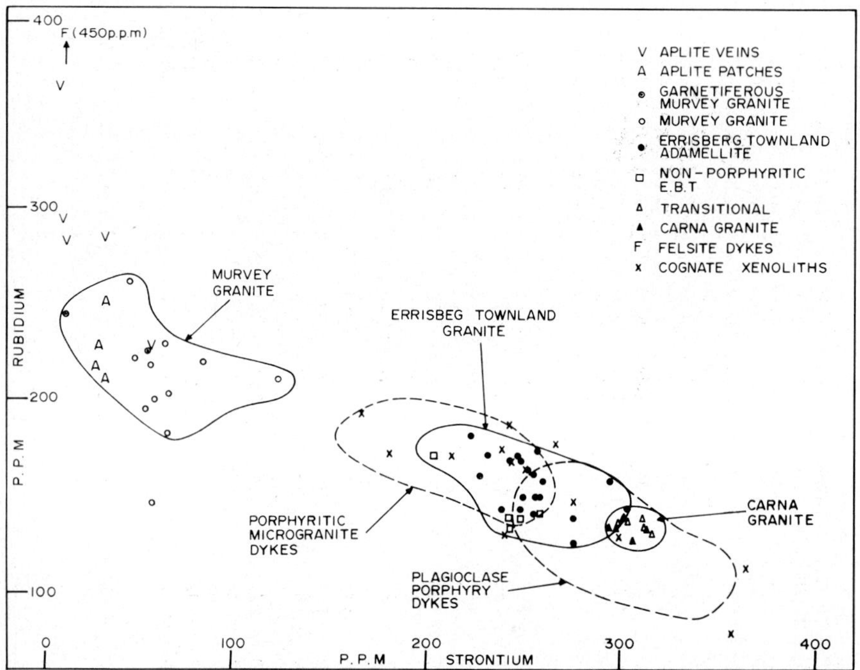

Fig.5. Variation of rubidium with strontium in the granites, dykes and xenoliths of Lettermullan.

zinc. An anomalous enrichment of copper, chromium, and several other metals, relative to their host leucogranite characterizes the aplite veins (Fig.6), although the absolute concentrations of these metals in the Galway Granite are well within published ranges for normal "granites" (Table III).

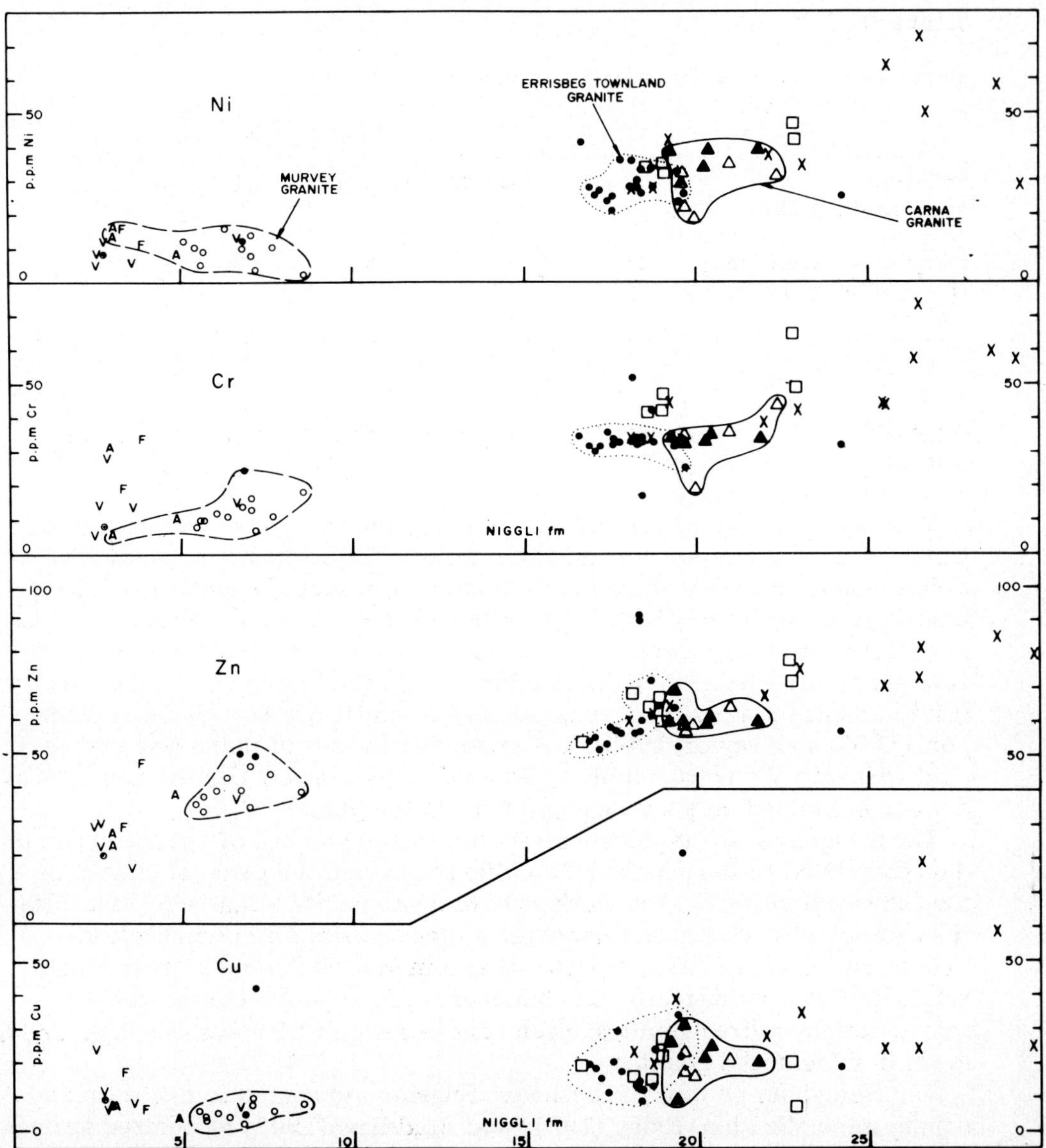

Fig.6. Variation of nickel, chromium, zinc and copper with Niggli fm in the granites of Lettermullan. Symbols as in Fig.5.

$$\text{Niggli fm} = \frac{\text{wt.\% (Fe}_2\text{O}_3 + \text{FeO} + \text{MgO} + \text{MnO})}{\text{formula wt. (Fe}_2\text{O}_3 + \text{FeO} + \text{MgO} + \text{MnO})} \times 1000$$

TABLE III

Mean base metal concentrations (ppm) in the Galway Granite

	Cu	Pb	Zn	Mo	U
Published (see Garrett, 1973)	10—30	15—35	39—64	< 2	0—6
Means of Galway Granite (Pb from Coats and Wilson, 1971)	18	57	55	< 2	2—4

GEOCHEMICAL FACTORS RELATING TO MINERALIZATION IN ACID IGNEOUS ROCKS

Molybdenum mineralization of the Galway pluton has been precipitated from a very minor residual magmatic fraction. This ultimate magmatic fraction is characterized by anomalous enrichment in several elements besides base metals, and by very high Rb/Sr ratios in rocks crystallized from the fraction. Overall the Galway Granite has distinctly lower mean sulphur, chlorine, zirconium, rare-earths, P_2O_5 and H_2O than published mean concentrations for granitic rocks (compared to Nockolds and Allan, 1953; Turekian and Wedepohl, 1961; and Taylor, 1965). Lower concentrations of these elements accord well with the petrographic evidence that the Galway Granite magma was dry and contained only a very restricted volatile phase.

The meagreness of sulphide mineralization and absence of alteration can be directly related to the paucity of volatile fraction in the parental magma of the Galway Granite. Several workers have demonstrated that large concentrations of volatiles in parent magmas are a pre-requisite for mineralization. Tauson and Kozlov (1973) reported that mineralized "granites" had a high volatile content indicated by much higher Li, B, Sn and F concentrations than in unmineralized granites; mean concentration of Rb was also high, and Sr lower in mineralized granites.

The Hercynian plutons of southwest England are related to major tin and copper mineralization. Exley (1957) and Bradshaw (1967) recognized strong increases in Li, Sn, F and Rb in these granites and their minerals. The most striking geochemical feature is a remarkable Rb enrichment and Sr depletion. The 100-km^2 St. Austell Granite, described by Exley, has major Sn—W—Cu mineralization peripheral to the stock. The whole granite shows high Rb/Sr ratios. The stock is zoned with an outer biotite—muscovite granite with a Rb/Sr × 100 ratio of 900 forming the bulk. This grades into Li-mica granite and fluorite granite by progressive increase of albite, the fluorite granite occupying an inner core. The volatile-rich inner granite has a Rb/Sr × 100 ratio of 3000.

The high Rb/Sr ratios over wide areas of the southwest England granites are highly diagnostic of both alteration and mineralization. The behaviour of base metals is less consistent. In biotite and feldspar, for example, lead and tin show distinctly high concentrations compared with those from unmineralized granites (Bradshaw, 1967); zinc, however, is higher in feldspar only, and copper, the second most important ore metal, shows no consistent difference in rock-forming minerals from mineralized and unmineralized granites. The variation about the mean concentration of the base metals in the mineralized granites is wide, however, and Tauson and Kozlov (1973) proposed that this greater dispersion is a better guide to ore potential than absolute concentrations.

The inherently complex patterns of base metal variation are difficult to recognize in regional low-density geochemical sampling of unknown plutons. It is considered that the Rb/Sr ratio offers a more consistent guide to mineralization potential. In the Galway Granite the mean Rb/Sr × 100 ratio is 81, although over the bulk of the pluton, comprising adamellite, the ratio is 53. In the marginal Murvey Granite Rb/Sr × 100 ratio increases to 353, whilst in the garnetiferous Murvey Granite and the sparse aplite veins, the values range between 417 and 4271; mineralization, although very limited, is entirely restricted to the rocks with highest Rb/Sr ratio. The highly mineralized St. Austell Granite (Exley, 1957) exhibits Rb/Sr × 100 ratios in excess of 1000 for the whole stock. The amplitude and areal extent of an Rb/Sr ratio anomaly is thus a distinct guide to ore potential in acid igneous rocks. Tauson and Kozlov (1973), for example, quote mean rubidium of 400 ppm and mean strontium of 100 ppm for mineralized granites, giving a mean Rb/Sr × 100 ratio of 400.

Recent data published by Sheraton and Black (1973) suggest that 400 may be a useful division between well mineralized and non-mineralized granites. Eight major granitic rock units in the northeast Queensland tin fields, ranging in age from 1475 to 262 million years, have average Rb/Sr × 100 ratios as follows: four granites mineralized in tin, etc.: 370—1640; one granite moderately mineralized in copper, lead and tungsten: 230; one granite weakly mineralized in copper, lead and zinc: 74; two unmineralized granites: 6 and 34. The most extensive mineralization (the Herberton tin field) is associated with the Elisabeth Creek Granite; the avearge Rb/Sr × 100 of this granite is 1640, almost double the next highest value. Increasing Rb/Sr ratio in the eight granites corresponds exactly to the increase in importance of their associated mineralization, whereas tin and other metals, although *generally* concentrated in the mineralized granites, are unevenly distributed.

MINERAL EXPLORATION OF ACID IGNEOUS ROCKS USING ROCK GEOCHEMISTRY

Only base or precious metals are normally used in regional geochemical exploration of acid igneous rock areas, even though Coope (1973) observed that no consistent scheme of variation of these metals in rocks and minerals exists for porphyry copper mineralization. Complex patterns and wide dispersion of base or precious metals may be a guide to mineralization, but the lower concentrations, and more restricted occurrence of these metals in ferromagnesian and accessory minerals, makes anomalous pattern recognition difficult unless mineralized rocks themselves are sampled. In regional assessment of acid igneous rock environments the analysis, by atomic absorption or X-ray fluorescence, of rubidium and strontium in whole rocks, in addition to the base or precious metals sought, will yield substantially more information for little extra cost. Rubidium and strontium are widely dispersed in acid igneous rocks, and vary continuously and sensitively with fractionation of the parent magma, and with alteration processes in which K_2O is increased. Rb/Sr ratio variation aids the recognition and mapping of acid igneous rock facies, and areas of anomalous high Rb/Sr ratio can be assessed according to amplitude, areal extent and relation to environment and other geological and geochemical features. Areas of high Rb/Sr ratio but with no base or precious metal concentration may relate to areas of alteration over buried ore deposits.

The use of trace elements to diagnose the presence of volatiles in the parental magma, and the relationships of these volatiles to mineral genesis, is not a new concept, and like the use of Rb/Sr ratios in academic geochemistry, has been frequently reported over the last 10 to 15 years. Lithium, the halides, and sulphur seem to indicate the presence of volatiles important in ore genesis. The analysis of these elements, in whole rocks and fluid inclusions may also prove to be important indicators in rock geochemical exploration programmes.

CONCLUSIONS

(1) A detailed comparison of geology, petrography and element variation has shown that Rb/Sr ratio allows accurate assessment of the degree and trend of fractionation in the Galway Granite magma. The behaviour of several elements indicates that the development of volatiles in the magma was severely restricted. Molybdenum mineralization has been precipitated only from this volatile phase, and is therefore predictably of a minor nature in the Galway pluton. The mineralization is associated with rocks yielding a Rb/Sr × 100 ratio in excess of 400, and as high as 4300.

(2) The use of Rb/Sr ratios in rocks is a reliable diagnostic tool in the regional assessment of the ore potential of acid igneous rocks. High Rb/Sr ratios over wide areas correlate with extreme fractionation of parent magmas and K_2O rich alteration zones, both of which are commonly associated with sulphide mineralization. The use of Rb/Sr in addition to base or precious metals, which have inherently more complex distribution patterns, is recommended for all regional rock geochemical programmes, especially those employing low-density sampling. Rb/Sr in drainage sediment and soils is also likely to be a useful prospecting tool.

ACKNOWLEDGEMENTS

Most of the data presented here were collected as part of a Ph.D. research programme, concluded in 1967, whilst the author was in possession of a N.E.R.C. award at the University of Bristol, England. The programme was supervised by Dr. B.E. Leake, and colleagues in the department, especially Dr. P.K. Harvey and Dr. G.L. Hendry, gave valuable assistance in analysis and computer data handling. Subsequent work, and preparation of this paper, has been under the auspices of Hunting Geology and Geophysics Limited, Hertfordshire, England, and the presentation has benefited from discussion with several colleagues.

REFERENCES

Aucott, J.W., 1968. Relationship between gamma-activity, geochemical variation and layering in the Galway Granite, Republic of Ireland. 23rd. Int. Geol. Congr., Prague, 6: 185—204

Bradshaw, P.M.D., 1967. Distribution of selected elements in feldspar, biotite and muscovite from British Granites in relation to mineralization. Trans. Inst. Min. Metall., Sect. B, Appl. Earth Sci., 76: B137—B148

Claxton, C.W., 1968. Mineral layering in the Galway Granite, Connemara, Eire, Geol. Mag., 105: 149—159

Coats, J.S. and Wilson, J.R., 1971. The eastern end of the Galway Granite. Mineral. Mag., 38: 158—161

Cole, G.A.J., 1922. Memoir and map of localities of minerals of economic importance and metalliferous mines in Ireland. Mem. Geol. Survey Ireland, Dublin (reprinted 1956)

Coope, J.A., 1973. Geochemical prospecting for porphyry-copper type mineralization — a review. J. Geochem. Explor., 2: 81—102

Exley, C.S., 1957. Magmatic differentiation and alteration in the St. Austell Granite. Q. J. Geol. Soc., Lond., 114: 197—230

Garrett, R.G., 1973. Regional geochemical study of Cretaceous acidic rocks in the northern Canadian Cordillera as a tool for broad mineral exploration. In: M.J. Jones (Editor), Geochemical Exploration 1972. Institution of Mining and Metallurgy, London, pp.203—219

Leake, B.E., 1974. The crystallisation history and mechanism of emplacement of the western part of the Galway Granite, Connemara, Western Ireland. Mineral. Mag., 39: 498—513

Leake, B.E., Hendry, G.L., Harvey, P.K. et al., 1969. The chemical analysis of rock powders by automatic X-ray fluorescence. Chem. Geol., 5: 7—86

Leggo, P.J., Compston, W., and Leake, B.E., 1966. The geochronology of the Connemara Granites and its bearing on the antiquity of the Dalradian Series. Q. J. Geol. Soc., Lond., 122: 91—118

Mursky, G., 1973. Fractionation of major, minor and trace elements in a zoned intrusive. 24th Int. Geol. Congr., Montreal, 10

Nockolds, S.R. and Allan, R.S., 1953. The geochemistry of some igneous rocks series, 1. Calc-alkali igneous rock trends. Geochim. Cosmochim. Acta, 4: 105—156

O'Brien, M.V., 1959. The future of non-ferrous mining in Ireland. In: The Future of Non-Ferrous Mining in Great Britain and Ireland. Institution of Mining and Metallurgy, London, p.16

Sheraton, J.W. and Black, L.P., 1973. Geochemistry of mineralized granitic rocks of northeast Queensland. J. Geochem. Explor., 2: 331—348

Tauson, L.V. and Kozlov, V.D., 1973. Distribution functions and ratios of trace element concentrations as estimators of ore-bearing potential of granites. In: M.J. Jones (Editor), Geochemical Exploration 1972. Institution of Mining and Metallurgy, London, pp.37—44
Taylor, S.R., 1965. Application of trace element data to problems in petrology. Phys. Chem. Earth, 6: 133—213
Turekian, K.K. and Wedepohl, K.H., 1961. Distribution of the elements in some major units of the earth's crust. Geol. Soc. Am. Bull. 72: 175—192
Wright, P.C., 1964. The petrology, chemistry and structure of the Galway Granite of the Carna area, Co. Galway. Proc. R. Irish Acad., 63B: 239—264

COPPER AND ZINC IN PROTEROZOIC ACID VOLCANICS AS A GUIDE TO EXPLORATION IN THE BEAR PROVINCE*

ROBERT G. GARRETT

Geological Survey of Canada, Ottawa, Ont. (Canada)

ABSTRACT

A lithogeochemical survey of the acid porphyry rocks of the Proterozoic Bear province has been undertaken. The data reveal patterns of copper and zinc which are relatable to known mineral occurrences. Additional areas of anomalous metal patterns are outlined which are possibly related to as yet undiscovered mineral occurrences. The behaviour of copper is essentially that of a chalcophile element whilst zinc behaves in an oxyphile manner, on average some 60% of the copper is in a sulphide form whilst only some 20% of the zinc is in that form. The use of a partial sulphide-selective leach is recommended for mineral exploration programmes and shown to be particularly important in the case of zinc. Significant regional patterns of zinc distribution are described which are believed to be related to the metallogenetic development of the Proterozoic acid volcano-sedimentary belts, and in particular their potential for stratabound volcanogenic sulphide deposits.

INTRODUCTION

During the summer of 1973 a bedrock geochemical sampling programme was undertaken in the southern half of the Wopmay subprovince of the Bear structural province of the Canadian Shield. The geochemical study was undertaken to assess the applicability of lithogeochemical techniques developed for use in the much younger granitoid terrane of the Canadian Cordillera (Garrett, 1971, 1973a, 1974a) to the more complex Proterozoic Shield terrane. The particular test area was chosen as it overlaps part of the Operation Bear-Slave lake sediment geochemical survey area (Allan et al., 1973) and is currently undergoing a mapping and geological revision programme (Garrett, 1974b; Hoffman, 1974; McGlynn, 1974). Additionally, it was known that a variety of mineral occurrences were present in the field area ranging from the typical Bear province Ag—U association, through porphyry copper like occurrences to stratabound copper sulphides.

The general geology of the field area has been reviewed by Stockwell and McGlynn (1970) and Fraser et al. (1972). The study area lies in Wopmay subprovince of the Bear province, which lies to the west of the Wopmay fault (a meridional fault at approximately 116°30′W) along the western margin of

---

*Published by permission of the Director, Geological Survey of Canada.

372

the Shield. This area west of the Wopmay fault is also known as the Great
Bear batholith which forms an element of Fraser's Wopmay Orogen (Wopmay
Belt of Stockwell and McGlynn, 1970). The complex of the Great Bear
batholith, of Late Aphebian age, consists mainly of high-level granites and
granodiorites. The roof of the batholith is best preserved in the north where
great thicknesses of ignibritic tuffs and some acid flows are found. The
volcano-sedimentary pile also includes basalts, andesites, conglomerates and
red bed sediments. These rocks are most common in the Echo Bay and
Camsell River areas in the west around the shores of Great Bear Lake.

The acid flows and pyroclastics make up some 30% of the batholith and
compositions vary from dacite to rhyolite; however, in the past they have
been grouped and mapped as quartz-feldspar porphyries. These porphyries

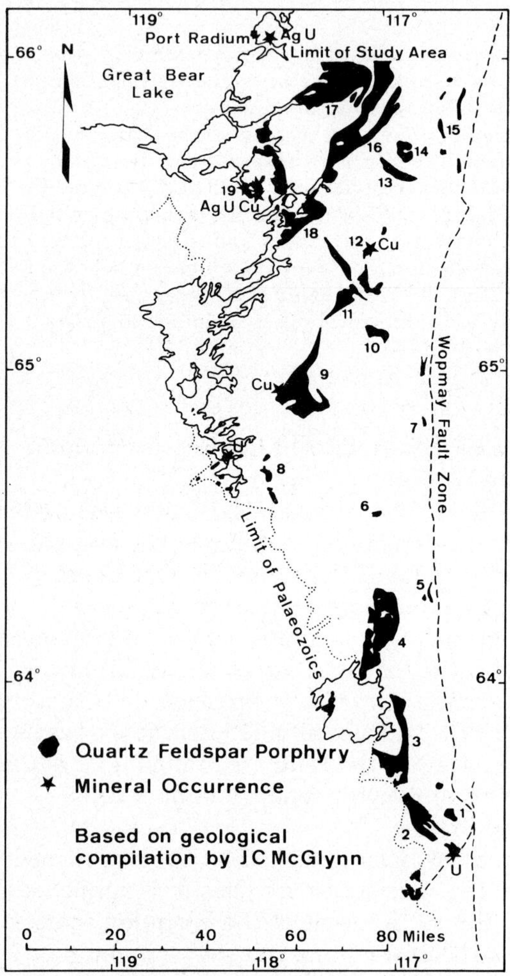

Fig.1. General geology and location map. Numbers refer to the numbers in square brackets
in text. (Based on geological compilation by McGlynn, 1974.)

are both extrusive and intrusive, and in the southern part of the study area gradational phases are found between fine-grained intrusive porphyries and the granitoid rocks. It seems likely that the entire assemblage forms a comagmatic pile with acid volcanics being outpoured onto a basement, possibly similar to the Hepburn batholith on the east side of the Wopmay fault. Great thicknesses of volcanics were extruded and younger granitoid rocks, possibly parts of the feeder system, were intruded into the pile. Locally, sedimentary rocks, shales, cherts and sandstones accumulated in transitory lakes.

The most important mineral occurrences lie in the Camsell River area where silver and copper are currently produced from the Terra mine ([19], see Fig.1). The silver occurs in cross-cutting veins with associated Cu, Co, Ni and U minerals. The major source of copper is from stratabound bodies within the mine which are related to iron formation units within the volcano-sedimentary pile. Nearby on the Norex property is a large area of altered andesites where there is pervasive pyrite and chalcopyrite mineralization. The general Calder River area also contains sulphide occurrences related to magnetite iron formation and cherts, and breccia pipes in the porphyries. At Tommie Lake [12] 25 miles to the southeast chalcopyrite is found associated with magnetite, chert and jaspilite horizons as well as in coarse-grained vein deposits which are interpreted by the author as gash veins into which quartz and chalcopyrite have been mobilized. Similar occurrences characterized by cobalt and arsenous minerals occur some 120 miles to the south near Lou Lake [2] (Shegelski and Thorpe, 1972). In the central part of the field area at Bode Lake [9] occurs copper mineralization which is interpreted by R.V. Kirkham (personal communication, 1973) as having many of the attributes of a porphyry copper. Other mineral occurrences are known to occur near Mazenod [3], DeVries [5], Hardisty [8] and Blackjack Lakes [15].

FIELD PROGRAMME

The quartz-feldspar porphyries were systematically sampled with aid of a helicopter and boat and foot traverses. 669 sites in the volcano-sedimentary rocks were sampled, only about 3% of these sites were obviously sedimentary and the remainder were either extrusive or intrusive porphyries. Differentiating between the intrusive and extrusive phases was often difficult due to the general fine grain and lack of small-scale structures in the rocks. In no case was material included in the data analysis if it was collected directly from the mineral occurrences described, only peripheral material was included for the data analysis. Field data was recorded on field cards which served as keypunch documents and computer technology was used extensively in the compilation and interpretation of the data. At each sample site two separate samples were collected with the aid of a sledgehammer. Care was taken to collect the freshest material available, discarding outer lichen-covered chips, from two points some 20 ft apart. The two samples from each site were bagged separately and shipped to the Geochemistry Section of the Geological Survey of Canada in Ottawa for sample preparation and analysis.

TABLE I

Summary of data statistics ($n = 1338$)

	Range		Arithmetic			Geometric mean	Log_{10} units					F
	min	max	$\overline{X}$	σ	C.V. (%)		$\overline{X}$	σ^2	σ^2_1	σ^2_2	σ^2_{SA}	(σ^2/σ^2_{SA})
Cu	1.0	60000.0	155	2093	1350	11	1.031	0.345	0.351	0.339	0.104	4.30
Cu px	1.0	50984.0	84	656	781	6	0.774	0.485	0.490	0.480	0.164	2.96
Zn	3.0	9500.0	124	424	342	72	1.857	0.142	0.137	0.147	0.024	5.91
Zn px	1.0	8200.0	47	340	723	16	1.209	0.172	0.167	0.178	0.043	4.02
Cu px/Cu	0.012	1.0	0.61	0.24	40							
Zn px/Zn	0.032	1.0	0.26	0.15	58							

Note: σ^2_1 is $\log_{10}$ variance of the first samples of each site pair and σ^2_2 is the variance of the second samples. σ^2 is the associated total data variance and σ^2_{SA} is the combined sampling and analytical variance.

SAMPLE PREPARATION AND ANALYSIS

The samples, weighing between 2 and 3 lbs. each, were reduced to minus 60 mesh, all the sample was crushed except for a hand specimen large enough for thin-section work. A split of this crushed material was ground in a ceramic ball mill to minus 100 mesh (Lavergne, 1965).

Total copper and zinc were determined by atomic absorption spectrophotometry after a HF/HClO$_4$ decomposition. A 400-mg sample was fumed to dryness three times and the residue taken up in 2M HCl and diluted to 20 ml to yield a 1M HCl concentration. Partial copper and zinc were also determined by atomic absorption spectrophotometry after a sulphide selective leach. This attack uses H$_2$O$_2$ in the presence of ascorbic acid and is identical to that described by Lynch (1971) for the ultramafic litho-geochemical study carried out by Cameron et al. (1971). A 100-mg sample was attacked in 10 ml of the leach solution, where results were in excess of the top standard, equivalent to 400 ppm in the rock, a 1/10 dilution was effected by using a 10-mg sample weight for attack.

DESCRIPTION OF DATA STATISTICS AND DATA PRESENTATION

The summary statistics of the total and partial copper and zinc determinations are given in Table I. As the sampling was of a replicate nature it is possible to carry out an analysis of variance to determine if the combined sampling and analytical variance is significantly smaller than the overall regional variability (Garrett, 1973b).

In all cases the standard deviations of the data are greater than their associated means, leading to coefficients of variation in excess of 100%; this indicates a skewness in the data which can be deduced as being positive from an inspection of the means and ranges. Therefore, a logarithmic transform was applied to the data before further computations in order to render the data into a more normal distribution. Additionally, the logarithmic transform can be justified on theoretical grounds. The analytical method is based on an atomic absorption spectrophotometric determination which like all spectrographic methods is based on Beer's Law which is a logarithmic relationship. Thus the overall distribution of the errors inherent in the determinations are log distributed and a transform should be applied to the data.

In Table I σ^2 is the total data variance of the 1338 samples and σ^2_1 and σ^2_2 are respective variances of the first and second samples from each of the 669 sampling sites. The combined sampling and analytical variance σ^2_{SA} is estimated from 669 paired samples and is used in computing an F ratio, σ^2/σ^2_{SA}. With so large a number of degrees of freedom, $N-1$ and N, and also with the two degrees of freedom being numerically close it is difficult to accurately evaluate the critical value of Fisher's F; however, at the 99% confidence level and 668 and 669 degrees of freedom it is approximately 1.2. In all cases the calculated value of F is in excess of this critical value from which we can deduce that the sampling and analytical variability is significantly less than the regional variability. This leads to the conclusion that the displayed elemental patterns are real and not due to local sampling or analytical variability.

The arithmetic data for the copper and zinc ratios of partial to total contents of the element are presented in arithmetic form, the means showing the different behaviour of these two elements.

Geochemical data are characterized by two features, level and relief. For any one group the level is easily expressed by the mean, and in the case of apparently lognormally distributed data the geometric mean. The appropriate parameter with which to describe the relief poses more of a problem and no one is generally accepted. In the present study the arithmetic coefficient of variation has been used. As data from the positive skew section of a distribution is included in the computation for standard deviation the computed value rises very steeply. This raw standard deviation could be used; however, the standard deviation generally increases with increasing mean so the coefficient of variation, i.e. $\sigma/\bar{X}$, has been used. Thus, for each of the 4 determinations two maps are presented, one depicting the geometric means of the 40 groups of volcano-sedimentary rocks and the other their arithmetic coefficients of variation, these maps of statistical parameters are considered analogous to the geochemical concepts of level and relief.

Two additional maps are presented each showing the average proportion of sulphide-selective leach metal to total metal for copper and zinc in each group.

The 1338 samples from the 669 sampling sites were divided in 40 groups. Each group contains an average of some 17 sites. In some instances the groups represent several small outcrop areas of porphyry all geographically close, and in others a much larger contiguous area of porphyry was divided up into convenient subareas. The 17 sites lead to 34 samples which are sufficient to estimate the mean metal content with a reasonable degree of confidence and also define the general nature of the frequency distribution. The relief of geochemical data is considered to be a more important criterion in interpretation than absolute level as mineralized areas are ideally reflected by high relief due to local erratic high values related to metal concentration processes. Thus, in interpreting the present data the two maps for each determination should be studied simultaneously and in the writer's opinion more weight should be given to groups exhibiting high coefficients of variation (relief) and lower means (levels) than groups having high means (levels) but low coefficients of variation (relief) which indicate uniform enrichment but no local concentration processes.

DESCRIPTION OF RESULTS

Total copper data covers a range of 1 ppm to 6%, a group of samples from Tommie Lake [12] contain the exceptionally high values, these rocks are

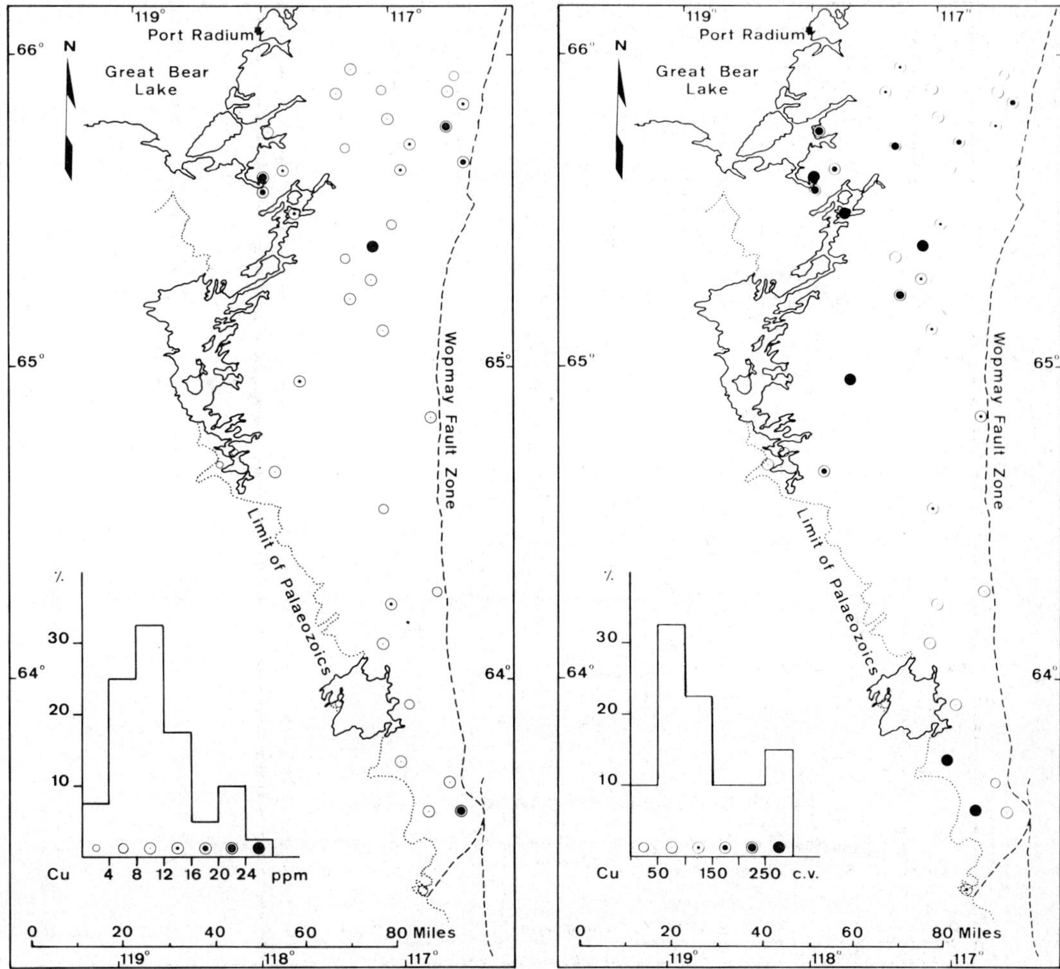

Fig.2. Geometric mean copper in volcano-sedimentary groups.

Fig.3. Arithmetic coefficient of variation for copper in volcano-sedimentary groups.

quartz-feldspar porphyries and other rocks high in magnetite which are of
sedimentary origin. The overall geometric mean of 11 ppm is within the
normal range of copper in acidic rocks, and 85% of the data falls between
5 and 50 ppm. The areal distributions for copper are shown in Figs.2 and 3.
The data distribution shows a bimodal distribution with the second population
being in excess of 20 ppm. Four groups fall into this population, three of
these are known to have mineralization in the vicinity, they are Tommie
Lake [12], Rainy Lake [19] and Blackjack [15]; the fourth, Hump Lake [1]
near which no mineralization is known has a mean of 21 ppm. The coefficient
of variation map, however, clarifies the matter. Hump Lake exhibits a normal

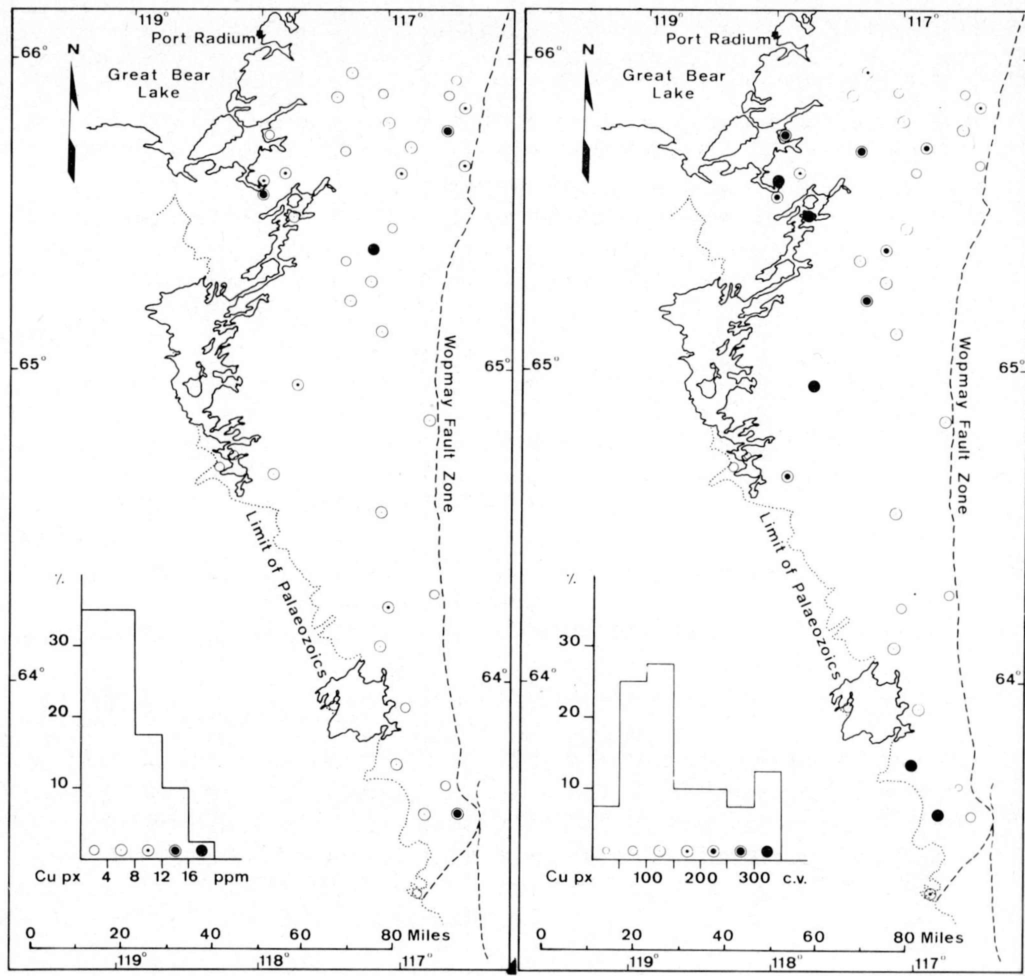

Fig.4. Geometric mean partial copper in volcano-sedimentary groups.

Fig.5. Arithmetic coefficient of variation for partial copper in volcano-sedimentary groups.

coefficient and on checking the range of the data there it lies between 7 and 56 ppm. Blackjack also reveals a somewhat unimportant coefficient. Thirteen groups show coefficients in excess of 150% and of these eight are known to be associated with mineralization; the Lou [2], Mazenod [3], Hardisty [8], Bode [9], Tommie [12], and Clut [18] Lakes and three groups in the Rainy Lake—Dinosaur Lake area [19]. The remaining five groups all have low associated means except Hansen Lake [14] and one group northeast of Blackjack. The group some 10 miles north of Rainy Lake on the shore of Conjuror Bay and the group on the Calder River [16] have high coefficients of variation but modal and low means, respectively, with respective maximum values of 322 and 400 ppm. These two groups lie in volcano-sedimentary belts extending from Rainy Lake and Clut Lake, and thus may be of interest.

The data for partial copper have a unimodal J-type distribution and the maps of the data (Figs.4 and 5) show that the regional patterns exhibited by the geometric means are essentially identical to those discussed for total copper. The coefficients of variation show a modified pattern in comparison and attention is drawn to the eleven groups with coefficients in excess of 250%. Five of these coefficients are in excess of 300% and these groups are all known to have sulphide mineralization associated with them. Of the remaining six, three are known to be associated with mineralization, Hardisty Lake [8], Tommie Lake [12] and one of the Rainy Lake groups [19]. The three remaining groups, Hansen Lake [14], Calder River [16] and Conjuror Bay were also outlined by total copper coefficients of variation but the author has no knowledge of mineralization reported in these areas.

The total zinc distribution is bimodal, with the second mode having a marked positive skew. This distribution was studied using a technique and algorithm for linear clustering described by Spath (1973). This method allows a decomposition of the distribution into a specified number of clusters using the criteria of minimizing the sum of the sums of squared deviations from the cluster mean for each cluster. A three-cluster model leads to boundaries at 60 and 110 ppm with respective frequencies and cluster means of 40%, 42 ppm; 42.5%, 85 ppm; and 17.5%, 128 ppm. In contrast the overall geometric mean is 72 ppm. The generally accepted levels for zinc in acid rocks are 60 ppm for granodioritic types and 40 ppm for granites (Levinson, 1974). The two low clusters with means of 42 and 85 ppm correspond reasonably with the means for acidic and intermediate acidic rocks. The third is believed related to mineralized rocks. The data range from 3 to 9500 ppm with some 85% falling between 20 and 200 ppm. The areal distribution of the zinc data is presented in Figs.6 and 7. The most noticeable feature of the total zinc areal distribution is the very marked gradient towards higher levels in the north and northwest of the field area. The groups corresponding to the third suggested population lie in the Rainy Lake [19], Conjuror Bay, Clut Lake [18], Cruikshank River [17] and Wylie Lake [13] areas. The last two of these plot into the 100—120-ppm histogram interval on Fig.6. Only the Rainy Lake and Clut Lake groups are known to contain mineralization. The coefficients of variation reveal six groups with variations in excess of 100%. Half of these exceed 200% and are all related to areas of mineralization, they are the Tommie [12], Clut [18] and Rainy [19] Lakes. The remaining three are more enigmatic and only the first of these is in an area of known mineralization, they are the Lou [2], Rae [4] and Hansen [14] Lakes.

The data for partial zinc are presented in Figs.8 and 9. The overall distribution of partial zinc, unlike that of total zinc, is unimodal, but is somewhat positively skewed. The tail of the distribution starts around 20 ppm and 10 groups have geometric means in excess of 20 ppm. Only three of these have known mineralization nearby, Tommie Lake [12], Clut Lake [18] and

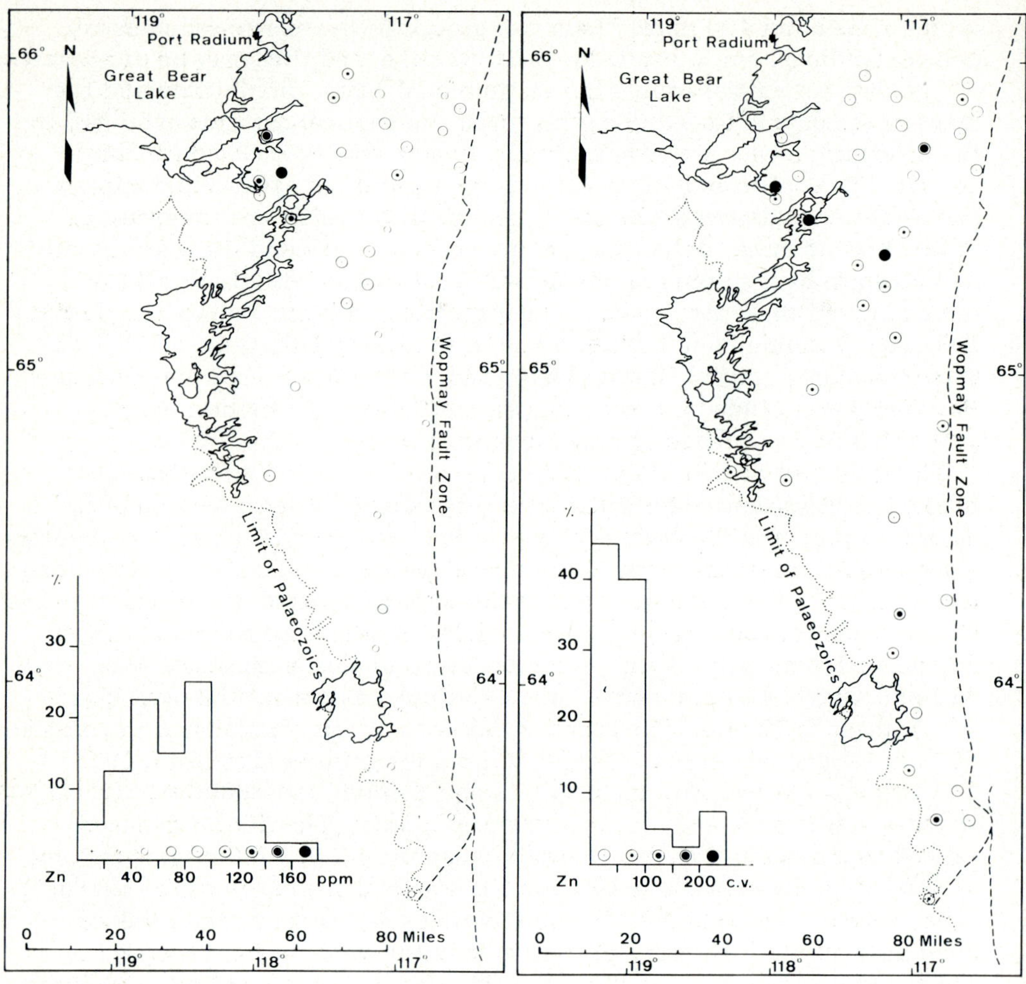

Fig.6. Geometric mean zinc in volcano-sedimentary groups.

Fig.7. Arithmetic coefficient of variation for zinc in volcano-sedimentary groups.

Rainy Lake [19], three others are of particular interest as they have means
in excess of 25 ppm but are not associated with known mineralization, Wylie
Lake [13], Hansen Lake [14] and Cruikshank River [17]. The coefficients
of variation reveal 11 groups with variations in excess of 100%, seven of these
are known to have mineralization in the area, Lou Lake [2], Hardisty Lake
[8], Bode Lake [9], Tommie Lake [12], Clut Lake [18] and two from the
Rainy Lake area [19]. The remaining four are at Rae Lake [4], Angle Lake
[10], Hansen Lake [14] and Cruikshank River [17]. The general regional
trend to higher levels of zinc in the north and northwest is also revealed in
the partial zinc data; however, the steady increase from the south is not as
well depicted by the partial data.

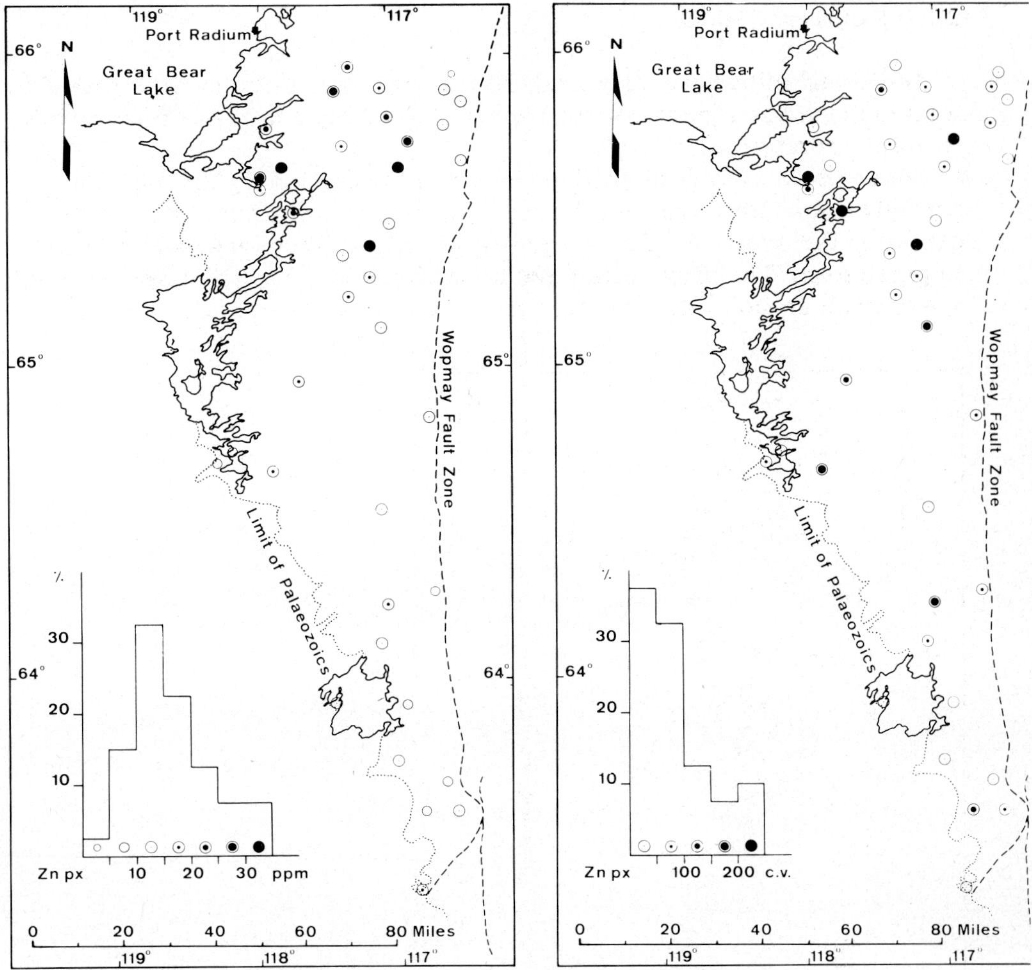

Fig.8. Geometric mean partial zinc in volcano-sedimentary groups.

Fig.9. Arithmetic coefficient of variation for partial zinc in volcano-sedimentary groups.

Summarizing the conclusions for both copper and zinc, seven groups show anomalous patterns, these are ordered in a general decreasing order of prominence; the Rainy [19], Tommie [12], Clut [18], Lou [2], Hansen [14], Hardisty [8] and Bode [9] Lakes. Only at Hansen Lake is there no known mineralization. Anomalous copper patterns were observed at four groups, Mazenod Lake [3], Self Lake [11], Blackjack [15] and Calder River [16], only at two of these, Mazenod Lake and Blackjack is mineralization known. Lastly, anomalous zinc patterns were found in four groups where mineralization is unknown at present, Rae Lake [4], Angle Lake [10], Wylie Lake [13] and Cruikshank River [17].

DISCUSSION OF RESULTS

   Additional plots have been made of the areal distribution of the partial to total metal ratios and also of the relationship of these ratios to total metals to aid interpretation (Figs.10—13).

   Copper is on average divided in the molar proportion 61/39 between partially extractable and total forms. This would indicate that 61% of the copper in the quartz-feldspar porphyry volcano-sedimentary belts is in a sulphide form. The distribution of the ratios reveal a negative skew with only a very small group of data falling above the mode; these data are for the

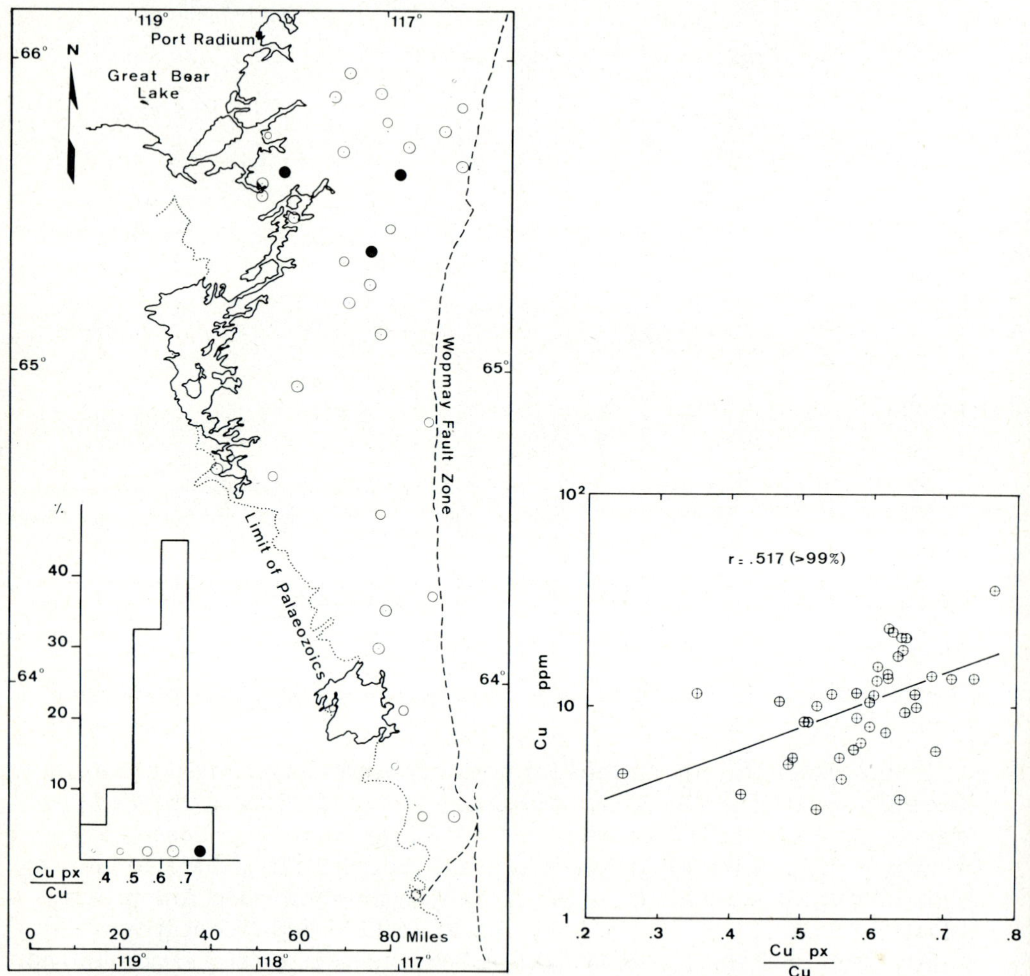

Fig.10. Average partial copper to total copper in volcano-sedimentary groups.

Fig.11. Plot of partial copper to total copper ratio versus total copper.

Tommie Lake [12], Wylie Lake [13] and Dinosaur Lake [19]. Of these three, the first and the last are from areas already selected as anomalous for copper. Wylie Lake has slightly elevated levels of both total and partial copper but modal or low coefficients of variation. It may be concluded that the Wylie Lake group has uniformly slightly elevated levels of copper, a higher than usual proportion of which is in a sulphide form, but which also shows no local concentration features which would be indicative of mineralization.

The total copper versus partial to total ratio plot reveals a positive correlation in the data, the line is the least-squares linear regression fit. The correlation is significant and indicates that as the total amount of copper in the rock

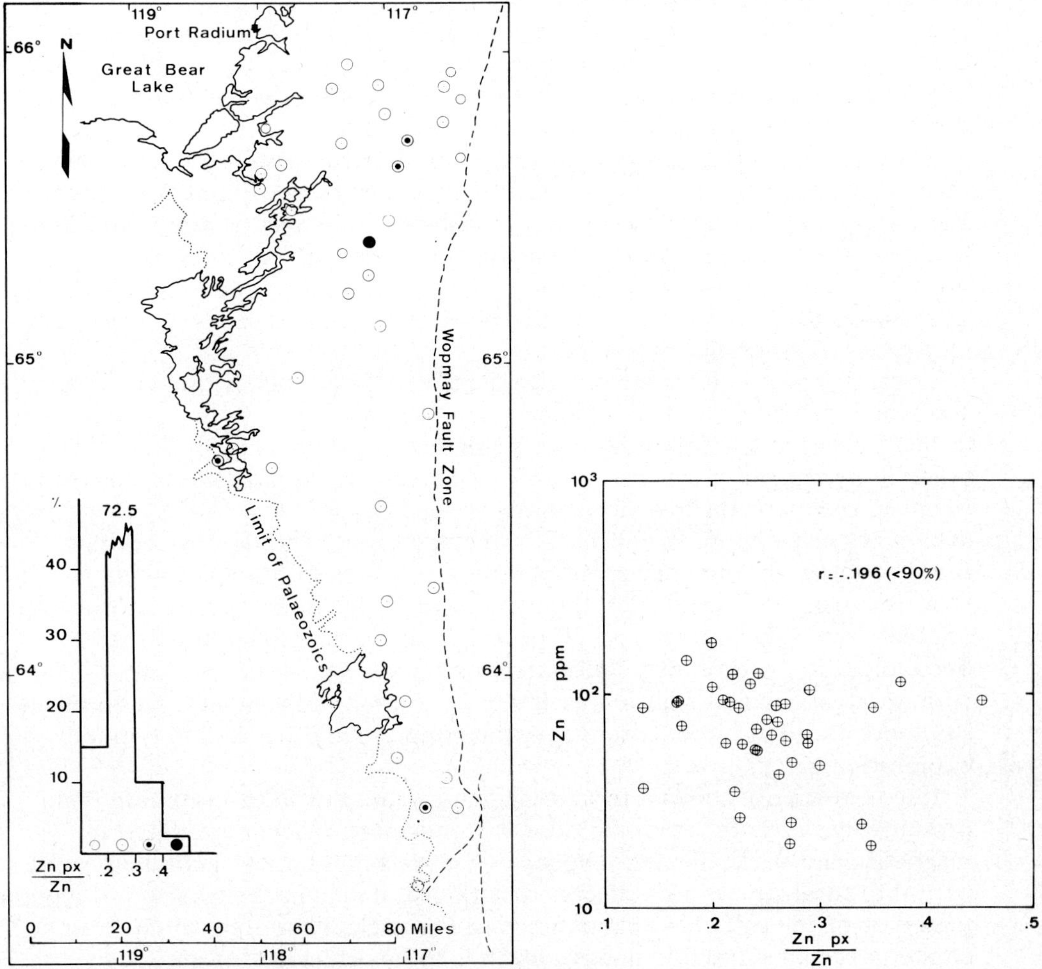

Fig.12. Average partial zinc to total zinc in volcano-sedimentary groups.

Fig.13. Plot of partial zinc to total zinc ratio versus total zinc.

increases a larger proportion goes into the sulphide phase, this could be regarded as typical behaviour for a chalcophile element.

The behaviour of zinc is in contrast to that of copper; firstly, the molar proportion of partial and total zinc is reversed and in the ratio 26/74, and secondly, the data are distributed with a positive skew. Five volcano-sedimentary groups show increased average levels of zinc in the sulphide phase, of these four are already selected as anomalous; two are from areas of known mineralization, these are respectively Lou [2] and Tommie [12] Lakes and Wylie [13] and Hansen [13] Lakes. The fifth is a group of scattered small volcano-sedimentary outcrops around Beaverlodge and Hottah Lakes 10 miles west of Hardisty Lake [8]. As with the Wylie Lake group copper data neither the means or coefficients of variation for the total and partial data are particularly unusual in the Beaverlodge group. It is concluded that the generally less than modal levels are simply associated with a uniformly higher proportion of zinc in the sulphide phase which shows no great tendency to local concentration indicative of high relief and therefore mineralization.

The plot of total zinc versus partial to total metal ratio shows that the relation between the two is random and that the effect of increasing the total zinc in the rock has an unpredictable effect on the partial to total zinc ratio. It is most probable that the increased levels of zinc are accommodated by substitution in magnetite or dark silicates such as biotite and hornblende. In this respect the behaviour of zinc in the Bear province acid volcano-sedimentary rocks is essentially oxyphile.

Thus, two particular features contrast the different behaviour of copper and zinc in the rocks of the study area. Firstly, almost two-thirds of the copper in the rocks is in the sulphide form, probably in pyrite, chalcopyrite and possibly pyrrhotite if it is present. This observation would tend to confirm the proposal made on thermodynamic grounds by Putnam (1973) that biotite at unit activity and in equilibrium with chalcopyrite should only contain 6 ppm copper. Putnam extends the 6-ppm figure to 30 ppm, allowing for the thermodynamic uncertainties, and considers it very unlikely that biotite can contain any more copper than 30 ppm in true isomorphous substitution. Secondly, the positive correlation between total copper and partial to total copper ratio indicates that increasing copper levels are accommodated in the sulphides and that either total or partial copper could be used in copper exploration programmes.

In contrast only about a fifth of the zinc is present in the sulphide form, presumably in pyrite, the remainder it is suggested is accommodated in magnetite and dark silicates. The lack of correlation between total zinc and partial to total zinc ratio indicates that the total zinc in the rock is not a good predictor of the sulphide zinc content of the rock. Therefore, in contrast to copper a partial sulphide-selective leach is to be preferred for zinc if an estimate of the sulphide zinc potential of the rocks is required.

In studying the data from a broad regional geochemical viewpoint the most important feature is the gradient seen in both total and partial zinc from low levels in the south to high levels in the north and northwest. One possible interpretation based on the bimodal nature of the total zinc data is that dacitic rocks should be more common in the north whilst rhyolitic types would predominate in the south. The partial zinc data indicate increased levels of sulphide held zinc in the north, these may be a contributory factor in the total zinc patterns. Equally well one could argue that the increased partial levels are to be expected due to the postulated prominence of intermediate dacitic rocks in the north. However, of the six groups exhibiting low partial to total zinc ratios, $\ll 0.2$, five lie north of 65°N. It is of interest that as one proceeds north the proportion of tuff and extrusive porphyries increases relative to intrusive phases. Thus, the northern parts of the area are represented by higher-level porphyries and the south by deeper-level rocks. The higher levels of partial zinc in the north indicate the areas of greatest sulphide zinc potential. As a generalization of the volcanogenic ore genesis model one might state that copper can occur either in high level or extrusive phases, e.g. porphyry coppers, pipe breccias and volcanogenic stratabound massive sulphides. Zinc in contrast is concentrated in the extrusive environment where zinc is found in the stratabound massive sulphides. The lack of any marked gradient in the copper patterns could correlate with lack of any exclusive site for copper deposits in the model, whereas the increased zinc levels in the north correlate with the increased abundance of extrusive rocks which are the preferred host for zinc deposits in the general volcanogenic model. This hypothesis is open to criticism as the typical Shield volcanogenic model does not truly apply to the field area. In Archean areas basic and intermediate rocks are far more common than acidic varieties. In the field area the opposite is true with basic and intermediate rocks only accounting for a small proportion of the rocks. In this respect the field area resembles more a basin and range situation (Hoffman, 1974). However, the author considers the hypothesis stimulating and therefore worthwhile.

It is also interesting to note that the regional lake sediment survey carried out by the G.S.C. in 1972 revealed a large area of elevated zinc levels in the northwest of the survey area (Allan and Cameron, 1973; Allan et al., 1973). At the time of the initial interpretation of the lake sediment data it was not clear what the source of the zinc in the sediments might be. Now it is proposed that as the areas of high partial zinc in the porphyries and areas of high zinc in lake sediments coincide that the zinc in the lake sediments has moved there in solution following the weathering of the porphyries of the volcano-sedimentary belts and the release of zinc from sulphides. This correlation and proposal, which seems very reasonable, can only increase one's faith in the applicability of low density reconnaissance geochemical programmes based on the secondary environment for delineating major geochemical features related to the bedrock.

CONCLUSIONS

The reconnaissance lithogeochemical survey data has been presented using the statistical parameters of geometric mean and coefficient of variation which are analagous to the geochemical criteria of level and relief. The known mineral occurrences containing significant amounts of copper and zinc sulphides present in the volcano-sedimentary porphyry belts have been outlined by anomalous metal patterns in the enclosing rocks. Additionally, attention has been drawn to certain areas which exhibit similar metal patterns but where no mineral occurrences are known; these areas are considered to have an increased mineral potential.

In contrast to copper zinc is best depicted by maps of the selectively leached sulphide phase if the objective of the survey is mineral potential estimation. This is largely due to the fact that copper behaves in an essentially chalcophile manner whilst zinc behaves in an oxyphile manner. Related to this fact is that the molar ratios for sulphide held metal to total metal are 61/39 and 21/79 for copper and zinc, respectively.

From a mineral exploration viewpoint the sulphide-selective leach is recommended. Although the total metal in this study was determined after a $HF/HClO_4$ attack which is both expensive in capital and operating costs it is considered probable that a less rigorous attack based on aqua regia, which would be less expensive, would remove almost as much copper into solution and probably less zinc, due to that element's oxyphile tendencies. Work by Brabec and White (1971) on granitoid rocks from the Guichon Creek batholith indicates somewhat similar conclusions as to metal location in the rock. Using their data from 37 samples some 79% of the copper is aqua regia extractable over a range of 2—155 ppm, as the extractable copper increases from an average of 15 to 132 ppm the extractable percentage increases from 71 to 91%. Their data for zinc shows less spread with a mean extractable ratio of 51% over a range of 9—45 ppm aqua regia extractable zinc, again as extractable zinc increases from 20 to 34 ppm the extractable percent increases from 48 to 55%. However, the author of this paper is not convinced that this 7% increase is significant. Although the rock types in the present study and Guichon Creek are not identical, both are acid and igneous and some comparisons may be drawn. Aqua regia extracts some 80% of the copper from the rock, whereas the sulphide selective leach removes 60%, in the case of zinc the extractable percentages are 50% and 20%, respectively. It is quite obvious that the differences are most severe for zinc, the more problematical of the two elements. It is proposed that the aqua regia removes significant amounts of zinc from the dark silicates and magnetite. The general increase in extractable copper to total copper ratio with increasing copper levels is similar to that described for the Proterozoic volcanics and indicates the chalcophile nature of copper.

On the basis of these observations it is proposed that the method of analysis for copper is relatively unimportant. Copper is chalcophile and total, aqua regia or H_2O_2-ascorbic acid extractable copper all yield data of use in estimating mineral potential. Ideally, the author would choose the latter, but for exploration work would settle for aqua regia due to ease of analysis and cost. In the case of zinc the aqua regia attack doubles the amount of zinc extracted relative to the H_2O_2-ascorbic acid attack. Thus the patterns of zinc distribution in sulphides are clouded by zinc from silicate and oxide sources. On these grounds it is proposed that the weak sulphide-selective attack is the preferred method of analysis if, as in exploration, the sulphide mineral potential is the major topic of interest in the survey area.

Lastly, the zinc data reveal broad regional patterns in agreement with reconnaissance lake sediment survey data. It is proposed that these patterns are related to the level of the present surface in relation to the original volcano-sedimentary pile, and that they reflect the metallogenetic development of the area, particularly the potential for massive stratabound volcanogenic sulphide deposits. One interesting aside is that if such deposits also occurred in the south they have been eroded and presumably their weathering products were dumped into the western sedimentary basin. Much of this metal would have been in solution or in a relatively soluble form. Could it have been a source for some or all of the zinc deposits now found in the sedimentary basin?

ACKNOWLEDGEMENTS

The author would like to express his thanks to R.F. Benson and N.G. Lund who assisted in the field work. Mr. Lund also assisted in the compilation of the data prior to data processing. Particular thanks are due to the sample preparation and analytical staffs who worked with great diligence in providing the analytical data which forms the basis of this paper. Finally, the interest of fellow members of the Geochemistry Section and Geological Survey who provided a sounding board for discussion of the ideas presented in this paper is acknowledged.

REFERENCES

Allan, R.J. and Cameron, E.M., 1973. Zinc content of lake sediments, Bear-Slave Operation. Geol. Survey Can. Map 10-1972, Sheet 1

Allan, R.J., Cameron, E.M. and Durham, C.C., 1973. Reconnaissance geochemistry using lake sediments of a 36,000 square mile area of the northwestern Canadian Shield. Geol. Survey Can. Paper 72-50: 70 pp.

Brabec, D. and White, W.H., 1971. Distribution of copper and zinc in rocks of the Guichon Creek batholith, British Columbia. Can. Inst. Min. Metall., Spec. Vol., 11: 291—297

Cameron, E.M., Siddeley, G. and Durham, C.C., 1971. Distribution of ore elements in rocks for evaluating ore potential: nickel, copper, cobalt and sulphur in ultramafic rocks of the Canadian Shield. Can. Inst. Min. Metall., Spec. Vol., 11: 298—313

Fraser, J.A., Hoffman, P.F., Irvine, T.N. and Mursky, G., 1972. The Bear Province. In: R.A. Price and R.J.W. Douglas (Editors), Variations in Tectonic Styles in Canada. Geol. Soc. Can. Spec. Paper, 11: 454—503

Garrett, R.G., 1971. Molybdenum, tungsten and uranium in acid plutonic rocks as a guide to regional exploration, S.E. Yukon. Can. Min. J., 92: 37—40

Garrett, R.G., 1973a. Regional geochemical study of Cretaceous acidic rocks in the northern Canadian Cordillera as a tool for broad mineral exploration. In: M.L. Jones (Editor), Geochemical Exploration 1972. Institution of Mining and Metallurgy, London, pp.203—219

Garrett, R.G., 1973b. The determination of sampling and analytical errors in exploration geochemistry — a reply. Econ. Geol., 68: 282—283

Garrett, R.G., 1974a. Mercury in some granitoid rocks of the Yukon and its relation to gold-tungsten mineralization. J. Geochem. Explor., 3: 277—289

Garrett, R.G., 1974b. Bear Province lithogeochemical survey. In: Report of Activities, Part A. April to October 1974. Geol. Survey Can. Paper 74-1A: 63 pp.

Hoffman, P.F., 1974. Volcanism and plutonism, Sloan River map area (86K), Great Bear Lake, District of Mackenzie. In: Report of Activities, Part A. April to October 1974, Geol. Survey Can. Paper 74-1A: 173—176

Lavergne, P.J., 1965. Field and laboratory methods used by the Geological Survey of Canada in geochemical surveys, No. 8. Preparation of geological materials for chemical and spectrographic analysis. Geol. Survey Can. Paper 65-18: 23 pp.

Levinson, A.A., 1974. Introduction to Exploration Geochemistry. Applied Publishing Ltd., Calgary, Alta., 612 pp.

Lynch, J.J., 1971. The determination of copper, nickel and cobalt in rocks by atomic absorption spectrometry using a cold leach. Can. Inst. Min. Metall., Spec. Vol., 11: 313—314

McGlynn, J.C., 1974. Geology of the Calder River map-area (86F), District of Mackenzie. In: Report of Activities, Part A. April to October 1974. Geol. Survey Can. Paper 74-1A: 383—385

Putnam, G.W., 1973. Biotite—sulphide equilibria in granitic rocks: a revision. Econ. Geol., 68: 884—886

Shegelski, R.J. and Thorpe, R.I., 1972. Study of selected mineral deposits in the Bear and Slave Provinces. In: Report of Activities, Part A. April to October 1971. Geol. Survey Can. Paper 72-1A: 93—96

Spath, H., 1973. Algorithm, clustering of one-dimensional ordered data. Computing, 11: 175—177

Stockwell, C.H. and McGlynn, J.C., 1970. Geology of the Canadian Shield. In: R.J.W. Douglas (Editor), Geology and Economic Minerals of Canada. Geol. Survey Can. Econ. Geol. Rept. 1: 838 pp.

Part 5
EXPLORATION IN LATERITIC TERRAIN

Chairmen:

J.A. HANSULD
AMAX Exploration Incorporated, Toronto, Ont., Canada
M. SHIIKAWA
Akita University, Akita, Japan

GEOCHEMICAL EXPLORATION PROBLEMS IN WESTERN AUSTRALIA EXEMPLIFIED BY THE MT. KEITH AREA

C.R.M. BUTT and N.R. SHEPPY

Division of Mineralogy, C.S.I.R.O., Wembley, W.A. (Australia)
Newmont Proprietary Limited, Mount Isa, Qld. (Australia)

ABSTRACT

The Archean shield of southwestern Australia has had a long and complex weathering history which has given rise to many exploration problems. Extensive and poorly understood chemical redistribution of elements occurred during mid-Tertiary lateritization and was followed by dissection under arid conditions which continue to the present. The laterites have often been partly truncated and subsequently covered by a variety of aeolian or water-lain sediments. Areas of outcropping residuum are relatively uncommon. Geochemical prospecting is thus hindered by problems of obtaining adequate samples and subsequently by difficulties of interpreting exploration data.

Exploration of the Mt. Keith area exemplifies many of these problems. Mineralization of economic significance at Mt. Keith consists predominantly of pentlandite disseminated in serpentinized dunite, and the main body has reserves of 290 million tons assaying 0.60% Ni. The ultramafic units occur in a variety of geomorphological situations, with complete or partly truncated laterites either overlain by up to 20 m of transported overburden or exposed as outcropping residuum. Orientation over the main body showed that thresholds of 4000 ppm Ni and 300 ppm Cu in highly weathered, oxidized zones over the ultramafics were indicative of disseminated sulphides, although locally intense silicification can cause some dilution of anomalies. Rotary drilling located several such anomalies throughout the property but follow-up diamond and percussion drilling met with varied success. Although disseminated sulphides were encountered on many occasions, other Ni—Cu anomalies, although essentially similar to those occurring over the main deposit, overlie ultramafics containing only rare or no sulphides. These enrichments of the weathered zone are considered to be due to concentration during lateritic weathering and are examples of one of the main interpretive problems confronting geochemical exploration in this terrain. Parameters other than simple copper and nickel concentrations and ratios, therefore, are necessary for a more accurate definition of anomaly significance, even when searching for targets of these dimensions.

INTRODUCTION

The history of weathering and erosion of the Yilgarn Block of southwestern Australia has been long and complex. It has given rise to a number of physiographic features which are individually peculiar and which together form a unique landscape (Mulcahy, 1967). In consequence, the terrain presents some unusual exploration problems and it is significant that prior to 1965, most mineral discoveries in Western Australia were of weathering resistates, such as gold and tin, or weathering products, such as iron ore and bauxite. During exploration for sulphide ores, the deep weathering and transported overburden have hindered geological mapping and the highly conducting, oxidized surface layers inhibit electromagnetic surveys. Geochemical techniques have been particularly important, but they are limited by problems of adequate sampling and interpretative procedures.

In this paper, some of the important features of the terrain are described and their influence on geochemical prospecting discussed. The exploration history of the disseminated nickel sulphide deposits at Mt. Keith exemplifies some of the typical problems.

WEATHERING HISTORY, GEOMORPHOLOGY AND GEOCHEMISTRY

The Archean shield of Western Australia has probably been exposed to terrestrial conditions since the Mesozoic and possibly since the Palaeozoic. The salt lake chains are probably some of the oldest landforms. They appear to be disconnected remnants of a pre-existing drainage system which is still being followed by the present subsurface drainage. The main components form riverine systems either side of a broad continental divide which approximately bisects the Archean shield (Fig.1, after Mulcahy and Bettenay, 1971, and Beard, 1973). Beard (1973) argues that these systems date from the Mesozoic, whilst, on sedimentary evidence, Johnstone et al. (1973) suggest more specifically, early Jurassic or even early Triassic ages.

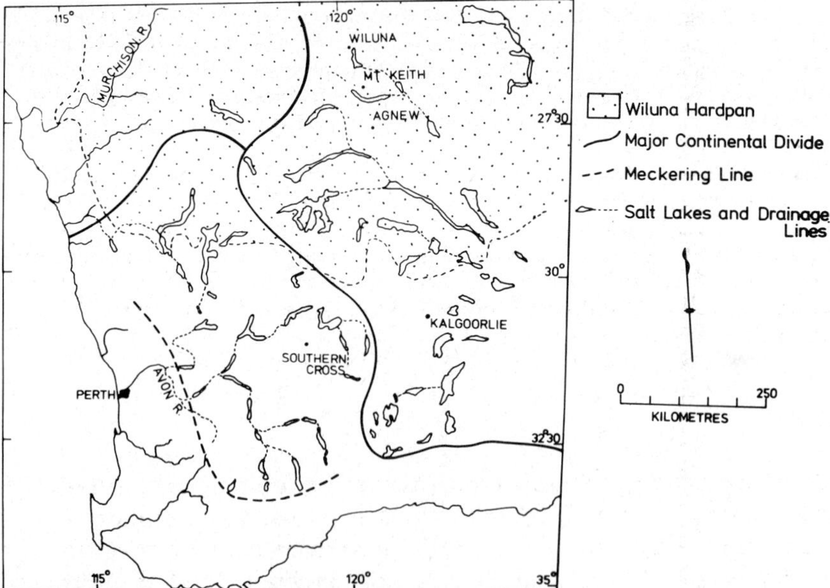

Fig.1. Location map depicting the principal drainage systems and some major geomorphological boundaries in southwestern Australia.

During this time span, climatic changes have been considerable and the present landforms and surficial materials often have properties inherited from previous weathering cycles. There are two main climatic and weathering phases of particular importance:

(1) Periods of lateritization, involving deep weathering and peneplanation. Lateritization here refers to the formation of weathering profiles up to 100 m in depth which, among other characteristics, have, or have had, an upper horizon of aluminous or ferruginous composition. The term "laterite" refers to the whole profile, not to any individual horizon.

(2) Subsequent periods of aridity and semi-aridity, continuing to the present.

This is only a very broad generalization and the two phases have overlapped in place and time.

Lateritization

The main period of lateritization was possibly during the Oligocene and Miocene (Beard, 1973; Johnstone et al., 1973), and is considered to have operated on a tectonically stable land mass with subdued relief. Sedimentary evidence given by these authors suggests that a warm, probably humid climate prevailed, permitting deep weathering over extensive areas. "Lateritization" was not a single event, however, as laterites have developed on surfaces of different ages (Mulcahy and Hingston, 1961).

The predominantly granitic rocks of the shield have weathered to give a characteristic profile 45—60 m deep which shows surface iron oxide accumulation over a pallid, kaolinitic saprolite zone (Walther, 1915). Over the more basic rocks the same principle is followed although mineralogy is more complex with more exaggerated ferruginous expressions and carbonate development.

Aridity and dissection

With a change to arid conditions, dissection of the lateritic surface has taken place. Most base-metal exploration activity has been in the inland region, and the landforms here are of particular importance. The geochemical profiles vary according to the geomorphic situations in which they occur. To the west, this region extends to the Meckering line (Fig.1) which marks a change from the irregular, disconnected drainage of the salt lake systems to the closer network of the seasonal drainage near the coast (Mulcahy, 1967).

The geomorphology of the inland region has been described by Jutson (1950), with some detailed studies by Mabbutt et al. (1963) and Bettenay and Hingston (1964); the main features are shown diagrammatically in Fig.2 and described below.

Old Plateau. Lateritic residuals, most widespread near the divides, probably represent the peneplain of the main lateritic phase. Profiles may be more or less complete and capped by the upper, iron oxide horizon, or partly trun-

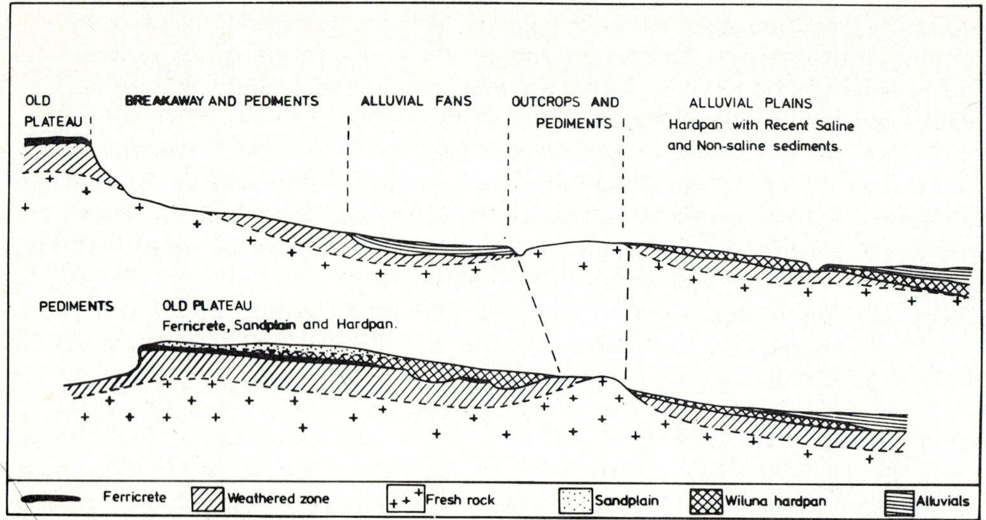

Fig.2. Idealized section showing some of the major landforms developed over Archean rocks in inland Western Australia.

cated and capped by silcrete or hardened saprolite. Frequently, there is a transported cover of colluvium, gravels and sand, which can become very deep in undulations of the pre-existing surface. The sandplains often have aeolian features and although probably derived from the breakdown of lateritic material (Brewer and Bettenay, 1973) are not residual.

Breakaways, pediments and outcrop. The residuals are often bounded by escarpments (breakaways), below which, on the pediments, the various horizons of the laterites are exposed, occasionally to fresh rock. More recent residual soils can be developed on pediments, mostly from pre-weathered material. Thin alluvial fans form the lower pediment slopes.

Alluvial, colluvial and outwash plains. Erosion of the lateritized surface is to the base level of the salt lakes, and extensive sedimentary deposits have accumulated on the valley sides and floors. An important component is a rudaceous colluvium, termed the Wiluna Hardpan (Litchfield and Mabbutt, 1962, Bettenay and Churchward, 1974), which is widespread in the Murchison district of Western Australia, extending some 400 km south of Wiluna (Fig.1). It consists of poorly sorted silicified sandstones, breccias and conglomerates composed of quartz, chert and weathered rock fragments, in sheet-like deposits up to 30 m in depth. The hardpan can overlie complete or truncated laterites, and either outcrop or be itself eroded or overlain by later deposits.

Chemical deposits occur in the "sumps" of extensive basin-like valleys, with calcrete development (the site of patchy U—V mineralization) in tributary drainages. Evaporites occur in the salt lakes themselves and evaporite materials are added to nearby soils by deflation.

Geochemical exploration problems

The major chemical and mineralogical changes that have occurred during lateritization are responsible for many complexities in the interpretation of geochemical data from the region. The formation of low-grade nickel laterites over ultramafics, for example, impedes exploration for nickel sulphides and the concentration of various metals in "ironstones" leads to problems of identifying gossans (Clema and Stevens-Hoare, 1973). Weathering profiles are deep, the horizons poorly characterized and element variability considerable. Only broad differences between or within horizons can be recognized and it is difficult to distinguish geochemical variability produced by weathering from bedrock effects.

The nearly ubiquitous cover of transported overburden developed during the lateritic and, particularly, the arid phases are an obvious exploration problem. Conventional geochemical procedures involving the sampling of rocks, gossans, stream sediments and soils are only applicable in pedimented areas where "outcrop", whether of fresh rock, saprolite or residual soil, occurs. Most of the major nickel discoveries in Western Australia, for example, have been in such areas and geochemical techniques have been applied there quite successfully (Mazzucchelli, 1972; Hall et al., 1973; Dalgarno and Knowles, in prep.). Such areas, however, comprise less than 40% of the greenstone belts and perhaps less than 15% of the total landscape. Elsewhere sampling has to be done by drilling to residuum. There is some evidence of hydromorphic dispersion of ore metals into transported cover (Dalgarno and Knowles, in prep.) but this is not necessarily a common phenomenon. Biogeochemical methods have yet to demonstrate their applicability away from areas of residuum.

LOCATION AND PHYSIOGRAPHY OF THE MT. KEITH AREA

Mt. Keith is 80 km south-southeast of the old gold mining town of Wiluna and approximately 400 km north-northwest of Kalgoorlie, Western Australia (Fig.1). The claims of Metals Exploration Limited and its joint venture partners extend nearly 26 km along strike of a narrow, north-northwest trending greenstone "belt" which is proving to be most productive of nickel mineralization, particularly from Mt. Keith southwards. The deposits include a number of disseminated sulphide occurrences at Mt. Keith and the adjoining Yakabindie claims of Anaconda Australia Inc., and the massive sulphide deposits of Western Selcast, 60 km south at Perseverance, near Agnew.

Mt. Keith is situated near the highest part of the interior plateau and the area has been subjected to the periods of lateritic weathering and subsequent dissection described previously. Between Wiluna and Agnew, the greenstone belt passes through a wide range of geomorphological situations, varying from the oldest peneplain which is represented by plateau-like divides to lowlands of the salt lake drainages (Fig.3). The lateritic profiles developed can be complete, or partly or wholly truncated. At Mt. Keith, the greenstone belt crosses

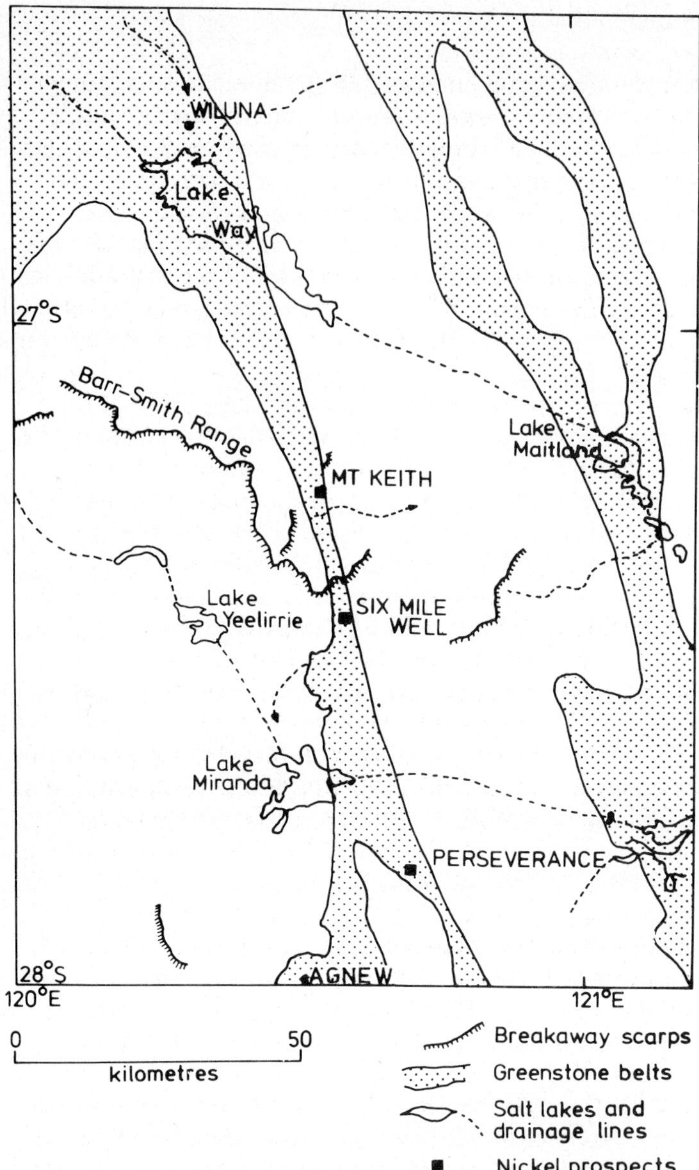

Fig.3. Location map, Wiluna—Agnew area, Western Australia.

the divide between the Lake Way—Lake Maitland drainage to the north and east, and the Lake Yeelirrie—Lake Miranda drainage to the west and south. North and west of Mt. Keith the divide is the Barr-Smith Plateau, developed on the porphyritic western granite. A major feature is a west-facing "breakaway" escarpment which, at Mt. Keith, swings east and marks the divide as it crosses the greenstone belt about 14.5 km south of the main mineralized body. To the north of the breakaway, relief is gentle with lateritic profiles complete or partially truncated, and overlain in places by up to 25 m of transported overburden of various types. Areas of residuum are uncommon, and one of the most extensive of these has developed on the western strata-bound peridotitic ultramafic, which is eroded to nearly fresh rock. South of the breakaway, relief is more pronounced, erosion is exposing the various lateritic horizons and about 70% of the area is outcrop and residuum.

Residual soils are mostly very shallow, red, gravelly sands, red earths and lithosols, commonly with ferruginous or siliceous fragments. Soils derived from transported materials, such as sand plain, are generally deeper, but profile development is weak. Most soils are alkaline and can have thin carbonate accumulation horizons, especially over basic and ultramafic rocks.

The Mt. Keith area has a semi-arid climate. Rainfall is erratic, with an annual mean of about 210 mm. Summers are long and hot, and maximum temperatures commonly exceed 38°C. Winters are mild with occasional severe frosts.

GEOLOGY AND MINERALIZATION AT MT. KEITH

The geology of Mt. Keith, shown on Fig.4, and its regional setting are discussed fully elsewhere (Sheppy and Burt, 1974). The greenstone belt here varies from 2.6 to 6.5 km in width, between bordering granitic rocks. The porphyritic western granite appears to be the oldest part of the local sequence and is unconformably overlain by an east facing series of partly metamorphosed boulder conglomerates and arkoses of the Jones Creek Conglomerates (Durney, 1972). This unit becomes argillaceous towards the top and is succeeded by a steeply dipping sequence of dominantly intermediate to felsic volcanogenic rocks, with subordinate mafic and ultramafic flows. One of the largest mafic units occurs at the contact with the eastern granite gneiss. Occasional pyritic carbonaceous shales and ferruginous cherts are found representing periods of quiescence. The whole sequence displays characteristics of lower greenschist facies metamorphism. The eastern granite gneiss contains rafts of amphibolites and banded magnetite quartzites and shows evidence of the metamorphism and *lit-par-lit* intrusion of a pre-existing stratigraphy by granite.

There are two main groups of ultramafic rocks: the stratabound group mentioned above and thought to be gravity differentiated flows, and a later, intrusive metadunite which is interpreted as a dyke system. These dyke-like metadunites are the hosts for the disseminated sulphide mineralization so characteristic of the Mt. Keith—Yakabindie area and, to date, the strata-bound metaperidotites and metapyroxenites have proved to be barren of nickel sulphides. The ultramafic flows often show "spinifex" textures and are located near the base of the intermediate-felsic volcanogenic sequence. The dykes of typically "cumulate" textured serpentinized dunite apparently occupy a fissure zone located centrally within the greenstone belt. Regional aeromagnetics and geological observation suggest that this fissure zone is a fundamental structural lineament which may have tapped mantle material.

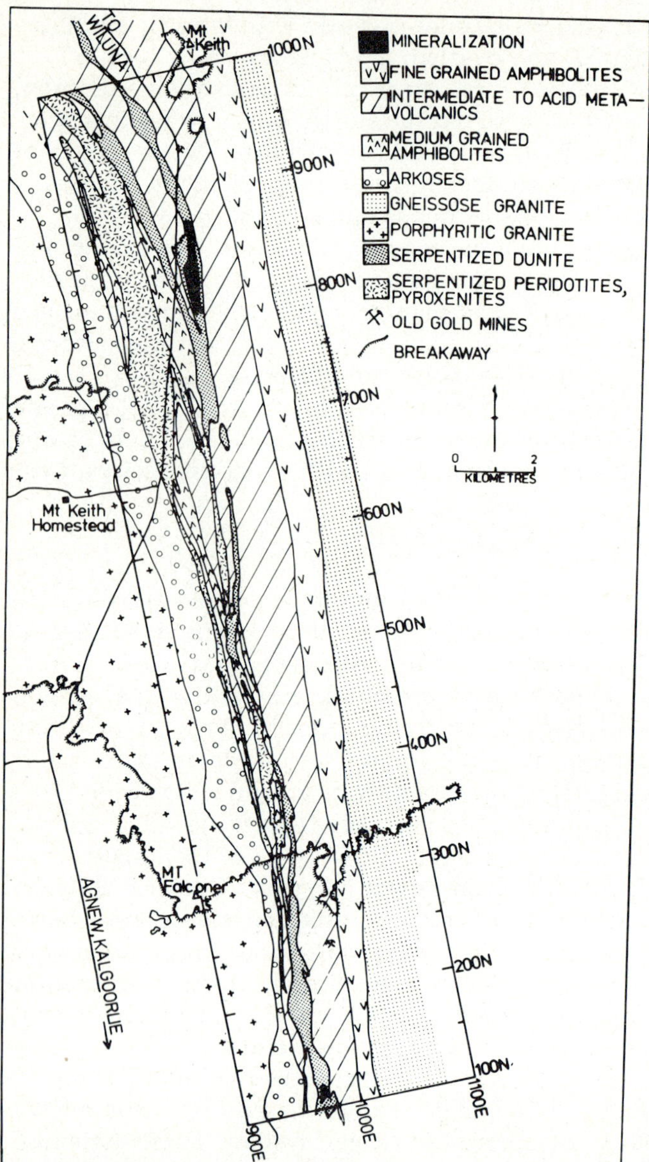

Fig.4. Regional geological map, Mt. Keith. Grid locations in text refer to the exploration grid depicted on this diagram.

The main mineralized body at Mt. Keith is hosted by a serpentinized dunite with narrow peridotitic margins (Fig.5) and dipping steeply to the west. The mineralization is essentially disseminated pentlandite, which displays considerable uniformity over large tonnages and clear-cut natural limits. Mineralization of economic significance extends continuously for some 1675 m, with the sulphides confined to a central zone within the inner serpentinized dunite

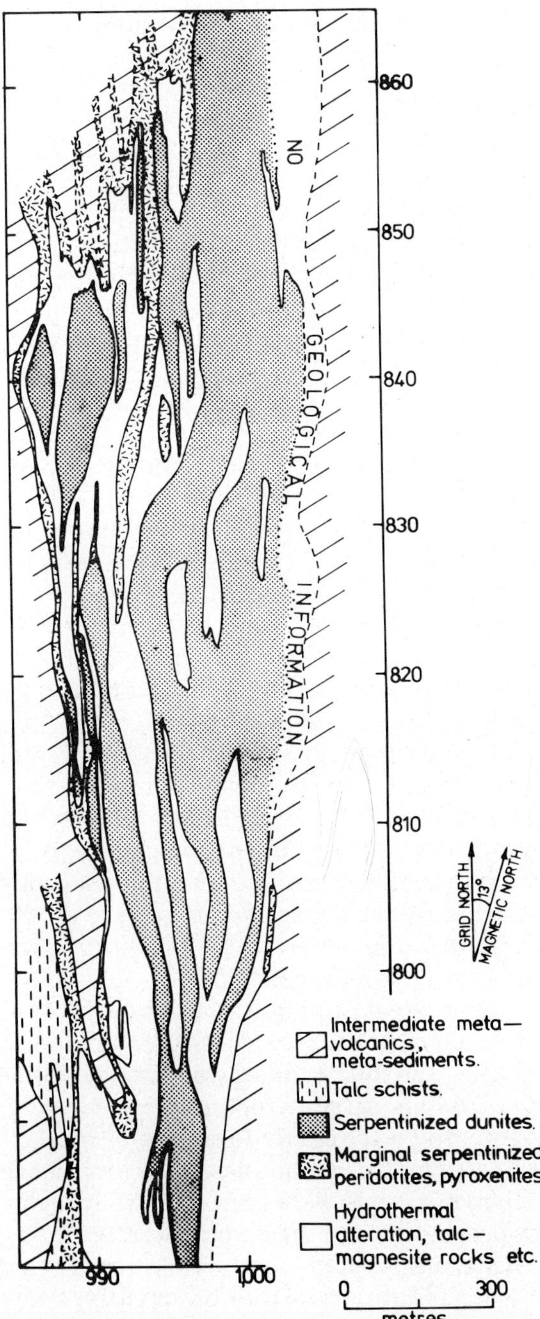

Fig. 5. Simplified geological plan of the main mineralized zone, Mt. Keith, projected to the 150-m level.

and with the pyroxene-bearing margins remaining barren. The sulphides are located interstitially between serpentinized olivine grains and in the primary zone comprise predominant pentlandite, subordinate pyrrhotite, and some pyrite in fissures. Millerite, violarite, chalcopyrite, and arsenides occur only as accessories. Primary zone oxides include intergranular and interstitial magnetite of an early mode, and rare interstitial chromite. A later magnetite mode occurs as intergrowths with sulphides and in fissures. The non-opaque assemblage consists of serpentine (dominantly lizardite), hydrotalcite varieties (pyroaurite and stichtite), magnesite, and minor brucite and talc. Hydrothermally altered zones near the margins and centrally within the hosting body show various stages of carbonatization, talc formation and magnetite remobilization. About 10% of the body is affected.

Reserves of the main body are 263 million tons indicated, 27 million tons inferred, at a grade of 0.60% Ni. Non-sulphide nickel approximates 0.11% for the assay range 0.40—0.80%. Ni/Co and Ni/Cu ratios are 34/1 and 60/1, respectively, and are thus very different from the ratios of about 45/1 and 11/1 (Ewers and Hudson, 1972) in massive ores of the Kambalda type. This latter point is of significance in considering the use of Ni/Cu ratios of weathering products during exploration, and is discussed later.

Disseminated nickel sulphide mineralization at Mt. Keith was first discovered by the local pastoralist in 1969, in a speculation hole drilled about 2.5 km south of the main body. The hole was sited close to a ferruginous outcrop assumed — correctly — to be derived from ultramafic rocks.

During subsequent exploration, the main target areas, namely the ultramafic units, were delineated by airborne and ground magnetometry. These are the most rapid discriminators of ultramafic rocks, particularly at Mt. Keith where thick transported overburden is present over at least 70% of the claim area. For more detailed exploration of the ultramafic zones, the technique adopted was vertical, i.e. approximately down dip, rotary drilling to an average depth of about 30 m. Holes were sited at variable intervals from 15 to 100 m apart on lines traversing the magnetic feature at 305- and locally 150-m (i.e. 1000 and 500 ft) spacings. Later, some closer interval drilling was done over ultramafic contacts. This rather costly method, which involved a programme of 1300 holes, was considered justified since the initial target was already defined. Drilling continued until cuttings of clearly identifiable rock materials were obtained, though drilling sometimes had to be terminated because of hard, meteorically silicified rocks. In most cases up to ten samples, each representing intervals of 1.53 m (5 ft) from principal horizons, were collected from each hole, crushed and analysed by nitric-perchloric acid digestion and atomic absorption spectrophotometry for nickel and copper. Analytical precision was generally ± 5—10% for nickel and ± 15—20% for copper. Subsequently, deep percussion and diamond drilling was used to test anomalies. Over the main body, a shaft was sunk to 153 m to obtain pilot plant samples.

To a large extent this exploration technique obviated a need for careful appraisal of geomorphology and soils, or for more than a general appreciation

of nickel and copper distributions in lateritic weathering profiles. It should
be emphasized, however, that the programme was aimed at discovering further
disseminated sulphides and was not designed to locate massive sulphides.

GEOCHEMISTRY OF THE MAIN ZONE OF MINERALIZATION

The discovery of the main body of disseminated sulphides followed closely
after the initial discovery to the south. Results from the exploration of this
deposit were used for orientation, to determine some parameters for exploration for similar bodies in the remainder of the property.

The main body lies in a slight topographic low, now a shallow, easterly
draining wash, occupying a rather deeper feature, also draining to the east.
There is an apparently complete laterite profile preserved, overlain by clays
and hardpan. The main features of the profile are given in Table I.

Detailed rotary drilling and follow-up diamond drilling were both carried
out on traverses at 150-m intervals along strike. The results are summarized
in Figs.6 and 7, which compare nickel and copper distributions at a depth of
30 m with those of the primary zone, projected to 150 m.

The highest copper and nickel contents in the primary zone tend to occur
together both centrally and close to the western margin, in black serpentinized
dunite. A comparison of these distributions shows that, in general, the main
zone for which reserves have been calculated is broadly delineated by the
4000-ppm Ni and 300-ppm Cu contours. In subsequent exploration, these
values were chosen as arbitrary threshold values to indicate the presence of
disseminated nickel sulphides.

The nickel anomaly is displaced slightly to the east, this being due to the
westerly dip of the structure, although possibly in part to hydromorphic dispersion down drainage. Rotary drilling has generally been at 30-m intervals
along the east—west traverses and this spacing is inadequate to determine
accurately the degree of lateral dispersion of nickel, if any, or the nature of
any halo. However, comparison of nickel and copper contents of weathered
metasediments no more than 30 m "downslope" of the mineralized body,
and of apparently equivalent contact rocks to barren ultramafics about 1.5
km south along strike, suggests that at the scale of the mineralization being
sought, background variations are greater than any such enrichment (Table II).

There has been a general twofold concentration of copper in the weathered
zone, relative to primary zone contents, but for the most part the copper
distribution delineated the mineralization quite specifically.

There are two situations over the main body itself in which the nickel and/
or copper contents at the 30-m level are below the arbitrary 4000-ppm and
300-ppm threshold:

(1) In the vicinity of the shaft, nickel content is 1200—3000 ppm although
copper can exceed 1000 ppm. This area is in the centre of the drainage wash,
where overburden is deeper. Hence, the samples have been taken from the
ferruginous zone rather than the lower, Ni-enriched horizons.

TABLE I

Principal weathering horizons developed over mineralized, serpentinized dunite, Mt. Keith
(the description is based on the profile exposed in the exploration shaft)

Depth (m)	Description
0—26	*Overburden* — probably transported
0—12	Hardpan, weakly bedded red (2.5YR 4/7)[*] to reddish yellow ((7.5YR 7/8) breccia, with quartzose chalcedonic and ferruginous fragments and sands. Iron oxide pisolites and accretions developed below 5 m. MnO_2 on partings.
12—26	Montmorillonitic swelling clays with interstitial secondary silica. Pale grey-green (10Y 8/1 to 2.5GY 8/2), mottled yellow (10YR 7/6) and dark red (10R 3/6), with haematitic pisolites and accretions in upper 2 m and lower 2.5 m.
26—45	*Ferruginous zone*
26—29	Pisolitic haematite with some interstitial clay.
29—39	Green (to 7.5GY 5/4) and brown clays with yellow and dark haematitic red mottling with variable development of pisolites and massive ferruginous accretions.
39—45	Massive haematite accretions, some pisolites in strongly mottled clays. Some silicification.
45—62	*Silicified zone* Secondary silica as pans to 30 cm thick and impregnating and replacing weathered serpentinized dunite with original textures retained until replacement almost complete.
62—80	*Supergene zone* Fresh to partly weathered serpentinized dunite. Pentlandite completely replaced by violarite; subordinate marcasite, pyrite and millerite. Some silica and magnesium mobile.
80—89	*Transition zone* Serpentinized dunite; primary pentlandite and pyrrhotite partly replaced by violarite and marcasite, respectively.
Below 89	*Primary zone* Serpentinized dunite with pentlandite and subordinate pyrrhotite.

Water table: 45 m.
[*] Colours: Munsell notation.

(2) In several areas, notably 998—1005E, 840—850N, both nickel and copper levels are depressed. The 30-m samples are here of silicified weathered ultramafics, in which the absolute enrichment of silica has essentially diluted the metal contents.

Both of the above situations simply exemplify the necessity of choosing a constant horizon for surveys of this type. However, logging of drill cuttings of such highly weathered materials is difficult and, at best, results in only

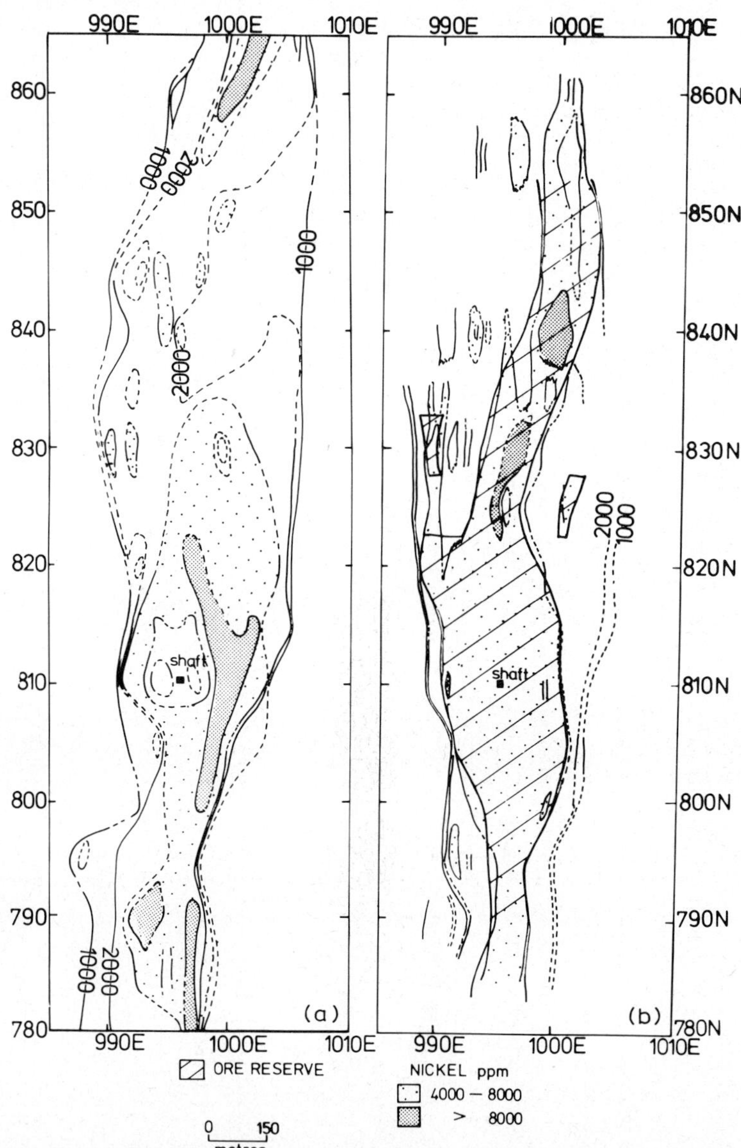

Fig. 6. Nickel distributions in weathered and primary zones over the main mineralized body, Mt. Keith: (a) nickel distributions in the weathered zone at 30 m depth, determined by vertical rotary and percussion drilling; (b) nickel distributions in the primary zone projected to the 150-m level, determined by diamond drilling.

broad subdivisions in which considerable variability exists. Consequently, an alternative adopted here was to consider the maximum value attained within a depth zone, assuming in part that this may represent a constant, enriched horizon. In general this has the tendency to broaden and intensify the anoma-

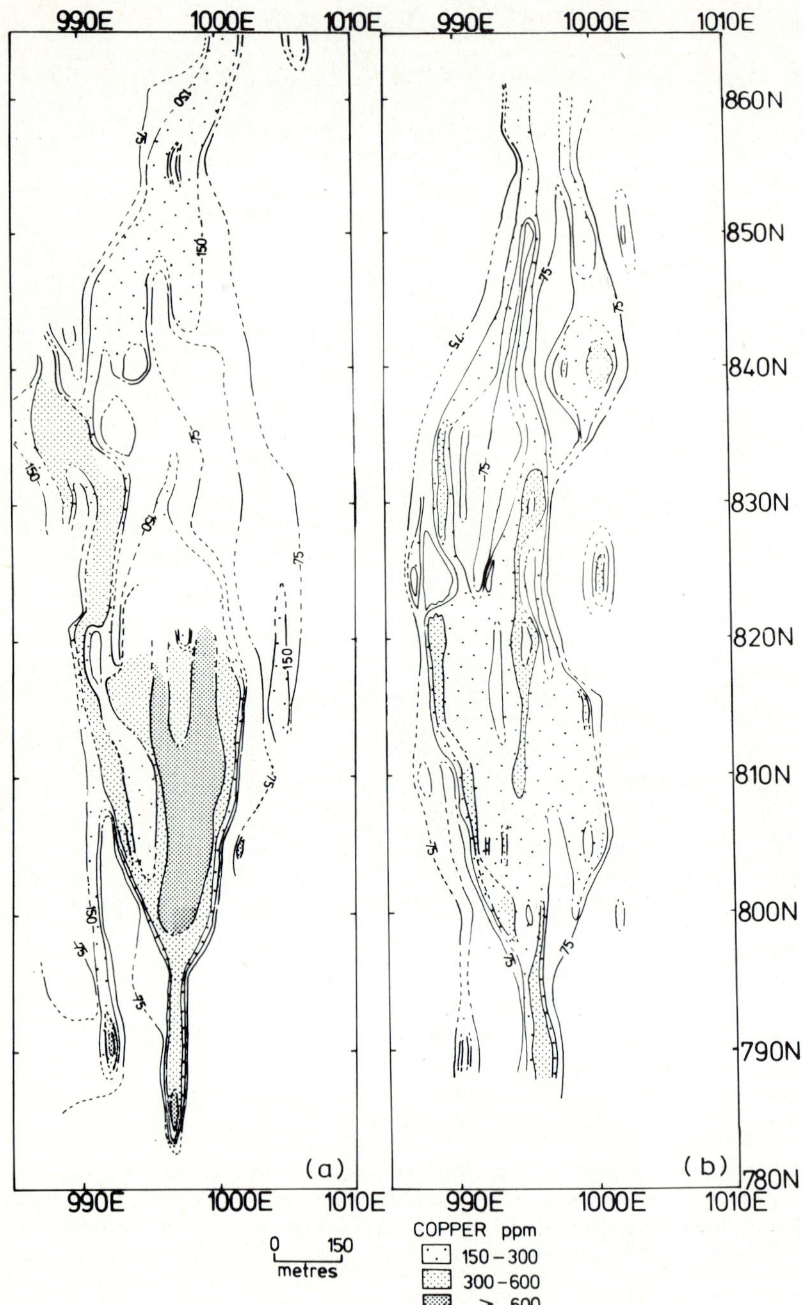

Fig. 7. Copper distribution in weathered and primary zones over the main mineralized body, Mt. Keith: (a) copper distributions in the weathered zone, at 30 m depth, determined by vertical rotary and percussion drilling; (b) copper distributions in the primary zone, projected to the 150-m level, determined by diamond drilling.

TABLE II

Nickel and copper contents of contact rocks, development area, Mt. Keith

		Ni (ppm)	Cu (ppm)	No. holes
Metasediments by	range	35—270	15— 95	4
mineralized body	mean	165	45	
Metasediments by	range	75—300	15—150	3
barren ultramafic	mean	195	100	

lies. Thus, near the shaft, maximum values of 1.5—2.5% Ni replace the lower
values shown on Fig.6a. In the other example cited, the silicification was so
intense that drilling had to be terminated without obtaining less diluted samples. Despite this, the main anomaly is so large that these areas of below threshold values have no great significance. Similarly, elsewhere, unless intense
silicification is very widespread, this empirical approach to data appraisal is
justified, particularly where the anomalies and deposits sought are of this
magnitude.

METAL DISTRIBUTION IN WEATHERING PROFILES

Metal distributions in the weathering profiles over the mineralized zone are
complex. Many of the features are illustrated by the shaft profile and preliminary results from some detailed studies are described below. The mean Fe,
SiO_2, Ni, Cu, Cr and V contents of the principal horizons are shown in Table
III. The characteristic iron and silica contents of these horizons reflect that
the presence or absence of iron oxides and silica were the main distinguishing
criteria during macroscopic examination. Similarly, these same horizons plot
into distinct fields on Ni—SiO_2 and Ni—Fe correlation diagrams (Fig.8).
None of these six elements show much variation in the supergene zones,
and those that do may reflect sampling problems rather than true variations
in metal contents. Nevertheless, a slight nickel enrichment between 73 and
80 m depth (7000—9000 ppm Ni) may reflect the reconcentration of the
metal leached from violarite at the top of the supergene zone. The main enrichment of nickel is in the silicified zone. Here, fluctuations in nickel, iron,
chromium and, in part copper contents are inversely related to that of silica.
The negative correlation between nickel and silica is clearly shown in Fig.8;
the resultant correlation between nickel and iron is only weak. The residual
nickel variance, therefore, is not merely a function of total iron content.
Maximum nickel enrichment occurs at the transition to the ferruginous
zone. Concentrations of 2.47, 2.55% Ni occur with maxima in manganese
(2.11%, 1.37%), cobalt (0.54%, 0.50%) and copper (0.14%, 0.13%), the accumulations presumably being the result of a water-table effect. In the remainder of
the ferruginous zone, nickel contents are generally depleted relative to the

TABLE III

Partial compositions of the main horizons of the residual weathering profile, Mt. Keith shaft

Horizon	No. of samples		Fe (%)	SiO$_2$** (%)	Ni (ppm)	Cu (ppm)	Cr (ppm)	V (ppm)
Pisolitic	6	M*	36.40	11.90	815	150	11,345	390
		S	2.63	8.10	328	110	3360	145
Pisolites	16	M	27.25	22.70	2020	500	8575	147
plus clay		S	5.40	7.60	675	210	4620	42
Accretions	7	M	37.90	20.70	9160	1175	5960	60
		S	2.30	6.40	4900	175	2210	22
Silicified	28	M	10.72	54.00	12,235	475	1015	9
zone		S	2.04	14.40	3935	120	325	8
Supergene	30	M	4.80	31.40	6825	230	850	12
zone		S	0.68	6.10	1455	73	190	12
Transitional	13	M	4.44	33.40	6145	230	960	54
zone		S	0.45	2.70	785	27	97	18

* M = arithmetic mean, S = standard deviation.
** = X-ray fluorescence analysis.

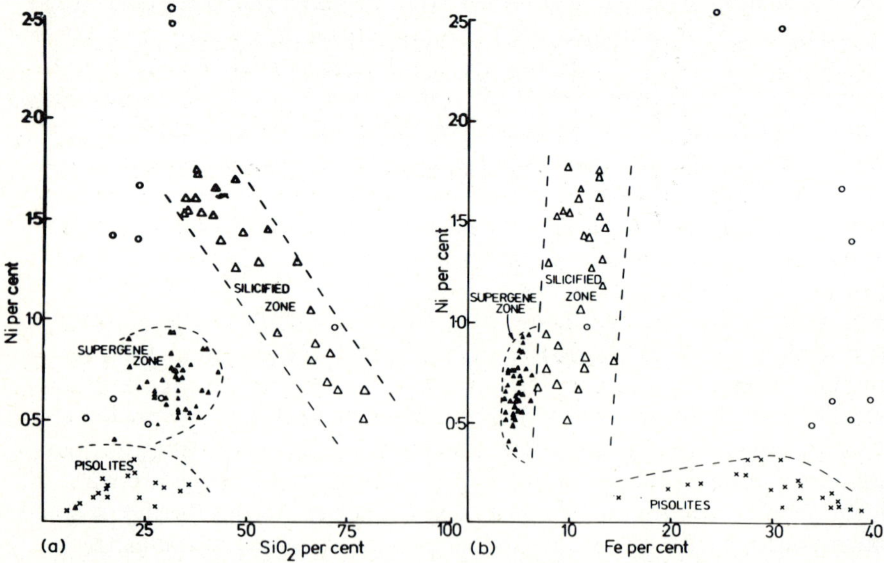

[× Pisolites, ○ Accretions] Ferruginous zone, △ Silicified zone, ▲ Supergene and Transitional zone

Fig.8. Scatter diagram showing correlation between nickel contents and silica and iron contents in the residual profile, Mt. Keith shaft: (a) Ni—SiO$_2$ diagram; (b) Ni—Fe$_2$O$_3$ diagram.

primary and supergene zones, particularly towards the top. Other elements, such as Cu, Cr and V are, however, concentrated in parts of this zone. Copper values are particularly high in horizons with massive haematite accretions and replacements, whereas chromium (to 1.6%) and vanadium (to 600 ppm) have accumulated in the overlying pisolitic horizons.

These high concentrations are presumably residual, the results of the leaching of mobile constituents such as magnesium and silicon. Subsequently, during reworking of the iron oxides to form pisolites, nickel, copper and some iron are further leached, concentrating residual chromium and vanadium still further. The residual concentration of vanadium during pisolite formation has been described by Taylor and Giles (1970).

The Ni/Cu ratio in the mineralized serpentinite is approximately 60/1, although in the region of the shaft it is closer to 30/1. Since, during weathering, nickel and copper tend to concentrate in different horizons, a direct comparison between ratios in fresh and weathered rocks is hard to make. In general, the ratio appears to decrease, presumably because of the lower mobility of copper. However, the high background copper contents of the pyroxenitic margins of the ultramafic and of amphibolitic country rocks reduces the value of Ni/Cu ratios as discriminators of mineralization.

EXPLORATION OF THE REMAINDER OF THE PROPERTY

The exploration techniques of ground magnetometry and rotary drilling were applied throughout the property. Promising areas, in which nickel and copper exceeded 4000 ppm and 300 ppm, respectively, were further investigated by vertical percussion drilling and/or inclined diamond drilling. As a result, disseminated sulphides were discovered at several sites, but, in general, the technique met with varied success and a number of anomalies were found to be related to only very low-grade mineralization or even barren ultramafics.

An attempt will be made to illustrate comparisons with the main mineralized body in the following examples from other areas.

Area "A"

This is the only location north of the main mineralized body where there is any surface expression of the bedrock. Southwards there are increasing depths of alluvial cover, reaching a maximum thickness of 25 m at the shaft. The nickel anomaly (to 1%) is up to 325 m wide, with the horizon of maximum enrichment close to the surface. In the centre of the anomaly, however, extreme silicification has diluted nickel contents to below 3500 ppm. Copper anomalies are restricted to zones near the ultramafic-metavolcanic contact, possibly, but not convincingly, because elsewhere here a pre-existing upper horizon of copper enrichment may have been eroded away.

Subsequent percussion and diamond drilling has established no consistent relationships between bedrock and weathered zone geochemistry. One percus-

sion hole intersected disseminated pentlandite assaying 0.65—0.75% Ni and
1000 ppm Cu in the primary zone where silicification had diluted the anomaly
to below 3500 ppm Ni and 100 ppm Cu. Another percussion hole, drilled in a
similarly intensely silicified area (2000—3000 ppm Ni, 5—10 ppm Cu) found
primary zone material of only 0.35% Ni and 10 ppm Cu.

A diamond drill hole was sited to test a coincident Cu—Ni anomaly on the
eastern ultramafic-metavolcanic contact, where concentrations of up to 1.5%
Ni and 1100 ppm Cu persisting to depths of 75 m suggest the possible occurrence
of contact-type massive sulphides (Fig.9). However, the drill hole only encountered disseminated pyrrhotite in the primary zone assaying up to 0.29%
Ni and 160 ppm Cu. This is one of the most perplexing results encountered
during exploration of the property and remains unexplained. A dioritic intrusion on the contact may be of significance, but there is no direct evidence.

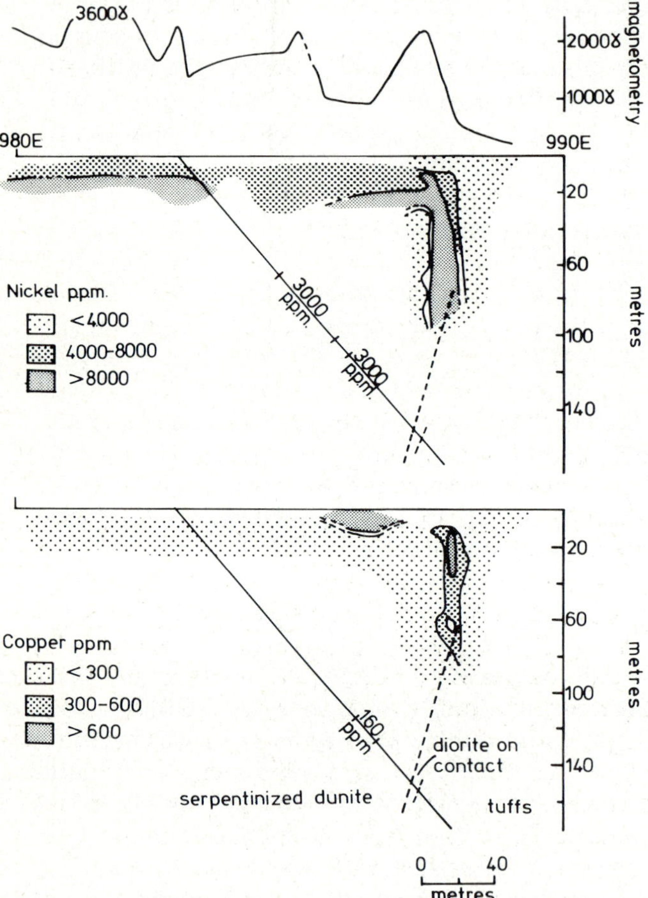

Fig.9. Section showing nickel and copper anomalies in the weathered zone on the eastern
contact of a serpentinized dunite, area "A", and the maximum intersection encountered
by diamond drilling. Magnetometry by vertical field, flux gate magnetometer.

Area "B"

Nickel anomalies occur over the metadunites throughout this area beneath the transported material. Follow-up drilling has established the existence of varying widths of centrally located disseminated pentlandite or pyrrhotite mineralization (grade 0.40—0.60% Ni) over the entire strike length of about 3000 m. Diamond drilling results gave fairly consistent correlations between weathering and primary zone metal contents (Table IV). The geomorphological situation is very similar to that encountered at the main mineralized zone, and here the same parameters of 4000 ppm Ni and 300 ppm Cu are applicable.

TABLE IV

Comparison of nickel and copper contents in primary and weathered zones, area "B", Mt. Keith

Drill Hole	Zone	Ni (%)	True width (m)	Cu (ppm)	True width (m)
III	weathered	0.50—1.00	152	100—200	60
	primary	0.53	36.5	100—200	15.2
IV	weathered	0.50—1.00	165	50—300	165
	primary	0.41	150	75	150
		0.50	91	100—200	91

Area "C"

This area lies on the backslope from the point at which the breakaway crosses the greenstone belt, about 14.5 km south of the main mineralized body. The ultramafic units are mostly narrow, discontinuous meta-pyroxenites or peridotites altered to talc magnesite rocks. The area is relatively elevated, and the sand plain more readily compared with the older Tertiary peneplain than the surface at the main body. The lateritic profiles differ from those described previously: silicification is generally only minor, and beneath the ferruginous horizons soft green, red-brown and yellow clays occur, becoming harder with depth and merging to fresh rock at about 75 m. In some profiles a lower ferruginous zone is found, often sufficiently hard to prevent rotary drilling. The laterite profiles are mostly complete and overlain by 2—20 m of sandstone hardpan and sand plain. The relief prior to deposition was stronger than at present, and the sandstone infills a gullied valley, expressed now as a shallow depression.

Rotary drilling showed a number of anomalies, mostly weaker than those found elsewhere. Three, with maximum expressions of 4000—5500 ppm Ni and 125—250 ppm Cu, were tested and each found to overlie barren, spinifex-textured meta-pyroxenites containing 1200—2400 ppm Ni and 20—40 ppm Cu.

Where stronger anomalies occur (0.4—1% Ni, 90—555 ppm Cu), patchy disseminated mineralization has been encountered.

Area "D"

Southwards from the breakaway, dissection has exposed and truncated the weathering profiles, and surficial materials are mostly residual. Fresh rocks are nowhere exposed. The location of this example is 400—1300 m south of the breakaway, down a fairly gentle pediment slope. It is surmised that the horizon of maximum enrichment for both nickel and copper is close to the surface in a geomorphic situation quite similar to that seen at area "A", which also occurs on a pediment slope.

A small, but well developed nickel anomaly assaying about 1% achieves a maximum width of 150 m. Copper content is less than 100 ppm except for the eastern contact where it achieves 650 ppm. The test diamond drill hole into the primary zone intersected green serpentinized dunite containing no economically interesting mineralization; nickel approximating 3000 ppm for the whole 190-m intersection 107 m beneath the anomaly, and copper rarely exceeding 20 ppm (90 ppm on the eastern contact). This is an instance where the nickel parameter established at the main body failed entirely, but where low copper may have been assessed as a negative indication.

Area "E"

In this area, well south of example "D" but located still in a comparable geomorphic environment, there is an interesting comparison to be made: essentially similar nickel and copper anomalies have arisen from very different bedrock geochemistry. The anomaly is shown in Fig.10. Diamond drill hole I intersected 180 m of disseminated pentlandite in serpentinized dunite similar to that of the main mineralized body, and assaying 0.67% Ni and 200—400 ppm Cu. The expression in the weathered zone, 10 m above, is 0.3—1% Ni and 1180 ppm Cu over the same width. Drill hole II, 305 m north, tested an anomaly assaying 1.5% Ni and 830 ppm Cu, but encountered only barren talc-magnesite rock beneath it. It is possible that this inconsistency has been caused by the mineralization at the drill site II having no appreciable extent, although elsewhere on the property such rapid vertical changes have not been observed.

DISCUSSION

Although at Mt. Keith the main targets have been large bodies of low-grade disseminated nickel sulphides, the results indicate some of the difficulties which may face a search for massive sulphides. In the case of nickel, exploration is concentrated in areas where the rocks and their weathering products already have high background contents of the metal. This, in itself, has a ten-

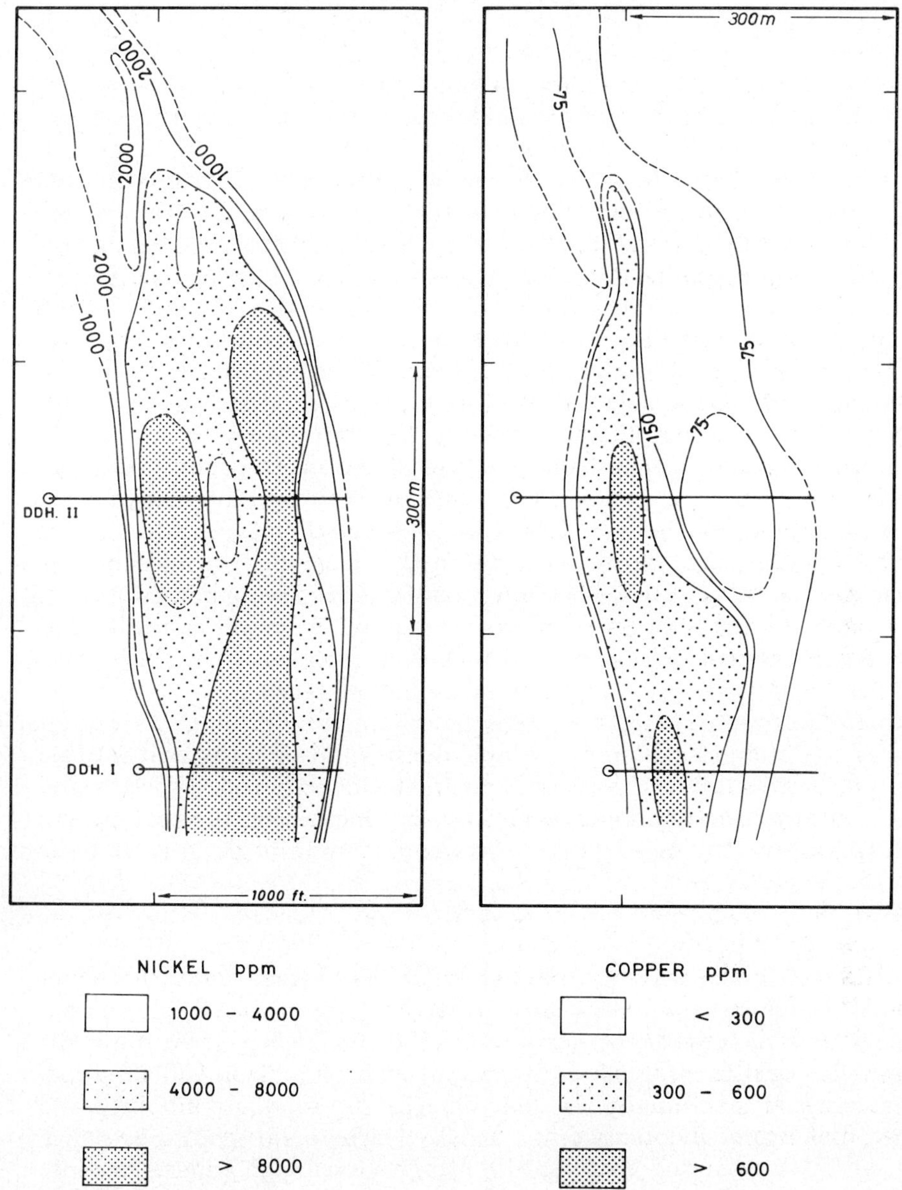

Fig.10. Nickel and copper anomaly in weathered serpentinized dunite, area "E", Mt. Keith plan

dency to reduce anomaly contrast. Exploration requirements are those of detailed follow-up and evaluation of targets already delineated by magnetometry. The geochemical problems can be considered as falling into two categories: sampling and interpretation.

Sampling

Transported overburden is widespread and surface sampling is frequently not worthwhile. Even in the few restricted areas of residuum at Mt. Keith there is often a thin cover of transported material, and "soil" sampling can only give an incomplete survey. There can be considerable difficulty in distinguishing between residuum and alluvial materials. In one of the few soil sampling exercises carried out, at area "A", the results indicate that the highest nickel contents and greatest contrasts can be expected in the coarse fractions. Thus, for 24 samples, the minus 80-mesh fraction contains 110—950 ppm Ni (mean 195 ppm), whereas the whole rock samples have 180—1550 ppm Ni (mean 715 ppm). There was no copper anomaly. This confirms the results of Mazzucchelli (1972) and Dalgarno and Knowles (in prep.) at Kambalda and Redross where surface sampling has wider application. The usual result of chemical weathering is for any concentration of cations to occur in fine-grained secondary minerals, in fine soil fractions. However, later silicification and ferruginization re-cements much of the weathered material into massive forms which are, perhaps, protected from further leaching. During subsequent arid periods, in which clastic dispersion predominates, the cemented material only partly disintegrates and remains coarse grained. Fine soil materials, however, are well dispersed, for example, during sheet wash erosion. Much that remains can be diluted by the addition of aeolian material.

Elsewhere, adequate samples can only be obtained by drilling. It is, of course, very costly to obtain a satisfactory sample density, particularly as it is often necessary to drill well into the residual profile to identify rock types. In the absence of alternative techniques, therefore, any increase in the cost effectiveness of exploration drilling relies on improvements in interpretative techniques.

Interpretation

The deep lateritic weathering completely alters the trace and major element geochemistry, mineralogy and structure of the original bedrock. In general, mobile elements such as magnesium and calcium are leached from the system and immobile constituents such as vanadium, chromium, iron and aluminium are left as residual accumulations. Elements with intermediate mobility, such as copper, nickel and silicon are only partially leached and tend to be redistributed within the weathering profiles, often to give local high concentrations. The mineralized serpentinite at Mt. Keith contains 40% MgO, 36% SiO_2, 7.5% FeO, and the weathering products mostly contain less than 1.5% MgO with different horizons concentrating up to 90% redistributed SiO_2, 80% residual and redistributed Fe_2O_3, and even 24% Al_2O_3, presumably residual. The primary serpentine-magnesite assemblage is replaced by secondary assemblages dominated by various forms of silica, iron oxide and clays. The dunitic texture is ultimately destroyed by silica and iron oxide replacement, and pisolite formation.

The mechanisms governing these transformations are poorly understood, and so the interpretation of the incomplete geochemical data used for exploration is very difficult. These interpretative problems exist whatever the sampling technique, or sample location. Original abundance in bedrock is only one factor controlling the metal content of a specific horizon in a weathering profile, and simple concentration limits seem inadequate to distinguish anomalies. Although concentrations of 4000—15,000 ppm Ni in weathered rock were a guide to economically significant intersections of disseminated sulphides in some areas at Mt. Keith, essentially similar anomalies elsewhere overlie rocks with rare or no sulphides. In other situations, silicification can suppress anomalies by simple dilution to values of 2000—3000 ppm Ni.

The association of copper with nickel is often used as a guide to nickel sulphide mineralization in the Archean of Western Australia since chalcopyrite commonly occurs in massive sulphides of the Kambalda type. However, chalcopyrite is rare in the Mt. Keith-type disseminated sulphides, and the copper content of the mineralized zones is usually insignificant. Copper appears to be rather less mobile than nickel during weathering and there is some evidence at Mt. Keith that it may concentrate as a residual. Thus, although pronounced copper anomalies are associated with the principal mineralized bodies at Mt. Keith, the presence of copper is not itself an indicator of sulphides. For example, the strong Cu—Ni anomaly on the ultramafic-metavolcanic contact illustrated in Fig.9 has no related sulphide accumulation. Nevertheless, from the evidence of the two main bodies which have disseminated sulphides at grades of 0.60% Ni and above, the occurrence of copper anomalies on a similar areal scale to those of nickel appears to be a possible indicator of the richer deposits.

Problems of data interpretation are a considerable hindrance to the efficient exploration of this terrain and require further research. A major requirement is to be able to distinguish, in all residual materials, the effects of weathering from those due to bedrock geochemistry. In the first instance, it is necessary to determine criteria for defining various horizons within the lateritic profiles and for describing the variance of element concentrations within such horizons. Ideally, it should be possible to use the shallowest obtainable residuum, whether this is sampled as a surface soil or by drilling, thereby reducing drilling costs or allowing greater sample densities.

The study of the geochemical mechanisms involved in the deep weathering of this terrain, and the determination of criteria for accurate data interpretation are the objects of a continuing research programme by C.S.I.R.O at Mt. Keith and elsewhere. Multi-element geochemical studies are being carried out over several lithologies in a variety of geomorphological situations with the ultimate aim of improving the efficiency of geochemical exploration techniques. Results of this work will be subject of future papers.

SUMMARY

The Mt. Keith area demonstrates many of the exploration problems en-

414

countered in the Archean Yilgarn Block of Western Australia. These are the result of the considerable chemical redistributions that have occurred in the deep lateritic profiles, and the extensive cover of transported overburden. A programme of magnetometry followed by exploration drilling successfully delineated a number of disseminated sulphide bodies. Orientation over the main zone of economically significant mineralization suggested that threshold contents of 4000 ppm Ni and 300 ppm Cu in weathered ultramafics were suitable exploration criteria. Exploration of the 25 km strike length of the dyke-like ultramafic units on the property tended to confirm this result. However, it was found that some anomalies overlie essentially barren ultramafics. Considerable difficulties were experienced in interpreting the geochemical data, as the tenor of the surface expression was no real guide to bedrock metal content. The presence of anomalous copper concentrations in overburden was not always a useful distinguishing criterion for either disseminated sulphides or the possible occurrence of massive sulphides. The metal is not an essential constituent of the disseminated mineralization and, like nickel, can become enriched in weathering profiles over apparently barren ultramafics. Nonetheless, the most significant results were obtained when Ni—Cu anomalies are coextensive, even when the maximum concentrations of the metals occurred in different horizons.

ACKNOWLEDGEMENTS

The authors wish to acknowledge the considerable contributions made by the staff of Metals Exploration Ltd. and its Joint Venture partners (Freeport of Australia Ltd., Australian Consolidated Minerals N.L.) to the work described in this paper. Members of the C.S.I.R.O. Divisions of Mineralogy and Land Resources Management are thanked for valuable discussions and for their assistance in the preparation of the manuscript.

REFERENCES

Beard, J.S., 1973. The elucidation of palaeodrainage patterns in Western Australia through vegetational mapping. In: Vegetation Survey of Western Australia. Vegmap Publications, Perth, W.A., Occ. Paper 1, 17 pp.

Bettenay, E. and Churchward, H.M., 1974. Morphology and physiographic relationships of the Wiluna Hardpan in arid Western Australia. J. Geol. Soc. Aust., 21: 73—80

Bettenay, E. and Hingston, F.J., 1964. Development and distribution of soils in the Merredin area, Western Australia. Aust. J. Soil Res., 2: 173—186

Brewer, R. and Bettenay, E., 1973. Further evidence concerning the origin of the Western Australian sand plains. J. Geol. Soc. Aust., 19: 533—541

Clema, J.M. and Stevens-Hoare, N.P., 1973. A method of distinguishing nickel gossans from other ironstones on the Yilgarn shield, Western Australia. J. Geochem. Explor., 2: 393—402

Dalgarno, C.R. and Knowles, A.R., in preparation. Near-surface geochemical expression of the the Redross nickel deposit, Western Australia

Durney, D.W., 1972. A major unconformity in the Archean, Jones Creek, Western Australia. J. Geol. Soc. Aust., 19: 251—260

Ewers, W.E. and Hudson, D.R., 1972. An interpretive study of a nickel-iron sulphide ore
    intersection, Lunnon Shoot, Kambalda, Western Australia. Econ. Geol., 67: 1075—1092
Hall, J.S., Both, R.A. and Smith, F.A., 1973. A comparative study of rock, soil and plant
    chemistry in relation to nickel mineralization in the Pioneer area, Western Australia.
    Proc. Australas. Inst. Min. Metall., 247: 11—22
Johnstone, M.H., Lowry, D.C. and Quilty, P.G., 1973. The geology of southwestern Australia
    — a review. J.R. Soc. W.Aust., 56: 5—15
Jutson, J.T., 1950. The physiography (geomorphology) of Western Australia. Bull. Geol.
    Surv. W.Aust., 95: 366 pp.
Litchfield, W.H. and Mabbutt, J.H., 1962. Hardpan soils of semi-arid Western Australia. J.
    Soil. Sci., 13: 148—159
Mabbutt, J.H., Litchfield, W.H., Speck, N.H., Sofoulis, J., Wilcox, D.G., Arnold, J.M.,
    Brookfield, M. and Wright, R.L., 1963. General Report on the Lands of the Wiluna—
    Meekatharra Area, Western Australia, 1958. C.S.I.R.O. Aust. Land Research Series
    No.7, Melbourne, Vic., 215 pp.
Mazzucchelli, R.H., 1972. Secondary geochemical dispersion patterns associated with the
    nickel sulphide deposits at Kambalda, Western Australia. J. Geochem. Explor., 1: 103—116
Mulcahy, M.J., 1967. Landscapes, laterites and soils in south western Australia.
    In: J.N. Jennings and J.A. Mabbutt (Editors), Landform Studies from Australia and
    New Guinea. Australian National Univ. Press, Canberra, A.C.T., pp.211—230
Mulcahy, M.J. and Bettenay, E., 1971. Soils and landscape studies in Western Australia, 1.
    The major drainage divisions. J. Geol. Soc. Aust. 18: 349—357
Mulcahy, M.J. and Hingston, F.J., 1961. The Development and Distribution of the Soils of
    the York—Quairading Area, Western Australia, in Relation to Landscape Evolution.
    C.S.I.R.O. Aust. Soil Publication No.17, Melbourne, Vic., 47 pp.
Sheppy, N.R. and Burt, D.R.L., 1974. The Mt. Keith nickel sulphide deposits. In: The
    Economic Geology of Australia and Papua-New Guinea. Australasian Institution of
    Mining and Metallurgy, Melbourne, Vic. (in press)
Taylor, R.M. and Giles, J.B., 1970. The association of vanadium and molybdenum with
    iron oxides in soils. J. Soil Sci., 21: 203—215
Walther, J., 1915. Laterit in West Australien. Z. Dtsch. Geol. Ges., 67B: 113—124

THE WEATHERING PRODUCTS OF NICKELIFEROUS SULPHIDES AND THEIR ASSOCIATED ROCKS IN WESTERN AUSTRALIA

J.R. WILMSHURST

C.S.I.R.O. Minerals Research Laboratories, North Ryde, N.S.W. (Australia)

ABSTRACT

Oxidized and leached outcrop derived from nickel sulphide-bearing ultramafics may be interpreted by chemical methods. In outcrop, certain of the minor and trace elements can be used in assessing the probable nature of the parent rock. Sulphide nickel, generally present in fresh rock in a ratio of ~ 15/1 with copper, tends to absolute and relative depletion during oxidation so that a higher Ni/Cu ratio suggests a non-sulphide or exotic nickel source. The presence of zinc at higher levels is likewise unfavourable, suggesting in the first instance, that the copper is not derived from ultramafic sulphide.

Of the persistent trace elements, palladium and platinum are promising indicators, in outcrop, of Ni—Cu sulphides, with qualitative and some possible semi-quantitative applications. It is noteworthy that while palladium and nickel show an "acceptable" correlation, platinum and nickel do not, and in some instances platinum may be virtually absent.

The presence of arsenic, at high levels, can indicate derivation from arsenic-bearing Ni—Cu sulphides, which may carry minor gold; anomalous but lower values of arsenic may derive from metasediments in the sequence. Zinc levels may be elevated when arsenic is present.

INTRODUCTION

The Ni—Cu sulphide mineralization of the Eastern Goldfields region of Western Australia occurs in a deeply weathered yet slightly dissected terrain. Since neither sulphides nor sulphidic rock are accessible to surface exploration methods, the interpretation of weathered outcrop plays an important part in the earlier stages of exploration. Surface rock "after" Ni—Cu sulphide, however, is frequently not sufficiently distinctive to allow immediate and simple classification. This is a consequence both of the nature of the mineralized and associated rocks and of the manner in which they have weathered.

This paper examines some of the characteristics of the oxidized surface zone rocks which might be used in interpreting outcrop derived from sulphide-bearing ultramafics. The sulphide mineralization of the ultramafic rocks of the region has been described specifically and generally by a number of workers: Woodall and Travis (1969), Hudson (1972, 1973), Martin and Allchurch (1974), Robinson and Stock (1974), Ross and Hopkins (1974), Sheppy and Rowe (1974).

The Spargoville region (Fig.1) to which much of this present work refers has

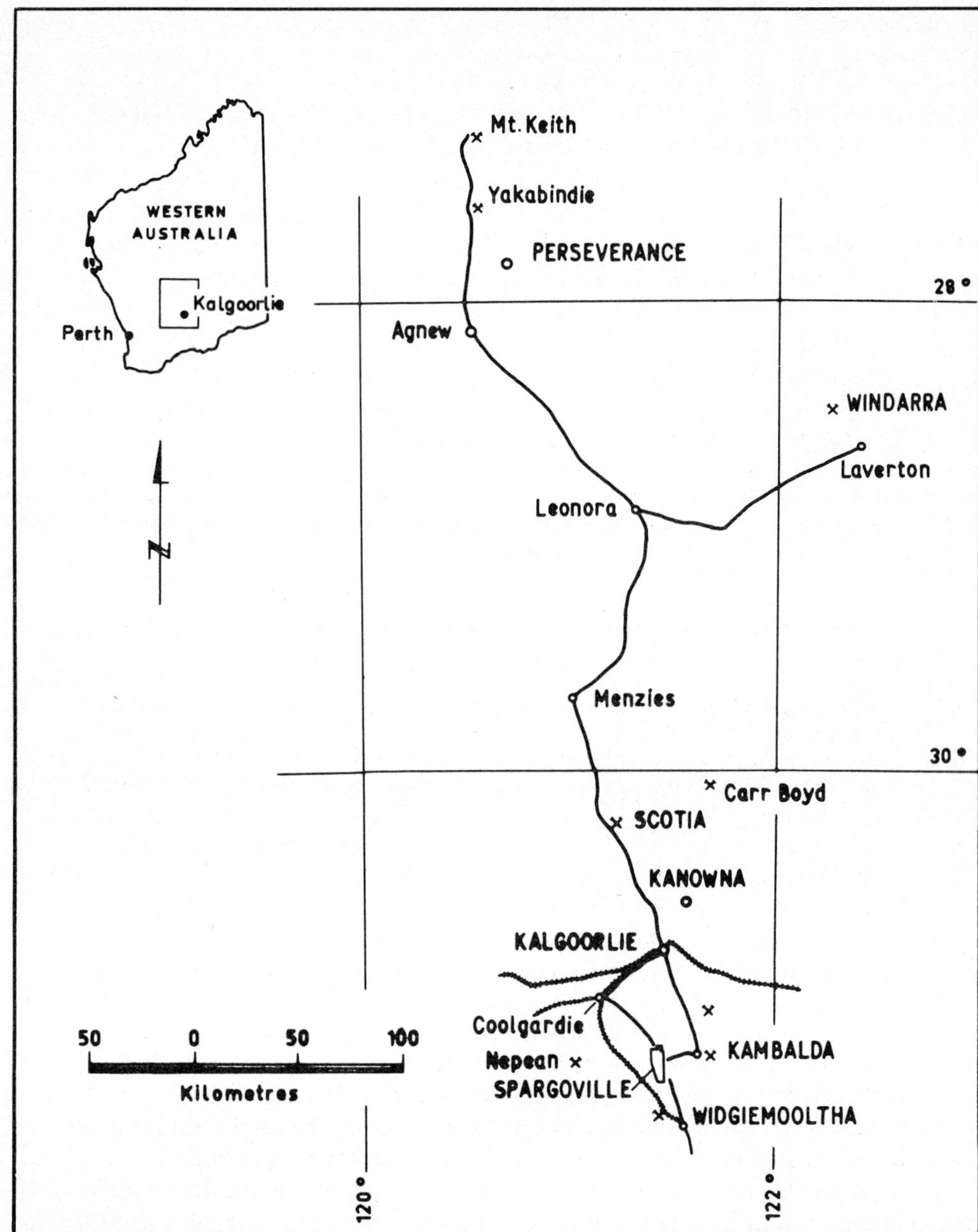

Fig.1. Location plan.

been studied by Hancock et al. (1971) and by Andrews (1974). In these deposits sulphides are disseminated throughout the ultramafic rock and concentrate as lenses or pods at or towards the lower contact of the unit. In each instance the ultramafic rocks are now usually composed of associations of talc, serpentine, amphibole and carbonate. Although rare on a regional basis, olivine is common in the Spargoville rocks.

Accessory magnetite, chromite and ilmenite occur together with the primary sulphides pentlandite, chalcopyrite, pyrrhotite and pyrite. Secondary sulphides, principally violarite, are present in the alteration zones and are "economic" sulphides in Western Australia.

Weathered outcrop derived from sulphide-bearing rock is frequently preserved although the exposure may be limited to a few metres in an essentially ferruginous and frequently "lateritized" terrain.

Generally, a number of factors act against a simple and conclusive geochemical approach to outcrop interpretation. In essence these arise from an immediate inability to determine whether the nickel in the outcrop was derived from in-situ sulphide, from silicates or is exotic. Similarly copper may have been derived from indigenous sulphide, but it may also have come from metasediment, from metabasalt or from remote ultramafic sulphide. Doubt as to the source of the copper is due to the retentive and collective properties of the iron "oxides" generally and of those of the lateritic oxide zone in particular.

Moreover, as already noted, since the Ni—Cu sulphides may vary from low to high grade and from massive bodies to disseminations, there is in practice no sharp discontinuity between barren and mineralized rock, and correspondingly, between the oxidized outcrop of barren and mineralized rock.

While it is self-evident that major concentrations of sulphide are not found at the contact(s) of all the ultramafic units in the sequence, our work has shown that, at Spargoville, basal Ni—Cu concentrations do occur elsewhere in the ultramafic sequence although they are insignificant in terms of those of the mineralized units. These minor concentrations which have a form suggesting gravitational settling of the sulphide can give rise to geochemically significant outcrop, especially as sample collection has been directed to contact zones by "theory" and by geochemical survey.

The problem, then, is more complex than it at first appears. It is necessary to distinguish "laterite" from gossan due to massive or disseminated sulphides and to detect outcrop derived from rock initially low in sulphide (e.g. 0.5% Ni). It is also desirable to distinguish outcrop that represents minor (contact) sulphides. Finally, there is the matter of the interpretation of the replacement silicified rock which can be anomalously *low* in nickel and copper even when derived from a highly sulphidic precursor.

METHODS OF ANALYSIS

In this work the major elements have been determined by X-ray fluorescence (XRF) analysis and by atomic absorption spectrometry (AAS). Fused borate buttons were used for the XRF determinations. For AAS determinations of the major and minor elements, samples were dissolved either by hydrofluoric-perchloric acid attack or by hydrochloric acid after sintering (350—400°C) with sodium peroxide (Corbett et al., 1974).

Arsenic was determined by the arsine-molybdenum blue method and at the higher concentrations by XRF. Results for the elements vanadium, silver and lead were obtained by atomic emission spectrometry. Palladium, platinum and gold were determined by AAS (carbon rod, background correction) after concentration into a silver prill by conventional fire assay.

RESULTS AND DISCUSSION

Geochemistry of the major rock units

A present lack of available and definitive information on the major rock units stems from a number of causes, but principally is due to the relatively short time since discovery and exploitation as well as poor exposure and the highly altered nature of many of the rocks in the Eastern Goldfields.

The outline geological map of the Spargoville property (Fig.2) shows three apparently isolated north-trending ultramafic sequences separated by meta-

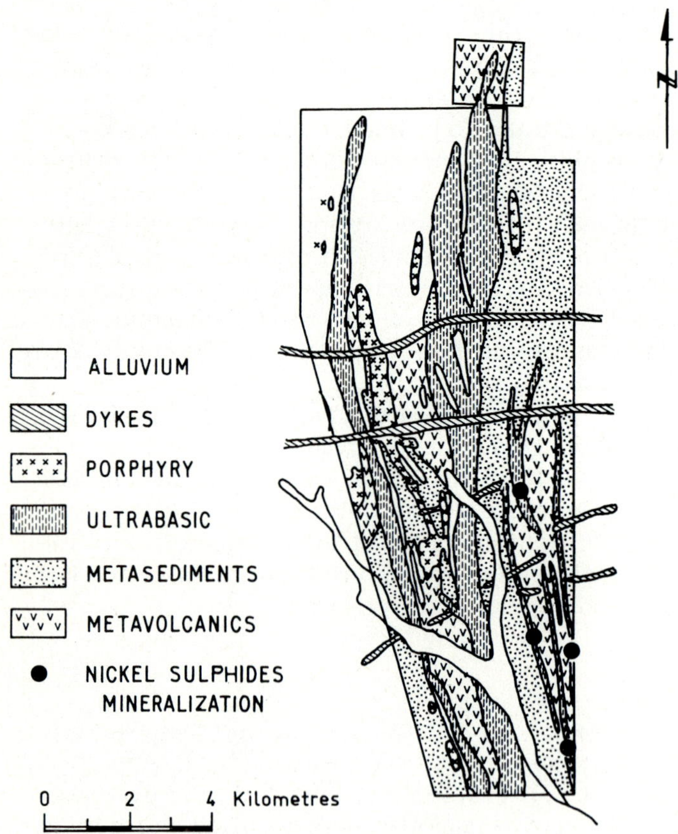

Fig.2. Geological map of Spargoville area.

TABLE I

Geochemistry of mineralized and unmineralized rock units, Spargoville deposit

Sample	Depth (m)	SiO$_2$ (%)	Al$_2$O$_3$ (%)	MgO (%)	Fe (%)	Cr (%)	Ni (%)	S (%)	Co (ppm)	Cu (ppm)	Zn (ppm)	As (ppm)	Ag (ppm)	Au (ppm)	Pd (ppm)	Pt (ppm)
5DB	97.4	47.1	6.1	21.8	8.5	0.30	0.10	0.28	120	30	40	2.0	<1	0.03	0.02	0.03
tremolite-chlorite unit																
5DB	178.0	39.3	5.8	22.4	9.5	0.25	0.69	2.0	180	540	80	5.0	1	0.02	0.33	0.30
internal contact																
5DB	189.9	41.5	2.7	41.5	6.8	0.25	0.25	0.45	130	50	20	0.6	<1	0.01	<0.01	0.02
dunite																
5DB	196.6	28.2	15.4	33.1	6.8	0.02	0.02	1.3	50	190	100	1.3	<1	0.01	0.01	0.01
dunite-porphyry contact																
5DB	225.9	38.6	1.54	38.8	10.7	0.30	1.34	1.95	360	1200	30	10	1	0.03	0.21	0.03
dunite-disseminated sulphide																
5DB	234.7	27.2	0.57	33.2	17.0	0.01	4.8	9.7	820	4700	80	27	4	0.03	0.66	0.13
dunite-disseminated sulphide																
5DB	240.6	17.0	0.38	17.0	30.2	0.06	9.5	21.0	1600	2600	10	36	4	0.02	1.00	0.84
dunite-massive sulphide																
S1	graphitic sediment					0.01	0.008	10.0	60	60	1000	25	2	0.03	<0.01	0.03
S2	sulphidic sediment					0.01	0.03	12.0	200	2000	20,000	60	4	0.06	<0.01	<0.01
5DB	263.4	52.2	13.8	6.2	7.2	0.01	0.005	0.42	80	140	60	5	<1	0.01	<0.01	<0.01
metabasalt																
CB	153.3	41.3	2.1	39.6	5.9	0.17	0.23	0.23	110	20	140	20	<1	<0.01	<0.01	<0.01
altered peridotite																
CB	117.3	31.2	2.5	22.6	16.1	0.32	0.45	4.5	470	2900	610	30	1	0.03	0.05	0.04
altered peridotite																
CB	153.6	38.3	3.4	36.8	8.2	0.26	0.93	1.07	280	310	170	40	<1	0.03	0.02	0.02
altered peridotite — contact																
CB	235.7	45.1	13.2	16.5	10.4	0.18	0.04	3.0	130	1800	4800	30	2	0.04	0.02	0.03
hybrid(?) metasediment																
5DR	315.0				47.7	0.55	11.9	37.6	2200	5500	40	30	4	0.06	0.68	0.03
remobilized massive sulphide																

422

basalt and metasediment. Significant mineralization has been intersected
along the Eastern ultramafics at the locations shown on the map.

The Central Belt is taken to represent an unmineralized ultramafic sequence
and selected material has been studied (core ddhCB). The mineralized unit of
the Eastern Belt at the 5D location has already been described by Hancock et
al. (1971).

Geochemical data for the relevant rock types of these two sections and for
associated metasediments are given in Table I. Because of the special significance of the precious metals, the discussion of their geochemistry has been
separated from and follows on after that of the other elements.

General geochemistry

The unmineralized ultramafic rocks.
The log of the "type" unmineralized ultramafic section (of the Central
Belt) is designated ddhCB and is shown in Fig.3.

Although it seems that the ultramafics of the section were originally
peridotitic in nature, the present mineralogy is essentially a metasomatic
alteration assemblage of serpentine, talc, chlorite, amphibole and generally,
minor relict olivine. Carbonate alteration is pervasive although some carbonate
zones occur, principally at contact or shear zones.

The individual ultramafic units are relatively thin, and show, in varying de-

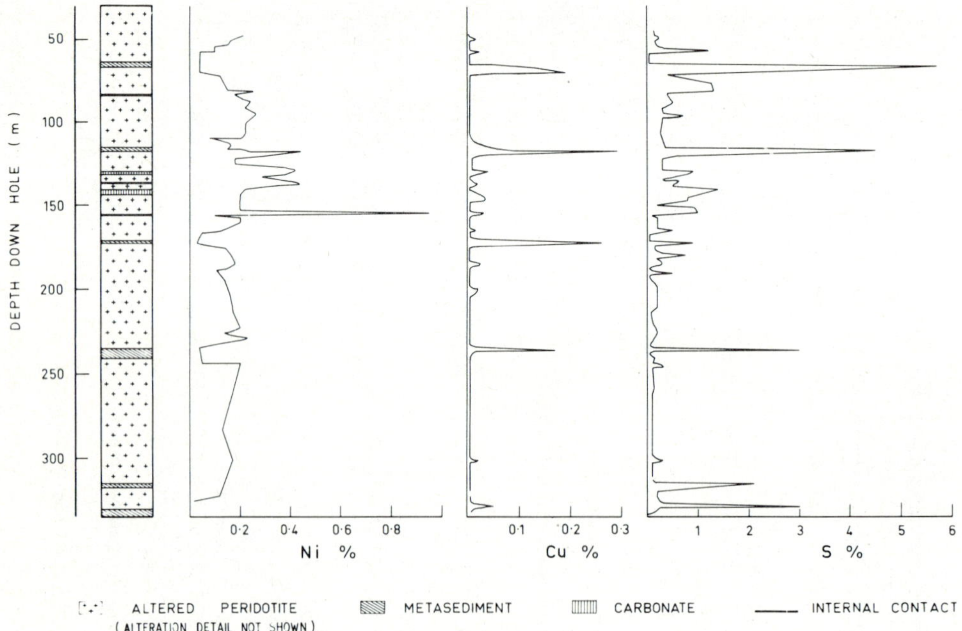

Fig.3. Outline log of "unmineralized" core ddhCB.

gree, minor contact sulphide concentrations which can be significant in outcrop, as will be shown later. Geochemically, the average levels of the indicator elements, nickel (1500 ppm) and cobalt (120 ppm) are of the expected order for peridotites as is zinc at 70 ppm. On the other hand, sulphur (1500 ppm) and arsenic (25 ppm) are quite anomalous and copper (~50 ppm) is high. It is difficult, however, to be certain that the latter figures are generally representative or that they really represent the original and not the altered rock.

Both nickel and copper show at least some degree of concentration at the contacts, with the highest nickel concentration (in this section) approximately 1%. The log also shows that copper (together with S and Zn) is associated with the intercalated metasediments and reaches relatively significant concentrations although still minor in amount. Very little movement of this copper is necessary to cause an apparent ultramafic Ni—Cu geochemical anomaly at the surface. The upper ultramafic unit(s) of the mineralized section (Figs.4 and 5) should also be considered. Although it is probable that these units were genetically related to the magma which gave rise to the mineralized

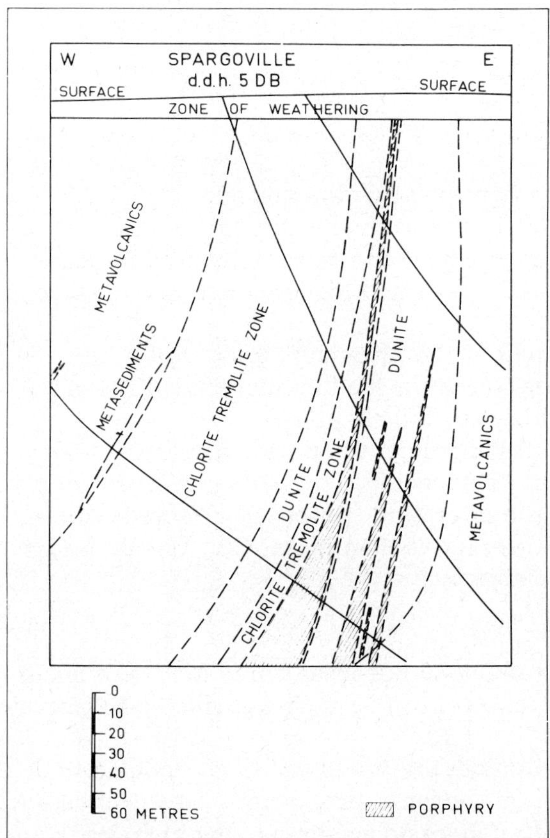

Fig.4. Simplified geological section through the ultramafic unit of Spargoville ddh5DB.

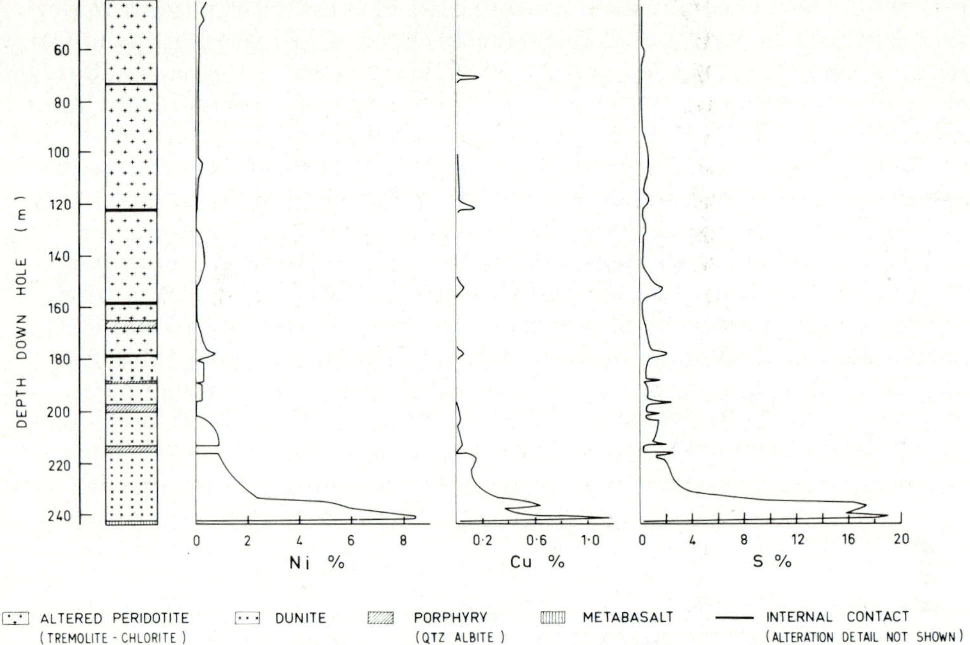

Fig.5. Outline log of mineralized core ddh5DB.

dunitic unit they are not themselves significantly mineralized.

Nickel (800 ppm) generally has lower values than in the Central Belt peridotite. This is possibly an alteration effect. However, a marked increase (2500 ppm) is found in the olivine-rich zones and higher concentrations are also present in the contact sulphides.

Copper varies considerably throughout the ultramafic units, but has a mean of some 40 ppm which is close to that in the "unmineralized rocks" of ddhCB and, as with nickel, it is highest at the contacts.

Cobalt is fairly uniformly distributed through the ultramafics except for enrichment in the sulphides as might be expected.

Zinc shows locally elevated levels in alteration (porphyry-affected) zones. Overall, it is present at marginally lower levels than in ddhCB. Arsenic is also lower and is not anomalous in these rocks.

The mineralized rocks

The Ni—Cu sulphides exhibit, for practical purposes, three modes of occurrence and all three may be found in one deposit as at Spargoville and Perseverance.

(1) Disseminated nickeliferous sulphides are present in most deposits and occur in the upper part of the sulphide concentration profile. However cases are known, for example Mount Keith, where no accompanying high-grade sulphide has been intersected. There is no gross compositional discontinuity

apparent in passing into disseminated sulphide from the barren host rock. There is a gradual increase in sulphide concentration with concomitant decrease in the silicate matrix, which in the Spargoville mineralized unit is now a partially altered assemblage after dunite that has residual olivine locally in excess of 50% (Hancock et al., 1971).

The primary sulphide assemblage consists of pyrrhotite, pentlandite, chalcopyrite and pyrite, infrequently subject to alteration to a Ni-rich assemblage, as for example pentlandite-heazlewoodite noted in Perseverance serpentinite. Supergene oxidation is important, although not necessarily complete, and produces essentially a two-violarite, pyrite-marcasite and chalcopyrite mineralogy.

The sulphide content of the rock results in increased concentrations of the elements nickel, iron, copper, cobalt and their trace congeners; however, since the silicates themselves carry relatively high levels of nickel, cobalt and iron, the lower concentrations of sulphide can be more readily detected from the copper content and that of other trace elements.

(2) Primary massive sulphides are usually found at the lower contacts of the ultramafic units, but are usually only a minor constituent of the zone, which may be only a few centimetres thick.

The mineralogy of the massive sulphide is not greatly different from that of the disseminated sulphides, although it may show mineralogical zoning as is reported for the Kambalda Lunnon Shoot (Ewers and Hudson, 1972). Such zoning is not apparent in the Spargoville mineralization although the distributions of both copper (in chalcopyrite) and cobalt (in pentlandite-pyrite) are distinctly non-uniform (Hancock et al., 1971). This might be due to metamorphism, nonetheless it is real and relevant to outcrop.

Oxidation again changes the primary sulphide assemblage in a general pattern to the two-violarite, pyrite-marcasite and chalcopyrite "system". Oxidation may extend to a considerable depth, probably as the result of a coupled electro-chemical reaction (Nickel et al., 1974). Mineralogically the sulphides, and thus the elements iron, nickel, cobalt and copper are dominant, however, the minor and trace elements may be non-uniformly distributed and may show vertical zoning, as for example copper, cobalt, palladium and platinum. The low concentrations recorded for silver are noteworthy.

(3) Remobilized massive sulphides occur as intrusive shoots of massive sulphide in country rock, usually metabasalt, and may carry more nickel and copper than in-situ sulphide. The massive sulphide may be almost entirely free of silicate and contain trace to minor amounts of carbonate.

The sulphide minerals consist essentially of generally well-crystallized pentlandite and pyrrhotite together with chalcopyrite and pyrite. However, remobilized sulphide does in places carry arsenides in significant concentration.

Chemically the make-up of the remobilized sulphide reflects the general absence of silicate phases and also of chromite and ilmenite. There is a major

variability in arsenic, which may be accompanied by trace Sb and may reach
a level of greater than 0.5%. Trace Pb (10 ppm) has been detected whether
arsenic is present or not. Silver levels do not seem to be elevated above those
of the primary sulphide, but the values of the other precious metals are of
interest.

The metasediments

The metasediments are relevant to the present discussion, not in their surface expression per se but rather in their minor and trace element contents,
the distributions of which may explain some anomalous ultramafic-derived
outcrop. The sediments range in type from the well-bedded pyrrhotitic
"syngenetic-sulphide" to those which contain a considerable proportion of
mafic silicates. They are usually fine grained, consisting of quartz, plagioclase
and variable amounts of "tremolite", chlorite and carbonate; graphitic carbon
is often present. The sulphide, which ranges from a trace to major constituent,
consists principally of pyrrhotite with trace to minor chalcopyrite and/or
sphalerite. The material examined contained nickel at the trace level only.

The metasedimentary rocks may thus be a geochemically significant source
of copper during oxidation and weathering. The association of zinc with the
metasedimentary copper, however, allows the use of the former as a negative
indicator of mineralization of interest.

The geochemistry of the precious metals

The precious metals are an important group of trace elements in the ultramafic sulphides from both the economic and geochemical viewpoint.

In this present work our immediate interest in the platinoids arose from
an initial supposition that they would be relatively immobile during the
weathering processes and would therefore be reliable indicators. The figures
and data which we obtained bear out this supposition but showed that the
initial concentrations and interelement distributions were not quite as anticipated.

The platinum group of elements includes Pt, Pd, Ru, Rh, Os and Ir. We
have concerned ourselves essentially with palladium and platinum. Other
members of the group are certainly present, notably iridium, and might be
expected to have diagnostic differential behaviour similar to that which we
suggest for palladium and platinum.

The unmineralized rocks

Palladium and platinum are strongly associated with the Ni—Cu sulphides
in the various rocks considered here. The "unmineralized" ultramafic rocks
themselves contain very low concentrations of both palladium and platinum,
around 0.01—0.02 ppm (for example Table I, CB/153.6 m, 103.3 m and
5DB/97.4 m), except at basal contacts where they show enrichment sympa-

thetic with that of nickel and copper. In the examples given by the Central
Belt rocks in Table I, palladium and platinum reach values of 0.05 and 0.04
ppm, respectively, in CB/117.3 m. This contrasts with values of 0.33 and 0.30
ppm in the contact sulphide zone (5DB/178 m) above the mineralized unit of
the Eastern Belt. However, it is difficult to make a strict comparison because
of alteration and the probable leaching of the sulphides which is general in
the contact zones and which could result in absolute and relative changes in
the concentrations of nickel and copper in the sulphides.

The mineralized rocks

In the fresh rocks we can recognize three situations which are relevant to
the use of palladium and platinum in interpreting outcrop: the disseminated
to massive sulphide "profile", remobilized sulphide and "hydrothermally"
altered/leached sulphide. Fig.6 shows the distribution of palladium and platinum in the Spargoville core 5DB. The diagram includes the mineralized
dunitic unit and the base of the overlying ultramafic unit and shows that
palladium and platinum exhibit differential behaviour.

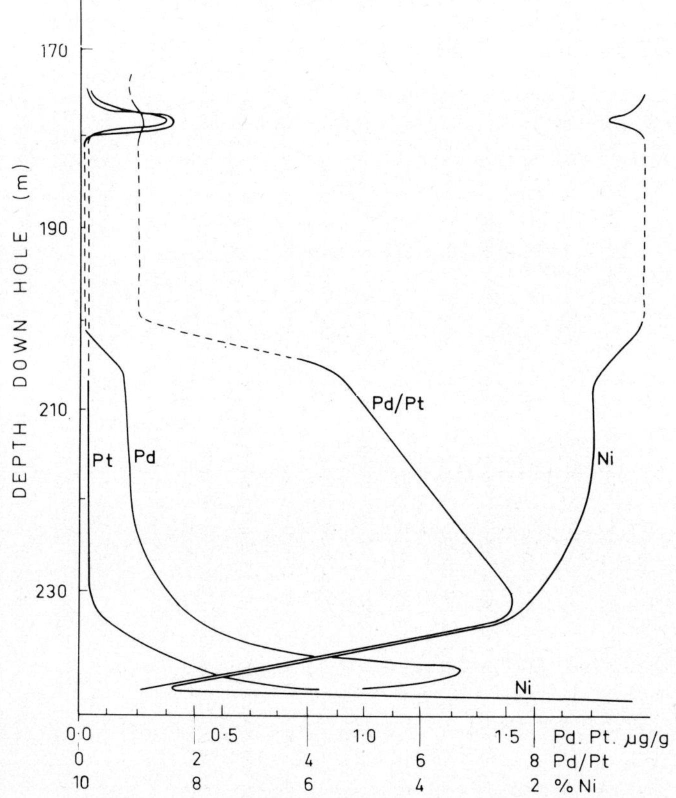

Fig.6. Distribution of palladium and platinum in the core 5DB.

TABLE II

Geochemistry of weathered rocks

Sample	MgO	SiO$_2$	Fe	Ni	Cu	Co	Zn	As	Pd	Pt	Au	Al$_2$O$_3$	Ca	Na	Mn	Ti	Cr	V	Ag	Ni/Cu
1	0.1	30.3	38.6	1.27	0.20	60	<20	7	0.41	1.0	0.02	<0.1	<0.01	150	0.05	0.01	0.006	<10	—	6.4
2	0.13	6.0	51.0	1.90	0.15	400	<20	5	0.28	1.0	0.10	0.24	<0.01	600	0.13	0.02	0.02	<10	—	12.7
3	0.10	8	51.1	0.32	0.093	240	4	0.049	0.22	0.80	0.03	3.5	<0.01	800	<0.01	0.01	0.58	150	—	3.4
4	0.11	8	51.3	0.118	0.188	340	20	0.086	0.11	0.27	0.12	3.2	0.01	200	0.01	0.08	0.09	80	—	0.63
5	0.2	9.4	51.9	0.59	0.165	120	90	0.004	0.40	<0.04	0.04	0.21	0.02	200	0.01	0.02	0.01	<10	—	3.6
6	0.1	40	32.2	0.09	0.06	60	40	0.009	0.15	0.38	<0.02	0.18	0.04	600	0.06	0.006	0.20	60	—	1.5
7	0.1	20	42.2	0.23	0.19	290	120	0.41	0.49	0.19	9.2	0.47	<0.01	<100	0.02	<0.01	0.02	10	—	1.2
8	0.23	15	41.5	3.52	0.50	1050	50	0.044	0.16	0.40	0.03	0.88	0.01	150	0.04	0.01	0.03	150	1	7.4
9	0.13	18	35.4	0.11	0.18	270	20	0.044	0.16	0.40	0.03	0.9	<0.01	600	0.01	0.11	0.03	40	4	6.1
10	0.03	93	2.0	0.01	0.016	20	20	0.005	4.6	0.62	1.4	0.06	0.05	150	<0.01	<0.01	0.15	60	—	0.62
11	0.95	4.6	51.5	2.06	0.60	400	20	0.003	1.8	0.10	0.01	0.20	0.13	100	0.08	0.11	0.02	<10	—	3.4
12	0.36	6.0	50.0	2.42	0.60	1500	20	0.002	2.6	0.12	0.04	0.1	<0.01	300	0.17	0.11	0.02	<10	—	4.0
13	0.31	4.5	54.4	1.32	0.40	600	20	0.002	2.6	0.13	0.01	0.1	<0.01	100	0.20	0.12	0.02	<10	—	3.3
14	0.19	10.3	50.4	1.61	0.25	200	20	8	0.68	0.32	0.12	0.19	<0.01	100	0.09	0.01	0.004	<10	—	6.4
15	12.4	63.2	8.9	0.61	0.40	80	150	0.043	1.1	0.04	5.7	2.9	0.03	200	0.06	0.10	0.30	400	—	1.5
16	0.23	75	12.3	0.24	0.22	100	20	0.014	4.5	0.16	0.13	0.54	0.12	150	0.03	0.01	0.15	25	2	1.1
17	9.7	67.1	7.6	0.27	0.03	200	240	0.016	0.01	0.02	0.01	2.4	0.16	200	0.05	0.08	0.40	150	—	9
18	4.7	20.6	34.2	0.41	0.04	150	250	0.69	0.02	0.12	4.2	5.3	0.89	400	0.09	0.14	0.60	100	<1	10
19	5.5	56.9	16.8	0.39	0.04	100	60	0.005	0.13	0.04	0.06	2.4	0.04	800	0.06	0.08	0.20	60	—	9.7
20	11.8	52.1	13.0	0.52	0.04	150	130	0.005	0.10	0.02	0.06	3.8	0.22	4000	0.07	0.11	0.40	150	—	13
21	2.1	60.7	22.1	0.66	0.04	40	70	0.017	0.33	0.03	0.34	0.38	1.27	100	0.05	0.006	0.15	150	—	16.5
22	1.4	20.2	49.5	0.99	0.04	150	80	0.017	0.02	<0.03	<0.02	0.24	0.60	300	0.05	0.04	0.10	20	—	25
23	0.12	29.9	45.5	0.78	0.04	200	60	0.037	0.14	0.14	0.02	0.10	0.06	200	0.02	0.006	0.20	25	—	19.5
24	3.3	70	6.2	0.13	0.04	240	20	0.017	0.09	0.05	0.25	1.5	0.01	100	0.02	0.06	0.035	150	—	3.3
25	3.5	70	4.9	0.10	0.03	140	20	0.015	0.06	0.04	0.16	1.3	0.01	300	0.01	0.04	0.03	150	4	3.3
26	4.6	30	35	0.076	0.092	210	50	0.025	0.10	0.14	0.05	0.69	0.01	150	0.05	0.08	0.05	200	8	0.83
27	0.68	10	54.8	0.31	0.19	290	170	0.48	0.09	0.09	50.0	1.5	0.02	<100	0.04	0.04	0.07	15	2	1.6
28	0.24	8	54.9	1.36	0.069	190	220	0.40	0.06	0.03	0.13	2.5	0.03	150	0.03	0.04	0.03	15	1	19.7
29	7.8	45	16	1.66	0.11	270	50	0.005	0.17	0.07	0.11	1.8	0.01	400	0.03	0.06	0.03	25	—	15.1
30	2.3	70	9.3	0.10	0.077	310	20	0.008	0.16	0.06	0.01	2.1	<0.01	600	0.04	0.09	0.04	80	10	1.3
31	1.27	72.5	13.6	0.24	0.015	20	230	9	0.02	0.02	0.04	0.20	<0.01	100	0.19	0.01	0.10	20	—	16
32	0.43	10	45.4	0.087	0.208	210	520	0.047	0.19	0.27	0.04	2.2	0.01	600	0.02	0.12	1.3	100	—	0.4
33	0.36	15.8	45.5	0.27	0.04	60	500	0.26	0.21	0.04	3.1	3.6	0.48	200	0.06	0.12	0.30	60	—	6.8
34	0.1	4	51.9	1.19	0.076	230	280	0.46	0.40	0.04	0.27	3.3	0.04	100	0.02	0.05	0.04	15	—	15.7
35	0.1	20	38.1	0.17	0.11	220	250	0.49	2.2	0.30	2.6	0.72	<0.01	100	<0.01	0.01	0.06	15	1	1.6

TABLE II (continued)

Sample	MgO	SiO$_2$	Fe	Ni	Cu	Co	Zn	As	Pd	Pt	Au	Al$_2$O$_3$	Ca	Na	Mn	Ti	Cr	V	Ag	Ni/Cu
37	1.4	88	4.1	0.25	0.036	160	50	3	0.07	0.08	0.02	0.34	0.01	250	0.05	0.01	0.20	80	—	6.9
38	1.1	87	4.4	0.24	0.039	200	50	3	<0.01	0.08	<0.01	0.24	0.01	200	0.08	0.01	0.15	80	—	6.2
39	9.7	79	3.4	0.16	0.01	150	30	3	0.21	0.06	0.03	0.40	0.01	100	0.12	0.02	0.06	150	1	16
40	1.7	56.4	18.3	0.37	0.055	410	130	0.004	0.06	0.06	0.02	<0.1	3.1	100	1.0	1.0	0.02	200	2	6.7
41	1.34	79	4.0	0.056	0.013	220	40	0.043	0.09	0.05	0.07	0.44	0.01	100	0.05	0.05	0.03	100	2	4.3
42	0.08	80	5.7	0.78	0.061	360	40	0.016	0.04	0.03	0.03	0.08	<0.01	600	0.11	0.01	0.01	80	1	12.8
43	2.0	70	5.8	0.051	0.035	190	20	0.022	0.09	0.05	0.02	1.2	<0.01	600	0.02	0.07	0.31	80	—	1.5
44	2.3	86.9	2.8	0.18	0.025	50	60	4	0.03	0.06	0.02	0.9	0.05	200	0.04	0.03	0.20	40	—	7.2
45	0.38	94	1.4	0.01	0.003	10	20	4	0.01	0.06	0.02	0.9	0.05	200	0.04	0.03	0.20	10	—	3.3
47	0.1	60	22	0.021	0.017	140	30	0.056	0.09	0.03	10.0	0.06	<0.01	100	<0.01	<0.01	0.01	15	4	1.2
48	0.10	6	53.0	0.11	0.007	76	145	0.017	0.10	0.04	0.01	2.1	0.1	<100	0.02	0.27	0.14	60	—	15.7
49	<0.1	4	48	0.40	0.028	70	310	0.041	0.10	0.04	0.01	2.1	0.15	<100	0.01	0.14	0.70	40	—	14.2
50	<0.1	4	57.2	0.62	0.006	150	470	0.042	0.02	0.05	0.02	1.5	0.03	<100	0.01	0.05	0.14	10	—	103
51	0.1	30	38.4	0.20	0.12	200	250	0.12	0.04	0.14	0.04	0.5	<0.01	300	0.01	0.01	0.05	100	4	1.7
52	0.3	10	45.3	0.58	0.071	20	1670	0.11	0.04	0.05	0.05	0.5	<0.01	200	0.02	0.02	1.9	150	—	8.3
53	<0.1	8	45.5	0.60	0.02	170	500	0.036	0.02	0.02	0.01	2.5	0.15	150	0.03	0.12	1.5	40	—	30
54	0.1	8	52.3	0.089	0.053	122	850	0.003	0.02	0.02	0.02	2.8	0.20	<100	0.03	0.12	0.02	10	—	1.7
55	<0.1	8	46.9	0.049	0.035	310	120	0.007	<0.04	0.04	0.04	8.6	0.01	<100	0.08	0.14	0.06	60	—	1.4
56	0.04	90	2.8	0.02	0.001	< 10	7	0.006	0.01	0.05	0.67	0.04	0.06	<100	0.11	<0.01	0.03	20	—	20
57	21.3	50	6.1	0.057	0.008	150	30	0.015	0.02	0.03	0.59	1.7	0.10	800	0.02	0.08	0.03	30	—	7
58	2.78	13.7	41.6	0.56	0.030	1000	390	0.06	0.05	0.06	0.73	5.7	0.31	400	0.91	0.41	0.25	150	1	18.7
59	0.60	38	31.6	0.20	0.0015	20	360	0.002	<0.01	0.02	4.7	4.7	<0.01	100	0.09	0.15	0.60	60	—	133
60	3.6	60	5.5	0.056	0.008	110	20	0.043	0.01	0.04	0.01	18.0	0.12	600	0.02	0.07	0.03	100	6	7
61	0.22	11.6	50.9	0.06	0.002	20	20	0.06	0.02	<0.01	0.01	6.1	0.13	100	0.06	0.10	0.20	<10	—	30
62	0.1	19.6	45.3	0.31	0.002	600	700	0.011	<0.01	0.02	0.02	0.32	<0.01	150	0.71	0.01	0.04	100	—	150
63	0.1	4	58.6	0.104	0.094	300	2260	0.001	0.05	0.06	0.07	2.6	<0.01	100	0.01	0.22	0.05	40	—	1.1
67	0.48	30.7	37.6	0.03	0.015	20	3000	0.008	0.01	0.04	0.01	3.7	0.11	6000	0.09	0.14	0.04	150	—	2
68	0.63	33.5	35.0	0.01	0.015	20	800	0.03	0.02	0.02	0.01	3.9	0.09	4000	0.06	0.19	0.06	200	—	1
70	<0.1	45.6	22.6	<0.01	<0.01	20	300	1	0.02	0.04	0.02	9.1	0.64	8000	<0.01	0.30	0.006	30	—	0.5
71	11.0	55	10.5	0.14	0.096	160	40	0.010	0.08	0.06	0.02	7.8	0.22	1.5	0.04	0.50	0.16	150	1	1.5
72	6.5	75	4.4	0.047	0.031	190	20	0.020	0.07	0.05	0.04	1.2	0.10	1000	0.03	0.09	0.04	60	—	1.5
73	<0.2	70.8	7.8	0.01	0.006	< 20	< 20	6	0.02	<0.02	<0.01	10.8	<0.01	4	<0.01	0.23	0.04	60	—	1
74	<0.2	92	6.0	0.01	0.02	20	20	0.002	<0.01	0.02	0.27	0.14	<0.01	100	<0.01	0.13	0.08	20	—	0.5
75	0.42	8.6	50.4	0.12	0.01	20	1050	0.015	0.01	<0.02	0.01	2.6	0.39	100	0.11	0.10	0.06	<10	—	12

Figures italicized are in ppm; all other values are given as %.

From a background value of some 0.02 ppm, palladium increases sympathetically with nickel through the disseminated sulphide and reaches a maximum in the massive sulphide at the base of the unit. It is interesting and possibly of significance that a similar value is found for the ratio Pd/Ni in a Perseverance section and also in material from another deposit.

Platinum, in contrast, remains near the background value until the sulphide concentration reaches matrix level when it increases abruptly, maximizing at the base of the ultramafic unit.

Contrasting with the primary massive sulphide, remobilized massive sulphide carries negligible platinum while containing variable, but high, palladium (for example 5DR/315 m, Table I).

In the "hydrothermally" altered/leached sulphide zones we have found platinum to be dominant over palladium in certain non-oxidized leached and silicified zones, now very low in sulphide.

These data suggest that palladium is substantially more mobile than platinum under the conditions of alteration that obtained in this region and which led to the formation of the massive remobilized sulphides. The data of Fig.6 also suggest that the possibility of primary magmatic differentiation or partitioning of palladium and platinum should be considered.

It should be noted at this point, however, that although the data presented below would suggest quite restricted mobility for palladium and platinum during oxidation and weathering of the sulphidic rock, the work of Cousins (1969) and Fuchs (1972) indicates that this might not always be the case.

Geochemistry of the weathering products

Samples were taken from outcrop or from exposures in shallow trenches at the Spargoville property and a number of other localities by Dr. W. Hancock and the author. Geochemical data are given in Table II.

It should be noted that specimens 3, 8 and 9 were contiguous and undisturbed. Although differing morphologically they are indicative of the variation in the indicator elements which may be observed between samples. The pair 51, 52 is also interesting in that although they were closely spaced and undisturbed surface samples, one is apparently a modified ultramafic contact "gossan" while the other is largely residual and has a substantial contribution from adjacent non-ultramafic sulphide.

In the following section the products of weathering are dealt with in terms of an admittedly imprecise "weathering profile". This has some virtue in allowing a preliminary decision as to the degree and manner in which the surface rock may have varied from its original composition.

The weathering profile. It is not easy to define a generalized and coherent pattern for weathering alteration in the presence of sulphides, particularly in respect of silicification. In practical terms we are really concerned with the mobility of the indicator elements in the several zones of the weathering profile. In the present context the main difficulty is in knowing whether one is

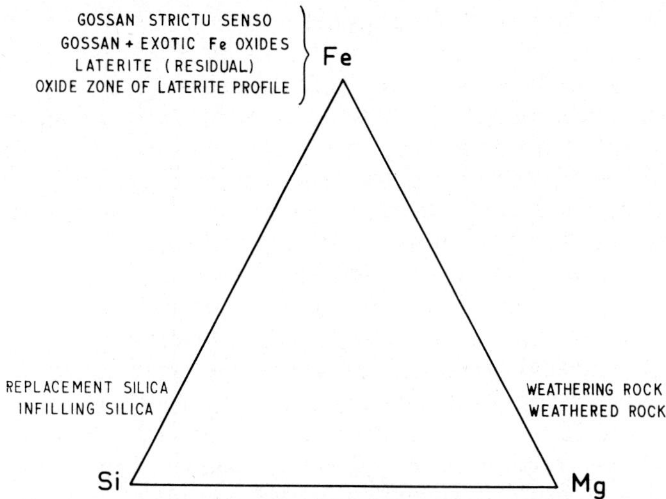

Fig.7. Diagrammatic representation of weathering products.

dealing with the present profile, or an earlier and probably "compound" profile. If the latter, the interpretative problems are somewhat greater.

It seems necessary to recognize that an apparently similar outcrop may arise in a number of ways and that the several pathways may, and almost certainly will, affect the geochemical make-up and that each will record the nature of the original rock in varying degree. Thus, as is shown diagrammatically in Fig. 7, one finds:

(a) At the iron apex:
    the strict gossan
    the gossan + exotic iron oxides
    residual laterite
    the oxide zone of the laterite weathering profile, both in situ and where erosion has removed the covering.

(b) At the silica apex:
    replacement silica
    spatial infilling silica
    perhaps residual silica related to sulphide oxidation.

(c) Toward the MgO apex:
    both the presently weathering rocks and that which was part of an earlier cycle profile.

Notwithstanding the above imprecision it will be apparent from the following discussion that the classification of the outcrop specimens in terms of a nominal weathering type is in fact of distinct value in the initial selection and sorting of field samples.

Weathered rocks

These show much of the texture of the original rock: there is only minor silicification and/or ferruginization. They are considered here in relation to their unweathered equivalents.

Weathered unmineralized ultramafic. To a first approximation the major elements follow the parent rock in terms of aluminium, calcium and sodium, and thus help distinguish basaltic, sedimentary, porphyritic, and tremolite-chlorite rocks from the dunitic (serpentinized and talc-carbonate) rocks.

Similarly for nickel, the higher levels in the dunitic rocks tend to persist as against the somewhat lower levels in tremolite-chlorite. However, it is well known that nickel (and Co) are sensitive to the position of the rock in the weathering profile and we cannot make much of such minor variations.

In weathered outcrop representing original unmineralized ultramafic, nickel and cobalt may be ~0.2% and 150 ppm, respectively. More significantly, however, copper remains close to the initial level, giving a high Ni/Cu ratio. Zinc also shows little or no increase. However, there is the practical problem, already alluded to and discussed further below, of the possibility of unmineralized material being affected, not infrequently, by exotic copper.

Palladium and platinum show little or no enrichment at perhaps 0.02—0.04 ppm.

Weathered ultramafic with disseminated sulphide. The effect of the presence of sulphide is to add Ni,Co, Cu, Fe, Ag, Pd, Pt, S, Se and Te to the system. Since nickel and cobalt are present in the silicates of the fresh rocks, and are profile dependent in weathering, it is difficult to be positive, for lower concentrations, as to their source in considering the weathered rock.

One can be more certain about copper. With a Cu/Ni ratio of e.g. 1/15, 5000 ppm sulphide Ni (perhaps 7000 ppm total) corresponds to some 300 ppm Cu which is significantly above background and which is largely retained by the weathered rock. The corresponding levels for zinc tend to be low. Nonetheless, where the initial sulphide concentration was low, and hence the available copper was minimal, elevation of the zinc concentration, as may occur adjacent to porphyry, will raise doubts as to the source of the residual copper. This is more of a problem with ferruginous and siliceous outcrop.

As has been described, in the Spargoville example, platinum does not increase much above background until nickel reaches a relatively high concentration. However, palladium follows nickel and with the latter at 0.5% will be present at the level of about 0.06 ppm which although low, is persistent during weathering.

Weathered massive sulphide. This is the classic gossan in the general sense of Blanchard (1968). Boxworks are clearly evident, although the pathway through oxidation and leaching is apparently complex as shown by Nickel (1973) and Nickel et al. (1974).

The outcrop specimens carry high concentrations of both nickel and copper, but show variable cobalt which is a reflection both of variation in the initial concentration in sulphide and variable loss during weathering. Zinc levels are low.

Absolute levels for Pd + Pt are high, as may be expected from the fresh rock data, and the anticipated variation in the ratio of the two elements is also observed. Allowing that there is some uncertainty in the nominal fresh rock-weathered rock correlation and that the sulphides have been affected by hydrothermal alteration in varying degree, the *primary sulphides* show high platinum (and Pd) whereas the remobilized sulphides have moderate to high palladium but low platinum.

High arsenic figures indicate the prior existence of arsenides, particularly in remobilized sulphide. Gold is frequently detectable in the outcrop and may show some enrichment.

Ferruginized rocks

In rocks of this nature, the major problem is that of discriminating against "laterite" and "indurated laterite". Simple distinction using copper is not necessarily sufficient because of the general regional abundance of this element and the low levels at which it may be geochemically significant. As has already been noted, zinc may be negatively indicative in this sense, but this is not altogether unambiguous.

Ferruginized unmineralized ultramafic. The nickel content is generally high and may depend on the level in the precursor and on the nature of the weathering. The copper content of the strictly unmineralized rock is low, giving a high value for Ni/Cu that also applies to uncontaminated "laterite". However, this is not so in other outcrops derived from nominally unmineralized ultramafic rock where the copper levels may be high enough to suggest mineralization. When accompanied by high zinc, and with low or only slightly elevated arsenic concentrations, it is probable that the copper is exotic. However, with high arsenic the situation is ambiguous. The arsenic and zinc may have come from sediment or porphyry and may not have been originally associated with mineralization or they may have derived from porphyry contaminated ultramafic and, hence, from arsenical Ni—Cu sulphide.

Palladium and platinum are less ambiguous; both elements exhibit close to background values in this class of outcrop.

High chromium also can be suggestive of residual concentration, however, there are marked variations in the concentration of the element in the fresh rocks and it may also be present in amphibole and chlorite in addition to chromite. Hence deductions based on the nature of the chromium in the outcrop may be open to question.

Ferruginized ultramafic with disseminated sulphide. Both nickel and copper vary over a considerable range depending on the initial concentrations and on the nature of the weathering. Arsenic may be present where the rocks were affected by porphyry intrusion; arsenic is generally accompanied by zinc and at times by gold. Palladium and platinum are found to be present at significant levels in keeping with the proposed derivation.

Ferruginized massive sulphide. The outcrop shows variable but usually high nickel and higher copper values, hence giving a low Ni/Cu ratio. Cobalt is quite variable and is clearly mobile under the conditions of alteration and of secondary oxide deposition in the region.

434

The usual low zinc levels suggest limited retention of the original sulphide zinc and subsequent minimal fixation under these conditions of alteration and weathering.

Values for palladium and platinum are again in agreement with the nominal sulphide type and the degree of mineralization. The high gold values recorded do in fact relate to higher concentrations in the sulphide, but the presence of "paint" gold in several instances shows that its mobility is quite significant.

Siliceous weathering products

There is some difficulty in deciding initially what one is dealing with in some of the more highly silicified material. It is necessary to make a decision as to whether a rock is silicified by replacement and hence bears some relationship to the precursor or whether the silica has been deposited as massive spatial infilling and possibly coloured with iron oxides and therefore has at best an obscure relationship with the "host" rocks. Textural investigation would seem to be especially relevant where the more siliceous rocks are concerned.

Silicified unmineralized ultramafic. Silicification results in a lowering of the indicator levels. Nickel is readily removed, perhaps more so than copper, resulting in a movement of the Ni/Cu ratio to lower values. There may be little distinction left between mineralized and unmineralized rock based on either absolute or relative nickel and copper figures. It is possible that there can be some enrichment in gold, but it is important to note that this does not seem to be so with palladium and platinum, although there is some slight uncertainty in the background values.

Silicified ultramafic with disseminated sulphide. Highly silicified rock after disseminated sulphide is the most difficult to assess using normal analytical data. As above, the process or conditions of silicification reduce what may be already marginally anomalous concentrations of the key elements. In our material the retention of even moderate nickel seems to be linked with the retention of a few percent iron, and the nickel decreases rapidly as the iron is removed. Titanium also is leached and hence cannot be used to indicate the nature of the replaced rock as might be suggested by its "retention" in other types of siliceous replacement rock.

Based on the known relationship to mineralization, palladium and platinum are present at the expected anomalous levels: the projected relationship with the fresh rocks, however, is not sufficiently certain to establish the degree of retention of palladium in particular.

Silicified massive sulphide. Although a higher silica content in the "gossan" tends to give lower indicator concentrations, there is the suggestion in this instance that the process of silicification might be self-limiting, i.e., that the silicification of the highly oxidized (ferruginous) rock gives a product immune to further leaching. It is, in fact, not easy to correlate changes in the concentrations of nickel and copper with the degree of silicification since both their absolute and relative values can be seen to vary markedly in the weathered sulphides.

It is clear, however, that the minor elements give little indication of the sulphide content of the original rock when the replacement by silica is nearly complete and there may even be doubt as to whether sulphide had been present originally. In our material, the high palladium and platinum concentrations are, however, quite distinctive of massive sulphides.

CONCLUSIONS

The overall interpretation of the weathering products of nickeliferous sulphides and associated rocks is frequently not straightforward, for geological and geochemical reasons which have been discussed above. An interpretative scheme can, however, be based on the geochemical characteristics of the outcrop, with reasonable expectation of success.

The most useful information is that for the elements Ni, Cu, Zn, As, Pd and Pt. Of these nickel, copper and zinc are readily determined by AAS. Arsenic is less readily determined but is necessary if one is to consider high As—Ni—Cu sulphides as well as "contamination" both of sulphidic and non-sulphidic rock.

There are instances, however, where this short-form geochemical data is inconclusive because of either barely anomalous indicator concentrations or a suggestion of exotic derivation. Data for palladium and platinum can be helpful in these cases since it has been shown that these elements resist movement during weathering and leaching and show a definite association with the Ni—Cu sulphides.

It is suggested that an integrated approach using geochemical, textural and field data is desirable.

ACKNOWLEDGEMENTS

The author is indebted to the Directors of Australian Selection (Pty) Limited for making available the material for this investigation. Thanks are also due to Dr. W. Hancock of Selection Trust for many useful discussions during the course of the work. The author wishes to thank Mr. J. Corbett (C.S.I.R.O.) and Mr. I. Hodges (N.S.W. Mines Department) for the determination of the precious metals and Mr N. Morgan (C.S.I.R.O.) for the AES determinations.

REFERENCES

Andrews, P., 1974. Nickel deposits of Spargoville Western Australia. In: C.L. Knight (Editor), Economic Geology of Australia and Papua New Guinea. Australasian Institute of Mining and Metallurgy, Melbourne, Vic. (in press)

Blanchard, R., 1968. Interpretation of leached outcrops. Nevada Bur. Min. Bull., 66, 196 pp.

Corbett, J.A., Godbeer, W.C. and Watson, N.C., 1974. The application of sodium peroxide sintering techniques to the analysis of minerals and rocks by atomic absorption spectroscopy. Proc. Australas. Inst. Min. Metall., 250: 51—54

Cousins, C.A., 1969. The Merensky Reef of the Bushveldt Igneous Complex. In: H.D.B. Wilson (Editor), Magmatic Ore Deposits. Econ. Geol. Monogr., 4: 239—251

Ewers, W.E. and Hudson, D.R., 1972. An interpretive study of a nickel—iron sulphide ore intersection, Lunnon Shoot, Kambalda, Western Australia. Econ. Geol., 67: 1075—1092

Fuchs, W.A., 1972. Geochemical behaviour of platinum, palladium, and associated elements in the weathering cycle in the Stillwater Complex, Montana. M.Sc. Thesis, Pa. State Univ., University Park, Pa.

Hancock, W., Ramsden, A.R., Taylor, G.F. and Wilmshurst, J.R., 1971. Some ultramafic rocks of the Spargoville area, Western Australia. Geol. Soc. Aust., Canberra, A.C.T., Spec. Publ., 3: 269—280

Hudson, D.R., 1972. Evaluation of genetic models for Australian sulphide nickel deposits. Australas. Inst. Min. Metall. Conf., Newcastle, N.S.W., pp.59—68.

Hudson, D.R., 1973. Genesis of Archaean ultramafic-associated nickel-iron sulphides at Nepean, Western Australia. Papers presented at Annu. Conf. Australas. Inst. Min. Metall., Perth, W.A., pp.99—109

Martin, J.E. and Allchurch, P.E., 1974. Geology of Perseverance nickel deposit, Western Australia. In: C.L. Knight (Editor), Economic Geology of Australia and Papua New Guinea. Australasian Institute of Mining and Metallurgy, Melbourne, Vic. (in press)

Nickel, E.H., 1973. Violarite-A: key mineral in the supergene alteration of nickel sulphide ores. Paper presented at Annu. Conf. Australas. Inst. Min. Metall., Perth, W.A., pp.111—116

Nickel, E.H., Ross, J.R. and Thornber, M.R., 1974. The supergene alteration of pyrrhotite-pentlandite ore at Kambalda, Western Australia. Econ. Geol., 69: 93—107

Roberts, D.E. and Travis, G.A., 1974. Textural evaluation of nickel sulphide gossans. In: C.L. Knight (Editor), Economic Geology of Australia and Papua New Guinea. Australasian Institute of Mining and Metallurgy, Melbourne, Vic. (in press)

Robinson, W.B. and Stock, E.C., 1974. The discovery and evaluation of the Windarra Nickel Deposits, Western Australia. In: C.L. Knight (Editor), Geology of Australia and Papua New Guinea. Australasian Institute of Mining and Metallurgy, Melbourne, Vic. (in press)

Ross, J.R. and Hopkins, G.M.F., 1974. The nickel sulphide deposits of Kambalda, Western Australia. In: C.L. Knight (Editor), Economic Geology of Australia and Papua New Guinea. Australasian Institute of Mining and Metallurgy, Melbourne, Vic. (in press)

Sheppy, N.R. and Rowe, J., 1974. The nickel sulphide deposits of Nepean. In: C.L. Knight (Editor), Geology of Australian Ore Deposits (in press)

Woodall, R. and Travis, G.A., 1969. The Kambalda nickel deposits, Western Australia. Proc. Ninth Commonwealth Min. Metall. Congr., Paper 26, pp.517—535

GEOCHEMICAL SOIL SURVEYS IN EXPLORATION FOR NICKEL—COPPER SULPHIDES AT PIONEER, NEAR NORSEMAN, WESTERN AUSTRALIA

R. COX

Placer Prospecting (Australia) Proprietary Limited, Sydney, N.S.W. (Australia)

ABSTRACT

Ni—Cu sulphides at Pioneer are localized at or near the base of a serpentinized ultramafic body forming part of the Norseman-Kalgoorlie "Greenstone Belt" of the Archean Shield of Western Australia. Geochemical and petrographic investigations have shown that the ultramafic body consists of several repetitions of differentiated cycles, each cycle consisting of a basal layer of meta-dunite overlain by serpentinized peridotite passing upwards into meta-pyroxenite.

The primary sulphide bodies consist essentially of a pyrrhotite—pentlandite— chalcopyrite assemblage within structurally controlled loci. In-situ alteration of pyrrhotite to marcasite and of pentlandite to violarite is associated with supergene processes. As a result of weathering these sulphides have been oxidized to a limonitic goassanous material, extending down to 40 m below surface. The gossans retain essentially the geometrical form of the sulphide bodies, and also retain anomalous nickel and copper values.

Outcrop in this strongly dissected, flat terrain is negligible and the only visible surface expression of the Ni—Cu mineralization is the sporadic occurrence of gossan float.

The typical soils of the area are solonized brown soils showing a characteristic profile development. The distribution of nickel and chromium in these soils over unmineralized ultramafic, mafic and metasedimentary rocks accurately reflects bedrock nickel and chromium distribution and delineates lithologic boundaries. In areas of sub-outcropping mineralization the soils show strong anomalies in the distribution patterns of nickel, copper, cobalt and Ni/Cr ratio which are associated with secondary iron and manganese oxides.

Experience has shown that gossan searching and geochemical soil sampling are the most effective, rapid, and economical techniques for positively detecting sub-outcropping Ni—Cu mineralization.

This paper describes a case history of successful geochemical exploration at Pioneer in the Western Australian "nickel belt".

INTRODUCTION

Since the discovery of nickel at Kambalda in 1965 (Woodall and Travis, 1969) at least 29 medium- to high-grade massive and disseminated nickel sulphide deposits have been outlined in the Yilgarn Block of the Archean Shield of Western Australia (Fig.1, Table I). These ore bodies, totalling in excess of 86 million tons, contain approximately 2.09 million tons of nickel. In addition, several other large, low-grade, disseminated occurrences have

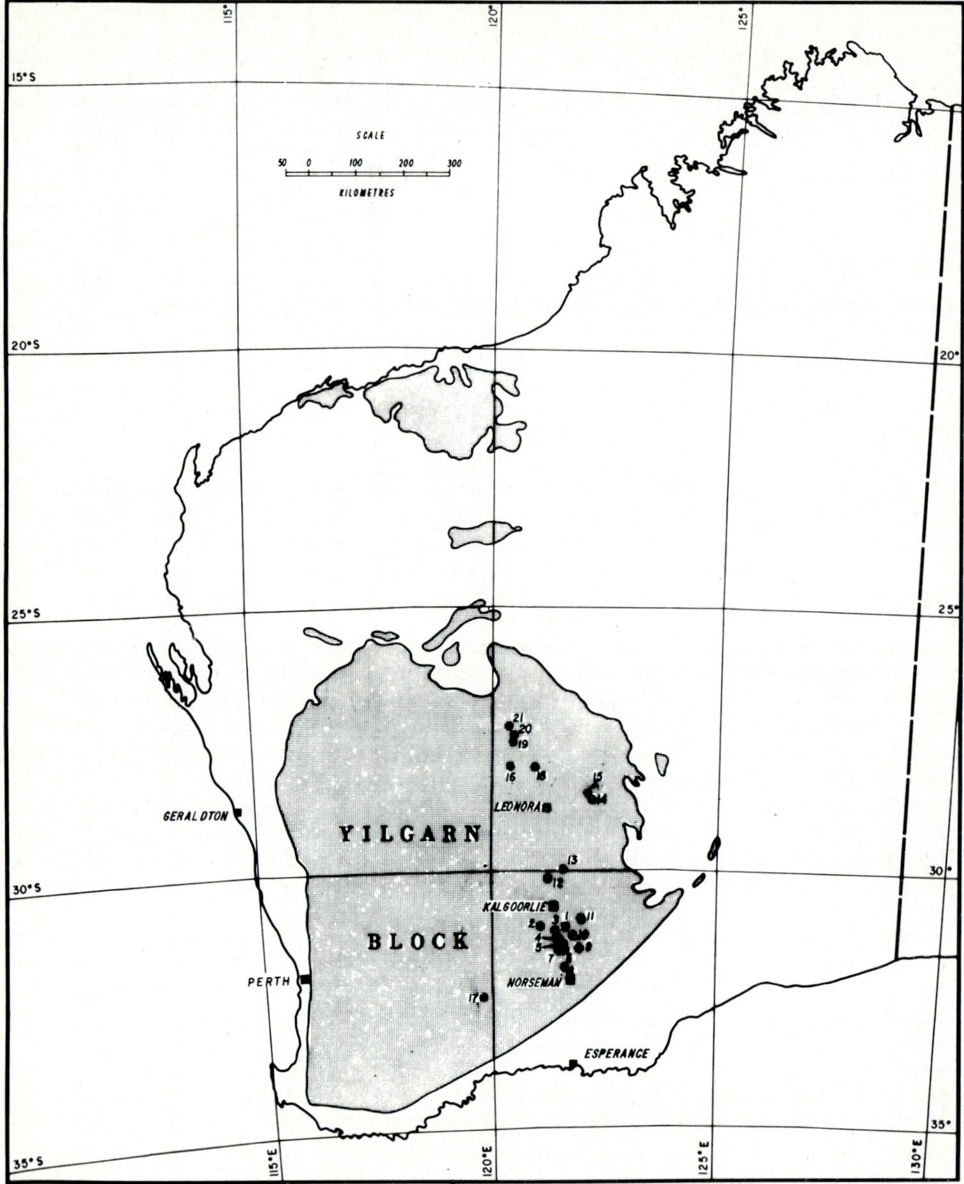

Fig.1. Location plan, nickel deposits of the Yilgarn Block (Archean) of Western Australia.

been discovered, most notable of which is Mt. Keith. Thus the Yilgarn Block has emerged as one of the great nickel fields of the world.

The Yilgarn Block underwent intense tropical weathering during Tertiary times producing characteristic lateritic and siliceous deposits from the Archean rocks by intense leaching to depths up to 100 m. The resulting silica cap and

TABLE I

Nickel sulphide deposits* of the Yilgarn Block, Western Australia

Property	Published reserves		
	Ton	% Ni	% Cu
Medium- to high-grade deposits			
(1) Kambalda (at least 10 deposits)	24,051,000	3.24	—
(2) Nepean	590,000	3.58	—
(3) Spargoville (No.2)	110,000	2.32	0.19
(No.3)	650,000	2.47	0.23
(4) Mt. Edwards	1,540,000	2.22	0.16
(5) Widgiemooltha (No.3)	907,000	1.20	—
(6) Redross	907,000	3.50	—
(7) Wannaway	3,630,000	1.23	0.10
(8) Pioneer	not published		
(9) Tramways	not published		
(10) St. Ives	not published		
(11) Mt. Monger	114,000	2.75	—
(12) Scotia	1,029,000	1.75	0.25
(13) Carr Boyd (3 pipes)	1,536,000	1.34	0.45
(14) South Windarra	3,500,000	2.02	0.09
(15) Mt. Windarra	5,400,000	2.18	0.20
(16) Perseverance	40,000,000	2.22	—
(17) Diggers Rocks	2,300,000	1.40	—
Total	86,264,000	2.42 wt.av.	
Low-grade disseminated deposits			
(18) Weebo Bore	not published		
(19) Yakabindie	50,000,000	0.5	—
(20) Mount Sir Samuel	not published		
(21) Mt. Keith	272,000,000	0.6	—

*Locations shown in Fig.1.

laterites have subsequently been partly eroded and covered by a variety of weathering products. Such weathering conditions obviously present severe obstacles to exploration for nickel sulphide deposits in this area.

It was realized at a fairly early stage of exploration that: (a) anomalous nickel and copper values in limonitic gossans represented the oxidation products of Ni—Cu sulphide mineralization, and that (b) many of the deposits are located at or close to the basal contact of an ultramafic body. This early knowledge and experience was largely responsible for subsequent successful exploration which relied heavily on geological mapping and magnetic surveys (to delineate ultramafic bodies) combined with gossan searching and geochemical soil sampling.

The Ni—Cu sulphide deposits at the Pioneer prospect (Cox and Tyrwhitt, 1974) occur at latitude 31°59′S, longitude 121°38′E, some 30 km north of Norseman and 85 km south of Kambalda (Fig.1) and were discovered in September 1968 by gossan searching along ultramafic bodies.

The mineralization is located within a narrow belt of north-trending ultramafics, mafics and metasediments comprising part of the Norseman-Kalgoorlie "Greenstone Belt" (Horwitz and Soufoulis, 1965; Soufoulis, 1966; Gemuts and Theron, 1974) in the predominantly granitic terrain of the Archean Yilgarn Block. Ni—Cu sulphides occur at and near the base of a serpentinized ultramafic body. The succession has been folded into a domal structure (the Pioneer Dome) so that the basal contact of the ultramafic body outcrops in a roughly elliptical pattern (Fig.2). The sulphide bodies consist essentially of a pentlandite/violarite—pyrrhotite/marcasite—chalcopyrite assemblage occurring within structurally controlled loci at or close to the basal contact of the ultramafic body. The contact zone of massive sulphides up to 1 m thick is overlain by a zone of disseminated sulphides of variable thickness which grades upwards into unmineralized ultramafic rocks. The outcropping bodies are oxidized to a depth of approximately 40 m below surface and a zone of supergene sulphides is developed between the oxidized and primary sulphides.

The Pioneer area is covered by essentially residual soils. Deep weathering of the rocks, to 25—100 m below surface, is ubiquitous, and the remnants or "roots" of a Tertiary laterite profile are locally preserved. In such an environment geochemical soil sampling provides a very effective, rapid and cheap method of defining anomalous Ni—Cu mineralization related to sulphides, and of delineating lithologic units.

Orientation surveys were carried out to determine the distribution of nickel, copper, cobalt, chromium and zinc, both laterally and vertically, within the soil profiles and weathered bedrocks characteristic of the Pioneer area. On the basis of these studies, detailed soil surveys were completed over most of the ultramafics of the Pioneer Dome, which have an aggregate strike-length of approximately 65 km (Fig.2). Initially 400 ft × 50 ft (122 m × 15.2 m) or 200 ft × 50 ft (61 m × 15.2 m) grids were used reducing to 200 ft × 25 ft (61 m × 7.6 m) and even 100 ft × 25 ft (30.5 m × 7.6 m) where initial encouraging results warranted further and more detailed follow-up. The results were computer-profiled and contoured and all anomalous situations, in terms of coincident nickel, copper, cobalt and Ni/Cr ratio values associated with ultramafic bodies, were followed up by further investigations involving costeaning and, where warranted, percussion and diamond drilling.

Three major nickel, copper, cobalt and Ni/Cr ratio soil anomalies were located on, or close to, the basal contact of the favourable ultramafic body and these coincide exactly with gossan exposed in costeans. Two of the anomalies are at the JH prospect, and one at the BB prospect, on the southeast flank of the Pioneer Dome (Fig.2). The detailed results from these two prospects are presented in this paper.

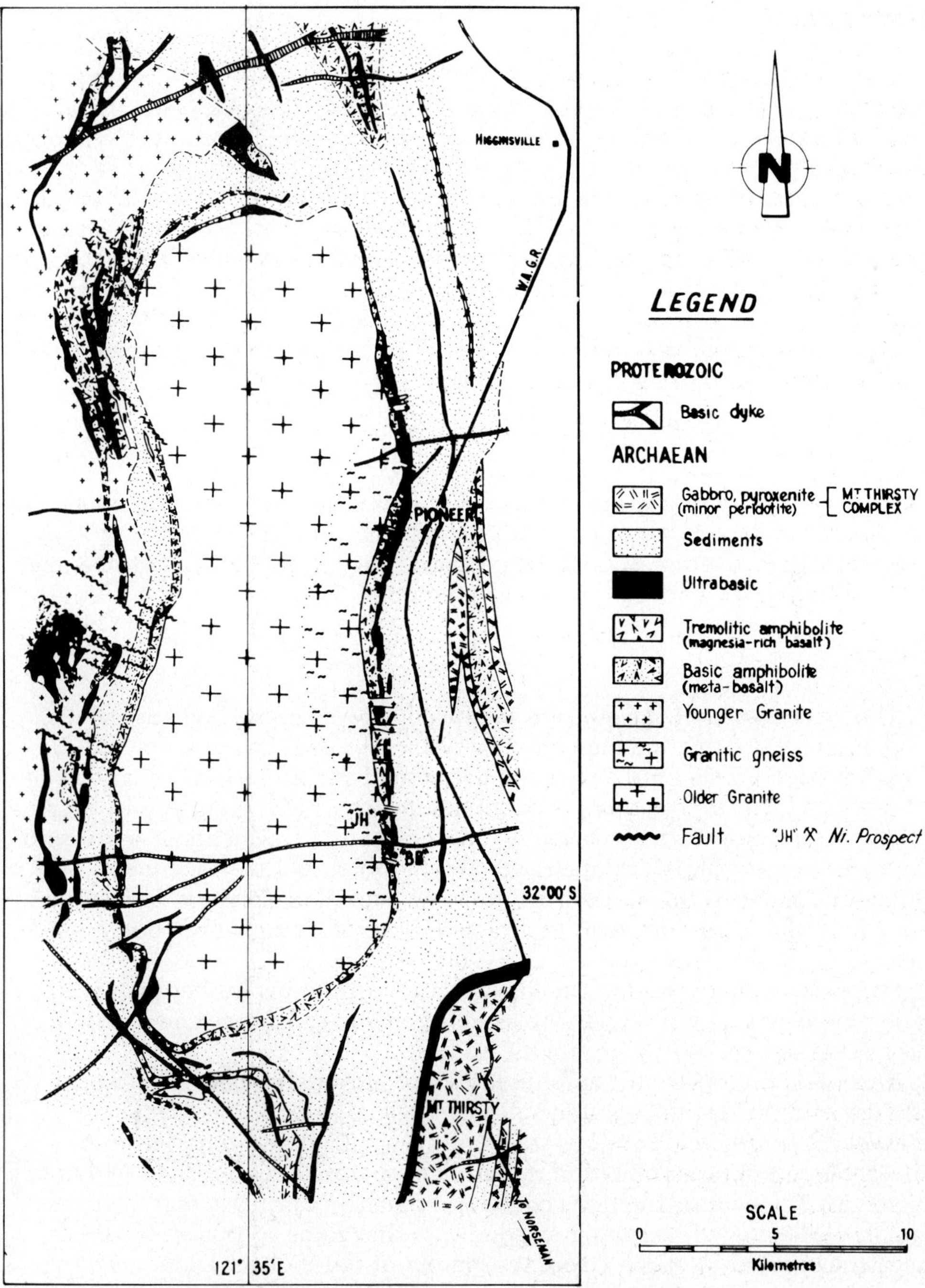

Fig.2. Pioneer Dome regional geology survey plan. (Reproduced with acknowledgement from Cox and Tyrwhitt, 1974.)

Physiography

The Pioneer area is approximately 300 m above sea level, on the strongly dissected arid interior plateau of Western Australia. The valleys are broad, alluvium filled and contain many large dry salt lakes; no perennial rivers occur in the area. Laterite remnants of the former sandy plateau are present, mainly as minor detritus with only occasional outcrops on slopes. At the JH and BB prospects the area is almost flat (1—1½° slope to the east) with shallow soil cover (10 cm to 1 m, averaging 20 cm) and almost no outcrop. Transportation of colluvium is limited in this area of flat topography, and weathering is essentially in situ. The topsoil, however, contains considerable "float" material of fresh and weathered bedrock which, due to its largely residual character, facilitates some form of geological mapping.

Climate

An arid climate prevails in the area. Rainfall is light and sporadic, averaging 27.66 cm per annum, and significant temperature changes occur both daily and seasonally. Average annual temperature range is 10°—24°C, but maxima and minima of 43° and - 2°C have been recorded.

Soils

The typical soils of the area are solonized brown soils (Stephens, 1962), which have a strongly alkaline reaction with pH values in the range 8.0—9.5. Characteristic profiles are developed over the ultramafic rocks, consisting of a brownish loam topsoil grading down at a depth of 20 cm or so into a lighter-coloured calcareous zone containing calcrete-coated fragments of weathered bedrock. The profiles developed over amphibolites and metasediments are generally similar to the above with the loam topsoil usually containing more abundant rock fragments, and the calcrete horizon being more poorly developed.

Soil colour often aided in the identification of the parent bedrock and, with experience, colour aerial photography was found to be a most useful tool in the delineation of ultramafic bodies.

The soils developed over amphibolites and ultramafics have a hard clay surface with dormant lichens and surface organic litter derived from *Melaleuca sheathiana* in various stages of decay. There is no herbaceous cover. Microscopic examination reveals little material coarser than ½ mm in diameter. The coarser fractions consist dominantly of quartz and feldspar grains; well-rounded, frosted grains indicate that some of this material has been transported by wind, (from the granite of the Pioneer Dome). The trace element values (Ni, Cu, Zn, Co, Cr) are lower in the coarser fractions due to the larger proportion of transported quartz and feldspar present. Hall (1971) records that the silt and clay fractions are composed almost entirely of montmorillonite.

Vegetation

Vegetation in the Pioneer prospect area is largely an open sclerophyllais woodland (Beard, 1969) dominated by *Eucalyptus salmonophloia, Eu. oleosa* and *Eu. dundasii* (12—15 m tall) with an under-storey of *Melaleuca sheathiana, Casuarina* Spp. *Santalum spicatum* and *Casia cremophila.* Spinifex is not uncommon but other shrubs and plants are virtually absent from the prospect area. However, saltbush (*Atriplex vesicaria*) and blue bush (*Cratystylis conocephala*) are locally present over saline or basic soils.

Casuarina Spp. is well developed on highly weathered granite and its abrupt change in density of growth from granite to amphibolite provides a clear indication of the contact between these two rock types.

Gossanous areas are densely populated by *Melaleuca sheathiana* and biogeochemical investigations by Hall (1971) have shown that this species growing over the gossans contains abnormal quantities of nickel compared to background plant analyses. Biogeochemical methods were not used as standard exploration techniques at Pioneer but an interesting review of the results obtained from a research programme has recently been published by Hall et al. (1973).

GEOLOGY

The Pioneer Dome (Fig.2 — Cox and Tyrwhitt, 1974; Gemuts and Theron, 1974) is the most southerly of several similar bodies in the Norseman-Kalgoorlie "Greenstone Belt" of the Yilgarn Block (Archean Shield). Nickel sulphide bodies have been discovered within this belt at Kambalda, Nepean, Spargoville, Mt. Edwards, Widgiemooltha, Redross, Wannaway, Pioneer, Tramways and St. Ives (Fig.1 and Table I). The Pioneer Dome is a complex of two sub-domes, having an hour-glass shape, with a central core of weakly foliated leucocratic granite (Older Granite) and quartz-feldspar biotite gneiss. These rocks, representing granitized basement, are surrounded by and overlain unconformably by a Lower and Upper Greenstone Group separated by quartizitic metasediments.

MINERALIZATION

Detailed exploration of the Pioneer Dome area has located several small bodies of Ni—Cu sulphides all of which lie at, or close to, the basal contact of the *main ultramafic body* within the Lower Greenstone Group. Many of these minor occurrences are localized within narrow shear zones of chlorite schist. These probably formed during late-stage metamorphism, with the nickel and copper sulphides moving into favourable structural traps from the adjacent ultramafics.

Only two significant bodies of Ni—Cu mineralization warranted delineation by pattern drilling, the JH and BB prospects, located on the southeastern flank of the Pioneer Dome (Figs.2 and 3).

444

JH Prospect. At the JH prospect mineralization occurs as two discrete lenses
at or near to the basal contact of the *main ultramafic body*. Both lenses
occur within a minor embayment of this basal contact. This contact strikes
0—030°, averaging 022° and dips 40°E, (Figs.3 and 4). The two lenses are separated by, and displaced on, a subvertical transverse fault, striking northwest,
along which quartz-feldspar pegmatite has been intruded.

The northern lens has a strike length of 115 m, is up to 6 m wide, and
lenses out at approximately 110 m below surface. The southern lens has a
strike length of 60 m, is up to 1.6 m wide, and also lenses out at approximately 110 m below surface.

BB Prospect. At the BB prospect, 1.5 km south along strike from JH prospect,
one lens of Ni—Cu mineralization occurs 15 m above the basal contact of the
main ultramafic body, which here strikes 030° and dips 30°E (Fig.3). This
lens has a strike length of 90 m, is up to 3 m wide and plunges 35° NNE. Due
to its small dimensions it has been drill tested to only 60 m below surface.

The *main ultramafic body* is quite lenticular, lensing out along strike
(Fig.3) and in depth (Fig.4). In addition, this body is much disrupted by
cross-cutting quartz-feldspar pegmatite dykes, and by sills particularly along
the basal contact.

At both the JH and BB prospects the mineralization is irregular, occurring
in the form of a series of discrete lenses and pods, and generally is predominantly in the form of massive veins and fracture fillings of marcasite (after
pyrrhotite) and violarite (after pentlandite). Finely disseminated pyrrhotite,
pentlandite and chalcopyrite are closely associated with the massive sulphides.
Drilling of these lenticular, pod-like bodies has indicated average grades of
1.1% Ni and 0.1% Cu.

During weathering the sulphides oxidize to limonitic gossanous material
which extends from surface to 40 m below surface. The only surface expression of the Ni—Cu mineralization at the JH and BB prospects is the occurrence of scattered small fragments of gossan float which assay up to 1.17%
Ni and 0.68% Cu. Costeaning and drilling have shown that the gossans retain
essentially the geometrical form of the original sulphide body, although a
considerable lateral spread of ferruginous silcrete associated with Tertiary
lateritization, particularly over the ultramafic, has carried with it Ni—Cu
values originating from the weathering sulphide zone producing a mushroom-
style dispersion pattern.

The Pioneer gossans are leached of nickel and show copper values slightly
higher than the underlying massive sulphides. Most gossans contain nickel and
copper in the range of 2000—5000 ppm Ni/500—5000 ppm Cu, with a few
over 1% Ni/0.6% Cu. These values compare with sulphide mineralization in the
range 0.5% Ni/0.02% Cu (disseminated sulphides) up to 6% Ni/0.4% Cu (massive sulphides). The distribution and weighted average assay values (from horizontal groove samples) of gossans exposed in shallow costeans are shown in
the accompanying gossan location plan (Fig.5). These gossans are lateritized

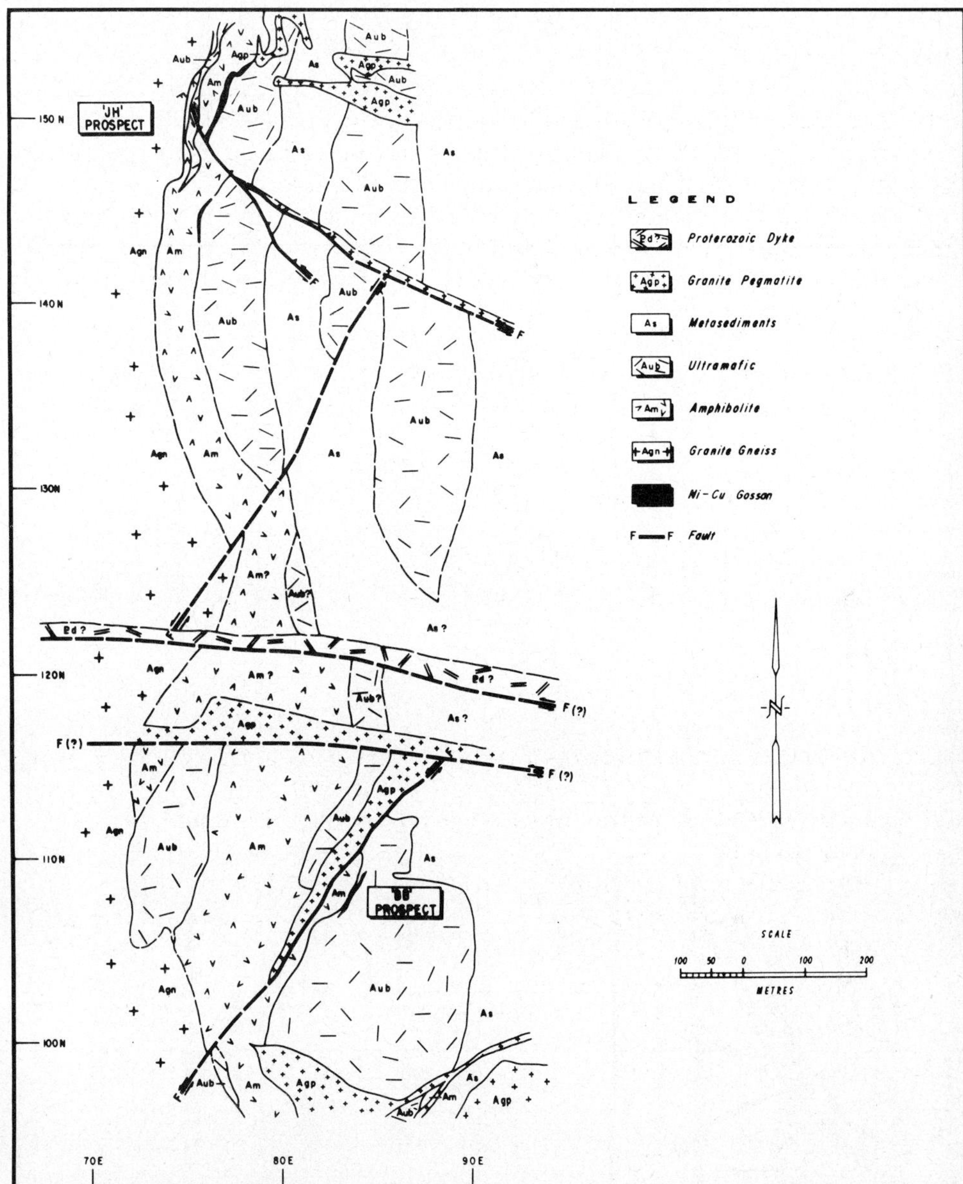

Fig.3. Geology plan, JH-BB prospect area.

and silicified, with chromium values generally in excess of 500 ppm (Fig.6). Recent work by Cochrane (1973) on the geochemistry of nickeliferous gossans has included samples from the Pioneer JH and BB prospects. He notes that most gossans from the nickel fields of the Yilgarn Block are partially lateritized or silicified, and that the chromium level can possibly be correlated

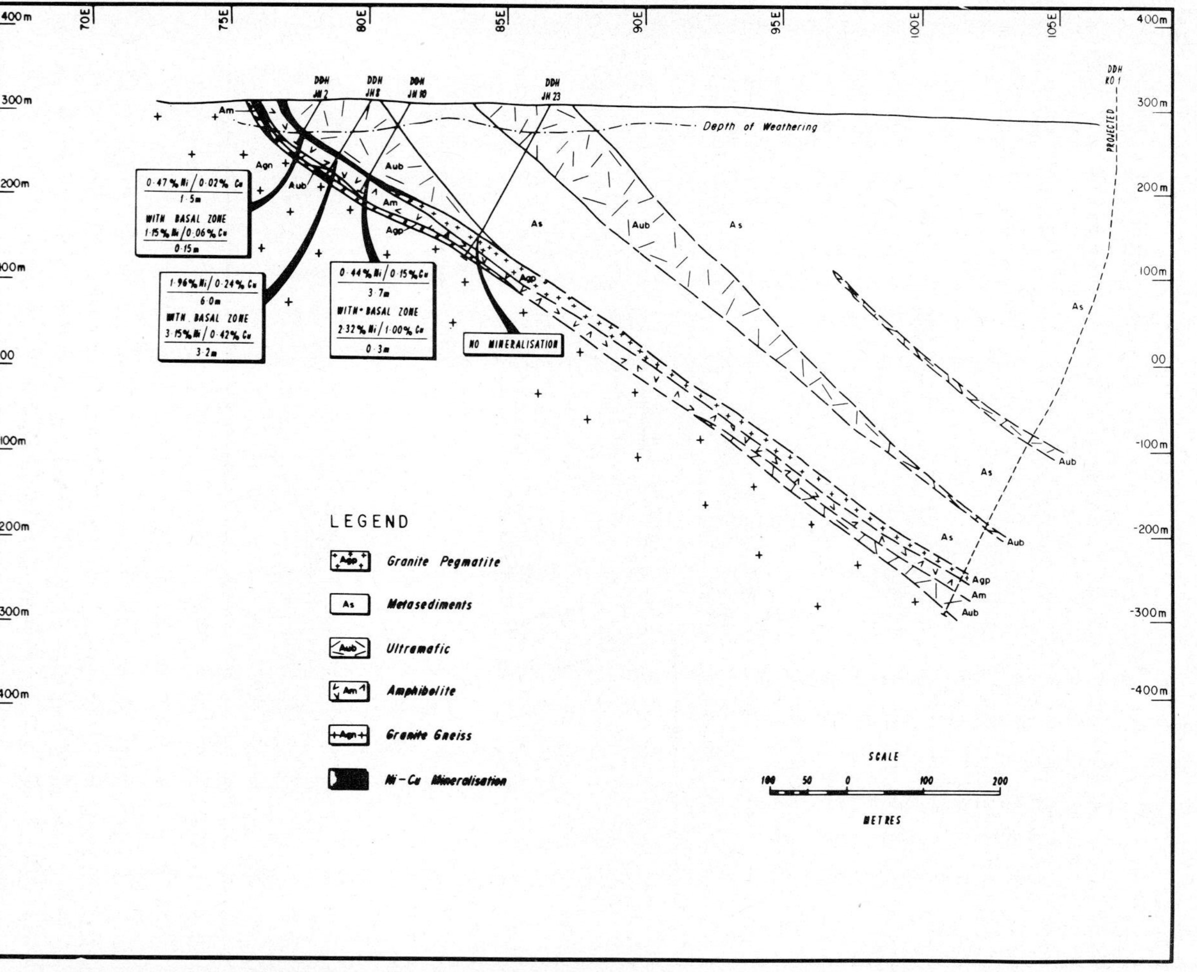

Fig.4. Geology cross-section 150.5N, JH prospect, looking north.

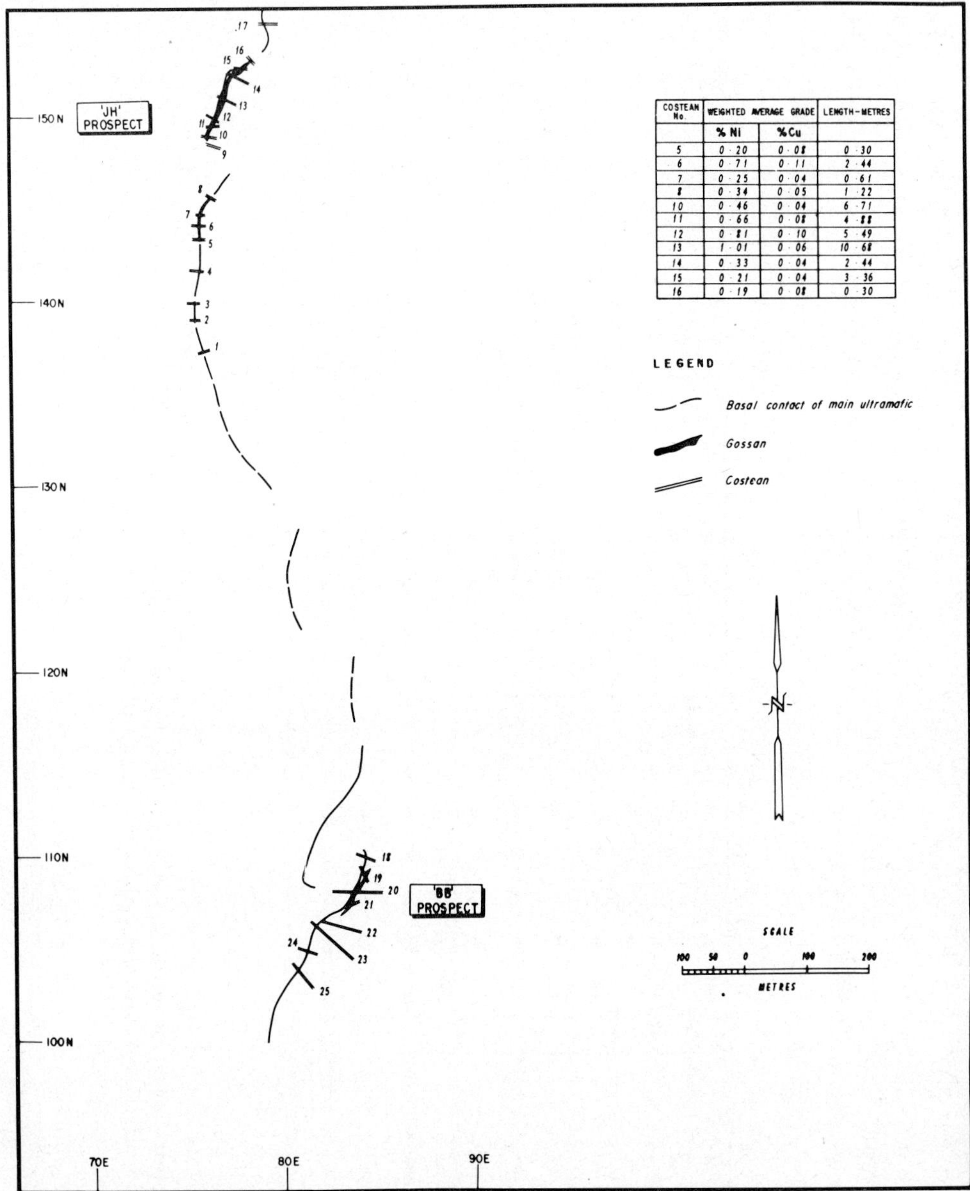

COSTEAN No.	WEIGHTED AVERAGE GRADE		LENGTH—METRES
	% Ni	% Cu	
5	0·20	0·08	0·30
6	0·71	0·11	2·44
7	0·25	0·04	0·61
8	0·34	0·05	1·22
10	0·46	0·04	6·71
11	0·66	0·08	4·88
12	0·81	0·10	5·49
13	1·01	0·06	10·68
14	0·33	0·04	2·44
15	0·21	0·04	3·36
16	0·19	0·08	0·30

Fig.5. Gossan location plan, JH-BB prospect area, showing weighted average grades from horizontal groove samples cut through gossan.

directly with the degree of lateritization. His investigations have shown that chromium levels greater than 500 ppm are typical in lateritized material in this environment.

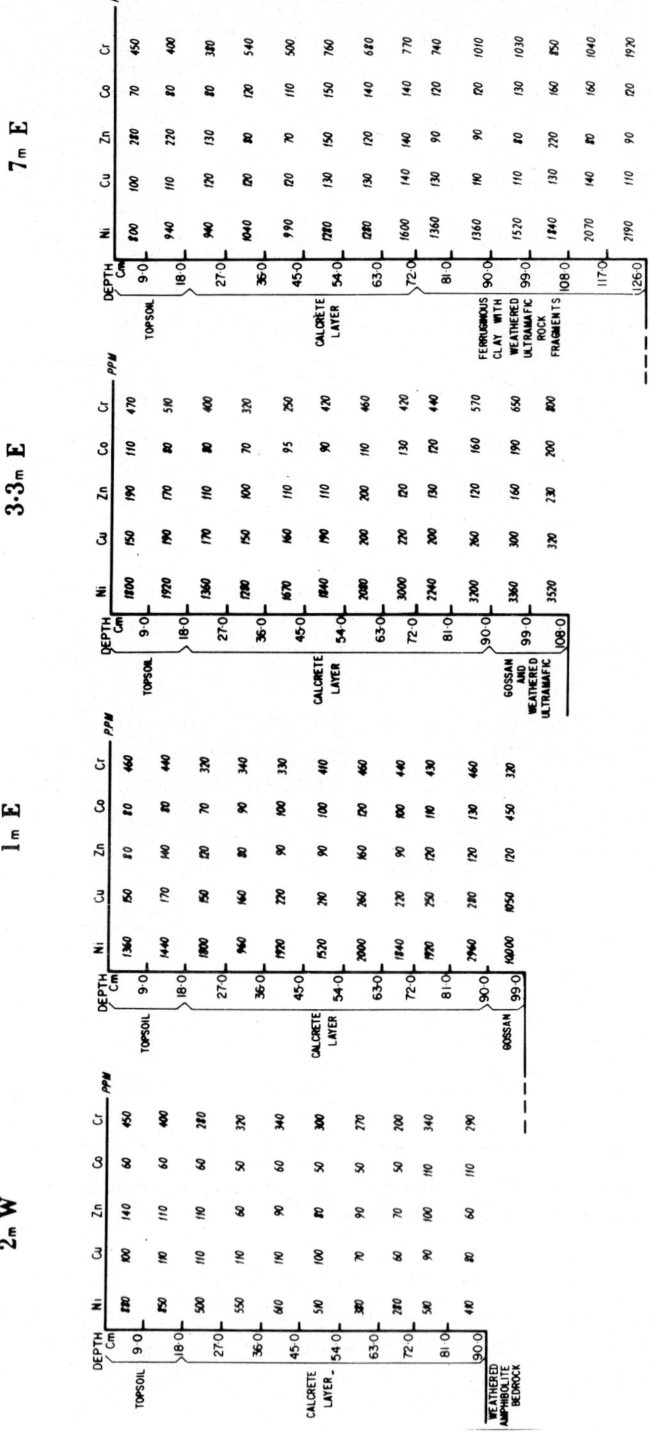

Fig.6. Geochemical orientation survey, JH prospect, showing vertical groove sample results from north wall of costean No.10 at 2 m west and 1, 3.3 and 7 m east of basal contact of main ultramafic body.

GEOCHEMICAL PROGRAMME

Orientation surveys

Although deep weathering of the rocks, to 25—100 m below surface, is ubiquitous and outcrop is almost totally absent, the Pioneer area is covered by essentially residual soils of shallow depth (10 cm to 1 m), averaging approximately 20 cm. In order to assess the applicability of geochemical soil sampling in the exploration for Ni—Cu mineralization associated with ultramafic rocks in this environment, geochemical orientation surveys were undertaken to analyse qualitatively and quantitatively the distribution of metal values (Ni, Cu, Co, Zn, Cr), both laterally and vertically, within the soil profiles and weathered bedrocks characteristic of the Pioneer area.

Costeans were dug with a backhoe, up to 3 m depth, in order to map and sample the soil/weathered bedrock profiles. The data obtained from a costean (No.10, Fig.5) dug across a mineralized (gossanous) section of ultramafic/amphibolite contact at the JH prospect serves to illustrate the techniques employed and results obtained.

Horizontal and vertical groove samples were taken from one wall (north wall) of each costean for orientation purposes. Fig.6 presents the results obtained from vertical groove samples taken on the north wall of costean No.10 at selected distances from the amphibolite/ultramafic contact (2 m west in amphibolite; 1 m east, 3.3 m east and 7 m east in ultramafic), from surface down to weathered bedrock. Figs.7 and 8 present the results, for nickel and copper respectively, obtained from horizontal groove samples taken on the north wall of costean No.10 at three selected horizons: (1) weathered bedrock, (2) calcrete, at 10 cm below base of topsoil, and (3) topsoil.

Bedrock

From analyses of the percussion chips from exploratory drillholes at the JH prospect, background metal values in weathered ultramafics (5—40 m below surface approximately) are as follows: 50—1400 ppm Ni, 25—200 ppm Cu, 10—120 ppm Co, 10—150 ppm Zn, and 100—1500 ppm Cr. The horizontal groove samples of weathered bedrock from costean No.10, taken at a depth of approximately 1 m below surface show:

(1) Nickel values of less than 200 ppm Ni in amphibolite at 6 m west of contact, increasing uniformly to 600 ppm Ni at 1.3 m west of contact, then rising dramatically to a maximum of 12,000 ppm Ni at 2.3 m east of contact (i.e., in ultramafic), and falling irregularly to 1500—2000 ppm Ni between 7 and 17 m east (Fig.7).

(2) Copper values of 50—100 ppm Cu at 6 m west to 1.3 m west of contact in amphibolite, rising dramatically to a maximum of +1200 ppm Cu at 1.3 m east of contact (i.e., in ultramafic), and falling irregularly to 150—200 ppm Cu between 5 and 17 m east (Fig.8).

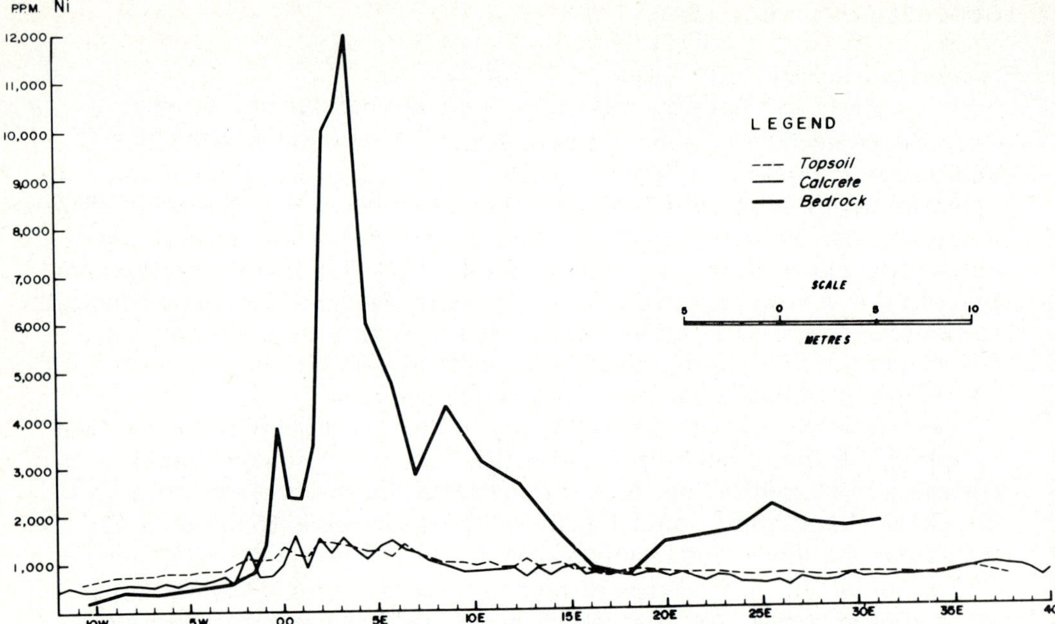

Fig.7. Geochemical orientation survey, JH prospect, showing nickel content of topsoil, calcrete and weathered bedrock from horizontal groove samples taken on north wall of costean No.10.

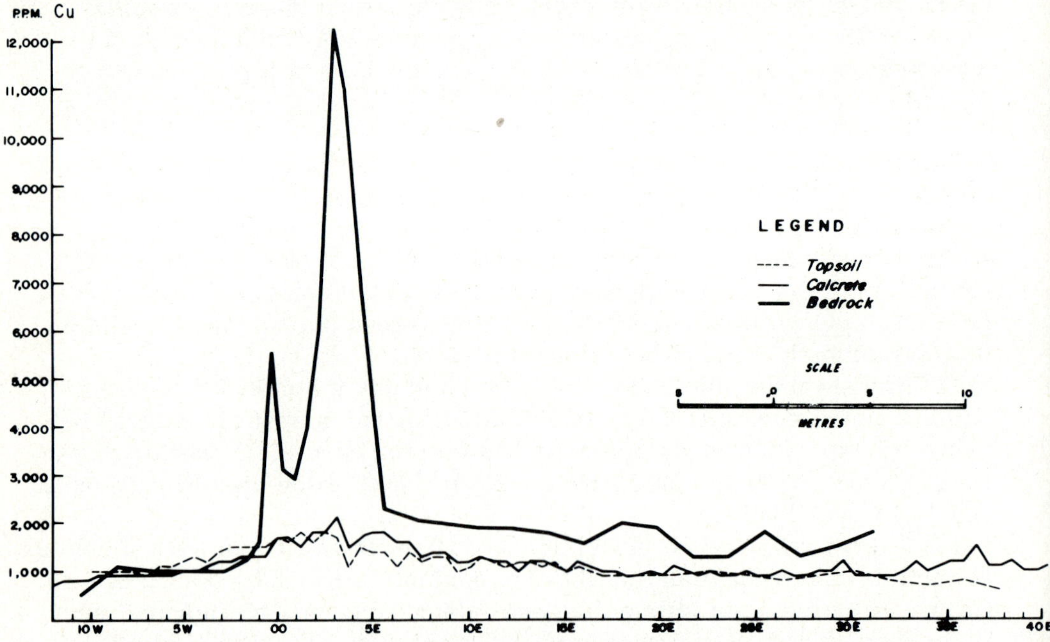

Fig.8. Geochemical orientation survey, JH prospect, showing copper content of topsoil, calcrete and weathered bedrock from horizontal groove samples taken on north wall of costean No.10.

(3) Cobalt content rises in sympathy with copper, from 110 ppm Co in amphibolite at 2 m west of contact to a maximum of 450 ppm Co at 1 m east (i.e., in ultramafic), falling to 120 ppm Co at 7 m east (Fig.6).

(4) Similarly zinc rises from 80 ppm Zn in amphibolite at 2 m west of contact to a maximum of 230 ppm Zn at 3.3 m east (i.e., in ultramafic), falling to 90 ppm Zn at 7 m east (Fig.6).

(5) The Ni/Cr ratio rises from 1.41 in amphibolite at 2 m west of contact to 31.25 at 1 m east (in ultramafic), dropping to 4.40 at 3.3 m east and 1.14 at 7 m east (Fig.6). (Note: the chromium value is *acid-soluble* chromium, not total chromium.)

Calcrete horizon

A calcrete horizon 0.5—1.5 m thick, is developed over both amphibolite and ultramafic bedrock, and is overlain by the topsoil. The calcrete horizon, which has a white, bleached appearance, consists of fragments of weathered bedrock, (with a *skin* of calcrete) in a groundmass of soft white calcareous clay (montmorillonite). Over amphibolite bedrock the fragments are commonly of relatively fresh rock, whereas over ultramafics the fragments are usually deeply weathered, often ferruginous clays.

Vertical groove samples (Fig.6) enable the following observations to be made: (1) nickel, copper and cobalt values decrease upwards over both amphibolite and ultramafic bedrock, whilst zinc values show a generally erratic vertical distribution pattern; (2) acid-soluble chromium values show no obvious vertical distribution pattern over amphibolite, whereas over ultra-mafic they show an irregular decrease upwards.

Topsoil

The shallow residual topsoils of the Pioneer area reflect the flat topographic relief (gentle over-all slope of 1- 1½° eastwards) and extremely arid climate of the interior plateau. The top few centimetres contain a large proportion of fine wind-blown and sheet-wash material (quartz and feldspar grains from the granite of the Pioneer Dome) which decrease in quantity with depth. Top-soils are masked only by regional drainage channels (which have little topo-graphic relief in this mature peneplained plateau but which stand out extremely well on colour air photos), which contain saline gravels and alluvium below a thin surface cover of wind-blown and sheet-wash material.

In the initial orientation survey the distribution of the various elements in different size fractions of soil was investigated and the minus 80-mesh fraction found to give the best degree of correlation between metal values in soil and weathered bedrock. The minus 80-mesh results from costean No.10, shown in Figs.6, 7 and 8, may be summarized as follows:

(1) Nickel values of 600 ppm Ni at 5.2 m west of contact (i.e., over amphibolite) increase to 850—880 ppm Ni at 2 m west of contact, then rise rapidly to 1360—1400 ppm Ni at 1 m east (i.e., over ultramafic), reaching a maximum of 1800—1920 ppm Ni at 3.3 m east and decreasing to 800—940

ppm Ni at 7 m east and 600—700 ppm Ni between 11.6 and 18.3 m east.

(2) Copper values of 100—110 ppm Cu between 5.2 and 2 m west of contact (i.e., over amphibolite) increase to a maximum of 150—190 ppm Cu at 3.3 m east (i.e., over ultramafic) and decrease to 100—110 ppm Cu at 7 m east and 60 ppm at 18.3 m east.

(3) Cobalt values increase in sympathy with copper values, from 60 ppm Co at 2 m west of contact (over amphibolite) to reach a maximum of 80—110 ppm Co at 3.3 m east (over ultramafic), decreasing to 70—80 ppm Co at 7 m east.

(4) Zinc values increase systematically from 110—140 ppm Zn over amphibolite at 2 m west of contact to 220—280 ppm Zn at 7 m east over ultramafic.

(5) The Ni/Cr ratio increases from 1.95 at 2 m west (over amphibolite) to a maximum of 3.83 at 3.3 m east (over ultramafic) decreasing to 1.77 at 7 m east.

The close association of the sympathetic nickel, copper, cobalt and zinc soil anomalies over the gossan is taken as evidence of essentially residual soils. The concentration of nickel, copper, cobalt, zinc and chromium in the finer (silt and clay) soil fractions indicates that these anomalies developed during an earlier period of chemical weathering. Hall (1971) subsequently used electron probe and chemical extraction techniques to show that silts, clays and Fe—Mn oxides associated with the fine fractions contain the most significant proportion of trace elements in the soil. It appears that manganese is able, because of its slight mobility, to separate from the iron oxide in the bedrock and incorporate trace elements, mainly nickel and cobalt, whereas copper and zinc are controlled largely by the silicate lattice of montmorillonite.

These results at Pioneer are in contrast to those obtained at Kambalda (Mazzuccheli, 1972) where nickel is preferentially concentrated in the coarse fraction of the soils associated with sulphide mineralization, with a corresponding depletion in the minus 80-mesh fraction. (Both nickel and copper nevertheless display strong anomalies in the minus 80-mesh fraction of soils in the immediate vicinity of sub-outcropping mineralization at Kambalda.)

In summary, from the initial orientation survey results it was apparent that sub-outcropping Ni—Cu sulphide mineralization of the type closely associated with the basal contact of ultramafic bodies had a marked surface geochemical expression. In particular the nickel, copper and cobalt values show strong and sympathetic vertical and lateral dispersion patterns. Zinc values generally show a more irregular and diffused dispersion pattern. It was noted in particular that maximum values for nickel, copper, cobalt, zinc and Ni/Cr ratio in residual topsoil are generally coincident with maximum values for these metals in weathered bedrock. The metal values in the topsoil show a greater lateral spread (mushroom-style dispersion pattern) and the maxima are generally offset 2—4 m east of the weathered bedrock maximum values, reflecting a gentle eastward-sloping topographic gradient.

Acid-soluble chromium values show a strong vertical distribution pattern above weathered ultramafic at 7 m east (Fig.6) and irregular patterns above weathered gossanous ultramafic at 3.3 and 1 m east, and also above weathered amphibolite at 2 m west of contact. In the weathered bedrock and calcrete horizon, chromium values decrease fairly systematically across the contact from east (ultramafic) to west (amphibolite). In the topsoil, chromium values show a very weak lateral migration.

The broadest dispersion pattern (laterally) for nickel, copper and cobalt values occurs in topsoil. The calcrete horizon contains higher metal values but these show less lateral dispersion than in the topsoil; i.e., the anomalies are *sharper*. Similarly, the metal values in the weathered bedrock are considerably higher than in the soil profile above but these values show a minimum lateral dispersion.

As over 90% of the strike length of ultramafic bodies around the Pioneer Dome is covered by shallow residual soils, systematic sampling of the topsoil is likely to be the cheapest and most rapid method of searching these bodies for associated Ni—Cu sulphide mineralization. In order to minimize the dilution caused by wind-blown sand and sheet-wash in the top few centimetres of topsoil, it would be preferable to sample the base of the topsoil horizon at 10—20 cm below surface.

Analytical procedures

Soil samples

After the initial size analysis tests were completed, procedures were standardized so that all soils were sieved to minus 80 mesh through a nylon screen. Base-metal analyses were carried out by digesting 0.5 g of sample with mixed perchloric acid/aqua regia acid to dryness in a test tube. The residue was taken up with concentrated hydrochloric acid and diluted to a volume mark on the test tube using distilled water. The solutions were read on an atomic absorption spectrophotometer using matched matrix standards. Precision was ± 10% at the 200-ppm level. The chromium values are of acid-soluble chromium using an AAS air-acetylene flame.

Costean samples

These samples (from the calcrete horizon and weathered bedrock) were dried at 120°C, rolled and cut to produce a 50—100-g sample for analysis. This sample was then pulverized in a Braun pulverizer. Base-metal analyses were carried out on 1.000 g by a perchloric acid/aqua regia leach to dryness in a 150-ml beaker. The residue was taken up with concentrated hydrochloric acid and made up to volume in a measuring cylinder with distilled water. The solutions were read as for soils by AAS. Results were quoted to ± 5% precision at the 200-ppm level.

454

Systematic soil surveys

Following the discovery of gossan float at the JH and BB prospects and the completion of orientation surveys as described above, a systematic soil survey was undertaken covering the two prospects on a 200 ft × 25 ft (approx. 61 m × 7.6 m) grid, reducing to 100 ft × 25 ft (30.5 m × 7.6 m) where warranted by geological complexities. Samples were taken from the base of the topsoil, in order to obtain a good spread of nickel, copper and cobalt values related to mineralization, and a better distribution pattern of chromium (and hence Ni/Cr ratio) than could be obtained from sampling the top of the topsoil. On this last point it should be noted that, due to the almost total lack of outcrop, geochemical sampling and systematic ground magnetic traversing were particularly valuable tools for confirming the geology of the prospect area as determined by float mapping, (viz. Figs.3 and 9—12 inclusive).

The results of the systematic soil survey over the JH and BB prospect area are of particular interest.

From a comparison of the contour plan of soil sample nickel values (Fig.9) with the surface geology float plan (Fig.3) it can be seen that the boundaries of the major lithologic units (ie., ultramafics, amphibolites, metasediments, granite gneiss and pegmatite) can be accurately delineated on the basis of the nickel content of the soil.

Over the *main ultramafic body* the highest nickel values are associated with the basal (western) contact, reflecting an overall change in composition of the layered body (as determined from drilling, D.M. Hausen, private communication, 1970; Franzen, 1971) from a dunitic base to a pyroxenite top. Superimposed on this overall trend are three major nickel anomalies associated closely with the basal contact at 109N 84.10E (BB prospect) and at 144N 76E and between 149N 76.25E and 151N 77.25E (JH prospect). These three soil anomalies coincide *exactly* with gossans exposed in the costeans (Fig.5).

The contour plan of soil copper values (Fig.10) indicates three major copper anomalies which coincide *exactly* with the three nickel soil anomalies/gossans described above. Copper values are of little practical value, however, in delineating lithologic boundaries. Cobalt behaves in a manner similar to copper.

The purpose of determining the Ni/Cr ratio for soil samples is to define anomalies associated with: (1) the contact zones of ultramafic bodies, and/or (2) with nickel anomalies related to suspected ultramafic bodies, e.g., in areas of deep soil cover.

In such areas Ni/Cr ratios greater than 1.0—1.5 are related to nickel sulphide mineralization. In unmineralized areas the Ni/Cr ratio is less than 1.0. Three major Ni/Cr ratio anomalies (Fig.11) satisfying the above conditions coincide *exactly* with the nickel, copper and cobalt soil anomalies described above.

CONCLUSIONS

The Ni—Cu sulphide mineralization at Pioneer gives rise to well-defined

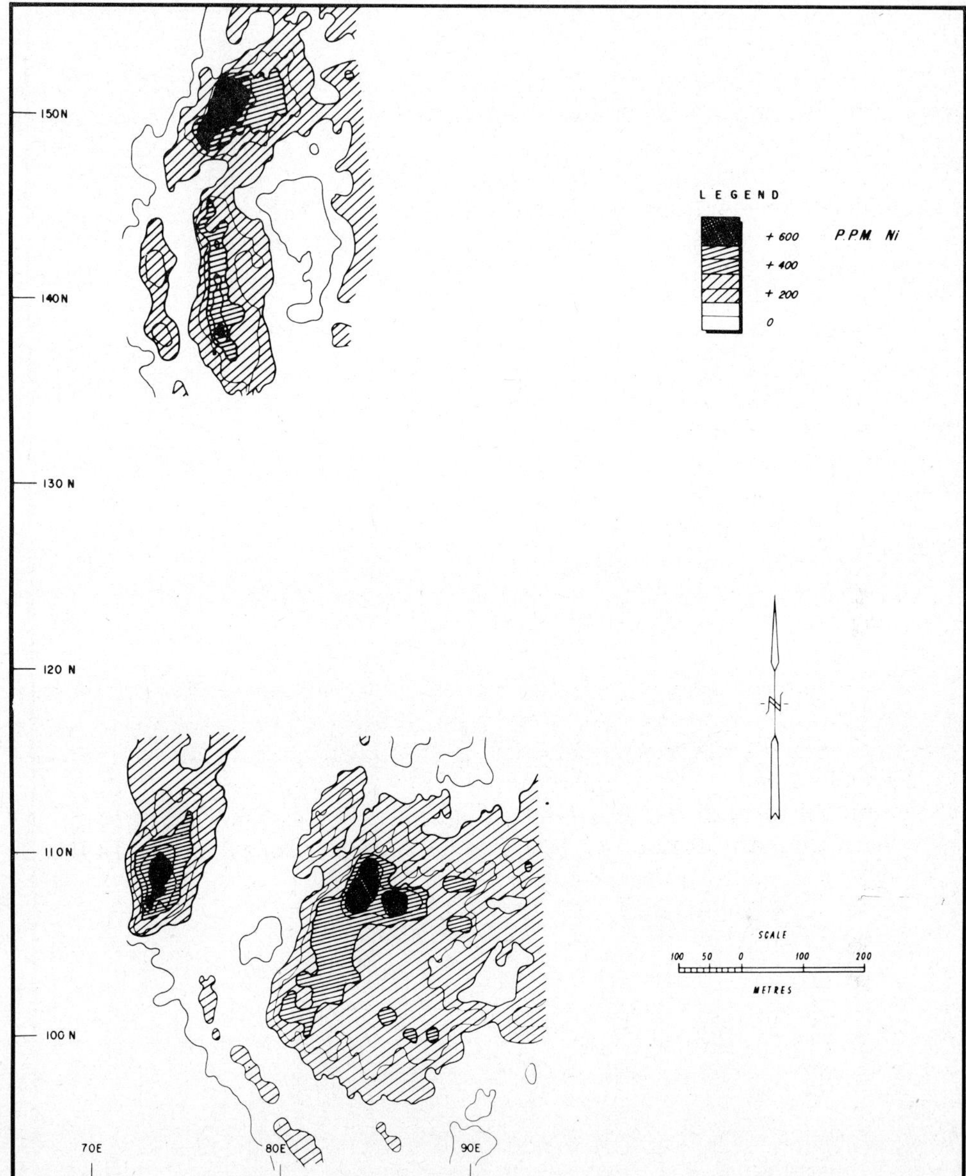

Fig.9. Geochemical soil survey, JH-BB prospect area, nickel contour plan (minus 80-mesh fraction).

nickel, copper, cobalt and Ni/Cr ratio anomalies in the calcrete and residual soil overlying the ultramafic and amphibolitic bedrock. Soil sampling, therefore, gives reliable indications of sub-outcropping mineralization and, although not responsible for the initial discovery, played a very significant part in the subsequent exploration and evaluation of the whole Pioneer Dome area.

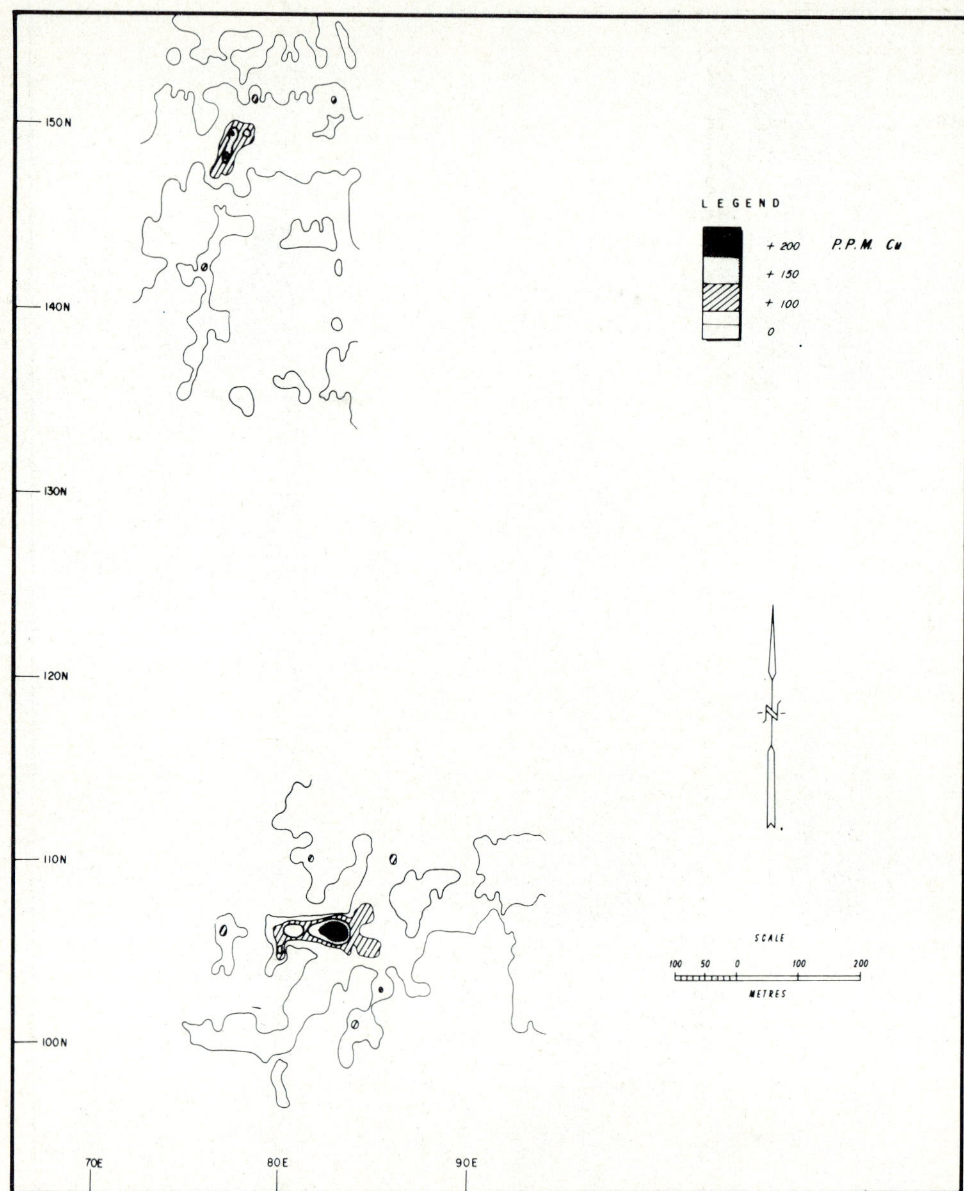

Fig.10. Geochemical soil survey, JH-BB prospect area, copper contour plan (minus 80-mesh fraction).

Geochemical orientation surveys at Pioneer showed:

(1) The development of a calcrete horizon above weathered ultramafics and amphibolites, this horizon being overlain by soils which are essentially residual in character.

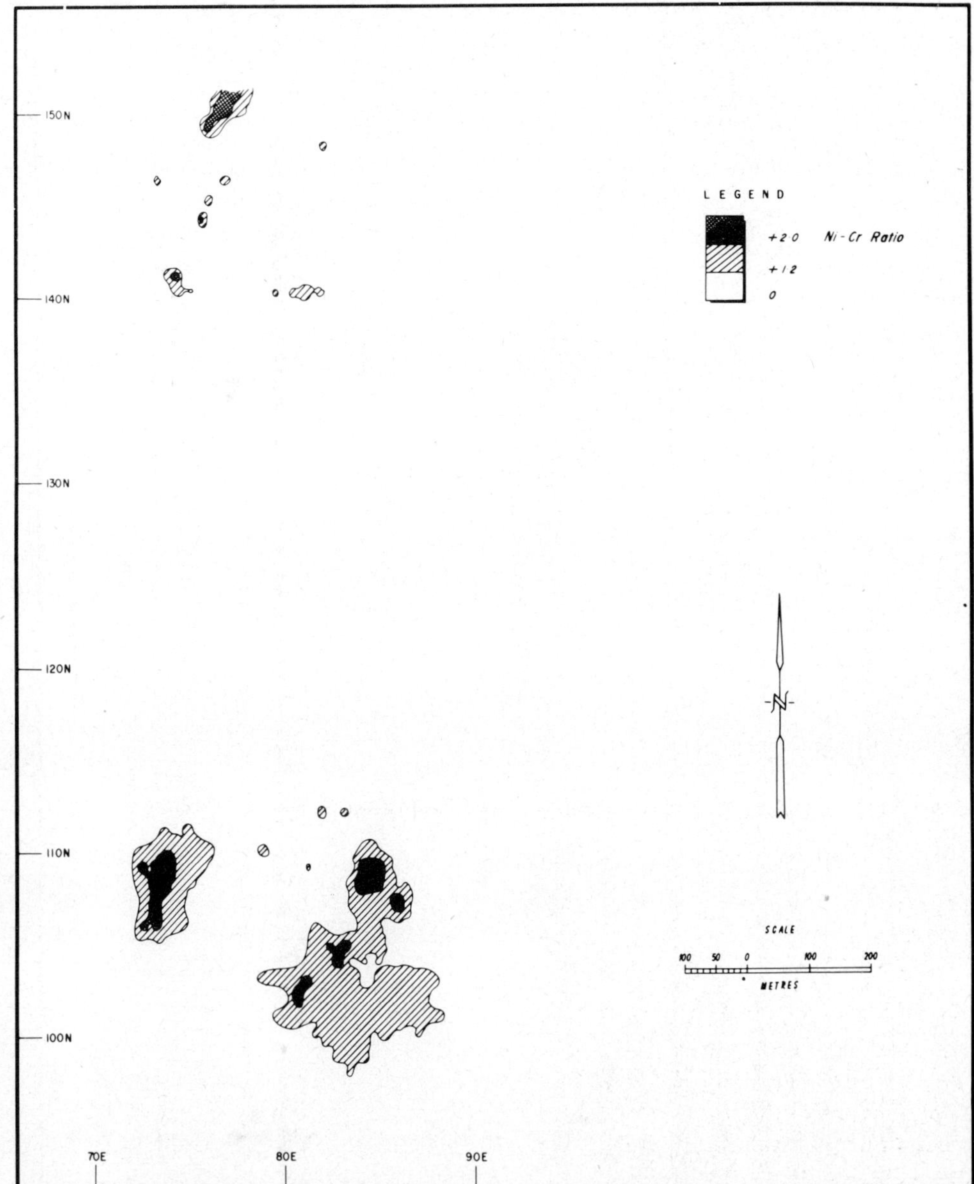

Fig.11. Geochemical soil survey, JH-BB prospect area, Ni/Cr ratio contour plan (minus 80-mesh fraction).

(2) The development of gossans (which were exposed by costeaning) above Ni—Cu sulphide mineralization.

(3) The development of strong vertical and lateral dispersion patterns (mushroom-style) for nickel, copper, cobalt, chromium and Ni/Cr ratio values

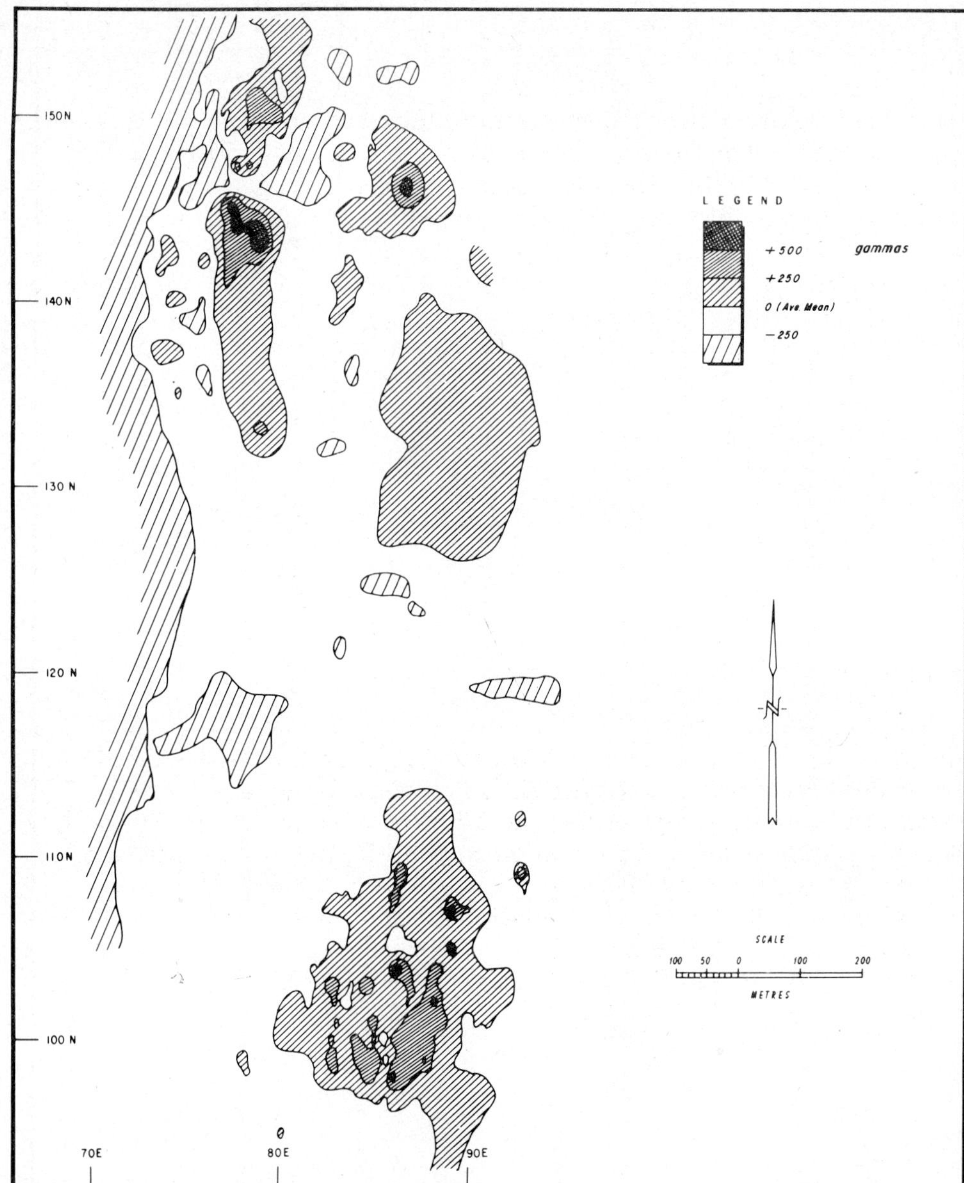

Fig.12. Ground magnetic survey, JH-BB prospect area.

in the calcrete horizon and soil above weathered ultramafic and amphibolite
bedrocks associated with contact-type mineralization. Zinc values are generally too irregular to be of direct value in locating a sub-outcropping mineralized body.

(4) Sympathetic anomalies of nickel, copper and cobalt values are present
in all size fractions of the residual soils and coincide *exactly* with the gossans

exposed by costeaning. The coarser soil fractions generally show a dilution of these metal values due to the presence of particles of wind-blown sand and some sheet-wash material.

Geochemical orientation surveys were followed up by a detailed soil sampling programme. The three major nickel, copper, cobalt and Ni/Cr ratio anomalies present in the JH-BB prospect area coincide *exactly* with gossans exposed by costeaning. Away from mineralization the distribution of nickel and chromium in the calcrete horizon and soil accurately reflects the distribution in bedrock of these elements and facilitates the delineation of lithologic units, particularly the boundaries of ultramafic bodies. The trend of amphibolites, metasediments and cross-cutting granite pegmatite dykes can similarly be traced on the basis of lower nickel and chromium values.

Experience with a whole range of geological, geochemical and geophysical exploration techniques in the Pioneer area has shown that gossan searching and geochemical soil sampling are the most effective, rapid and economical techniques for positively locating sub-outcropping Ni—Cu mineralization in this environment. A full understanding of the limitations, as well as the advantages, of the techniques is required before the results are applied to another area, however close that area may be geographically.

ACKNOWLEDGEMENTS

This paper briefly presents some of the geochemical results obtained at Pioneer during an extensive and integrated exploration program carried out by a team of geologists under the writer's supervision on behalf of the Newmont Pty Limited/Norseman Gold Mines N.L. joint venture. The author gratefully acknowledges the permission to publish given by these two companies. Acknowledgement is made of the teamwork of Newmont colleagues D.S. Tyrwhitt, M.G. Mason, R.G. Adamson, M.A. Ward, J. Hastie, G.O. Dickson, D.M. Hausen and J. Pasovsky, and of Norseman Gold Mines N.L. consultant W.N. Thomas. Geochemical analyses were carried out by the Norseman Gold Mines N.L. laboratory under the supervision of Messrs J. Dowson and V. Garnys.

REFERENCES

Beard, J.S., 1969. The vegetation of the Boorabbin and Lake Johnston areas, Western Australia. Proc. Linnol. Soc. N.S.W., 93: 239—269
Cox, R. and Tyrwhitt, D.S., 1974. Pioneer nickel prospects, near Norseman. In: C.L. Knight (Editor), Economic Geology of Australia and Papua New Guinea. Metals Volume, Australasian Institute of Mining and Metallurgy, Melbourne, Vic. (in press)
Cochrane, R.H.A., 1973. A quide to the geochemistry of nickeliferous gossans and related rocks from the Eastern Goldfields. In: Report of the Department of Mines, Western Australia, for the year ending December 1972. Div. IV, pp.115-121

Franzen, N.W., 1971. An investigation into the petrology of the ultramafic and related rocks at Pioneer, Western Australia. B. Sc. (Honours) Thesis, Dep. of Geology, Univ. of Western Australia, Perth, W.A. (unpublished)

Gemuts, I. and Theron A., 1974. Stratigraphy and Mineralization of the Archaean between Coolgardie and Norseman. In: C.L. Knight (Editor), Economic Geology of Australia and Papua New Guinea. Metals Volume, Australasian Institute of Mining and Metallurgy, Melbourne, Vic. (in press)

Hall, J.S., 1971. A comparative biogeochemical investigation in the semi arid/arid zone of West Australia. B.Sc. (Honours) Thesis, Dep. of Economic Geology, Univ. of Adelaide, Adelaide, S.A. (unpublished)

Hall, J.S., Both, R.A. and Smith, F.A., 1973. A comparative study of rock, soil and plant chemistry in relation to nickel mineralization in the Pioneer area, Western Australia. Proc. Aus. Inst. Min. Metall., 247: 11—22

Horwitz, R.C. and Soufoulis, J., 1965. Igneous activity and sedimentation in the Precambrian between Kalgoorlie and Norseman, Western Australia. Proc. Aus. Inst. Min. Metall., 214: 45—49

Mazzucchelli, R.H., 1972. Secondary geochemical dispersion patterns associated with the nickel sulphide deposits at Kambalda, Western Australia. J. Geochem. Explor., 1: 103—116

Soufoulis, J., 1966. Widgiemooltha, Western Australia, Explanatory Notes for 1:250,000 Geological Series sheet SH/51-14. Geol. Survey, W.A.

Stephens, C.G., 1962. A Manual of Australian Soils. C.S.I.R.O., Melbourne, Vic. 61 pp.

Woodall, R. and Travis G.A., 1969. The Kambalda nickel deposits, Western Australia. Proc. 9th Comm. Min. Metall. Congr. (London), pp.517—533

GEOCHEMISTRY IN THE EXPLORATION OF NICKELIFEROUS LATERITE

PABLITO M. ONG and ARTURO C. SEVILLANO

A. Soriano Corporation, Makati, Rizal (Philippines)

ABSTRACT

Applied geochemistry was used in prospecting and in the preliminary exploration of four nickeliferous laterite deposits in Palawan, Philippines.

Near-surface soils were collected from a grid with a 300-m square sampling pattern. Based on the evaluation of geochemical results from the soil samples, each laterite prospect was subdivided into "geochemical areas", namely: (a) highly anomalous, (b) anomalous, (c) slightly anomalous, and (d) background areas.

At Long Point, Palawan, the first laterite prospect, the geochemical areas were probed by test pitting. The majority of test pits in the slightly anomalous to highly anomalous areas, particularly in the latter, penetrated possible nickel ore (>0.9% Ni) in laterite and/ or in the subjacent decomposed ultramafic rocks. By contrast, most of the test pits sunk to check the background areas encountered nickel deposits which are below the cut-off grade.

Using the resulting geochemical anomaly maps as guides in the exploration of other laterite prospects, test pitting was largely confined to the slightly to highly anomalous portions of the nickeliferous laterite prospects. Sinking of unnecessary test pits in areas containing low-grade nickel was thus avoided resulting in substantial savings in exploration costs.

INTRODUCTION

It is well known in the mining industry that applied geochemical techniques have been extensively employed in prospecting and in the preliminary exploration for copper, zinc, lead and other common base-metal deposits. This paper presents the first attempt to employ geochemical techniques in prospecting and in the initial exploration of nickeliferous laterite deposits in the Philippines.

In September, 1967, the Exploration Division of A. Soriano Corporation discovered nickeliferous laterite deposits at Long Point (Fig.1) and nearby areas in Palawan. Together with geological mapping, geochemical techniques were employed in preliminary appraisal of the laterite deposit at Long Point with successful results. The same geochemical techniques were subsequently employed as an aid in the search for other economically promising laterite deposits in Palawan. They were effective in delineating the Berong and Moorsom prospects in Palawan.

The geochemical survey was carried out primarily to: (1) delimit economi-

462

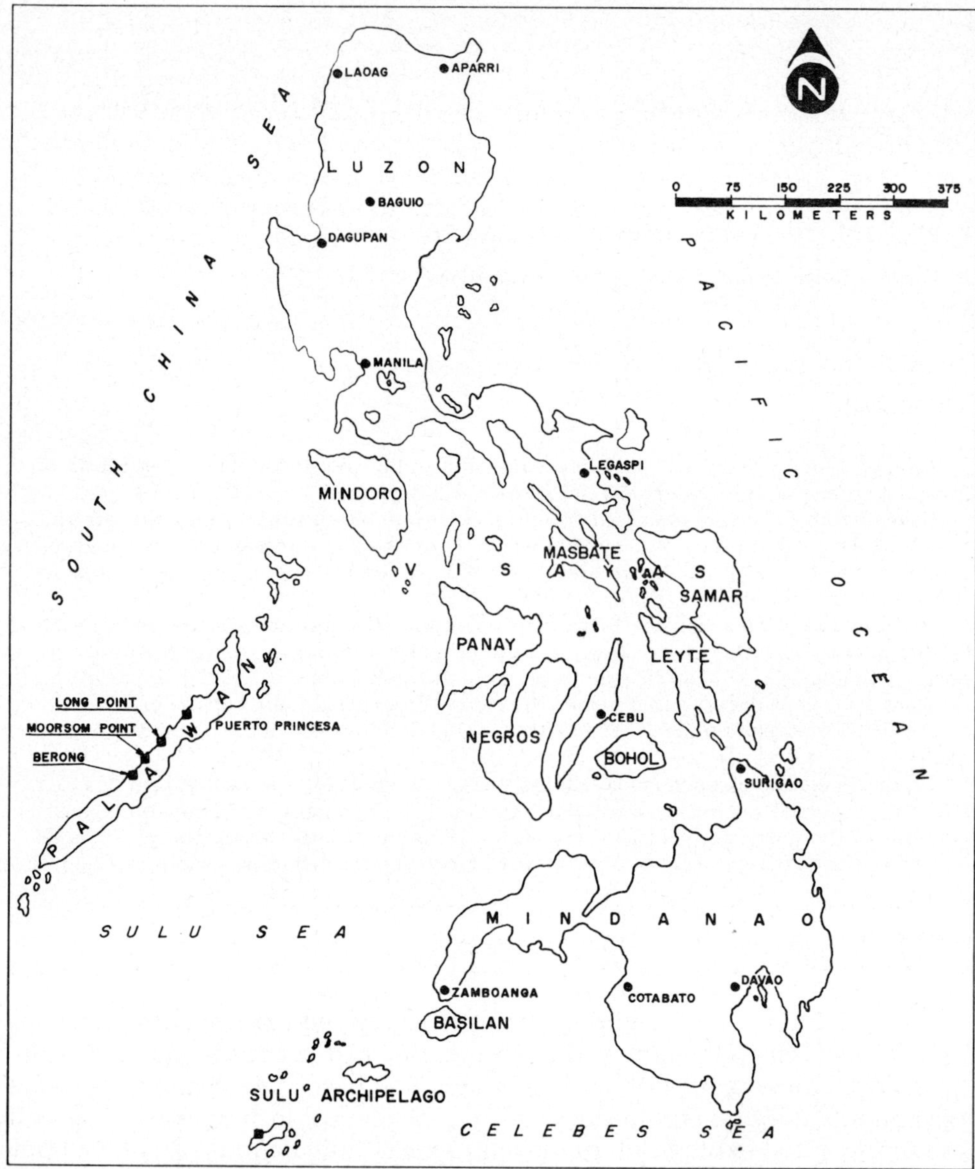

Fig.1. Map of Philippines showing study areas.

cally promising deposits of nickeliferous laterite, and subjacent decomposed peridotite and dunite from similar types of deposits which are less promising for economic exploitation, and (2) delineate the concealed contacts of the ultramafics with gabbro, diabase, metavolcanic and other rock types which have comparatively low nickel and cobalt contents.

Four terms are frequently referred to in this paper, they are: (a) typical nickeliferous laterite, (b) atypical laterite, (c) geochemical areas, and (d) possible nickel ore. Typical nickeliferous laterite is the reddish-brown to yellowish-brown ferruginous soil, which is the end-product of the chemical weathering of peridotite and/or dunite. An atypical laterite is the dark brown soil derived from the chemical weathering of gabbro and other non-ultramafic rocks. A geochemical area is an area defined by a given range of nickel or cobalt geochemical values, for example the area representing the 1000—5000 ppm nickel values. Possible nickel ore is a 1-m or more thickness of laterite and/or decomposed ultramafic rock with an average nickel content of at least 0.9% Ni.

GENERAL DESCRIPTION OF THE STUDY AREAS

Long Point area

Long Point forms a rectangular promontory on the western coast of Palawan Island. Within the claims area the topography is somewhat rugged, with elevations reaching 501 m above sea level (a.s.l.).

The most common rock type on the claims is a peridotite with dunite facies which intrudes metavolcanic rocks (Fig.2). Both are slightly serpentinized in places. Pyroxenite, gabbro and diabase are also present and generally occur as dykes in the peridotite.

Along ridges and spurs, the ultramafic rocks are largely overlain by typical nickeliferous laterite. The metavolcanic rocks and gabbro, on the other hand, are blanketed by atypical laterite.

Berong and Moorsom areas

The Berong and Moorsom groups of claims adjoin each other on the western side of Palawan about 25 km southwest of Long Point. The Berong group constitutes the southern section of the claims area and the Moorsom group comprises the northern portion. Berong is generally characterized by subdued to moderate topography, whereas Moorsom is moderate to rugged. On the whole, the topography of the Berong and Moorsom areas, particularly the latter, is more rugged than Long Point; the highest elevation within the claims area is 1164 m a.s.l.

General geology and the development of laterite of the Berong and Moorsom areas is essentially the same as that of Long Point (Fig.3).

FIELD TECHNIQUES

Prospecting

Ultramafic areas were first selected from the 1:1,000,000 scale geological map of the Philippines, and from the existing geological literature. This was followed by careful study of the selected areas using 1:50,000 scale topo

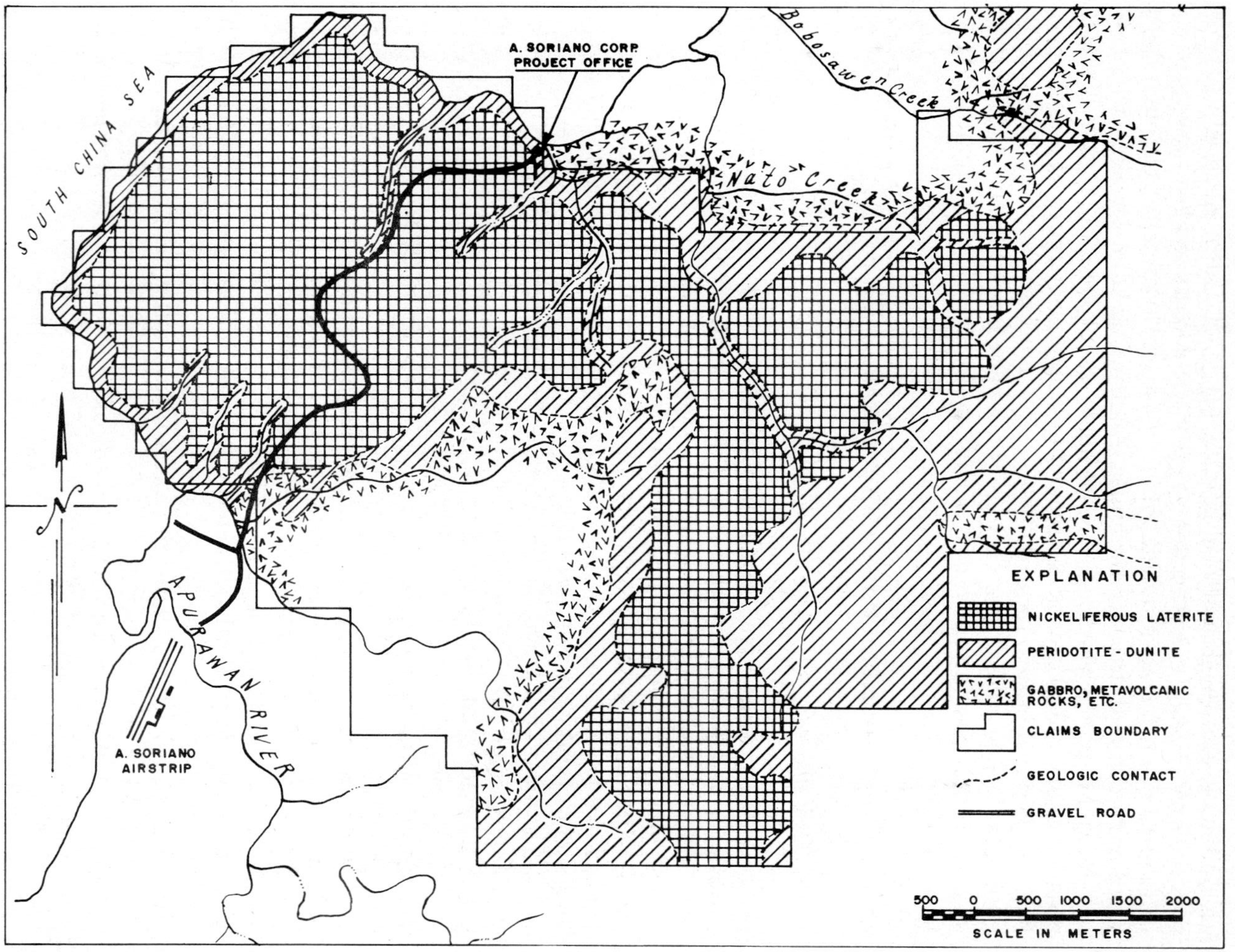

Fig.2. Geological map of Long Point area, Palawan.

graphic maps. Localities characterized by subdued to slightly subdued topography were delineated, particularly mountains with broad nearly flat ridges and spurs and gentle slopes where thick laterite and decomposed ultramafic rocks are usually found.

Typical laterite has both a characteristic colour and a distinctive vegetation and can easily be recognized from the air in sparsely forested places. An air-

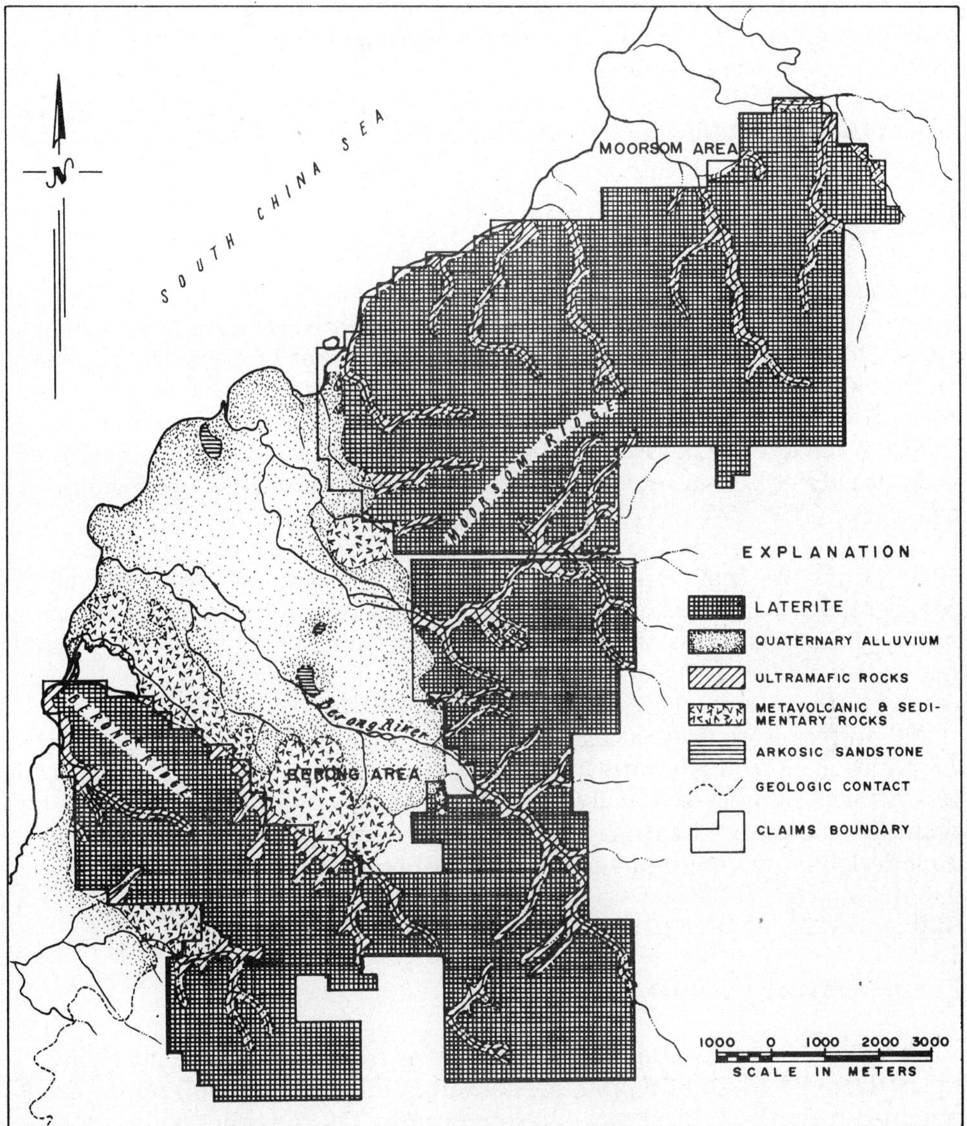

Fig.3. Geological map of Berong and Moorsom areas, Palawan.

borne reconnaissance was made over the possible lateritic areas before the ground survey was started. When the results of the pre-ground work studies were completed, a geological team was dispatched to the areas delineated to undertake combined geological and geochemical surveys.

Initial prospecting mainly involved cutting reconnaissance lines across possible lateritic areas. These lines were purposely run along spurs, ridges and gentle slopes in order to encounter residual soils. Soil samples were collected at an average depth of 35 cm from each sampling point spaced at about 250 m along the lines. No stream sediment samples were collected. In places where rock outcrops or boulders were encountered, rock types were mapped. Prospecting test pits were dug at selected points in order to obtain a general idea of the characteristics of subsurface geology — particularly the occurrence of nickel.

Preliminary exploration

In the course of staking claims in the areas indicated by the initial geological-geochemical prospecting as economically promising, soil samples were collected from a depth of about 35 cm at each corner of every claim. This established a 300-m square grid geochemical sampling pattern over the claims. The different rock types encountered along the survey lines were mapped.

Where the laterite deposit looked promising test pits were dug at selected sites, usually at 600-m intervals. Channel samples were collected from these pits.

Upon completion of the geochemical survey and the preliminary test pitting work, the main exploration programme was started. This involved pitting at closer intervals concentrated mainly within slightly to highly anomalous geochemical areas. A second series of test pits were dug at 300-m intervals to coincide with claim corners. A number of test pits were also dug, for check purposes, in background areas and in the atypical laterites.

All samples were dried at base camp before sending them to the A. Soriano Corporation central laboratory in Manila. The minus 80-mesh fraction of the geochemical samples was analysed colorimetrically for nickel and cobalt by Stanton's methods (Stanton, 1966). Test pit samples were assayed by the dimethylglyoxime method and later by atomic absorption.

GEOCHEMICAL AND TEST PIT DATA

Long Point geochemical results

The nickel content of near-surface samples from the Long Point claims ranges from 80 to 24,000 ppm. Nickel values higher than 1000 ppm were obtained from the typical nickeliferous laterite. On the other hand, values less than 1000 ppm came from the atypical laterite which had been verified by geological mapping to overlie gabbro and metavolcanic rocks.

Nickel values of 1000 ppm and greater were statistically evaluated. As interpreted from the statistical evaluation and a visual scan of the distribution of values, the threshold level was established at 5000 ppm. All the nickel results from typical nickeliferous laterite areas have been grouped into four geochemical ranges, namely:

(a) highly anomalous	greater than 15,000 ppm,		
(b) anomalous	10,000	—	15,000 ppm,
(c) slightly anomalous	5000	—	10,000 ppm,
(d) background	1000	—	5000 ppm.

Laterite containing highly anomalous nickel values (Fig.4) occurs as scattered patches in the northwestern section of the claim area, where dunite predominates. A similar irregular distribution of laterite with anomalous nickel content occurs in the northwestern, eastern and southern parts of the area. Laterite containing slightly anomalous nickel values covers a large area extending from the northwestern section of the claims to the eastern and southern portions.

By contrast, the low-nickel laterite or laterite-containing background nickel values is located mainly in the easternmost portion of the claims where bedrock is pyroxene-rich peridotite with gabbro and diabase dykes, and where the topography is somewhat more rugged compared to Long Point proper. It also occurs as narrow zones near the ultramafic/gabbro and ultramafic/metavolcanic rock interfaces. Low-nickel laterite covers only one-fifth of the original prospect area.

Nickel values below 1000 ppm were mostly obtained from the middle section of the southwestern half of the original prospect area where there is widespread occurrence of atypical laterite overlying metavolcanic rocks. Based mainly on the geochemical data and on results obtained from eight check test pits, this area was excluded in the main exploration work.

Cobalt concentrations in both typical and atypical laterites vary from 30 to 2800 ppm. Values less than 100 ppm came mostly from the atypical laterite. Typical nickeliferous laterite generally contained more than 100 ppm cobalt. Using the same method as for the determination of the nickel threshold, the cobalt threshold was established at 500 ppm. Like the nickel values, cobalt values above 100 ppm have been grouped into four geochemical ranges, as follows:

(a) highly anomalous	greater than 1500 ppm,		
(b) anomalous	1000	— 1500 ppm,	
(c) slightly anomalous	500	— 1000 ppm,	
(d) background	100	— 500 ppm.	

The slightly anomalous to highly anomalous (plus 500 ppm) cobalt values have distribution patterns which correspond to nickel "highs" (Fig.5). The areal distribution of the cobalt values below the threshold is strikingly similar to that of background nickel values.

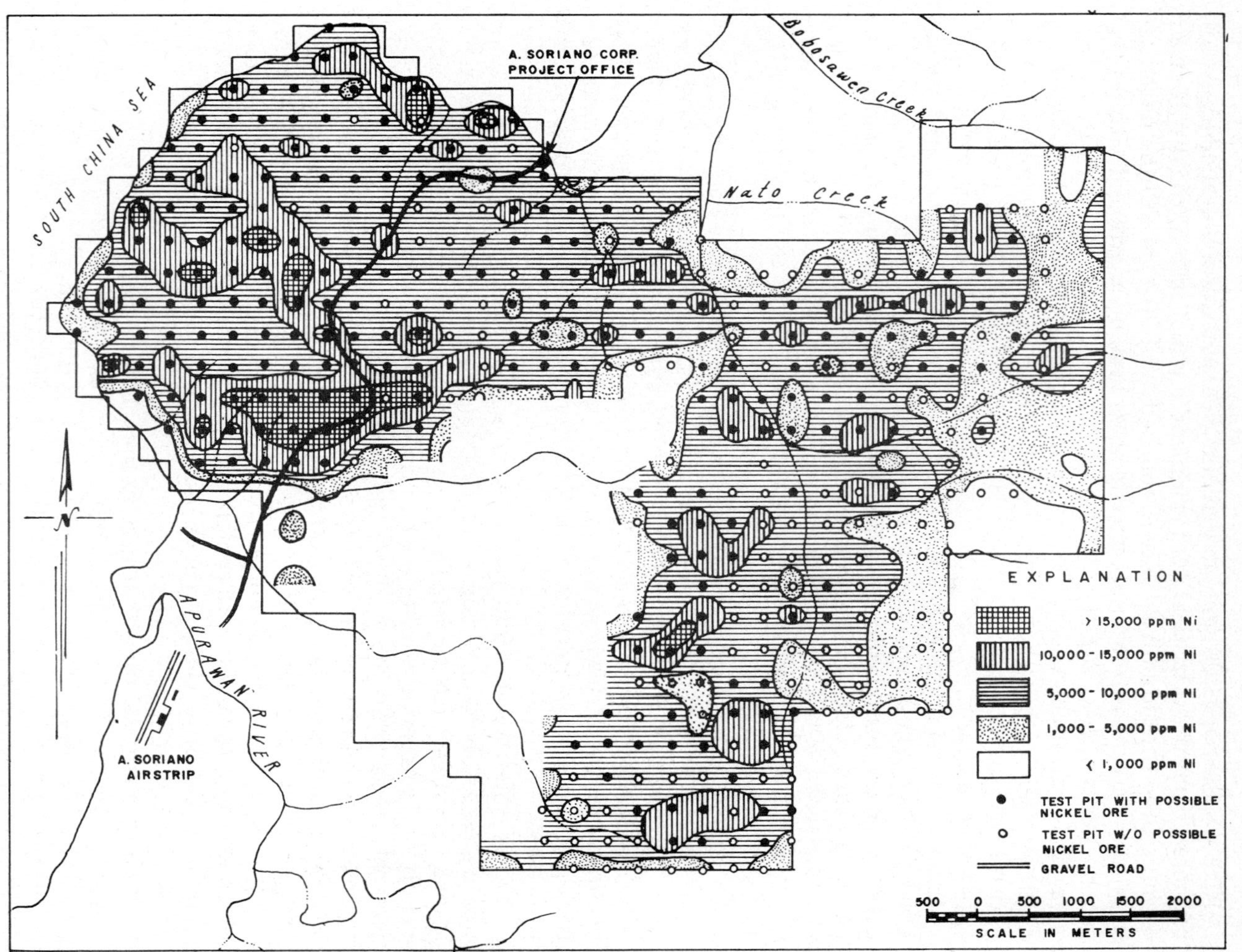

Fig.4. Nickel geochemical map of Long Point area, Palawan.

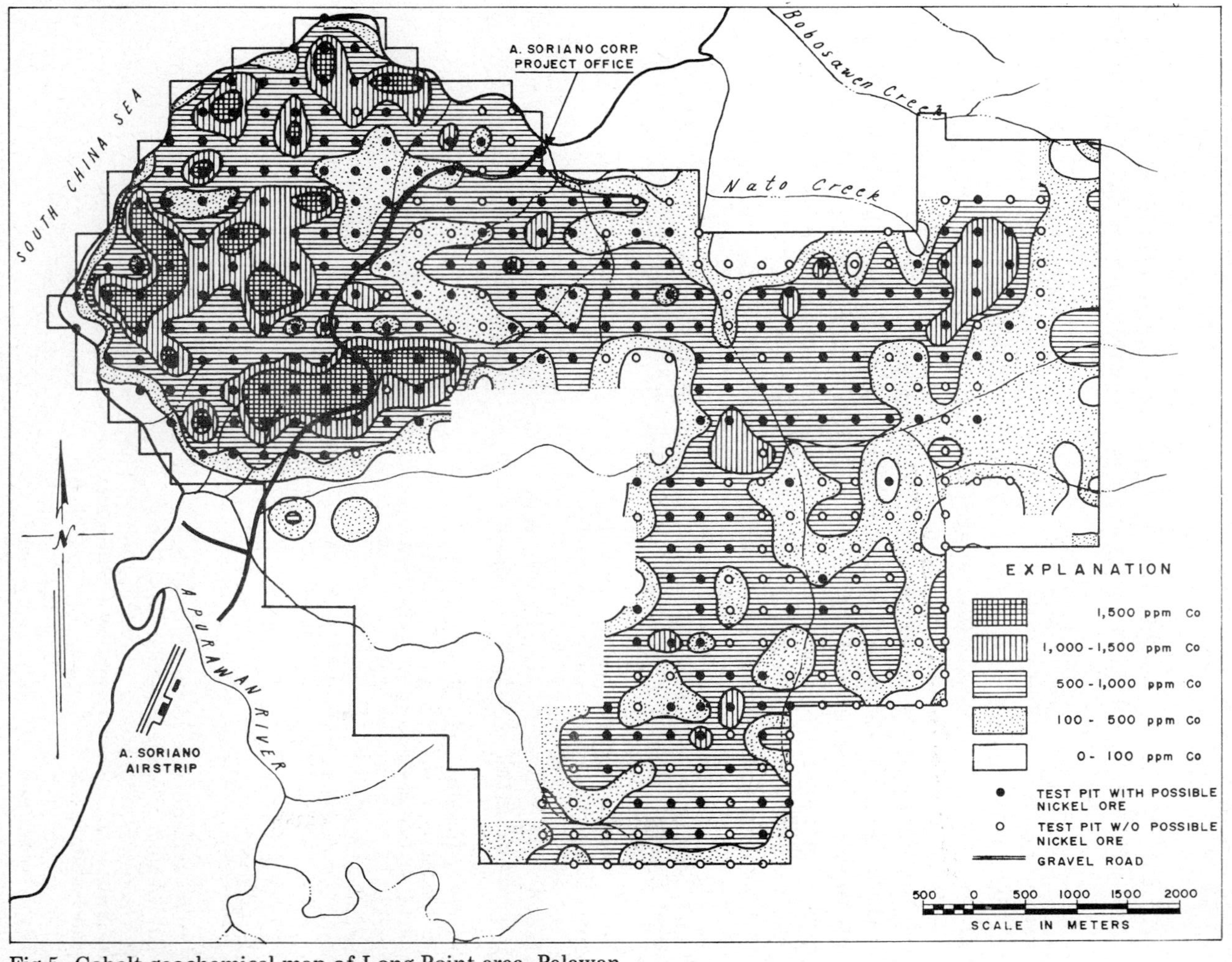

Fig.5. Cobalt geochemical map of Long Point area, Palawan.

Long Point test pit results

Three hundred and eighty-four test pits were dug at 300-m intervals on the Long Point claims. Table I shows the test-pit distribution in the different nickel geochemical areas.

96% of the test pits in the highly anomalous nickel areas penetrated possible nickel ore compared with 86% of the pits in anomalous nickel areas and 70% of the pits in slightly anomalous areas. Only 27% of the test pits in the background areas penetrated possible ore. Most of these pits are located near the boundaries of the slightly anomalous areas, and probably intersected the subsurface lateral extension of possible nickel ore bodies.

It is significant to note that, in the claims covering atypical laterite where all samples yielded nickel values less than 1000 ppm, none of the eight test pits hit possible ore. The nickel contents of subsurface samples of atypical laterite including weathered bedrock are all less than 0.1%.

In the case of cobalt Table II indicates that 100% of test pits in highly

TABLE I

Total number of test pits and the percentage of pits with possible nickel ore in the different nickel geochemical areas at Long Point

Nickel geochemical areas	Total no. of pits	Pits with possible ore	Percent of pits with possible ore
Highly anomalous	23	22	96
Anomalous	63	54	86
Slightly anomalous	219	153	70
Background	71	19	27
Atypical laterite	8	0	0

TABLE II

Total number of test pits and the percentage of pits with possible nickel ore in the different cobalt geochemical areas at Long Point

Cobalt geochemical areas	Total no. of pits	Pits with possible ore	Percent of pits with possible ore
Highly anomalous	28	28	100
Anomalous	39	35	90
Slightly anomalous	192	140	73
Background	110	47	43
Atypical laterite	15	1	7

anomalous areas penetrated possible nickel ore. In anomalous and slightly
anomalous areas, the ore intersection rate of the test pits is somewhat lower
at 90% and 73%, respectively. By comparison, only 43% of test pits in the
background areas were in ore-grade material. Only one of the fifteen test
pits in the atypical laterite penetrated possible ore.

Fig.6 shows the areal distribution of indicated nickel ore reserve in laterite
and in the underlying decomposed ultramafic rocks. By using a nickel cut-off
of 0.9%, 87 million dry metric tons averaging more than 1% Ni can be outlined. The overall areal distribution pattern of the indicated nickel ore reserves
conforms well with the slightly to highly anomalous nickel and cobalt geochemical areas.

The foregoing data indicate that, by and large, near-surface concentrations
of nickel and cobalt are reliable geochemical guides in the search for nickel
ore in laterite and in the subjacent decomposed ultramafic bedrock.

Berong and Moorsom geochemical results

The geochemical data from the Berong and Moorsom areas were treated in
the same manner as those for Long Point.

The range of nickel values encountered is quite comparable with Long Point
with ranges of 100—20,500 ppm and 100—17,600 ppm for Berong and
Moorsom, respectively. This enabled the use of the same anomaly categories.
The distribution of highly anomalous values is, however, much more limited
and erratic than at Long Point (Fig.7). Nickel values in the slightly anomalous
class (5000—10,000 ppm) cover almost the whole claim area whilst concentrations classed as anomalous (10,000—15,000 ppm) occur mainly on the
ridges and spurs of the southwestern part of the Berong claims.

Background nickel levels (<5000 ppm) are mostly to be found in the east
of the Berong area and the southeast of the Moorsom claims.

Cobalt values in the Berong soils are comparable with those at Long Point
ranging from 40—2400 ppm. In the Moorsom area the highest values are only
1600 ppm. Despite this difference the same threshold value of 500 ppm has
been used throughout. Using the same anomaly categories as for Long Point
it can be seen in Fig.8 that highly anomalous (>1500 ppm) and anomalous
values (1000—1500 ppm) for cobalt are very sparsely distributed whilst the
slightly anomalous values (500—1000 ppm) like the same category of nickel
values, occur over almost all the claim area.

Background cobalt values closely follow background nickel values in their
distribution.

Berong and Moorsom test pit data

One thousand and sixty-four test pits were dug at 300-m intervals on the
Berong and Moorsom claims.

Table III shows a comparison of test pit distribution in different nickel

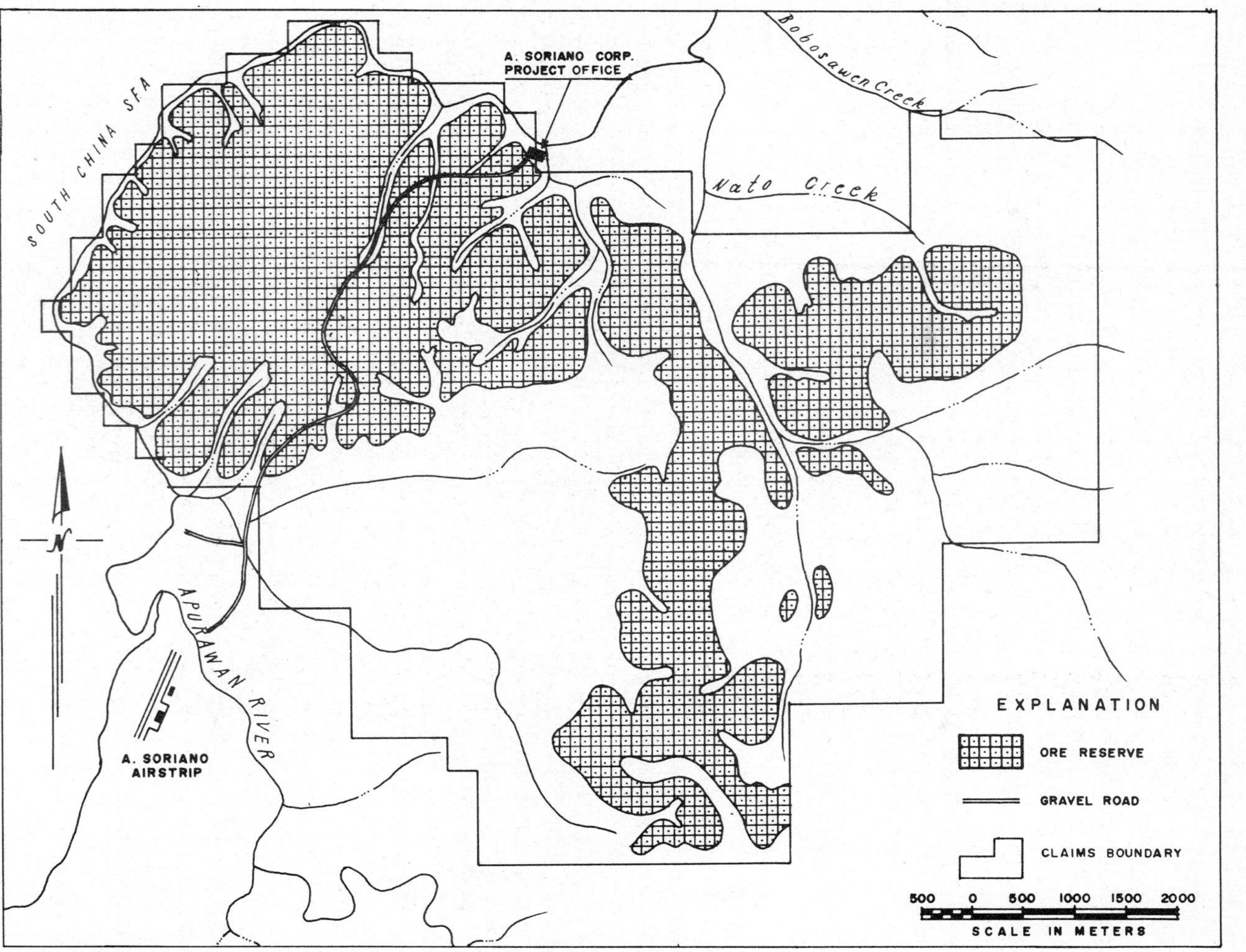

Fig.6. Map showing indicated ore reserve, Long Point, Palawan.

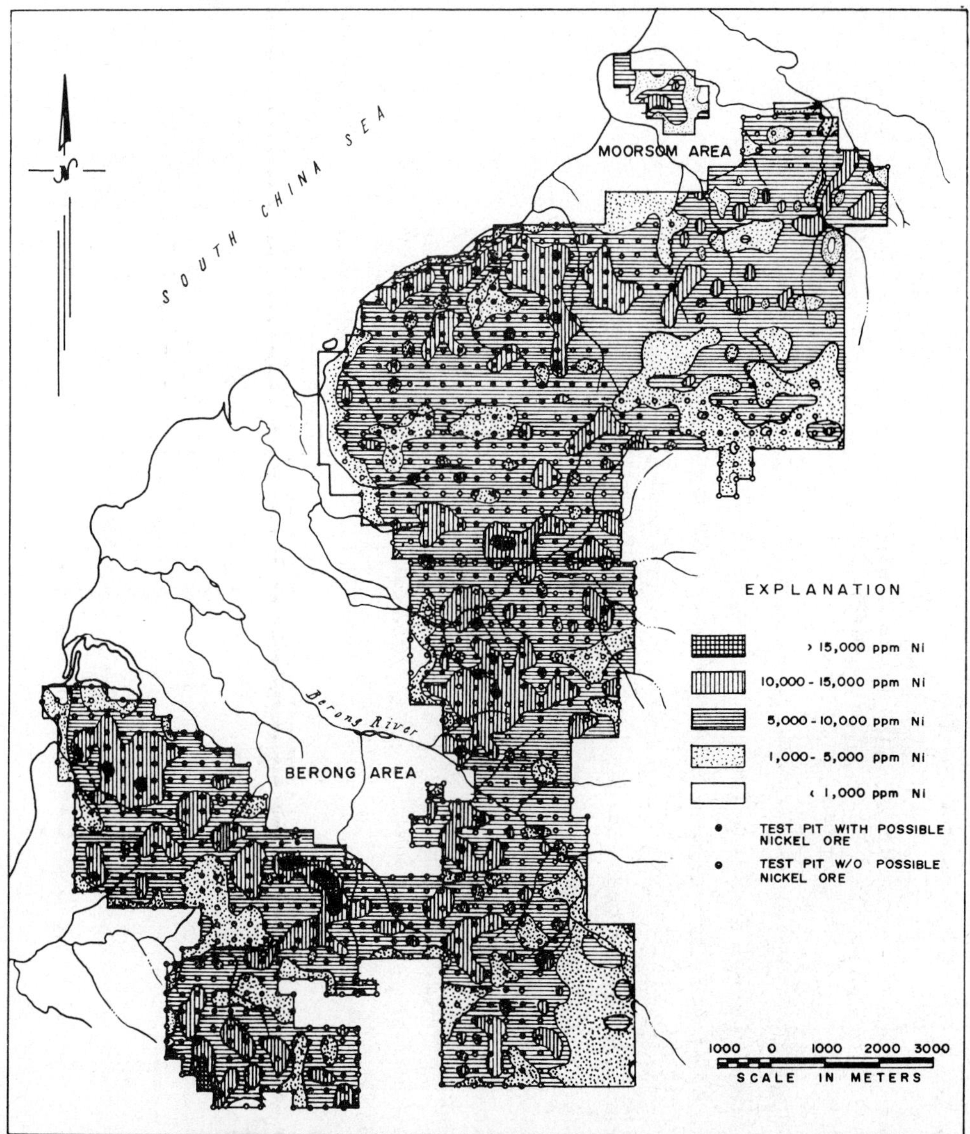

Fig.7. Nickel geochemical map of Berong and Moorsom areas, Palawan.

geochemical areas at Berong and Moorsom. Slightly anomalous to highly anomalous nickel areas in Berong have higher percentages (45—70%) of pits penetrating possible nickel ore as compared to similar areas in Moorsom where it varies from 37% in the slightly anomalous to 70% in the highly anomalous areas. Background nickel areas at Berong have a lower percentage (20%) of pits intersecting ore in contrast to that of Moorsom (33%). To some extent, the difference in the pit results between Berong and Moorsom may

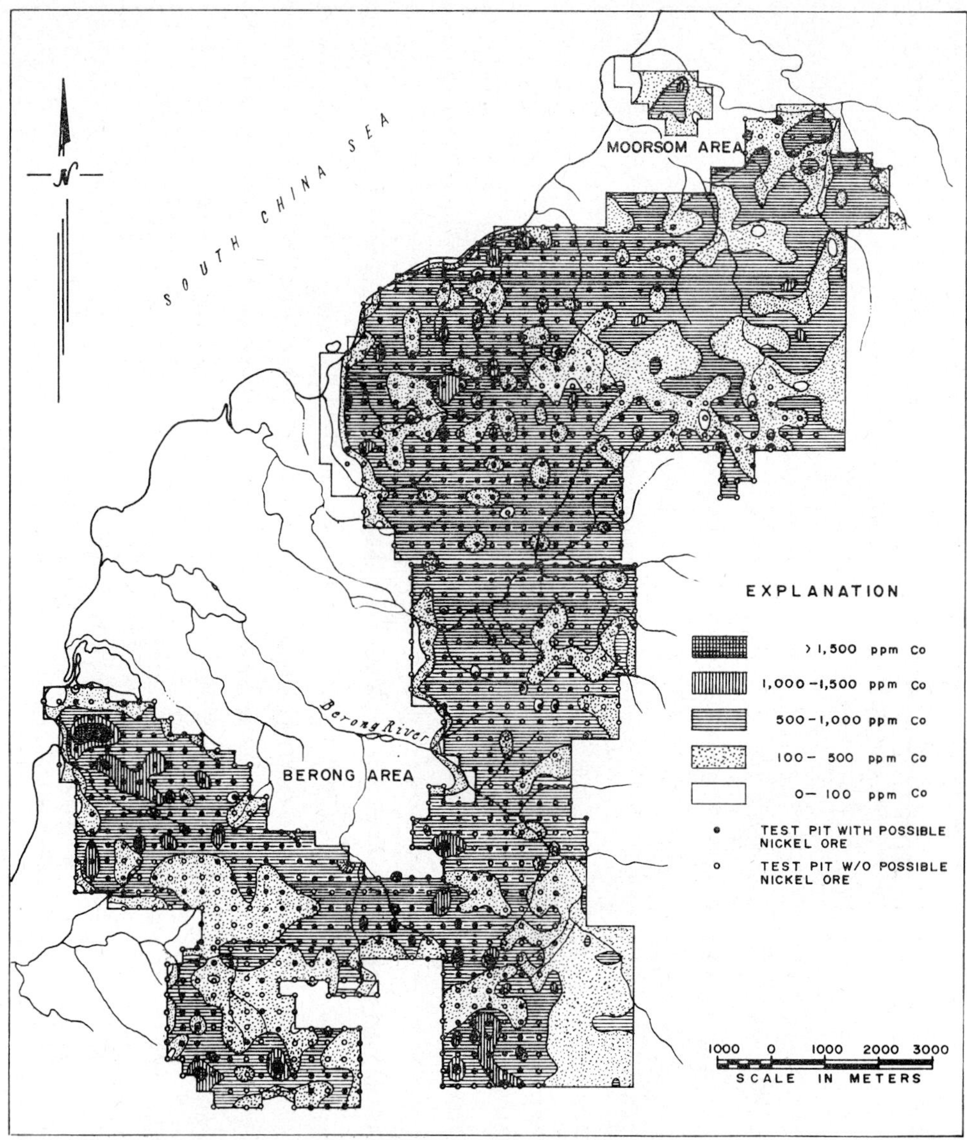

Fig.8. Cobalt geochemical map of Berong and Moorsom areas, Palawan.

be explained by the differing tenor of ore between the two places. However, the disparity is thought to be more likely to be a reflection of the contrast in topography between the two areas. Within the atypical laterite overlying non-ultramafic rocks at Berong and Moorsom, none of the test pits bottomed in possible nickel ore.

The cobalt results for the Berong and Moorsom test pits substantially follow those for nickel; the minor differences are not critical for exploration.

TABLE III

Total number of test pits and percentage of pits with possible nickel ore in the different nickel geochemical areas in Berong and Moorsom

Nickel geochemical areas	Total no. of pits		Pits with possible ore		Percent of pits with possible ore	
	Ber.	Moors.	Ber.	Moors.	Ber.	Moors.
Highly anomalous	20	10	14	7	70	70
Anomalous	152	66	99	39	65	59
Slightly anomalous	346	219	155	82	45	37
Background	140	92	28	30	20	33
Atypical laterite	15	4	0	0	0	0

As compared to Long Point, Berong and Moorsom areas as a whole have a lower success ratio of pits encountering ore in the slightly to highly anomalous nickel and cobalt geochemical areas. This, again, can be largely attributed to the dissimilarity in topography among the three study areas — Long Point having the least rugged topography of the three.

In spite of the minor discrepancies in the percentage of pits penetrating ore for the various categories of nickel and cobalt values, it is clear that the chances of encountering ore-grade material at depth increases as the intensity and extent of the surface geochemical anomaly increases.

Fig.9 shows the areal distribution of the indicated nickel ore reserve in Berong and Moorsom. Using a cut-off grade of 0.9% Ni, Berong contains an inferred ore reserve of 132 million dry metric tons averaging well over 1% Ni compared to the Moorsom reserve of 35 million dry metric tons averaging over 1%.

Almost all the indicated ore reserve in the Berong and Moorsom claims lie within the slightly to highly anomalous nickel and cobalt geochemical areas.

GENERAL DISCUSSION

Although the 5000-ppm Ni threshold value proved to be reliable in the Long Point, Berong, and Moorsom areas, this value should not be considered, without verification, as an accurate threshold for all nickeliferous laterite deposits. Geochemical results from each laterite prospect should be carefully evaluated to determine the local threshold. On the whole, however, the chance of finding nickel deposits with economic potential is much better in laterite prospects where the geochemical nickel results are consistently well above 5000 ppm than in prospects where the majority of the nickel values are below 5000 ppm.

The overall results of exploration work in the study areas clearly indicate

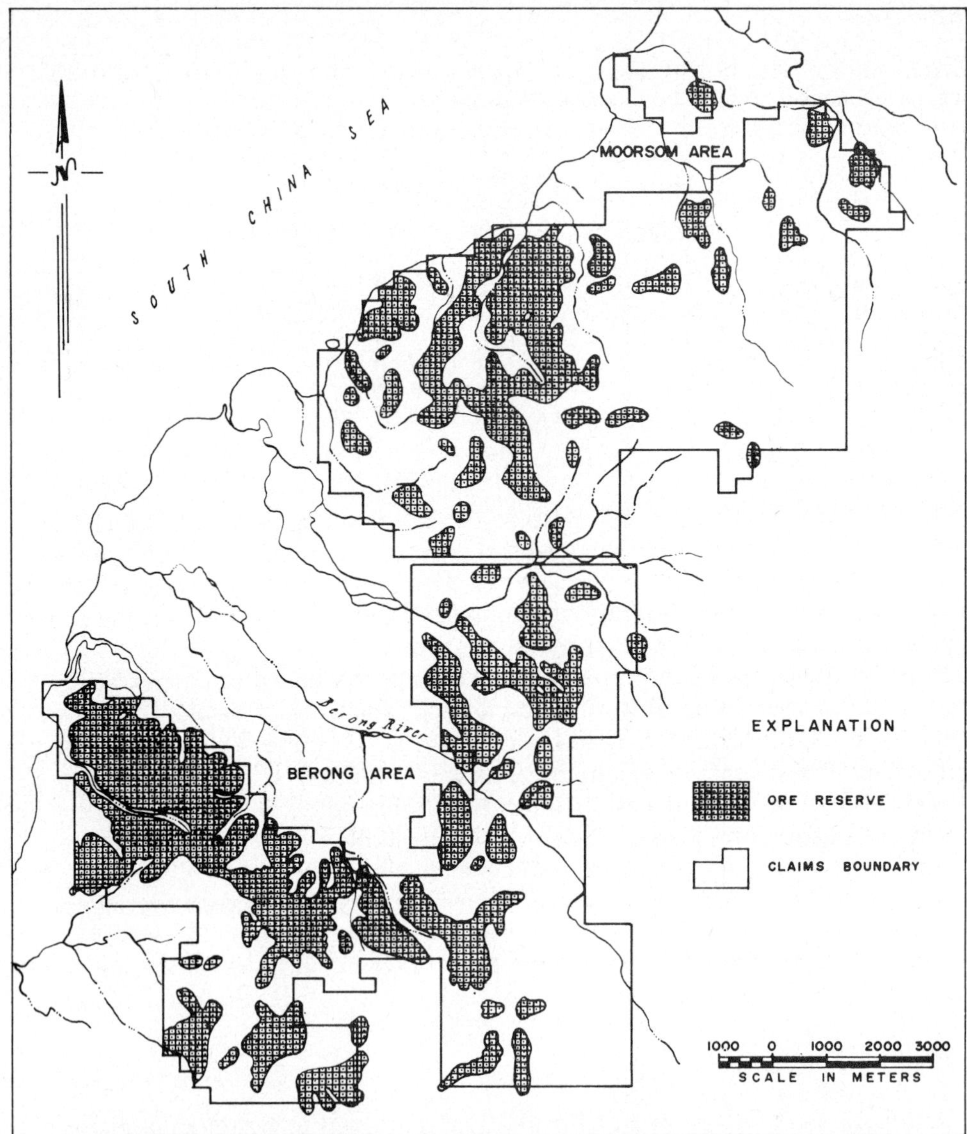

Fig.9. Map showing indicated ore reserves in Berong and Moorsom areas, Palawan.

that geochemical surveys have value in exploration for nickel in laterite and in the underlying decomposed peridotite. Geochemical surveys are particularly useful in delimiting priority areas for test pitting, or drilling. On the other hand such surveys can also delimit lateritic areas with no potential for nickel mineralization. Hence, a geochemical survey reduces exploration expenses because extensive exploration work can be avoided in the background geochemical areas of laterite deposits.

The nature of the topography must be taken into consideration when interpreting geochemical results.

As observed from the results of test pitting in the three study areas and from previous experience with the laterites in the Surigao Mineral Reservation, Mindanao, Philippines (Ong, 1963), the concentration of nickel and cobalt in an in-situ laterite mantle increases with depth. The concentration of cobalt is usually highest near the base of the laterite column and, by comparison, the highest concentration of nickel usually occurs in the decomposed peridotite or dunite underneath the laterite.

The rate of erosion in areas with gentle to moderate topography is relatively low. Near-surface geochemical samples from such areas represent the top portion of the laterite mantle. When anomalous nickel and cobalt values are obtained from these samples, nickel-bearing deposits with a nickel content higher than those in surface samples are likely to be found in the lower portion of the laterite columns and in the underlying decomposed peridotites.

Areas with rough topography, on the other hand, are characterized by immature or truncated soil profiles, due to a relatively high rate of erosion. In most cases, the lower portion of the profile is exposed at surface and in some places, the laterite profile may even be completely eroded to expose decomposed peridotite. Geochemical samples from such terrain most often represent different horizons of laterite profile or decomposed peridotite. Geochemical results from these samples are somewhat erratic. If the samples represent the bottom horizon of the laterite they may yield anomalous values which do not reflect underlying higher nickel contents, because they may already represent the near-maximum, if not the maximum, nickel concentration developed in the profile. As a result, near-surface geochemically anomalous values in rugged terrain do not necessarily indicate ore-grade deposits at depth.

CONCLUSIONS

The conclusions of importance which can be drawn from this study are summarized below:

(1) Geochemical prospecting can delineate nickeliferous laterite areas with good potential for nickel mineralization.

(2) Systematic geochemical sampling, coupled with proper interpretation of results, can delineate economically promising nickel deposits in laterite and in the subjacent decomposed ultramafic rocks from similar deposits which are less promising for nickel exploitation.

(3) The nature of the topography should be taken into consideration when determining the significance of the geochemical anomalies. Considering other factors equal, geochemical techniques appear to be more effective as a tool in exploration for nickeliferous laterite in areas with subdued to moderate topography compared to places with rugged terrain.

(4) Cobalt is as effective as nickel in delineating economically promising nickel deposits in laterite and/or in the subjacent decomposed peridotite rocks.

(5) Geochemical surveys based on nickel and cobalt are also useful in tracing the concealed contacts of peridotites and non-ultramafic rocks.

(6) Geochemical surveys can reduce the exploration cost for nickel deposits in laterites.

ACKNOWLEDGEMENTS

The authors wish to thank A. Soriano Corporation, particularly Messrs. C.D. Clarke and D.E. Lewis, Senior Vice-President of the Mining Department and Vice-President for Mineral Exploration, respectively, for the permission to present this paper. Sincere gratitude is extended to the staff of the Exploration Division who were largely responsible for the geological-geochemical surveys. Likewise, sincere thanks are also conveyed to the staff of the Division's Geochemical and Assay Laboratories for the analyses of samples, and to the office staff for drafting the maps and typing the manuscript.

REFERENCES

Ong, P.M., 1963. Geochemical and biogeochemical studies of the Surigao mainland laterite, Mindanao, Philippines (unpublished)

Stanton, R.E., 1966. Rapid Methods of Trace Analysis. Edward Arnold, London, 96 pp.

Part 6
EXPLORATION IN HOT DESERT AND TROPICAL TERRAIN

Chairmen:

P.R. DONOVAN
McPhar Geophysics Proprietary Limited, Manly, N.S.W., Australia
M.A. CHAFFEE
United States Geological Survey, Denver, Colo., U.S.A.

CASE HISTORIES FROM A GEOCHEMICAL EXPLORATION PROGRAMME — WINDHOEK DISTRICT, SOUTH WEST AFRICA

M.J. SCOTT

Johannesburg Consolidated Investment Company Limited, Windhoek (South West Africa)

ABSTRACT

The arid South West African climate promotes the formation of immature alkaline soils which inhibit the dispersion of copper. In regional exploration this limited dispersion capacity is overcome by collecting samples at very close intervals. Compositing the samples prior to assay reduces their numbers without significantly reducing the rate of ground coverage or complicating the interpretation.

Soil sampling programmes over areas of limonitic outcrop resulted in the discovery and subsequent definition of one important and two minor ore bodies during 1970 and 1971. The deposits are of the Cu—Zn-bearing massive sulphide type and are considered to be genetically related to submarine volcanic activity in the late Precambrian. The surface expressions, where exposed, take the form of heavily leached gossans and magnetite-rich quartzitic rocks.

An empirical study of the near-surface distribution of copper and zinc in soil and weathered rock revealed widespread sub-soil dispersion of copper and almost complete dispersion of zinc as a result of the introduction of sulphide-derived acid into the groundwater. The significance of these results in geochemical soil sampling is discussed.

INTRODUCTION

Geochemical soil and, to a lesser extent, stream sediment sampling are the most widely used base-metal prospecting techniques in South West Africa. Johannesburg Consolidated Investment Company has been actively engaged in geochemical exploration in the Territory since early 1970; by the end of 1973, 1.2 million samples resulting in 2.04 million assays had been collected and processed by this Company. During this four-year period a total of twenty-six drilling programmes based on these results have been either completed or are in progress. Of these twenty-six drilling targets, eight were previously unknown and were discovered as a result of regional soil sampling programmes.

In April 1970, gossan samples from an area lying 20 km to the northeast of Windhoek were brought to the Company by a local prospector. Assays of the visibly unmineralized material revealed significant copper and zinc values and subsequent reconnaissance prospecting in the area showed three limonite-rich localities to be present. Drilling based on soil sampling programmes outlined economic concentrations of copper and zinc. Production from one of

482

these prospects, Otjihase, will commence in 1975 at the initial milling rate of
100,000 tons per month.

This paper is intended to provide a descriptive outline of the successful use
of geochemistry in exploration. The results of an investigation into the mobility of copper in the near-surface environment are presented. They represent
an attempt to solve a prospecting problem rather than an in-depth scientific
study.

GEOLOGY, CLIMATE AND SOILS OF SOUTH WEST AFRICA

Nearly 50% of the total surface area of 825,000 km^2 is underlain by sedimentary or metasedimentary formations (Fig.1a). Granites, basement schists,
and wind-blown sands occupy the remainder. The bulk of the sediments are
of late Precambrian age and are correlated with the Katanga System of the
Zairian and Zambian Copperbelts (Martin, 1961).

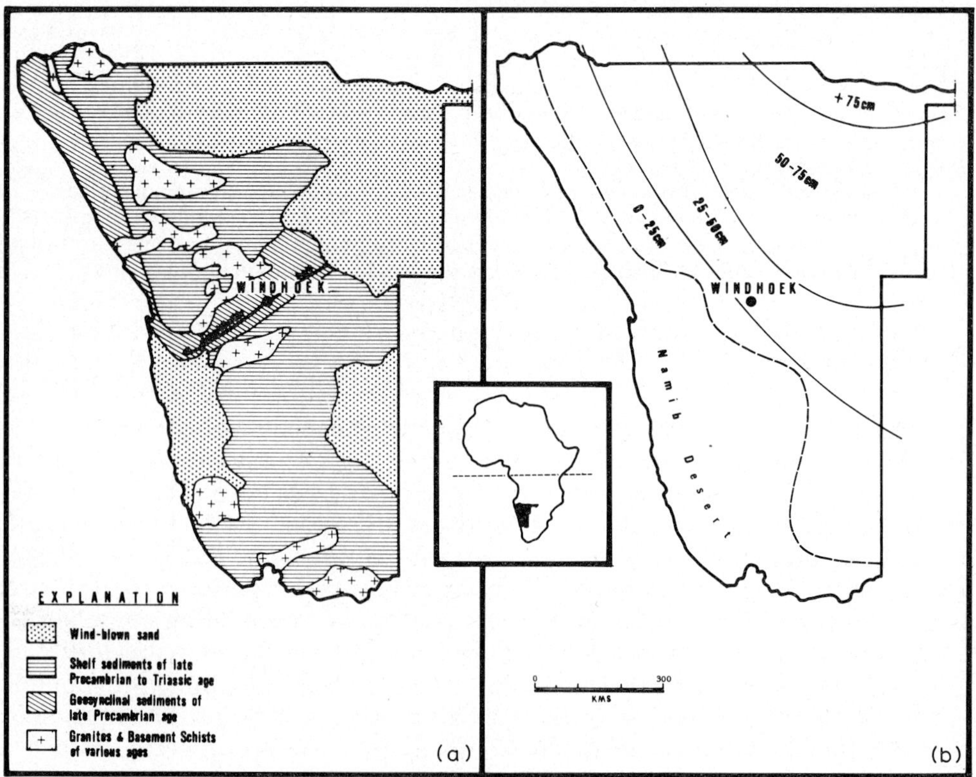

Fig.1. Simplified geology (a) and annual rainfall (b) of South West Africa.

Economic interest in deep-water sediments is generated by Cu—Zn deposits associated with volcanic activity during sedimentation. The more important of these deposits lie along the Amphibolite Belt trending northeast through Windhoek. Within the shelf facies, Pb—Zn (Cu) mineralization in dolomites, and sedimentary copper deposits in shales and sandstones, provide the main interest. The environmentally dictated target metals are thus copper, lead and zinc.

Throughout the area of geological interest, the climate is arid to semi-arid, the rainfall increasing from almost nothing at the coast to 50 cm annually near the edge of the sand-covered region in the northeast (see Fig.1b). The climate is hot as well as dry, with average temperatures between 15 and 25°C in July, and 20 and 30°C in the middle of the five-month long wet season in January. The average potential evaporation rate at Windhoek is 320 cm annually.

Outside the sand-covered areas the soils are mainly residual, although a variable proportion of wind-blown material appears to be a universal component. They typically show little profile development and grade from lithosols in the west to immature light-brown calcareous soils in the regions of higher rainfall. In the more arid regions and particularly in areas of arenaceous bedrock the mineralogical composition of the surface soil conforms closely to that of the underlying rock. Most soils give an alkaline reaction in the pH range 8.0—9.5. However, this is quite variable, and in areas where no calcrete is developed, pH values can be as low as 6.0. This is especially true of the upland areas of locally higher rainfall.

REGIONAL EXPLORATION METHOD

In view of the good results obtained using geochemical soil sampling elsewhere in southern Africa (Cornwall, 1961; Sharpe, 1964; Philpott, 1975) it was natural to select this method as a prospecting tool in the residual soils of South West Africa. The relative immobility of copper is a predictable problem in the more alkaline soils of this environment (Hawkes and Webb, 1962; Hansuld, 1966). In South West Africa, consequently, any regional reconnaissance soil sampling programme cannot be based on sample intervals as large as the 61-m (200 ft) and 50-m spacings used in Zambia and Rhodesia (Cornwall, 1961; Philpott, 1975).

In reconnaissance exploration soil samples are collected at 10-m intervals from just above or within the C horizon, along traverses spaced 200—500 m apart. Since the resulting volume of samples exceeds reasonable laboratory capacity, equal amounts of the material from neighbouring sites are mixed in the field to give 20-m sample points. Following digestion of the sample in a hot 18/1/1 concentrated $HClO_4/HNO_3/HF$ mixture both copper and zinc contents of the minus 80-mesh fractions are determined by atomic absorption spectroscopy. The values are plotted as profiles together with the available lithological information and areas of abnormally high metal content are visually identified. Mixing the samples gives a smoothing effect which does not significantly alter the background value. Single-point anomaly sites exceeding the arbitrary threshold of 2.5 times the modal value are examined in the field and resampled. The visually selected areas of significant metal values are resampled along traverses 25 or 50 m apart and individual sample determinations made. Values from lithologically homogeneous areas are presented in contour-plan form while the results from heterogeneous environments are subjected to statistical analysis. These latter values are grouped by lithological units and threshold values determined, usually graphically, for each unit. A log-normal distribution is assumed (Ahrens, 1954) and lower anomalous limits taken at either a multiple of modal value or at the base of the upper 2.5% of the value range. This figure corresponds closely to the modal value plus two standard deviations. The resulting anomalous sites are investigated for the presence of mineralization.

Follow-up work on potentially economic discoveries takes the form of trenching, percussion drilling and diamond drilling, aided by geological mapping, photogeological interpretation, and ground geophysical techniques such as I.P., E.M. and magnetometer surveys.

GEOLOGICAL SETTING OF DISCOVERIES

The prospects, Otjihase and Ongombo, are genetically related to vulcanism within eugeosynclinal sediments of the Precambrian Khomas Series. They lie along the Amphibolite Belt at positions 15 km (Otjihase) and 40 km (Ongombo) to the northeast of Windhoek (Fig.1a). The sediments of the Khomas Series comprise a monotonous sequence of northwesterly dipping quartz-biotite schists, within which lie the amphibolites, quartz-sericite schists, and talc schists making up the 2 km wide Amphibolite Belt.

The deposits can best be described as bedded magnetite-rich quartzites carrying pyrite-chalcopyrite-sphalerite (Otjihase) and pyrite-chalcopyrite (Ongombo). Otjihase is a massive sulphide body, while Ongombo ore is disseminated. Actual sulphide mineralogies at Otjihase and Ongombo are 7% chalcopyrite, 1.9% sphalerite, 30% pyrite, and 6.5% chalcopyrite and 11% pyrite, respectively. The ore is contained within shoots having copper contents of ten or more times the magnetite-quartzite host-rock level of 0.1—0.2%. At Otjihase these shoots are relatively narrow and plunge in a west-northwesterly direction, obliquely across the maximum foliation dip direction of 16° to the northwest.

Despite the relatively thin soil cover, the proportion of bedrock exposed does not exceed 15%. The more resistant magnetite quartzites show stronger surface expressions and these outcrops are frequently limonite stained, grading in places into gossans. Secondary copper and zinc minerals are almost entirely absent and are rarely seen even in trench and stream bed exposures.

The discovery area lies at a height of 2000 m (6600 ft) above sea level and, as a result of this elevation, receives an average annual rainfall of 37 cm. Soils consequently show some profile development and have neutral to mildly acidic reactions in the pH range 7.0—6.0.

SOIL SAMPLING

Almost identical exploration procedures were employed at both prospects. Soil sampling, percussion drilling to 40 m vertical depth, and diamond drilling formed the basic stages. Soil sampling served to define the mineralized horizons, particularly in areas of no outcrop, and percussion drilling outlined the near-surface positions of the ore shoots. Down-dip definitions and ore-reserve estimations were obtained from diamond drill information.

The soil samples were collected at 20-m intervals along traverses spaced 50 m apart and orientated perpendicular to the strike. They were taken at 15 cm depth, just below the A and usually above the C horizon. The minus 80-mesh fraction was analysed for total copper and zinc contents. Copper results are shown on Figs.2 and 3 where the values are contoured in multiples of the background value. In these figures the primary ore grade is diagrammatically represented by metre-percent copper contours, the product of true thickness and grade. Zinc values have been omitted.

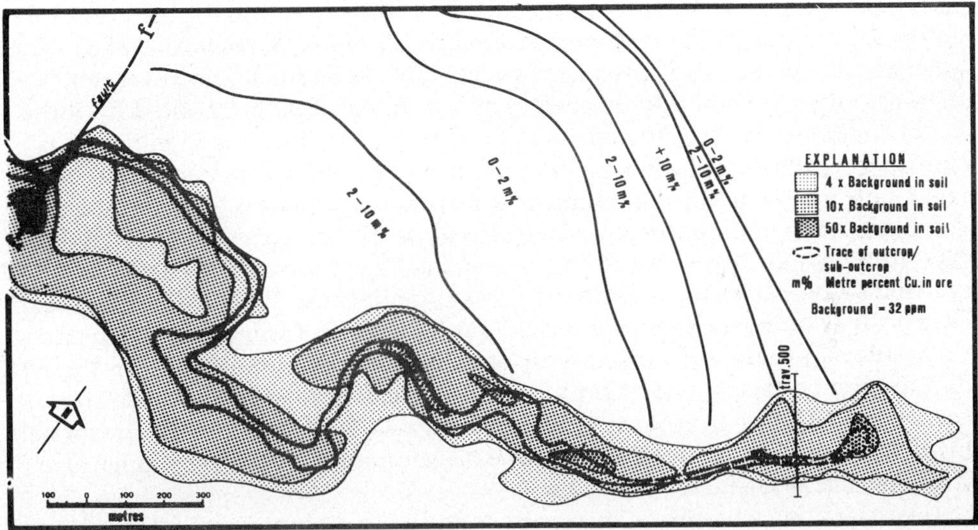

Fig.2. Otjihase prospect. Copper distribution in ore body and soil.

At both prospects the soil copper anomaly accurately outlines the surface or near-surface position of the magnetite-quartzite host rock. The broken pattern towards the west at Otjihase is due entirely to topography causing a sinuous outcrop pattern and preservation of the horizon in some places on the footwall side.

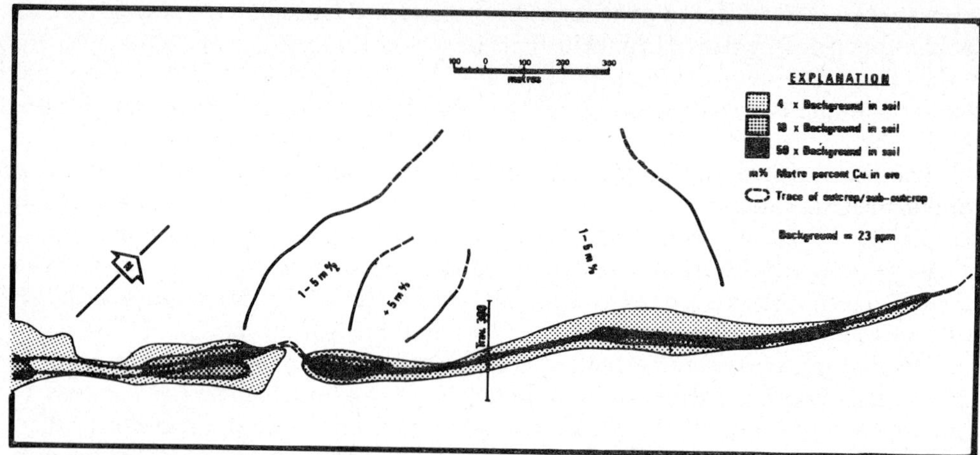

Fig.3. Ongombo prospect. Copper distribution in ore body and soil.

COPPER AND ZINC DISPERSION AT OTJIHASE AND ONGOMBO

Numerous authors have described the solution and transportation of copper and, less often, zinc, by sulphuric acid released from oxidizing iron-rich sulphides (Bateman, 1951; Park and MacDiarmid, 1964; Blanchard, 1968). Few quantitative, in-situ studies of the process have been made, however, and there is only a general understanding of the physical path travelled by these metals between release from the sulphide and their appearance in the soil. Considering the local problems experienced with restricted trace-metal movement in the soils, and the present prospecting emphasis on Cu—Zn-bearing pyrite deposits, an empirical investigation into the dispersion of metals from the Otjihase and Ongombo ore bodies seemed appropriate. These two deposits are well suited to an investigation of this type because (1) the country rock is chemically homogeneous for a considerable distance from the hanging and footwalls; (2) there is no primary dispersion of copper and zinc from the well-defined ore bodies as evidenced by diamond drill core assays; (3) the pyrite content at Otjihase is 30%, at Ongombo only 11%; (4) a comparison can be made between the subdued Ongombo soil anomaly and the very much stronger one at Otjihase.

Investigations were made into copper and zinc distribution in both the overlying soils and the near-surface rock around the ore bodies. Two traverses over the ore shoots were selected (Figs.2 and 3) and soil and percussion drill samples collected as follows. The soil samples were taken at 10-m intervals from 15 to 20 cm depth and sieved through a nest of screens of 10, 20, 35, 80 and 140 mesh sizes, each fraction being analysed for total copper and zinc contents. The vertical percussion holes were drilled at 25-m intervals to a depth of 30 m and sampled every 50 cm. These drill-pulp samples were analysed for total copper and zinc contents and pH determinations made on

10-g splits in 25 ml distilled water suspensions. The accuracy of these pH measurements must be viewed in the light of comments made on this method of determination (Garrels and Christ, 1965). Nevertheless, the values obtained show recognizable patterns and a high degree of reproducibility was obtained from the Radiometer pH meter used. The very dry environment probably renders changes in sample pH during the period between collection and measurement insignificant.

In Fig.4 the copper distribution in the country rock is contoured in multiples of background value and, for reasons of clarity, only three of the six soil fraction copper contents determined are shown. It is convenient also, to separate high and low pH areas by only one contour, drawn arbitrarily at pH 8.0 particularly as the values tend to be erratic in the lower pH zone. The average pH of the soil is 6.2, and of the country rock away from the reef 9.1; just above the oxide/sulphide interface of the ore body it reaches a minimum of 3.5, rising to 7.0 in the oxidized reef material and to 5.0 in the deeper sulphide ore. The maximum pH effect thus surrounds the point at which the sulphides abruptly disappear, decreasing rapidly towards surface and more gradually with depth.

The position of the zone of pH depression moves down the ore body as the sulphides retreat during oxidation and its shape will be modified by seasonal and other changes in the direction of groundwater movement. Dilute acid solutions emanating from the oxidizing ore move the acid-soluble copper into the country rock where, once fixed by rising pH, it is unlikely to be remobilized by encroaching alkaline groundwater. This process results in dispersion of copper to the limits of wall-rock penetration achieved by circulating acid solutions. Considering the five times vertical-scale exaggeration used in drawing this section, the movement of copper released from the ore body is largely vertically downwards.

From the average copper grades in the oxidized and unoxidized ore, and assuming only minor loss of copper from the latter, the quantity of ore body copper moved into the wall rock can be calculated. It is found that 7.1% of the original copper remains in the oxidized reef zone indicating that 92.9% has been lost into the surrounding rock. A similar calculation can be made for the amount of copper reaching surface between the four times background contour. This reveals that only 23% of the original copper present in depth in the ore body reaches the surface between these limits. Presumably the missing copper has been widely dissipated along joints and bedding planes and is now present as small, erratic, additions to country-rock background. The sectional area studied is unfortunately too small to explore this possibility.

It is also possible to calculate the relative amounts of copper entering the soil directly from the leached reef material and via the exotic oxide-bearing wall rock. Here it is found that, of the copper reaching the soil between the four times background limits, 30% is contributed by the reef at surface and 70% through weathering of the surrounding schists. Although the grade of the country rock schist is lower than that of the reef at outcrop, the 100-m

488

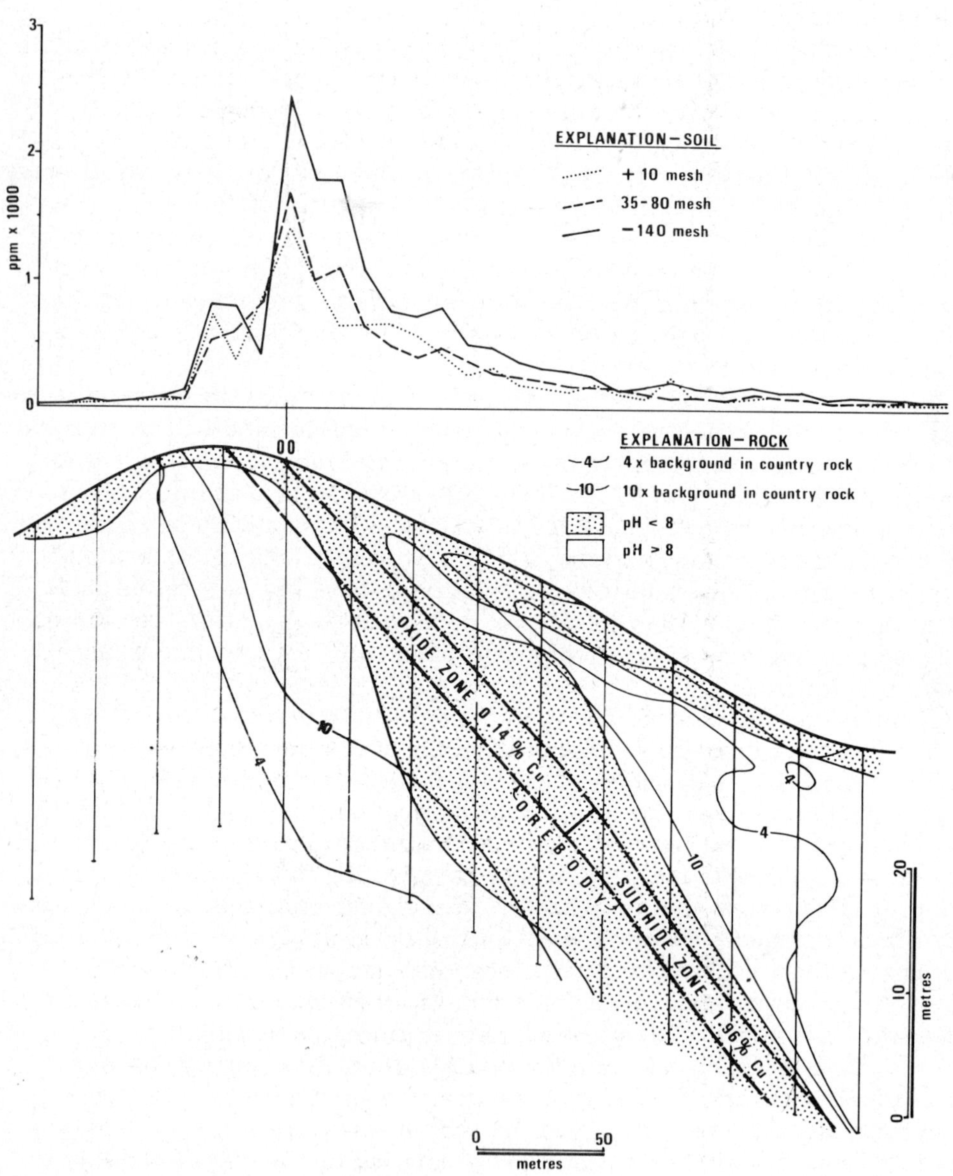

Fig. 4. Otjihase Prospect, Traverse 500. Copper distribution in soil and rock.

surface width of this zone is more than four times greater. There is thus a built-in hydromorphic dispersion pattern, similar in form and intensity to an undisturbed soil anomaly, in the rock beneath the soil.

Downslope movement is apparent in the soil anomaly and has had the effect of extending significant values as far as 200 m from the reef outcrop and 150 m from the nearest significant site of country-rock copper contribution. From the distribution of copper in the various size fractions, the influence of physical weathering, even in this higher rainfall area, can be seen. Movement of soil copper into the finer material (minus 140 mesh) with time (or downslope position from source) is not marked (Table I). Despite this limited movement of ions the minus 140-mesh soil fraction yields better contrast and dispersion than the coarser fractions.

TABLE I

Percentage of total soil copper in minus 140-mesh fraction at Otjihase

At outcrop (%)	Downslope position (%)			
	50 m	100 m	150 m	200 m
30	39	34	43	38

Zinc is relatively more mobile than copper under alkaline conditions (Mason, 1958; Hawkes and Webb, 1962) and this is well illustrated in Fig.5 where the twice background zinc contour outlines the distribution.

A very subdued soil anomaly results from nearly complete leaching and removal of the zinc by acidified groundwater. Only 6.9% of the zinc originally present in the ore appears at surface within the two times background limits, and of this 1.9% (or 27% of the zinc in the soil) is contributed directly from reef material. The figure of 93% of zinc lost in the sub-soil zone of weathering could possibly be higher where less alkaline groundwaters are associated with zinc-bearing massive sulphides.

The Ongombo study was complicated by the discovery that the apparently residual soil here contains a sufficient proportion of aeolian material to explain the subdued soil anomaly. In addition the ore body was found to be poorly developed under the section chosen for study. In spite of these factors the results are considered to be of interest and are shown in Fig.6, in which, as the country rock is 0.6 pH unit more alkaline than Otjihase, the pH 8.5 contour has been used.

The sharply restricted area of lower pH, contrasting strongly with that at Otjihase, is immediately apparent and is considered to be a reflection of the different sulphide content of these two bodies.

A minimum value of pH 5.0 was obtained at the top of the sulphide zone and the country rock and soil averages are 9.7 and 7.3 pH units, respectively. Consequently movement of copper into the country rock is much less pro-

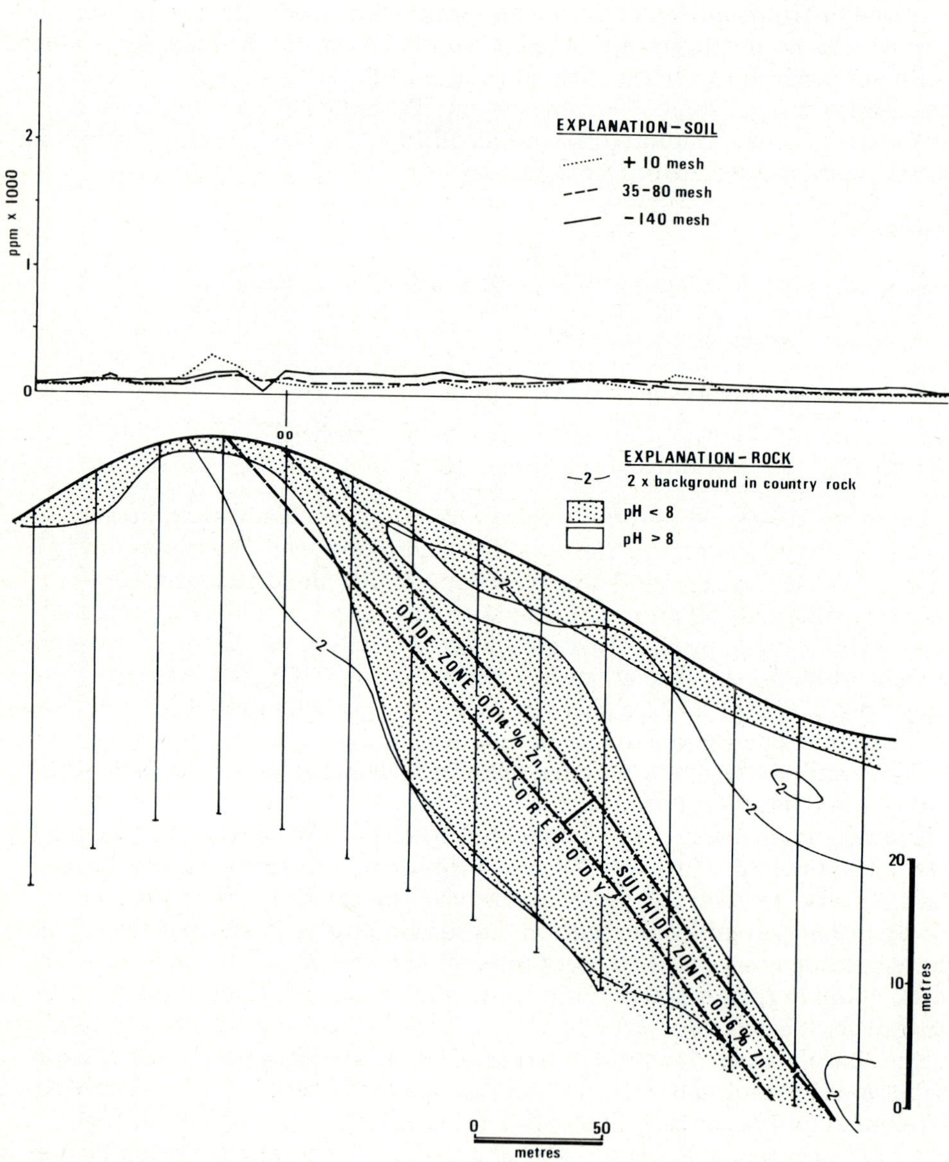

Fig.5. Otjihase Prospect, Traverse 500. Zinc distribution in soil and rock.

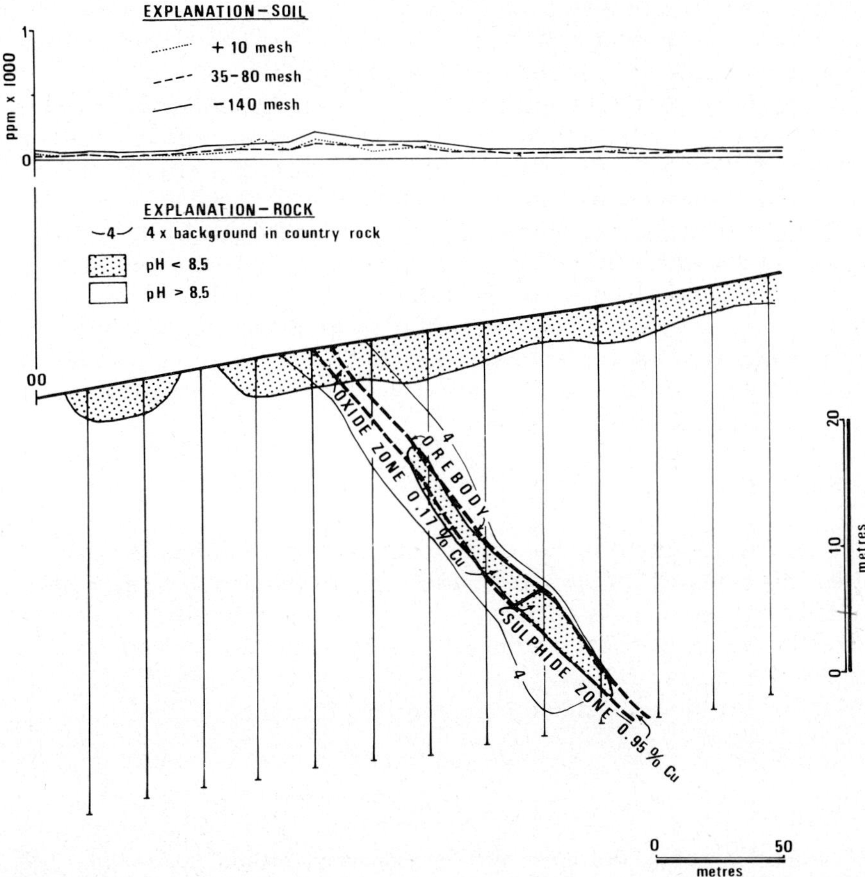

Fig.6. Ongombo Prospect, Traverse 300. Copper distribution in soil and rock.

nounced than at Otjihase and the amount of sulphide-derived copper which would appear as an identifiable anomaly at surface has increased slightly to 28%. Of the total copper which would appear at surface between the four times background limits at Ongombo, however, 64% is contained in the oxidized reef sub-outcrop.

CONCLUSIONS

Soil copper dispersion is adversely affected by arid conditions but, in the special case of massive sulphide bodies, a sub-surface hydromorphic dispersion pattern is introduced into the soil from weathering bedrock. This dispersion into the wall rock results from solution and transport of copper in dilute sulphuric acid solutions formed by oxidation of the sulphide ore. The pattern shape is a function of groundwater movement and may result in surface anomaly patterns unrelated to topography. In the case of zinc, which is more mobile

492

under alkaline conditions, a very subdued profile, or possibly no detectable
anomaly at all, may be present at surface. It appears unlikely, however, that
sub-surface copper could be completely dissipated by this process.

In the discovery area environment the minus 140-mesh soil fraction gives
better copper and zinc contrast than the other sizes used and the anomalous
values extend further from the outcrop in this size range. For copper at least
this appears to be applicable even in largely wind-blown material.

The percentages of identifiable ore-body copper, which reach the surface
at Otjihase (23%) and Ongombo (28%), are surprisingly alike and could be
applicable to similar deposits elsewhere. If this can be proved and the reef
width is known it may be possible to calculate an approximate ore-body
grade from the excess of copper above background in the residual soil. Surface
distortion of the soil anomaly and the rate of removal of soil copper in solution
would, however, be complicating factors.

ACKNOWLEDGEMENTS

Grateful acknowledgement is due to the management of Johannesburg
Consolidated Investment Company for permission to publish this paper and
to all the staff members who assisted in data collection. In addition special
thanks are due to Colin Bleach for draughting the figures used, to Johann
Vermeulen for the assays, to Naomi Lehnerdt for typing the manuscript and
to Mr. C. Cousins for his helpful and constructive criticism.

REFERENCES

Ahrens, L.H., 1954. Log-normal distribution of the elements. Geochim. Cosmochim. Acta,
    5: 49—73
Bateman, A.M., 1951. The Formation of Mineral Deposits. John Wiley and Sons, New
    York, N.Y., 371 pp.
Blanchard, R., 1968. Interpretation of leached outcrops. Nevada Bur. Min. Bull., 66: 195 pp.
Cornwall, F.W., 1961. Geochemistry — Chartered Exploration. In: F. Mendelsohn (Editor),
    The Geology of the Northern Rhodesian Copperbelt. MacDonald, London, pp.187—208
Garrels, R.M. and Christ, C.L., 1965. Solution, Minerals, and Equilibria. Harper and Row,
    New York, N.Y., 450 pp.
Hansuld, J.A., 1966. Eh and pH in geochemical exploration. Can. Inst. Min. Metall. Bull.,
    59(647): 315—322
Hawkes, H.E. and Webb, J.S., 1962. Geochemistry in Mineral Exploration. Harper and
    Row, New York, N.Y., 415 pp.
Martin, H., 1961. The Damara System in South West Africa. 4th Meet. 5th Reg. Com.
    Geol., Pretoria, C.C.T.A. Bull., 80: 91—95
Mason, B., 1958. Principles of Geochemistry. John Wiley and Sons, New York, N.Y., 329 pp.
Park, C.F. and MacDiarmid, R.A., 1964. Ore Deposits. W.H. Freeman and Co., San
    Francisco, Calif., 475 pp.
Philpott, D.E., 1975. Shangani — a geochemical discovery of a nickel—copper sulphide
    deposit. In: I.L. Elliott and W.K. Fletcher (Editors), Geochemical Exploration 1974.
    Elsevier, Amsterdam, pp.503—510
Sharpe, J.W.N., 1964. The Empress nickel—copper deposit, southern Rhodesia. In: S.H.
    Haughton (Editor), The Geology of Some Ore Deposits of Southern Africa, II. Geol.
    Society of South Africa, Johannesburg, pp.497—508

GEOCHEMICAL BEHAVIOR OF GALENA UNDER SEMI-ARID CLIMATIC CONDITIONS: AN EXAMPLE FROM UPPER VOLTA, WESTERN AFRICA

PIERRE J. GOOSSENS

Department of Geology and Geological Engineering, Michigan Technological University, Houghton, Mich. (U.S.A.)

ABSTRACT

A small but high-grade galena vein has been discovered using geochemical exploration methods in northwestern Upper Volta. The climate of the area is transitional between north sudanian and sahelian, with an evaporation/precipitation ratio of about 5/1. This unusual western African mineralization occurs in an alluvial fan containing transported fragments of lateritic material in a clay matrix. Bedrock is composed of schists and quartzites of the Precambrian Birrimian system. Following discovery of cerussite fragments, a soil geochemical survey was carried out in an area covering 1.2 km² using a 50-m square grid. A 10-m square grid was used in anomalous zones. Samples were analysed by AAS for lead, copper, zinc, nickel and silver. Plots of concentration against cumulative frequency and computer facilities were used for the statistical treatment of the results.

Geochemical background levels for lead, copper, zinc and nickel are respectively 45, 60, 42 and 30 ppm; thresholds are respectively 200, 150, 95 and 80 ppm. The area of anomalous lead values seems to trend parallel to the vein strike along a zone where trenching led to the discovery of a massive vertical, 1 m thick, argentiferous galena vein a few meters below the surface. The vein occurs within an amphibolite unit.

Sampling on both sides of the vein, in deep trenches and in shallow drill-holes, was carried out in order to trace the geochemical dispersion of the lead and other metals (Cu, Ag, Zn, Ni) which occur in trace amounts in the galena. Lead has been traced in the soil for distances of up to 15 m laterally from the vein. Silver is present within the lead zone but in very small concentrations. Copper can be traced over a larger area. Zinc and nickel are sporadically distributed and apparently not related to the zones containing lead anomalies.

Minerals identified in the deposit include Pb- and Cu-phosphates, Fe-hydroxides, anglesite, cerussite and malachite, covellite and various clay minerals. The primary sulphide-oxidation envelope shows a characteristic mineral zoning with development of covellite at the contact with the galena. The covellite is surrounded by an anglesite zone which is surrounded by cerussite and malachite zones. The carbonate zone is surrounded by a Pb- and Cu-phosphate zone. This zoning (sulphides → sulphate → carbonates → phosphates) indicates an adjustment of the lead and copper minerals to increasing alkalinity in the soil.

The use of the secondary lead (and Cu) dispersion patterns in regional soil geochemical exploration is limited by morphologic and climatic conditions. Other methods of prospecting for massive sulphide mineralization might be more successful in this sector of western Africa.

INTRODUCTION

In 1969, the "Direction Générale des Mines" (DGM) of Upper Volta identified cerussite in a piece of rock obtained near Gan, a small village at latitude 13°15′30″N and longitude 2°53′00″W (Fig.1). Because pieces of cerussite and anglesite were found in the area during a subsequent field examination, a limited soil geochemical survey was undertaken. This survey led to the discovery of a concealed small, but massive, vein of galena (Goossens, 1970).

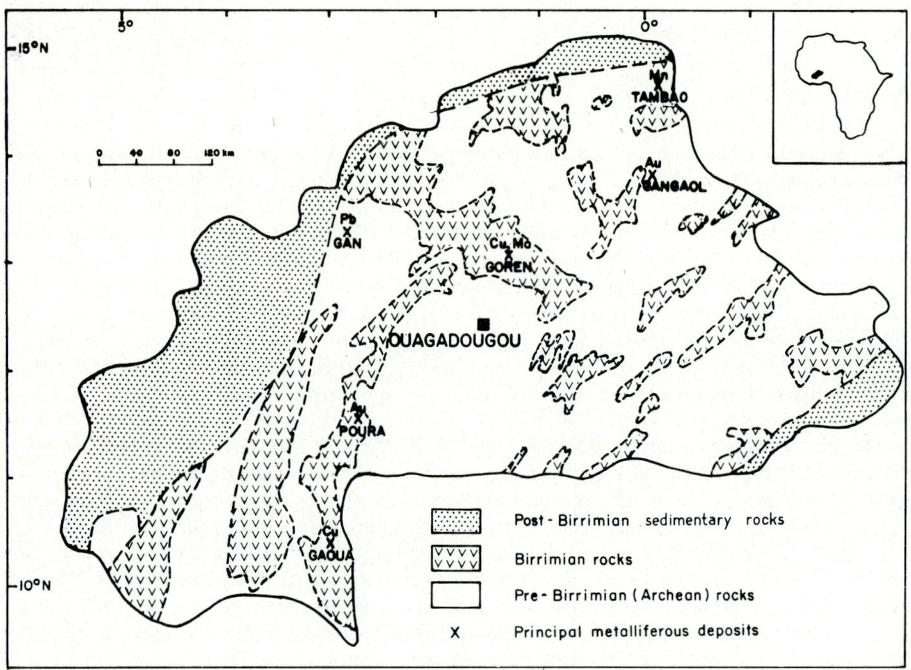

Fig.1. Simplified general geology of Upper Volta with the location of the principal metallic mineralizations (Gaoua and Goren: porphyry copper; Tambao: manganese oxide deposit; Poura: Homestake-type gold deposit; Gangaol: isolated gold-quartz vein) and the location of Gan lead prospect. Note the "greenschist belts" (in chevrons) are generally trending north-northeast, except, north of Ouagadougou, where they intersect another trend east-southeast (modified after Goossens, 1971).

The area of interest is located on a large, flat, granitic plain where small hills of Birrimian rocks capped by laterite form the only topographic relief. Intermittent streams have produced a large alluvial fan which slopes gently to the south-southwest. Vegetation is scarce savanna type in the dry season (October to May) and is essentially composed of small, thorn trees and a few baobabs. During the rainy season, the area is cultivated for millet.

The climate is transitional between north sudanian and sahelian with a rainfall ranging between 500 and 700 mm per year. Temperature ranges between 19° and 40°C during the dry season and 20° and 35°C during the wet season. The evaporation/precipitation ratio is high (5/1) and the streams are active only during the 3-month rainy season.

REGIONAL GEOLOGY

The oldest rocks in the Gan area are pre-Birrimian granites and migmatites with general lineations trending N10°E to N50°E (Fig.1). This Archean shield is covered and/or intruded by granitoid and gabbroic rocks and a series of volcano-sedimentary rocks forming part of the Birrimian system. The Birrimian orogeny is assumed to have occurred at the end of early Proterozoic time (Brunnschweiler, 1974). Birrimian metavolcano-sedimentary rocks occur in elongated belts and resemble in many aspects the Archean greenschist belts of the Canadian Shield. Near Gan, Birrimian rocks are exposed and apparently overlie the Archean basement. Lead mineralization is found within these metavolcanic and metasedimentary rocks which have a schistosity trending N65°E. All metallic mineralization in Upper Volta is found within the Birrimian rocks.

In the vicinity of lead mineral occurrences, the metamorphosed volcano-sedimentary sequence is composed of partially bleached amphibolites and black quartzites called by the French geologists "quartzites à minerai" (Ducellier, 1963). In thin section this quartzite is seen to be composed of almost 50% manganese oxides; the rest being quartz, muscovite and light brown biotite. Minor amounts of chalcopyrite and digenite are sometimes present. The texture suggests a metamorphosed manganiferous volcanic tuff. Encrustations of malachite and dioptase and other unidentified variegated oxides are commonly developed on "quartzite" outcrops near Gan. The newly discovered galena vein is found within the amphibolites and trends N110°E making an angle of 50° with the schistosity.

Birrimian rocks normally form the only topographic features in this part of Upper Volta, where they are capped by laterite. Erosion of these lateritic caps provides fragments of iron oxides and hydroxides, quartz, and clay minerals, which accumulate in topographic depressions. In the mineralized area, the amphibolite is covered by a thick overburden that is partly residual in the lower section and is eluvial in the upper section. A typical soil profile in the mineralized area shows, in the C horizon, bleached altered schist fragments in a grey clayey matrix. Clay minerals become more abundant upwards and grade into a saprolitic horizon. The C horizon and the saprolite are probably the only residual type of soil. The upper section of the soil profile is progressively enriched in lateritic gravels with less abundant clay. This upper horizon may represent the eluvial soil arising from the destruction of the fossil lateritic cappings. Overburden thickness varies greatly but is usually more than 3 m. Silty alluvium accumulates in shallow depressions.

In conclusion, the soil profile in the mineralized area near Gan can be summarized as a typical C horizon and saprolite, both residual, which are overlain by transported eluvial and alluvial material.

LEAD OXIDE MINERALOGY

Because no other galena occurrence had been reported in this part of western Africa, a mineralogical study of the different encrustations and oxidation products collected around the galena vein was undertaken in order to determine the geochemical behavior of the lead under the prevailing climatic conditions. The main objective was to find which element or elements besides lead could be used effectively in reconnaissance exploration for galena mineralization.

A typical zoning of oxidation products has been found around the unaltered galena (Fig.2). The presence of copper in the galena has been reported from qualitative spectrographic analyses. A thin zone of covellite encloses the primary lead sulphide. The covellite zone is surrounded by anglesite which is the first oxidation product of galena to be formed. The lead sulphate is surrounded by a larger zone of cerussite and copper carbonates, malachite and azurite. Cerussite and copper carbonate-bearing gravels are very common in the area. Around this carbonate zone a wide lead-phosphate zone is developed. Lead phosphates are very common in the area and easy to find due to their variegated colours.

This zoning (Pb sulphide → Cu sulphide → Pb sulphate → Pb and Cu carbonates → Pb and Cu phosphates) clearly indicates an adjustment of the lead (and Cu) minerals to an increase in the alkalinity of the soil, both laterally and vertically, from the primary sulphide. Garrels and Christ (1965, pp.236 and 240) investigated the stability fields of the lead and copper minerals under variable conditions of pH and Eh, at 25°C, with a total dissolved sulphur species = 10^{-1} and a partial pressure of CO_2 between $10^{-3.5}$ and 10^{-4}. At an Eh between 0.0 and +0.1 (oxidation conditions) and with an increasing pH, the mineral stability fields progress from galena to covellite to anglesite to cerussite + malachite (Fig.3). Although partial pressure of phosphorus has not been taken into account in this diagram it is known that the phosphate precipitates under alkaline conditions (Krauskopf, 1967, p.90). It is assumed the lead-phosphate field of stability overlaps the right side of the cerussite and malachite field of stability, i.e., at higher pH. This will depend largely on the availability of $(PO_4)^{3-}$ and its interaction with $(CO_3)^{2-}$ (Gulbrandsen, 1969). Thus the mineral zoning observed at the Gan galena occurrence agrees with the calculated Eh—pH variation diagram.

GEOCHEMICAL SURVEY

The geochemical soil survey centred around the cerussite occurrences and covered an area of 1.2 km^2. Samples were collected every 50 m along north—south and east—west lines on a square grid pattern. Following this first survey,

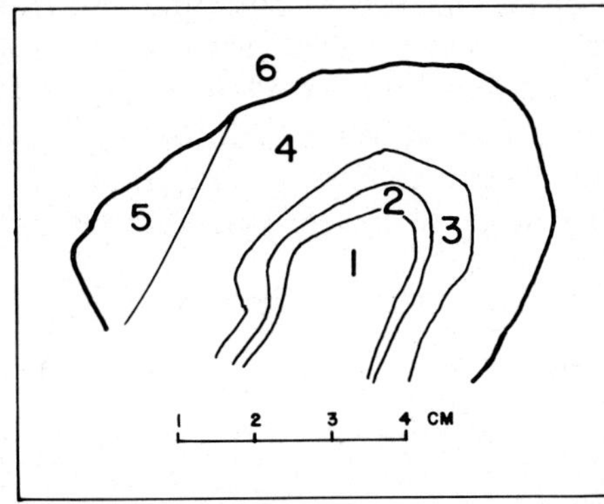
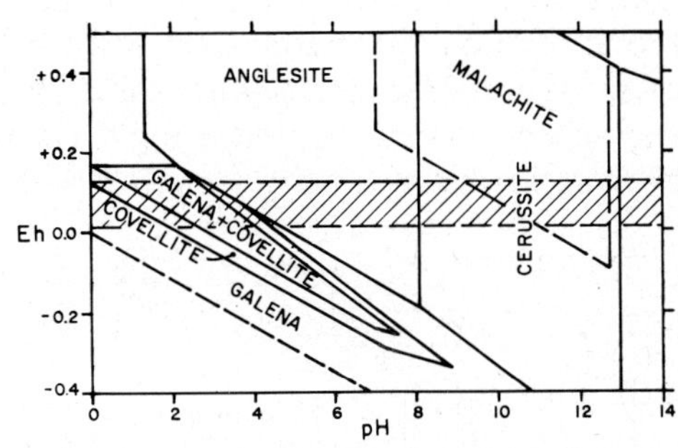

Fig.2. Oxidation zoning of a galena vein, near Gan, Upper Volta. This sketch is made from an ore hand specimen. The thickness of individual zones can vary but generally increases towards the phosphate zone; 1 = galena, 2 = covellite, 3 = anglesite, 4 = cerussite, 5 = malachite, 6 = lead and copper phosphates.

Fig.3. Stability relations among lead and copper compounds in water at 25°C and 1 atm total pressure. Total dissolved sulphur = 10^{-1}, P_{CO_2} = $10^{-3.5}$ to 10^{-4} (Garrels and Christ, 1965, pp.236 and 240). The hatched area is assumed to represent the Eh and pH conditions in the oxidation zones of the galena vein near Gan, Upper Volta.

two areas were selected for more detailed work, using a 10-m square grid. One area centred around a concentration of cerussite fragments whilst the other covered the lead zones in the southeast corner of the original area. In order to study the vertical and lateral dispersion of lead and other metals, soil profiles were sampled in trenches near the galena vein and in auger drill holes in non-mineralized areas.

Soil samples were taken deep enough to avoid the alluvium and plant roots (about 30 cm). The soil samples were shipped to Ouagadougou where the U.N. project has its laboratory facilities. There the samples were crushed to 200 mesh and analysed for lead, silver, copper, zinc, and nickel by atomic absorption spectrophotometry following dissolution in concentrated nitric acid (Pb, Ag) or aqua regia (Cu, Zn, Ni).

The chemical data for lead, copper, and zinc have been interpreted using graphical plots of concentration versus cumulative frequency (Lepeltier, 1969) and also using a computer program written by G. Carlson (Geology Department, M.T.U.). Nickel and silver do not show any significant trends and were therefore not entered into the program.

Discussion of the results

Lead has a background of 45 ppm which is unusually high and is due to the derivation of the samples from a lead-bearing area. For the soil in Upper Volta, Borucki and Zeegers (1970) estimated lead regional background to be between 4 and 24 ppm. The lead threshold (Lepeltier, 1969, p.545) is 200 ppm and these values are contoured with a heavy line in Fig.4. As a result of the contouring two anomalous lead zones are well defined; one in the west of the prospected area and another in the southeast. The area in the southeast with the highest lead values was resampled using a 10-m grid. Auger holes were drilled on a 25-m grid; and two deep trenches were excavated. Although extremely high

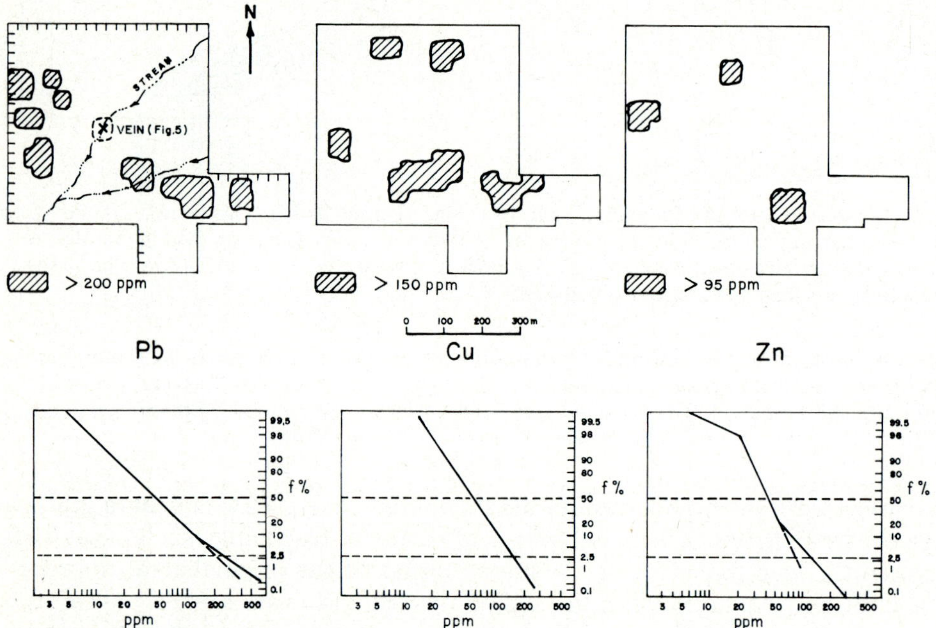

Fig.4. Lead, copper and zinc in soil, Gan area. Cumulative frequency versus element concentration diagrams for lead, copper and zinc.

lead values were reported (up to 3000 ppm), no sulphide mineralization was found. The vertical distribution of the lead in the drill holes is erratic with some high values being found in the upper part of the soil whilst others are found grouped close to the bedrock.

After a detailed soil geochemical survey was carried out using a 10-m grid in the centre of the prospected area a galena vein was discovered by trenching (Fig.4). A concentration of cerussite fragments was found in this area. Although the soil survey using a 50-m sampling interval did not reveal any particularly anomalous zone, the soil survey using a 10-m interval recorded values up to 1600 ppm Pb. Trenches were excavated below these high values and the galena vein was found. Fig.5 shows the distribution of lead, copper, and zinc in three vertical sections. Section 1, 0.30 m away from the sulphide vein, returned values up to 8000 ppm Pb. Section 2, 3 m away from the sulphide vein, contained lead values only up to 1000 ppm. Section 3 is 15 m away from the sulphide vein and no anomalous lead values were recorded.

According to these data, the migration of lead is restricted, both laterally

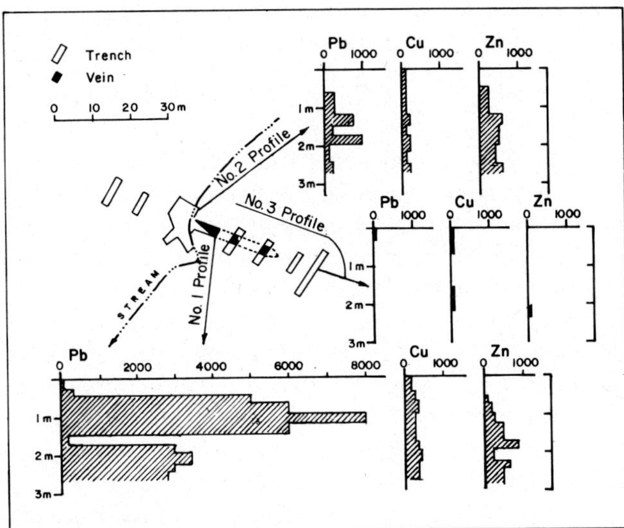

Fig.5. Vertical distribution of lead, copper and zinc (ppm) in trenches made at different distances from the Gan galena vein (section 1: 0.30 m distant, section 2: 3 m distant, section 3: 15 m distant).

and vertically, to a very small envelope less than 15 m in extent around the primary sulphides. Within this secondary lead halo the values decrease sharply with distance from the vein to background levels. In this zone lead is fixed as sulphates, carbonates and, to a larger extent, phosphates.

It should be pointed out that the areas of anomalous soil are topographically below the known galena vein. Therefore, a certain amount of mechanical transport is assumed. This mechanical transport is probably due to the migration of the upper soil material during the rainy season. If this assumption is true, more mineralization might be found to the north and northeast and above the areas containing lead anomalies.

Copper has a mean background concentration of 60 ppm. The threshold is estimated (Fig.4) at 150 ppm. The distribution of values $\geqslant$ 150 ppm shows a relatively good correlation with the anomalous lead areas, especially in the southeastern corner of the prospected area. The distribution of copper in drill hole samples from the southeastern area shows extremely good correlation with lead. The same can be said for the distribution of copper in the three profiles of Fig.5.

Zinc has a background value of 42 ppm. According to the cumulative frequency distribution (Fig.4) and the method of data interpretation suggested by Lepeltier (1969, p.545), the zinc values reflect two geochemical populations. This multiple population will not be explained here. The threshold level for zinc is estimated at 95 ppm. It is considered that the zinc distribution simply reflects dispersion associated with the lead sulphide mineralization and that no zinc mineralization can be expected to occur.

Silver was only determined in a few samples and values did not exceed 8 ppm, although the galena contains as much as 500 ppm Ag.

CONCLUSIONS

In the immediate vicinity of the Gan galena occurrence lead and copper show relatively good correlation and their secondary dispersion patterns seem to be well developed. The area covered by this secondary dispersion pattern appears large enough to be used as pathfinder in local soil geochemical sampling programs. Under the prevailing alkaline soil conditions, the mobility of these two metals is restricted by their fixation in supergene carbonate and phosphate minerals. The dispersion of these elements is largely controlled by the degree of mechanical transport, i.e., topography and erosion. Topographic conditions must, therefore, be taken into consideration when planning soil geochemical surveys. Topography as well as the climatic environment will also control soil development. In the Gan area, residual soil is covered by transported soil. This upper, transported overburden can be moved over large distances and can completely conceal the secondary dispersion pattern developed in the residual soil. If the residual soil is well developed and not covered by transported overburden, lead and copper dispersion patterns will be developed but will be restricted to a very small area. Under the climatic conditions such as obtain near Gan and in the major part of Upper Volta regional soil surveys using 1000 m x 1000 m or even 500 m x 500 m square grids will probably not locate any anomalous samples.

The Precambrian geology of Upper Volta, particularly the Birrimian metavolcano-sedimentary series, is similar in many aspects to the Archean, greenschist belts of the Superior province of the Canadian shield (Hutchinson, 1974). Therefore, there exists a good possibility that volcanogenic massive sulphide deposits might be present, both in the Archean and Proterozoic systems. Although intensive geochemical surveys have been carried out in Upper Volta (Borucki and Zeegers, 1970), only sub-economic disseminated copper sulphide mineralization has been discovered by geochemical exploration to date. This unsuccessful result is considered to be largely due to the lack of large secondary halo development above massive mineralization as described above.

Stream sediment geochemical surveys are inappropriate in this part of Upper Volta. The alluvium is poorly developed and, during the dry season, dust wind and storm accumulate large amounts of sand and silt in the creeks.

It is suggested here that mineral exploration programs in Upper Volta should emphasize other ways of prospecting such as geochemical analysis of bedrock samples, surface geological investigations (with special attention to variegated rock fragments produced by the oxidation of sulphide ores), and geophysical exploration techniques.

ACKNOWLEDGEMENTS

The author is grateful for the information and technical help provided by
G. Carlson, J. Moreau, A. Ruotsala, N. Scofield, and H. Zeegers. The manuscript has been reviewed by C. Lepeltier and A. Ruotsala; their comments and
interest were helpful and appreciated. The drawings were performed by
P. Ostlender.

REFERENCES

Borucki, J. and Zeegers, H., 1970. Champ d'Application de la Géochimie comme méthode
  d'exploration du Birrimien en Haute Volta. Projet Tambao, Nations Unies, Rapp. No.10
  (unpublished)
Brunnschweiler, R.O., 1974. New K—Ar age determinations from the West African Shield
  in the Niger Republic. Geology, 2(1): 17—20
Ducellier, J., 1963. Contribution à l'étude des formations cristallines et métamorphiques du
  Centre et du Nord de la Haute Volta. Mém. BRGM, No.10 (Paris)
Garrels, R.M. and Christ, C.L., 1965. Solutions, Minerals, and Equilibria. Harper and Row,
  New York, N.Y.
Goossens, P.J., 1970. Reconnaissance Géochimique de la Zone Plombifère de Gan. Projet
  Tambao, Nations Unies, Rapp. No.7 (unpublished)
Goossens, P.J., 1971. Les gisements du type porphyry copper: géologie, prospection,
  économie et découvertes récentes en Afrique de l'ouest. Min. Métall. (Paris), October,
  pp.227—228; November, pp.257—261
Gulbrandsen, R.A., 1969. Physical and chemical factors in the formation of marine apatite.
  Econ. Geol., 64: 365—382
Hutchinson, R.W., 1974. Volcanogenic sulfide deposits and their metallogenic significance.
  Econ. Geol., 68: 1223—1246
Krauskopf, K.B., 1967. Introduction to Geochemistry. McGraw-Hill, New York, N.Y.,
  706 pp.
Lepeltier, C., 1969. A simplified statistical treatment of geochemical data by graphical
  representation. Econ. Geol., 64: 538—550

SHANGANI — A GEOCHEMICAL DISCOVERY OF A NICKEL—COPPER SULPHIDE DEPOSIT

D.E. PHILPOTT

Johannesburg Consolidated Investment Company Limited, Salisbury (Rhodesia)

ABSTRACT

Ni—Cu mineralization occurs in the Archean rocks of the Shangani area generally within sequences comprising tholeitic basalts and ultramafics.

The discovery of the Shangani mine in February 1970 was the result of the pragmatic application of geochemical prospecting techniques.

INTRODUCTION

Rhodesia is situated between latitudes 15°30'S and 22°30'S and longitudes 25°15'E and 33°00'E. It is largely savannah country with an altitude range of 550—2590 m above mean sea level. Along latitude 19°00'S the annual rainfall varies from less than 600 mm to as much as 1600 mm.

During the period between 1969 and 1972 exploration in Rhodesia proceeded at a rapid and competitive rate. Emphasis was placed on a rapid appraisal of geological environments, securing ground and a thorough and systematic approach to exploration. All potentially mineralized environments were explored until they were proved negative by applying multiple techniques.

During this period, the J.C.I. exploration team comprising six geologists, eight semi-skilled and about 140 unskilled staff, collected 737,967 samples from an area of 5,562 km² and carried out over 2 million metal analyses.

In addition to the Shangani discovery nickel sulphides were located in four other areas.

REGIONAL GEOLOGY

The Archean rocks of the Shangani area occur in a well-defined east—west trending structure which is essentially a syncline having its axis aligned in a northwest—southeast direction (Fig.1). The stratigraphic succession is well exposed in the northern part of the area but further south exposures are discontinuous and in places entirely absent.

The rocks can be divided into two main groups as follows:

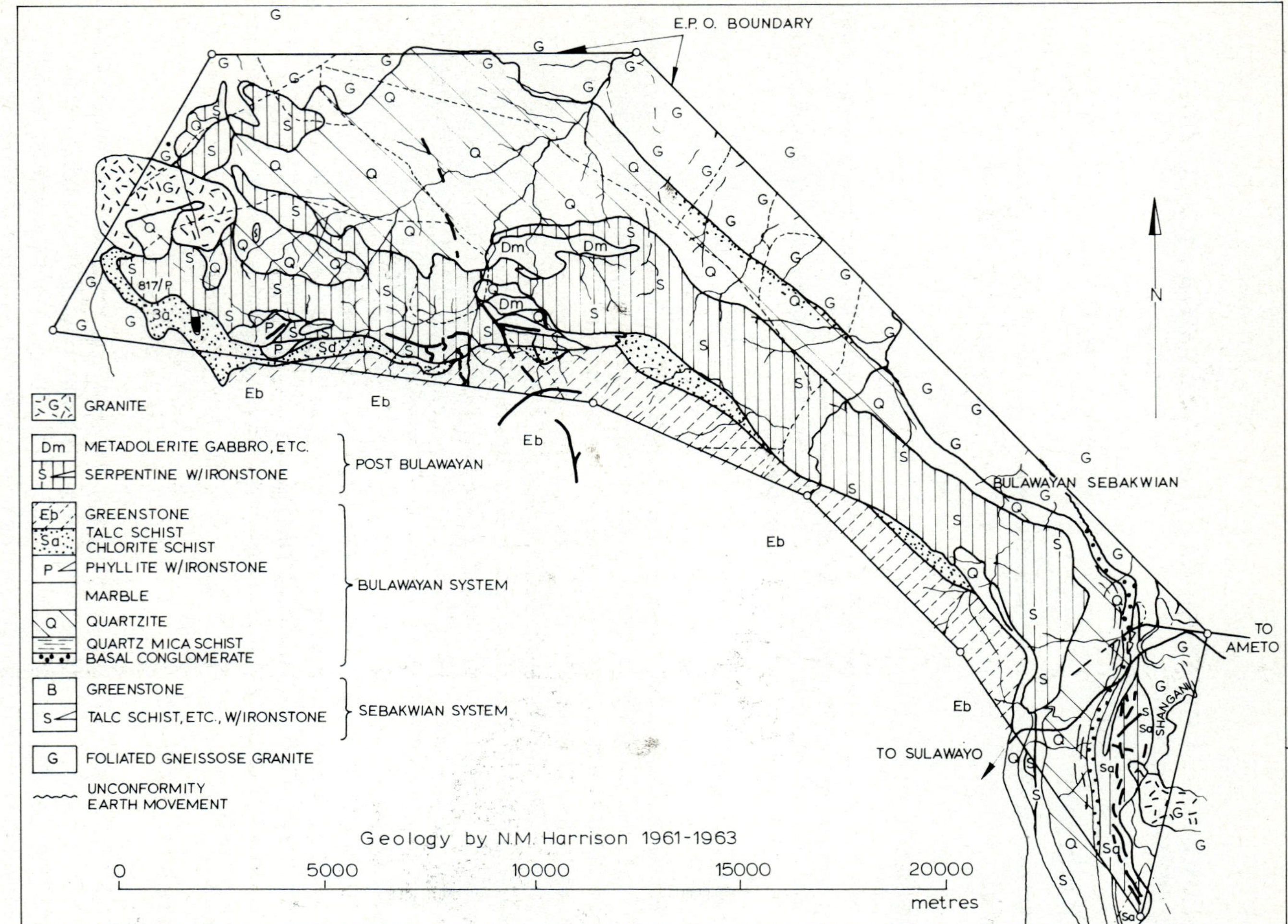

Fig.1. Shangani E.P.O. No.233 — geological plan.

(1) The early Archean Sebakwian rocks consisting of serpentinites and associated ultramafics.

(2) The middle Archean Bulawayan assemblage of metamorphosed tholeiitic basalts ("greenstones") and differentiated ultramafic rocks.

The Sebakwian rocks

Intermediate tuffs and agglomerates. An extended sequence of agglomerates and water-lain tuffs comprise the lowermost members of the succession. Metamorphic effects attributable to granitic intrusions cause local variations in lithology. The poorly exposed tuffaceous rocks are the principal rocks in the vicinity of the Shangani mineralization.

Carbonaceous shale.

This is a very distinctive horizon found along the entire strike length of the Shangani ultramafic belt. It is generally poorly exposed and often displaced by faults and dykes. The shale always contains pyrite and pyrrhotite and is recognizable in surface exposures by a limonitic or cherty opaline gossan containing low nickel and erratic copper values (Table I).

Ultramafic rocks. Closely associated with the shale are assemblages of serpentinite. These are often talcose and carbonated and have margins of amphibolite and pyroxenite.

The Bulawayan rocks

The rocks of this system although having some lithological similarities to the Sebakwian have a quite distinctive chemistry.

Extrusive rocks. The Bulawayan system consists mostly of a massive sequence of pillowed metabasalts of tholeitic affinity.

Intrusive ultramafics. The lavas are intruded by large differentiated ultramafic bodies comprising serpentinites, dunites, peridotites, pyroxenites and gabbros. These rocks occur throughout the Shangani area and have a distinctive geochemical character.

The serpentinites, which tend to form hills, are often weathered and altered

TABLE I

Carbonaceous shale values

Sample	Cu (ppm)	Ni (ppm)	Zn (ppm)	Pb (ppm)
1	520	100	70	50
2	400	100	70	50
3	280	200	250	50
4	125	100	20	20
5	100	90	15	25
6	125	200	20	20

to truly berberitic serpentinites. In these high relief areas they are characterized by cherty, opaline horizons containing high nickel values due to concentration during weathering.

EXPLORATION PROCEDURE AT SHANGANI

Stage one

After selection of the area on geological grounds and securing the right to
prospect a soil sample grid was laid out using an east—west baseline.

Samples were collected from a depth of 25—30 cm at 60-m intervals along
north—south traverse lines 300 m apart. Where no soil was present rock
samples were collected. Rock types, soil types and vegetation assemblages
were recorded during the sampling traverse leading to the rapid production of
a basic geological map. An area of about 500 km^2 was prospected in this way
in about six weeks.

The pre-numbered samples were sent to the base camp laboratory to be
crushed and pulverized to pass through an 80-mesh screen. After attacking the
samples with a mixture of concentrated perchloric acid and 5% nitric acid followed by dilution to 10 ml with 5% hydrochloric acid, the metals of interest
were determined by atomic absorption techniques. With two instruments the
laboratory had a capacity of 120,000 determinations per month permitting
analytical data to be returned to the field crews within seven to ten days from
their dispatch from the field.

When the analytical results were available they were plotted and anomalous
areas given a priority rating, A, B, C or D. The ratings are defined as follows:
C and D priority is assigned to those anomalies which, from a study of the
aerial photos and geology, can be confidently attributed to sources other than
mineralization, e.g. copper values as high as 150 ppm would be given a C—D
rating if they occurred in an area of mafic dykes. About 40% of such anomalies
would be re-examined to confirm the interpretation. Priority B anomalies are
those which cannot confidently be assigned to geological causes. All priority B
anomalies were re-examined. Anomalies which are obvious and distinctive,
especially when supported by a broad area of intermediate values, are given an
A priority rating and are investigated rapidly and intensively.

Ultramafic rocks of all types were usually indicated by erratic nickel and
cobalt values in soils. Serpentinites were characterized by very high nickel
values (> 2700 ppm) and low copper values compared to pyroxenites and
gabbros where lower nickel values were accompanied by significant levels of
copper. Some experience was required to distinguish low contrast Ni—Cu
anomalies due to carbonaceous shale from those due to unserpentinized ultramafics.

Frequently, erratic results were obtained from the survey due to differences
in the soil types encountered. For example, the basalts occur in low-lying areas
and are characterized by black, clayey, neutral (pH 6.0—6.5) soils whilst the

serpentinites have a variable cover of red slightly acid (pH 5.5—6.0) soils.

In the case of the Shangani area the initial survey resulted in the location of eight A anomalies, five B priority anomalies and thirty C—D anomalies (Fig.2). Only the A priority anomalies will be dealt with here. The initial survey revealed one particularly strong and extensive Ni—Cu anomaly near a small hill (Table II).

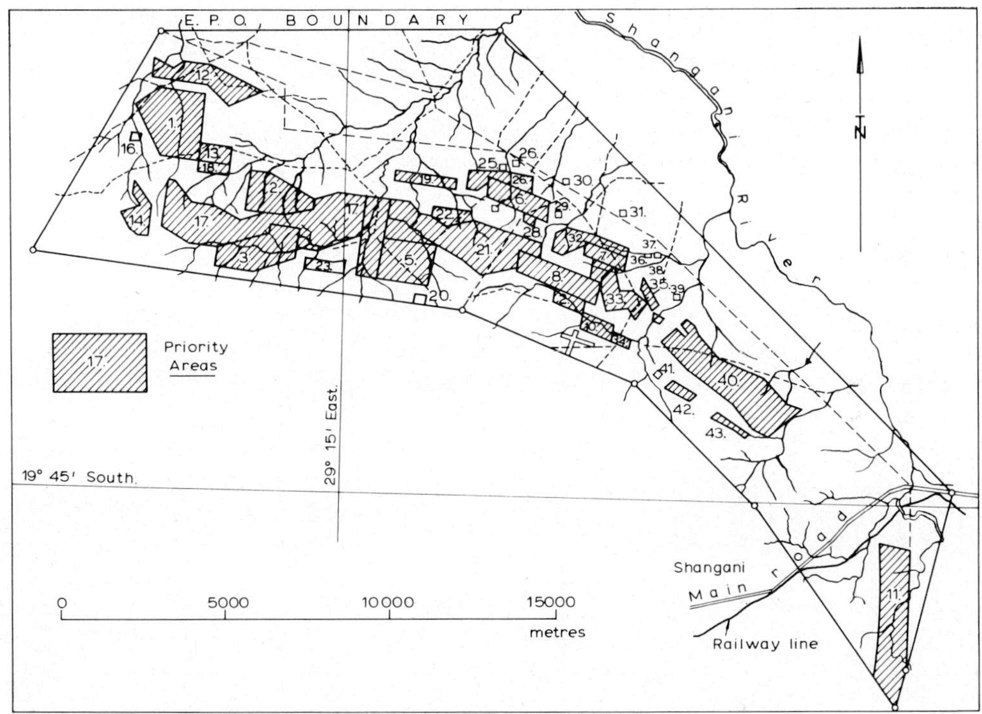

Fig.2. Shangani E.P.O. No.233 — priority plan

TABLE II

Soil analyses of discovery anomaly

	Traverse 330 (spread over 360 m x 300 m)	Traverse 340 (spread over 420 m x 300 m)
Background		
Cu (ppm)	50—80	50—80
Ni (ppm)	500	500
Anomaly		
Cu (ppm)	200	250
Ni (ppm)	3200	4000

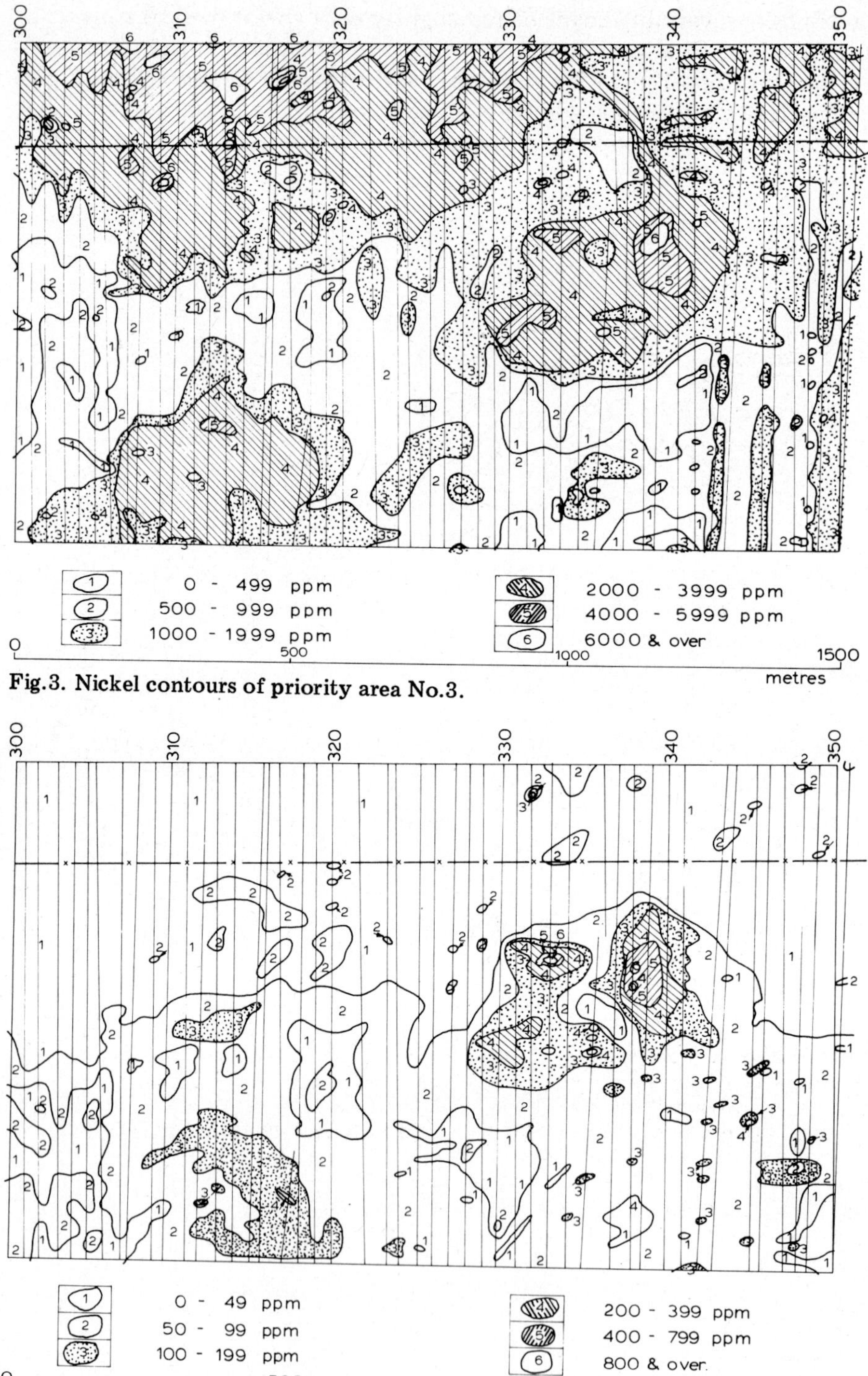

Fig.3. Nickel contours of priority area No.3.

1 0 - 499 ppm
2 500 - 999 ppm
3 1000 - 1999 ppm
4 2000 - 3999 ppm
5 4000 - 5999 ppm
6 6000 & over

0   500   1000   1500
metres

Fig.4. Copper contours of priority area No.3.

1 0 - 49 ppm
2 50 - 99 ppm
3 100 - 199 ppm
4 200 - 399 ppm
5 400 - 799 ppm
6 800 & over.

0   500   1000   1500
metres

TABLE III

Shangani gossan values

Sample	Cu (ppm)	Ni (ppm)	Zn (ppm)	Pb (ppm)
1	3100	5900	500	50
2	1350	1200	25	50
3	1950	1500	50	25
4	3600	1500	190	50
5	1800	1500	45	50
6	1300	1600	30	25
7	2300	2400	50	50
8	2800	1900	30	50
9	1600	3300	90	50

Stage two

Further investigation of the A priority anomalies was undertaken using a more detailed soil and rock sampling programme (Figs.3 and 4, Table II). The most important result of this follow-up work was the discovery of a small (1 m x 6.5 m) scree-covered gossan, on an ultramafic contact. When sampled the gossan returned encouragingly high nickel and copper values

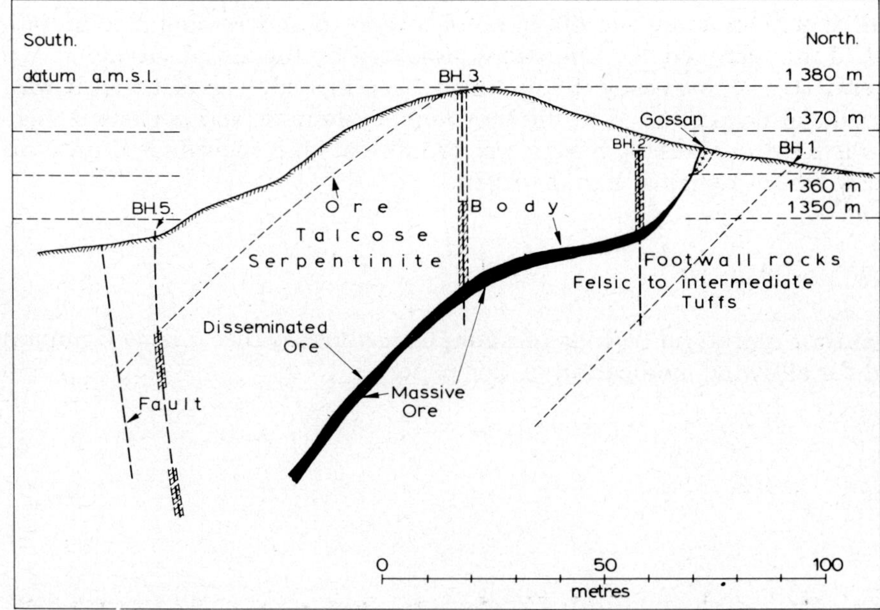

Fig.5. Cross section through hill — Shangani Ni—Cu sulphide deposit.

(Table III). The significance of this gossan was tested using all means available, i.e., trenching, geophysics (E.M. and I.P.), waggon drilling and diamond drilling.

The gossan was shown to be the surface expression of an ore body which at 122 m below surface is 160 m wide and has been proven to extend to a depth of at least 900 m (Fig.5).

TABLE IV

Analyses of Shangani sulphides

Average grade	Massive ore (%)	Disseminated ore (%)
Nickel	9.75	0.56
Copper	1.20	0.08
Cobalt	0.23	0.01
Iron	36.05	7.37
Sulphur	31.15	1.43
P.G.H. plus Au (g/ton)	5.58	0.89

The orebody itself is composed of both massive and disseminated Cu—Ni sulphides of economic grade (Table IV, Fig.6).

COMMENTARY

The Shangani anomaly was only one of several to undergo detailed investigation out of the many Ni—Cu anomalies indicated by the initial survey.

It is easy to see how easily it could have been missed. The fact that it was discovered can be attributed to the thorough, systematic and persistent manner in which every anomaly having any promise at all is re-evaluated until no reasonable doubt remains as to its origin.

ACKNOWLEDGEMENTS

The author is grateful to Johannesburg Consolidated Investment Company Limited for allowing publication of this paper.

RELATION BETWEEN COPPER CONTENT IN SOILS AND COPPER GRADE OF SOME PORPHYRY COPPER DEPOSITS IN HUMID TROPICAL REGIONS

MORITSUNA SAIGUSA

Mitsubishi Corporation, Tokyo (Japan)

ABSTRACT

Before conducting information and outline drilling of a porphyry copper deposit, it is considered possible that the grade of the target can be estimated by a detailed study of soil sampling data. To illustrate how this estimation may be carried out, the geochemical data and the copper content of three porphyry copper deposits in the humid tropics, Santo Niño, Philippines; Nungkok, Malaysia; and Chaucha, Ecuador; are compared.

The ratio of cold-extractable copper to total copper appears to be one of the important factors affecting the estimation of ore grade. This ratio may be influenced by the thickness of overburden. For example at Santo Niño where the overburden is very thin, the ratio is only 0.0055. On the other hand, at Chaucha, where the overburden is very thick, the ratio is 0.35. Depending on the value of this ratio, the copper content of a particular ore body will be from 1.5 to 3 times the total copper value in the soils above the ore body.

INTRODUCTION

The presence of a geochemical anomaly does not guarantee the presence of an ore deposit. To ensure recognition of an ore deposit by geochemical methods many problems, related to the geochemistry of soils and the weathering processes of mineralized rocks, need to be solved. This paper illustrates a method of interpreting geochemical anomalies which has been found useful in exploration for porphyry copper deposits in the humid tropics. The method is based on the relationship between the copper content of soils and of the ore bodies as indicated by drill core assays. The porphyry copper deposits of Santo Niño, Philippines; Nungkok, Malaysia; and Chaucha, Ecuador; will serve as examples.

SANTO NIÑO DEPOSIT

This deposit is located in the mountainous area of northern Luzon Island in the Philippines, 11 km northeast of Baguio City (Fig.1). The area is characterized by steep topography and distinct wet and dry seasons.

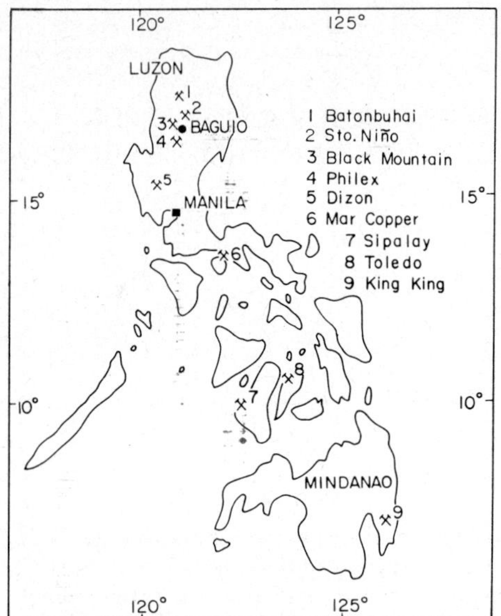

Fig.1. Porphyry copper deposits in the Philippines (2 = Santo Niño).

Geology and mineralization

The geology of the Santo Niño deposit was described by Bryner (1970).
The rock formations present include porphyries, gabbros, volcanoclastics and
metamorphic rocks. Porphyritic rocks, of Neogene age, have an aphanitic texture and a composition, based on the anorthite content of the plagioclase,
ranging from basalt through andesite and dacite to soda rhyolite. Andesite and
dacite varieties are most common. In many cases tectonic crushing and alteration mask the primary characteristics of these rocks.

The volcanoclastic unit, which is the youngest group of rocks in the area,
is confined almost entirely to what appears to be a downfaulted block that
lies to the east of the porphyritic rocks. Most of the volcanoclastics have a
distinctive clastic structure and greenish-white colour. The clast size ranges
from merely coarse grain to cobble conglomerate with little sorting in evidence.
The lithic fragments of the volcanoclastic units are mostly derived from the
porphyritic rocks.

Copper mineralization is preferentially associated with fracture zones in
dacitic porphyries, although some mineralization is found in other rock
types.

The main mineralized zone at Santo Niño has horizontal dimensions of the
order of 330 m × 200 m using a cut-off grade of 0.4% Cu. Chalcopyrite occurs
as open space fillings in shattered and granulated vein quartz and pyrite. Sel-
vedges of molybdenite are observed along faults but molybdenite is not as
abundant as chalcopyrite. Pyrite and magnetite occur as veinlets and disseminations generally replacing mafic minerals.

Soil sampling and analysis

Soil samples were collected from the upper part of the B horizon along a line of drill holes running through the centre of the ore body from east to west (Fig.2).

The minus 80-mesh fraction of the air-dried samples was digested in perchloric acid and analysed for total copper by atomic absorption. Cold-extractable copper was extracted with a mixture of citric acid and hydroxylamine hydrochloride followed by atomic absorption analysis.

Total copper and cold-extractable copper values

Over the mineralized zone total copper values range from 2000 to 4580 ppm compared to less than 2000 ppm in adjacent areas. Very little of this copper is cold extractable. Over the ore body the ratio of cold-extractable copper to total copper (Cx.Cu/T.Cu) averages only 0.005 (Fig.2; Table I).

The overburden in the area is relatively thin due to the steep topography and high rate of erosion. There are differences in thickness, however, and the ratio of copper in drill core to copper in the overlying soil (DH.Cu/T.Cu)

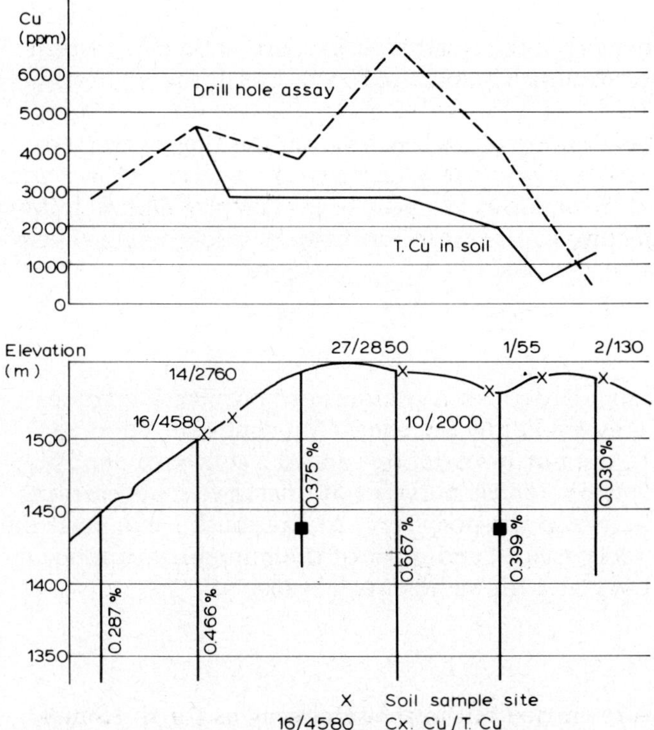

Fig.2. Geochemical sections, Santo Niño deposit.

TABLE I

Geochemical data, Santo Niño, Philippines

	T.Cu (ppm)	Cx.Cu (ppm)	Cx.Cu/T.Cu	DH.Cu (ppm)	DH.Cu/T.Cu
Outside	130	2	0.015	300	2.3
ore zone	55	1	0.018		
Inside	2000	10	0.005	3990	2.0
ore zone	2850	27	0.009	6670	2.3
	2760	14	0.005		
	4580	16	0.003	4660	1.0
Average	3048	16.8	0.005	4770	1.56

seems to be related to these differences. In general, overburden is thicker on the flatter hilltops and gives rise to DH.Cu./T.Cu ratios greater than 2, whereas on the hillsides where there is almost no superficial cover the ratio is close to 1. The overall DH.Cu/T.Cu ratio for this thinly overburdened ore body is 1.5 (Table I).

NUNGKOK DEPOSIT

This copper deposit is located in the northwestern part of Sabah, eastern Malaysia, just to the west of Mount Kinabalu (4101 m a.s.l.), the highest mountain in Borneo (Fig.3).

The mineralization occurs 1300 m above sea level on the rugged western slope of Mount Nungkok (1674 m). In this area there are no distinct wet and dry seasons; heavy rains fall throughout the year to give an annual precipitation of about 3500 mm. In spite of the steep topography thick jungle covers the mountains.

Geology and mineralization

In the vicinity of the Nungkok deposit a granodiorite porphyry intrudes mudstones, siltstones and greywackes of Eocene to Miocene age. There are two areas of mineralization, the northern deposit and the southern deposit. The northern deposit consists of a dense network of quartz veinlets carrying chalcopyrite and pyrrhotite within the porphyry. At the southern deposit the quartz veinlets are more widely spaced and most of the mineralization occurs at the contact of the porphyry and the sedimentary rocks.

Soil sampling and analysis

Sampling and analysis were carried out in the same way as for the Santo Niño deposit.

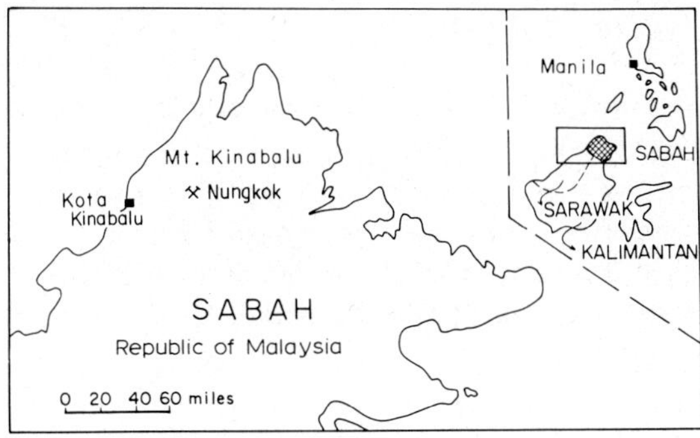

Fig.3. Location map, Nungkok deposit, Sabah, Malaysia.

Total copper and cold-extractable copper values

At the southern ore body total copper in the overlying soils ranges from 1200 to 2800 ppm and cold-extractable copper from 200 to 450 ppm. At the northern ore body total copper ranges from 1400 to 2620 ppm and cold-extractable copper from 150 to 600 ppm. Although some high values, up to 2600 ppm Cu, are found in the area between the two main deposits, they are very sporadic (Table II).

The average Cx.Cu/T.Cu ratios for the southern, northern and intermediate areas are 0.14, 0.13 and 0.16, respectively; substantially higher than the ratios obtained at Santo Niño. The difference is attributed to the presence of thicker overburden and higher precipitation at Nungkok.

TABLE II

Geochemical data, Nungkok, Malaysia

	T.Cu (ppm)	Cx.Cu (ppm)	Cx.Cu/T.Cu	DH.Cu (ppm)	DH.Cu/T.Cu
Southern	1200	230	0.19	1040	0.87
ore body	2800	230	0.08	1360	0.49
	2570	450	0.17	4500	1.75
	1720	200	0.11	4310	2.50
	1280	230	0.18		
Average	1914	268	0.14	2750	1.43
Northern	2580	150	0.06	4790	1.85
ore body	1400	250	0.18	4160	2.97
	2620	150	0.05	1800	0.70
	2450	600	0.24	5390	2.20
	1794	313	0.17	5160	2.90
Average	2168	293	0.13	4260	1.96
Overall average	2041	281	0.14	3505	1.71

Relation between total copper in soil and drill core

Six holes have been drilled in the southern deposit, five of which extend to depths of about 70 m and one to 430 m (Fig.4). Examination of the cores shows that mineralization is continuous throughout their length. For this deposit the ratio of copper in core to copper in the overlying soils averages 1.43 (Table II). For the northern ore body the ratio averages 1.96. For both ore bodies the average value of the DH.Cu/T.Cu ratio is 1.71 which is of the same order as that for Santo Niño.

CHAUCHA DEPOSIT

This copper deposit is located in the southwestern part of Ecuador 70 km southeast of the port of Guayaquil (Fig.5). The mineralization was found during a United Nations Development Project (U.N.D.P.) and, since April

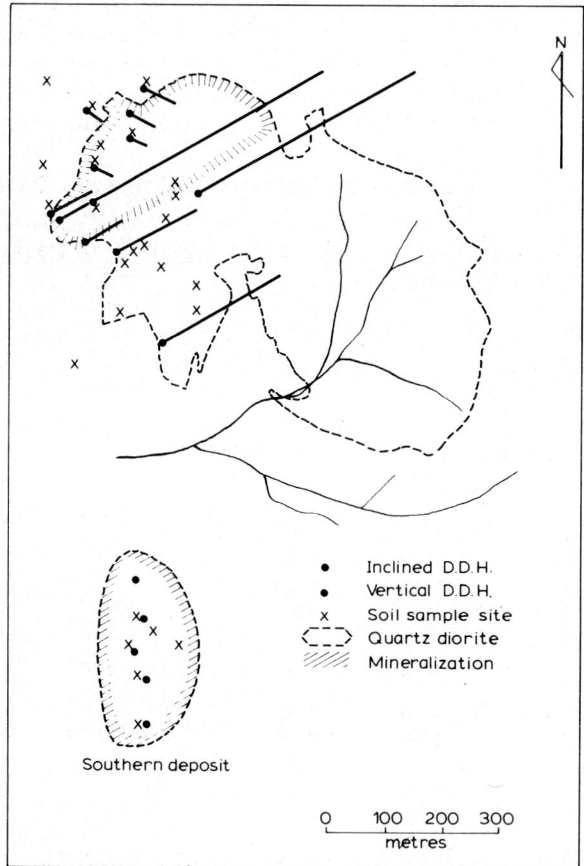

Fig.4. Location map, soil samples and drill holes, Nungkok deposit.

Fig.5. Location map, Chaucha deposit, Ecuador.

1970, has been further investigated by the Overseas Mineral Resources Development Company of Japan (O.M.R.D., 1972). Soil sampling and I.P. surveys delineated what is now called the Naranjos ore body at an altitude of about 1700 m above sea level. The climate of the area is temperate, despite the proximity to the equator, with a dry season which lasts from May to October. In the rainy season rain falls daily and averages 1300 mm annually. At levels below 1000 m elevation the area is heavily forested but above this level trees are virtually absent.

Steep topography characterizes the area with the metamorphic rocks giving rise to more abrupt relief than the more easily weathered intrusives.

Geology and mineralization

Cretaceous submarine volcanic and pyroclastic rocks of the Andean geosyncline cover a basement of presumed Paleozoic metamorphics and are intruded by mineralized quartz diorites (Fig.6). Chalcopyrite mineralization is widely distributed together with pyrite, molybdenite and magnetite. Chalcocite, crysocolla and various oxides and carbonates occur as secondary minerals. Generally unstratified Quaternary sediments believed to be of glacial origin overlie the deposit. They consist of relatively unsorted angular to more or less rounded boulders from a few centimetres to 10 cm in diameter in a clay matrix.

518

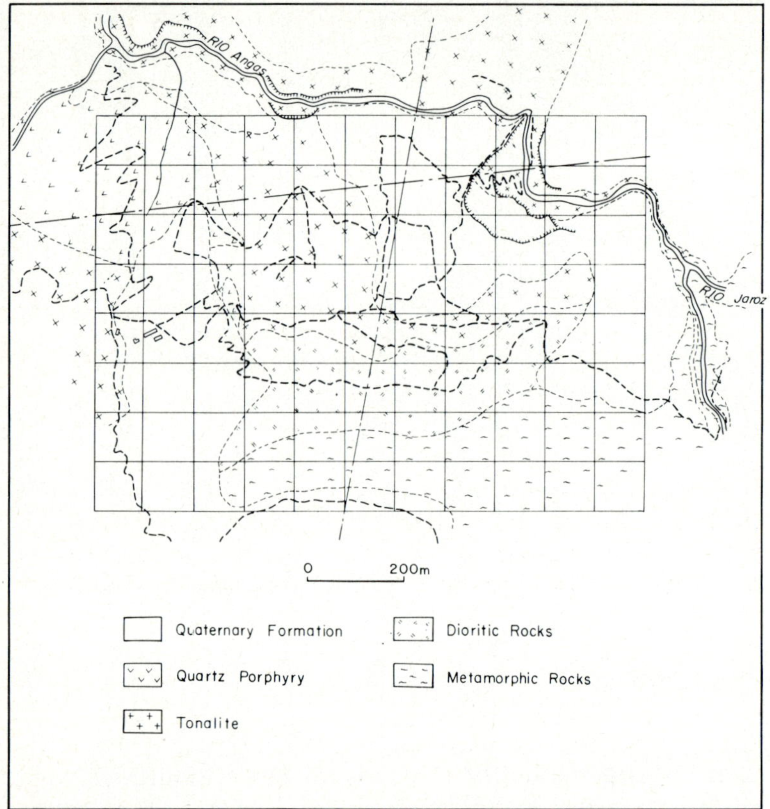

Fig.6. Geological map, Naranjos ore body, Chaucha.

Soil sampling and analysis

The B horizon soil was sampled in pits, 1—1.5 m in depth, sunk at 50-m intervals on lines 200 m apart. In this discussion the total copper data of both the U.N.D.P. and O.M.R.D. surveys are used because both data sets are in good agreement, according to O.M.R.D. checks, despite differences in the analytical methods used. The minus 80-mesh fraction of the O.M.R.D. samples was digested in nitric acid before determining total copper by atomic absorption.

Since U.N.D.P. did not do any cold-extractable copper determinations, estimated values were calculated by a moving average technique for their sample stations using adjacent O.M.R.D. data. Cold-extractable copper was determined by O.M.R.D. using atomic absorption following a 24-hour leach with dilute (1/19) hydrochloric acid.

Total copper and cold-extractable copper values

In the vicinity of the Naranjos deposit total copper in soils ranges from 100

to 8000 ppm and an extensive area of soil carrying over 2500 ppm Cu overlies part of the ore body (Fig.7).

The average value for the Cx.Cu/T.Cu ratio is 0.35 and actually gets as high
as 0.95. These values are very much higher than those obtained at Santo Niño
and Nungkok. When plotted the most extensive area of high values is seen to
occur along the steep slopes down to the Angas River with another, smaller
area immediately over the ore body (Fig.8). This is thought to be due to
downslope movement of soluble copper into the overburden at the base of
the slope.

It seems that some of the effects of topography in displacing the anomalies
can be minimized by plotting the relatively insoluble component of the soil
copper. This parameter was calculated by subtracting Cx.Cu values from T.Cu
values, i.e., (T- Cx)Cu. When these (T- Cx)Cu values are plotted the base of
slope anomaly is eliminated and the anomaly over the ore body emphasized
(Fig.9).

A section through an area of relatively deep overburden is shown in Fig.10
and, whilst the T.Cu values are not anomalous over the ore body, the Cx.Cu/
T.Cu values clearly are. In Fig.11, a section is illustrated where the overburden
is thin; total copper, the Cx.Cu/T.Cu ratio and (T- Cx)Cu are all high over the
mineralization.

Relation between total copper in soil and drill

The ratio of drill core copper to total soil copper, DH.Cu/T.Cu, for the

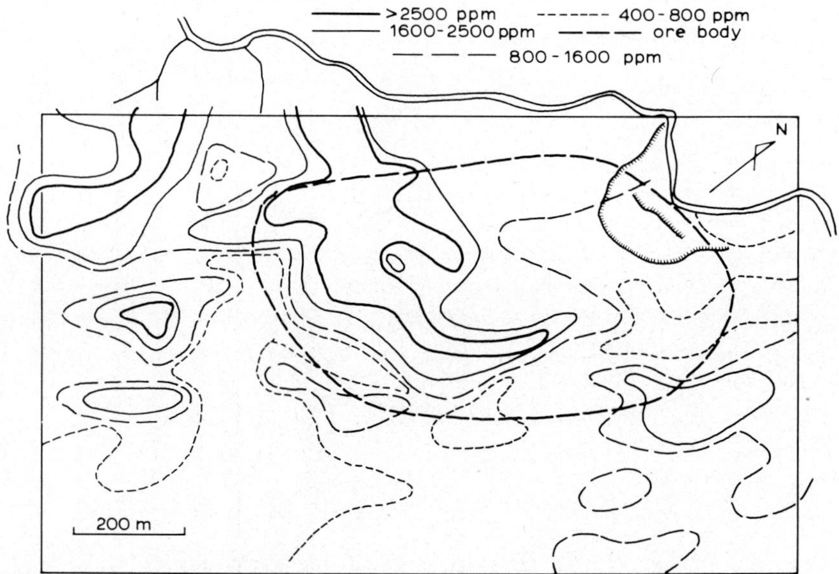

Fig.7. Total copper, T.Cu, in soils, Naranjos ore body.

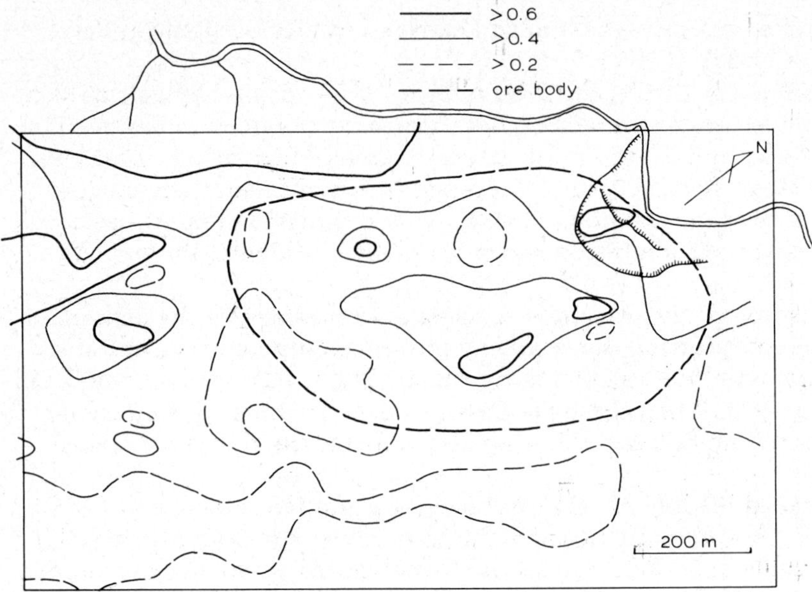

Fig.8. Ratio of cold-extractable copper to total copper, Cx.Cu/T.Cu, in soils, Naranjos ore body.

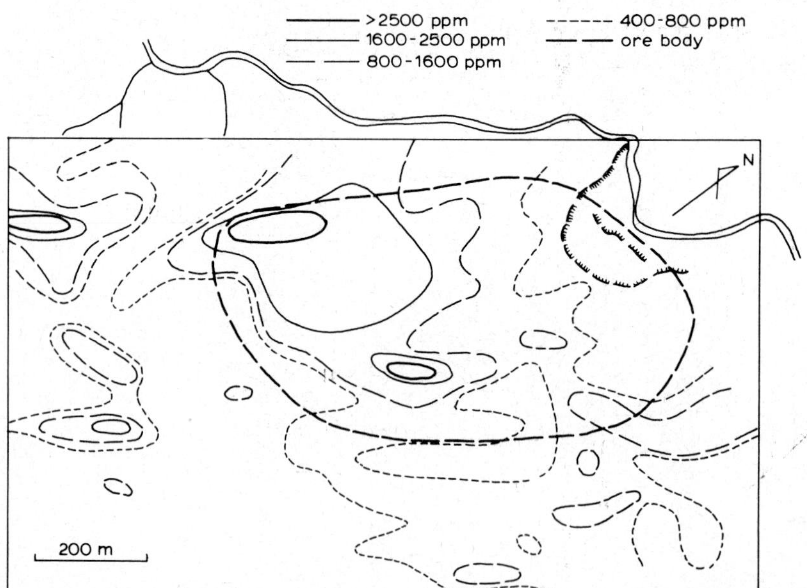

Fig.9. Relatively insoluble copper, (T- Cx)Cu, in soils, Naranjos ore body.

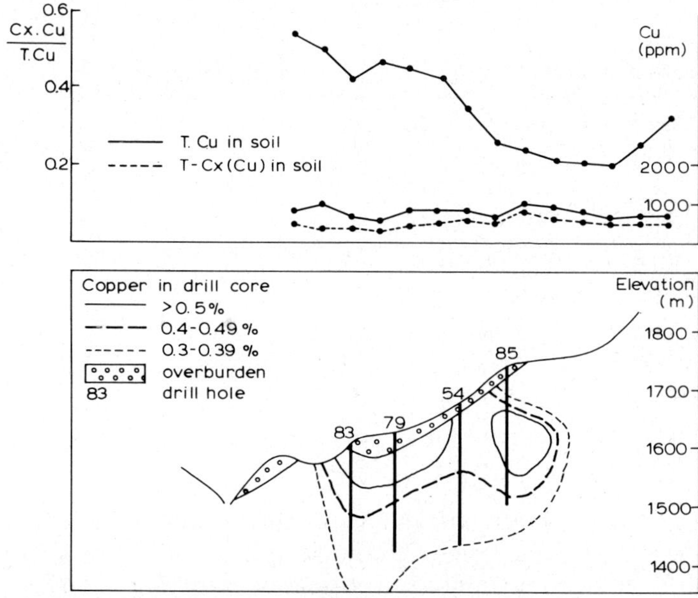

Fig.10. Section along line 20.2, Naranjos ore body.

Naranjos deposit averages 2.7, i.e., nearly twice the average values obtained at Santo Niño and Nungkok (Table III). In areas where the overburden exceeds 20 m the ratio averages 3.9, whereas in areas of thinner cover it falls to an average of 2.2.

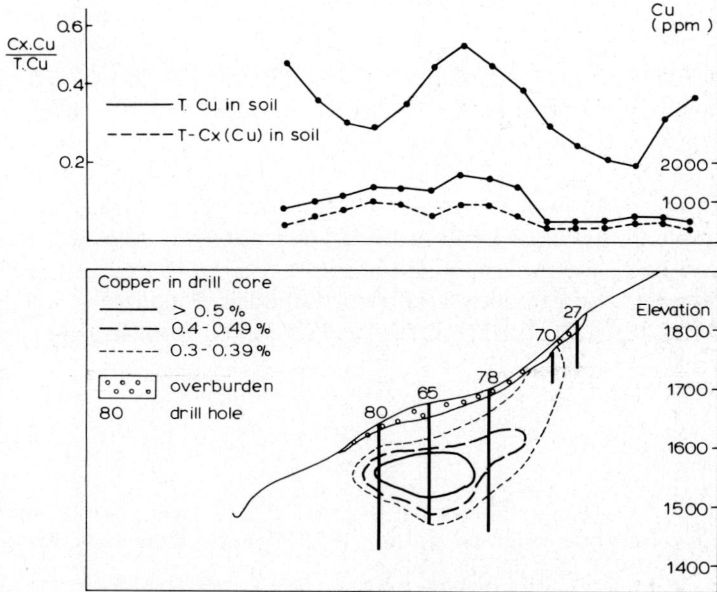

Fig.11. Section along line 18.2, Naranjos ore body.

TABLE III

Geochemical data, Chaucha, Ecuador

	Overburden thickness		Average
	< 20 m	> 20 m	
T.Cu (ppm, average)	1667	1100	1420
Cx.Cu/T.Cu	0.32	0.39	0.35
DH.Cu (ppm, average)	3600	4300	3890
DH.Cu/T.Cu	2.2	3.9	2.7

CONCLUSIONS

The three examples described are not sufficient to confirm the relationship between the soil geochemical data and the drill core assays, but it is considered that this approach could be used to estimate the copper content of porphyry-type deposits. Both the ratio of cold-extractable copper to total copper in soils and of drill hole copper to total soil copper are important in making this estimation and both ratios are clearly influenced by topography and overburden thickness. Depending on the level of the Cx.Cu/T.Cu ratio the copper content of a particular deposit will be from 1.5 to 3 times the total copper level in the soils overlying the deposit.

ACKNOWLEDGEMENTS

The author expresses his grateful thanks to Mr. H.A. Brimo, President of Philex Mining Corporation, for permission to publish the results of this study at Santo Niño. Thanks are also due to Mr. D.E. Lewis, Vice President of Atlas Consolidated Mining and Development Corporation, for his kind guidance·in the soil sampling at Nungkok and permission to present the paper. Grateful acknowledgement is made to the staff geologists of Overseas Mineral Resources Development Company, Ltd., for making available the data contained in their report on the Exploration of the Chaucha Deposits, Ecuador. Finally the author wishes to thank Mitsubishi Corporation for permission to publish this paper.

REFERENCES

Bryner, L.D., 1970. Geology of the Santo Niño copper prospect. J. Phil. Geol., 24: 33—40.
Overseas Mineral Resources Development Company, Ltd., 1972. Report on the exploration of the Chaucha deposits, Ecuador. 174 pp. (in Japanese)

GEOCHEMICAL PROSPECTING IN THE EASTERN DESERT OF EGYPT

V.A. BUGROV and I.M. SHALABY

United Nations Development Project, Cairo (Egypt)
Geological Survey and Mining Authority of Egypt, Cairo (Egypt)

ABSTRACT

This paper describes some of the results of the geochemical work carried out by the
U.N.D.P. Project "Assessment of the Mineral Potential of the Aswan Region" during
1968—1973. The first regional geochemical surveys in Egypt were carried out as part of the
Project in 1968 in the rocky Eastern Desert. The field conditions prevailing in the Project
area produced various problems in selecting types and techniques of geochemical surveys,
which had to be solved in the course of orientation work. Reconnaissance, regional and
detailed geochemical surveys were carried out. Stream sediment, colluvium and bedrock
sampling, often supplemented by heavy mineral panning, was used depending on the
physiographic and sampling conditions. Regional prospecting was found to be most successful when either the minus 0.075-mm fraction of stream sediment samples was analysed
by cold-extraction techniques or the minus 1-mm alluvial fraction was analysed by spectrographic analysis.

Panning of mineral concentrates was also applied with a fair degree of success, especially
when prospecting for gold and heavy minerals resistant to weathering (i.e., cassiterite,
wolframite, columbite, etc.). Bedrock sampling was used as a routine method and produced
good results. In areas where there was some unconsolidated material, eluvium, colluvium,
this was sampled during detailed surveys, mostly using the minus 1-mm fraction. In localities
covered with an overburden containing aeolian sand, the minus 1-mm plus 0.25-mm fraction
was sampled. The removal of the minus 0.25-mm (aeolian) material substantially enhanced
anomaly contrast.

Major successes of the Project's geochemical prospecting programme included the discovery of several occurrences of Ta—Nb albitite-type ores; tin mineralization in greisens;
stockwork-type molybdenum ores and Cu—Ni—Co sulphide ores in a gabbro-peridotite complex. These results drastically modified earlier views on the metallogeny and mineral potential of the Eastern Desert of Egypt.

INTRODUCTION

This paper describes geochemical work undertaken in the Eastern Desert
region of Egypt between 1968 and 1973 as part of the Project "Assessment of
the Mineral Potential of the Aswan Region", a joint venture of the U.N.D.P.
and Geological Survey of Egypt.

Geochemical surveys were carried out in an area covering 75,000 km²
situated in the southeast part of Egypt, extending from the Sudanese border
in the south to latitude 25° in the north, and from the River Nile in the west
to the Red Sea coast in the east (Fig.1).

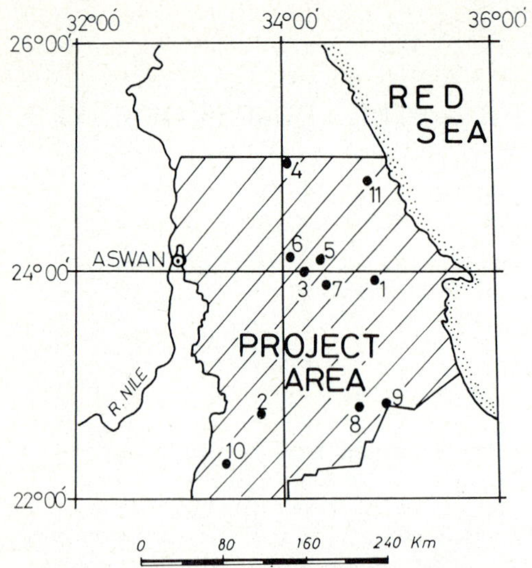

Fig. 1. Location of project area. 1 = Darhib; 2 = Abu Swayel; 3 = Gabbro Akarem; 4 = Muelha; 5 =-Homerit Mikpid; 6 = Homr Akarem; 7 = Nikeiba; 8 = El Naga; 9 = Nugrub—El Fogani; 10 = Gezira; 11 = Nugrus.

This region is a highly dissected stony desert with strong topographic relief. The maximum elevation is 1900 m above sea level and the average altitude of the areas studied is about 400 m. Rain falls only once every several years; consequently vegetation is extremely sparse and is found only in wadis. Rock outcrops are well exposed in the greater part of the area, although in places they may be covered with aeolian sand.

There is no detailed geological map of the region. However, there is a geological map at a scale of 1:1,000,000 and some photogeological maps at a scale of 1:100,000. Some localities (maximum 100—200 km² each) have been mapped geologically at a scale of 1:100,000 or 1:40,000. About one-third of the project area has been covered by aeromagnetic and radiometric surveys at a scale of 1:100,000.

The regional geochemical prospecting programme undertaken by this project was the first of its kind in Egypt. The main purpose was to provide a technical and economic evaluation of the mineral potential of this area of Egypt using airborne geophysical surveys and photogeological interpretation followed by combined ground geological, geophysical, and geochemical studies.

Geochemical operations were sub-divided as follows:

(1) Reconnaissance geochemical work to verify airborne-geophysical anomalies on the ground. In this case, bedrock was usually sampled on traverses crossing the geophysical anomalies to outline zones of hydrothermal alteration, contact alteration, and autometasomatism (i.e., greisenization and albitization), with the aim of finding mineralization.

(2) Reconnaissance work based on photogeological studies utilizing the

system described above. Areas selected for reconnaissance surveys were based on geological features and the experience gained during the ground checking of airborne geophysical anomalies. The success of this work was largely due to the high percentage of outcrop.

(3) Regional geochemical surveys that covered thousands of square kilometres and were based upon general geological features.

(4) Detailed geochemical prospecting undertaken in promising areas of up to several square kilometres, delineated during reconnaissance or regional surveys.

Geochemical sampling techniques varied widely, depending upon particular objectives, scope of operations, and sampling conditions. The principal techniques involved included: (a) wadi alluvium stream-sediment sampling for dispersion trains; (b) sampling of unconsolidated colluvium for secondary dispersion patterns; (c) bedrock sampling for primary dispersion patterns.

Generally, geochemical field operations were combined with geological and geophysical investigations, and were often accompanied by heavy-mineral concentrate sampling.

Between 1968 and 1973, a team of three to seven geochemists and geologists collected some 35,000 geochemical samples, undertook detailed geological mapping of selected localities, provided geological support to exploratory and drilling operations, and assisted in surveying and laying out survey grids. Depending on the geochemical method used, different analytical techniques were employed; including wet chemical, spectrographic, and atomic absorption methods.

SELECTION OF SAMPLING TECHNIQUES

In order to ensure successful application of geochemical methods, a series of experimental surveys was carried out at several selected localities using various techniques (Bugrov, 1974).

The principal results of these investigations may be summarized as follows:

(1) In stream sediments and colluvium, base metals are dispersed in both mechanical and chemical patterns. The former can be outlined using a coarse size fraction (minus 1.0-m plus 0.25-mm) and the latter by using a fine size fraction (minus 0.075-mm).

(2) The coarse material is particularly amenable for spectrographic analysis whilst wet chemical techniques and especially partial extraction methods are more appropriate for the finer fractions.

(3) One important benefit arising from the use of the coarse fraction is that aeolian sand, which acts as a diluent, is removed and anomaly contrast is enhanced.

(4) Deposits of resistate minerals can be reliably located using panned concentrates of heavy minerals from alluvium and colluvium.

(5) Because of the widespread occurrence of outcrop, bedrock sampling is a widely applicable and very valuable technique.

526

RESULTS OF INVESTIGATIONS

Several of the successful results of the project have already been reported (Bugrov, 1974). One further example of an integrated mineral survey will be presented here.

The project which led to the discovery, in March 1972, of Cu—Ni—Co sulphides in a gabbro-peridotite complex at Gabbro-Akarem (Figs.1 and 2) commenced with a study of aerial photographs which suggested that in one locality (marked on the available geological maps as schists and undifferentiated granodiorites) there was a massif of possible mafic-ultramafic type. Geochemical reconnaissance sampling of bedrock along three traverses was combined with geological observations across the area in order to find indications of mineralization. This work finally resulted in the discovery of a gabbro-peridotite massif, subsequently named Gabbro Akarem, containing traces of Cu—Ni—Co mineralization. The gabbro-peridotite massif is concordant with gneisses, schists, and amphibolites, and is probably a lopolith. The country rocks dip towards the massif in a basin-shaped manner with dips varying from 40° to 75°. The margins of the massif are composed of gabbro-amphibolite and amphibolitized gabbro. The central core is made up of gabbroic rocks including norite, olivine-gabbro and pyroxenite. Peridotites (mostly hartzburgite with some pyroxenite) occur as bands in the gabbros, especially in the central parts. Sulphide mineralization (impregnations of pyrrhotite and chalcopyrite) when present can easily be seen in the peridotite, pyroxenite and norite phases. In oxidized zones the mineralized rocks are distinguished by their intense ferruginization and the presence of secondary carbonates and silicates of copper

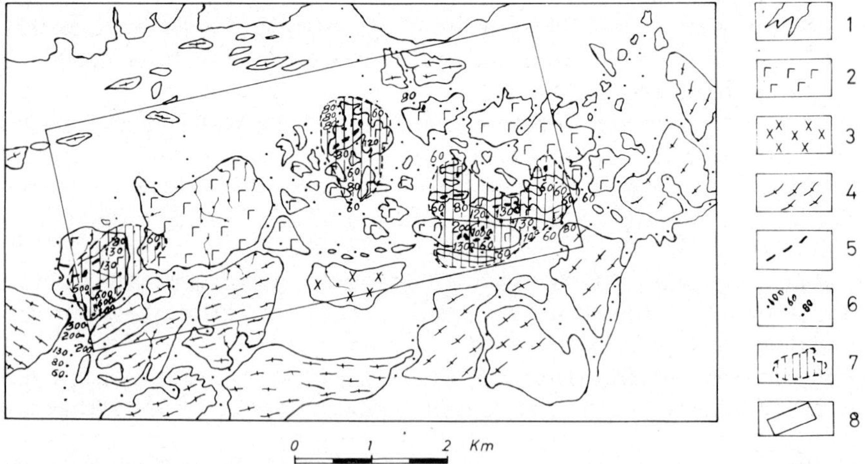

Fig.2. Results of stream sediment sampling for copper and nickel, Gabbro Akarem area. 1 = alluvium and colluvium; 2 = undifferentiated gabbro-peridotite intrusion; 3 = granodiorite; 4 = gneisses and schists; 5 = gossans; 6 = localities of samples with copper and nickel content; 7 = anomalous zones, 8 = locality prospected at scale 1:10,000.

and nickel. Gossans related to massive ore are generally found over peridotite-gabbro contacts.

The area of this massif measures 10 km x 3 km (Fig.2). Four zones of gossans related to Cu—Ni sulphides were located, with a maximum distance between them of about 7 km. A number of smaller gabbro outcrops containing sulphide mineralization (disseminated pyrrhotite and chalcopyrite) were also found. These finds led to the implementation of a regional geochemical prospecting programme for Cu—Ni sulphide ores utilizing drainage sampling within an area of about 800 km^2.

Stream sediment sampling at Gabbro Akarem was undertaken on a grid approximately 1000 m x 500 m (1000 m being the distance between wadis, and 500 m the distance between sampling stations along a wadi). Where necessary, a 500 m x 250 m grid was used on the mafic and ultramafic formations. The minus 1-mm fraction was collected and analysed spectrographically for 12 elements. As a result of the regional geochemical prospecting in this area, six anomalous dispersion trains of copper and nickel were delineated, each 1 to 2 km long. This paper deals only with the results obtained in a small part of the investigated area (about 50 km^2), which includes the largest gabbro-peridotite massif with the longest dispersion trains and highest concentrations of copper and nickel. The results of stream sediment sampling in this locality, given in Fig.2, show the anomalous values as the sum of the copper and nickel contents found in the samples. The sum of the copper and nickel content was found to produce high contrast and extensive anomalies. The dispersion trains found in the area of the Gabbro Akarem massif extend for 1.5—2 km using a threshold (for Cu + Ni) of 60 ppm, and have a maximum intensity of 1300 ppm (Fig.2). New occurrences of gossans, from 50 m to 2 km in extent, were found during regional geochemical sampling. Regional stream sediment sampling outlined three areas of high copper and nickel values related to gossan zones in mineralized gabbro and peridotite. Based on these results and supported by the occurrence of these gossans within the boundaries of the gabbro-peridotite massif, the whole area was re-sampled at a scale of 1:10,000 using a 100 m x 50 m grid covering an area of approximately 30 km^2.(Figs.2 and 3). The minus 1-mm fraction of the colluvium was sampled.

As was the case with the drainage samples, delineation of colluvium anomalies was based on the sum of the (Cu + Ni) contents of the samples. The anomaly threshold was taken as the mean metal content plus three standard deviations (100 ppm Cu + Ni). The dispersion patterns for (Cu + Ni) in colluvium at Gabbro Akarem are presented in Fig.4. The anomalies mainly reflect the peridotite zones. The highest concentrations of (Cu + Ni), from 500 to 4000 ppm, coincide with gossan zones or occurrences of intensely ferruginized rocks. The largest areas of mineralized rocks with associated gossans were found in the northern half of the central part of the massif and in its south-eastern part. These areas can be traced on the surface for 500—1000 m, with a visible width of 100—150 to 300—350 m. Follow-up drilling in the first geochemical anomaly investigated led to the discovery of disseminated Cu—Ni—Co

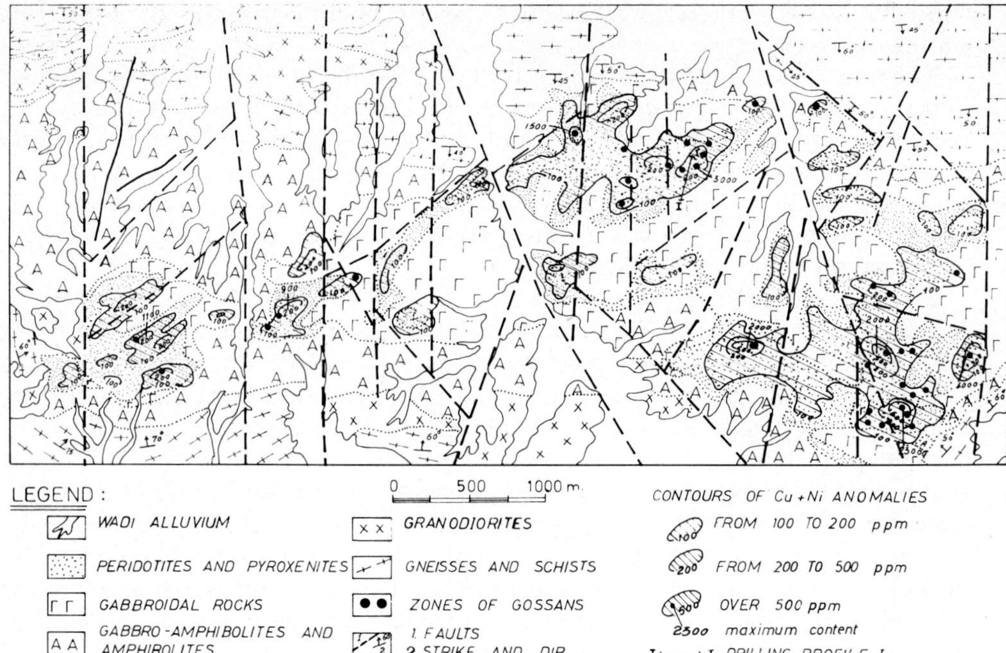

Fig.3. Gabbro Akarem: generalized geological map with results of colluvium sampling for copper and nickel.

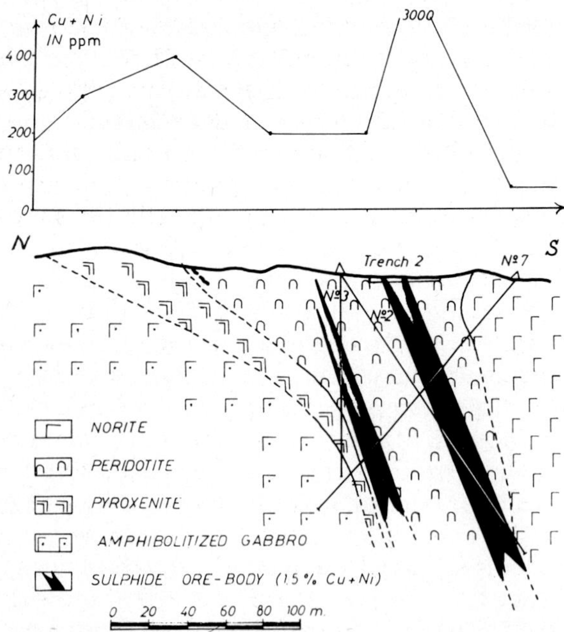

Fig.4. Geological cross-section through drill profile I—I and relationship with anomalies in the minus 1-mm fraction of colluvium sampling.

sulphides with bodies of massive sulphide composed predominantly of pyrrhotite, chalcopyrite, and pentlandite with rarer cubanite. This mineralization was encountered in three drill holes on profile I—I (Figs.3,4). The outline of the sulphides marked on the geological cross-section through drilling profile I—I represents the sum of copper and nickel contents above 1.5%. The true thickness of one body is about 13 m, whilst the other is about 4 m, with copper in both attaining a maximum of 1.52%, nickel 2.44%, and cobalt 0.136%.

Gold and the elements of the platinum group are very low; Au — up to 20 ppb, Pt — up to 30 ppb, and Pd — up to 135 ppb.

This discovery is of major importance, because it proves, for the first time, the presence in the Eastern Desert of Egypt of magmatic Cu—Ni sulphide ores in a differentiated gabbro-peridotite complex.

CONCLUSIONS

Stream sediment, colluvium, and bedrock sampling in combination with heavy-mineral concentrate panning, where necessary, can be applied in geochemical prospecting for ore deposits with a high degree of success in the conditions of the Eastern Desert of Egypt.

The wide application of geochemical techniques in prospecting for ore deposits initiated in 1968 by the U.N. Mineral Survey Project "Assessment of the Mineral Potential of the Aswan Region" revealed the favourability for mineral discovery in this area. Geochemical prospecting played an important role in discovering for the first time in Egypt Cu—Ni—Co sulphide ore in differentiated rocks of a gabbro-peridotite complex (Gabbro Akarem). In addition, geochemical methods were mainly responsible for discovering occurrences of Ta—Nb (Nugrus, Nikeiba) and beryllium (Homerit Mikpid, Homr Akarem). Tantalum, niobium, zirconium, and rare-earth mineralization was found in the nepheline syenites in some of the ring complexes, tin mineralization of the greisen type was discovered at Gabel Muelha and Homr Akarem, and molybdenum stockwork-type ores were located at Homr Akarem. These discoveries drastically modified previous views on the metallogeny of the Eastern Desert.

The experience of carrying out geochemical surveys accumulated by this U.N. Project, may be successfully employed in other areas of similar physiographic conditions.

ACKNOWLEDGEMENTS

Grateful thanks are due to Dr. R. Said, Chairman of the Egyptian Geological Survey and Mining Authority, and to Dr. M.S. Garson, Project Manager, for the attention and assistance that they extended to the geochemical team, and for their kind permission to publish the results of some surveys. We feel indebted to our colleagues — geologists and geochemists — in the project who carried out the bulk of the field operations. We acknowledge with gratitude the efforts of Mr. Anatoly P. Alexandrov who translated this manuscript from Russian into English.

REFERENCES

Bugrov, V., 1974. Geochemical sampling techniques in the Eastern Desert of Egypt. J. Geochem. Explor., 3: 67—75
Solovov, A.P., 1959. Theory and practice of metallometric surveys. Academy of Sciences of the Kazakh S.S.R., Alma-Ata, 226 pp. (in Russian)
Solovov, A.P., 1965. Application of methods of mathematical statistics in geochemical prospecting. A short methodological guide. Ministry of Geology of the Kazakh S.S.R., Alma-Ata, 122 pp. (in Russian)

INTERPRETATION OF MOLYBDENUM CONTENT IN GROUNDWATER UTILIZING Eh, pH AND CONDUCTIVITY

PAUL B. TROST and CHARLES TRAUTWEIN

1221 South Welch Circle, Lakewood, Colo. (U.S.A.)

ABSTRACT

A program of geochemical exploration for porphyry copper deposits in southeastern Arizona was conducted utilizing groundwater samples. The samples were generally collected from wells which had been pumping for at least ten minutes. The pH and Eh were immediately determined at the sample site. Samples were then collected in two polyethylene bottles, one of which was acidified, and sent to a commercial laboratory for analysis of conductivity, molybdenum and other trace elements.

Both correlation diagrams and theoretical calculations utilizing Eh—pH diagrams suggest that the molybdenum content is not primarily controlled by the Eh—pH conditions of the groundwaters.

Correlation diagrams showed a much higher correlation between the molybdenum content and conductivity than between the molybdenum content and Eh—pH.

Case studies conducted near porphyry copper deposits show little change in pH, or Eh, as the target is approached; however, the molybdenum content increases significantly. Thus, one might utilize a ratio of (molybdenum/conductivity) to suggest the proximity of an oxidizing porphyry copper deposit.

INTRODUCTION

Exploration for porphyry copper deposits in the southwestern United States is continually invoking new geochemical methods to cope with the frequent pediment and alluvial blankets covering these deposits. In the 1950's and 60's, some exploration companies began determining the molybdenum content of groundwater obtained from domestic wells. This exploration technique was probably based on the empirical and theoretical observations of molybdenum mobility in neutral to alkaline aqueous solutions. Since most major porphyry copper deposits are known to contain molybdenum in the range of 0.02% (Lowell and Guilbert, 1970) and since many groundwaters in southern Arizona are neutral to alkaline, then presumably groundwater being dispersed from an oxidizing or oxidized porphyry copper deposit would contain anomalous molybdenum (Bloom, 1966).

Huff (1970) conducted a case study of copper and molybdenum in groundwater around the Pima-Mission ore bodies and outward. His results showed that anomalous molybdenum was indeed present in the groundwaters around these mines, whereas the copper failed to suggest the nearby disseminated mineralization.

However, the major drawback of this method has been the failure to devise methods which will economically lead to the source of the anomaly. Thus, anomalous areas of 10—100 square miles may be suggested as interesting (Texas Gulf, Inc., Company Report) but other indirect methods are then necessary for further target delineation. A second drawback has been the apparent displacement and enrichment of molybdenum anomalies. This was shown by Ward et al. (1960) for an area in which water, presumably originating along a mountain margin fault, contained 18 ppb Mo and 700 ppm dissolved solids. However, by the time the spring water emerged near the valley floor, 4½ miles distant, the molybdenum content had increased to 340 ppb and the dissolved solids to 48,000 ppm.

New impetus was given to this type of geochemical exploration when numerous Russian authors began publishing results of their groundwater geochemical studies. Chernyaev et al. (1965) reported on the apparently successful employment of groundwater geochemical exploration in a semi-arid to arid area of the southern Urals. Goleva et al. (1969) reported on the use of Mo, Cu, F, Zn, W, Pb, Be, Li and As in groundwater for locating buried Mo—W mineralization in permafrost areas. Apparently, the Russians were also concerned by the dimensions of the anomalous area for Chernyaev et al. (1968) later reported on the use of statistical measures using models to aid in the interpretation.

A geochemical groundwater exploration program concentrated in pediment-covered areas of geologic interest, was initiated in 1971 by Trost for Texas Gulf, Inc. During the next 1½ years, 1392 samples were collected and analyzed for 3—7 elements per sample.

Values for pH ranged from 6.2 to 9.0, Eh from —200 to +300 mV, molybdenum from <2 to 540 ppb and conductivity from 100 to 10,000 micromhos/cm. Iron was not determined since Huff's study showed no variation in iron content of water from wells near ore bodies (containing anomalous Mo) and those at some distance away. Also a brief study of Eh—pH diagrams (Hansuld, 1966a) showed that very little iron would be interacting with the molybdenum in the pH—Eh ranges observed during the field studies.

METHODS

Sample collection

Groundwater samples were obtained primarily from three sources in southern Arizona: domestic wells, irrigation wells and windmill tanks.

Water samples were collected in two 500-ml polyethylene bottles; initially, both bottles were acidified immediately with 5 drops of 25% HCl (sample set A). For sample set B, however, only one bottle was acidified so that a true conductivity value could be obtained. Samples were acidified to prevent trace element absorption by container walls or formation of iron hydroxides.

Determination of Eh and pH

The pH and Eh were determined on a third sample immediately at each well site after the well had been pumping for at least 10 minutes to ensure a fresh sample. A portable Beckman Electromate pH/Eh meter was employed utilizing a Beckman combination pH electrode and a Beckman combination Eh electrode. The date, sample number, sample location, pH, Eh and any observations (well depth, visible iron oxide) were noted at each sample site. No pH or Eh measurements were made on samples passing through pressure tanks because of a possible pH shift due to changes in pCO_2. pH measurements were considered reliable to $\pm$ 0.1 pH unit; Eh measurements to $\pm$ 10 mV. Resampling of selected wells after an 8-week period showed the molybdenum content to be within 10%, and pH and Eh to be within 5% of the original value.

Sample analysis

Water samples were shipped to a commercial laboratory for trace element analysis and determination of conductivity. The determination of conductivity proved less costly and much more rapid than determination of total dissolved solids (TDS). The conductivity or specific electrical conductance is defined as the ability of a 1-cm cube to conduct an electrical current at 25°C and is reciprocal of resistance. Units of measurement are micromhos/cm. As noted by Davis and DeWiest (1966), the specific conductance, or conductivity, (K) can be used as an estimate of total dissolved solids in ppm by multiplying specific conductance by 0.7; or $K \times 0.7 \simeq$ TDS. The conductivity f potable subsurface waters ranges from 30 to 2000 micromhos/cm, ocean water from 45,000 to 55,000 micromhos/cm and oilfield brines may exceed 100,000 micromhos/cm.

Molybdenum was determined colorimetrically at a commercial laboratory by a modified version of the U.S.G.S. method (Ward et al., 1963). A sensitivity of 2 ppb was obtained with a reproducibility of 2 ppb at the lower limits of detection. Davis and DeWiest (1966) listed an average abundance of 0.1—100 ppb for molybdenum in potable groundwater, whereas Hawkes and Webb (1962) suggest a molybdenum content of 0.3—3.0 ppb.

Hydrologic considerations

If possible, the depth of each sampled well was determined by contacting the owner or examining the well logs maintained by the state of Arizona. In addition, the water levels and lithologic logs were examined when the data was available to determine the static water levels. Data from perched water tables was noted and discarded during the interpretation phase.

534

Rates of groundwater flow in one of the basins, typical of the area, represented by set A varied from 10 to 40 ft per year as determined by a hydrologist studying the basin.

RESULTS

Molybdenum distribution

During the geochemical exploration program, a total of 1392 samples were
collected by trained personnel throughout southern Arizona. Fig.1a shows a
histogram of the distribution of molybdenum for the 1392 samples. This histogram represents the cumulative data from many Arizona basins with diverse
basinal fill and geohydrologic parameters. Data henceforth discussed will be

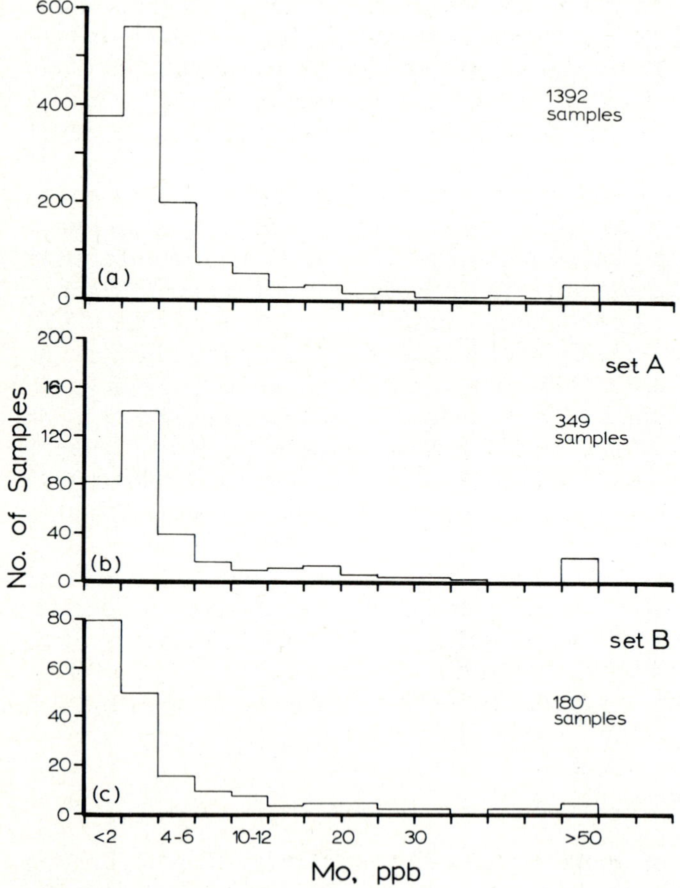

Fig.1. Histograms of molybdenum (ppb) in groundwaters of southern Arizona: (a) total
population 1392, mean 6 ppb; (b) data set A, total population 349 mean 15 ppb — the
high mean for this data set reflects the numerous samples collected in an anomalous area;
(c) data set B, total population 180, mean 6 ppb.

confined to three basins in southern Arizona. Data will be presented as sets A and B. As previously mentioned, set A had both bottles acidified, whereas set B had only one bottle acidified.

Fig.1b and 1c show the distribution of molybdenum in well waters for sets A and B, respectively. Set A, with a population of 349, shows a close correlation to the histogram of the total population. The histogram of set B, with a population of 180, shows a more log-normal distribution probably reflecting the fact that a major percentage of samples came from two different basins not included in set A. The total population and set B have a mean of 6 ppb Mo. Set A shows a mean of almost 15 ppb Mo reflecting numerous samples collected from an anomalous area.

Relationship of pH/Eh with molybdenum content

Huff (1970) determined the pH at some of his sample sites and reported pH values ranging from 7.1 to 7.7 for both samples taken near the Pima-Mission ore body and also for those taken some distance away. Fig.2a shows that very little correlation exists between log [Mo] content and pH. Similarly

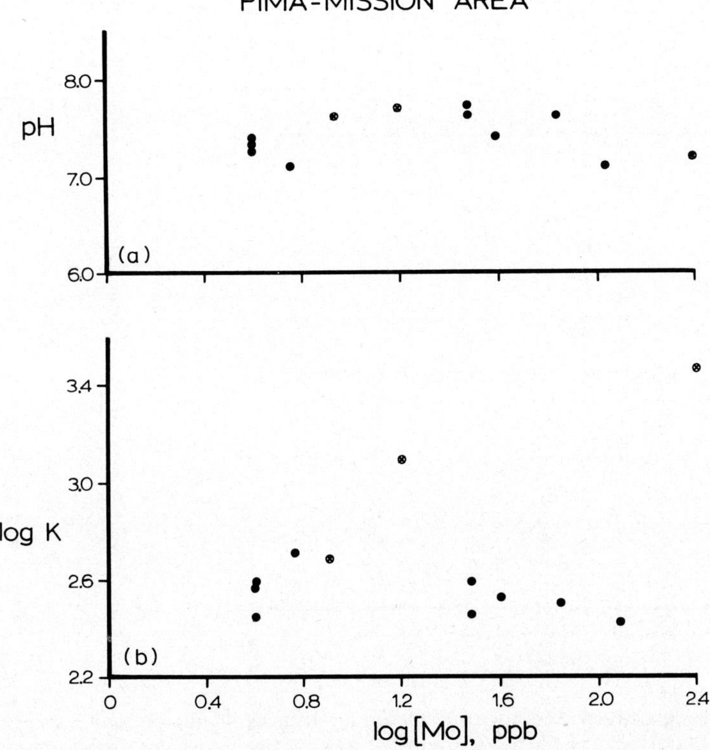

Fig.2. Correlation plots for groundwater within a radius of 9 miles of the Pima-Mission ore bodies: (a) pH vs. log [Mo], showing no correlation between pH and molybdenum content; (b) log K vs. log [Mo]. (Data of source: Huff, 1970.)

Fig.3a and b, plots of pH and Eh values vs. log [Mo] of samples located within four miles of the Tyrone ore body, show very little correlation between log [Mo] and pH or Eh. Sato (1960) determined the pH and Eh values of groundwaters around the Magma, San Manual, and the Cole and Campbell Mines (Bisbee area) in Arizona. His Eh and pH data for the groundwater within the oxidized portions of the ore bodies showed a pH range of 7.7—8.1 and Eh ranges from +340 to +380 mV. These values are thus quite similar to Huff's values, the Tyrone data and the values obtained during this study.

Eh—pH relationships potentially defining the stable molybdenum species in that portion of the aqueous environment studied in detail (see Fig.4a) were determined according to standard procedures as initially outlined by Pourbaix (1949) and notably popularized by Garrels and Christ (1965). Calculations involving thermodynamic evaluation of molybdenum, iron and sulfur species in the aqueous solution were substantially reduced by reference to previous investigations in the system Mo—Fe—S—H$_2$O, carried out by Titley and Anthony

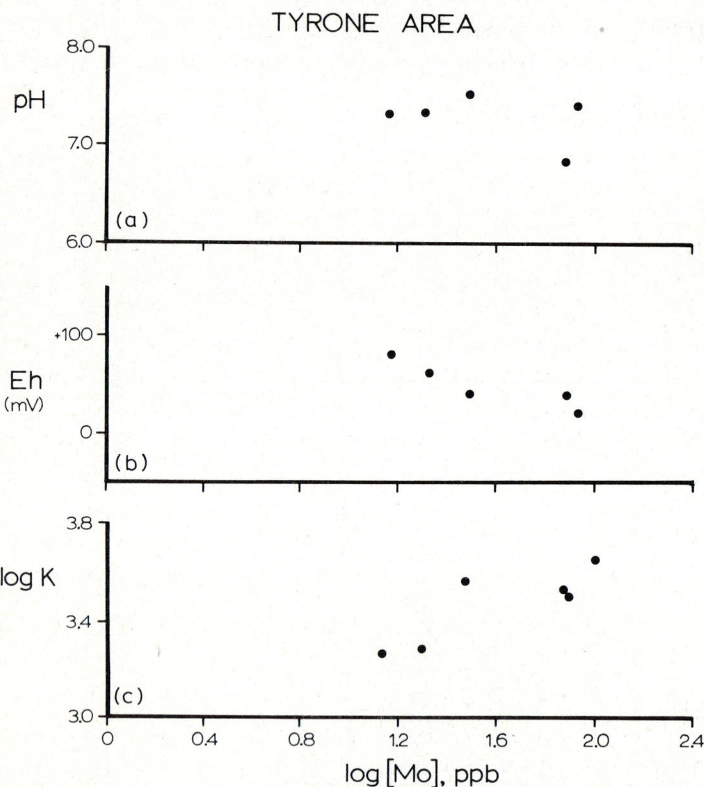

Fig.3. Correlation plots for groundwaters located within a radius of 4 miles of the Tyrone ore body, New Mexico: (a) pH vs. log [Mo], showing no correlation between increasing molybdenum content and pH; (b) Eh vs. log [Mo], showing an inverse correlation between Eh and molybdenum content; (c) log K vs. log [Mo], showing a good correlation of increasing molybdenum content with increasing conductivity.

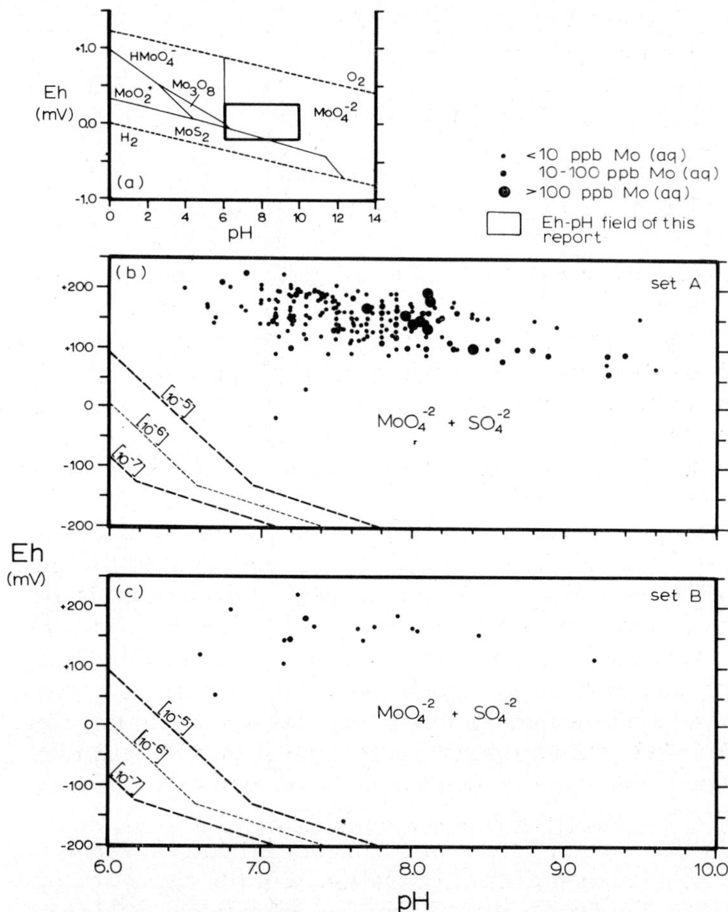

Fig.4. (a) Eh—pH diagram showing the species of potential interest to this study. (b) Expanded portion of Eh—pH diagram encompassing the Eh—pH parameters of this study and showing the sample values for data set A. (c) Same as Fig.4b, but for data set B. Conditions: $T = 25°C$, $P = 1$ atm; boundaries: $(S_{aq}) = 10^{-2}$ molal, $(Mo_{aq}) = 10^{-6}$ molal (Fig.4a), $(Mo_{aq}) = 10^{-5}, 10^{-6}, 10^{-7}$ molal (Fig.4b and c).

(1961), Titley (1963) and Hansuld (1966a,b).

Analyses of aqueous molybdenum and corresponding on-site pH and Eh values were determined for data set A; for data set B, however, aqueous sulfate was also determined. Both sample sets fall within a restricted field of Eh and pH values as illustrated in Fig.4b and c. Within the boundaries of this field, aqueous molybdenum concentrations were found to be theoretically controlled by two reactions:

538

$$Mo_3O_8 \text{ (ilsemanite)} + 4H_2O \rightleftharpoons 3MoO_4^{2-} + 8H^+{}_s + 2e^- \tag{1}$$
$$Eh = +1.955 - 0.2366pH + 0.0887 \log (MoO_4^{2-})$$

and:

$$MoS_2 \text{ (molybdenite)} + 12H_2O \rightleftharpoons MoO_4^{2-} + 2SO_4^{2-} + 24H^+ + 18e^- \tag{2}$$
$$Eh = +0.419 - 0.0787pH + 0.0033 \log (MoO_4^{2-}) + 0.0066 \log (SO_4^{2-})$$

The influence of metastable ferric hydroxide as a potential control on molybdate mobility by chemisorption seemed unlikely due to the absence of any suspended or colloidal matter in the samples analysed. Also, as previously noted, Huff (1970) showed no detectable difference in the iron content of waters from near mineralization or those farther away. It is conceivable, however, that absorption of molybdate may occur by iron hydroxides very close to the oxidizing ore body as the groundwaters are rapidly neutralized by the sediments through which the water disperses.

As illustrated in Fig.4b and c, the most notable characteristic in both sample sets is the degree of undersaturation in MoO_4^{2-}. As compared to the theoretically calculated activities, the molybdenum concentrations appear to be five orders of magnitude undersaturated with respect to either molybdenite or ilsemanite. Sample values appear to plot more nearly parallel to the MoS_2/MoO_4^{2-} field boundary rather than the Mo_3O_8/MoO_4^{2-} boundary; however, the concentrations of aqueous molybdenum in both sets of data do not conform to the predicted trends. Therefore, it was concluded that Eh—pH relationships do not appear to be major controls on MoO_4^{2-} activity based upon analogy with calculated stability relations.

Fig.5a and b show the observed field relationships of pH and Eh versus log [Mo] for data set A. Eh shows very little correlation with the change in molybdenum concentration (Fig.5b). A slight increase in molybdenum concentration may be present with an increase in pH for data set A (Fig.5a); however, the slope of the line does not correspond to the calculated slope defined by either the ilsemanite-molybdate or the molybdenite-molybdate reactions.

Fig.6a and b are similar correlation plots for data set B. Again, Eh shows very little correlation with molybdenum or to the calculated slope of the two potentially controlling reactions. As for data set A, however, there does appear to be a slight correlation between pH and log [Mo] in data set B. This slight correlation is not defined by the slopes of either of the two reactions, again suggesting that other factors may be the major controls on the molybdenum mobility.

Relationship of conductivity (K) with molybdenum content

Barakso and Bradshaw (1971) showed that molybdenum content in water increased with increase in pH. Their leach experiments also suggested, however, that the type and amount of cations present in a solution influences the solubility of molybdenum. In fact, their experiments showed an increase of

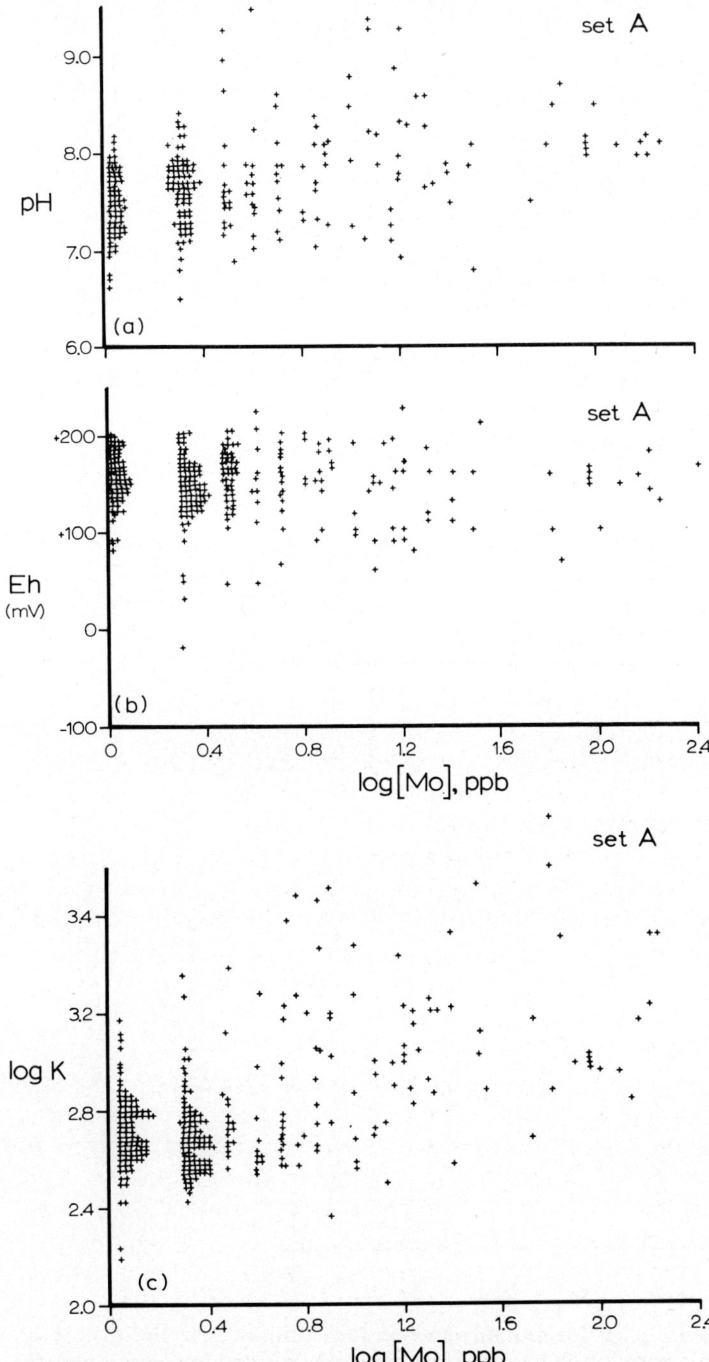

Fig.5. Data set A. Correlation plots for groundwaters in southern Arizona: (a) pH vs. log [Mo], suggesting a slight correlation between the molybdenum content and pH; (b) Eh vs. log [Mo], showing poor correlation between molybdenum content and Eh; (c) log K vs. log [Mo], suggesting a good correlation between increasing conductivity and increasing molybdenum content.

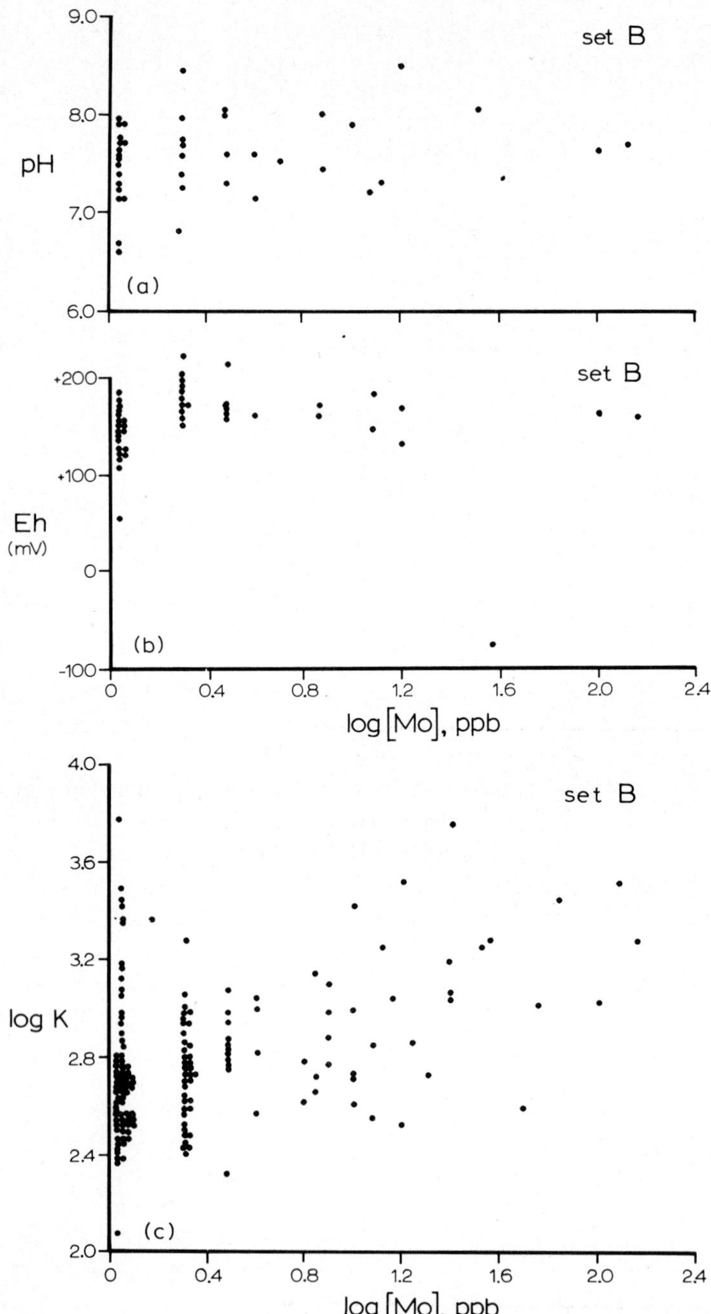

Fig.6. Data set B. Correlation plots for groundwaters in the southern Arizona: (a) pH vs. log [Mo], showing possible correlation between the increasing pH and increasing molybdenum content; (b) Eh vs. log [Mo], showing poor correlation between the increasing Eh and increasing molybdenum content; (c) log K vs. log [Mo], showing good correlation between increasing conductivity and increasing molybdenum content.

molybdenum in solution of 1—3 times when Na_2CO_3 is used to adjust to pH 8.0 as compared to $CaCO_3$. This strongly suggests that the chemistry of the solution plays a major role in determining the molybdenum content. Although the chemical makeup of a natural water is extremely complex, it may be simplified in this case by considering only those ionic species capable of interacting with molybdate, and also by considering the total amount of ionic species (i.e. conductivity, K) present in the natural waters.

Fig.2b shows that two possible relationships exist between the $\log K$ and $\log [Mo]$ values in groundwater samples collected within 9 miles of the Pima-Mission ore bodies. Samples closest to the mineralization did show a close correlation between the molybdenum content and conductivity, whereas those located 4—9 miles away showed no correlation. Fig.3c shows an excellent correlation between the molybdenum content and conductivity in groundwater samples located within 4 miles of the Tyrone ore body.

Fig.5c, a plot of $\log [Mo]$ vs. $\log K$ for data set A, shows a wide scatter of points. However, a non-linear relationship between molybdenum content and conductivity can be easily visualized. Fig.6c, a plot of $\log [Mo]$ vs. $\log K$ for data set B shows slightly less scatter and also a good, and probably non-linear, correlation between the increased molybdenum content and increased conductivity. Set B probably shows less scatter because the conductivity was determined on the unacidified sample.

Since similar correlations are observed for both sets of data, and for most case studies, this strongly suggests a direct relationship between increasing conductivity with increasing molybdenum content.

Plots of $\log [SO_4^{2-}]$ vs. $\log K$ for data set B shown in Fig.7a, suggests that a good, non-linear correlation also exists between those two species. This may be explained as due to the sulfate, along with chloride, ions accounting for a major portion of the anionic species present in solution.

Fig.7b, a plot of $\log [Mo]$ vs. $\log [SO_4^{2-}]$ also shows a good correlation although the data point scatter is much greater than for the plot of $\log [SO_4^{2-}]$ vs. $\log K$. This greater scatter may suggest that additional ionic species influence the molybdenum content of the groundwater. If the water samples were collected quite close to an oxidizing porphyry copper, then the correlation could be readily explained by the oxidation of molybdenite.

The relationships suggested by Barakso and Bradshaw's data, the data presented by $\log [Mo]$ vs. $\log K$, and $\log [Mo]$ vs. $\log [SO_4^{2-}]$, may be explained in one or more of the following manners: (1) that the solubility of a salt shows a proportional increase with increasing ionic strength; (2) that ion pairs are forming, such as $Ca^{2+}MoO_4^{2-}$ and $2Na^+MoO_4^{2-}$, in such low concentrations that precipitation does not occur; and (3) that the molybdate ion is forming a complex with sulfate (and possibly other ionic species).

Solubility related to ionic strength

Krauskopf (1967), and many others, have pointed out that as the concentration of dissolved ions increases in a solution, the solubilities of many salts will be proportionately increased. Qualitatively, this is due to ions entering

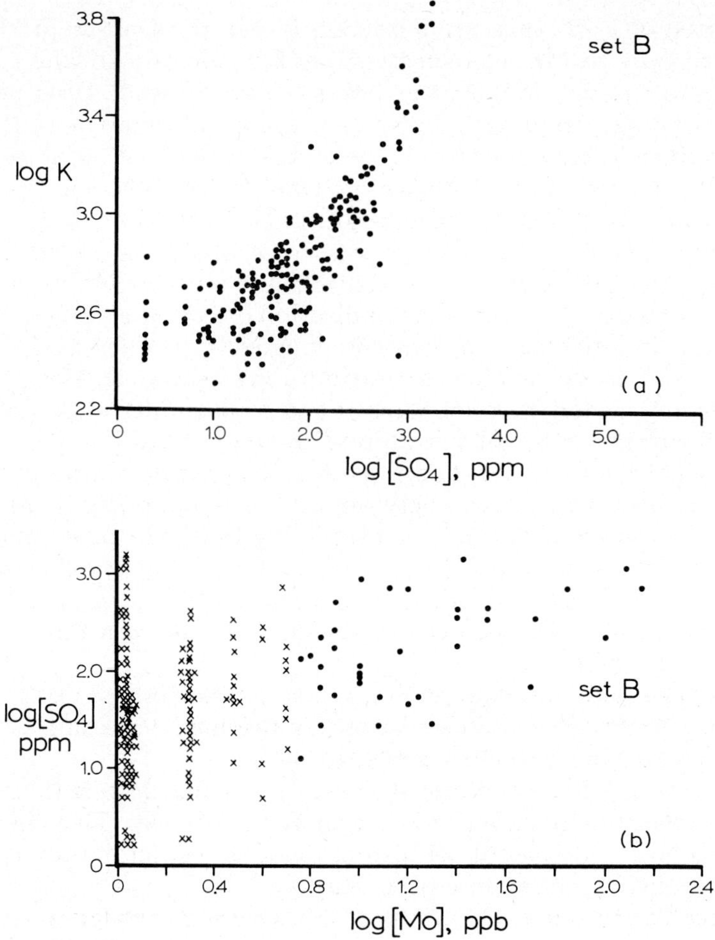

Fig.7. Data set B. Correlation plots of (a) log K vs. log [SO_4], showing close correlation between an increasing conductivity and increasing SO_4 content; and (b) log [SO_4] vs. log [Mo], showing good correlation between the increasing molybdenum content and increasing SO_4 content. Samples designated by ✕ correspond to samples with a molybdenum content at or below mean (6 ppb).

into the hydration sphere of oppositely charged ions and, thus, effectively competing with another ion that may tend to precipitate the hydrated species. Quantitatively, this is expressed as the ionic strength, where the ionic strength is a function of both the charge on an ion (Z) and its concentration in the solution (C). For numerous ions, the effect is cumulative, such that:

$$\mu = \tfrac{1}{2} \sum C Z^2$$

Thus, as the total dissolved solids (i.e. conductivity) increases, one would expect an increase in the solubility of the molybdate salts, such as $CaMoO_4$, due to the increase in ionic strength and corresponding decrease of the activity coefficients.

Formation of ion pairs

An ion pair can be considered as an uncharged complex ion which, if it were in greater concentration, would probably form a solid precipitate. The formation of ion pairs has been well documented for complex positive ions interacting with anionic species (Cotton and Wilkinson, 1966) and is also strongly suspected for complex anionic species (e.g. MoO_4^{2-}) interacting with simple cations (e.g. Na^+, Ca^{2+}). Bonding in ion pairs is probably of a coordinate-covalent type rather than covalent or ionic. Bonds are thus relatively weak and the ions may be separated without much difficulty. Since the compounds $CaMoO_4$, K_2MoO_4, $MgMoO_4$ and Na_2MoO_4 are well known, it is possible that these cations, common to groundwater, may indeed form ion pairs with the molybdate anion.

Formation of complexes with molybdate

According to Cotton and Wilkinson (1966) "the oxo-anions (of MoO_4^{2-}) also give complexes with sulfate (SO_4^{2-})..." This complex probably results through some type of polymerization of the sulfate and molybdate radicals. Thus, the field data substantiates the theoretical and laboratory-derived data, and probably is of major importance in explaining the observed relationships between log [Mo] and log [SO_4] in natural groundwaters.

CONCLUSIONS

An exploration program for porphyry copper deposits in southern Arizona was conducted utilizing trace element contents in groundwaters. In addition to the molybdenum content, the pH, Eh and conductivity were determined, as was sulfate on some of the samples. Molybdenum content ranged from <2 to 540 ppb, pH from 6.2 to 9.0; Eh from —200 to +300 mV, conductivity from 100 to 10,000 micromhos/cm and sulfate from 2 to 2400 ppm.

Based on theoretical calculations employing thermodynamic data, the only molybdenum species present at these conditions would be the molybdate anion, MoO_4^{2-}. Concentrations of molybdenum were found, however, to be 10^4—10^6 orders of magnitude undersaturated as compared to the theoretical Eh—pH calculations. Plots of log [Mo] vs. pH showed only a weak correlation, suggesting that the Eh and pH exert only a small influence on the amount of molybdenum present in solution. Similarly, log [Mo] vs. Eh plots showed only a poor correlation between the molybdenum content and Eh.

In contrast to the relatively poor correlation between pH, Eh and log [Mo], plots of log [Mo] vs. log K showed a scattered, but easily detectable correlation between the molybdenum content of the groundwater and the total dissolved solids. Furthermore, a good correlation was found to exist between the molybdenum and sulfate contents as shown by a plot of log [Mo] vs. log [SO_4]. This relationship may be explained by one or more factors involving ionic strength, formation of a molybdate-sulfate complex, and/or formation of ion pairs.

It thus appears that the pH—Eh environment may be important for the initial solubilization of molybdenum, presumably as molybdate; however, once molybdenum is in solution and being dispersed from the immediate environment of the ore body by the groundwater, there is little control exerted by pH or Eh as compared to the controls of the chemistry of the solution itself, especially the sulfate content.

Since the case studies suggest a correlation between increasing molybdenum and increasing conductivity as an ore body is approached, one might utilize the ratio of log [Mo/K] in conjunction with a plot of log [Mo/K] vs. log [Mo] to suggest the proximity of a buried oxidizing porphyry copper deposit.

ACKNOWLEDGEMENTS

The authors wish to thank Texas Gulf, Inc. and especially Dr. Leo Miller of that organization for allowing publication of these data.

REFERENCES

Barakso, J.J. and Bradshaw, B.A., 1971. Molybdenum surface depletion and leaching. In: R.W. Boyle and J.I. McGerrigle (Editors), Geochemical Exploration. Can. Inst. Min. Metall., Spec. Vol., 11: 78—84

Bloom, H., 1966. Geochemical exploration as applied to copper—molybdenum deposits. In: S.R. Titley and C.K. Hicks (Editors), Geology of the Porphyry Copper Deposits, Southwestern North America. Arizona Univ. Press, Tucson, Ariz., pp.111—119

Chernyaev, A.M., Chernyaeva, L.E. and Kovalev, V.F., 1965. Hydrogeochemical condition in prospecting for sulfide deposits in arid areas of the southern Urals, eastern Urals, and northern Kazakhstan. Chem. Abstr., 74: 79346b

Chernyaev, A.M., Kovalev, V.F., Chernyaeva, L.E. and Kozlov, A.V., 1968. Comparative evaluation of hydrochemical conditions of prospecting for sulfide deposits in arid regions of the Urals, areas east of Urals, and northern Kazakhstan. Chem. Abstr., 72: 102520y

Cotton, F.A. and Wilkinson, G., 1966. Advanced Inorganic Chemistry. Interscience, New York, N.Y., 1136 pp.

Davis, S.N. and DeWiest, R.J.M., 1966. Hydrogeology. John Wiley and Sons, New York, N.Y., 463 pp.

Garrels, R.M. and Christ, C.L., 1965 Solutions, minerals and equilibria. Harper and Row, New York, N.Y., 450 pp.

Goleva, G.A., Boguslavskii, M.E. and Fedosaeva, U.V., 1969. Water dispersion haloes of blind molybdenum and tungsten mineralizations. Chem. Abstr., 74: 144703g

Hansuld, J.A., 1966a. Eh and pH in geochemical exploration. Can. Min. Metall. Bull., March, pp.315—322

Hansuld, J.A., 1966b. Behavior of molybdenum in secondary dispersion media — a new look at an old geochemical puzzle. Min. Eng., December, pp.73—77

Hawkes, H.E. and Webb, J.S., 1962. Geochemistry in Mineral Exploration. Harper and Row, New York, N.Y., 415 pp.

Huff, L.C., 1970. A geochemical study of alluvium-covered copper deposits in Pima County, Arizona. U.S. Geol. Survey Bull. 1312-C, 31 pp.

Krauskopf, K.B., 1967. Introduction to Geochemistry. McGraw Hill, New York, N.Y., 721 pp.

Lowell, J.D. and Guilbert, J.M., 1970. Lateral and vertical alteration-mineralization zoning in porphyry ore deposits. Econ. Geol., 65 (4): 373—408

Pourbaix, M.F., 1949. Thermodynamics of Dilute Aqueous Solutions. Edward Arnold and Co., London

Sato, M., 1960. Oxidation of sulfide ore bodies, 1. Geochemical environments in terms of Eh and pH. Econ. Geol., 55: 928—961

Titley, S.R., 1963. Some behaviorial aspects of molybdenum in the supergene environment. Trans. Soc. Min. Eng., 15: 199—204

Titley, S.R. and Anthony, J.W., 1961. Some preliminary observations on the theoretical geochemistry of molybdenum under supergene conditions. Ariz. Geol. Sci., Digest, 4: 103—116

Ward, F.N., Nakagawa, H.M. and Hunt, C.B., 1960. Geochemical investigation of molybdenum at Nevares Spring in Death Valley, California: U.S. Geol. Survey Prof. Paper, 400B: 454—456

Ward F.N., Lakin, H.W. and Canney, F.C., 1963. Analytical methods used in geochemical exploration by the U.S. Geological Survey. U.S. Geol. Survey Bull., 1152, 100 pp.

Part 7
EXPLORATION IN GLACIATED TERRAIN

Chairmen:

W.W. SHILTS
Geological Survey of Canada, Ottawa, Ont., Canada
M.B. MEHRTENS
Rio Amex, Denver, Colo., U.S.A.

METAL DISPERSION IN LAKE SEDIMENTS RELATED TO MINERAL-IZATION AND ITS ROLE IN RECONNAISSANCE IN THE SOUTHERN CANADIAN SHIELD

W.B. COKER and IAN NICHOL

Department of Geological Sciences, Queen's University, Kingston, Ont., (Canada)

SUMMARY

In areas of glacial overburden with indefinite and disorganized drainage systems, as exist in large regions of the Canadian Shield, conventional geochemical exploration techniques involving the sampling of stream sediments and soils have found limited application. In such areas lake sediment composition could be an indicator of mineralization if either the host rock or the mineralization itself has some diagnostic geochemical feature which is transmitted to lake sediments. If these conditions are fulfilled, lake sediment sampling would offer a convenient procedure for reconnaissance scale exploration in Shield areas. The usefulness of the procedure has been previously demonstrated in permafrost areas of Canada (Allan, 1971; Allan et al., 1972). In the southern Shield, where organic activity is much greater, metal dispersion in lake sediments is controlled by a number of factors contributing to the existence of a complex relationship between lake sediment geochemistry and mineralization.

Orientation surveys were carried out over the Sturgeon Lake, Shebandowan Lake, Manitouwadge and Upper Manitou Lake greenstone belts within the Superior province to investigate the nature of metal dispersion in lake sediments associated with mineralization in order to evaluate the feasibility of utilizing lake sediments as a sample medium in reconnaissance surveys (Coker, 1974; Coker and Nichol, 1975).

The character of lake water and nature of lake sediment composition was investigated in four lakes in the Sturgeon Lake area associated with mineralization and different bedrock types.Lyon Lake is adjacent to a major Cu—Zn deposit while Darkwater Lake is over acid metavolcanics and granodiorite. Corsica Lake is associated with granitic gneisses and Lake "A" with basic metavolcanics. The above investigations indicated that the most representative and homogeneous sample of a single lake basin occurs in the central regions of a lake (Fig.1). However, it was difficult, if not impossible, to distinguish a lake adjacent to mineralization from one in barren terrain on the basis of the single element analyses of the organic-rich material from the lake centres. High concentrations of dissolved oxygen in the bottom waters of both Corsica Lake and Lake "A" resulted in the oxidation and precipitation of iron and manganese and coprecipitation of zinc to abnormally high levels in the sediments (Fig.1). In Lyon Lake, adjacent to an undeveloped Cu—Zn deposit, the waters are relatively alkaline (pH = 8.0) which limits anomalous zinc contents to the immediate area surrounding the inflow stream (Fig.1).

As the manganese distribution in lake sediments is closely related to the oxygen content and pH of the lake waters, the zinc contents attributable to these features may be screened out using Zn/Mn ratios, thus focussing attention on areas of high zinc related to mineralization. On this basis the lake centre sediments of Lyon Lake, which is adjacent to mineralization, have anomalous zinc >100 ppm associated with Zn/Mn × 100 values >80. In

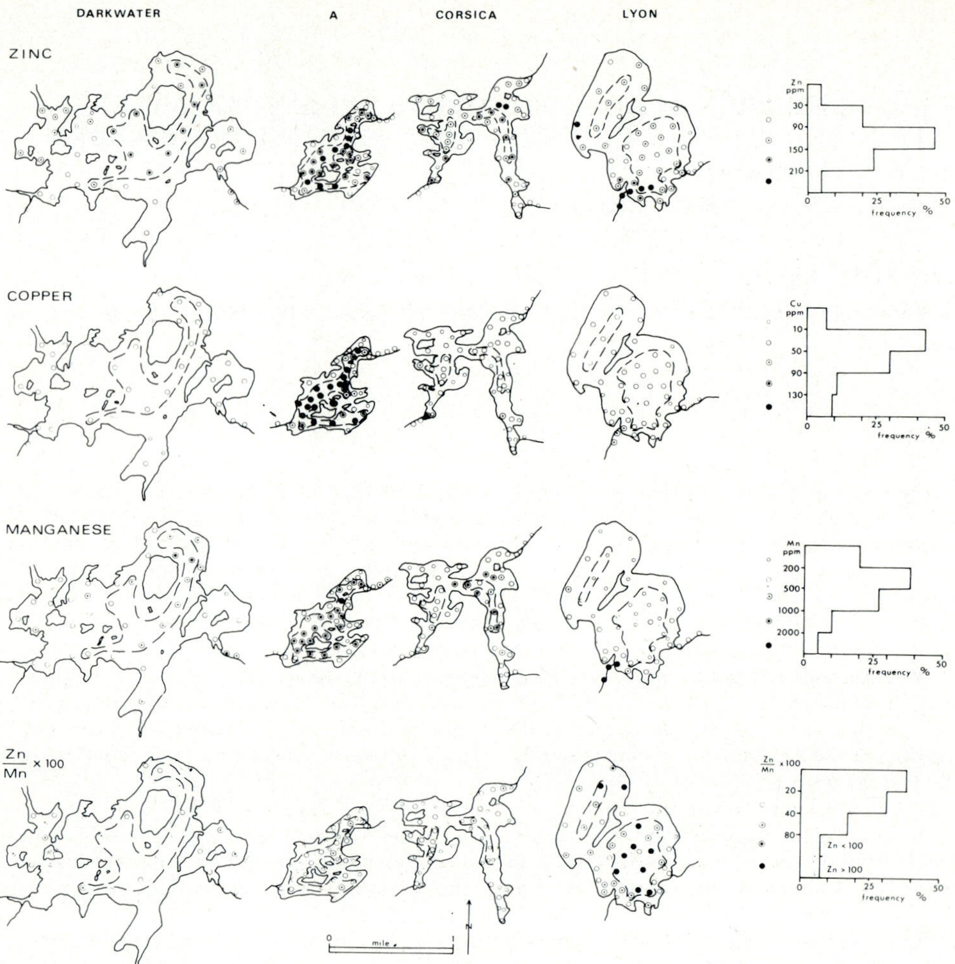

Fig.1. Dispersion of zinc, copper, manganese, Zn/Mn × 100 in lake sediments.

contrast the lake sediments with anomalous zinc contents related to coprecipitation in Corsica Lake and Lake "A" have Zn/Mn × 100 values of generally <40 (Fig.1).

On a regional basis lake sediments from lakes adjacent to the Mattabi Cu—Zn deposit do not have anomalous zinc contents (Fig.2) but do have anomalous Zn/Mn ratios (Fig.3). Lake sediments associated with the Shebandowan Cu—Ni deposit, although not distinguishable on the basis of nickel alone (Fig.4), are characterized by high Ni/Mn ratios (Fig.5).

A reconnaissance survey based on the analysis of organic-rich lake sediments from lake centres, and subsequent screening of the analytical data using Zn/Mn ratios, revealed the existence of prominent anomalous zones, one of which was comparable to that associated with the Mattabi mines deposit. Follow-up investigations confirmed the reconnaissance anomalies to be associated with favourable geology. Further geological and geophysical investigations are currently underway to evaluate the significance of the anomalous zones. Final assessment of the technique must await the results of the current programme to

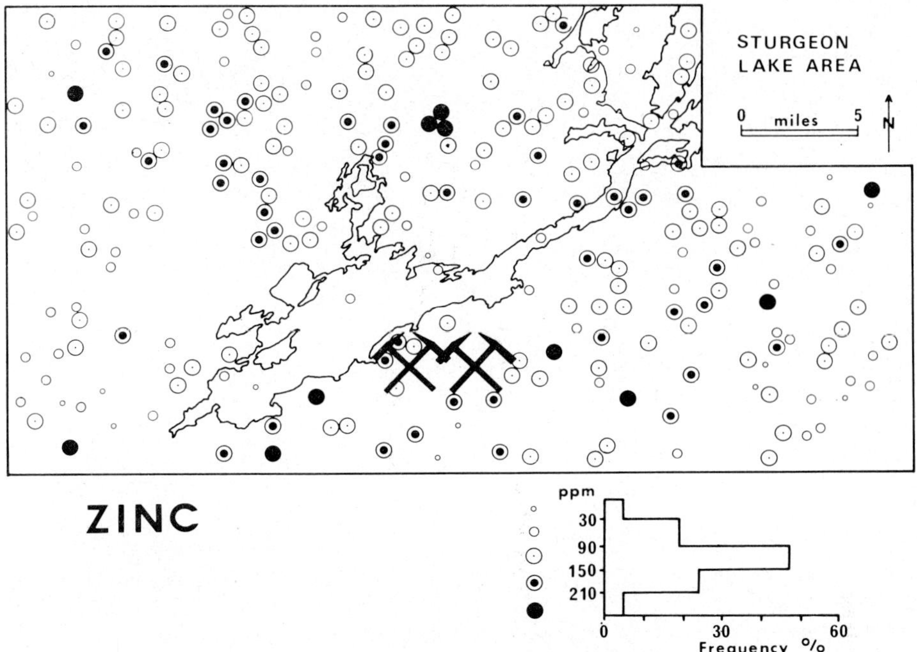

Fig.2. Dispersion of zinc in lake sediments, Sturgeon Lake area.

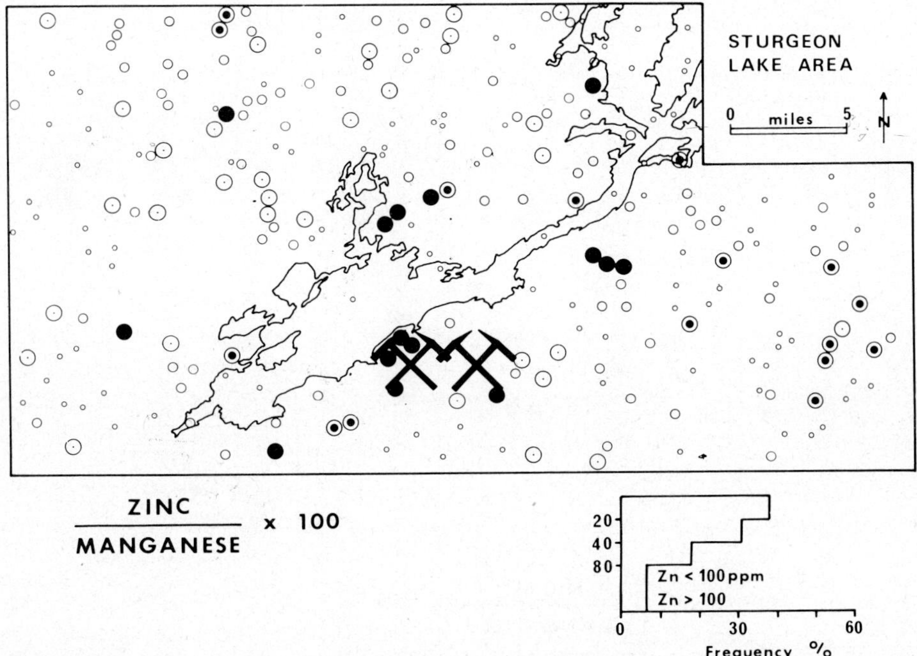

Fig.3. Dispersion of Zn/Mn × 100 in lake sediments, Sturgeon Lake area.

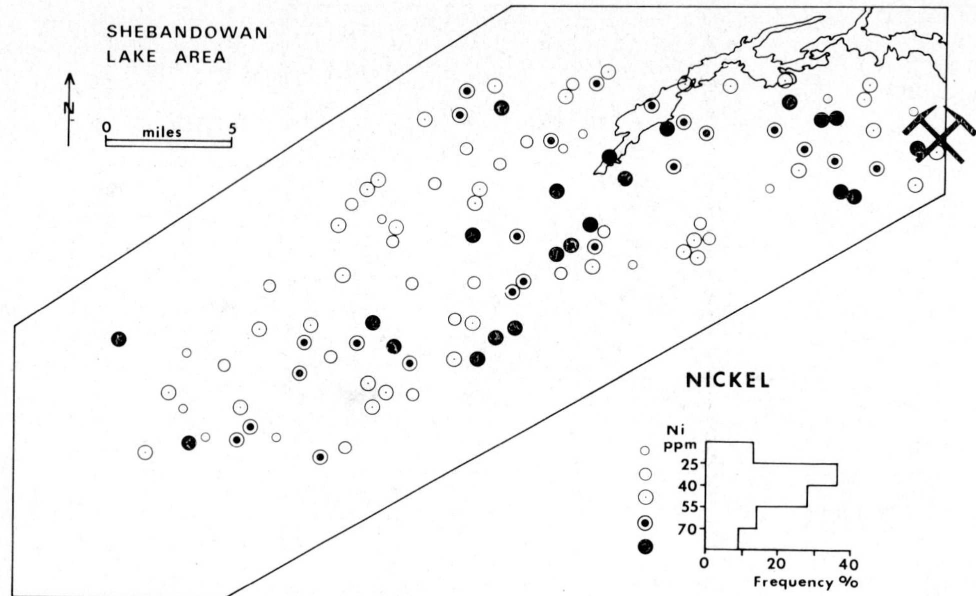

Fig.4. Dispersion of nickel in lake sediments, Shebandowan Lake area.

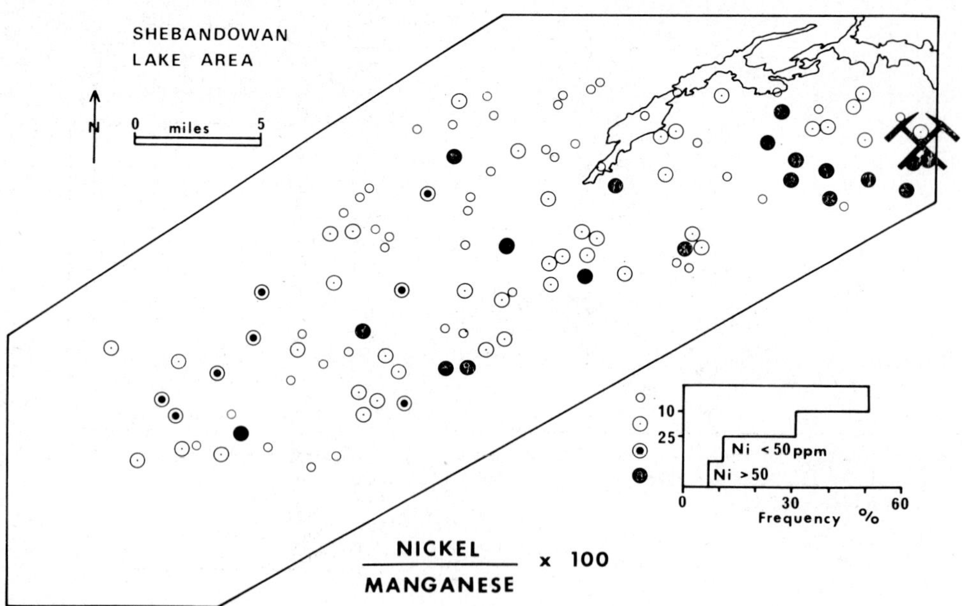

Fig.5. Dispersion of Ni/Mn × 100 in lake sediments, Shebandowan Lake area.

evaluate the significance of the anomalies revealed by the reconnaissance. Provisionally,
however, it is concluded that sampling and analysis of organic-rich lake sediments followed
by screening of the data may be a viable exploration procedure in the southern areas of
the Canadian Shield in the search for Cu—Ni and base-metal massive sulphide deposits.

ACKNOWLEDGEMENTS

The research formed part of the overall programme of research in exploration geochemistry in the Department of Geological Sciences, Queen's University, under the direction of I.N. and constituted the Ph.D. research programme of W.B.C. Rio Algom Mines
Limited provided generous financial support for this work and grateful appreciation is expressed to members of the Rio Algom group, in particular Dr. M.B. Mehrtens, Mr. J.A.
Sadler, Mr. R.C. Hart and Mr. C. Spence, for their keen interest in and enthusiastic support
of this research. In addition, the staff of the Department of Geological Sciences, Queen's
University, are thanked for assistance with analytical work and preparation of this paper.

REFERENCES

Allan, R.J., 1971. Lake sediment: a medium for regional geochemical exploration of the
    Canadian Shield. Can. Inst. Min. Metall. Bull., 64: 43—59
Allan, R.J., Cameron, E.M. and Durham, C.C., 1972. Lake geochemistry — a low density
    sample technique for reconnaissance geochemical exploration and mapping of the
    Canadian Shield. In: M.J. Jones (Editor), Geochemical Exploration 1972. Institution
    of Mining and Metallurgy, London, pp.131—160
Coker, W.B., 1974. Lake sediment geochemistry in the Superior province of the Canadian
    Shield. Thesis, Queen's University, Kingston, Ont. (unpublished)
Coker, W.B. and Nichol, I., 1975. The relation of lake sediment geochemistry to miner-
    alization in the southern Canadian Shield. Econ. Geol. (in press)

REGIONAL LAKE SEDIMENT GEOCHEMICAL SURVEY FOR ZINC MINERALIZATION IN WESTERN NEWFOUNDLAND

P.H. DAVENPORT[1], E.H.W. HORNBROOK[2] and A.J. BUTLER[1]

[1] *Newfoundland Department of Mines and Energy, St. John's, Nfld. (Canada)*
[2] *Geological Survey of Canada, Ottawa, Ont. (Canada)*

ABSTRACT

Lake sediment samples were collected from a region of about 3000 square miles
(7800 km^2) which is underlain by Lower Palaeozoic carbonate rocks containing several
showings of zinc and lead mineralization. A total of 2494 sites were sampled, providing an
average sample density of 1 sample per 1.2 square miles (3 km^2). Samples were collected
from the central basins of lakes, and organic-rich material was the preferred sample type.
An orientation study carried out in an area of significant zinc mineralization near Daniel's
Harbour indicated that the zinc content of organic-rich, centre lake-bottom sediments was
a reliable indicator of areas of mineralization within this terrain.

The samples were dried, sieved to minus 80 mesh (< 177 μ) and analysed for zinc, lead,
manganese and iron by atomic absorption spectrophotometry. The loss on ignition (L.O.I.)
of the samples was determined as an estimate of their organic content. Analysis of variance
was carried out on 128 pairs of duplicate samples, and the combined sampling and analytical
errors are significantly small at the 99% confidence level compared with the over-all
variability of zinc, lead, manganese, iron and L.O.I.

Most of the areas containing the more significant zinc showings in the region are
reflected in the zinc content of the lake sediments by anomalous or high background zinc
values. The zinc distribution is, however, strongly correlated with the distributions of iron
and loss on ignition, suggesting that zinc is coprecipitated with iron, and adsorbed or
chelated by organic material. Stepwise multilinear regression was carried out with zinc as
the dependent variable, and iron and L.O.I. as the independent variables. The regression
equation accounted for 52.7% of the zinc variability.

The residual zinc distribution after regression better defines the areas of known zinc
mineralization than does the actual zinc distribution. In addition, the lake sediments are
anomalous in residual zinc scores in some areas where no zinc showings have been recorded,
but where the geological setting is favourable for mineralization.

INTRODUCTION

Lake sediment as a low-density sample medium in regional geochemical
exploration programmes has received increasing use over the last several years.
Much of the early work was carried out by mining companies and remains
unpublished, but recently, published results have become available (e.g. Allan,
1971; Allan et al., 1972, 1973). In common with other sample media used in
geochemical exploration, the effectiveness of the trace element content of

556

lake sediments to delineate areas of mineralization will vary from area to area, depending on the physical features of the area and the type of mineralization being sought.

A regional lake sediment sampling programme was carried out over about 3000 square miles (7800 km^2) of Lower Palaeozoic carbonate rocks in western Newfoundland in 1973 (Fig.1). The programme was carried out to

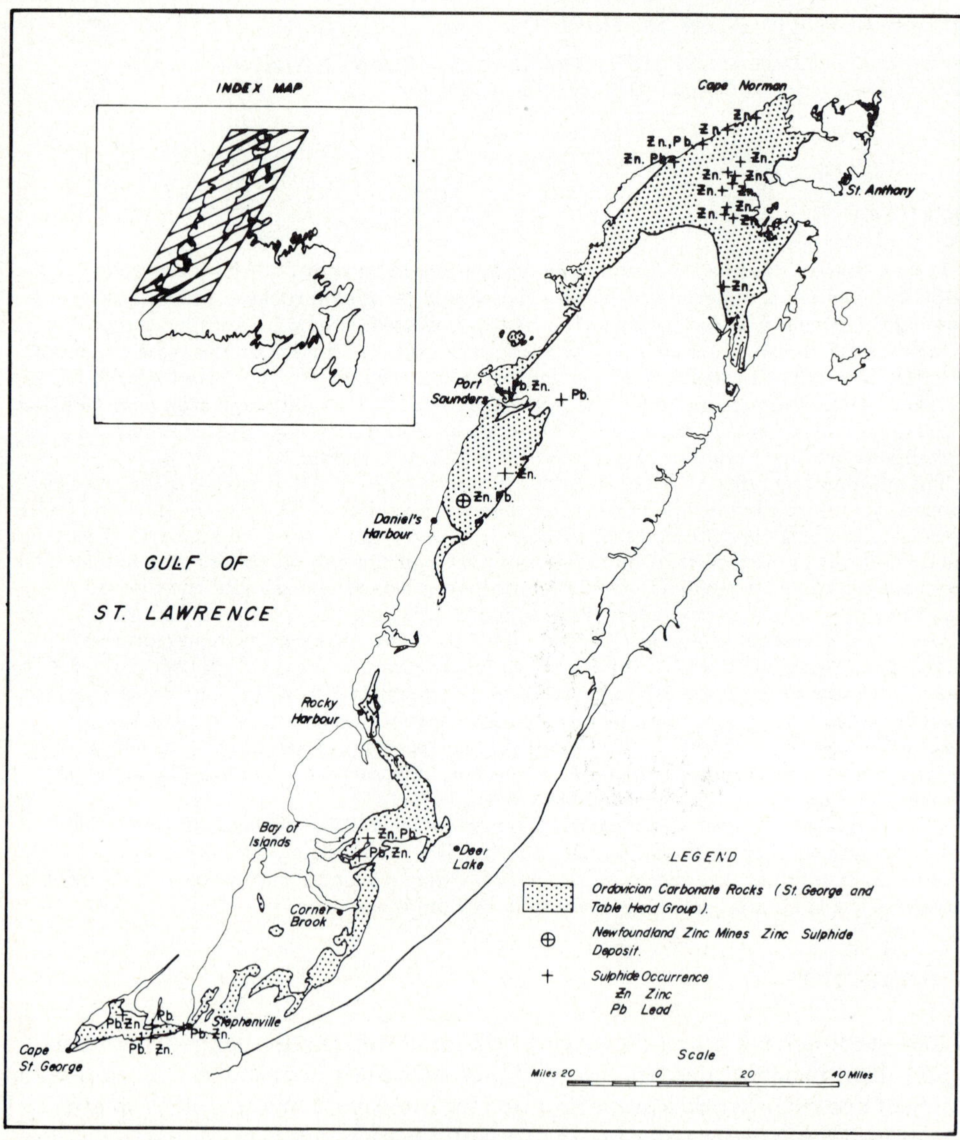

Fig.1. Distribution of Lower Palaeozoic carbonate rocks in western Newfoundland (from Williams, 1967) over which geochemical lake sediment survey was conducted.

delineate areas of potential zinc mineralization. The effectiveness of this type
of survey in detecting zinc mineralization in this terrain was demonstrated
in an orientation study carried out in an area containing significant zinc
mineralization near Daniel's Harbour (Fig.1) (Hornbrook et al., 1974).

Whilst the interpretation of the data is not yet completed, the survey has
been successful in detecting most of the significant known zinc showings in
the belt, and has indicated new areas of probable mineralization, many of
which are geologically favourable.

GEOLOGY OF STUDY AREA

The lake sediment sampling programme was confined almost exclusively
to the area underlain by the St. George and Table Head Groups of carbonate
rocks in western Newfoundland (Fig.1). These groups are comprised of limestone and dolomite, with very minor shale, and are of Middle or Upper
Cambrian to Middle Ordovician age (Williams, 1967). They represent plat-
formal carbonate deposition. The contact between the underlying St. George
Group and the Table Head Group is disconformable over most, if not all, of
its extent in the belt (Cumming, 1968).

Base-metal mineralization in the area may be subdivided conveniently into
three main types. The first type is stratabound sphalerite mineralization
occurring in the upper part of the St. George Group, a few hundred feet
stratigraphically below the disconformity with the Table Head Group. This
type is best exemplified by the Newfoundland Zinc Mines' deposit near
Daniel's Harbour (Fig.1). Here sphalerite occurs in a number of pods, the
largest of which contains 4.4 million tons of 8.8% Zn (Anonymous, 1974).
The setting of these deposits is described by Collins and Smith (1972a, b). A
second type of deposit contains both sphalerite and galena as ore minerals.
Available details of this type of deposit are few, but they occur stratigraphically lower in the St. George Group. In general they are stratabound, but
sulphide veinlets are associated in places. They are known to be present north
of Corner Brook (Fig.1), and are described by Lilly (1963). A third type is
the galena-sphalerite veins found in the Table Head Group rocks, particularly
on the Port au Port Peninsula. Little data concerning these deposits are available.

The following brief description of the Quaternary geology is taken from
Brookes (1972), Vanderveer (1973) and Grant (1974). During the glacial
maximum Labrador ice affected only the northern tip of the Great Northern
Peninsula, flowing generally southeastward to south-southeastward across the
most northerly part of the area (Fig.2) and southwestward down the Strait
of Belle Isle towards St. John Bay. As the glacial maximum passed, Labrador
ice retreated and the area was affected only by the late Wisconsin ice cap that
existed on the Long Range Mountains even during the Wisconsin glacial
maximum. Flow from this cap was generally west over the area underlain by
the St. George and Table Head Groups (Fig.2) and radially off the high lands

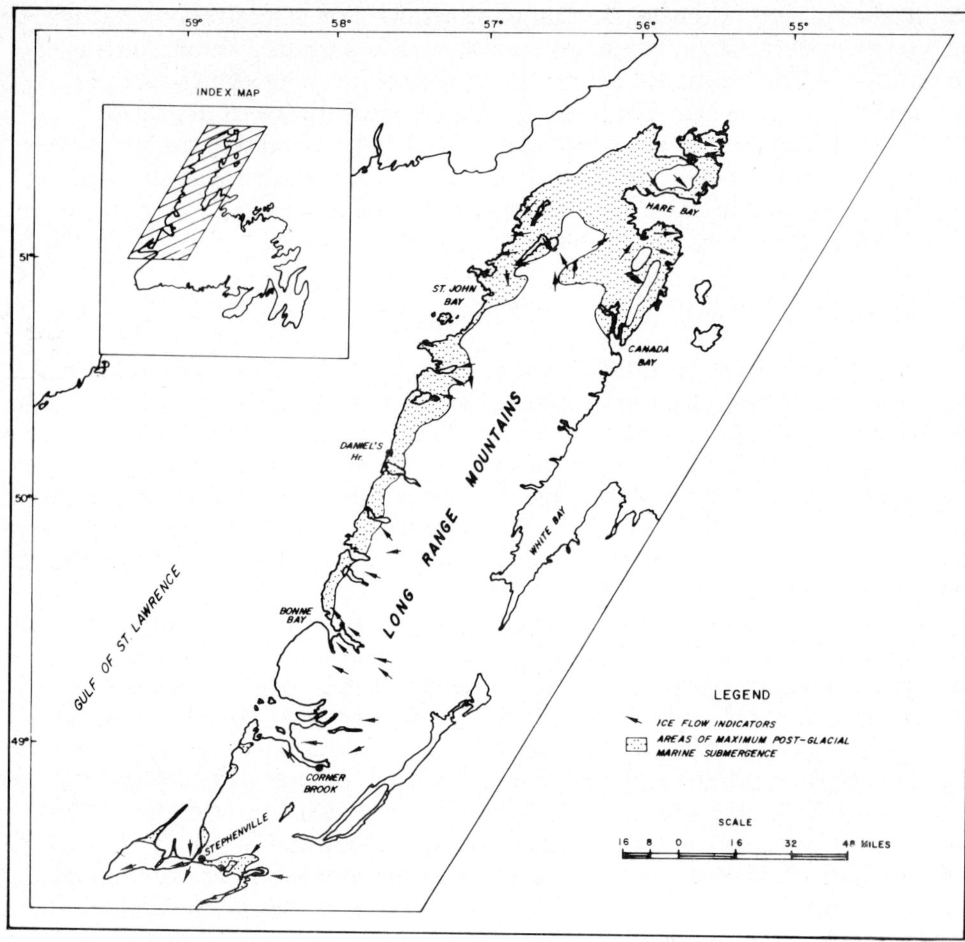

Fig.2. General ice flow directions and area of maximum marine overlap during Wisconsin Glaciation in western Newfoundland (compiled from Brookes, 1972; Vanderveer, 1973; and Grant, 1974).

and across the low-lying northern end of the Great Northern Peninsula. During this stage ground moraine was deposited over much of the area.

The sea advanced as this ice front retreated, and DeGeer moraines and marine deposits were laid down. In the late stages of glaciation the sea was held out from parts of the study area by glacial lobes fed from the Long Range Mountains. Marine deposits and shell fragments are found up to 450 ft above the present sea level in the north of the area and there is a general decrease in the maximum marine overlap from Hare Bay southwards to Stephenville. The maximum marine limit is shown on Fig.2, but in fact the areas actually inundated are much more restricted and were controlled by the position of the wasting ice front. Continued isostatic uplift and retreat of the sea led to some marine deposition along the coast (Grant, 1972) and deposition

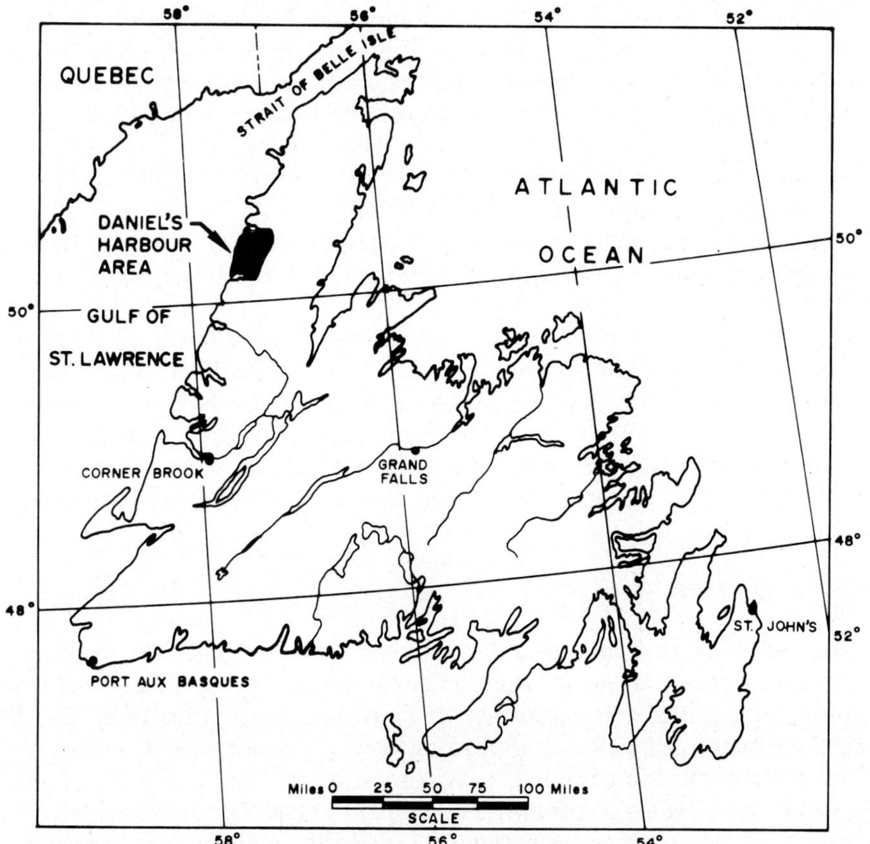

Fig.3. Location of Daniel's Harbour orientation study area.

of ablation moraine accompanied the wastage of the ice back into the highlands.

Marine deposition is rather minor, and marine clays are restricted to near the present coastline. Marine deltas and beach deposits are present in places, but more extensive areas are covered by lag-gravel deposits, the result of wave action on moraine. DeGeer moraines are most common in the northern part of the area.

In the northern area the drift is generally thin, ranging from 0 to 30 ft (9 m) in thickness. From Bonne Bay to Stephenville the drift is thin except in low-lying plains where thicknesses of the order of 200 ft (60 m) may be encountered, e.g. the area southeast of Stephenville.

ORIENTATION STUDY

To determine the optimum procedures for carrying out a regional geochemical exploration programme for zinc mineralization in the St. George and Table Head Group carbonate rocks, an orientation study was carried out in 1972. An area of approximately 350 square miles (900 km^2; see Fig.3), was selected which included the Newfoundland Zinc Mines Limited deposits which are currently being brought into production. The showings have been evaluated by drilling and one has been trenched but were otherwise undisturbed. Stream sediment, organic lake centre-bottom sediment and lake water samples were collected over this area, in addition to a detailed soil and basal till sampling study close to the zinc mineralization. The results of this study are discussed more fully by Hornbrook et al. (1974).

Briefly stated, the zinc distribution in the stream sediments, lake sediments and lake waters all reflect the presence of the major zinc mineralization. The potential of lake water samples was reduced because of the insufficiently sensitive analytical method employed, which allowed the reliable identification only of highly anomalous zinc concentrations. The distribution of streams in the orientation study area is erratic, owing to the development of underground drainage in places. Thus it is not possible to maintain a uniform sample density over the area, and this situation is also true in many places within the area underlain by the St. George and Table Head Groups outside the orientation study area.

Zinc distribution in lake sediments

The distribution of zinc in lake sediments is shown in Fig.4. The average sample density is 1 sample per 2 square miles (1 sample per 5.2 km^2), and the sample points have a fairly uniform distribution throughout the area. The frequency distribution of zinc is log-normal, and the contour intervals in Fig.4 are arbitrarily chosen at 0.5, 1.5, 2.5 and 3.5 standard deviations above the mean (150, 400, 1000, 2700 ppm Zn, respectively). The major zinc deposits and related showings are reflected by a multi-station anomaly, where the lake sediment samples contain zinc concentrations ranging from 6250 to 14,500 ppm. Only one of these sampled lakes may be contaminated by trenching over an adjacent showing. Other lower order anomalies occur on the east margin of the sampled area, where zinc showings are known.

The lake sediments were analysed for Cu, Pb, Ag, Co, Ni, Mn, Fe and L.O.I., in addition to zinc. Only very minor concentrations of copper, lead and silver are associated with the zinc mineralization, and the cobalt and nickel contents of the mineralization are very low. The distributions of these elements did not, therefore, reflect the presence of the zinc showings. Zinc is weakly correlated with iron and L.O.I. Despite the statistical significance of these correlations, however, the distribution of zinc residuals after linear regression with these parameters is essentially similar to the actual zinc distribution. There is no correlation between zinc and manganese in the orientation study data.

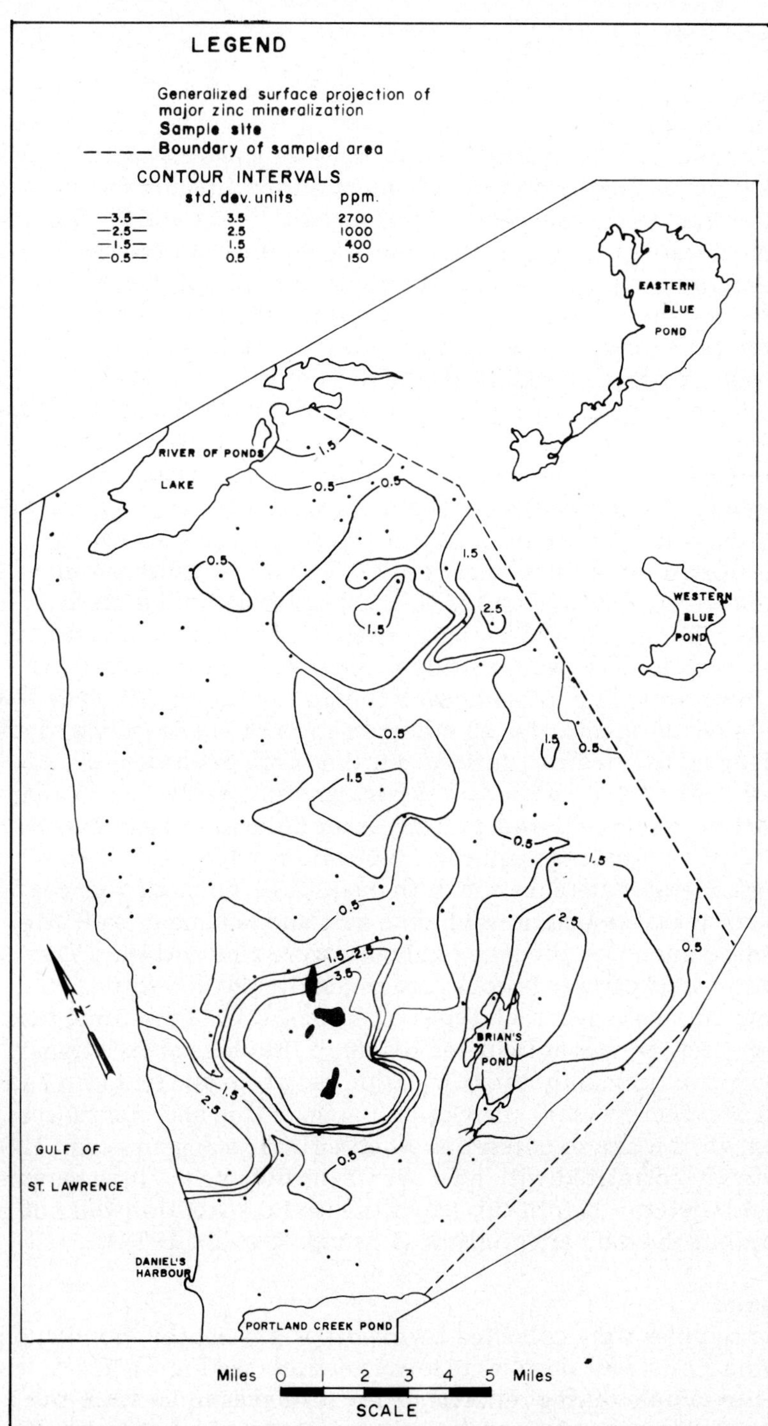

Fig.4. Distribution of zinc in lake sediments, orientation study area.

REGIONAL SURVEY

Design of survey

The results of the orientation study indicate that organic sediments from
the central parts of lakes are the best sample medium for a regional reconnaissance survey for zinc mineralization in this terrain. Coker and Nichol
(1975) have demonstrated that organic sediment samples from the central
parts of lakes are more homogeneous in both major and minor element
composition than near-shore samples. Organic gyttja-like material is the most
common sediment in the regional study area, but in some areas, particularly
in the northern part, no organic sediments are developed in the lakes.

A sample density of approximately 1 sample per square mile (2.6 km^2) was
chosen for the regional survey. Over most of the area this density is easily
achieved, but in places the sample interval is greater. The zinc and lead
mineralization known in the belt occurs typically as clusters of pods or veins,
which may occupy an area of several square miles. In general, however,
significant dispersion of metals will occur only from mineralization which
intersects the bedrock surface. Several of the zinc sulphide bodies on the
Newfoundland Zinc Mines Limited property near Daniel's Harbour fortuitously
suboutcrop in or close to lakes, thus readily supplying zinc to the drainage
system. A fairly close sampling interval was adopted to minimize the possibility of failing to detect anomalies due to mineralization of which only a small
proportion intersected the bedrock surface. Furthermore the host rocks of
the mineralization, although commonly a particular stratigraphic unit, are not
geochemically distinct in trace elements from other units in the succession
(Sangster, 1968, J.A. Collins, personal communication, 1972).

The known base-metal mineralization in the belt of carbonate rocks contains
the three important metal associations of zinc, zinc and lead, and lead. The
only ore metals determined on the lake sediments were zinc and lead. The
lead and zinc contents of carbonate rocks are generally low and show little
variation between the two major rock types of dolomite and limestone; thus
little background variation was anticipated owing to lithological variations.
Although in the orientation study the correlation between zinc and iron and
L.O.I. was weak, and there was no correlation between zinc and manganese,
it was considered worthwhile to determine whether iron, manganese and L.O.I.
are more significantly correlated with zinc over the whole belt. These parameters,
therefore, were also determined. In this paper the lead distribution will not be
discussed further, but the data are available (Davenport et al., 1974).

Sampling method
Lake sediment samples were collected by means of a weighted, hollow, pipe-
like sampler having a one-way valve in its lower section (see Fig.5). The valve
prevents loss of the sample during retrieval of the loaded sampler from the
lake bottom. Sampling was carried out from the pontoon of a helicopter by

Fig.5. Lake sediment sampler.

allowing the sampler to free-fall through the water to penetrate the organic-mineral soil accumulation at the lake bottom. After retrieval of the sampler, the sample was shaken out of the lower section, examined, and transferred to paper sample bags. Features of the sample and sample site such as water depth, colour and composition of the sample and nature of the surrounding terrain were recorded.

Sample preparation

The samples were partially air dried in the field, and oven dried in the laboratory. During the orientation study, it was found that highly organic samples tend to form a hard, brittle cake when dry. Therefore, such samples collected during the regional survey were disaggregated in a food blender with stainless steel blades and glass jug. All samples from the regional study were sieved to minus 80 mesh ($<177\ \mu$), through stainless steel sieves.

Analytical methods

Both the samples from the orientation study and those from the regional survey were analysed by atomic absorption spectrophotometry for zinc, lead, manganese and iron. An aliquot of 0.2 g of the sample was digested in 3 ml of a $4M$ $HNO_3/0.1M$ HCl mixture at 100°C for 90 minutes and when cool the solution was made up to 10 ml with water. The analytical precision was determined by running a series of 10 control samples in each analytical batch. The precision of zinc, manganese and iron at the 95% confidence level were ±14%, ±12% and ±14%, respectively.

The percentage weight loss on ignition (L.O.I.) was determined on all samples as an estimate of their organic content. Approximately 0.2 g of sample was heated in an oven overnight at 110°C to drive off excess moisture, and weighed. The sample was then ignited in a muffle furnace on a 3 hour long, time-temperature controlled rise to 500°C, which temperature was maintained for 1 hour. The sample was then reweighed, and the weight loss calculated from the ratio of the difference in weight before and after ignition to the weight of the dried sample before ignition, expressed as a percentage. Timperley and Allan (1974) report that no carbonate decomposition occurs when ashing is carried out below 550°C.

Sample reproducibility

To determine the sample reproducibility and representativity, approximately 5% (128) of the 2494 sites were sampled in duplicate. These duplicate samples were collected throughout the study, and the means and standard deviations of the compositions of the subsets are similar to those of the whole data set. The method proposed by Garrett (1973) was used to determine whether the over-all variance of zinc, manganese, iron and L.O.I. in the data subsets is significantly greater than the variance due to sampling and analysis at each site. The F values of these variance ratios are given in Table I. It can be seen that the variance ratios for zinc, manganese, iron and L.O.I. all greatly exceed the critical F ratio (1.51) at the 99% confidence level.

TABLE I

Analysis of sampling and analytical variance for Zn, Mn, Fe and L.O.I. from 128 pairs of duplicate lake sediment samples

	F ratio*
Zn	13.01
Mn	8.59
Fe	17.05
L.O.I.	6.74

*F ratio is the ratio of the variance of the distribution of each element in the data subset to the variance within the each sample pair (sampling and analytical variance). The critical F ratio with 128, and 127 degrees of freedom at the 99% confidence level is 1.51 (Garrett, 1973).

Zinc distribution

The distribution of zinc is shown in Figs.6, 7 and 8. The northern area (Fig.6) is arbitrarily separated from the central area to the south (Fig.7) for ease of presentation. The southern boundary of the central area is the approximate northern boundary of the Gros Morne National Park, and the northern boundary of the southern area (Fig.8) corresponds with the southern boundary of the park. All of the 2494 sample points are shown, and it can be seen that the sample site distribution in the northern and central areas are quite even, but in the southern portion of the southern area, especially on the Port au Port Peninsula, shown as an inset, the sample points are less evenly distributed, because of the uneven lake distribution. The zinc frequency distribution is log-normal, with a geometric mean of 55 ppm. Contour levels were arbitrarily chosen at 0.5, 1.5, 2.5, 3.5 standard deviations above the mean (107, 393, 1439, 5284 ppm Zn, respectively), based on the whole data set. Figs.6, 7 and 8 are hand contoured and represent the actual zinc distribution. In general the zinc distribution is quite smooth, with anomalous sites surrounded by sites containing high background zinc values. There are few high single point anomalies.

The northern area

From Fig.6 it is apparent that the eastern portion, east of lines joining A to C and C to E contains higher zinc values than the area to the west. The zinc showings (Fig.1) to the west and southeast of B (Fig.6) are reflected only by high background zinc values in the lake sediments. The high background zinc values from A to the area east of C approximately follow the upper part of the St. George Group near its contact with the overlying Table Head Group, and may indicate that the zinc mineralization is more extensive than shown in Fig.1. The highest values of zinc in the lake sediments occur immediately to the north of D, and southwards to the area south of E. Two anomalous samples

566

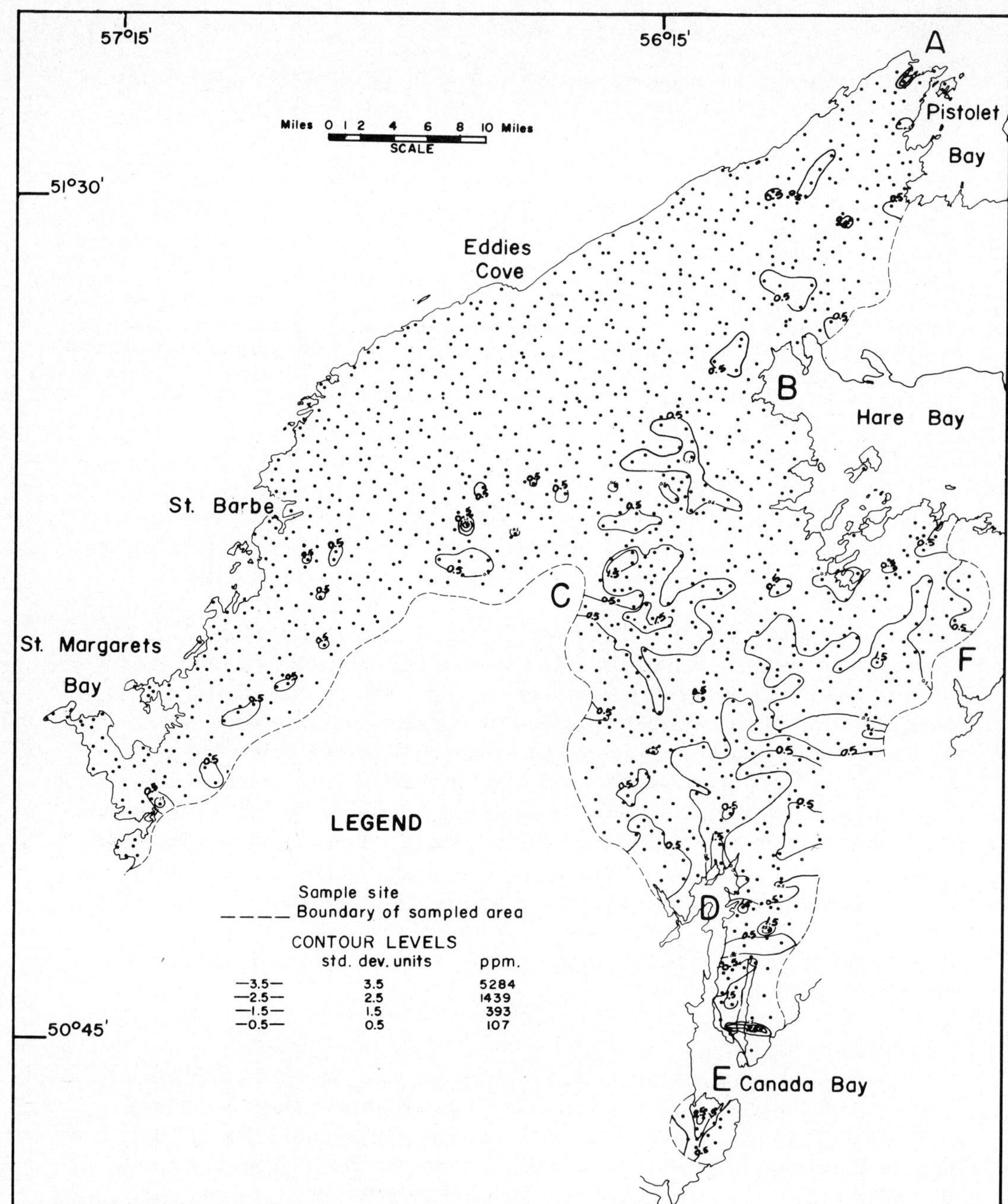

Fig. 6. Zinc distribution, northern area.

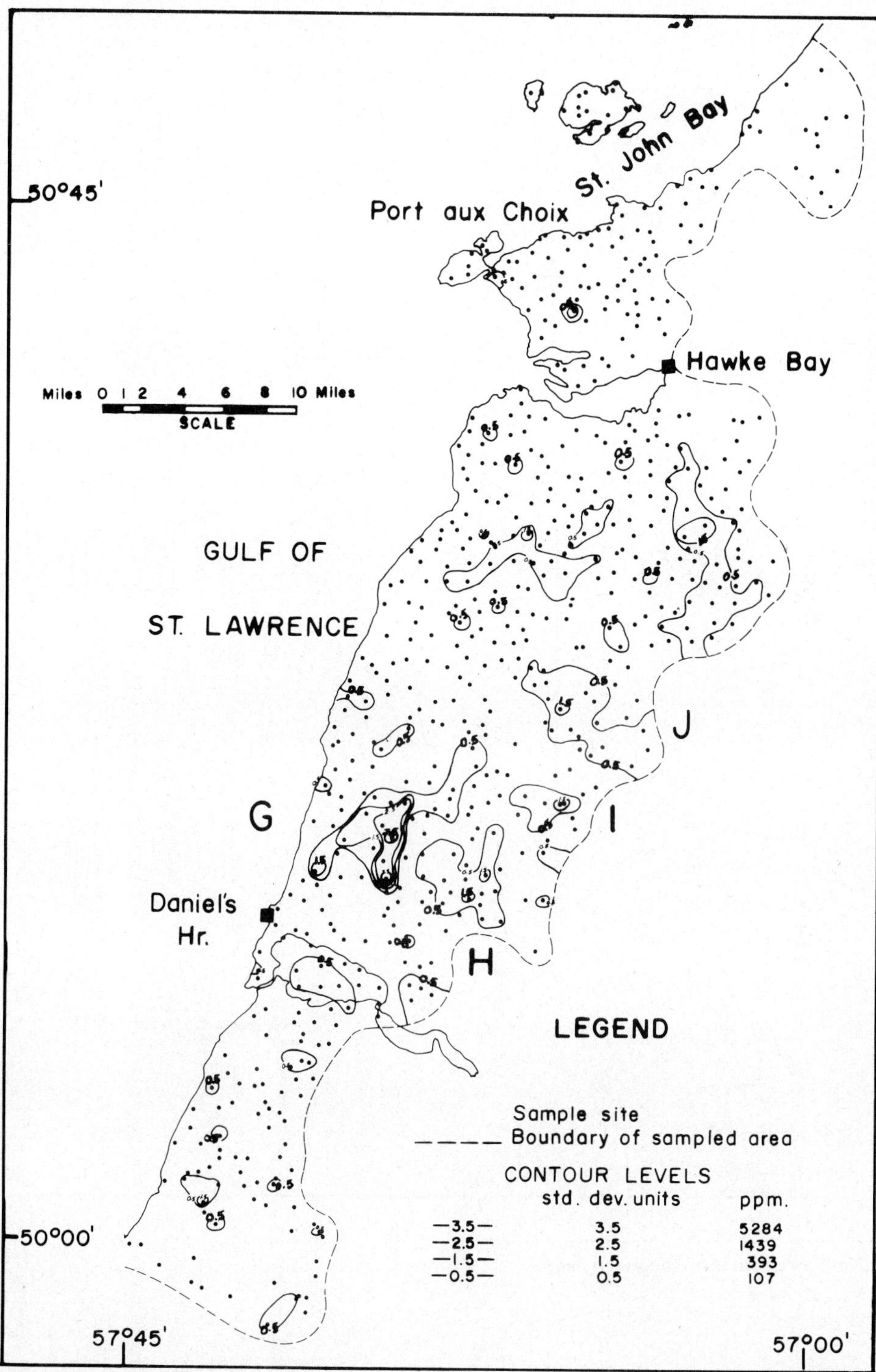

Fig.7. Zinc distribution, central area.

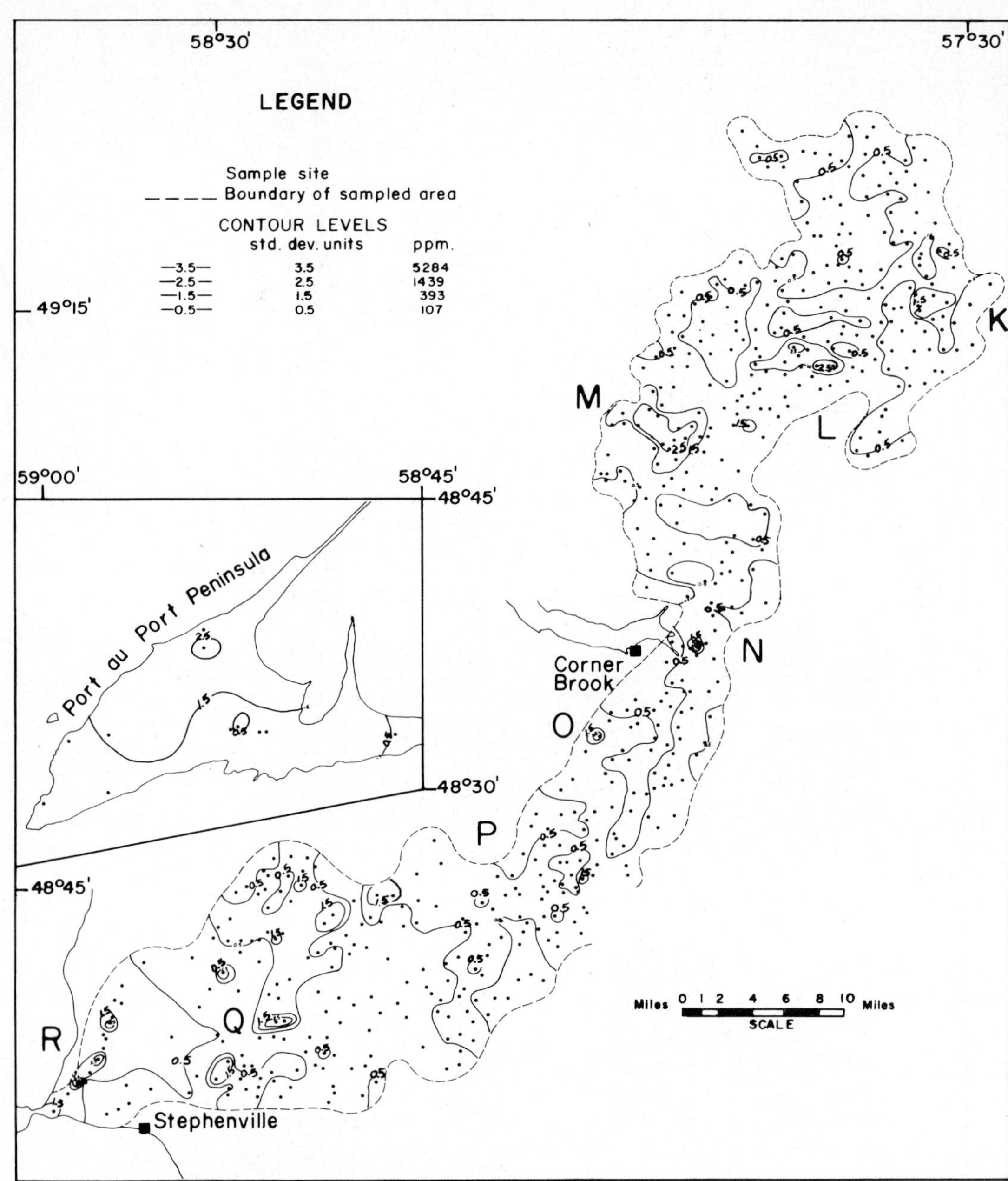

Fig.8. Zinc distribution, southern area.

between D and E occur to the east of the area underlain by carbonates, and are in lakes overlying predominantly clastic rocks of the Hare Bay Allochthon (Smyth, 1973). The area generally to the west of F is predominantly underlain by allochthonous clastic rocks which contain black shale units (Smyth, 1973). This area was sampled because windows of Table Head Group carbonates are present within the allochthon. This area is also characterized by high background zinc levels in lake sediments which may be attributable to high zinc concentrations in the shale units rather than to zinc mineralization. The showings indicated in Fig.1 on the northwest coast of the northern area are all minor, and the lack of any high background or anomalous zinc values in this area suggests that no significant showings of zinc occur inland.

The central area

The zinc distribution in this area (Fig.7) reflects the presence of the zinc deposit and related showings of Newfoundland Zinc Mines Ltd., east of G. As in the results of the orientation study (Fig.4) a multi-station anomaly is developed in the lake sediments from the area containing these deposits. The high background values north of H and west of·I and J (Fig.7) also correspond to high background or anomalous values in Fig.4. It should be noted that the contour levels of zinc in ppm differ between the orientation study and the regional study.

The southern area

The zinc distribution is shown in Fig.8. High background to anomalous zinc values occur to the west of K and north of L. There are no published data on showings in this area, but it is believed to be favourable from past geochemical surveys carried out by mining companies in the general area. The area to the east of M contains lead and zinc showings (Lilly, 1963) although there is little published information as to their grade or extent. The other significant anomalies occur to the east of Q and to the east of R. To the writers' knowledge no mineralization has been recorded from these areas. The northwestern part of the Port au Port Peninsula (inset, Fig.8) also contains high background to anomalous zinc contents. Some minor Pb—Zn mineralization is known to occur in the Table Head Group in this area.

Coprecipitation and adsorption of zinc

The actual zinc distribution identifies most of the more economically significant zinc showings, and identifies areas of high zinc where there is no known mineralization. To determine whether the distribution of zinc was correlated with those of manganese, iron and L.O.I. a correlation matrix was calculated (Table II). The distributions of manganese, iron and L.O.I. are all approximately log-normal, so that all the data are log-transformed. All these coefficients are significant at the 99% confidence level. Thus the zinc distribution is correlated significantly with iron, L.O.I. and manganese. Manganese

TABLE II

Correlation matrix of Zn, Mn, Fe and L.O.I. from lake sediments

	Zn	Mn	Fe	L.O.I.
Zn	1.0			
Mn	0.35	1.0		
Fe	0.60	0.71	1.0	
L.O.I.	0.47	−0.06	0.12	1.0

All the data have been transformed into logarithms. Number of samples is 2494.

and iron are themselves very closely correlated. Thus it appears that on a regional scale that zinc is subject to coprecipitation with both iron and manganese oxides and hydroxides, and adsorption or chelation by organic material. Stepwise multilinear regression was carried out with zinc as the dependent variable, and iron, L.O.I. and manganese successively as the independent variables. It was found that iron accounted for 36%, L.O.I. for 16% and manganese for only 0.1% of the variance of zinc. Thus the regression was carried out again with only iron and L.O.I. as the dependent variables.

Residual zinc distribution

The normalized zinc residuals are plotted in Figs.9, 10 and 11. The data are again hand contoured from the residual scores plotted at the sample sites. The contour intervals of +0.5, +1.5, +2.5 and +3.5 standard-deviation units are chosen to correspond with those of the actual zinc distribution, shown in Figs.6, 7 and 8.

The northern area

Areas of high background or anomalous zinc residual values (Fig.9) are better defined than is the case with actual zinc values (Fig.6). Three well-defined anomalous areas are present between A and C, and the known zinc showings (Fig.1) west of B (Fig.9) are fairly well defined. In a similar way the residual zinc anomalies immediately north of D, and extending discontinuously to the south of E over rocks stratigraphically favourable for zinc mineralization are enhanced relative to the actual zinc values. The residual zinc values to the west of F over allochthonous rocks are background, suggesting that the high background actual zinc values are not related to mineralization. The two high actual zinc anomalies furthest to the east from D and E are also anomalous in residual zinc. The cause of these anomalies occurring outside the St. George and Table Head Groups is unknown. The area to the west of a line joining A and C is rather featureless in residual zinc, as it was in actual zinc. Two anomalous clusters of residual zinc scores are present immediately to the north and northwest of C, which were not well defined in the actual zinc data (Fig.6).

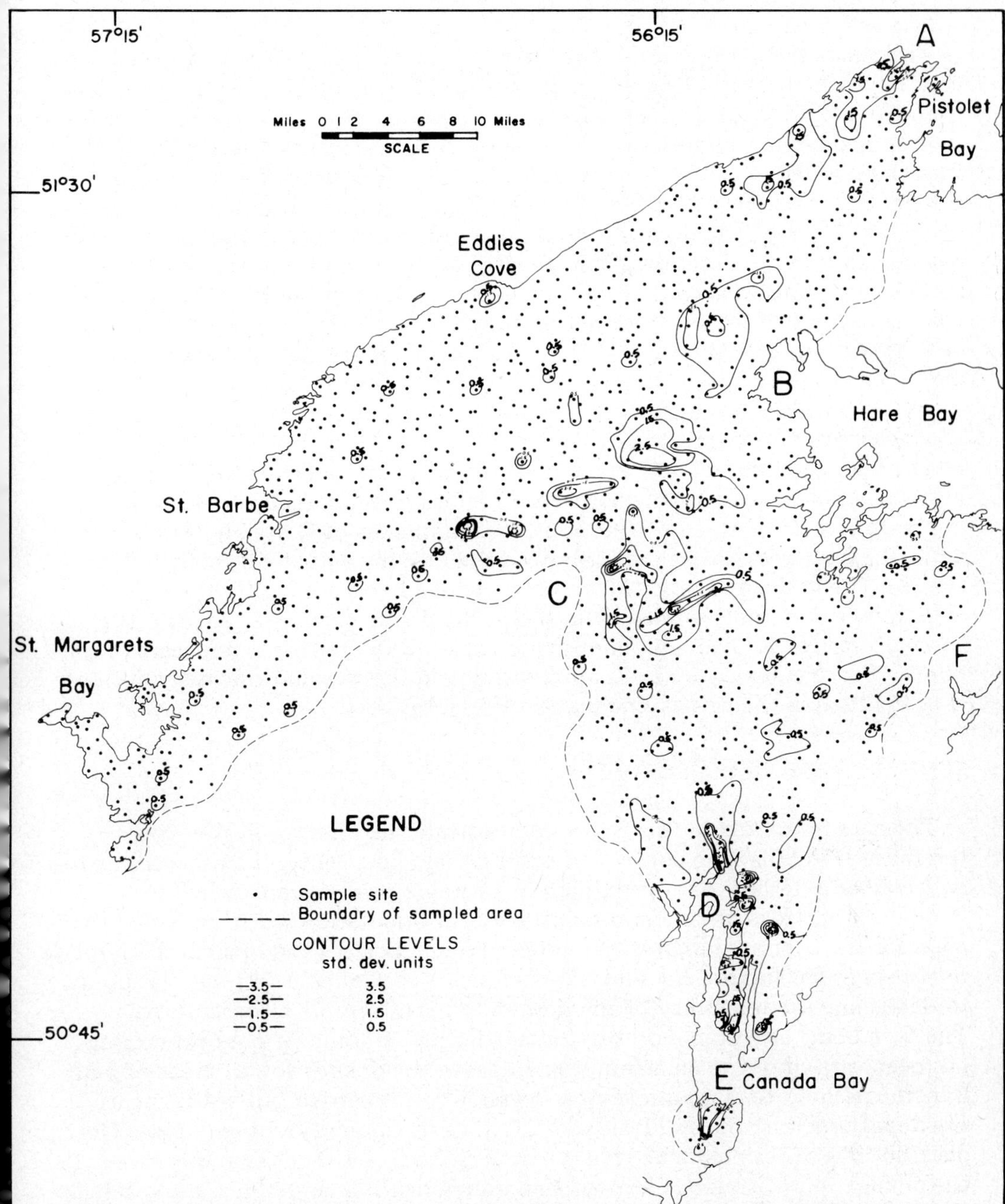

Fig.9. Residual zinc distribution after multilinear regression against iron and loss on ignition, northern area.

572

The central area

The most striking feature of the residual zinc distribution in this area (Fig.10) is the multi-station anomaly east of G, over the zinc deposits of Newfoundland Zinc Mines Ltd. (Fig.1). The anomaly is more clearly defined by the residual zinc distribution than by the actual zinc distribution (Fig.7). Elsewhere in this area the residual zinc pattern differs from the actual zinc distribution pattern, with generally a better clustering of anomalous and high background areas. The discontinuous strip of high background and anomalous residual zinc values describing the arcuate pattern from H to H' (Fig.10) includes two known showings of zinc mineralization and follows the favourable upper part of the St. George formation. This pattern was scarcely apparent in the actual zinc distribution (Fig.7). This area may contain substantially more zinc mineralization than has been found to date.

The southern area

The high background and anomalous residual zinc values (Fig.11) cluster better than in the case of the actual zinc values. The anomalies west of K, north of L, west of N, east of O, south of P and west of Q are all better defined by the residual zinc distribution than by the actual zinc distribution (Fig.8). The anomaly east of M, associated with known Zn Pb showings (Fig.1) is considerably enhanced in the residual zinc data (Fig.11), as is the anomaly east of R. The anomalous zinc values in the northwestern part of the Port au Port Peninsula are even more striking in the residual zinc distribution (Fig.11) than in the actual zinc distribution (Fig.8).

SUMMARY AND CONCLUSION

The data presented in this paper demonstrate the effectiveness in this terrain of using the zinc content of organic-rich, lake centre-bottom sediments, collected at a density of approximately 1 sample per 1 square mile (2.6 km^2), in delineating areas of zinc mineralization. An orientation study provided the basis for the over-all design of the programme indicating the optimum approach for the regional survey. An analysis of variance of replicate samples was used to determine the reliability of the data in terms of sample representativity. The data being acceptable on this basis, a limited amount of data processing has been carried out in an attempt to improve the delineation of areas of zinc mineralization. Stepwise multilinear regression was carried out with zinc as the dependent variable and iron and L.O.I. as the independent variables. The distributions of residual zinc values resulting from this regression improves the delineation of zones of known zinc mineralization, and hence the residual zinc distribution is believed to be a more reliable indicator of new areas of potential zinc mineralization.

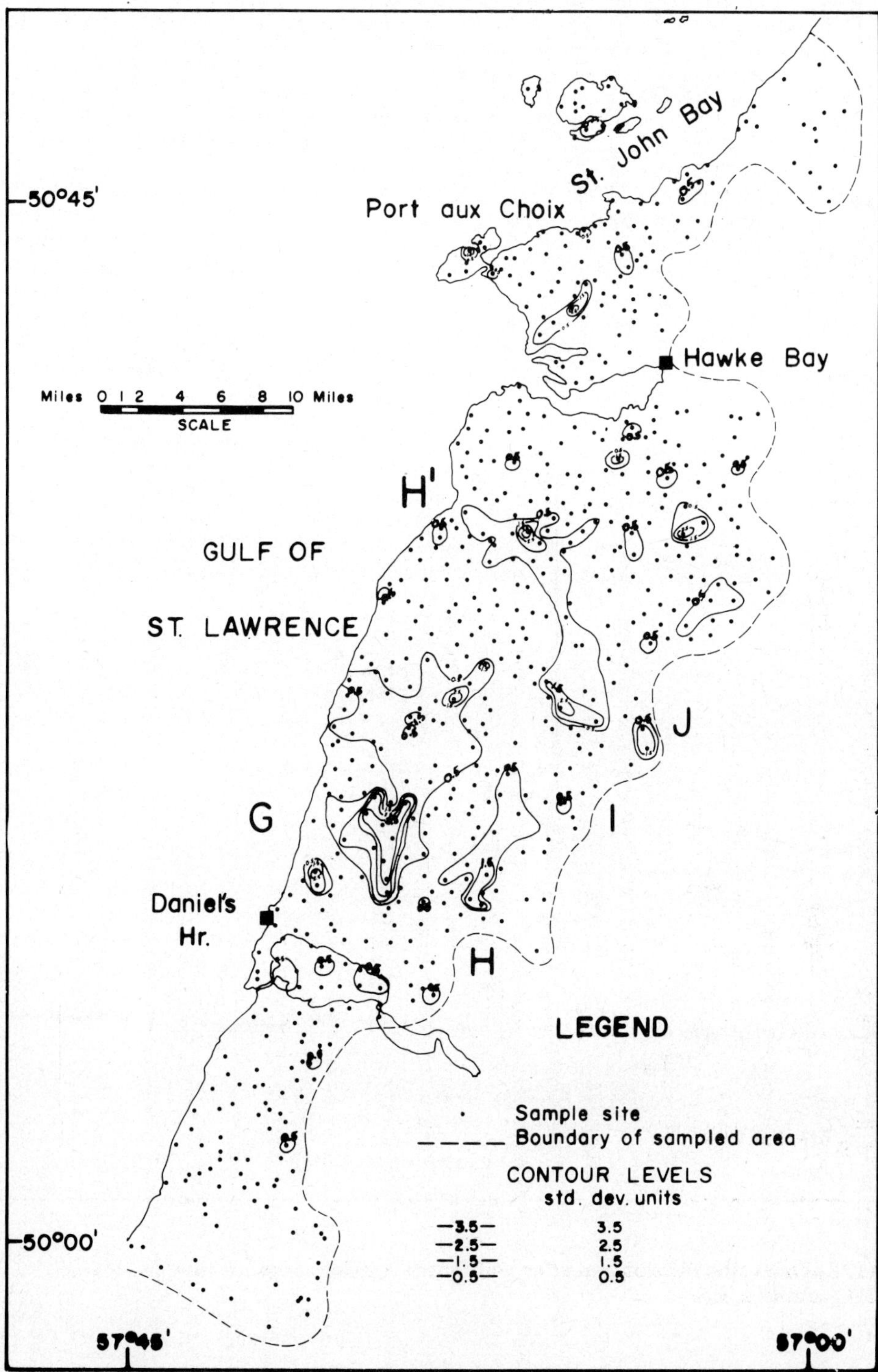

Fig.10. Residual zinc distribution after multilinear regression against iron and loss on ignition, central area.

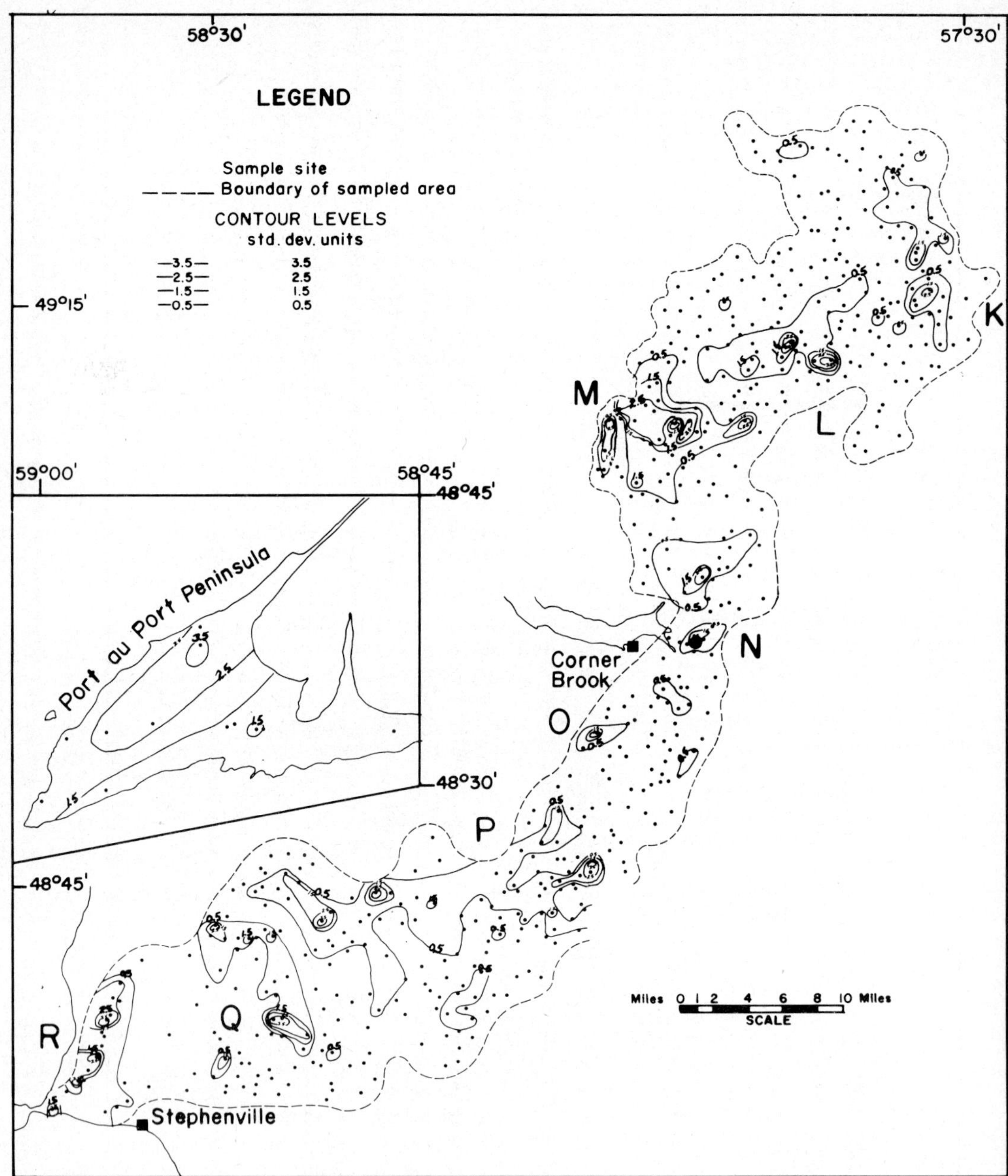

Fig.11. Residual zinc distribution after multilinear regression against iron and loss on ignition, southern area.

Regional background variation

From Figs.6, 7 and 8, a general increase in the average zinc content is apparent progressively from the northern area to the central area, and from the central area to the southern area. This regional effect is also apparent, although less noticeable, in the normalized residual zinc scores (Figs.9, 10 and 11). Furthermore, iron and L.O.I. also show this same trend, and manganese to a lesser extent, being roughly equivalent in mean level in the northern and central areas, but substantially higher in the southern area. These regional variations are summarized in Table III.

Table IV compares the means of the actual zinc and normalized residual zinc scores expressed in terms of standard deviation units from the means of each variable in the whole data set. Thus it can be seen that although the means of the residual zinc scores in the three areas are more similar than the means of the actual zinc values, the residual zinc score means are significantly different.

For each sample site, the lake area, lake water depth at the sample site, and the topographic relief — in a circular 1 square mile (2.6 km^2) search area centred on the sample site — were measured. The correlation coefficients between the lake area, lake depth (at sample site), topographic relief, zinc, manganese, iron and L.O.I. are given in Table V. Zinc is quite strongly correlated with both lake depth and topographic relief, in addition to iron, L.O.I. and manganese. It is also noteworthy that manganese and iron are correlated

TABLE III

The means and standard deviations of Zn, Mn, Fe, L.O.I. and normalized residual Zn score (after stepwise multilinear regression against Fe and L.O.I.) distributions in lake sediments from all the data and from the northern, central and southern areas

		All data ($N^* = 2494$)	Northern area ($N = 1420$)	Central area ($N = 493$)	Southern area ($N = 581$)
Zn	geometric mean (ppm)	55	39	61	115
	standard deviation	0.524	0.505	0.447	0.481
Mn	geometric mean (ppm)	170	142	133	325
	standard deviation	0.561	0.437	0.510	0.748
Fe	geometric mean (%)	0.924	0.782	0.955	1.353
	standard deviation	0.472	0.414	0.511	0.523
L.O.I.	geometric mean (%)	15.1	11.5	18.5	24.8
	standard deviation	0.600	0.644	0.551	0.436
Zn scores	mean	0	−0.165	0.020	0.387
	standard deviation	1.000	0.899	1.006	1.115

Data have been transformed into logarithms.
$*N$ = number of samples in each data set.

TABLE IV

Comparison of the means of actual Zn and residual Zn scores (after stepwise multilinear regression with Fe and L.O.I.) in the northern, central and southern areas, expressed as standard-deviation units from the mean of the whole data set

	Northern area	Central area	Southern area
Actual Zn means	−0.285	0.086	0.611
Residual Zn means	−0.165	0.020	0.387

TABLE V

Correlation matrix between lake area, lake depth, topographic relief, Zn, Mn, Fe and L.O.I. for 2494 lake sediment samples

	Lake area	Lake depth	Topographic relief	Zn	Mn	Fe	L.O.I.
Lake area	1.00						
Lake depth	0.53	1.00					
Topographic relief	0.00	0.36	1.00				
Zn	0.11	0.38	0.38	1.00			
Mn	0.37	0.48	0.26	0.35	1.00		
Fe	0.33	0.44	0.23	0.60	0.71	1.00	
L.O.I.	−0.28	0.09	0.28	0.47	−0.06	0.12	1.00

All data have been transformed into logarithms.

quite strongly with lake depth. This tends to suggest that in the deeper lakes, chemical precipitation of manganese and iron, possibly with the coprecipitation of zinc is an important process. The addition of these three parameters to the regression equation in addition to iron and L.O.I. accounted for only a further 2% of the variation of zinc. Thus the significant differences between the mean zinc residual scores between the three areas cannot be accounted for by the parameters measured.

There is no apparent reason to expect the background zinc content of the carbonate rocks to change significantly from north to south along the belt, nor is there any reason considering the present level of knowledge, to suggest that the mineral potential increases in this direction. The belt of Cambro-Ordovician carbonate rocks in western Newfoundland does extend, however, between the latitudes of 51°30′ north and 48°30′ north. Thus climatic conditions are rather different at the two extremes of the belt, with consequent differences in such factors as weathering and erosion processes and vegetation type. The vegetation surrounding each sample site was noted as being either predominantly forest, mixed forest and swamp, swamp or barren rock. The proportion of samples collected in forested areas increases from 51% in the northern area, to 65% in the central area, to 91% in the southern area. In each of the three areas, the mean zinc content of the sample from forest areas is higher than the mean zinc contents of samples from the other three categories. Thus one major component of the background variation is caused by the increase in forest cover from north to south within the area. Whilst this is partly a function of climate, the nature of the drift and topography are important factors in controlling vegetation type. In the northwest part of the northern area, the drift cover is thin or absent, the topography very flat, and forest coverage minimal.

A final interpretation of these lake sediment data must take into account all the relevant parameters that have been recorded. The use of stepwise multilinear regression is helpful in removing local variations in the environment of sedimentation, and to some extent in reducing the background variation in zinc content on a regional scale The correspondence between the loci of known zinc mineralization and residual zinc anomalies in the lake sediments is good, and the residual zinc distribution is a considerably more reliable guide to areas of known and potential zinc mineralization than the actual zinc distribution. A still more reliable picture will emerge when the data are subdivided into more homogeneous subsets on the basis of the available and relevant qualitative parameters, such as vegetation and physiography, before carrying out stepwise multilinear regression and other data processing as appropriate on each subset.

ACKNOWLEDGEMENTS

This work was carried out by the Mineral Development Division, Newfoundland Department of Mines and Energy, as part of the Canada‑ Newfoundland Mineral Exploration and Evaluation Agreement, funded jointly by the Department of Regional Economic Expansion and the Department of Energy, Mines and Resources.

We are greatly indebted to Dr. R.G. Garrett, Geological Survey of Canada, for carrying out the data processing at some inconvenience to his own work. The advice and information provided by Mr. D.G. Vanderveer, Mineral Development Division, on the Quaternary geology of the study area is gratefully acknowledged.

Finally we wish to thank Mr. D.G. Vanderveer, and Mr. J.M. Fleming, Director, Mineral Development Division, for critically reading the manuscript.

REFERENCES

Allan, R.J., 1971. Lake sediment: a medium for regional geochemical exploration of the Canadian Shield. Bull. Can. Inst. Min. Metall., 64(715): 43—59

Allan, R.J., Lynch, J.J. and Lund, N.G., 1972. Regional geochemical exploration in the Coppermine River area, District of Mackenzie; a feasibility study in permafrost terrain. Geol. Survey Can. Paper 71-33, 52 pp.

Allan, R.J., Cameron, E.M. and Durham, C.C., 1973. Reconnaissance geochemistry using lake sediments of a 36,000-square-mile area of the northwestern Canadian Shield. Geol. Survey Can. Paper 72-50, 70 pp.

Anonymous, 1974. Higher prices raise Teck net subsiduaries show potential. Northern Min., 59(47)

Brookes, I.A., 1972. The glaciation of southwestern Newfoundland.Ph.D. Thesis, McGill University, Montreal, Que., 208 pp.

Coker, W.B. and Nichol, I., 1975. Metal dispersion in lake sediments related to mineralization and its role in reconnaissance in the southern Canadian Shield. In: I.L. Elliott and W.K. Fletcher (Editors), Geochemical Exploration 1974. Elsevier, Amsterdam, pp. 549—553

Collins, J.A. and Smith, L., 1972a. Sphalerite as related to the tectonic movements, deposition, diagenesis and karstification of carbonate platform. Proc. 24th Int. Geol. Congr. Montreal, Sect.6, pp.208—215

Collins, J.A. and Smith, L., 1972b. Lithostratigraphic controls of some Ordovician sphalerite. In: G.C. Amstutz and A.J. Bernard (Editors), Ores in Sediments. Springer-Verlag, New York, N.Y., pp.721—725

Cumming, L.M., 1968. St. George's—Table Head disconformity and zinc mineralization, western Newfoundland. Bull. Can. Inst. Min. Metall., 61: 721—725

Davenport, P.H., Hornbrook, E.H.W. and Butler, A.J., 1974. Geochemical lake sediment survey for Zn and Pb mineralization over the Cambro-Ordovician carbonate rocks of western Newfoundland. Dept. of Mines and Energy, Govt. of Newfoundland, St. John's, Open File Rept.

Garrett, R.G., 1973. The determination of sampling and analytical errors — a reply. Econ. Geol., 68(2): 282—283

Grant, D.R., 1972. Postglacial emergence of northern Newfoundland. Geol. Survey Can. Paper 72-1, Part B: 100—102

Grant, D.R., 1974. Surficial geology maps, western Newfoundland. Geol. Survey Can., Open File Rept. 180

Hornbrook, E.H.W., Davenport, P.H. and Grant, D.R., 1974. Regional and detailed geochemical exploration studies in glaciated terrain in Newfoundland. Dept. of Mines and Energy Publ., Govt. of Newfoundland St. John's (in press)

Lilly, H.D., 1963. Geology of the Hughes Brook—Goose Arm area, west Newfoundland. Memorial Univ. of Newfoundland Geol. Rept. No.2, 123 pp.

Sangster, D.F., 1968. Some chemical features of lead—zinc deposits in carbonate rocks. Geol. Survey Can. Paper 68-39

Smyth, W.R., 1973. Stratigraphy and structure of the southern part of the Hare Bay allochthon, northwestern Newfoundland. Ph.D. Thesis, Memorial Univ., St. John's, Nfld., 172 pp.

Timperley, M.H. and Allan, R.J., 1974. The formation and detection of metal dispersion halos in organic lake sediments. J. Geochem. Explor., 3: 167—190

Vanderveer, D.G., 1973. A terrain analysis in western Newfoundland, from I.R. color photography (1:110,000). Newfoundland "ERTS" Project. On file with Dept. of Mines and Energy, Govt. of Newfoundland, St. John's

Williams, H., 1967. Geology of Newfoundland. Geol. Survey Can., Map 1231A

A COMPARISON OF STREAM SEDIMENT SAMPLING METHODS IN PARTS OF GREAT BRITAIN

R.C. LEAKE and R.T. SMITH

Geochemical Division, Institute of Geological Sciences, London (Great Britain)

ABSTRACT

Studies of the distribution of elements in three different size fractions of alluvium are described, based on surveys in north and southwest Scotland and the Cheviot area of northern England. Known mineralization in these areas is sub-economic and varies from veins with galena and baryte to a complex leached tabular cupriferous structure and weak disseminations of chalcopyrite, cupriferous pyrite and molybdenite associated with acid intrusive rocks. Analyses for economically interesting elements in the minus 100-mesh fraction of sediment, initially wet-screened, are compared with results obtained on panned heavy mineral concentrates and on samples of a water-suspended fraction.

The water-suspended material is easily collected and is particularly valuable in the detection of gossan, fine-grained or friable mineralization and stockworks with molybdenite and chalcopyrite concentrations which are amenable to rapid weathering. The "fines" are a natural complement to panned concentrates which are most useful in the location of vein deposits under active erosion. However, the broad size spectrum of the sieved sediment makes it the best single sample for use in detecting the complete range of mineralization types.

INTRODUCTION

Studies of element partitioning in drainage sediments (Gibbs, 1973) have shown that many metals occur in detrital mineral grains and rock fragments and are also coprecipitated with hydrous manganese oxides and other grain coatings, adsorbed onto clay-size and colloidal material, or complexed with organic matter. In regions of low geochemical contrast it is therefore important to know the distribution of elements in different size fractions in order to select the most appropriate sampling technique to meet the objectives of a survey. The areas of investigation on which this paper is based are in regions where known mineralization is sub-economic and where there is a variety of mineral occurrences including veins with lead and barium, complex tabular cupriferous structures, and weak copper and molybdenum disseminations in acid intrusive rocks.

Earlier investigation in northern Scotland (Plant, 1971) had shown that the optimum size fraction for regional geochemical reconnaissance was minus 100 mesh (<150 μm). However, they also showed that for most elements studied the geochemical contrast increased with decreasing particle size. More recently, Leake and Aucott (1973) demonstrated the value of panned concentrates in geochemical reconnaissance, particularly in regions of crystalline rocks subjected to glaciation. Both sieved sediments and panned concentrates are now compared with a third sample type, namely water-suspended material or "fines" comprising mainly silt and clay grade alluvium.

SAMPLE COLLECTION AND PREPARATION

Heavy-mineral concentrates were obtained by panning 2—3 kg of sediment screened through a 0.25-cm sieve to constant volume, producing on average 25—30 g of concentrate. This was prepared for analysis as described by Leake and Aucott (1973). The normal sieved sediment samples were collected and prepared for analysis in the manner described by Plant (1971). Fines were collected by decanting suspended silt and clay-grade material from a mixture of unsieved sediment and water and collecting it in polyethylene bags. After sieving through 1 mm mesh to remove coarse organic matter, and filtration through Whatman No.41 filter paper, the residues were dried at 110°C and finally dispersed by light manual crushing.

Analytical methods

Methods used for the analysis of the three sample types are shown in Table I. Standardization between the analytical techniques was carried out to enable variations of metal concentration between different sample types to be compared.

TABLE I

Analytical methods employed

Sample type	Location	Element	Analytical technique	Subsample weight (g)
Panned concentrate	all	Cu, Zn, Ba, Pb	XRF	8
Sieved sediment	Shin	Mn, Fe, Ni, Cu, Zn, Mo, Ba, Pb	XRF	8
	Galloway	Ni, Cu, Zn, Pb	AAS	0.5
		Ba	XRF	8
	Cheviot	Cu, Zn, Pb	AAS	0.5
		Ba	OES	0.05 (0.01—0.015)*
Fines	all	Mn, Fe, Co, Ni, Cu, Zn, As, Br, Mo, Ba, Pb	XRF	8

XRF = X-ray fluorescence spectrometry, AAS = atomic absorption spectrophotometry, OES = optical emission spectroscopy.
* Weight of sample volatilized.

The AAS technique and the OES method 2, adopted by Plant (1971) were used for the analysis of minus 100-mesh sediments. Panned concentrates and sieved sediments from the Shin area were analysed by the XRF method of Leake and Aucott (1973). Use was made of the relatively high content of organic matter in the fines to introduce a matrix correction by means of scattered background measurement in the XRF analysis of these samples. An acid mixture of $3HNO_3 + 1HClO_4$ was employed for metal extraction prior to AAS determination and the good agreement between copper values obtained by AAS and XRF (Fig.1), except at low levels, indicates efficient extraction from the sieved sediment samples. Generally, there is also good agreement between AAS and XRF results for zinc, though the latter are consistently somewhat higher. This trend is less pronounced when a $HF + HClO_4$ mixture (Leake and Aucott, 1973) is used, which shows that some zinc in minerals is resistant to $HNO_3 + HClO_4$ attack. The lower correlation between AAS and XRF results for lead is probably due to the lower analytical precision of the AAS method over the concentration range encountered. Over a wider range the scatter is much less apparent (Leake and Aucott, 1973). The XRF values for molybdenum, determined separately, compare favourably with results obtained by a colorimetric method based on solvent extraction of the molybdenum-dithiol complex after carbonate fusion.

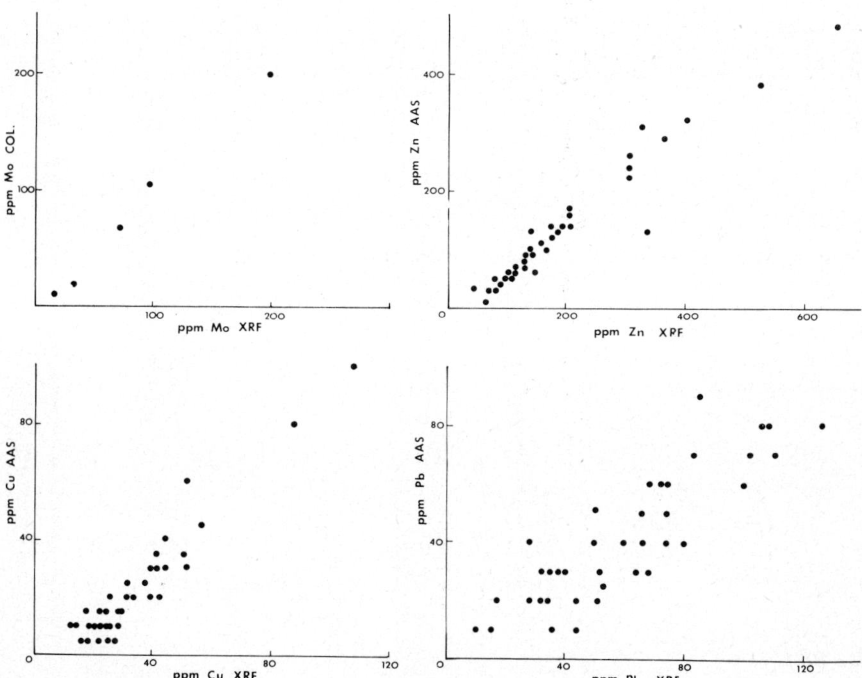

Fig.1. Comparison between determinations of the copper, zinc and lead content of minus 100-mesh (150 μm) sieved sediments by AAS and XRF and the molybdenum contents by a colorimetric method and XRF.

Sampling and analytical precision

Sampling and analytical precision obtained in the geochemical survey of the Galloway region has been estimated using the method of Garrett (1969, 1973) on results from duplicate panned concentrates and sieved sediments collected from 40 sites over a period of about 1 year. The equations used to determine the parameters shown in Table II, that is regional variance, variance in the difference between duplicates, total sampling and analytical precision (% P), and Snedecors F ratio, are given in Michie (1973). All the F values exceed the value of $F = 1.7$ which, on the basis of duplicate sampling at 40 sites, Garrett (1973) would regard as indicating detectable regional variation in element concentrations at the 95% confidence limit. It can be seen from Table II that the total analytical and sampling precision for copper and lead is appreciably better for sieved sediments than panned concentrates but the contrast shown by the lead is greater in the concentrates. In the case of zinc the difference in precision between the two sample types is small but the sieved sediment does exhibit greater regional contrast. The above statistics for copper and lead are influenced by the presence of a significant number of samples in the population, particularly of concentrates with element contents below the detection limit (i.e. < 10 ppm).

TABLE II

Estimates of regional variance and total sampling and analytical precision (P) based on analyses for copper, zinc and lead in drainage samples from the Galloway region (logarithmic data)

Mean	Regional		Difference between duplicates		$P(\%)$	F
	Mean	variance	mean	variance		
Panned concentrate						
Cu	0.95	0.51	0.35	0.088	62	5.8
Zn	2.05	0.073	0.13	0.013	24	5.7
Pb	1.44	0.66	0.44	0.102	67	6.5
Stream sediment						
Cu	1.29	0.136	0.21	0.011	22	12.4
Zn	2.31	0.176	0.17	0.011	22	16.0
Pb	1.88	0.177	0.22	0.037	40	4.8

Particle-size analysis

Twelve concentrates were sized by dry sieving and the same number of sediments and fines wet-sieved at 240 mesh (63 μm) for size analysis using an Andreasen pipette (Piper, 1950). The data in Fig.2 show that the concentrates contain only a very small proportion of material with grain size less than 100 mesh (150 μm), so that overlap between the sample types is minimal. Sieved sediments are generally more than twice as coarse as corresponding fines. Variable loss of silt- and clay-grade material during field collection of wet-screened sediments accounts for the difference in the shape of the size-distribution curve between 50 and 10 μm shown by the two sample types.

THE DISTRIBUTION OF BASE-METAL ANOMALIES IN DRAINAGE SEDIMENT

Cheviot area, northern England

This area is underlain by Lower Devonian andesites with some pyroclastics, intruded by the Cheviot granite which varies in composition from granite to pyroxene granodiorite (Carruthers et al., 1932; Jhingran, 1943). West of the lavas there is a small area of Wenlockian greywackes and shales overlain unconformably to the south and east by Lower Carboniferous mudstones and sandstones. Ablation moraine is widespread in the valleys, while much of the

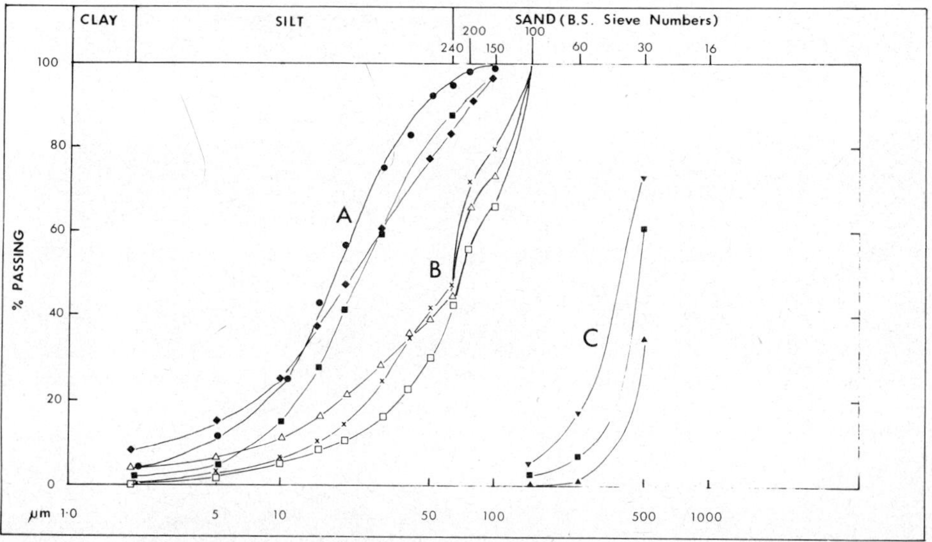

Fig. 2. Comparison of the size distribution of fines samples (A), sieved sediments (B) and panned concentrates (C) from Galloway, southwest Scotland.

high ground, particularly over the granite, is covered with peat. Fracture-bound mineralization up to 2 m in thickness, containing barytes accompanied either by malachite or by galena and its alteration products, occurs sparsely within the volcanic rocks. Sphalerite, in association with vein barytes and calcite, occurs in boulders derived from a northwesterly trending crush zone along the southwestern contact of the granite. Chalcopyrite and malachite are associated with weak uranium mineralization in the Wenlockian sediments

584

and to the north of the granite a zone several metres wide is impregnated with malachite (noted as Cu on Fig.3).

The contents of copper and zinc in panned concentrates and sieved sediments, collected simultaneously, are shown in Fig.3. Summary statistics and anomalous metal concentrations identified by cumulative frequency curves from data relating to 164 sites are given in Table III, those referring to sieved sediments being derived from Haslam (1974).

The number of sites at which anomalies occurred in panned concentrates and sieved sediments, or in both sample types, is given in Table IV, which also includes the number of those anomalies that could be attributed to base-metal mineralization from upstream outcrop, to the presence of mineralized alluvial boulders, or to the presence of ore minerals in the heavy-mineral concentrates. In 50 panned concentrates examined microscopically, the following ore minerals were identified in decreasing order of abundance: barytes, malachite,

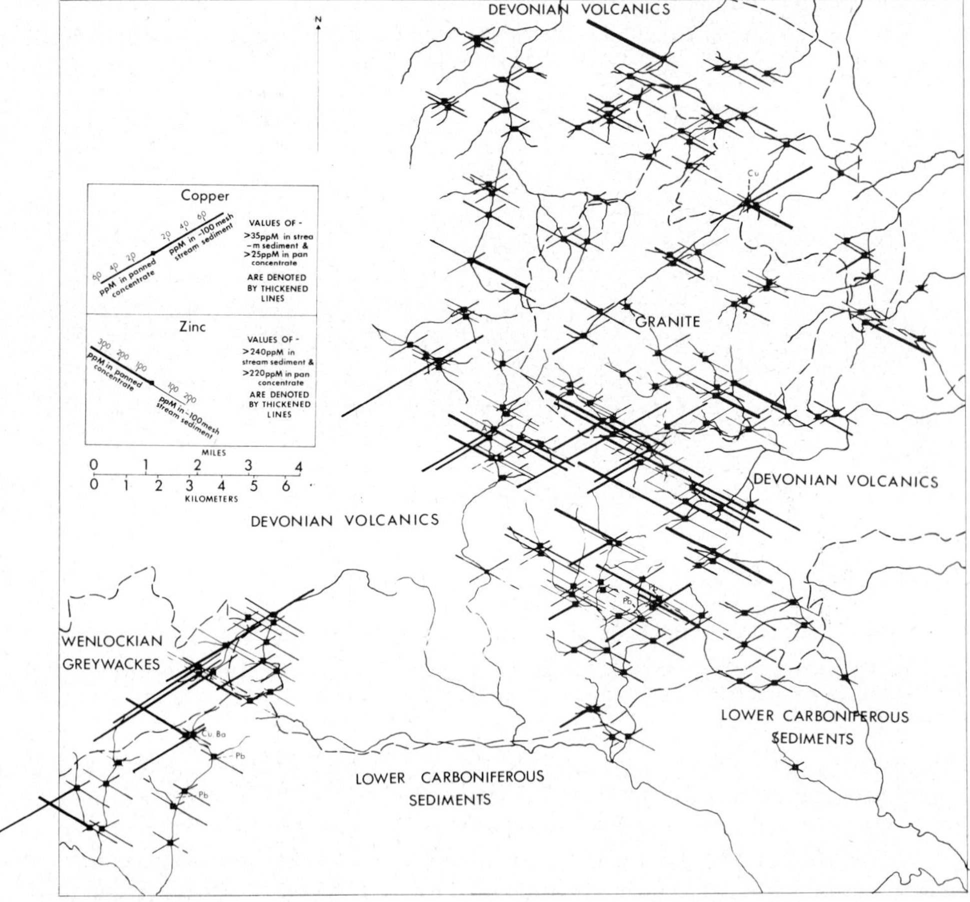

Fig.3. Copper and zinc contents of minus 100-mesh (150μm) stream sediments and panned concentrate collected simultaneously from the Cheviot area, northern England.

TABLE III

Summary statistics and anomalous metal concentrations for concentrates and sediments from 164 sites in the Cheviot area, northern England

| | Panned concentrate | | | | | Sieved sediment | | | | |
	$\overline{X}_A$	$\overline{X}_L$	σ_L	range	A	$\overline{X}_A$	$\overline{X}_L$	σ_L	range	A
Cu	21	13	0.39	<10— 125	25	18	14	0.23	<5— 70	35
Zn	135	117	0.22	35— 590	220	161	145	0.19	40—500	240
Ba	3410	998	0.52	135— 7.1%	900	1010	813	0.24	75— 1.0%	1700
Pb	117	33	0.57	<10—6760	140	53	43	0.28	<10—350	130

$\overline{X}_A$ = mean arithmetic data (ppm), $\overline{X}_L$ = mean logarithmic data (ppm), σ_L = standard deviation logarithmic data, A = values (in ppm) above which samples are anomalous.

sphalerite, chalcopyrite, cerussite and anglesite. Galena and pyromorphite also occur in isolated samples.

It is apparent from Table IV that there are many more copper, lead and barium anomalies in concentrates than in sieved sediments, but that the number of zinc anomalies is very similar for both sample types. From a total of 18 copper anomalies in panned concentrates not observed in the corresponding sediment, seven are correlated with fracture-bound barytes mineralization in the lavas and Lower Carboniferous sediments, three with calcite and barytes veins in the crush zone at the south contact of the granite, and three with chalcopyrite mineralization within the Wenlockian sediments.

Of the 7 zinc anomalies in the concentrate alone, four can be correlated with barytes mineralization and two with the fracture-bound mineralization of the crush zone. In contrast, alluvium at six of the 8 sites showing anomalous zinc only in the sediments, is derived from the vicinity of the crush zone or the line of its extension along strike. The poorer analytical precision obtainable for barium in sieved sediments by the optical emission method, as against

TABLE IV

The number of sites at which anomalies occur in either or both panned concentrates and sieved sediments

| | Anomalous in sediment (non-anomalous in panned concentrate) | | Anomalous in panned concentrate (non-anomalous in sediment) | | Anomaly in both samples | |
	total	related to mineralization	total	related to mineralization	total	related to mineralization
Cu	3	2	18	16	5	5
Zn	14	8	10	7	3	3
Ba	5	0	38	36	11	11
Pb	0	0	9	7	4	4

barium in concentrates determined by the higher-precision XRF method, invalidates a comparison of barium values in the two sample types. Nevertheless, the much greater number of barium anomalies in concentrates indicates that barytes occurs in alluvium predominantly as fragments greater than 150 μm in size. Of those concentrates containing anomalous lead which is not reflected in the corresponding sieved sediments, three can be correlated with fracture-bound barytes mineralization and one with copper-bearing vein mineralization in the crush zone at the southern contact of the granite.

The relative efficiency of concentrates and sieved sediments as a means of detecting mineralization also depends on distance from source. There are four instances in which both sample types successfully identify copper, lead or barium mineralization which can be traced to a lower order tributary. In addition, there are seven similar cases (1 copper, 1 zinc, 2 lead and 3 barium) in which only the concentrate is anomalous and one (zinc) in which only the sediment is anomalous. Barium dispersion in concentrates from one source persists for at least 5 km downstream compared with a dispersion distance of less than 1 km detectable in the corresponding sieved sediment.

In the Cheviot area, the dispersion of copper, barium and lead from the predominantly fracture-bound mineralization takes place more readily as particles with dimensions in excess of 150 μm. The panned concentrate is the most effective sample for the detection of this dispersion. Sphalerite mineralization, however, is reflected in stream sediment over a greater particle-size range.

Galloway, southwest Scotland

In this region Lower Palaeozoic greywackes and grits with belts of black shale, chert and spilitic volcanic rocks are intruded by large granitic masses varying in composition from granite to diorite (Phillips, 1956; Parslow, 1968). Several types of mineralization are developed of which four are considered here. A linear fracture zone trending north-northwest, containing barytes veins up to 3 m wide with minor amounts of chalcopyrite, intersects the eastern margin of a large granite mass and its contact aureole. Evidence of the barytes mineralization and associated chalcopyrite is less readily detected in the sieved sediment than the concentrate but the occurrence of sphalerite in the central part of the area is clearly indicated by anomalous zinc concentrations in both sample types. The weaker zinc anomalies found only in the sieved sediment samples from the northern part of the area have so far been attributable to small fractures with calcite infillings.

In a second area, means and standard deviations for copper in sieved sediments and the corresponding concentrates from 23 sites are 26 ppm and 12 ppm, and 53 ppm and 108 ppm Cu, respectively. The stronger contrast afforded by the concentrates can be attributed to the local mechanical dispersion of fresh cupriferous pyrite and chalcopyrite seen to be disseminated through porphyritic acid intrusive rocks in grains up to 1 cm in diameter.

All three sample types were collected from an area in the west of the region. The lead contents of sieved sediments and panned concentrates collected simultaneously from one drainage basin in the area are shown in Fig.4. The anomalous levels (> 140 ppm Pb for sieved sediment, > 50 ppm Pb for panned concentrates and > 200 ppm Pb for fines) are based on the interpretation of cumulative frequency plots of data from the whole of the western half of the Galloway region. The lead anomalies, the majority of which are apparent in the first two sample types, can be attributed to the presence of cerussite in alluvium derived from a Lower Palaeozoic black shale, chert and spilitic lava sequence. Furthermore, close agreement exists between the lead contents of sieved sediments and fines, the correlation coefficient being significant at greater than the 99.9% confidence limit for samples from 52 sites in the above and adjacent drainage basins. Lead concentration at the most anomalous site (A in Fig.4) is 980 ppm in fines, showing that dispersion occurs in a wide range of particle sizes.

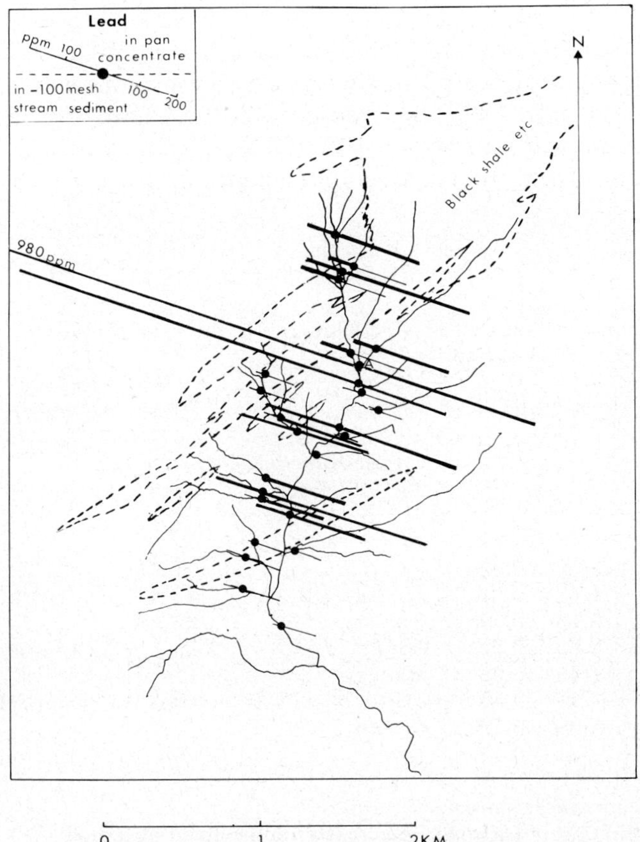

Fig.4. The lead contents of minus 100-mesh (150 μm) stream sediments and panned concentrates collected simultaneously from area D in the Galloway region of southwest Scotland. Anomalous values are shown as thickened lines.

The degree of correlation between the zinc contents of the sieved sediments and fines (correlation coefficient significant at the 99.9% level) is high, but it is appreciably lower (correlation coefficient significant at the 98% level) between the zinc contents of sieved sediments and panned concentrates.

Zinc anomalies (> 180 ppm in panned concentrates, > 400 ppm in sieved sediments and > 500 ppm in fines), though few in number, are generally coincident in the three sample types except in the vicinity of thin (up to 1 m) quartz-arsenopyrite veins adjacent to a large granite intrusion, where only the sieved sediments and fines are anomalous.

Precipitation of hydrous oxides of manganese and iron on surfaces in stream sediment has been demonstrated to have an important influence on trace metal content of alluvium in certain environments (Nichol et al., 1967). Early orientation work on stream sediments from the Galloway area (Gallagher et al., 1971) showed that levels of manganese and other elements increased markedly at particle sizes less than 100 μm. A series of fines samples from 38 sites was therefore analysed for Mn, Fe, Co, Ni, As, Br, Zn and Pb. Carbonaceous matter was also determined from weight loss on ignition at 450°C. Summary statistics and correlation coefficients significant at the 95% level are given in Table V.

The association of high concentrations of lead with organic-rich soils and the upper layers of peat bogs (Salmi, 1967) has also been noted where organic matter is in contact with stream water in northern Sweden (Brundin and Nairis, 1972). In contrast, fines from the Galloway area containing up to 35%

TABLE V

Summary statistics and correlation coefficients for selected metals in fines samples from 38 sites in the Galloway region of southwest Scotland

| | Arithmetic | | Logarithmic | | correlation coefficient significant at: | | |
	$\overline{X}$	σ	$\overline{X}$	σ	95—99% level	99—99.9%	>99.9%
Mn	0.96%	1.14%	−0.16	0.32	As		Fe, Co
Fe	5.37%	1.25%	0.72	0.39		—Br	Mn, Co
Co	37	30	1.48	0.27	Zn, As, —Br, Pb		Mn, Fe
Ni	144	35	2.15	0.11		Zn, —Carb	
Zn	458	249	2.60	0.24	Co, Pb	Ni	As
As	57	46	1.66	0.29	Mn, Co		Zn
Br	35	20	1.44	0.37	—Co	—Fe	Carb
Pb	105	45	1.98	0.19	Co, Zn		
Carb	17.4%	6.6%	1.21	0.17		—Ni	Br

All arithmetic values in ppm except where shown.
$\overline{X}$ = mean, σ = standard deviation, Carb = carbonaceous matter determined by weight loss on ignition at 450°C.

organic matter show no significant correlation with lead. This absence of marked enrichment of metals in the organic phases is presumed to be due to the coarseness of the organic material (essentially fibrous and shredded vegetation derived from coniferous forest and moorland) and to the high erosion rates, a factor suggested by Dall'Aglio (1971) to account for similar observations in the case of uranium.

The only element which shows a significant positive correlation with organic matter is bromine, which is also known to have a strong association with organic matter in soils and peats (Leake, 1971). Nickel, on the other hand, shows a significant negative correlation, indicating dilution by organic matter.

Dispersion of copper from a gossan outcrop of a linear breccia zone at the western margin of a granite mass results in anomalous copper (> 30 ppm) persisting in the sieved sediment for 1.5 km downstream of the outcrop. Fines also contain five times the background value at a site 1.5 km from the outcrop. In the concentrates, however, copper is at the background level only 0.6 km below outcrop. At exposure, copper occurs predominantly in a brown Fe-rich limonite cementing rock fragments, and as green amorphous Cu—As secondary mineral coatings. Minor chalcopyrite associated with quartz occurs in other parts of the structure. Comparison of the three sample types indicates that copper from this source is dispersed chiefly in the finest size fractions because of the friable nature of the gossan and the secondary mineral coatings.

Shin area, northern Scotland

The geology of this area is dominated by rocks of the Moinian assemblage (Read, 1931) comprising sparsely micaceous, siliceous or feldspathic psammite with subsidiary bands of micaceous psammite and semipelite. The country rock is intruded by major post-tectonic bodies of adamellite and the distribution of barium and copper in the area, based on the analysis of panned concentrates and sieved sediments, is illustrated in Fig.5. Anomalous amounts of barium and copper in concentrates correlate closely with the known widespread distribution of barytes, with notably high concentrations along crush zones. High geochemical contrast is afforded by the panned concentrates although dispersion trains are generally shorter relative to those of the sieved sediment, except where locally steeper gradients facilitate mechanical dispersion. A greater number of anomalies (Table VI) are detected by sieved sediment, although where second- or lower-order streams have been sampled there is difficulty in locating a bedrock source and false anomalies due to secondary environmental effects are frequently recognized.

Weak copper anomalies are detected most effectively by sieved sediments, particularly in areas of molybdenum mineralization in the Moinian rocks where copper occurs as chalcopyrite and cupriferous pyrite in fine-grained coatings on joint surfaces. Whereas panned concentrates are less effective under these circumstances, owing to the low efficiency for the recovery of fine-grained copper minerals, they are more successful than sieved sediments

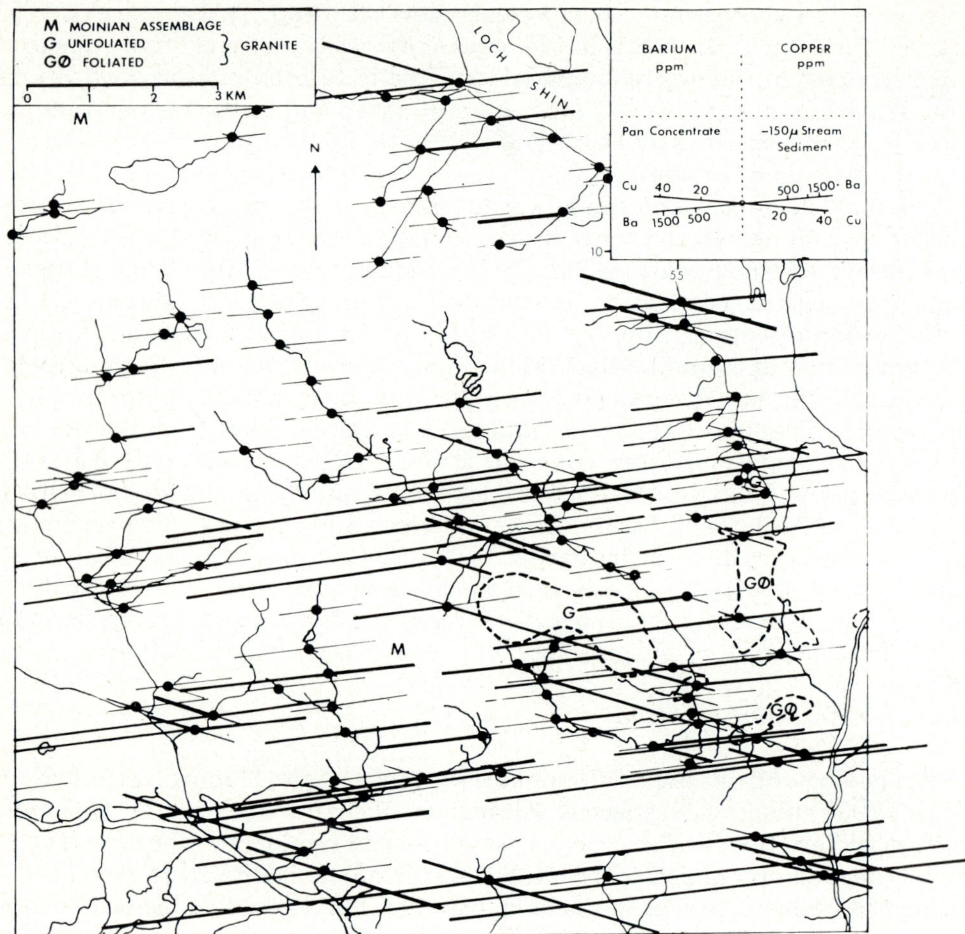

Fig.5. Copper and barium contents of sieved sediments and panned concentrates collected simultaneously from the Shin district of north Scotland. Anomalous values are shown as thickened lines.

in detecting copper associated with fracture-bound barytes mineralization. The mean and standard deviation for all sites anomalous in either or both sample types are compared in Table VI.

Because of analytical difficulties in the determination of molybdenum in the presence of appreciable amounts of zircon in panned concentrates from the area, the alluvial dispersion of molybdenum was investigated in sieved sediment and fines. Correlations between elements and between sample types significant at the 95% confidence limit are given in Table VII and, from these, the following observations are made:

(1) Zinc, molybdenum, manganese and barium are highly correlated in both sample types; copper, lead and nickel are moderately correlated, but for iron

TABLE VI

Summary statistics of copper and barium contents in sieved sediment and panned concentrate from all sites anomalous in either or both sample types ($N = 81$)

$\overline{X}_A$	$\overline{X}_L$	σ_L	Range	No. of anomalies
Sieved sediment				
Cu 51	46	0.20	5—105 ppm	14
Ba 3150	2400	0.26	240—7500 ppm	40
Panned concentrate				
Cu 53	40	0.40	5—110 ppm	17
Ba 9695	1000	0.93	80 ppm to 5.8%	29

No. of anomalies in both sample types: Cu 7, Ba 21

$\overline{X}_A$ = mean of arithmetic data (ppm), $\overline{X}_L$ = mean of logarithmic data (ppm), σ_L = standard deviation of logarithmic data.

in the two sample types there is no significant correlation. Examination of the fines samples indicates that the highest concentrations of iron occur in the vicinity of limonitic precipitates related to oxidising pyrite. Iron values in the corresponding sieved sediment sample occasionally indicate limonitic iron, but the effect is obscured by the greater abundance of detrital hematite, magnetite and mafic silicates.

(2) Barium and manganese are highly correlated in the fines samples, a feature attributable to coprecipitation, and not apparent in sieved sediments where this effect is probably masked by the presence of fine particles of detrital baryte.

(3) In contrast to the Galloway area, zinc and manganese are highly correlated in both sample types, suggesting a strong coprecipitation influence on the distribution patterns of zinc. Although no more than traces of sphalerite or smithsonite have been found in this area, zinc enrichment is known in peaty soils and till associated with galena-fluorite veins at one locality (Gallagher et al., 1971).

(4) The distribution of molybdenum in the two sample types is related to manganese as shown by a significant correlation at the 95% level, though the effect is less important than for iron, nickel, and zinc in the sieved sediment and for lead, iron, barium and zinc in the fines.

Molybdenite mineralization was detected by both sample types. The fines, however, yield more strongly anomalous concentrations but similar background values in areas of subdued relief mantled by waterlogged organic-rich overburden. Copper anomalies, though scarce, are more readily identified in the fines and there are three instances of Mo—Cu mineralization being detected only in this type of sample.

TABLE VII

Correlation coefficients for selected metals in sediment and "fines" sample pairs collected from 24 sites in the Shin district of northern Scotland

	Arithmetic		Logarithmic				
	$\overline{X}$	σ	$\overline{X}$	σ	correlation coefficient significant at:		
					95—99% level	99—99.9%	>99.9%
Cu	34	22	1.47	0.23	*Cu*, Ni		
Cu	51	41	1.60	0.31	Cu, *Pb*	*Ni*	
Pb	59	49	1.72	0.18	—Ni		
Pb	113	114	1.89	0.50	*Cu*, Zn, *Zn*, Mo	*Mn*	
Zn	153	88	2.11	0.28	*Pb*, *Ni*, *Mo*, *Ba*	Ni, Mo, Fe, *Fe*, Ba	*Zn*, Mn, *Mn*
Zn	248	148	2.33	0.26	*Pb*, *Mo*	*Ni*, Mo, Mn, Ba	Zn, *Mn*, *Ba*
Ni	124	90	1.99	0.31	—Pb, *Ni*, *Fe*, *Mn*	Zn, Mn	
Ni	115	61	2.61	0.21	Cu, Zn, Ni, Ba, *Ba*	*Cu*, *Zn*	
Mo	59	50	1.52	0.60	*Pb*, Mn, Ba, *Ba*	Zn, *Zn*	*Mo*, *Mn*
Mo	93	84	1.77	0.48	Zn, *Zn*, *Fe*, *Mn*, *Ba*		Mo
Fe	7.75%	6.19%	0.74	0.39		Zn, Mn	
Fe	16.69%	8.40%	1.17	0.23	*Ni*, Mo	Zn, *Mn*	Mn
Mn	0.76%	0.70%	—0.38	—0.57	Mo	*Zn*, Ni, Fe	Zn, *Fe*, *Mn*
Mn	0.89%	0.48%	—0.18	—0.44	Ni, *Mo*, Ba	*Pb*, Mo, *Fe*, *Ba*	Zn, *Zn*, Mn
Ba	1642	693	3.18	0.18	*Ni*, Mo, *Mn*	Zn, *Zn*	*Ba*
Ba	1899	1228	3.20	0.27	Zn, *Ni*, Mo, *Mo*		Zn, *Mn*, Ba

All arithmetic values in ppm except where shown. Data for "fines" samples are shown in italics throughout.

$\overline{X}$ = mean, σ = standard deviation.

CONCLUSIONS

The results presented here confirm the marked differences which could be expected in the distribution of elements such as copper, lead, zinc, molybdenum and barium in three different drainage sample types from areas in Scotland and northern England. Generally, the degree of contrast in the geochemical distribution of certain elements can be enhanced by collecting and analysing heavy-mineral concentrates, or by employing samples of the water-suspended fractions of clay- to silt-size alluvium. Panned concentrates are best suited to the detection of coarse-grained sulphides and baryte in deposits under active erosion, such as veins in parts of the Cheviot and Galloway regions. The fines fraction is most useful for locating mineralization that is readily weathered and dispersed from source, for example the fine-grained Mo—Cu mineralization occurring on joint coatings in the Shin area of Northern Scotland and the heavily leached copper mineralization in the Galloway region.

Panned concentrates and fines complement each other and give effective coverage of the alluvial size spectrum except in the fine sand size range. In the areas examined, the collection of a panned concentrate and a fines sample at each site permits the largest number of discrete mineral occurrences to be detected. However, sieved stream sediment provides the best single sample, at a similar sampling density. This sample type covers a wide particle size range, is characterized by relatively high sampling and subsampling precision and is well suited to geochemical mapping of a wider range of elements than have been considered here. The panned concentrate and "fines" samples have their best application in the follow-up stage of metalliferous mineral prospecting for the more precise delineation of metal concentrations in different modes.

ACKNOWLEDGEMENTS

The work was carried out during a uranium reconnaissance undertaken on behalf of the United Kingdom Atomic Energy Authority and a programme of mineral reconnaissance being carried out on behalf of the Department of Industry, both managed by Dr. S.H.U. Bowie. The paper draws freely on sampling and analytical data obtained by colleagues in the Geochemical Division and is published by permission of the Director, Institute of Geological Sciences, London.

REFERENCES

Brundin, N.H. and Nairis, B., 1972. Alternative sample types in regional geochemical prospecting. J. Geochem. Explor., 1: 7—46

Carruthers, R.G., Burnett, G.A. and Anderson, W., 1932. The geology of the Cheviot Hills (explanation of one-inch geological sheets 3 and 5). Mem. Geol. Survey England and Wales.

Dall'Aglio, M., 1971. A study of the circulation of uranium in the supergene environment in the Italian Alpine Range. Geochim. Cosmochim. Acta, 35: 47—59

Callagher, M.J., Michie, U.McL., Smith, R.T. and Haynes, L., 1971. New evidence of uranium and other mineralization in Scotland. Trans. Inst. Min. Metall., Sect. B, Appl. Earth Sci., 80B: 140—171

Garrett, R.G., 1969. The determination of sampling and analytical errors in exploration geochemistry. Econ. Geol., 64: 568—569

Garrett, R.G., 1973. The determination of sampling and analytical errors in exploration geochemistry — a reply. Econ. Geol., 68: 282—283

Gibbs, R.M., 1973. Mechanisms of trace metal transport in rivers. Science, 180: 61—73

Haslam, H.W., 1974. Geochemical survey of stream waters and stream sediments from the Cheviot area. Rep. Inst. Geol. Sci. (in press)

Jhingran, A.G., 1943. The Cheviot granite. Q. J. Geol. Soc. Lond., 98: 241

Leake, R.C., 1971. The use of soil A-horizon trace element surveys in geochemical prospecting at Bradwell Moor, near Castleton, Derbyshire. Rep. I.G.S. Analyt. Ceram. Unit, No.68

Leake, R.C. and Aucott, J.W., 1973. Geochemical mapping and prospecting by use of rapid automatic X-ray fluorescence analysis of panned concentrates. In: M.J. Jones (Editor), Geochemical Exploration 1972. Institution of Mining and Metallurgy, London, pp.389—400

Michie, U. Mcl., 1973. The determination of sampling and analytical errors in exploration geochemistry. Econ. Geol., 68: 281—282

Nichol, I., Hornsail, R.F. and Webb, J.S., 1967. Geochemical patterns in stream sediments related to precipitation of manganese oxides. Trans. Inst. Min. Metall., Sect. B, Appl. Earth Sci., 76B: 113—115

Parslow, G.R., 1968. The physical and structural features of the Cairnsmore of Fleet granite and its aureole. Scott. J. Geol., 4: 91

Phillips, W.J., 1956. The Criffell-Dalbeattie Granodiorite Complex. Q. J. Geol. Soc. Lond., 112: 221

Piper, C.S., 1950. Soil and plant analysis. Interscience, New York, N.Y.

Plant, J., 1971. Orientation studies on stream-sediment sampling for a regional geochemical survey in northern Scotland. Trans. Inst. Min. Metall., Sect B, Appl. Earth Sci., 80B: 324—345

Read, H.H., 1931. The geology of central Sutherland (explanation of one-inch geological sheets 108-109). Mem. Geol. Survey Scotland

Salmi, M., 1967. Peat in prospecting: applications in Finland. In: A. Kvalheim (Editor), Geochemical Prospecting in Fennoscandia. Interscience, New York, N.Y., pp.113—126

MORAINE SAMPLING UNDER VARVED CLAYS IN THE ROUYN–VAL D'OR AREA: GENERAL RESULTS*

P. LA SALLE, B. WARREN, P. GILBERT, H.L. JACOB and Y.R. LA SALLE

Quebec Department of Natural Resources, Quebec, Que. (Canada)

ABSTRACT

A project to employ miners laid off by the mines of the Cadillac area (in the Rouyn–Val d'Or mining belt) was started in July 1971. Ground moraine (lodgement till), or what was thought to be ground moraine, was sampled along or between roads wherever access was possible. Most of the samples were collected under varved clays with a piston sampler. It is estimated that 80—90% of the samples were collected at the bedrock contact. The minus 80-mesh fraction was analysed for Cu, Zn, Ni, Pb, Co, Mn, Cd and Ag. The coarse fraction was passed through bromoform, ground and analysed for the same elements. 7000 samples were so collected, and the coarse fraction of samples showing anomalous results examined under binoculars. Results are presented as maps.

INTRODUCTION

This project was undertaken in June 1971, by the Quebec Department of Natural Resources, with the financial support of ARDA. The main purpose was to reduce unemployment caused by the closing down of molybdenite mines in the Cadillac area and, at the same time, to have useful work done for money expended.

Thus, it was suggested that licensed miners, working in crews of 4, sample the till under the varved clays of former glacial lake Barlow-Ojibway, within the Rouyn–Val d'Or mining belt (Fig.1). Samples were collected with percussion drills of the Cobra or Pionjar type. Similar equipment was used by Gleeson and Cormier (1971) in outlining the Louvicourt ore body. Similar surveys have also been carried out by Skinner (1972) and Gunton and Nichol (1974).

METHODS

Sampling procedure

First a hole is drilled with a sharp-pointed steel rod to reconnoitre the depth to bedrock. The closed sampler is then sunk into the hole. This proce-

---

* Published with permission of the Quebec Minister of Natural Resources.

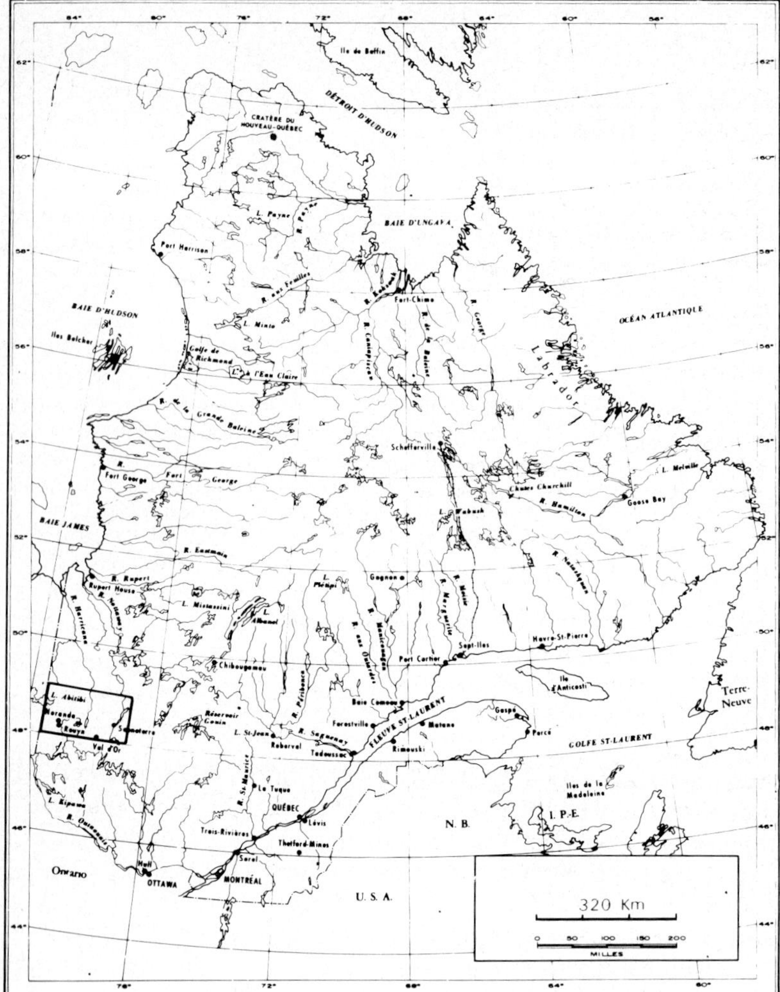

Fig.1. Location of study area.

dure minimizes breakdown of the sampling tool when attempting to reach bedrock through the till but, in some cases, may contaminate the potential sample by pushing down material from above. Once the sampling depth is reached the sampler is opened and then sunk to its full length to fill the sample chamber.

The sample is usually taken as close as possible to the contact between bedrock and till, because this is where transport by glacial movement is assumed to be minimal. However, boulders may be mistaken for bedrock.

Despite the advantages of this sampling device, it does not work well at depths exceeding approximately 25 m (82.5 ft). Firstly, because of inertia and friction, the sampling spoon does not open easily at greater depths. Secondly, it does not function well when the overburden is sandy; in such ground the sampler is sunk with difficulty, and is also difficult to withdraw because of sand-locking.

A third limitation for such large-scale sampling is that the geologist has to rely, to some degree, on the drilling crew to appraise the quality of the sample, correctness of depth and other types of control. Consequently, 10—15% of the identifications made in the field had to be rejected. However, control as to the identification of the material sampled was exercised in the laboratory where every sample was examined for grain-size distribution and possible contamination.

Sampling area and field-data recording

The area sampled is shown in Fig.1. Roughly, it is the extent of the Rouyn—Val d'Or mining belt. At first, samples were taken along all roads at an interval of half a mile. After this had been completed, the ground between was sampled at the same scale and then certain areas were resampled at an interval of one-quarter mile. In this way almost all parts of the area were covered, although not always on a regular grid, to give approximately 7000 samples of till. In the field the crews recorded the following data on a printed computer card, with a preset number for the sample printed on a detachable stub: soil zone, depth of sampling, position of sample with respect to bedrock, type of trees, type of overburden, type of sample, number of the sampling crew, date, and project number.

Laboratory procedure

After drying, samples were submitted to the procedures summarized in the flow-chart (Fig.2). Samples were analyzed by X-ray fluorescence for arsenic, molybdenum, yttrium, rubidium, and strontium, because it was impossible to get reliable results for the first two elements by atomic absorption. Results for the last three elements may not have any economic interest at present. However, as the source used to irradiate the sample, ^{109}Cd, gave reproducible readings for all five elements, it was decided to also record the results for yttrium, rubidium, and strontium. They may prove useful later on in mapping bedrock using till chemistry. As they become available, all the field data and laboratory results are registered on magnetic tapes. This data bank has been validated for possible inconsistencies and errors. When completed, it is hoped that it will be made available for purchase by exploration personnel.

THE FIELD AREA

Bedrock geology

According to the most recent Quebec government map, most of the bed-

598

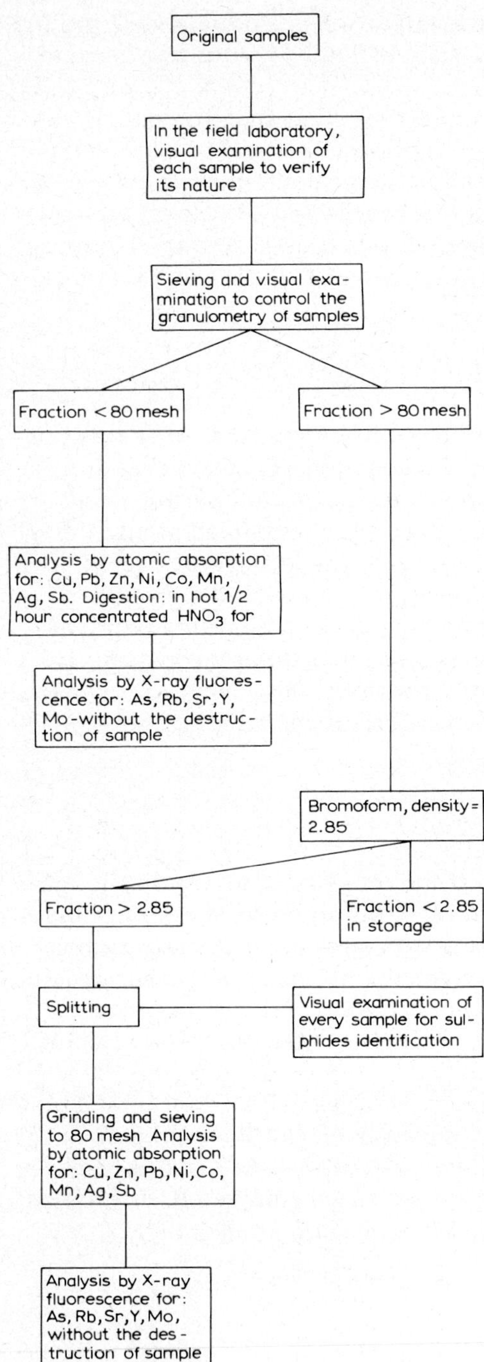

Fig. 2. Analytical flow-chart.

rock of the Rouyn Val d'Or area can be ascribed to the following rock categories:

(1) Volcanic rocks: rhyolite, dacite, andesite, basalt, and accompanying tuff.
(2) Metasedimentary rocks: graywacke, conglomerate, argillite.
(3) Plutonic rocks: large granitic masses, some minor mafic bodies.

However, according to Descarreaux (1973), the normative classification of volcanic rocks in the Rouyn Val d'Or area indicates some discrepancies with the government map, and normative basalt seems to be the most abundant rock type.

The structural trend of both the main rock units and major faults is generally east—west. This feature, coupled with the roughly north—south movement of glacier flow, presents an interesting possibility for prospecting using glacial sediments.

Quaternary geology

Quaternary geology of the area is not well known, and it is certainly a simplified picture to show the stratigraphy of the unconsolidated sediments as in Fig.3. Results must be interpreted with caution since in some areas, such as the James Bay lowlands, two tills and interglacial sediments are present. In the region sampled, deep natural sections are absent, and stratigraphy is inferred but not known.

Mineralogy of samples

Fifteen samples of till and eleven samples of varved clays were analyzed by X-ray diffraction (Fig.4). It appears that in the <2-μ fraction tills contain up to 25% illite and up to 15% vermiculite, whereas the varved clays contain up to 45% illite and up to 20% vermiculite. Feldspar and quartz are the other most abundant minerals. Feldspar goes up to 35% and quartz to over 15% in

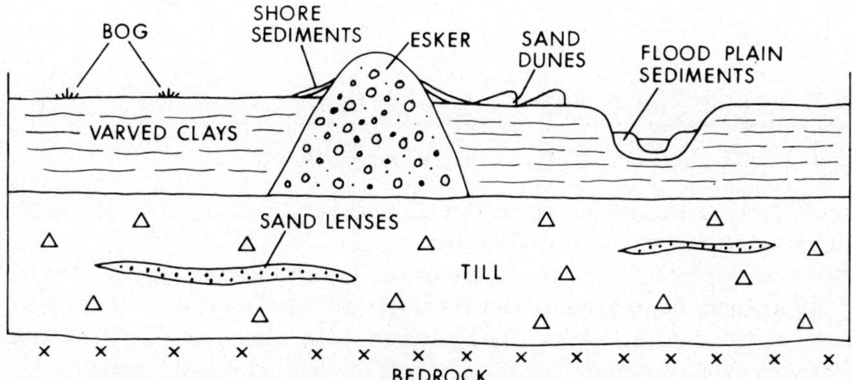

Fig.3. Simplified sketch of Quaternary geology.

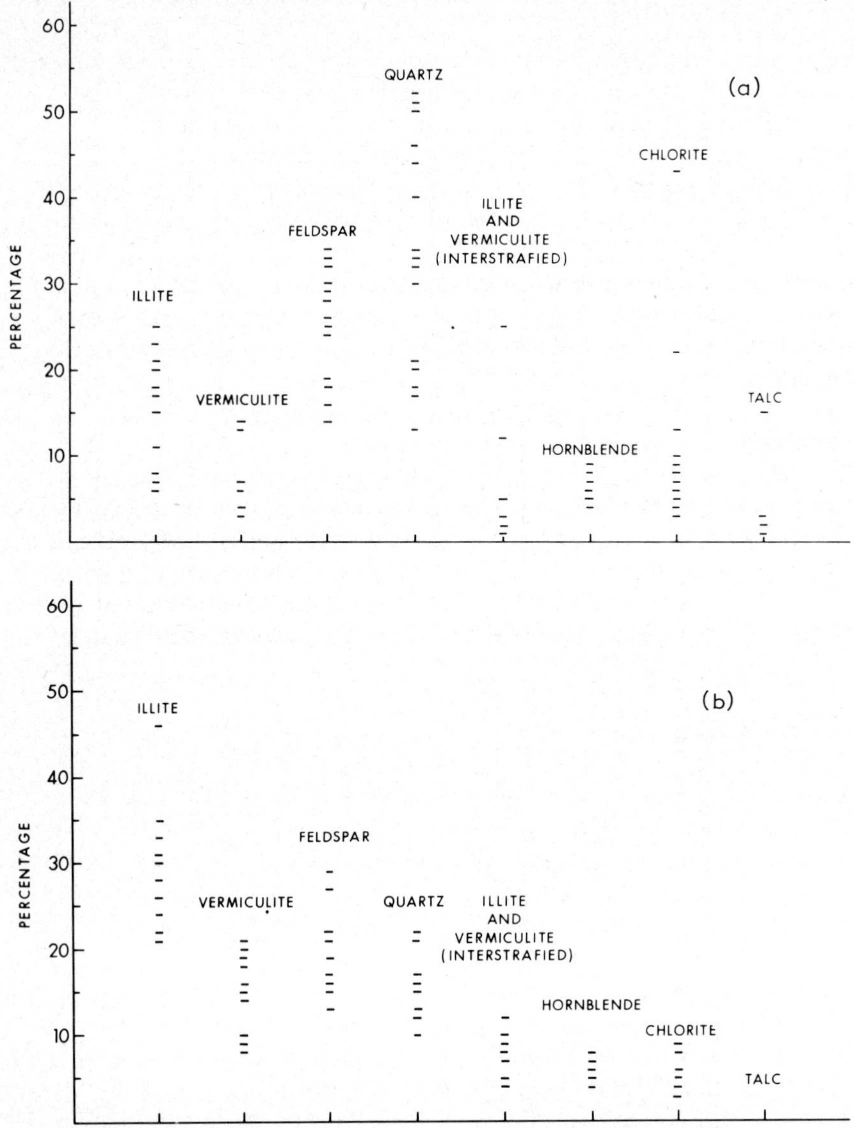

Fig.4. Mineralogy of clay fraction ($<2\,\mu$): (a) tills, (b) varved clays.

the moraines. This is not unusual, since the till and varved clays are both products of glacial attrition of the regional bedrock.

However, the presence of large percentages of clay minerals suggests that either there might have been long-distance transport from an area of sedimentary rocks such as the James Bay Lowlands, or that the clays are derived from an interglacial soil. It is possible that the clay minerals may adsorb certain elements (Samama, 1973) thereby producing anomalies or background variations unrelated to sulphide mineralization.

RESULTS

Reliability and statistical distribution

Table I summarizes results of repeated analysis, on different days, of the same sample. For copper, zinc, nickel, cobalt, and manganese, the results are reliable within 15%.

Population distributions show a tendency towards log-normality. Programmes for contours and trend surfaces (e.g. G.P.C.P.-II, Calcomp, 1972) assume a gaussian distribution (Davis, 1973; Hall, 1973) for the independent variable and also a regular distribution of sampling sites. The latter criterion was impossible to attain in the present case. As for the first criterion, log-transformation of the results would, in most cases, normalize the distributions: this had not, however, been possible at the time of writing.

EXAMPLES OF MAPS PRODUCED

As it is impractical to present here a large number of maps we have chosen the LaMotte sheet (32D/8W) to illustrate the relationship between chemical composition of till and bedrock. Fig.5 shows distribution of nickel values as well as the location of ultrabasic bodies. Generally, the high values for nickel correlate with the zones of ultrabasic rock. The high values are not, however, necessarily located above the zones of ultrabasic rocks, but sometimes show a short down-glacier transport. There is also some dilution in the till sample since analysis of bedrock shows values for nickel as high as 1500 ppm. The same data have been plotted using an advanced programme (G.P.C.P.-II, Calcomp, 1972) to give a contoured map of the original values, the third-order trend surface and residuals of the third-order trend (Figs.6—8). If it is assumed here that the trend surface represents a regional background (Davis, 1973) it follows that by subtracting the trend surface from the original data, local anomalous values will appear as residuals. However, the geometric

TABLE I

Reproducibility of atomic absorption analyses

Element	Mean	Standard deviation	Coeff. of variation
Cu	35.3	4.7	13.3
Zn	27.3	2.6	9.3
Pb	6.1	1.1	19.3
Ni	51.6	3.7	7.2
Co	13.7	1.2	9.3
Mn	175.4	8.0	4.5
Ag	3.0	0.95	31.0
Sb	4.5	0.93	20.5

shape of the residuals and their displacement with respect to sample locations poses some problems. Blanking-out of areas where there are no samples would be possible. However, this does not resolve the difficulties encountered here, since blank areas are considered by the programme in calculating contours.

CONCLUSIONS

For this project the price of sampling ranged between $0.50 and $2.50 per linear foot, depending on overburden thickness. This figure can be compared with the cost for other similar projects (see Gunton and Nichol, 1974, p.74).

Results show that if samples are taken close to contact with bedrock, till chemistry reflects bedrock chemistry. This suggests that by determining a large number of elements — especially those capable of differentiating rock types most specifically — it may be possible, using factor analysis, to supplement bedrock mapping.

For exploration the method rests on the assumption that mineralized outcrops below the till will be sufficiently large that material incorporated into the till will produce an anomalous fan detectable by point sampling. Isolated anomalous values must be treated with care. They may indicate either that the mineralized outcrop is very small, or at least that the width and length of the fan that it produced is smaller than the grid interval. Obviously, mineralized deposits that do not outcrop below the till will not be detected. Computer programmes for contours, trend surfaces and residuals must be used cautiously as results depend on grid size and homogeneity of the sampling.

REFERENCES

Calcomp, 1972. G.P.C.P.-II, a general purpose contouring program. Calcomp, Anaheim, Calif.
Davis, J.C., 1973. Statistics and data analysis in geology. Wiley, Toronto, Ont., 550 pp.
Descarreaux, J., 1973. A petrochemical study of the Abitibi volcanic belt, and its bearing on the occurrences of massive sulphide ores. Can. Inst. Min. Metall. Bull., 66: 61—69
Gleeson, C.F. and Cormier, R., 1971. Evaluation by geochemistry of geochemical anomalies and geological targets using overburden sampling at depth. In: R.W. Boyle (Editor), Geochemical Exploration. Inst. Min. Metall., Spec. Vol., 11: 159—165
Gunton, J.E. and Nichol, I., 1974. Delineation and interpretation of metal dispersion patterns related to mineralization in the Whipshaw Creek area. Can. Inst. Min. Metall. Bull., 67: 32—41
Hall, A., 1973. The median surface — a new type of trend surface. Geol. Mag., 110: 467—472
Samama, J.C., 1973. Ore deposits and continental weathering — a contribution to the problem of geochemical inheritance of heavy-metal contents of basement areas and of sedimentary basins. In: G.C. Amstutz and A.J. Bernard (Editors), Ores in Sediments. Springer-Verlag, New York, N.Y., pp.247—266
Skinner, R.G., 1972. Drift prospecting in the Abitibi clay belt. Overburden drilling program methods and costs. Geol. Survey Can. Open File No.116, p.27

SEMI-REGIONAL GEOCHEMICAL STUDIES DEMONSTRATING THE EFFECTIVENESS OF TILL SAMPLING AT DEPTH

C.F. GLEESON and E.H.W. HORNBROOK

C.F. Gleeson and Associates Limited, Ottawa, Ont. (Canada)
Geological Survey of Canada, Ottawa, Ont. (Canada)

ABSTRACT

In Canada till sampling has been carried out by different techniques for many years primarily to detect and define metal dispersion halos associated with metallic sources.

The technique described in this paper involves the sampling of till at or near bedrock by the use of a light percussion drill and piston sampler. Successful studies were carried out in areas underlain by ultramafic bodies and gold occurrences in the Timmins—Val d'Or region of Ontario and Quebec.

The results of the lake-sediment till sampling programme in the region show that lake sediment geochemistry will not detect underlying ultramafic rocks and gold occurrences where these targets are covered with glaciolacustrine sediments. However, geochemical results from till samples taken at the bedrock/till interface show that, in this environment, overburden sampling at depth is a useful detail and semi-reconnaissance exploration technique for outlining large geological targets such as ultramafic intrusives, as well as smaller targets such as gold-bearing structures.

Generally stronger contrast and better anomaly definition can be obtained from the geochemical analyses of the heavy-mineral concentrates, specific gravity greater than 2.96, from the till than with the minus 80-mesh ($<177\mu$) or minus 230-mesh ($<63\mu$) fractions from the till. By using nickel and copper as indicator elements the presence of ultramafic intrusives, overlain by up to 44.4 m (148 ft) of glacial sediments, were detected by sampling at 800-m (½ mile) centres. Smaller targets, such as gold-bearing structures, were outlined using arsenic as a pathfinder element and by sampling at 400-m (¼ mile) centres.

INTRODUCTION

From the conceptual and experimental stages at which geochemical overburden sampling at depth was in Canada when first studied by Ermengen (1957) in 1956, Gleeson (1960) in 1957 and later by such workers as Boniwell and Dujardin (1964), Lee (1963), Fortescue and Hornbrook (1969), Van Tassel (1969), Gleeson and Cormier (1971) and Garrett (1971), it has rapidly moved into an accepted geochemical exploration technique which is being used now throughout glacial terrains and in nearly all geological environments in Canada.

The major impetus which finally helped establish this geochemical approach in Canada came partially as a result of regional studies undertaken by the Geological Survey of Canada during the winter 1971—1972 in the Abitibi area of Quebec and Ontario, Canada. A programme of drift sampling at depth in 11 localities of the region was carried out by Skinner (1972) involving 394 rotary drill holes and 140 percussion holes. In addition, a programme of regional lake sediment sampling and till sampling at depth under lakes was carried out by Hornbrook and Gleeson (1972, 1973) over some 20,000 km² (12,000 square miles) of the Precambrian shield. The till sampling at depth was done over four lakes in the Abitibi area; 30 sites were sampled on Pelletier Lake, 77 on Macamic Lake, 54 on Abitibi Lake and 60 on Nighthawk Lake (Fig.1).

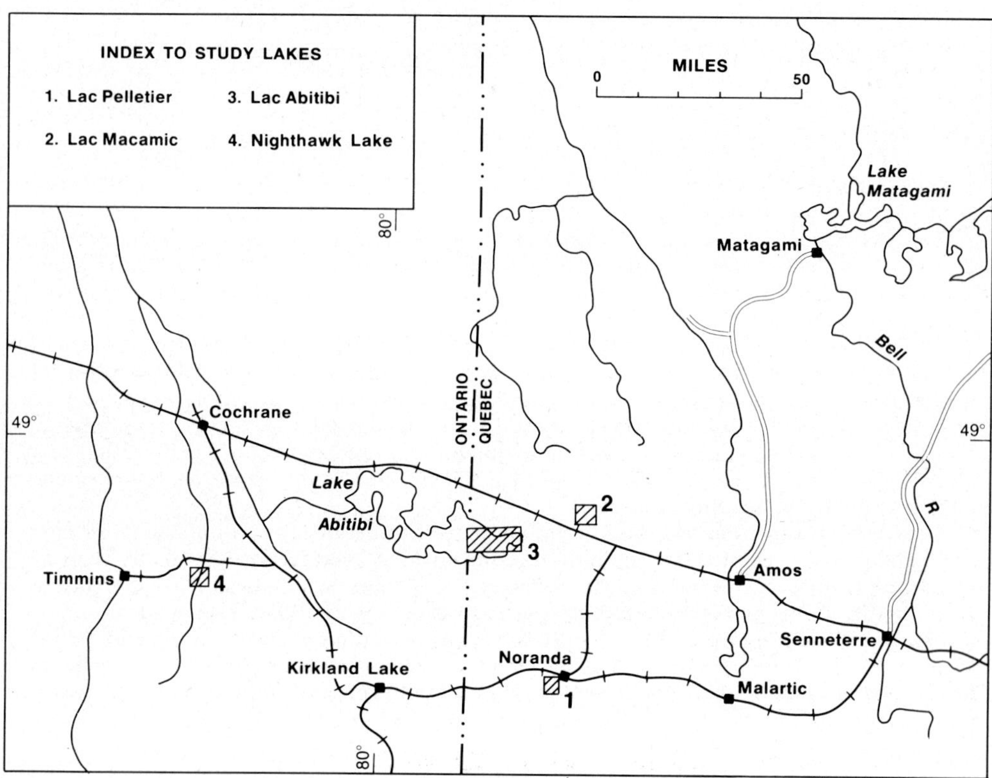

Fig.1. Location map of Abitibi area, Ontario/Quebec.

The purpose of the study by Hornbrook and Gleeson was to test the applicability of the semi-reconnaissance geochemical overburden sampling technique at depth in the Abitibi Clay belt. Prior to these studies the technique had proven useful to the mineral exploration companies in detailed surveys involving evaluation of geophysical and geological targets. This paper will discuss some of the pertinent highlights of the geochemical results from the vicinity of a gold-bearing structure on Pelletier Lake, Quebec and adjacent to gold-bearing structures and ultramafic rocks on Nighthawk Lake, Ontario.

Field methods

Basically, the drilling and sampling equipment and the drilling techniques employed were the same as those originally outlined by Gleeson and Cormier (1971). The programme was carried out using two drill crews equipped with Pionjar-type portable percussion drills. In the Abitibi programme (4 lakes) a total of 221 sites were sampled and 3637 m (12,124 ft) of overburden were drilled at an average cost of $6.30 a metre ($2.10 a foot) or about $10 per sample.

At each site, a lake sediment sample and an overburden sample from atop bedrock was taken. However, on Pelletier Lake duplicate till samples were taken at each site so that studies could be carried out on various size fractions.

On Pelletier Lake, stations were sampled every 150 m (500 ft) on lines 300 m (1000 ft) apart and on Nighthawk Lake samples were taken at 400- and 800-m (¼ and ½ mile) centres.

Sample preparation and analysis

To test the metal dispersion patterns in the various size fractions, the lake sediment samples from Pelletier Lake were dried, split and one portion was passed through an 80-mesh (177 μ) stainless steel screen and the other portion was put through a 230-mesh (63 μ) stainless steel screen. The Nighthawk lake sediments were sieved through a 230-mesh (63 μ) stainless steel screen. The minus 80-mesh ($<$177 μ) and minus 230-mesh ($<$63 μ) fractions were analysed geochemically.

To study metal dispersion in various size fractions of the till two samples were taken at each drill site over Pelletier Lake. One sample was sieved to obtain minus 10-mesh ($<$2 mm), plus 80-mesh ($>$177 μ) and minus 80-mesh ($<$177 μ) fractions and the second sample was sieved to obtain minus 50-mesh ($<$297 μ), plus 230-mesh ($>$ 63 μ) and minus 230-mesh ($<$63 μ) fractions. In the Nighthawk Lake study only the latter separations were done.

Heavy-mineral concentrates from the tills were obtained from the minus 10-mesh ($<$2 mm), plus 80-mesh ($>$ 177 μ) and minus 50-mesh ($<$297 μ), plus 230-mesh ($>$ 63 μ) fractions by separation with tetrabromoethane (specific gravity 2.96). The heavy-mineral fractions were split and one part pulverized to minus 100-mesh ($<$149 μ) for geochemical analysis. The second portion was retained for mineralogical examination and the light fraction was stored for possible future studies. In addition, geochemical analyses were done on the minus 80-mesh ($<$177 μ) and minus 230-mesh ($<$63 μ) fractions of the tills from Pelletier Lake and on the minus 230-mesh ($<$63 μ) fraction of the tills from Nighthawk Lake.

All samples were analysed geochemically for Cu, Pb, Zn, Mo, Ni, Mn, Ag, and As. The first seven elements were analysed using a Techtron AA-5 atomic absorption spectrophotometer after digesting 0.5 g of sample in hot aqua regia for 2 hours. Arsenic was determined colorimetrically utilizing silver diethyl-

TABLE I

Generalized stratigraphy and inferred events, Abitibi Clay Belt (after Skinner, 1972)

Units	Description	Events	Age (^{14}C yr B.P.)
Recent deposits	Peat, humus and alluvium Wind-blown sand, especially along eskers.		
Glacio-lacustrine silt and clay	Thin silt-clay varves; clay oxidized to chocolate brown. Not more than 5—10 ft thick.	Rapid dissipation of ice sheet; drainage of small post-Cochrane lakes.	by 7800 yr
Cochrane till	Clayey, silty till with few pebbles; essentially reworked Barlow-Ojibway sediments. Locally, a thin bed of sand and gravel at base of till. Southern limit shown on maps by Hughes (1960), Boissonneau (1965), and Prest et al.(1967). Unit < 1—35 ft thick. Difficult to recognize in drilling; essentially clay with a few pebbles.	Surge of lobe of ice sheet; affected only part of area.	8300 yr
Glacial Lake Barlow-Ojibway deposits	Grade from sand and gravel at bottom up through sand, then thick sandy silt varves, then into thin clayey varves at top of unit. Can be over 150 ft thick. Includes esker facies.	Glacier margin retreated perhaps as far north as James Bay lowlands; eskers formed during this retreat.	
Till	Bouldery, sandy to silty compact till, in places overlain by a relatively loose, sandy ablation facies. Locally has lenses of sand and gravel up to 10 or 20 ft thick. Unit can be as much as 75 ft thick.	Retreat / Glaciation	10,000 yr
Bedrock			

dithiocarbamate (Vasak and Sedivec, 1952) after 0.5 g of sample was digested
cold with a mixture of nitric and perchloric acids; then in a sand bath at
$170°-200°$C for 4 hours or until the nitric acid was expelled.

This report will deal only with the arsenic results from Pelletier Lake and
the Cu—Ni—As results from Nighthawk Lake. The total data from all lakes
sampled have been released as an open file report by the Geological Survey
of Canada (Hornbrook and Gleeson, 1972).

Glacial stratigraphy

A generalized glacial stratigraphic sequence for the Abititi area has been
drawn up by Skinner (1972) and it is presented here in Table I. The areas in
the Abititi region are underlain by deposits from glacial lake Barlow-Ojibway
and till.

PELLETIER LAKE

This lake is located in Rouyn Township about 3.2 km (2 miles) southwest
of the towns of Rouyn-Noranda, Quebec, at latitude $48°13'$N and longitude
$79°03'$W (Fig.1). The depth of water in the area sampled varies from 0.9 to
5.4 m (3—18 ft) and the thickness of the glacial overburden varies from 2.9
to 34.8 m (9.5—116 ft). Thickness of the lodgement till varies from 0 to 3 m
(0—10 ft) but generally it is less than 0.3 m (1 ft). Here and there, on top of
the till, is a gravelly layer 0.3 to 7.2 m (1—24 ft) thick and this is overlain by
a layer of fine sand 0.15—9 m (0.5—30 ft) thick and above the sand are glaciolacustrine silts and clays which vary from 2.7 to 22.5 m (9—75 ft) in thickness. The direction of the last glacial advance in this area was approximately
south 20° east.

In general there is a bedrock high along the easterly edge of the sampled
area and from there the bedrock forms a ridge about 120 m (400 ft) wide
which more or less follows the base line. North and south of this ridge the
bedrock is at depths greater than 7.5 m (25 ft).

Geological setting of Lake Pelletier

This lake was chosen as a test site because the east-northeast trending gold-
bearing Pelletier Lake fault underlies the portion of the lake sampled. A former gold producer, Stadacona Mines Limited, is located at the east end of
this fault where it joins with the main northeast- striking ore-bearing Stadacona fault. The faults cut Precambrian volcanic and dioritic rocks.

The Stadacona mine was in production from 1936 to 1958 and during
that time some 3,053,420 tons grading 0.16 oz. Au per ton were mined. The
ore zones comprise parallel and cross-cutting quartz-ankerite veins in the
Stadacona fault. The maximum width of the ore zone was 6 m (20 ft) but
rarely exceeded 2.4 m (8 ft). Besides quartz and ankerite the zone contained

tourmaline, pyrite, free gold, petzite, arsenopyrite, chalcopyrite, galena, talc and chrome-bearing mica (Wilson, 1962).

The rocks underlying Pelletier Lake are predominantly interbedded flows of Precambrian pillowed andesite, rhyolite agglomerate and tuff into which masses and dykes of diorite have been intruded. The volcanic and dioritic rocks are intersected by the Pelletier Lake fault and the southwestern extension of the Stadacona fault. Diamond drilling on the Pelletier Lake fault has revealed the presence of a gold-bearing pyritic, carbonatized (ankerite) shear zone veined with quartz. The zone appears to be 7.5—12 m (25—40 ft) wide.

Geochemical results

The Pelletier Lake study was carried out to determine the effectiveness of till sampling at depth in outlining gold-bearing structures. Also to demonstrate the blanketing effect of impervious glacial lake clay sequences on the vertical migration of trace metals from underlying anomalous tills, the lake sediments were taken. Because of the presence of arsenopyrite in the gold-bearing zones, arsenic has been used as the pathfinder element.

Figs. 2, 3 and 4 respectively show the results for arsenic in the lake sediments (minus 80 mesh and minus 230 mesh), tills (minus 80 mesh and minus 230 mesh) and heavy-mineral concentrates from the tills (minus 10-mesh, plus 80 mesh and minus 50-mesh, plus 230 mesh).

Lake sediments (minus 80 and minus 230-mesh fractions)

Arsenic in the minus 80-mesh fraction of the lake sediments varies from 3 to 18 ppm (Fig.2a). The Pelletier Lake fault zone has no anomaly associated with it. High values of 10 and 15 ppm in the southeast corner of the grid may be related to arsenic from the Stadacona fault zone which outcrops on the eastern shore of the bay. The 18-ppm value in the southwest part of the grid could also be related to sources of arsenic on shore as outcrops are abundant here. The high value (10 ppm) in the northeast part of the grid could be caused by contamination from mine tailings in the adjacent bay. A corresponding high value of 60 ppm also occurs here in the minus 230-mesh fraction (Fig.2b). Except for this value, arsenic in the minus 230-mesh lake sediment material varies from 3 to 9 ppm and no significant increase in arsenic is obvious over or near the gold-bearing structures.

Therefore, the distribution of arsenic in lake sediments from Pelletier Lake does not indicate the presence of the gold-bearing structures underlying the glaciolacustrine deposits.

Till (minus 80 and minus 230-mesh fractions)

The arsenic values in the minus 80-mesh fraction of the tills show a marked increase over and in the vicinity of the Pelletier Lake fault zone (Fig.3a). Arsenic in the minus 230-mesh fraction does not appear to increase immediately over the gold-bearing fault (Fig.3b). However, an increase does occur in the

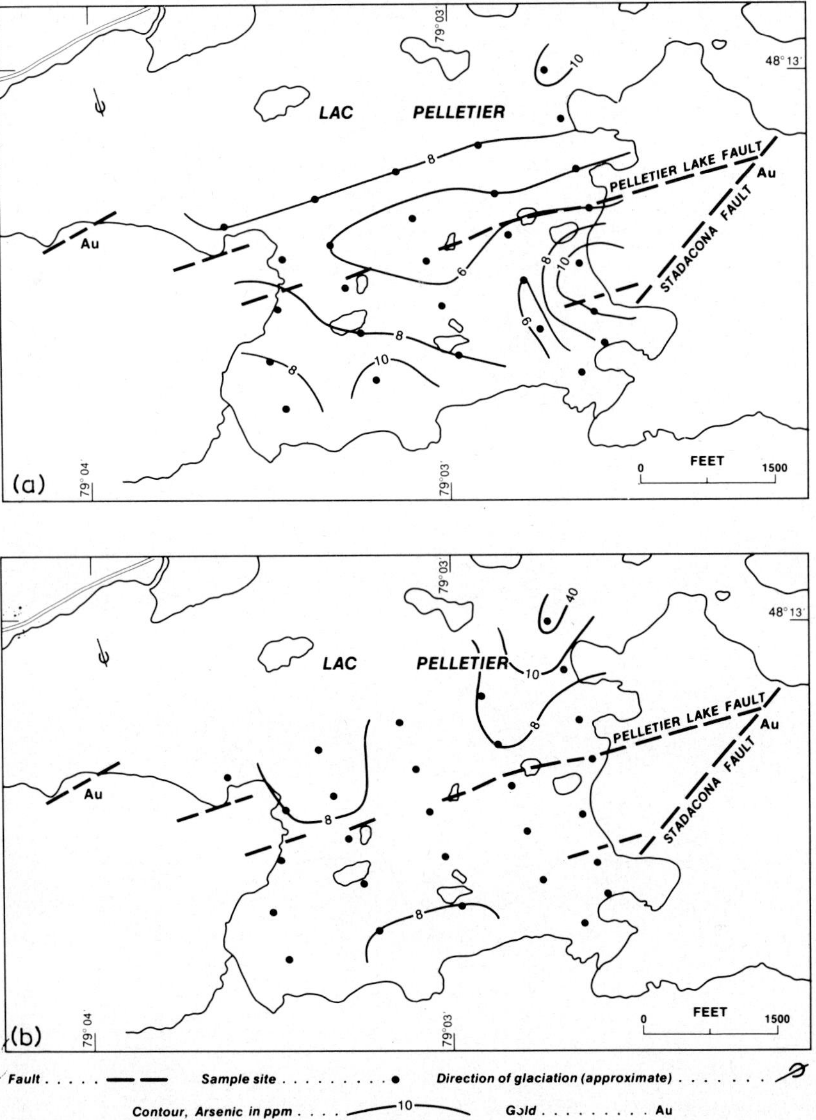

Fig.2. Arsenic distribution in lake sediments — Pelletier Lake, Quebec; (a) minus 80-mesh fraction, (b) minus 230-mesh fraction.

minus 230-mesh fraction of the till southeast of the Pelletier Lake fault. Overburden here is rather thin and this anomaly could be glacially transported from the northwest or it could represent hydromorphic dispersion, probably by groundwater, from the gold-bearing faults on shore. The latter is probably more likely because a southeasterly glacial dispersion fan from the fault is not evident from the results.

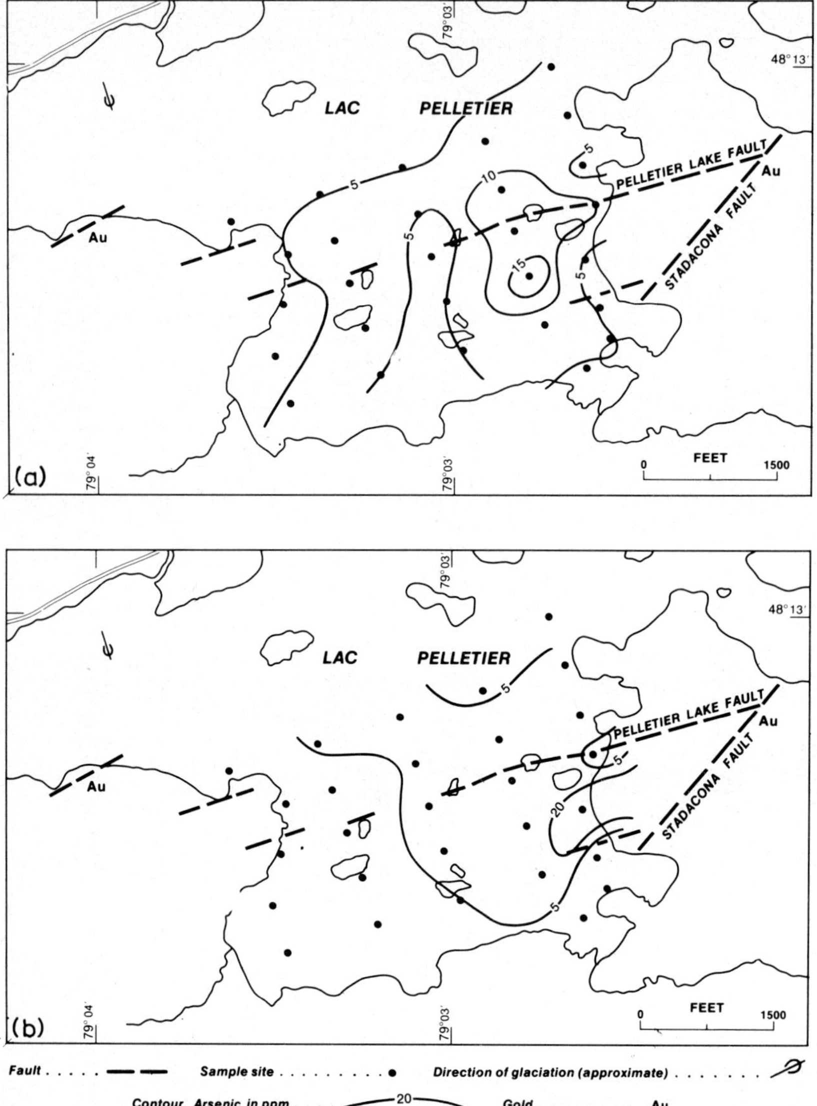

Fig.3. Arsenic distribution in till — Pelletier Lake, Quebec; (a) minus 80-mesh fraction, (b) minus 230-mesh fraction.

Till (heavy-mineral fractions)

The minus 10-mesh, plus 80-mesh heavy-mineral fraction shows a good anomaly in excess of 20 ppm As over and down glacial ice-direction movement from the Pelletier Lake fault zone (Fig.4a). Several unexplained anomalies occur in the northeast and southwest part of the sampled area.

The minus 50-mesh, plus 230-mesh heavy mineral fraction (Fig.4b) is also

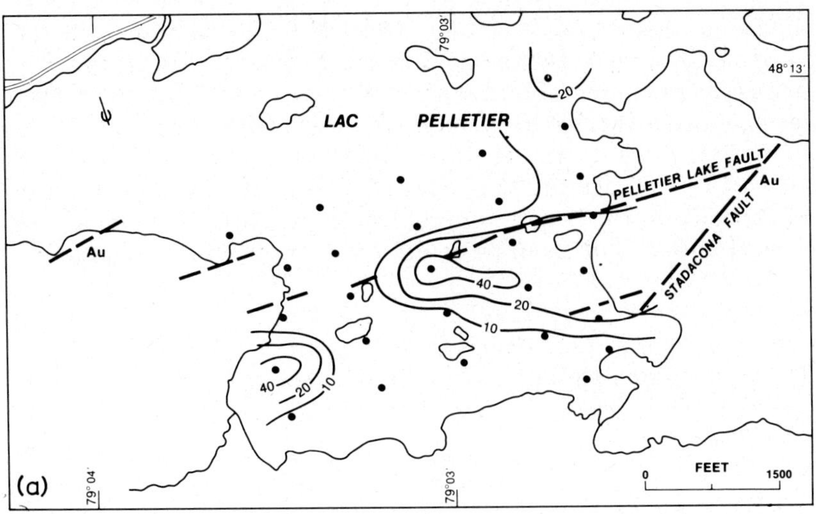

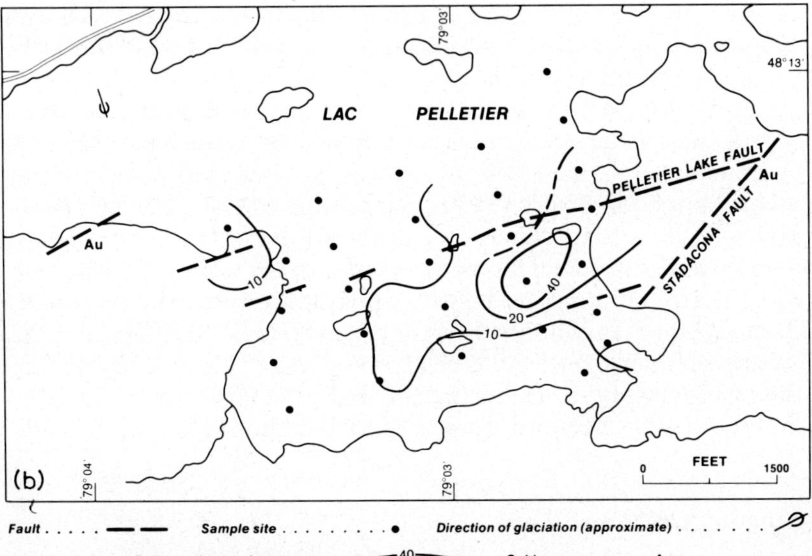

Fig.4. Arsenic distribution in heavy-mineral fraction of till — Pelletier Lake, Quebec; (a) minus 10-mesh, plus 80-mesh fraction, (b) minus 50-mesh, plus 230 mesh fraction.

anomalous over and down ice direction from the Pelletier Lake fault zone. However, the anomaly is less intense and more restricted than that found using the minus 10-mesh, plus 80-mesh material.

Summary

The Pelletier Lake gold-bearing fault zone can be delineated by sampling the tills atop bedrock on a 300 m × 150 m (1000 ft × 500 ft) grid. Using arsenic as the pathfinder element for the gold-bearing structure the minus 10-mesh, plus 80-mesh heavy-mineral material shows a more distinct anomalous pattern. However, more down-ice dispersion is evident with the minus 50-mesh, plus 230-mesh heavy-mineral fraction. The arsenic anomaly in the minus 80-mesh fraction of the till overlies Pelletier Lake fault. Its dispersion is more restricted than those from the heavy-mineral fractions. Arsenic in the minus 230-mesh till fraction does not seem to be effective in outlining the gold-bearing fault.

Lake sediments are ineffective in delineating the fault zones; however, there appears to be some hydromorphic dispersion of arsenic into the lake sediments possibly from the outcrops along the shore line and/or groundwaters from the fault zones. This creates single station anomalies in lake sediments taken near the shore.

NIGHTHAWK LAKE

Nighthawk Lake is located 30.4 km (19 miles) east of the town of Timmins, Ontario (Fig.1) at latitude 48°30′N and longitude 80°55′W. In Northeast Bay of Nighthawk Lake the water varies in depth from 0.3 to 3 m (1—10 ft). The overburden is from 2.7 to 44.4 m (9—148 ft) thick and most of the north half of the bay and the extreme south part is covered by more than 24 m (80 ft) of glacial overburden.

Glacial striae measured by Leahy (1971) vary from 145° to 175°T. The glacial stratigraphy here consists generally of the following: an upper layer of glaciolacustrine clay and silt 0.3—18.6 m (1—62 ft) thick, which is underlain by fine sand and silt 1.5—30.6 m (5—102 ft) thick; underlying the sand and silt is 0.6—10.2 m (2—34 ft) thick layer of sand, gravel and boulders. In places the sand and/or gravel layer overlies 0.3 m (1 ft) or less of dense lodgement till. Some of the bouldery layers according to Hughes (1965) form part of the basal till. In most places samples were taken from the till on top of bedrock.

Geological setting of Nighthawk Lake

This lake was chosen to test the effectiveness of reconnaissance geochemical till sampling in outlining general geological targets of possible economic interest. The two targets of interest here are the ultramafic intrusions and gold occurrences.

Northeast Bay of Nighthawk Lake is underlain by several bodies of serpentinized peridotite (Fig.5). These ultramafics intrude volcanic rocks which in places have been altered to chlorite-carbonate schists. On the peninsulas to the south the rocks are mainly volcanic and these have been intruded also by peridotite.

In the south part of the bay, gold occurrences are known to be present in

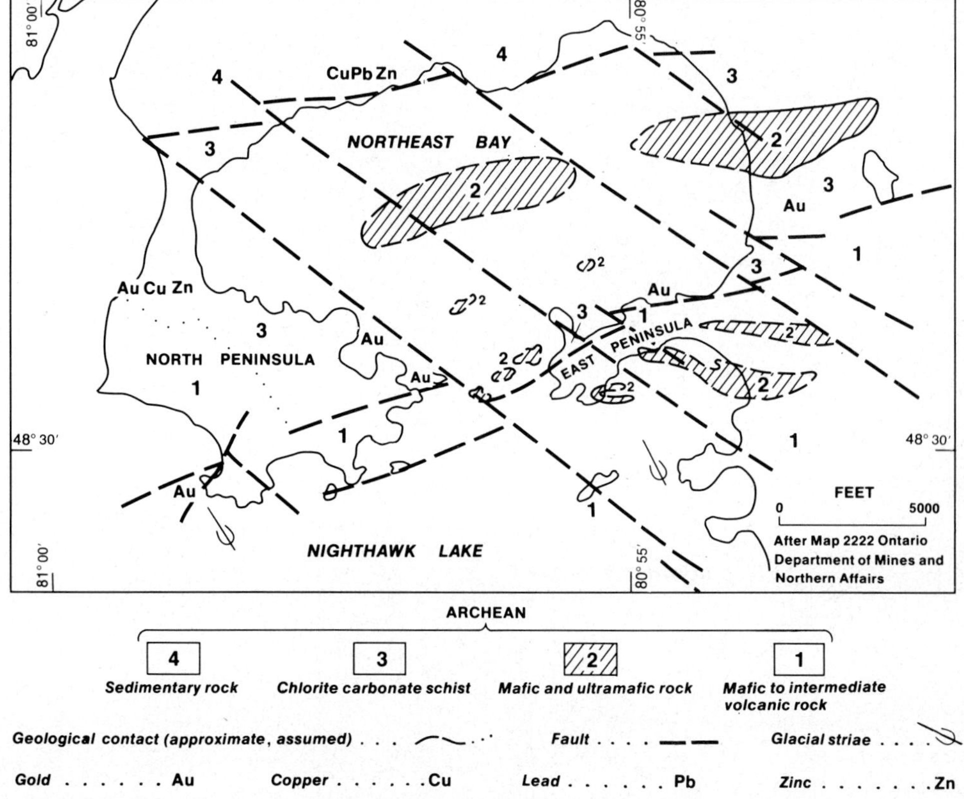

Fig.5. General geology of Northeast Bay, Nighthawk Lake, Ontario.

pink to grey felsite or porphyry dykes and quartz veins and stringers. These occur in strongly schistose volcanic rocks associated with grey and brown carbonates and green chrome-micas. The dykes and veins are frequently associated with fault zones which trend N70° E. The mineralized carbonate zones may contain about 5% pyrite and traces of chalcopyrite, sphalerite, pyrrhotite, galena, and free gold.

Gold occurrences have been described also as occurring in quartz veins cutting serpentine-chlorite-carbonate schist which contain abundant pyrite (Leahy, 1971).

Between 1924 and 1944 Porcupine Peninsula mine, which is located at the southwestern tip of North Peninsula, produced about 100,000 tons of ore from which 27,416 oz. of gold was recovered (Leahy, 1971).

Geochemical results

The Nighthawk Lake study was done to determine the effectiveness of systematic semi-reconnaissance till sampling at depth in outlining large geo-

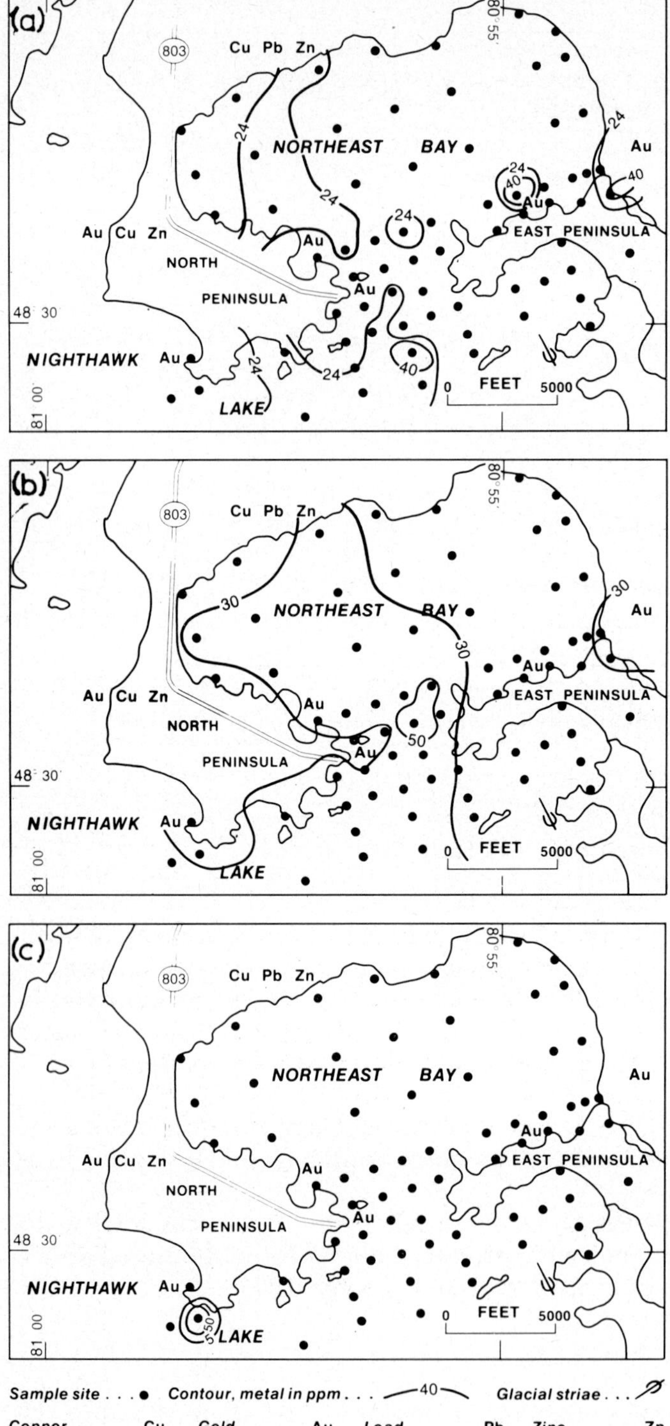

Fig.6. Metal distributions in minus 230-mesh lake sediments — Nighthawk Lake, Ontario; (a) copper, (b) nickel, and (c) arsenic.

logical targets of economic interest (ultramafic intrusions) and gold-bearing structures. Also lake sediments were sampled to demonstrate the ineffectiveness of this technique in areas underlain by thick glaciolacustrine deposits.

The distribution of copper, nickel, and arsenic in the lake sediments, tills and heavy-mineral concentrates from the tills are shown in Figs.6, 7 and 8, respectively.

Lake sediments (minus 230-mesh fraction)

The lake sediment values for copper (Fig.6a) vary from 4 to 63 ppm and average 15 ppm, for nickel (Fig.6b) they vary from 8 to 105 ppm and average 25 ppm, for arsenic (Fig.6c) they range from 0.5 to 100 ppm and average 2 ppm.

All arsenic values but one are less than 5 ppm. The one high result (100 ppm) occurs near a small island (Pine Island) in the southwest part of the grid. This single high value may be due to contamination from the mine tailings of Porcupine Peninsula mine.

There is no apparent relationship between the distribution of nickel, copper and arsenic in the lake sediments and the geology of the area.

Till (minus 230-mesh fraction)

The distribution of copper, nickel and arsenic in the minus 230-mesh fraction of the tills are shown in Fig.7a, b and c, respectively.

Copper values vary from 8 to 50 ppm and average 15 ppm. There is little apparent correlation between the known geology and copper in the fine fraction of the tills.

Nickel values vary from 12 to 370 ppm and average 52 ppm. Anomalous values in nickel are common in areas underlain by ultramafic rocks. The northeasterly trend of the nickel anomalies (Fig.7b) between the two peninsulas coincides with the attitude of some of the ultramafic bodies (Fig.5). Similarly the easterly nickel trend on the north side of East Peninsula corresponds with the strike of the serpentinized ultrabasic rocks there. Ultrabasic rocks or their altered equivalents are in close proximity to the gold-bearing structures. The northeast part of the bay is underlain by an anomalous nickel zone which coincides with the location of a peridotite intrusion. The anomaly is open to the north and probably represents dispersion of the fine fraction of the till from a source to the north.

Arsenic in the minus 230-mesh fraction of the till (Fig.7c) varies from 1 to 33 ppm and averages 2 ppm. There does not seem to be any regional trend for arsenic in this fraction. Local circular anomalies coincide with anomalous nickel on the north side of East Peninsula and at two locations in the narrows between and south of East and North Peninsulas.

Till (heavy-mineral fraction)

The distribution of copper, nickel and arsenic in the minus 50- plus 230-mesh heavy-mineral fraction (specific gravity > 2.96) is shown in Fig.8a, b and c, respectively.

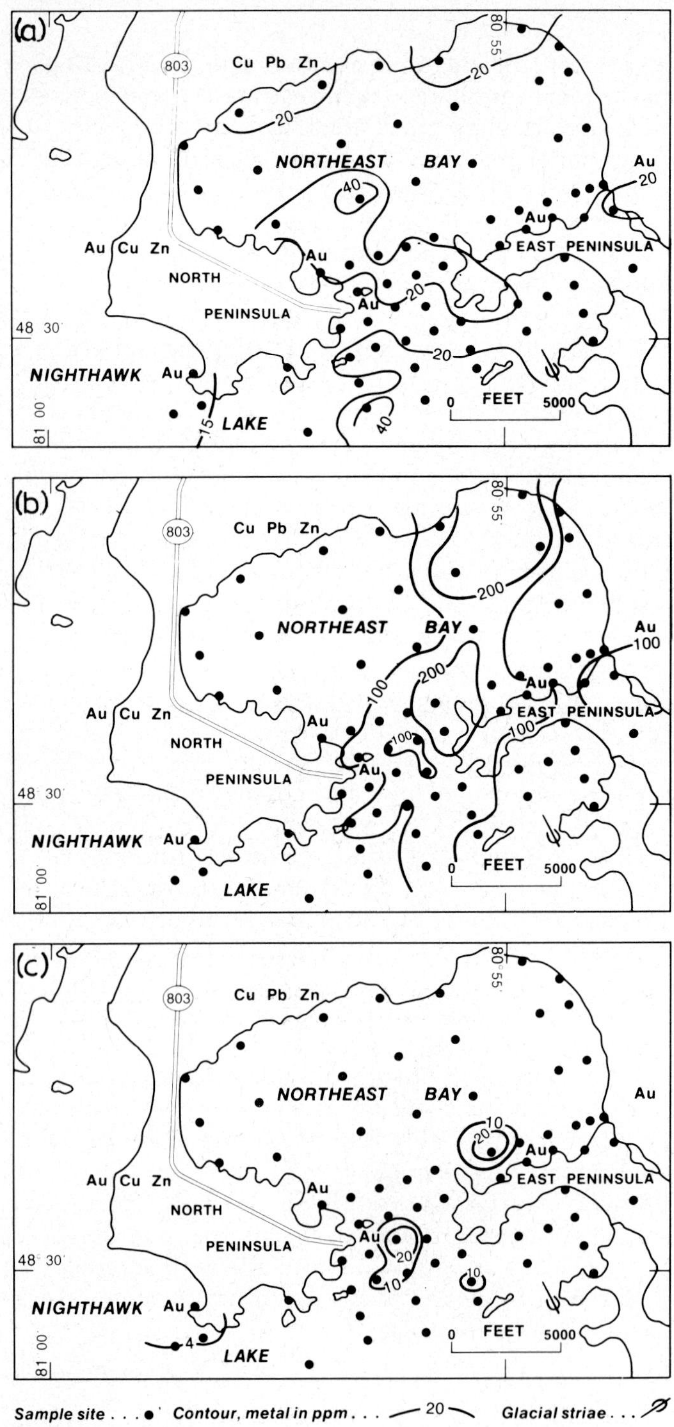

Fig. 7. Metal distribution in minus 230-mesh till — Nighthawk Lake, Ontario; (a) copper, (b) nickel, and (c) arsenic.

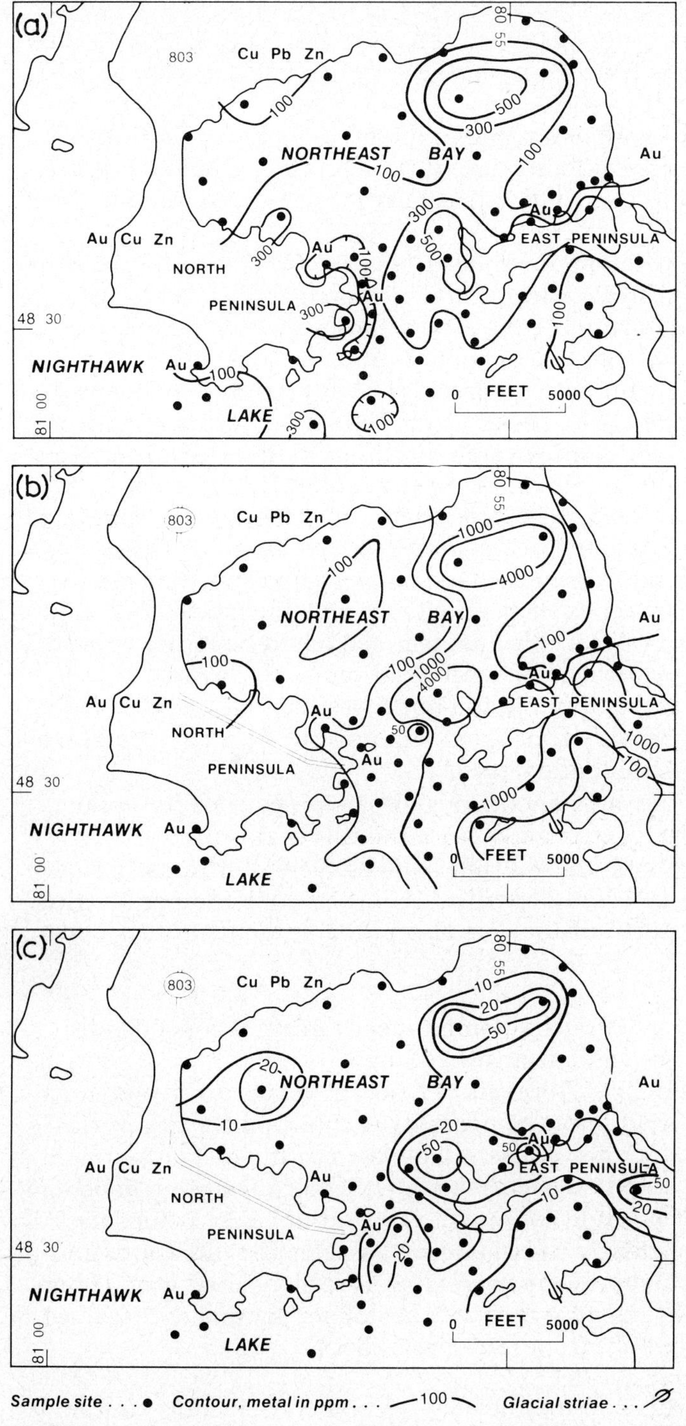

Fig.8. Metal distribution in minus 50-mesh, plus 230-mesh heavy-mineral fraction of till — Nighthawk Lake, Ontario; (a) copper, (b) nickel, and (c) arsenic.

Copper. Copper values vary from 14 ppm to 2200 ppm and the average is 88 ppm.

Strong easterly trending copper anomalies underlie the north half of the bay, these are related to ultramafic rocks and talc-chlorite-carbonate schists. Traces of chalcopyrite were identified in the heavy-mineral concentrates from here.

On East Peninsula strong copper anomalies are present. Gold deposits occur here in carbonated and schistose volcanic and ultramafic rocks. Also copper anomalies underlie a large portion of the area between the two peninsulas and south of them. This regional increase in copper probably is related to ultramafic rocks and to an area of intense hydrothermal alteration that has produced large bodies of carbonatized rocks. Carbonatization occurs in areas intruded by albite-syenite and albite-syenite porphyry dykes. According to Leahy (1971) the carbonates are ankerite, calcite, dolomite and magnesite. Other common minerals in these rocks are sericite, fuchsite, pyrite, quartz, albite and leucoxene. Trace amounts of chalcopyrite have been identified in the heavy-mineral concentrates from the area between and south of the two peninsulas and carbonate minerals such as ankerite and magnesite are common.

Another source of the copper in the area sampled could be minor amounts of chalcopyrite which are known to occur in gold veins.

An increase of copper to 164 ppm in the northwest part of the sampled area could be part of a dispersion train from a minor Cu—Pb—Zn occurrence immediately north in the sedimentary rocks.

Hence, it appears that the large anomalous dispersion pattern for copper found in the heavy-mineral concentrates from the tills in this part of Nighthawk Lake are related probably to multiple sources such as ultramafic rocks, altered carbonated zones and gold deposits. Minor base-metal mineralization in the sedimentary rocks north of the lake also probably contributes copper to the tills.

Nickel. Nickel in the minus 50- plus 230-mesh heavy-mineral concentrates varies from 20 to 11500 ppm and averages 163 ppm.

The strongest nickel anomaly (7000 and 11,500 ppm) occurs in the northeast part of the bay. Minor amounts of pyrrhotite, chalcopyrite and pyrite in the heavy-mineral concentrates indicate that some of the nickel found in the tills here is related to sulphide occurrences. Other significant nickel anomalies are present on East Peninsula and north and south of East Peninsula. All these nickel anomalies are closely associated with ultramafic intrusions and/or their altered equivalents. Many of the nickel highs are associated with anomalies in copper, which indicates that most of the copper anomalies discussed previously are derived probably from ultramafic rocks.

Arsenic. The minus 50- plus 230-mesh heavy-mineral concentrates contain 0.5 to 160 ppm arsenic and average 5 ppm.

High arsenic values are coincident with zones high in nickel. In the vicinity of the gold-bearing structures on East Peninsula high arsenic values occur. The broad distribution of arsenic anomalies suggests that gold mineralization could be wider spread than previously known. The coincidence with nickel probably results where gold-bearing structures intersect ultramafic rocks such as on East Peninsula.

Although no arsenopyrite was seen in the heavy-mineral concentrates many of the samples contain from a trace amount to 7% pyrite. Pyrite is commonly associated with the gold deposits and the arsenic in the heavy-minerals is probably derived from pyrite.

Summary

Lake sediment sampling in this area is ineffective in outlining broad geological targets such as ultramafic intrusions or more restricted targets such as gold-bearing structures underlain by up to 44.4 m (148 ft) of glacial sediments.

However, till sampling at depth on 400- and 800-km (¼ and ½ mile) centres has proven effective in outlining both of the above types of targets. The most definitive results have been obtained from the minus 50- plus 230-mesh heavy-mineral concentrates. Coincident anomalies in copper, nickel and arsenic have been obtained over the ultramafic intrusions and over the gold structures where they cut the ultrabasic rocks. The results suggest that the gold mineralization could be wider spread than previously known, this is particularly true north and southwest of East Peninsula.

GENERAL DISCUSSION

Although Cu, Pb, Zn, Mo, Ni, Mn, As and Ag content were determined in all samples from four different lakes in the Abitibi area, only selected elements over two lakes have been discussed here. To compare the type of results obtained in lakes that are apparently devoid of mineral deposits (Abitibi and Macamic Lakes) with those obtained from lakes underlain by mineral occurrences and/or ultramafic rocks (Nighthawk and Pelletier Lakes) a summary of background values obtained from all four lakes is presented in Table II.

This table shows that, except for arsenic and manganese, slight increases in element background values are apparent in the minus 230-mesh fraction of the lake sediments from Pelletier Lake as compared to the minus 80-mesh fraction. All metals, except arsenic in the tills from Pelletier Lake, have lower average metal values in the minus 80-mesh fraction than in the minus 230-mesh fraction. Also, except for arsenic and manganese, the averages of the metal values in the minus 10- plus 80-mesh heavy-mineral fraction of the till from Pelletier Lake generally are lower than the metal averages in the minus 50- plus 230-mesh heavy-mineral fraction.

The mode of element dispersion in the till is by particulate transport and this increase in arsenic in the coarser fractions of the till may be a reflection

628

TABLE II

Geometric means of trace elements from lake sediments and tills, Abitibi region

Lake	Fraction and material	Geometric mean (ppm)							
		Cu	Pb	Zn	Mo	Ni	Mn	As	Ag
Pelle-tier	−80 mesh lake sediments	50	18	93	2.1	42	576	7.1	1.2
	−230 mesh lake sediments	52	25	96	2.5	44	552	6.2	1.2
Pelle-tier	−80 mesh tills	35	11	44	1.3	27	330	4.5	1.0
	−230 mesh tills	52	18	65	1.8	34	409	3.7	0.7
Pelle-tier	−10+80 mesh hvs tills	115	20	76	2.4	57	597	9.2	1.4
	−50+230 mesh hvs. tills	173	29	84	2.4	58	430	7.9	1.5
Night-hawk	−230 mesh lake sediments	15	13	38	1.5	25	362	1.6	1.3
	−230 mesh tills	15	13	27	1.8	52	263	2.0	1.3
	−50+230 mesh hvs tills	88	21	45	6.9	163	396	4.7	1.0
Macamic	−230 mesh lake sediments	25	19	66	2.6	39	596	2.5	1.2
	−230 mesh tills	10	9	26	2.0	18	145	1.2	0.7
	−50+230 mesh hvs tills	22	16	33	2.2	36	254	2.3	1.0
Abitibi	−230 mesh lake sediments	15	15	64	1.5	24	360	2.8	0.9
	−230 mesh tills	14	10	28	1.4	20	190	1.2	0.6
	−50+230 mesh hvs tills	33	10	31	1.7	32	295	2.1	0.7

Note: hvs = heavy-mineral concentrate, specific gravity > 2.96

of the proximity of the source of arsenic (i.e. Lake Pelletier fault). The increase of arsenic in the coarser fraction on the lake sediments also suggests that some of the dispersion of arsenic in the lake sediments may be in the form of minute particles. One possible source of such dispersion would be contamination from mine tailings.

Background values in Table II for most elements in the heavy fraction of the tills show marked increases over the lakes containing gold mineralization and ultramafic rocks, (Pelletier and Nighthawk Lakes) as compared to those under which no ultramafic rocks or gold deposits are known (Abitibi and Macamic Lakes).

From the Pelletier Lake study it is concluded that analyses of the coarser heavy minerals (i.e. minus 10-mesh, plus 80 mesh) is as effective in outlining gold-bearing structures as the heavy concentrates from finer fractions. When sampling close to the source of the metals the minus 10-mesh, plus 80-mesh fraction is preferable.

In future studies of this type it would be interesting to compare the element contents of whole till samples with the element contents of the heavy-mineral fractions from the till. The whole till samples would be prepared for analysis by pulverizing them to minus 100 mesh much in the same manner as one would prepare a rock sample for geochemical analysis. It is suggested that where lodgement till is sampled, meaningful results possibly could be obtained by analyzing the whole till sample. Such an approach is economically justified because expensive heavy-mineral separations would be avoided.

To establish optimum sample densities for detailed and semi-reconnaissance work additional investigations involving till sampling are required over and near various types of mineral deposits and geological targets.

Recent work by Gunton and Nichol (1974) has demonstrated that overburden sampling at depth can be effective in evaluating surface geochemical anomalies that are suspect of being transported to a position remote from their bedrock sources.

Also in Finland (Wennevirta, 1973) sampling of the bedrock/till interface not only has been used as an effective geochemical exploration technique, but by examining the rock fragments in the coarse fraction of the till they have been able to estimate the nature of the underlying geology. This aspect of the technique has not been stressed in Canada but it could prove to be a valuable geological mapping technique in extensively overburdened areas such as the Abitibi Clay Belt.

GENERAL CONCLUSIONS

In Canada sampling of till at the bedrock/till interface by means of light portable percussion drills is rapidly becoming one of the most frequently used geochemical exploration techniques for evaluating geophysical and geological targets in glacial terranes. Although, in the past, the technique has been restricted to detailed studies of specific anomalies, it is now apparent that the method is applicable on a semi-reconnaissance basis. The results presented have clearly demonstrated that larger geological targets such as ultramafic intrusions can be outlined by sampling at 800-km (½ mile) centres, and analysing the heavy-mineral fraction for such elements as copper and nickel. Smaller targets, such as gold-bearing structures, have been delineated by sampling at 400-km (¼ mile) centres and analysing the heavy-mineral fraction for arsenic. The technique has been proven effective in areas underlain by up to 44.4 m (148 ft) of glaciolacustrine deposits and where surficial geochemistry does not work.

REFERENCES

Boissonneau, A.N., 1965. Algoma-Cochrane; surficial geology. Ont. Dept. Lands Forest, Map S365

Boniwell, J.B. and Dujardin, R.A., 1964. Discovery and exploration of the Poirier ore deposit. Can. Inst. Min. Metall. Bull. 57 (629): 945—951

Ermengen, S.V., 1957. Geochemical prospecting in Chibougamau. Can. Min. J., 78 (4): 99—104

Fortescue, J.A.C. and Hornbrook, E.H.W., 1969. Progress report on biogeochemical research at the Geological Survey of Canada, 1963—1966. Geol. Survey Can. Paper 67-23, Part II

Garrett, R.G., 1971. The dispersion of copper and zinc in glacial overburden at the Louvem deposit, Val d'Or, Quebec. In: R.W. Boyle (Editor), Geochemical Exploration. Can. Inst. Min. Metall., Spec. Vol., 11: 157—158

Gleeson, C.F., 1960. Studies on the distribution of metals in bogs and glaciolacustrine deposits. Ph.D. Thesis, McGill Univ. (unpublished)

Gleeson, C.F. and Cormier, R., 1971. Evaluation by geochemistry of geophysical anomalies and geological targets using overburden sampling at depth. In: R.W. Boyle (Editor), Geochemical Exploration. Can. Inst. Min. Metall., Spec. Vol., 11: 159—165

Gunton, J.E. and Nichol, I., 1974. Delineation and interpretation of metal dispersion patterns related to mineralization in the Whipsaw Creek area. Can. Inst. Min. Metall. Bull., 67 (741): 66—75

Hornbrook, E.H.W. and Gleeson, C.F., 1972. Regional geochemical lake bottom sediment and till sampling in the Timmins—Val d'Or region of Ontario and Quebec. Geol. Survey Can. Open File Rept. No.112

Hornbrook, E.H.W. and Gleeson, C.F., 1973. Regional lake bottom sediment — moving average — residual anomaly maps in the Abitibi region of Ontario and Quebec. Geol. Survey Can. Open File Rept. No.127

Hughes, O.L., 1960. Surficial geology of Iroquois Falls, Cochrane District. Geol. Survey Can. Map 46-1959

Hughes, O.L., 1965. Surficial geology of part of the Cochrane District, Ontario, Canada. Geol. Soc. Am., Spec. Paper 84: 535—565

Leahy, E.J., 1971. Geology of the Nighthawk Lake area, District of Cochrane. Ont. Dept. Mines and Northern Affairs Geol. Rept. 96

Lee, H.A., 1963. Glacial fans in till from the Kirkland Lake fault: a method of gold exploration. Geol. Survey Can. Paper 63-45

Prest, V.K., Grant, D.R. and Rampton, V.N., 1967. Glacial map of Canada. Geol. Survey Can. Map 1253A

Skinner, R.G., 1972. Drift prospecting in the Abitibi Clay Belt, overburden drilling programme methods and costs. Geol. Survey Can. Open File Rept. No.116

Van Tassel, R.E., 1969. Exploration by overburden drilling of Keno Hill Mines Limited. In: 2nd Int. Geochemical Exploration Symposium. Q. Colo. School of Mines, 64 (1): 457—478

Vasak, V. and Sedivec, V., 1952. Colorimetric determination of arsenic. Chem. Listy., 46: 341—344

Wennevirta, H., 1973. Sampling of the Bedrock/Till Interface in Geochemical Exploration; Prospecting in Areas of Glacial Terrain. Institution of Mining and Metallurgy, London, pp. 67—71

Wilson, M.W., 1962. Rouyn-Beauchastel Map area, Quebec. Geol. Survey Can. Mem. 35

AN ELECTROCHEMICAL MODEL FOR ELEMENT DISTRIBUTION
AROUND SULPHIDE BODIES

B. BÖLVIKEN and Ö. LOGN

Geological Survey of Norway, Trondheim (Norway)
A/S Sydvaranger, Oslo (Norway)

ABSTRACT

It is suggested that the upper lithosphere can be considered as a primary redox potential field, where the groundwater represents an electrolyte, and the siliceous bedrock a system of membranes in the electrolyte. The redox potentials of this primary field decrease with depth from the surface and hence electrical currents flow — positive (e.g. H^+) downwards and negative (e.g. OH^-) upwards — in the lithosphere. A body of electrically well-conducting sulphide mineralization can be regarded as a dipole inert electrode in the electrolyte, the lower end constituting the anode and the upper the cathode. Secondary currents in addition to those of the primary field flow between cathode and anode; cations move upwards and anions downwards in the electrolyte, and electrons upwards in the electrode. At some places the direction of the currents are the same for both primary and secondary fields; at other places primary and secondary currents counteract each other. Consequently, patterns of varying current density and ion concentration develop around the electrode. If the environment of the electrode is homogeneous, these patterns will be quite regular; at a vertical electrode current density and ion concentration will be highest in the vertical extensions of the electrode. At a suboutcropping ore body the overburden represents an environmental heterogeneity, because the electrical conductivity of the overburden is higher than that of the country rock and lower than that of the ore.

Thus, when adapted to surface conditions, the model suggests high current density and high ion concentration in the vicinity of the boundaries of the upper part of the ore, and low values for these parameters in the overburden above the ore. Aspects of the possible influence of electrochemical dispersion on element distribution are discussed. Empirical data for self potentials, redox potentials, specific conductivity and pH in ground and surface waters, as well as contents of Li, Na, Cu, Zn, Pb, Cl and values for pH and specific conductivity of soil extracts at the Joma massive pyrite deposit are interpreted in terms of the presented model.

INTRODUCTION

The existence of natural self potentials (SP) at certain ore deposits has been known for more than a century, and SP measurements are widely used in prospecting (Sato and Mooney, 1960; Society of Exploration Geophysicists, 1966, 1967; Parasnis, 1966, 1967, 1970; Bitterlich and Wöbking, 1972). The origin of these potentials is not completely understood, although largely explained by Sato and Mooney (1960) who state that: "Self potentials associated

632

with a sulphide ore body result from the ohmic potential drop within the
country rocks. The electric current is produced by separate but simultaneous
reduction of oxidizing agents near the surface and oxidation of reducing
agents at depth. The ore does not participate directly in either reaction, but
serves as a conductor to transfer the electrons from the reducing agents to the
oxidation agents. The possibility for the above reaction to occur depends
upon differences in oxidation potential of groundwaters at different depths".
This electrochemical mechanism of self potentials is widely accepted in the
literature (Becker and Tilford, 1965; Habashi, 1966; Parasnis, 1966, 1967,
1970; Henriet, 1971; Malmqvist and Parasnis, 1972; Govett, 1973).

In another paper (Logn and Bölviken, 1974) the present authors have
reported some results of surface and drill hole self potential measurements at
the Joma pyrite deposit, Norway, which seem to agree well with the electrochemical theory for SP. Fig.1 shows a typical result from one of these drill
holes. The self potential curve indicates two distinct patterns: (1) abruptly
alternating potentials with minima down to −800 mV and (2) a generally
smooth potential trend, the variation of which is within some tens of milli-
volts. These two patterns can be recognized in all the drill holes investigated
as well as at the surface. The abrupt potential changes occur at boundaries of
sulphide horizons and country rock giving low potentials inside the ore. The

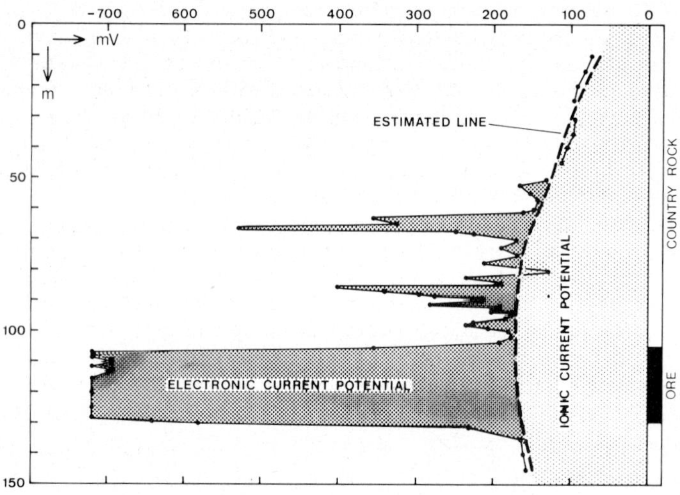

Fig.1. Self potentials in diamond drill hole No.119 Joma pyrite deposit, Norway. Potential
pattern can be separated into: (1) electronic current potentials (abruptly changing path),
and (2) ionic current potential (smooth trend) by graphical estimation.

potentials of the smooth trend decrease when approaching the upper part of the ore, and increase when approaching the lower part. The two types of potential pattern were interpreted in the following way:

(1) The abruptly alternating potentials are a consequence of the presence of electromotive forces at the country rock/sulphide interfaces. They indicate that electronic currents pass through the ore bodies. These potentials were termed "electronic current potentials".

(2) The smooth potential trend indicates that ionic currents, pass through the pore water of the country rock outside the ore body. These potentials were termed "ionic current potentials".

Electronic and ionic current potentials can be separated graphically as indicated in Fig.1. When this procedure is used on rather complicated drill hole SP data from sections through the ore (Fig.2), more simple patterns of both electronic and ionic current potentials appear; an example of the latter is reproduced in Fig.3. The distribution of ionic current potentials indicates

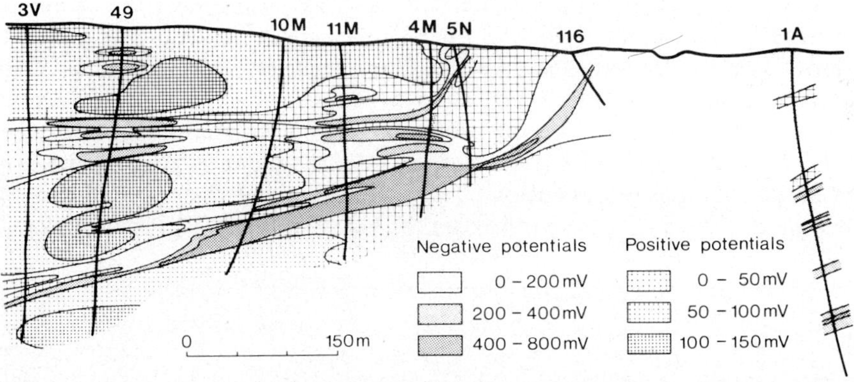

Fig.2. Self potentials in drill holes projected into a vertical section of the Joma pyrite deposit, Norway. (After Logn and Bölviken, 1974.)

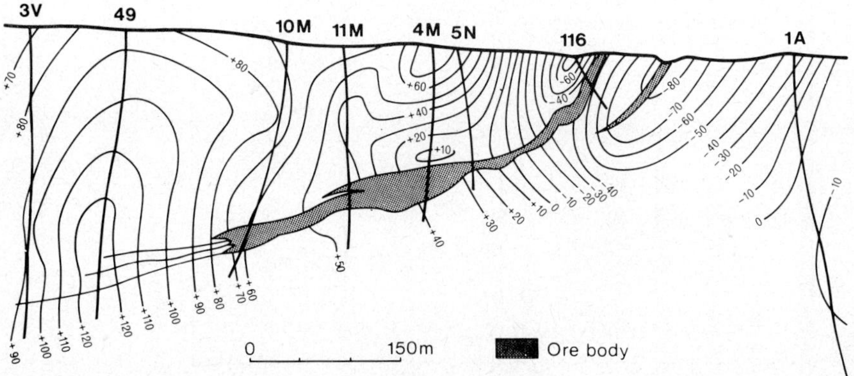

Fig.3. Ionic current potentials in drill holes estimated from data in Fig.2, Joma pyrite deposit, Norway. (After Logn and Bölviken, 1974.)

that the direction of positive electricity is upwards in the country rock. Electronic current potentials are highest at the upper end of the ore and lowest at the lower end, which seem to indicate that the direction of positive electricity is downwards in the ore. In other words, cations migrate upwards and anions downwards in the country rock, and electrons upwards in the ore body. These electrical currents might have geochemical consequences.

To elucidate some possible effects on the distribution of trace elements around an ore body, an electrochemical model of an ore deposit in country-rock groundwater is introduced in this paper. In geochemical prospecting any possibility for tracing electrochemical effects around ore bodies in surface material would be of particular interest. The model is, therefore, adapted to surface conditions. Towards the end of the paper some data from the Joma area, Norway, are discussed in relation to the presented model.

THE PRIMARY REDOX POTENTIAL FIELD OF THE LITHOSPHERE

All available information indicates that the groundwater near the surface of the earth normally has higher redox potentials and lower pH than the ground-waters at depth (Baas Becking et al., 1960). Consequently, systems of vertical redox potential and pH gradients must exist in the waters of the upper lithosphere. Such systems could be termed primary redox potential fields (Bölviken, in prep.). The primary redox potential field of a homogeneous part of the upper lithosphere could be outlined as follows (Fig.4): terrestrial waters with dissolved ions (e.g. Na^+ and Cl^-) and redox pairs (e.g. Fe^{3+}/Fe^{2+}) constitute an

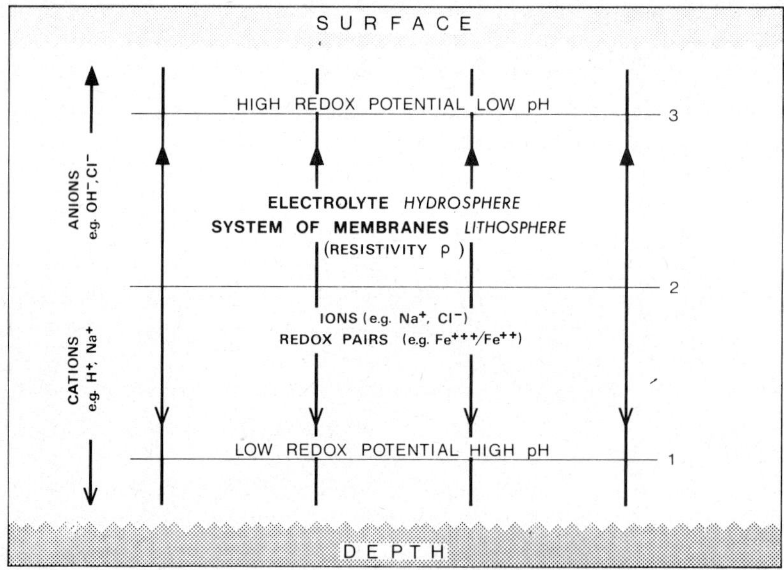

Fig.4. Diagrammatic representation of the primary redox potential field of a vertical section through a homogeneous part of the lithosphere.

indefinite electrolyte and the siliceous bedrock represents a porous medium of membranes in the electrolyte. The redox potential (and consequently the pH) vary regularly through the system, equipotential surfaces being horizontal Redox potentials represented, for example, by the ratio Fe^{3+}/Fe^{2+} increase upwards, whereas pH increases downwards.

Three types of transportation of material may be distinguished in this redox potential field: (1) groundwater movement, (2) diffusion of dissolved species caused by concentration differences, and (3) migration of ions due to electromotive forces.

Groundwater movement and the diffusion is to a great extent controlled by the permeability of the rocks; these types of transport shall not be further discussed here. Migration of ions due to electromotive forces depends on (1) potential differences, (2) electrical resistivity (ρ) of the media through which the currents flow, and (3) concentration (c) and equivalent conductivity (λ) of the ions in question. In a homogeneous part of the lithosphere, the main electrical current direction will probably be vertical and current densities uniform. Cations might migrate downwards and anions upwards, H^+ and OH^- being the outstanding current carriers with their ubiquity and high equivalent ionic conductance.

PRIMARY AND SECONDARY REDOX POTENTIAL FIELDS AROUND ORE BODIES

Let us consider a vertical sulphide ore body with negligible electrical resistance (R_0) occurring in the primary redox potential field of the earth. If oxidation of the sulphides is slight, the ore body could be considered as an inert electrode in an electrolyte, the composition of which varies between the ends of the electrode (Fig.5). Because of the redox potential differences originally present in the electrolyte (terrestrial waters) and the difference between the type of electrical conductance in electrode (ore) and electrolyte, a secondary redox potential field is developed locally around the electrode. The electrode takes on the character of a dipole, the lower end becoming an anode and the upper a cathode.

Secondary currents in addition to the primary ones will flow between the anode and the cathode. In the electrolyte (country-rock groundwater) near the electrode cations will move upwards from the anode towards the cathode, and anions downwards from the cathode towards the anode. Within the electrode (ore body) electrons will move upwards from the anode to the cathode in order to compensate for the electrons produced by the chemical reactions at the anode, and those consumed by the reactions at the cathode. Potential gradients will build up around cathode and anode and the system will tend to approach a state as schematically indicated in Fig.5.

In nature this state will be reached to a variable extent depending on the effects of groundwater movement, oxidation of ores and other parameters such as potential differences, velocity of redox- and diffusion processes, and electrical resistivity of the various media. Oxygen of the atmosphere and

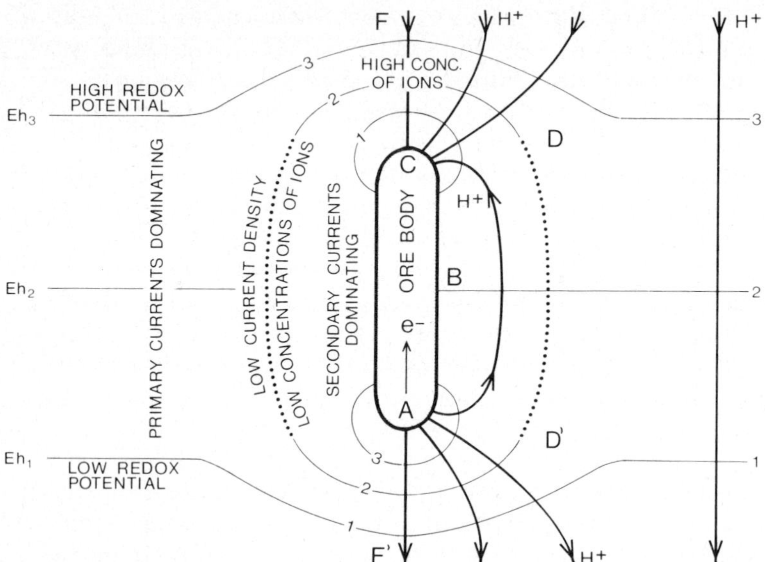

Fig.5. Schematic model of an ore body as electrode in a primary redox potential field. Eh_1, Eh_2 and Eh_3: selected equipotential surfaces. Heavy lines indicate path of primary and secondary currents, arrows indicate direction of positive electricity (cations) in the electrolyte. Arrow inside ore body: direction of electron flow in the ore. A = anode; B = electrical symmetry point at country rock/ore interface; C = cathode; D = limit of zone where secondary currents counteract primary currents; F = extension of the ore.

reducing agents in the inner of the earth would be the ultimate energy source of such natural galvanic cells. The maximum difference in redox potential between the electrolyte at the anode and that at the cathode could be considered as the electromotive force (EMF) of the cell.

At any point in the electrolyte the current densities resulting from the developed potential pattern will be the vectorial sum of the primary and secondary current densities. Far away from the electrode (ore) primary currents will dominate; close to the ore secondary currents are predominant. At a certain distance from the electrode (line D—D', Fig.5) the primary and secondary current densities will be approximately equal, and since they are of opposite direction the total current density approaches zero along line D—D'. Concentration of ions carrying the current will consequently be low in this zone. At the vertical extension of the ore (line F—F', Fig.5) the primary and secondary currents will both have the same vertical direction. Along the line F—F' the total current density is at a maximum, and ion concentration will consequently be high. In the space outside the lines D—D' and F—F' primary and secondary currents will either amplify or counteract to a variable degree, and the ion concentration at different places in the electrolyte would vary accordingly.

Various ions in the electrolyte will behave differently. Generally, cations will move towards the cathode where they will be reduced or kept in solution depending on their local concentration and on their position in the standard potential range. If the reduced form is a gas, it will diffuse away from the cathode; if it is a solid it will precipitate at the cathode. Oxidizing agents of the primary field will be in excess in the environment of the cathode. These agents will counteract the reduction processes at the cathode. They might also, in turn, oxidize the species reduced at the cathode when these diffuse away from the latter.

Protons have the highest equivalent conductance of all ions and will most probably be the main carrier of current in an acid environment. Hydrogen is more noble than those elements normally occurring in the highest concentrations in natural waters, and H^+ may be reduced at the cathode according to the equation:

$$2H^+ + 2e^- = H_2$$

Anions will move towards the anode where they will be oxidized or kept in solution according to their concentration and the position in the standard potential range. The oxidized form of the anions, which normally would be a gas, will diffuse away from the anode. The reducing agents of the primary field will be in excess in the environment of the anode. These agents will counteract the oxidation processes at the anode. They might also, in turn, reduce the species oxidized at the anode when these diffuse away from the anode.

Hydroxyl ions have the highest equivalent conductance of all anions and would be the main current carrier in a basic environment. Hydroxyl will probably be oxidized at the anode together with elements, for example Cl_2. The reactions at the anode can be represented as:

$$2OH^- \rightleftharpoons H_2O + \tfrac{1}{2}O_2 + 2e^-$$

In Fig.6 the patterns of redox- and self potentials as well as ion concentrations caused by the galvanic cell are diagrammatically illustrated along four selected traverses. On approaching the ore from a distant point, potential patterns will be quite regular, potentials decreasing towards the cathode and increasing towards the anode. Owing to the EMF of the cell, concentration of protons and cations more oxidizing than H^+ will tend to decrease towards the cathode (increasing pH), and increase towards the anode (decreasing pH). Similarly, concentrations of hydroxyl and anions more reducing than OH^- will tend to increase towards the cathode and decrease towards the anode.

Cations more reducing than H^+ and anions more oxidizing than OH^- will probably show a more complicated picture. At traverses outside the line D—D' (traverses 1 and 4, Fig.6) concentration of these cations will increase towards the cathode and decrease towards the anode. The anions will conversely decrease towards the cathode and increase towards the anode. At traverses between points D and D' (traverses 2 and 3, Fig.6) the concentration of these ions will be at minimum along the line D—D'.

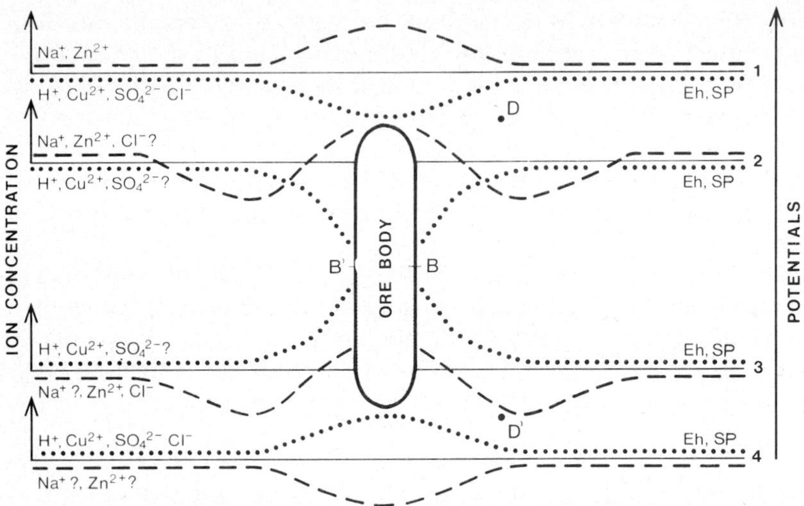

Fig.6. Estimated patterns of redox potential, self potential and concentration of ions along four traverses in the electrolyte (groundwater) around a schematic model of a vertical inert electrode (ore body) in a primary redox potential field (upper lithosphere). Traverses 1 and 4 cross the extension of the electrode, traverses 2 and 3.cross the electrode between points D and D' (see Fig.5). Question mark after symbol indicates particular uncertainty for that element.

For all ions in question the concentration at any point will be a result of the local redox potential, the current density and the necessity for electro-neutrality between cations and anions. The relative importance of the different effects cannot be estimated quantitatively at the present stage. Some of the indicated trends are therefore uncertain; the most questionable are indicated with a question mark in Fig.6.

ADAPTION OF THE ELECTROCHEMICAL MODEL TO SURFACE CONDITIONS

The processes described above presuppose homogeneous and isotropic conditions in the electrolyte, a situation which could never be fulfilled in nature. Of the many possible heterogeneities we shall here briefly discuss the nature and effect of the overburden, since this is of particular interest in geo-chemical prospecting. Fig.7 shows a model of the upper part of an ore body outcropping in the bedrock beneath glacial till. Ice movement is assumed to have taken place from right to left, and glacial debris from the ore is contained in the till. Post-glacial oxidation of this debris has occurred as well as post-glacial oxidation of the ore and country rock.

Owing to great differences in electrical resistivity (ρ) current directions will decline abruptly at the boundaries of bedrock (high ρ) and overburden (low ρ). In the overburden the resulting current density will be high in the vicinity of the outer edges of the ore, and low above the ore. The concentra-

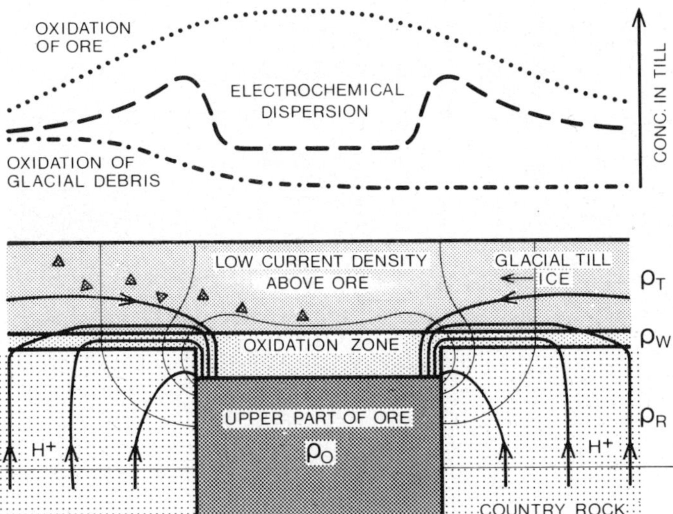

Fig.7. Schematic model of the upper part of a sulphide ore body outcropping under glacial till. Thin lines: equipotential surfaces. Heavy lines with arrows: current direction of positive electricity (H^+) in the electrolyte: ρ_R, ρ_W, ρ_T, ρ_O, : electrical resistivity in country rock, oxidized zone, glacial till and ore, respectively. Upper curves indicate paths to be expected from various dispersion mechanisms in the groundwater of the overburden sampled close to the bedrock.

tion of ions taking part in the electrochemical processes are supposed to be high where the current densities are high. Resulting concentration patterns in the electrolyte of the overburden are shown in the upper part of Fig.7.

Electrochemical dispersion will cause characteristic maxima at the edge of the ore and a minimum above the ore. For elements with an originally lower concentration in the ore than in the country rock this minimum would be more pronounced than for elements with similar concentrations in ore and country rock. For elements enriched in the ore the electrochemical dispersion pattern in the groundwater may be modified by dispersion caused by oxidation of ore minerals in situ or in the glacial debris. Clearly, the development of these distribution patterns would also be modified by groundwater movement in bedrock and overburden.

INTERPRETATION OF EMPIRICAL DATA FROM THE JOMA AREA FOLLOWING THE ELECTROCHEMICAL MODEL

The Joma deposit is situated in inner North Trøndelag, central Norway, in an area of cold climate approximately 600 m above sea level (a.s.l.). The economic mineralization consists of about 17 million tons of massive pyrite with varying amounts of chalcopyrite, sphalerite and galena, in a country rock of highly metamorphosed greenstones rich in carbonates (Fig.8). A

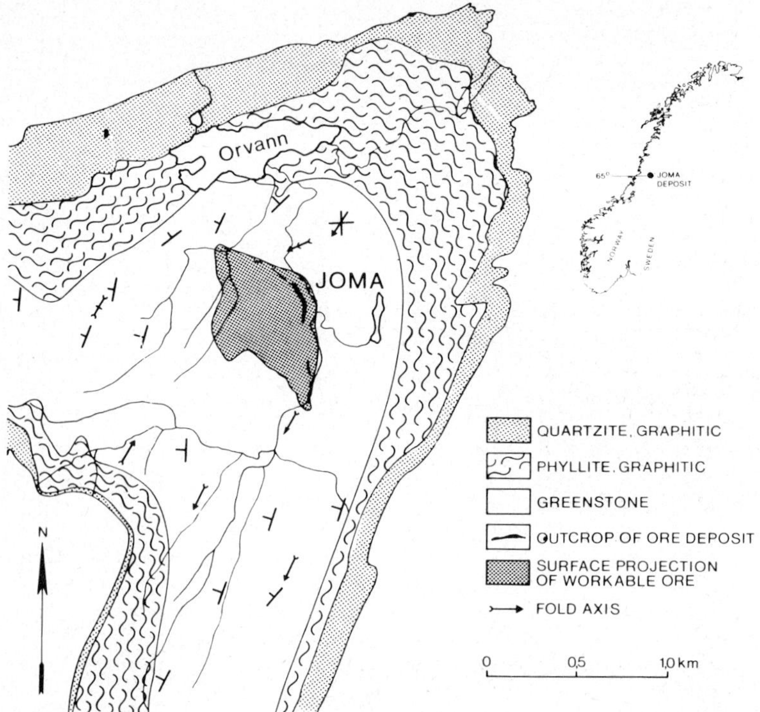

Fig.8. Geological map of the area surrounding the Joma pyrite deposit, central Norway. (After R. Kvien, personal communication, 1974.)

graphitic phyllite dipping under the greenstones is observed in the lower half of drill hole 1A and in the lowermost 2—3 m of drill hole 3V (Figs.2 and 3). The overburden, mostly 0.5—1.2 m thick above the deposit, consists of glacial till and also partly bog and weathered rock. A more detailed account of the deposit and its surroundings has been given earlier (Logn and Bölviken, 1974). Unless otherwise stated, the methods referred to in the following account have been described in an earlier paper about in-situ measurements of pH, Eh and SP in drill holes (Bölviken et al., 1973).

Drill hole data

Looking back to Fig.3 it is seen that the self potential trends from the drill holes at Joma and those outlined from the model agree quite well, taking into account the dip and complex geometry of the sulphide mineralization. Simultaneous in-situ measurements of pH, redox potentials (Eh) and self potentials (SP) have been carried out in drill hole 117, which intersects the thickest part of the ore near the outcrop. In this hole (Fig.9) pH decreases regularly with depth from 6.7 near the surface to a minimum of 6.0 in the middle of the ore at a depth of 50 m. From 50 m, pH increases with depth

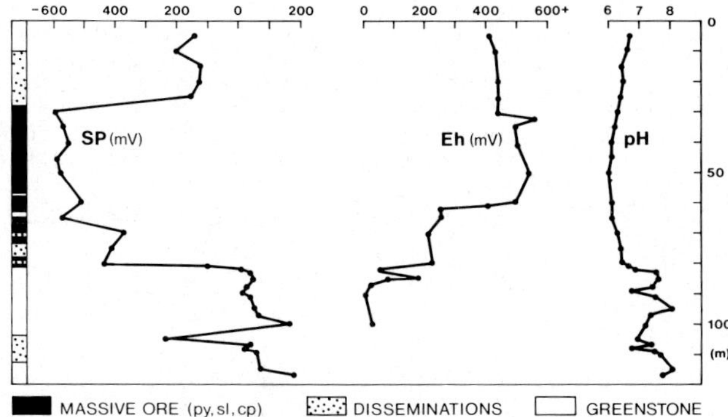

Fig.9. Self potentials (SP), redox potentials (Eh) and pH in drill hole No.117, Joma pyrite deposit, Norway. All parameters are measured in situ (Bölviken et al., 1973).

to 6.5 at the footwall boundary of the ore. In the footwall country rock the pH values are quite irregular, generally they increase with depth. Eh increases with depth in the upper part of the hole, from approximately 400 mV near the surface to a maximum of about 570 mV at a depth of 50 m where the pH is at minimum. Further down Eh decreases abruptly with depth reaching approximately 50 mV at a depth of 100 m. Local Eh maxima appear close to the ore boundaries both in the hanging- and footwall. Electronic current SP are constantly about −450 mV through the ore, while ionic current SP increase with depth from approximately −150 mV near the surface to +150 mV at a depth of 110—120 m.

The authors are not able to explain all these results at the present stage. The pH and Eh values seem to be greatly influenced by oxidation of sulphides. It is uncertain whether or not the local Eh maxima and other features of the Eh curve could be effects of a secondary redox potential field.

Judging from the results in another drill hole (Fig.1), the self potentials in hole 117 appear to be influenced by the geometry of the sulphide mineralization. The ore body dips approximately 30°, a reason why the potential pattern above its upper part must be deformed. When this is taken into account, the results obtained for ionic current self potentials appear to agree fairly well with what would be expected from the model in Fig.5.

pH values in waters from a section through the deposit and country rock are shown in Fig.10. The data are all from in situ measurements. In all the drill holes intersecting ore, pH seems to have a bimodal distribution in the country rock with an acid and a basic trend. In a drill hole located in the footwall (1A) such bimodality is less pronounced and has only been observed in the upper half of the hole. When the pH data are compared with the geological description of the drill cores (R. Kvien, personal communication, 1974) and the SP data (Fig.2), it seems quite clear that practically all the acid peaks on the pH curves coincide with occurrences of massive or disseminated sulphides. The sulphides appear to have been somewhat oxidized since the

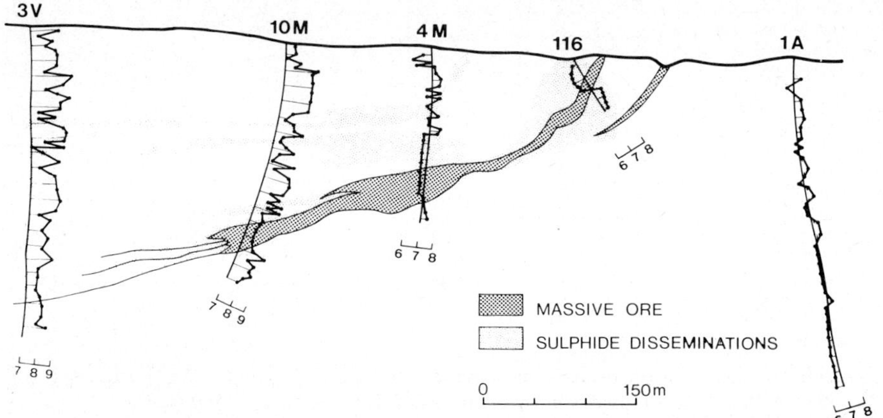

Fig.10. pH in drill-hole waters, from a vertical section through the Joma pyrite deposit, Norway. In-situ measurements with glass electrode.

pH in the massive ore decreases regularly from approximately 8.0 at depth to 6.0—6.5 near the surface.

The most basic trend probably reflects the pH in porewater within the country rock (greenstones rich in carbonates). pH values of this trend seem to decrease with depth in the deepest drill holes through the ore. Drill hole 1A in the footwall intersects greenstones in the upper half and more or less bituminous phyllites in the lower half (R. Kvien, personal communication, 1974). The change in pH pattern occurs halfway down the hole at the border between the two formations. The pH of the basic peaks in the upper half of hole 1A (greenstone) is nearly constant with depth (approximately 7.5), while the pH in the lower half (phyllites) decreases slightly with depth (from 6.9 to 6.5). It is also interesting to note that even though the low pH produced by the sulphides should be expected to lower the pH in the country rock in the vicinity of the sulphide mineralization, the pH in the country rock is higher nearer the ore body (3V, 10M, 4M, 117) than away from it (1A), although the greenstone in hole 1A does not seem to differ substantially petrographically from that in the other holes (R. Kvien, personal communication, 1974). The total picture of the distribution pattern of pH appears to be in agreement with what should be expected in a secondary redox potential field around the Joma deposit.

The authors believe that the pH patterns described above reflect conditions in the bedrock; we find it unlikely that the results are significantly influenced by the original composition of the water used during the diamond drilling because:

(1) The holes had been drilled at least 5 years before the pH measurements were made (1971).

(2) The water that had been used during the drilling must have been taken from nearby streams. These all have pH values between 6.5 and 7.5 and low

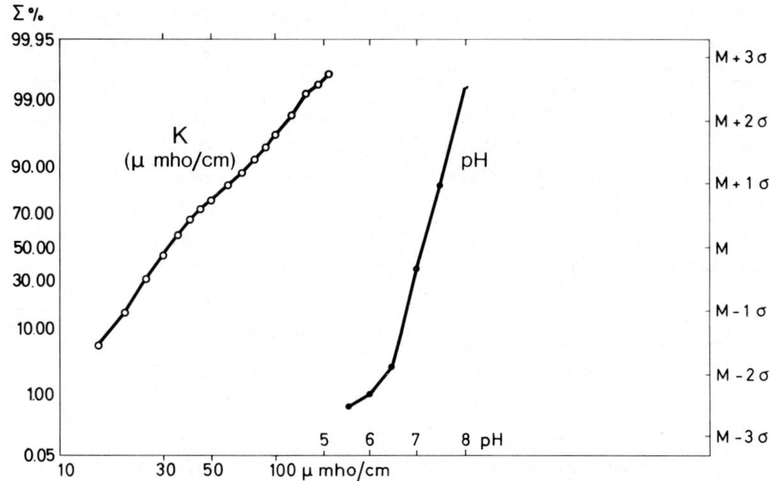

Fig.11. Frequency distributions of specific conductivity (K) and pH in 936 samples of stream waters from an area surrounding the Joma pyrite deposit, Norway.

contents of dissolved salts (Fig.11), which indicate low buffer capacity.

(3) The details indicated on the pH curves were reproducible, and results appeared to be very little influenced by stirring the drill-hole waters on moving the probe up and down (Bölviken et al., 1971).

Specific conductivity of the water in drill holes was measured (1971) in samples brought to the surface. Fig.12 shows the results from the drill holes sampled. The specific conductivity is rather uniform within the holes; in-situ measurements, like those of pH, would probably have given more diverse results. There are, however, clear differences between the holes. The conduc-

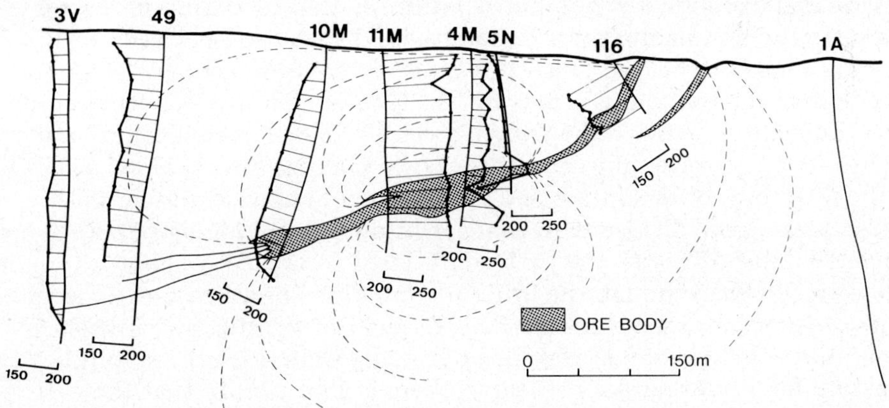

Fig.12. Specific conductivity in drill-hole waters from a vertical section through the Joma pyrite deposit, Norway. The conductivity was measured after ca. 1 hour in samples brought to the surface. Unit on curves: μmho/cm. Dashed lines indicate estimated paths of current flow in the country rock.

tivity is highest in the drill holes passing through the central part of the ore (11M, 4M), and lowest in those intersecting the ore near the surface (116) or at depth (3V). This must be due to a relatively low salt concentration in these last-mentioned holes. It cannot be an effect of pH variations since the water of drill hole 116 is acid and that of 3V basic; both these should produce better specific conductivities than the more neutral water of holes 11M and 4M, other things being equal. This pattern of specific conductivity can, however, be explained by reference to the electrochemical model. Drill holes 11M and 4M intersect the thickest part of the ore body. The ore is almost flat dipping in this part, the main direction of the secondary currents consequently being nearly horizontal and little influenced by primary currents. Due to the geometry of the ore body the current densities are probably high and rather uniform in the country rock above the central part of the ore (Fig.12). Local high current density imposed on the electrolyte in the rock would attract ions from the environment and promote local high specific conductivity.

Surface data

Samples of glacial till were taken as close as possible to the bedrock (maximum depth = auger length = 1.4 m) along a profile across the central part of the ore (north of the profiles shown in Figs.2 and 3). After drying and sieving to minus 180 micron, the lithium, sodium, copper, lead and zinc contents of these samples were determined (atomic absorption) after hot nitric acid digestion. The Cl^- (ion-selective electrode), pH (glass electrode) and specific conductivity (as introduced by Govett, 1973, 1975) were determined in water extracts of the soil. Results are shown in Fig.13.

The conductivity of soil slurries as well as copper, lead and zinc contents of the soil show similar patterns (Fig.13a): all are markedly high just above both massive and disseminated sulphides indicating that oxidation processes in this case are the dominant factor behind these distribution patterns.

pH, lithium, sodium and Cl^- (Fig.13b) show distribution patterns different from those of conductivity and heavy metals. Lithium has low concentrations just above the main ore, above the hanging wall sulphide disseminations and above the footwall massive mineralization. Local maxima are obtained at both sides of the main ore. Both sodium and Cl^- depict a somewhat similar path, but the trends are more diffuse than that for lithium. Lithium distribution seems to agree quite well with the pattern outlined from electrochemical dispersion (Fig.7). Even though the lithium content is known to be higher in the country rock than in the sulphides this should not significantly influence its content above the disseminations in the hanging wall, since the sulphide mineralization here is rather poor. Clearly, there is a possibility that the distribution pattern obtained for lithium in the till merely reflects primary distribution in the bedrock or syngenetic secondary distribution in the overburden. On the other hand, Li^+ is one of the cations with high ionic equivalent conductance which according to the electrochemical model should be one of the principal current carriers beside H^+.

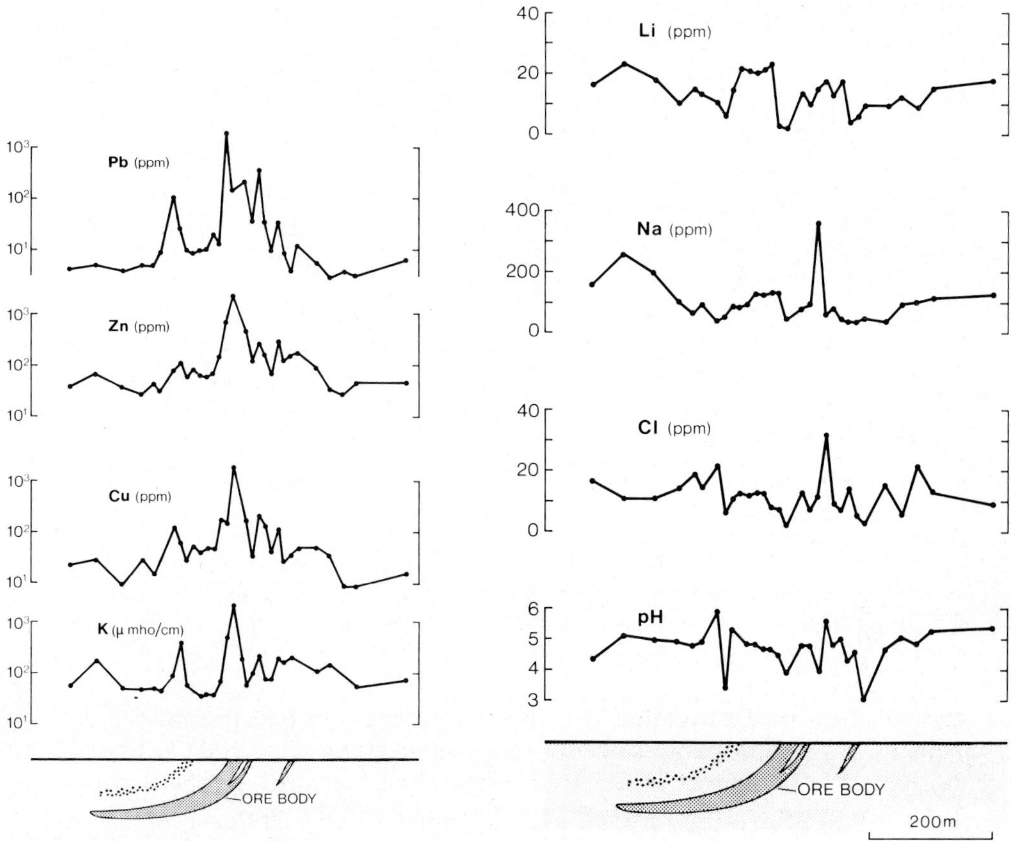

Fig.13. Specific conductivity, pH and chloride in water extracts and copper, zinc, lead, lithium and sodium in nitric acid extracts of glacial till from a profile crossing the Joma pyrite deposit, Norway.

pH is rather randomly distributed although some similarities with the lithium distribution might be recognized. Owing to oxidation of the ore indicated by the pH pattern in the ore and the heavy-metal distribution in the overburden, pH would be expected to be low above the ore which is the case only to a certain extent, indicating that one or more mechanisms counteract a drastic pH decrease near the ore. From the present data no conclusions can be drawn regarding the possibility that electrochemical processes may represent one of these mechanisms.

In an orientation survey around the Joma deposit (1964) surface waters were sampled at 200—500 m intervals along all streams in the area. Specific conductivity and pH were determined in these samples approximately 1 month after sampling. Frequency distributions of the results are shown in Fig.11, and

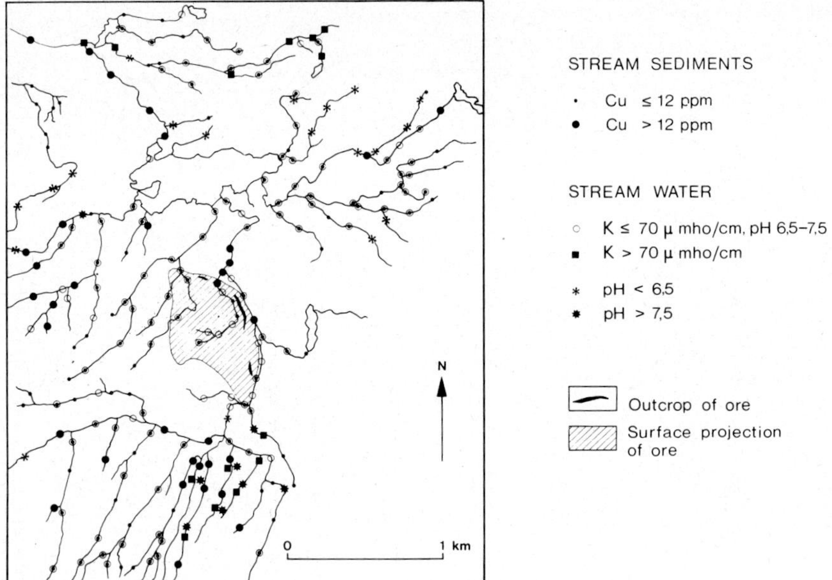

Fig.14. Anomalies (> 90 percentile) of cxCu in stream sediments and specific conductivity and pH in stream water of an area surrounding the Joma pyrite deposit, Norway.

anomalies in Fig.14 together with results for cxCu in stream sediments. A zone of low pH probably caused by phyllites of the same type as in drill hole 1A (Figs.8 and 10) was found approximately 1 km north of the deposit. The Joma ore body does not give clear pH or conductivity anomalies.

Two main conductivity anomalies including all samples with results higher than the 90 percentile were found outside the deposit. One is in a drainage system 1.5 km north of the Joma deposit, the other in several small streams at the southern extension of the main ore. The southern conductivity anomaly coincides with areas of high pH (> 8) stream waters and high metal contents in stream sediments. This stream sediment anomaly has been related to disseminated sulphides which are known to occur in this zone. Since the pH values are high, the high conductivity in these stream waters cannot be caused by oxidizing sulphides alone. Oxidizing sulphides in host rocks especially rich in carbonates could probably produce the combination of high conductivity and high pH. Coinciding high conductivity and high pH in stream water could, however, be caused by electrochemical processes and reflect high current densities due to natural galvanic cells in the drainage area. This last interpretation offers interesting possibilities for regional exploration through pH and conductivity measurements in stream waters and stream sediment slurries.

CONCLUSIONS

The electrochemical model presented in this paper is a tentative and highly simplified image of natural features and processes. In nature many factors influence the complex system of an ore deposit in its environment. More research is needed in order to prove whether or not electrochemical ion transport might really be one of the dominating factors contributing to the element distribution around an ore deposit. The natural current densities at Joma were estimated to be of the order of 5 mA/m^2 within the ore body and 0.1 μA/m^2 in the country rock (Logn and Bölviken, 1974). The authors consider that natural current densities of this magnitude may probably have some geochemical consequences beyond those of interest solely to pure academic discussions. Electrochemical and other mechanisms might in some instances interact. This seems to be the case when considering the effects on specific conductivity of natural interstitial waters. Since glacial overburden could be considered chemically as a more or less neutral blanket positioned over the bedrock ready to receive any pattern imposed upon it, the search for electrochemical dispersion at shallow depths in glacial till appears to be a favourable alternative to deep overburden sampling. Interesting too is the possibility that electrochemical mechanisms might influence the mineralogical compositions of ore and country rock, as earlier pointed out by Govett (1973), Govett and Whitehead (1974), and others.

REFERENCES

Baas Becking, L.G.M., Kaplan, J.R. and Moore, D., 1960. Limits of the natural environment in terms of pH and oxidation—reduction potentials. J. Geol., 68: 243—284
Bitterlich, W. and Wöbking, H., 1972. Geoelektronik. Springer-Verlag, Vienna, 349 pp.
Becker, A. and Telford, W.M., 1965. Spontanous polarization studies. Geophys. Prosp., 13: 173—188
Bölviken, B., in preparation. The redox potential field of the earth.
Bölviken, B., Logn, Ö., Breen, A. and Uddu, O., 1973. Instrument for in situ measurements of Eh, pH, and self potentials in diamond drill holes. In: M.J. Jones (Editor), Geochemical Exploration 1972. Institution of Mining and Metallurgy, London, pp.415—420
Govett, G.J.S., 1973. Differential secondary dispersion in transported soils and post-mineralization rocks: an electrochemical interpretation. In: M.J. Jones (Editor), Geochemical Exploration 1972. Institution of Mining and Metallurgy, London, pp. 81—91
Govett, G.J.S., 1975. Soil conductivities: assessment of an electrochemical exploration technique. In: I.L. Elliott and W.K. Fletcher (Editors), Geochemical Exploration 1974. Elsevier, Amsterdam, pp.101—118
Govett, G.J.S. and Whitehead, R.E.S., 1974. Origin of metal zoning in stratiform sulphides: a hypothesis. Econ. Geol. 69: 551—556
Habashi, F., 1966. The mechanism of oxidation of sulphide ores in nature. Econ. Geol., 61: 587—591
Henriet, J.P., 1971. Large self-potential anomalies in the North Ri basin (Ardennes, Belgium). Bull. Soc. Belge. Geol. Paléont. Hydrol., 80: 85—98
Logn, Ö. and Bölviken, B., 1974. Self potentials at the Joma pyrite deposit, Norway. Geoexploration, 12: 11—28

Malmqvist, D. and Parasnis, D.S., 1972. Aitik: geophysical documentation of a third-generation copper deposit in North Sweden. Geoexploration, 10: 149—200
Parasnis, D.S., 1966. Mining Geophysics. Elsevier, Amsterdam, 356 pp.
Parasnis, D.S., 1967. Principles of Applied Geophysics. Methuen, London, 176 pp.
Parasnis, D.S., 1970. Some recent geoelectrical measurements in the Swedish sulphide ore fields illustrating scope and limitations of the methods concerned. Min. Groundwater Geophys. 1967 (Ottawa), pp.290—301
Sato, M. and Mooney, H.M., 1960. The electrochemical mechanism of sulphide self potentials. Geophysics, 25: 226—249
Society of Exploration Geophysicists, 1966, 1967. Mining Geophysics, Vol.I, 492 pp.; Vol.II, 708 pp.

THE USE OF ANALYTICAL STANDARDS TO CONTROL ASSAYING PROJECTS

WALTER E. HILL, Jr.

Hazen Research, Inc., Golden, Colo. (U.S.A.)

ABSTRACT

Analytical standards or reference materials for the exploration and mining industry are finely ground, well homogenized, exhaustively analyzed rock, tailings, ore or concentrates used to provide a datum or data to collate the analytical results on previously unanalyzed exploration or mining samples. They are useful in checking analytical precision and accuracy, in developing or comparing analytical methods, in training technicians, and as a tool in monitoring and verifying the results of exploration and property evaluation projects. In the course of the last two and a half years AMAX Exploration has experienced no major assaying problems where standards were available and used. Manpower and monetary savings have been significant.

INTRODUCTION

In 1970 the few standards available for rock and ore analysis from government, professional organizations and commercial sources were not oriented toward geochemical exploration or ore-body evaluation. Nor were they available in quantity at reasonable prices. Since then the A.E.G.—U.S.G.S. and Canadian Mines Branch standards have become available and are a partial answer to the problem.

The AMAX standards program was conceived in 1970 as a proprietary answer to the availability problem. Where public and organizational standards programs have to be all things to all people, the AMAX program was tailored to the specific requirements of AMAX's mining operations and exploration.

As a proprietary program it was conceived and set up as a continuing production and certification unit to provide new reference materials for both exploration and operations each year. To date, in three years of operation over 175 different standards have been produced. For quality control in assaying, these materials are used to verify the precision and accuracy of analytical work of any laboratory and to help choose the best laboratory when several are available.

TYPES OF ERRORS

One use of standards is to recognize, isolate, and rectify analytical errors. These fall into two major categories: sporadic and systematic.

652

Sporadic errors occur now and then as aberrant individuals or small groups and may not reappear in hundreds or thousands of assays. They are usually due to forgotten factors, calculation errors, decimal errors, or failure to add a necessary reagent to one or two samples. Although difficult to identify, once found, the individual analyses can be corrected with relative ease.

Systematic errors result from a continuing problem in the analytical laboratory that biases, either high or low, a large contiguous suite of results, the extent of the error may be a single day's output from one analyst or the entire laboratory output for weeks or months. These are the serious errors that can affect ore calculations and decisions on a project's viability. Furthermore, once questionable assays have tainted an evaluation project or property, the stigma is there in management eyes for a long time. It is psychologically important to have the analyses right the first time and to be able to prove it. Consequently, it is important to identify the problem as soon as possible and rectify the affected analyses after eliminating the source of the problem. This usually involves extremely interesting investigation based on a knowledge of chemistry and use of common sense. Once located, the awareness generated will usually eliminate recurrence of the problem for a period of time.

AMAX'S STANDARDS PROGRAM

AMAX Exploration's initial development of proprietary standards was prompted by a large group of poor assays that jeopardized the confidence of a group of several thousand assays. Spot checks on the pulps at a second laboratory indicated large discrepancies. AMAX prepared 1-kg quantities of standard material from 16 different drilling rejects, covering the range of values anticipated. These materials were well homogenized and exhaustively analyzed. Significant portions of the questionable assays were reassayed again but this time with intercalated standards. The problem was quickly resolved but at a cost almost double for assaying, plus inestimable losses in manpower and delays on the project.

The sixteen standards were then successfully used for routine assay control on similar evaluations and drilling projects. Poor assays were immediately identified and rerun at no additional cost to AMAX.

However, in less than one year supplies of these standards began to dwindle and a replacement program was undertaken. A wide range of ores, rocks, tailings, and concentrates were available. The size of each standard was increased from 1 kg to approximately 14 kg, this being the ore weight needed to fit the optimum operating volume of the largest plexiglas twin-shell Patterson-Kelly blender made at that time. Crushing and screening equipment were upgraded to match blending capability.

Standard materials

Wherever possible, the materials used for standards production are either natural rock, minerals, ore or mill products. This gives an authentic mineral assemblage and matrix signature so that standards are virtually indistinguishable from routine samples.

Carefully selected and composited drill core rejects are excellent materials for standards preparation. These materials are usually available shortly after drilling is initiated and are characteristic of the ore intersected. Standards, matrix matched in this manner to the ore from the first year or two of drill samples, prove valuable in later evaluation and mine development studies for that property.

Production of standards

Selected material is prepared by grinding and screening until 100% passes a 200-mesh screen. Coarse plus 200 reject is recrushed and screened. Screening assures removal of tramp iron or alumina from the grinding cycle and minimizes subsampling variance. When sufficient fine material has been prepared it is blended in a Patterson-Kelly twin-shell blender for 4 hours.

The bulk homogenized material is then sampled for homogeneity. If subsampling, on ten or more samples shows satisfactory reproducibility, we proceed with development of the analytical program and determination of certification statistics.

Certification

The certification program is assembled from subroutines to give: (1) the mean, (2) standard deviation, (3) F test, (4) Student's t test, and (5) a regression analysis as the estimate of true metal content.

All results received from all laboratories are plotted. However, to prevent obvious errors such as incorrect factors, calculation or keypunch errors, from adversely affecting the statistics, gross errors are rejected for certification purposes.

The first or tentative value is obtained on 30 sets of analytical results aggregated from five or more laboratories over a period of time. This is used as a guide until 60 results are assembled and statistically processed to give an accepted value. The accepted value rarely differs by ± 0.01% from the tentative value. As a further check, a third statistical run was made on the first group of standards to reach 120 determinations each. The accepted values from 120 sets of analyses differed less than ± 0.003% from the accepted values with 60 analyses. This seems to be the point of diminishing returns in our certification program.

Applications

It is imperative to have assays right the first time or at least to know when they are within tolerance limits. This can be accomplished with standards routinely submitted for analysis with unknowns. The standards must be included in sufficient density to appear in all groups digested for analysis, and they should cover a range of values from high to low, comparable to the unknowns, for the element or elements sought.

With analyses from each digestion group and from each analyst we have an excellent datum to check analytical precision and accuracy and can either verify or reject each day's analytical results in whole or in part. If verified, results can be used and plotted for ore reserves and mine planning on the day received. If rejected, no engineering time is wasted and plotting waits until corrected and verified results are received.

In evaluating a laboratory's analyses, precision is initially more important than accuracy. Precise data can easily be corrected to provide accurate results. Lack of precision usually indicates a large number of interacting independent variables and a laboratory that cannot provide the required quality.

Logistics of shipping and handling samples often unduly influences the choice of laboratory. Another factor used in selection of a laboratory is turnaround time. How many projects have been assigned because, "I can get the answers back in three days to a week?" Turnaround time may be fast because no one else is using that laboratory, but conversely slow turnaround is no guarantee of good results. Cost of analysis is another selection criteria but again is a very poor yardstick of precision and accuracy. Even the appearance of the laboratory can be very misleading. Clutter and apparently poor organization can sometimes turn out excellent assays. Historically geologists and engineers have used the wrong criteria in choosing a laboratory for a project, primarily because there was no datum to obtain a rapid assessment of analytical quality.

The proprietary standards program developed by AMAX has provided us with the discriminant tools for choosing the best laboratory to do any particular job and the means to continuously monitor precision and accuracy until it is completed. The evaluation is made by sending a small suite of standards to each laboratory being considered for the project and requesting analyses for the element or elements of interest. When results are received they are plotted to show the deviation of reported analyses from accepted values (Fig.1).

Fig.1 exhibits reasonably good precision and accuracy and indicates that this laboratory could be used for assaying if it has the capacity available to handle the work. Results plotted in Fig.2 exhibit reasonably good precision but are about 10% low. By improving accuracy the laboratory concerned would also be satisfactory. Results in Fig.3 were obtained from a laboratory checked in two successive years with the same results. No precision and consequently no accuracy has been demonstrated and the laboratory would be a poor choice. AMAX Exploration routinely evaluates commercial labs before assigning large quantities of analytical work. The most recent large evaluation involving AMAX's standards program was The Banner acquisition.

The Banner—Twin Buttes evaluation in the fall of 1972 involved drilling of four deep core holes and 24 percussion drill holes in the northeast ore body.

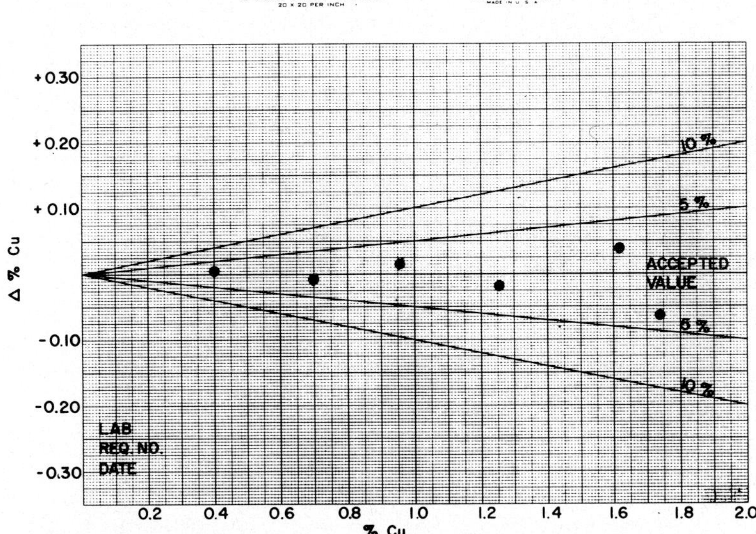

Fig.1. Results showing reasonable precision and accuracy.

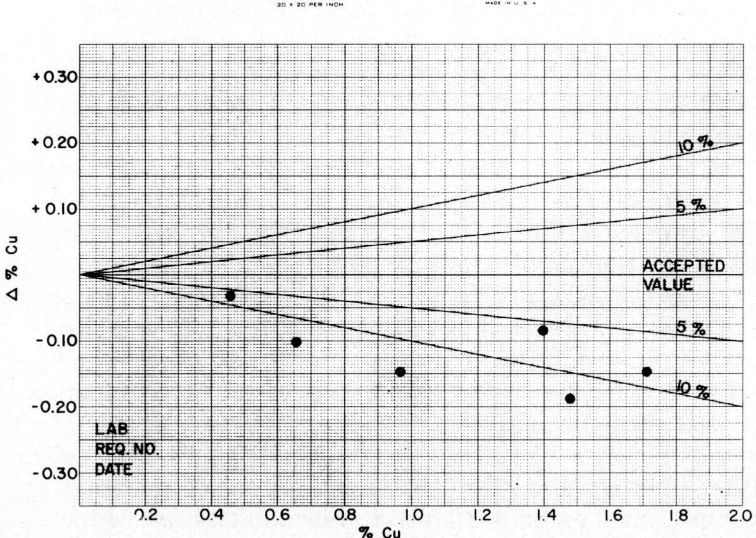

Fig.2. Results about 10% low but precision reasonable.

140 Becker drill holes were used to sample one of the major stockpiles to give 5124 assay samples. An accurate commercial analytical laboratory was chosen and samples delivered daily from the drills. To provide quality control, 750 pulp sacks of 32 different Cu—Mo standards (for a total weight of approximately 22 kg) were serially numbered and given to the laboratory for insertion at the rate of one standard for each ten unknowns. When results

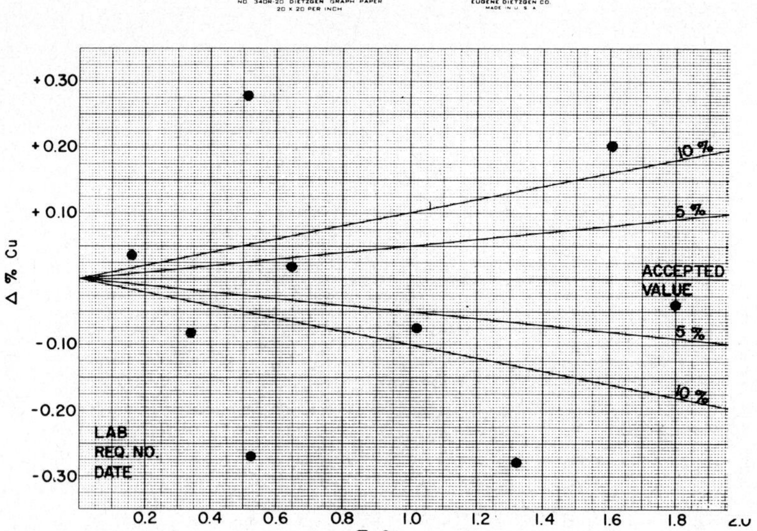

Fig.3. Results showing very poor precision and accuracy.

were received by AMAX the standards were uncoded and values plotted against accepted values. Assay results were either accepted or rejected on the basis of assays for the standards. Out-of-tolerance assays were called to the laboratory's attention immediately and corrected before a large mass of erroneous data had been generated.

During the 14-week assay evaluation AMAX was able to verify over 12,000 copper assays and 4000 molybdenum assays, with confidence limits of 98% and 95% respectively, as being within a ± 5% tolerance limit of the correct values. As an additional analytical check and to check sample preparation, when the pulps were retrieved 10% were randomly chosen for assaying at a different lab. Standards were also included with these samples. One group of check samples was reported over 10% lower than the original assays. The included standards indicated that the second laboratory was outside tolerance limits that day and, after spot-checking, the original results were used.

Cross-check assays also provide excellent quality information on the check laboratory in the event that the laboratory used for the primary assaying develops problems. Work can then be shifted to the second laboratory to provide continuity with demonstrated good quality.

The cost of the standards and replicate quality assurance program on the Twin Buttes evaluation was 21% higher than a single assay run on all samples. All assays were verified within a week of completion of the assaying. One group of molybdenum assays aberrant in the original results was resolved as a calculation problem on a new calculator purchased by the laboratory at the start of AMAX's assaying program.

The proven quality of the assays was contributory to successful acquisition of the Banner Mining Company and properties by AMAX several months later. The small amount of money spent for assay quality control was minimal compared to the 220 million dollars allocated to the purchase and associated new facility and work commitments connected with the acquisition.

CONCLUSIONS

The AMAX Exploration laboratories have been quality controlled for over two and a half years. Commercial laboratories are selected for projects on the basis of precision and accuracy on analyses of proprietary standards. The datum provided by routine submission of standards with assay samples permits quality to be judged and analyses verified before they leave in-house laboratories or as soon as results are received from commercial labs.

AMAX's reassay costs have dropped to virtually nothing. There have been no major reassay problems on any exploration or evaluation projects during the two and one-half years where standards were available and were used. AMAX's manpower savings in the laboratory, geological, and engineering staffs are difficult to calculate — but are significant.

ACKNOWLEDGEMENTS

I would like to thank Dr. P.D. Parker, Chief Geologist, and Dr. R.F. Horsnail, Chief Geochemist, of AMAX Exploration for their help and encouragement and for permission to publish the AMAX data used in the manuscript.

THE HOMOGENEITY OF SIX GEOCHEMICAL EXPLORATION REFERENCE SAMPLES

GLENN H. ALLCOTT and HUBERT W. LAKIN

U.S. Geological Survey, Denver, Colo. (U.S.A.)

ABSTRACT

U.S. Geological Survey analysts have made more than 46,000 determinations on six geochemical exploration reference samples for the purpose of furnishing data on the homogeneity of the samples. The samples are considered to be sufficiently uniform to be useful as geochemical exploration reference samples despite significant variance ($\alpha = 0.05$) in the data for some elements, as revealed by a hierarchical analysis of variance. Statistical parameters are presented, based on analyses of 50 randomly chosen samples, where operator, technique, and instrumental variables were controlled.

Non-Survey laboratories have also furnished data on various elements from the same six geochemical reference samples. A summary of these results is available from the authors upon request.

INTRODUCTION

A number of geological and geochemical variables were considered as criteria for selection of the samples. These variables include: (1) the type of sample media, (2) the variety of elemental constituents, (3) the range of concentration of the elemental constituents, (4) the variety of oxidation states, (5) the variety of types of chemical compounds, (6) the types of matrix (Si-rich, Fe-rich, etc.), (7) the types of mineral deposits, and (8) the types of geologic, geochemical, and climatic environment. The list is not exhaustive but based on such criteria and on a number of practical considerations, such as availability of personnel and knowledge of potential collection sites, three soil and three rock samples have been selected, prepared for analysis, and distributed to 176 laboratories.

The Committee on Geochemical Analysis envisioned two immediate goals with respect to the geochemical exploration reference samples. One of these goals was to establish the degree of homogeneity of the samples. This initial report deals with the overall variance as established by our analyses. A second goal was to establish interlaboratory variance based on various methods of analysis and sample treatment. To achieve this second goal, over 100 laboratories have cooperated by reporting their analytical results on randomly selected subsamples of the six geochemical exploration reference samples. Statistical analysis of this data is incomplete at this time and will be the subject of another report.

DESCRIPTION OF THE SAMPLES

Mineralogical composition

To determine their crystalline mineral composition, the milled samples were examined microscopically and by X-ray diffraction by Keith Robinson (written communication, 1974). The results are summarized in Table I, and include only those substances which could be identified in the finely ground samples.

TABLE I

Crystalline minerals identified in the milled reference samples

Sample	Major minerals	Minor minerals and contaminants
1	quartz, goethite	corundum, iron
2	quartz, feldspar	corundum, zircon, pyrite, rutile, tourmaline(?) muscovite, hydrobiotite-vermiculite, clay minerals, carbonaceous material
3	quartz, fluorite, calcite	corundum, barite, huebnerite(?)
4	quartz, alkali and plagioclase feldspar	corundum, sulfide minerals, andradite, muscovite-illite, biotite
5	quartz	corundum, chlorite, alkali feldspar
6	quartz	corundum, alkali and Na-rich plagioclase feldspar

Contamination

The reference samples contain contaminants and are not representative of the original composition of the soil or rock. The known contaminants were introduced during the sample preparation step. The rock samples contain iron alloy derived from the crusher. All samples contain variable amounts of aluminum oxide ceramic derived from the balls and liner of the grinding mill. The composition of the ceramic is about 85% aluminum oxide, 7—12% silicon dioxide, a few percent of alkaline earth oxides, and a trace of iron. A third contaminant, boron, was detected during the semiquantitative spectrographic analysis of sieve fractions. The analyses of the minus 150-mesh, plus 325-mesh fractions revealed high boron relative to the plus 150-mesh and the minus 325-mesh fractions. This contaminant was probably derived from a freight car of minus 150-mesh boron carbide being unloaded in the area during the grinding process.

Geochemical exploration reference sample one (GXR1)

The sample was collected from the Drum Mountains, Juab County, Utah, by J.H. McCarthy, Jr. The Drum Mountains are a typical faulted range of the

Basin and Range province of the southwestern United States. Gold, copper, and manganese deposits have been mined in the area. This sample is a composite of three samples from two outcrops of jasperoid "reefs" in Cambrian limestone. The jasperoid "reefs" range from crystalline to cryptocrystalline in internal structure and from gray to reddish brown in color.

Geochemical exploration reference sample two (GXR2)

The sample was collected from the Park City mining district, Summit County, Utah, by J.H. McCarthy, Jr. The Park City mining district on the eastern slope of the Wasatch Range has produced lead, zinc, silver, and copper from fissure veins and from replacements in limestone. The sample is a composite of residual, gray-brown loams (Munsell color chart 5/2 of hue 10YR) from four sites along a line approximately 0.8 km long. The shallow soil varies from 18 to 30 cm in depth and overlies a thick-bedded Weber Quartzite of Pennsylvanian age.

Geochemical exploration reference sample three (GXR3)

The sample was collected from a hot spring deposit in Humboldt County, Nevada, by S.P. Marsh. Tungsten was produced from these deposits during the 1940's. It is a composite of red-brown to black, earthy, Fe—Mn-rich material, cementing and replacing coarse alluvium on a bedrock surface of intensely deformed phyllitic shale of the Preble Formation of Cambrian age. Calcareous tufa commonly caps the mineralized material.

Geochemical exploration reference sample four (GXR4)

The sample was furnished by Kennecott Copper Corporation from their mine in Utah. It is a mill heads sample of unoxidized porphyry copper ore composed primarily of quartz and feldspar with minor amounts of andradite garnet, biotite, muscovite-illite, and sulfide minerals.

Geochemical exploration reference sample five (GXR5)

The sample was collected from Somerset County, Maine, an area subjected to continental glaciation during the Pleistocene, by F.C. Canney. It is a composite from three sample sites of the B zone of moderately well-developed podzol soils formed on a thin ($\leqslant 1$ m) basal glacial till. The bedrock underlying the collection sites is: (1) a highly silicified and sericitized quartz monzonite containing abundant disseminated chalcopyrite, molybdenite, and pyrite; (2) a mineralized norite containing abundant pyrrhotite, and minor chalcopyrite and pentlandite; and (3) peridotite and altered norite containing abundant pyrrhotite, chalcopyrite, and pentlandite.

Geochemical exploration reference sample six (GXR6)

The sample was collected from Davidson County, North Carolina, an area once active in gold and base-metal mining, by Henry Bell and A.A. Stromquist. It is a composite of three samples, obtained at depths of 15—45 cm, of resid-

662

ual, B-zone, yellowish-red (Munsell color chart 5/8 of hue 5YR) soil. Two samples are from within the Silver Hill—Gold Hill fault shear zone, where the soils are derived from sericitized mudstone and phyllite that are slightly anomalous in lead, zinc, silver, copper, barium, and molybdenum. The third sample of the composite is from a cross shear, underlain in part by rhyolitic rock and in part by andesitic basalt, which here are slightly anomalous in arsenic and gold.

PREPARATION OF THE SAMPLES

The first step in the preparation of the samples to be used for analysis was to reduce the particle size of the six approximately 450-kg bulk samples. It was desirable to reduce the sample to 100% minus 200-mesh, because it was shown by Smith et al. (1929) that a uniform sample that has been ground to pass a 200-mesh sieve cannot be made heterogeneous by jarring or by vibrational storage conditions, regardless of the difference in the density of the components. The bulk rock samples were crushed to pass a 5-mesh sieve. The bulk soil samples were passed through a 10-mesh sieve, and the plus 10-mesh fraction discarded. Each of the six bulk samples was then ground for 12 hours in an aluminum oxide ceramic-lined mill with aluminum oxide ceramic balls. During this period the sample was heated by friction to temperatures between 80° and 90°C, which may have caused a partial loss of some volatile elements. After opening, the mill was run an additional 1.5 hours to dump the sample into a receiving hopper. A wet mechanical analysis of each of the milled samples demonstrated that over 97% of every sample had been ground to pass a 44-μ (325-mesh) opening.

The second step in the sample preparation was the partitioning of the bulk sample into subsamples. Drums of about 45-kg capacity were filled from the receiving hopper, and the order in which the 80-g bottles were filled from the drum was recorded. Each of the bottles in an ordered set of bottles was assigned a random number using a permutation of numbers from one to the total number of bottles for each milled sample.

The overall variance for each geochemical exploration reference sample was obtained from analyses of a random subsample. The bottles from each reference sample were assigned random numbers. The first fifty bottles were used for analysis and the results of these analyses are presented in this report.

The subsamples for the nested design were obtained by randomly sampling within ordered partitions of the bulk sample. The bottles in their original order from each drum were arbitrarily divided into six groups, which in effect divided each drum into six layers of bottles. Two layers of bottles from each drum were selected randomly, and the two bottles were selected randomly from those two layers, resulting in the selection of four bottles from each drum. Duplicate analyses made in random order from each bottle establish a four-level nested analysis of variance model that will estimate variance components between drums, between layers within drums, between bottles within layers, and between determinations within bottles. The results of the analyses for this statistical model are incomplete at this time.

ANALYTICAL PROCEDURES

The procedure used for a six-step semiquantitative emission spectrographic analysis for 30 elements is described by Grimes and Marranzino (1968), which is a modification of a method by Myers et al. (1961). The elements thus determined are Fe, Mg, Ca, Ti, Mn, Ag, As, Au, B, Ba, Be, Bi, Cd, Co, Cr, Cu, La, Mo, Nb, Ni, Pb, Sb, Sc, Sn, Sr, V, W, Y, Zn, and Zr.

Atomic absorption spectrometric methods were used for gold (Thompson et al., 1968), cadmium (Nakagawa and Harms, 1968), tellurium (Nakagawa and Thompson, 1968), indium and thallium (Hubert and Lakin, 1973), mercury (Vaughn and McCarthy, 1964), and silver, cobalt, copper, nickel, lead, and zinc (Ward et al., 1969).

Colorimetric methods were used for molybdenum (Ward, 1951a), antimony (Ward and Lakin, 1954), tungsten (Ward, 1951b), cold-extractable copper (Canney and Hawkins, 1958) and cold-extractable heavy metals (Bloom, 1955).

Arsenic was determined by a confined-spot technique (Almond, 1953), and uranium by paper chromatography (Thompson and Lakin, 1957). The arsenic and uranium methods and the colorimetric methods were compiled by Ward et al. (1963).

RESULTS

A basic statistical summary of the results of analyses on each geochemical exploration reference sample is tabulated in Tables II—VII. Two sets of analyses are summarized in each table for each reference sample. One set of data is a summary of a group of 50 subsamples randomly selected from the entire set of available subsamples (bottles). The second set is a summary of data from four subsamples, randomly selected from each drum, analyzed in duplicate, and summed across the entire reference sample. This data is included to compare the overall variance for random samples and the set of nested samples selected randomly from each drum. In each table, the range, arithmetic mean, median, mode, standard deviation, confidence interval about the arithmetic mean at the 95% level, and the coefficient of variation (%) are presented by element and method of analysis.

Evaluation of the coefficients of variation for the data shows that the overall variation is within the limits expected for the methods of analysis. It can be shown that the semiquantitative spectrographic technique, as originally developed by Myers et al. (1961), has a general precision such that the expected coefficient of variation is about 50% when reading 1/3rd-order intervals. In the authors' experience with the 1/6th-order technique used in this study, the expected coefficient of variation is slightly greater than 30% for standard samples analyzed over a period of months.

The atomic absorption method for tellurium, developed by Nakagawa and Thompson (1968), was applied by those investigators to samples of conglomerate, limestone, phonolite, and breccia. Coefficients of variation calculated

TABLE II

Statistical summary for geochemical exploration — Reference sample one: jasperoid from
Drum Mountains, Juab County, Utah

GXR1 (50 random samples)

Element	Minimum	Maximum	Arithmetic mean	Median	Mode	Standard deviation	Confidence interval 95%	Coeff. of variation (%)
Semi-quantitative spectrographic method								
Fe	G20	G20	G20	G20	G20	—	—	—
Mg	0.15	0.7	0.24	0.20	0.20	0.12	0.35	50
Ca	0.30	1.0	0.68	0.70	0.70	0.15	0.044	22
Ti	0.02	0.05	0.035	0.03	0.03	0.0095	0.0027	27
Mn	300	1000	630	700	700	120	35	19
Ag	15	50	29	30	30	9.3	2.7	32
As	500	1000	670	700	700	130	38	19
Au	L10	L10	L10	L10	L10	—	—	—
B	L10	L10	L10	L10	L10	—	—	—
Ba	500	2000	1300	1500	1500	390	110	30
Be	1	2	1.5	1.5	1.5	0.22	0.064	15
Bi	500	1000	540	500	500	97	28	18
Cd	L20	L20	L20	L20	L20	—	—	—
Co	5	10	7.0	7.0	7.0	0.83	0.24	12
Cr	L20	L20	L20	L20	L20	—	—	—
Cu	1000	1500	1500	1500	1500	99	29	6.6
La	L20	L20	L20	L20	L20	—	—	—
Mo	5	15	10	10	10	2.2	0.64	22
Nb	L10	L10	L10	L10	L10	—	—	—
Ni	30	100	63	70	70	13	3.8	21
Pb	300	700	420	500	500	110	32	26
Sb	150	200	190	200	200	18	5.2	9.5
Sc	L5	L5	L5	L5	L5	—	—	—
Sn	30	70	57	50	50	10	2.9	18
Sr	300	500	420	500	500	100	29	24
V	70	100	96	100	100	9.8	2.8	10
W	150	300	200	200	200	16	4.6	8.0
Y	30	70	52	50	50	9.6	2.8	18
Zn	700	1500	1200	1500	1500	330	96	28
Zr	20	50	32	30	30	8.4	2.4	26
Atomic absorption and colorimetric methods								
Ag	25	31	29	29	29	1.3	0.38	4.5
As	200	300	290	300	300	24	7.0	8.3
Au	2.4	3.2	2.8	2.8	2.8	0.14	0.041	5.0
Cd	1.8	2.3	2.1	2.1	2.1	0.12	0.035	5.7
Co	—	—	—	—	—	—	—	—
Cu	900	1400	1100	1000	1000	100	29	9.1
Hg	3.2	4.7	4.0	4.1	4.1	0.28	0.081	7.0
In	0.60	1.3	0.86	0.90	1.0	0.19	0.055	22
Mo	6.0	12	7.7	8.0	8.0	1.6	0.46	21
Ni	—	—	—	—	—	—	—	—
Pb	660	860	750	750	710	46	13	6.1
Sb	75	200	110	100	100	21	6.1	19
Te	—	—	—	—	—	—	—	—
Tl	L0.02	L0.02	L0.02	L0.02	L0.02	—	—	—
U	L4	L4	L4	L4	L4	—	—	—
W	L20	L20	L20	L20	L20	—	—	—
Zn	530	810	670	670	630	58	17	8.7

All values are in g/ton except Fe, Ca, Mg, and Ti, which are in percent. G = greater than;
L = less than; — = no data.
Analysts: G.L. Crenshaw, M.S. Erickson, W.H. Ficklin, R.T. Hopkins, Jr., A.E. Hubert,
J.M. Motooka, J.H. Turner, and R.L. Turner.

GXR1 (56 nested samples)

Minimum	Maximum	Arithmetic mean	Median	Mode	Standard deviation	Confidence interval 95%	Coeff. of variation (%)	Element
Semi-quantitative spectrographic method								
G20	G20	G20	G20	G20	—	—	—	Fe
0.15	0.70	0.24	0.20	0.20	0.10	0.019	42	Mg
0.50	1.5	0.93	1.0	1.0	0.20	0.038	22	Ca
0.03	0.05	0.034	0.03	0.03	0.0081	0.0015	24	Ti
500	1000	770	700	700	140	27	18	Mn
15	70	32	30	30	11	2.1	34	Ag
500	1000	700	700	700	110	21	16	As
L10	L10	L10	L10	L10	—	—	—	Au
L10	L10	L10	L10	L10	—	—	—	B
700	2000	1200	1000	1000	300	57	25	Ba
1.0	1.5	1.4	1.5	1.5	0.20	0.038	14	Be
200	1000	610	500	500	240	46	39	Bi
L10	L10	L10	L10	L10	—	—	—	Cd
5	10	7.9	7	7	1.5	0.29	19	Co
L10	30	L20	20	20	—	—	—	Cr
1000	1500	1500	1500	1500	110	21	7.3	Cu
L20	L20	L20	L20	L20	—	—	—	La
3	15	10	10	10	2.8	0.53	28	Mo
L10	L10	L10	L10	L10	—	—	—	Nb
30	70	62	70	70	10	1.9	16	Ni
200	700	480	500	500	120	23	25	Pb
150	200	200	200	200	6.0	1.1	3.0	Sb
L5	L5	L5	L5	L5	—	—	—	Sc
30	70	53	50	50	9.0	1.7	17	Sn
300	700	440	500	500	96	18	22	Sr
70	100	96	100	100	11	2.1	11	V
150	300	200	200	200	14	2.7	7.0	W
20	70	48	50	50	8.8	1.7	18	Y
700	1500	1200	1000	1000	300	57	25	Zn
20	50	30	30	30	5.8	1.1	19	Zr
Atomic absorption and colorimetric methods								
22	31	28	29	29	1.7	0.32	6.1	Ag
200	300	300	300	300	16	3.0	5.3	As
3.1	4.8	3.8	3.7	—	0.47	0.089	12	Au
1.6	2.6	2.3	2.3	2.3	0.17	0.032	7.4	Cd
10	22	15	14	14	2.7	0.51	18	Co
700	1100	970	1000	1000	81	15	8.4	Cu
—	—	—	—	—	—	—	—	Hg
—	—	—	—	—	—	—	—	In
6	12	8.3	8.0	8.0	2.1	0.40	25	Mo
28	44	35	34	38	3.7	0.70	11	Ni
620	880	780	800	800	52	9.9	6.7	Pb
80	170	120	120	130	16	3.0	13	Sb
0.5	5.0	2.7	3.0	3.0	1.4	0.27	52	Te
L0.02	L0.02	L0.02	L0.02	L0.02	—	—	—	Tl
L4	L4	L4	L4	L4	—	—	—	U
L20	L20	L20	L20	L20	—	—	—	W
430	710	580	590	630	60	11	10	Zn

TABLE III

Statistical summary for geochemical exploration — Reference sample two: soil from Park City, Summit County, Utah

GXR2 (49 random samples)

Element	Minimum	Maximum	Arithmetic mean	Median	Mode	Standard deviation	Confidence interval 95%	Coeff. of variation
Semi-quantitative spectrographic method								
Fe	1.5	3.0	2.0	2.0	2.0	0.31	0.090	16
Mg	0.70	1.5	1.0	1.0	1.0	0.17	0.049	17
Ca	0.30	0.50	0.37	0.30	0.30	0.097	0.028	26
Ti	0.20	0.30	0.23	0.20	0.20	0.049	0.014	20
Mn	700	1500	900	1000	1000	190	55	21
Ag	10	20	15	15	15	1.7	0.49	11
As	L200	L200	L200	L200	L200	—	—	—
Au	L10	L10	L10	L10	L10	—	—	—
B	20	50	29	30	30	7.7	2.2	27
Ba	1000	2000	1500	1500	1500	180	52	12
Be	0.70	1.5	1.1	1.0	1.0	0.20	0.058	18
Bi	L10	L10	L10	L10	L10	—	—	—
Cd	L20	L20	L20	L20	L20	—	—	—
Co	5.0	10	7.0	7.0	7.0	0.84	0.24	12
Cr	20	50	27	30	30	8.4	2.4	31
Cu	20	70	62	70	70	11	3.2	18
La	20	20	20	20	20	0.0	—	—
Mo	L5	L5	L5	L5	L5	—	—	—
Nb	L10	L10	L10	L10	L10	—	—	—
Ni	10	30	19	20	20	3.7	1.1	19
Pb	150	500	310	300	300	84	24	27
Sb	L100	L100	L100	L100	L100	—	—	—
Sc	5.0	.10	7.0	7.0	7.0	0.73	0.21	10
Sn	L10	L10	L10	L10	L10	—	—	—
Sr	100	150	120	100	100	24	7.0	20
V	50	100	70	70	70	5.9	1.7	8.4
W	L50	L50	L50	L50	L50	—	—	—
Y	10	20	15	15	15	1.0	0.29	6.7
Zn	500	700	610	700	700	100	29	16
Zr	150	200	160	150	150	18	5.2	11
Atomic absorption and colorimetric methods								
Ag	11	15	13	13	13	1.1	0.32	8.5
As	L10	L10	L10	L10	L10	—	—	—
Au	0.02	0.06	0.038	0.040	0.040	0.0097	0.0028	26
Cd	2.7	4.5	3.4	3.3	3.3	0.39	0.11	11
Co	8	12	9.8	10	10	1.1	0.32	11
Cu	52	74	66	66	65	4.9	1.4	7.4
Hg	3.6	4.2	3.9	4.0	4.0	0.14	0.04	3.6
In	L0.2	L0.2	L0.2	L0.2	L0.2	—	—	—
Mo	L4	L4	L4	L4	L4	—	—	—
Ni	20	30	25	24	24	2.6	0.75	10
Pb	460	700	600	600	600	52	15	8.7
Sb	36	55	44	45	46	4.1	1.2	9.3
Te	L0.5	L0.5	L0.5	L0.5	L0.5	—	—	—
Tl	0.90	1.5	1.2	1.2	1.2	0.13	0.038	11
U	L4	L4	L4	L4	L4	—	—	—
W	L20	L20	L20	L20	L20	—	—	—
Zn	420	570	490	480	480	37	11	7.6

All values are in g/ton except Fe, Ca, Mg, and Ti which are in percent. G = greater than; L = less than; — = no data.
Analysts: M.S. Erickson, A.E. Hubert, R.T. Hopkins, Jr., J.M. Motooka, J.H. Turner, and R.L. Turner.

GXR2 (48 nested samples)

Minimum	Maximum	Arithmetic mean	Median	Mode	Standard deviation	Confidence interval 95%	Coeff. of variation (%)	Element
Semi-quantitative spectrographic method								
1.5	5.0	2.1	2.0	2.0	0.49	0.10	23	Fe
1.0	1.5	1.1	1.0	1.0	0.20	0.042	18	Mg
0.30	0.50	0.36	0.30	0.30	0.090	0.019	25	Ca
0.15	0.20	0.22	0.20	0.20	0.045	0.0094	20	Ti
700	1500	1000	1000	1000	190.	40	19	Mn
10	20	15	15	15	1.6	0.34	11	Ag
L200	L200	L200	L200	L200	—	—	—	As
L10	L10	L10	L10	L10	—	—	—	Au
20	50	29	30	30	4.3	0.090	11	B
1000	3000	1700	1500	1500	400	84	24	Ba
0.70	2.0	1.3	1.5	1.5	0.28	0.059	22	Be
L10	L10	L10	L10	L10	—	—	—	Bi
L20	L20	L20	L20	L20	—	—	—	Cd
5.0	10	7.4	7.0	7.0	1.3	0.27	18	Co
5.0	50	26	30	30	7.2	1.5	27	Cr
20	100	62	70	70	18	3.8	28	Cu
L20	L20	L20	L20	L20	—	—	—	La
L5	L5	L5	L5	L5	—	—	—	Mo
L10	L10	L10	L10	L10	—	—	—	Nb
10	50	23	20	20	9.4	2.0	41	Ni
150	1000	450	300	300	170	36	38	Pb
L100	L100	L100	L100	L100	—	—	—	Sb
3.0	10	6.9	7.0	7.0	0.68	0.14	9.9	Sc
L10	L10	L10	L10	L10	—	—	—	Sn
100	150	120	100	100	25	5.3	21	Sr
50	100	70	70	70	6.0	1.3	8.6	V
L50	L50	L50	L50	L50	—	—	—	W
10	20	15	15	15	1.8	0.56	12	Y
300	1000	580	500	700	160	34	28	Zn
100	200	160	150	150	25	5.3	16	Zr
Atomic absorption and colorimetric methods								
12	16	14	14	14	1.0	0.21	7.1	Ag
7.0	40	23	20	20	8.2	1.7	36	As
0.04	0.08	0.056	0.06	0.06	0.012	0.0025	21	Au
2.8	4.2	3.6	3.6	3.6	0.34	0.071	9.4	Cd
—	—	—	—	—	—	—	—	Co
50	80	64	64	65	5.7	1.2	8.9	Cu
3.6	4.4	3.9	3.9	3.9	0.18	0.038	4.6	Hg
L0.2	L0.2	L0.2	L0.2	L0.2	—	—	—	In
L4	L4	L4	L4	L4	—	—	—	Mo
—	—	—	—	—	—	—	—	Ni
480	770	630	630	660	59	12	9.4	Pb
28	49	41	40	37	3.9	0.82	9.5	Sb
L0.5	L0.5	L0.5	L0.5	L0.5	—	—	—	Te
1.0	1.9	1.4	1.3	1.3	0.19	0.04	14	Tl
L4	L4	L4	L4	L4	—	—	—	U
L20	L20	L20	L20	L20	—	—	—	W
390	570	500	500	540	46	9.7	9.2	Zn

TABLE IV

Statistical summary for geochemical exploration — Reference sample three: Fe—Mn—W-rich spring deposit from Humboldt County, Nevada

GXR3 (50 random samples)

Element	Minimum	Maximum	Arithmetic mean	Median	Mode	Standard deviation	Confidence interval 95%	Coeff. of variation (%)
Semi-quantitative spectrographic method								
Fe	10	20	15	15	15	1.7	0.49	11
Mg	0.50	0.70	0.70	0.70	0.70	0.028	0.0081	4
Ca	7.0	10	7.1	7.0	7.0	0.42	0.12	5.9
Ti	0.05	0.07	0.066	0.07	0.07	0.0078	0.0022	12
Mn	G5000	G5000	G5000	G5000	G5000	—	—	—
Ag	L0.5	L0.5	L0.5	L0.5	L0.5	—	—	—
As	1000	1500	1400	1500	1500	180	52	13
Au	L10	L10	L10	L10	L10	—	—	—
B	70	70	70	70	70	0.0	—	—
Ba	2000	5000	3000	3000	3000	400	120	13
Be	15	50	26	20	20	7.0	2.0	27
Bi	L10	L10	L10	L10	L10	—	—	—
Cd	L20	L20	L20	L20	L20	—	—	—
Co	10	30	29	30	30	4.0	1.2	14
Cr	10	20	15	15	15	1.7	0.49	11
Cu	5	10	9.1	10	10	1.5	0.44	16
La	L20	L20	L20	L20	L20	—	—	—
Mo	L5	7	L5	L5	L5	—	—	—
Nb	L20	L20	L20	L20	L20	—	—	—
Ni	20	50	31	30	30	6.1	1.8	20
Pb	L10	L10	L10	L10	L10	—	—	—
Sb	L100	L100	L100	L100	L100	—	—	—
Sc	10	15	15	15	15	0.71	0.21	4.7
Sn	L10	L10	L10	L10	L10	—	—	—
Sr	500	700	680	700	700	66	19	9.7
V	20	70	48	50	50	9.4	2.7	20
W	10000	G10000	G10000	G10000	G10000	—	—	—
Y	10	15	15	15	15	1.4	0.41	9.3
Zn	L200	L200	L200	L200	L200	—	—	—
Zr	50	100	55	50	50	11	3.2	20
Atomic absorption and colorimetric methods								
Ag	1.5	2.2	1.7	1.7	1.8	0.18	0.052	11
As	4000	6000	5900	6000	6000	400	120	6.8
Au	0.04	0.06	0.052	0.06	0.06	0.0098	0.0028	19
Cd	1.6	1.9	1.7	1.7	1.7	0.093	0.027	5.5
Co	30	44	36	36	36	3.0	0.87	8.3
Cu	8.9	11	9.6	9.5	10	0.45	0.13	4.7
Hg	0.25	0.33	0.30	0.30	0.30	0.018	0.0052	6.0
In	L0.2	L0.2	L0.2	L0.2	L0.2	—	—	—
Mo	6	12	7.4	8.0	8.0	1.3	0.38	18
Ni	38	50	43	42	42	2.6	0.75	6.0
Pb	92	110	100	100	100	3.7	1.1	3.7
Sb	28	45	35	35	33	4.4	1.3	13
Te	L0.5	L0.5	L0.5	L0.5	L0.5	—	—	—
Tl	3.3	4.6	3.8	3.8	3.7	0.33	0.096	8.7
U	L4	L4	L4	L4	L4	—	—	—
W	2000	6000	3800	4000	4000	740	210	19
Zn	110	180	150	150	150	13	3.8	8.7
CxCu	1	3	1.9	2	2	0.49	0.14	26
CxHm	100	220	110	100	100	24	7.0	22

All values are in g/ton except Fe, Ca, Mg, and Ti which are in percent. G = greater than;
L = less than; — = no data.
Analysts: C.W. Cole, G.L. Crenshaw, M.S. Erickson, W.H. Ficklin, R.T. Hopkins, Jr.,
A.E. Hubert, J.M. Motooka, J.H. Turner, and R.L. Turner.

GXR3 (76 nested samples)

Minimum	Maximum	Arithmetic mean	Median	Mode	Standard deviation	Confidence interval 95%	Coeff. of variation (%)	Element
Semi-quantitative spectrographic method								
10	15	15	15	15	1.4	0.22	9.3	Fe
0.30	0.70	0.69	0.70	0.70	0.059	0.0094	8.5	Mg
7.0	10	7.0	7.0	7.0	0.35	0.056	5.0	Ca
0.03	0.07	0.065	0.07	0.07	0.0097	0.0015	15	Ti
G5000	G5000	G5000	G5000	G5000	—	—	—	Mn
L0.5	L0.5	L0.5	L0.5	L0.5	—	—	—	Ag
1000	2000	1500	1500	1500	130	21	8.7	As
L10	L10	L10	L10	L10	—	—	—	Au
70	100	70	70	70	2.4	0.38	3.4	B
2000	5000	2900	3000	3000	420	67	14	Ba
15	30	23	20	20	5.0	0.80	22	Be
L10	L10	L10	L10	L10	—	—	—	Bi
L20	L20	L20	L20	L20	—	—	—	Cd
20	30	29	30	30	3.1	0.50	11	Co
7.0	20	15	15	15	2.2	0.35	15	Cr
5.0	20	8.2	7.0	7.0	2.0	0.32	24	Cu
L20	L20	L20	L20	L20	—	—	—	La
L5	20	L5	L5	L5	—	—	—	Mo
L20	L20	L20	L20	L20	—	—	—	Nb
20	50	29	30	30	4.5	0.72	16	Ni
L10	L10	L10	L10	L10	—	—	—	Pb
L100	L100	L100	L100	L100	—	—	—	Sb
10	15	15	15	15	0.89	0.14	5.9	Sc
L10	L10	L10	L10	L10	—	—	—	Sn
500	700	660	700	700	77	12	12	Sr
30	70	49	50	50	5.6	1.9	11	V
G10000	G10000	G10000	G10000	G10000	—	—	—	W
10	15	15	15	15	1.1	0.18	7.3	Y
L200	L200	L200	L200	L200	—	—	—	Zn
30	70	50	50	50	4.3	0.69	8.6	Zr
Atomic absorption and colorimetric methods								
1.3	2.0	1.6	1.6	1.6	0.16	0.026	10	Ag
6000	8000	6900	6000	6000	1000	160	14	As
—	—	—	—	—	—	—	—	Au
0.9	2.0	1.2	1.2	1.2	0.13	0.021	11	Cd
28	46	36	36	34	3.3	0.53	9.2	Co
7.8	14	10	10	11	0.99	0.16	9.9	Cu
0.19	0.34	0.27	0.27	0.27	0.025	0.004	9.3	Hg
L0.2	L0.2	L0.2	L0.2	L0.2	—	—	—	In
4	8	6.6	6	6	1.0	0.16	15	Mo
34	52	43	45	43	3.6	0.58	8.4	Ni
86	110	97	96	100	3.9	0.62	4.0	Pb
24	50	37	38	38	4.8	0.77	13	Sb
L0.5	L0.5	L0.5	L0.5	L0.5	—	—	—	Te
3.2	5.8	4.5	4.4	4.7	0.45	0.072	10	Tl
L4	L4	L4	L4	L4	—	—	—	U
2000	5000	3600	4000	4000	770	120	21	W
120	270	180	170	170	24	3.8	13	Zn
1	3	1.8	1.5	1.5	0.42	0.067	23	CxCu
—	—	—	—	—	—	—	—	CxHm

TABLE V

Statistical summary for geochemical exploration — Reference sample four: porphyry copper mill heads from Utah

GXR4 (50 random samples)

Element	Minimum	Maximum	Arithmetic mean	Median	Mode	Standard deviation	Confidence interval 95%	Coeff. of variation (%)
Semi-quantitative spectrographic method								
Fe	2.0	5.0	3.6	3.0	3.0	1.1	0.32	31
Mg	1.0	1.5	1.5	1.5	1.5	0.12	0.035	8.0
Ca	0.30	0.70	0.43	0.50	0.50	0.13	0.038	30
Ti	0.15	0.30	0.23	0.20	0.20	0.054	0.016	23
Mn	100	200	150	150	150	17	4.9	11
Ag	3.0	7.0	3.6	3.0	3.0	1.1	0.32	30
As	L200	L200	L200	L200	L200	—	—	—
Au	L10	L10	L10	L10	L10	—	—	—
B	L10	L10	L10	L10	L10	—	—	—
Ba	700	2000	1500	1500	1500	310	90	21
Be	1.5	2.0	1.7	1.5	1.5	0.24	0.070	14
Bi	10	15	14	15	15	1.9	0.55	14
Cd	L10	L10	L10	L10	L10	—	—	—
Co	7.0	15	13	15	15	2.6	0.75	20
Cr	30	100	53	50	50	17	4.9	32
Cu	3000	10000	8300	7000	10000	1800	520	22
La	50	70	68	70	70	5.5	1.6	8.0
Mo	200	500	290	300	300	67	19	23
Nb	L20	L20	L20	L20	L20	—	—	—
Ni	30	70	59	70	70	13	3.8	22
Pb	30	70	48	50	50	13	3.8	27
Sb	L100	L100	L100	L100	L100	—	—	—
Sc	7.0	7.0	7.0	7.0	7.0	—	—	—
Sn	L10	L10	L10	L10	L10	—	—	—
Sr	150	200	180	200	200	25	7.3	14
V	50	150	99	100	100	23	6.7	23
W	L50	70	L50	50	50	—	—	—
Y	10	15	11	10	10	2.4	0.64	20
Zn	L200	L200	L200	L200	L200	—	—	—
Zr	100	200	150	150	150	20	5.8	13
Atomic absorption and colorimetric methods								
Ag	3.0	4.0	3.6	3.6	3.6	0.20	0.058	5.6
As	40	80	62	60	60	9.3	2.7	15
Au	0.50	0.68	0.58	0.56	0.56	0.054	0.016	9.3
Cd	0.80	1.0	0.89	0.90	0.90	0.068	0.020	7.6
Co	16	20	18	18	18	1.2	0.35	6.7
Cu	5200	6700	6100	6100	6100	390	110	6.4
Hg	0.08	0.17	0.12	0.12	0.10	0.020	0.0058	17
In	L0.2	L0.2	L0.2	L0.2	L0.2	—	—	—
Mo	230	450	340	300	300	76	21	22
Ni	36	50	42	42	42	2.6	0.75	6.2
Pb	67	83	78	78	79	3.4	0.99	4.4
Sb	5.4	8.2	6.3	6.3	6.0	0.48	0.14	7.6
Te	0.2	2.0	0.87	0.80	0.80	0.33	0.096	38
Tl	3.1	4.2	3.5	3.5	3.5	0.24	0.070	6.9
U	L4	L4	L4	L4	L4	—	—	—
W	L20	L20	L20	L20	L20	—	—	—
Zn	46	67	59	60	60	5.7	1.7	9.6
CxCu	440	1800	890	880	880	210	61	24
CxHm	300	750	420	450	450	92	27	22

All values are in g/ton except Fe, Ca, Mg, and Ti which are in percent. G = greater than;
L = less than; — = no data.
Analysts: C.W. Cole, G.L. Crenshaw, M.S. Erickson, W.H. Ficklin, R.T. Hopkins, Jr.,
A.E. Hubert, J.M. Motooka, J.H. Turner, and R.L. Turner.

GXR4 (84 nested samples)

Minimum	Maximum	Arithmetic mean	Median	Mode	Standard deviation	Confidence interval 95%	Coeff. of variation (%)	Element
Semi-quantitative spectrographic method								
2.0	7.0	3.7	3.0	3.0	1.2	0.18	32	Fe
1.0	1.5	1.5	1.5	1.5	0.085	0.013	5.7	Mg
0.30	0.70	0.44	0.50	0.50	0.14	0.02	32	Ca
0.15	0.5	0.25	0.2	0.2	0.067	0.01	27	Ti
100	200	150	150	150	13	2.0	8.7	Mn
3.0	5.0	3.0	3.0	3.0	0.22	0.033	7.3	Ag
L200	L200	L200	L200	L200	—	—	—	As
L10	L10	L10	L10	L10	—	—	—	Au
L10	L10	L10	L10	L10	—	—	—	B
1000	2000	1500	1500	1500	110	17	7.3	Ba
1.5	3.0	1.9	2.0	2.0	0.24	0.036	13	Be
10	30	14	15	15	2.9	0.44	21	Bi
L10	L10	L10	L10	L10	—	—	—	Cd
7.0	70	12	10	10	5.2	0.78	43	Co
20	70	43	50	50	12	1.8	28	Cr
5000	15000	8100	7000	7000	1700	260	21	Cu
50	70	68	70	70	6.4	0.96	9.4	La
150	700	350	300	300	110	17	31	Mo
L10	L10	L10	L10	L10	—	—	—	Nb
30	70	55	50	50	12	1.8	22	Ni
30	70	46	50	50	9.8	1.5	21	Pb
L100	L100	L100	L100	L100	—	—	—	Sb
5.0	7.0	7.0	7.0	7.0	0.27	0.041	3.9	Sc
L10	L10	L10	L10	L10	—	—	—	Sn
150	500	180	200	200	37	5.6	21	Sr
50	150	97	100	100	27	4.1	28	V
L50	L50	L50	L50	L50	—	—	—	W
7.0	10	10	10	10	0.23	0.035	2.3	Y
L200	L200	L200	L200	L200	—	—	—	Zn
100	200	150	150	150	17	2.6	11	Zr
Atomic absorption and colorimetric methods								
3.0	4.0	3.6	3.7	3.7	0.20	0.03	5.6	Ag
120	160	150	160	160	19	2.9	13	As
—	—	—	—	—	—	—	—	Au
0.60	1.0	0.74	0.70	0.70	0.087	0.013	12	Cd
12	24	17	16	16	2.1	0.32	12	Co
4900	7000	6200	6200	6300	430	65	6.9	Cu
0.08	0.14	0.10	0.10	0.10	0.0095	0.0014	9.5	Hg
L0.2	L0.2	L0.2	L0.2	L0.2	—	—	—	In
240	360	350	360	360	30	4.5	8.6	Mo
32	48	39	38	38	3.2	0.48	8.2	Ni
70	91	79	78	73	5.6	0.84	7.1	Pb
4.0	7.0	5.7	6.0	6.0	0.70	0.11	12	Sb
—	—	—	—	—	—	—	—	Te
2.5	4.5	3.6	3.6	3.4	0.31	0.047	8.6	Tl
L4	L4	L4	L4	L4	—	—	—	U
L20	80	L20	L20	L20	—	—	—	W
50	81	64	64	63	6.8	1.0	11	Zn
1320	3080	2000	1980	1760	320	48	16	CxCu
—	—	—	—	—	—	—	—	CxHm

672

TABLE VI

Statistical summary for geochemical exploration — Reference sample five: soil from B
horizon from Somerset County, Maine

GXR5 (50 random samples)

Element	Minimum	Maximum	Arithmetic mean	Median	Mode	Standard deviation	Confidence interval 95%	Coeff. of variation (%)
Semi-quantitative spectrographic method								
Fe	1.5	7.0	5.8	5.0	5.0	1.3	0.38	22
Mg	0.50	1.5	1.5	1.5	1.5	0.18	0.052	12
Ca	0.20	0.50	0.33	0.30	0.30	0.082	0.024	25
Ti	0.10	0.30	0.21	0.20	0.20	0.04	0.012	19
Mn	100	300	250	200	200	54	16	22
Ag	L0.5	L0.5	L0.5	L0.5	L0.5	—	—	—
As	L200	L200	L200	L200	L200	—	—	—
Au	L10	L10	L10	L10	L10	—	—	—
B	10	10	10	10	10	0.0	—	—
Ba	1000	2000	1700	1500	2000	270	78	16
Be	1.0	1.0	1.0	1.0	1.0	0.0	—	—
Bi	L10	L10	L10	L10	L10	—	—	—
Cd	L20	L20	L20	L20	L20	—	—	—
Co	10	30	29	30	30	3.9	1.1	13
Cr	50	150	110	100	100	30	8.7	27
Cu	100	500	300	300	300	46	13	15
La	15	30	30	30	30	2.1	0.61	7.0
Mo	15	50	31	30	30	6.0	1.7	19
Nb	L10	L10	L10	L10	L10	—	—	—
Ni	30	100	78	70	70	16	4.6	21
Pb	10	15	14	15	15	2.1	0.61	15
Sb	L100	L100	L100	L100	L100	—	—	—
Sc	5.0	10	8.7	10	10	1.6	0.46	18
Sn	L10	L10	L10	L10	L10	—	—	—
Sr	50	100	98	100	100	9.1	2.6	9.1
V	30	100	76	70	70	14	4.1	18
W	L50	L50	L50	L50	L50	—	—	—
Y	5.0	20	15	15	15	2.2	0.64	15
Zn	L200	L200	L200	L200	L200	—	—	—
Zr	50	150	140	150	150	23	6.7	16
Atomic absorption and colorimetric methods								
Ag	1.2	1.5	1.3	1.3	1.3	0.067	0.019	5.2
As	10	20	16	20	20	4.8	1.4	30
Au	0.08	0.18	0.12	0.12	0.12	0.025	0.0072	21
Cd	L0.2	L0.2	L0.2	L0.2	L0.2	—	—	—
Co	28	40	33	32	32	2.6	0.75	7.9
Cu	320	400	360	360	350	17	4.9	4.7
Hg	0.09	0.18	0.15	0.14	0.14	0.019	0.0055	13
In	0.20	0.60	0.32	0.30	0.30	0.09	0.026	28
Mo	16	24	21	24	24	3.9	1.1	19
Ni	66	104	87	88	88	8.3	2.4	9.5
Pb	47	57	53	53	53	1.9	0.55	3.6
Sb	1	3	1.8	2	2	0.47	0.14	26
Te	L0.5	L0.5	L0.5	L0.5	L0.5	—	—	—
Tl	0.20	0.55	0.34	0.35	0.34	0.093	0.027	27
U	L4	L4	L4	L4	L4	—	—	—
W	L20	L20	L20	L20	L20	—	—	—
Zn	36	54	42	42	42	3.2	0.93	7.6
CxCu	264	330	280	264	264	26	7.5	9.3
CxHm	60	120	78	80	80	14	4.1	18

All values are in g/ton except Fe, Ca, Mg, Ti which are in percent. G = greater than;
L = less than; — = no data.
Analysts: C.W. Cole, M.S. Erickson, R.T. Hopkins, Jr., A.E. Hubert, J.M. Motooka,
J.H. Turner, R.L. Turner, and E.P. Welsch.

GXR5 (36 nested samples)

Minimum	Maximum	Arithmetic mean	Median	Mode	Standard deviation	Confidence interval 95%	Coeff. of variation (%)	Element
Semi-quantitative spectrographic method								
1.5	7.0	5.2	5.0	5.0	1.5	0.36	29	Fe
0.50	1.5	1.3	1.5	1.5	0.35	0.084	27	Mg
0.20	0.50	0.30	0.30	0.30	0.037	0.0088	12	Ca
0.15	0.20	0.19	0.20	0.20	0.023	0.0055	12	Ti
70	300	240	300	300	61	15	25	Mn
L0.5	0.7	0.5	0.5	0.5	—	—	—	Ag
L200	L200	L200	L200	L200	—	—	—	As
L10	L10	L10	L10	L10	—	—	—	Au
L10	L10	L10	L10	L10	—	—	—	B
1000	2000	1600	1500	1500	250	60	16	Ba
L1.0	1.0	L1.0	1.0	1.0	—	—	—	Be
L10	L10	L10	L10	L10	—	—	—	Bi
L20	L20	L20	L20	L20	—	—	—	Cd
10	30	24	20	20	5.6	1.3	23	Co
50	150	100	100	100	28	6.7	28	Cr
100	500	280	300	300	70	17	25	Cu
10	30	28	30	30	4.5	1.1	16	La
10	30	26	30	30	5.7	1.4	22	Mo
L10	L10	L10	L10	L10	—	—	—	Nb
20	100	70	70	70	12	2.9	17	Ni
10	20	15	15	15	2.6	0.62	17	Pb
L100	L100	L100	L100	L100	—	—	—	Sb
5	10	7.9	7.0	7.0	1.5	0.36	19	Sc
L10	L10	L10	L10	L10	—	—	—	Sn
100	100	100	100	100	—	—	—	Sr
20	100	69	70	70	11	2.6	16	V
L50	L50	L50	L50	L50	—	—	—	W
10	15	14	15	15	1.2	0.29	8	Y
L200	L200	L200	L200	L200	—	—	—	Zn
50	150	90	70	70	27	6.5	30	Zr
Atomic absorption and colorimetric methods								
0.90	1.4	1.2	1.2	1.3	0.10	0.024	8.3	Ag
20	50	23	20	20	5.4	1.3	23	As
0.04	0.08	0.058	0.06	0.06	0.012	0.0028	21	Au
L0.2	L0.2	L0.2	L0.2	L0.2	—	—	—	Cd
26	40	34	34	32	2.7	0.65	7.9	Co
340	440	370	360	360	20	4.8	5.4	Cu
0.10	0.26	0.15	0.14	0.14	0.028	0.0067	19	Hg
0.20	0.50	0.34	0.40	0.40	0.076	0.018	22	In
8	24	16	16	16	3.5	0.84	22	Mo
70	96	82	82	84	6.0	1.4	7.3	Ni
43	57	48	46	46	2.6	0.62	5.4	Pb
1	4	2.2	2.0	2.0	0.60	0.14	27	Sb
L0.5	L0.5	L0.5	L0.5	L0.5	—	—	—	Te
0.20	0.50	0.36	0.40	0.40	0.069	0.017	19	Tl
L4	L4	L4	L4	L4	—	—	—	U
L20	L20	L20	L20	L20	—	—	—	W
35	56	41	41	43	3.5	0.84	8.5	Zn
144	192	170	168	168	16	3.8	9.4	CxCu
—	—	—	—	—	—	—	—	CxHm

TABLE VII

Statistical summary for geochemical exploration — Reference sample six: soil from B horizon from Davidson County, North Carolina

GXR6 (50 random samples)

Element	Minimum	Maximum	Arithmetic mean	Median	Mode	Standard deviation	Confidence interval 95%	Coeff. of variation (%)
Semi-quantitative spectrographic method								
Fe	5.0	7.0	6.7	7.0	7.0	0.70	0.20	10
Mg	0.50	1.0	0.70	0.70	0.70	0.083	0.024	12
Ca	0.07	0.15	0.096	0.10	0.10	0.019	0.0055	20
Ti	0.30	0.50	0.36	0.30	0.30	0.091	0.026	25
Mn	700	1500	1000	1000	1000	110	32	11
Ag	L0.5	L0.5	L0.5	L0.5	L0.5	—	—	—
As	200	200	200	200	200	0.0	—	—
Au	L10	L10	L10	L10	L10	—	—	—
B	L10	L10	L10	L10'	L10	—	—	—
Ba	1000	1500	1300	1500	1500	250	73	19
Be	L1.0	1.0	L1.0	1.0	1.0	—	—	—
Bi	L10	L10	L10	L10	L10	—	—	—
Cd	L20	L20	L20	L20	L20	—	—	—
Co	10	15	10	10	10	1.4	0.41	14
Cr	30	70	59	50	70	12	3.5	20
Cu	50	70	65	70	70	8.9	2.5	14
La	L20	L20	L20	L20	L20	—	—	—
Mo	L5	L5	L5	L5	L5	—	—	—
Nb	L10	L10	L10	L10	L10	—	—	—
Ni	15	30	24	20	20	5.0	1.5	21
Pb	50	100	71	70	70	6.7	1.9	9.4
Sb	L100	L100	L100	L100	L100	—	—	—
Sc	20	30	26	30	30	5.0	1.5	19
Sn	L10	L10	L10	L10	L10	—	—	—
Sr	L100	L100	L100	L100	L100	—	—	—
V	150	200	180	200	200	24	7.0	13
W	L50	L50	L50	L50	L50	—	—	—
Y	10	15	13	15	15	2.5	0.73	19
Zn	L200	L200	L200	L200	L200	—	—	—
Zr	70	100	84	70	70	15	4.4	18
Atomic absorption and colorimetric methods								
Ag	0.90	1.3	1.1	1.0	1.0	0.071	0.021	7.8
As	200	300	270	300	300	46	13	17
Au	0.04	0.24	0.072	0.06	0.06	0.031	0.0089	43
Cd	L0.2	L0.2	L0.2	L0.2	L0.2	—	—	—
Co	—	—	—	—	—	—	—	—
Cu	55	83	67	67	66	5.4	1.6	8.1
Hg	0.03	0.05	0.039	0.04	0.04	0.0077	0.0022	20
In	L0.2	L0.2	L0.2	L0.2	L0.2	—	—	—
Mo	L4	L4	L4	L4	L4	—	—	—
Ni	—	—	—	—	—	—	—	—
Pb	120	160	140	140	140	7.2	2.1	5.1
Sb	4.0	6.0	4.9	5.0	5.0	0.45	0.13	9.2
Te	L0.5	L0.5	L0.5	L0.5	L0.5	—	—	—
Tl	2.1	3.2	2.5	2.4	2.5	0.22	0.064	8.8
U	L4	L4	L4	L4	L4	—	—	—
W	L20	L20	L20	L20	L20	—	—	—
Zn	89	120	100	100	110	7.3	2.1	7.3
CxCu	—	—	—	—	—	—	—	—
CxHm	—	—	—	—	—	—	—	—

All values are in g/ton except Fe, Ca, Mg, and Ti which are in percent. G = greater than;
L = less than; — = no data.
Analysts: C.W. Cole, G.L. Crenshaw, M.S. Erickson, R.T. Hopkins, Jr., A.E. Hubert,
J.M. Motooka, J.H. Turner, and R.L. Turner

GXR6 (64 nested samples)

Minimum	Maximum	Arithmetic mean	Median	Mode	Standard deviation	Confidence interval 95%	Coeff. of variation (%)	Element
Semi-quantitative spectrographic method								
3.0	10	6.3	7.0	7.0	1.3	0.23	21	Fe
0.30	1.5	0.81	0.70	0.70	0.22	0.040	27	Mg
0.05	0.15	0.089	0.10	0.10	0.019	0.0034	21	Ca
0.20	0.50	0.34	0.30	0.30	0.084	0.015	25	Ti
500	1500	950	1000	1000	230	41	24	Mn
L0.5	L0.5	L0.5	L0.5	L0.5	—	—	—	Ag
L200	L200	L200	L200	L200	—	—	—	As
L10	L10	L10	L10	L10	—	—	—	Au
L10	L10	L10	L10	L10	—	—	—	B
700	2000	1400	1500	1500	260	47	19	Ba
L1.0	1.0	L1.0	L1.0	L1.0	—	—	—	Be
L10	L10	L10	L10	L10	—	—	—	Bi
L20	L20	L20	L20	L20	—	—	—	Cd
5.0	15	10	10	10	2.5	0.45	25	Co
20	70	60	70	70	13	2.3	22	Cr
20	100	63	70	70	13	2.3	21	Cu
L20	L20	L20	L20	L20	—	—	—	La
L5	L5	L5	L5	L5	—	—	—	Mo
L10	L10	L10	L10	L10	—	—	—	Nb
10	70	23	20	20	6.6	1.2	29	Ni
50	100	70	70	70	8.0	1.4	11	Pb
L100	L100	L100	L100	L100	—	—	—	Sb
15	30	24	20	20	5.3	0.95	22	Sc
L10	L10	L10	L10	L10	—	—	—	Sn
L100	L100	L100	L100	L100	—	—	—	Sr
70	300	170	200	200	41	7.4	24	V
L50	L50	L50	L50	L50	—	—	—	W
10	20	15	15	15	1.4	0.25	9.3	Y
L200	L200	L200	L200	L200	—	—	—	Zn
50	100	76	70	70	13	2.3	17	Zr
Atomic absorption and colorimetric methods								
0.80	1.3	1.0	1.0	1.0	0.081	0.015	8.1	Ag
300	400	400	400	400	22	4.0	5.5	As
0.06	0.20	0.084	0.08	0.08	0.017	0.003	20	Au
L0.2	L0.2	L0.2	L0.2	L0.2	—	—	—	Cd
14	24	19	20	20	2.2	0.40	12	Co
55	78	69	69	70	4.9	0.88	7.1	Cu
0.01	0.10	0.04	0.04	0.04	0.011	0.0019	28	Hg
L0.2	L0.2	L0.2	L0.2	L0.2	—	—	—	In
L4	L4	L4	L4	L4	—	—	—	Mo
22	38	26	26	26	2.4	0.43	9.2	Ni
120	150	130	130	130	6.8	1.2	5.2	Pb
3.2	6.8	4.8	4.6	4.6	0.57	0.10	12	Sb
L0.5	L0.5	L0.5	L0.5	L0.5	—	—	—	Te
2.0	3.2	2.6	2.6	2.5	0.23	0.041	8.8	Tl
L4	L4	L4	L4	L4	—	—	—	U
L20	L20	L20	L20	L20	—	—	—	W
80	120	95	94	100	7.4	1.3	7.8	Zn
10	20	16	15	15	2.1	0.38	13	CxCu
10	100	45	40	40	13	2.3	29	CxHm

from data in their paper range from 9.4 to 28%. For two samples from the Cortez district, Nevada, the means of five replicates were 0.31 and 1.6 g/ton Te, and the coefficients of variation were 19 and 28%, respectively. In three samples from the Cripple Creek district, Colorado, the means of five replicates for each sample were 4.7, 11.1, and 25.7 g/ton Te, and the coefficients of variation were 9.4, 11, and 10, respectively. The coefficients of variation are consistently around 10 in concentration ranges from 5 to 26 g/ton Te, but increase by a factor of 2—3 at concentrations from 0.3 to 1.6 g/ton. These results are felt to be typical of the precision to be expected from the method under ideal conditions, and illustrate the general precision of the atomic absorption methods used in this study.

The data obtained by atomic absorption indicate that the overall variation is about what would be expected from analytical variation. The range in coefficients of variation for all six reference samples, for those elements analyzed by atomic absorption, are: Ag, 4.5—11; Au, 5.0—43; Cd, 1.8—12; Co, 1.9—18; Cu, 1.0—9.9; Pb, 3.6—9.4; Ni, 6.0—13; Te, 38—52; Zn, 7.3—11; In, 13—28; Hg, 3.6—28; and Tl, 6.9—27. In general, the higher coefficients of variation can be related to the lower concentrations.

The elements having a large range of coefficients of variation are gold, mercury, and thallium. It can be shown that for all three of these elements the large coefficients of variation are related to the low concentrations, which suggests that our analytical technique is inadequate for measuring at this level.

Tellurium has a fairly large coefficient of variation, but the amount of data available is sufficient only to suggest that this variation is related to low concentrations.

Colorimetric methods have been used to obtain data for arsenic, copper, molybdenum, antimony, tungsten, and cold-extractable heavy metals. The coefficients of variation range from 5.3 to 30%, which, in the experience of the authors and in the opinion of F.N. Ward (oral communication, 1974), is within the expected range of variation for the methods. Data are available from Ward and Lakin (1954) on the variability of the colorimetric technique for antimony, and this might be used to illustrate the precision of a typical colorimetric method. In developing their method, the authors applied the technique to four replicates each of seven samples of soil and two samples of rocks, to estimate the precision of the method. In concentration ranges of 2.3 to 9.5 g/ton the coefficient of variation varies from 7.4 to 14, while in the concentration ranges from 0.6 to 2.3 g/ton the coefficient of variation varies from 8.3 to 35.

COMPARISON OF METHODS OF ANALYSIS

Although exploration geochemists are probably more concerned with precision than accuracy, the data presented furnish some information on the differences between spectrographic and other analytical techniques. Silver concentrations are in reasonable agreement if the concentration is higher than a few g/ton, but atomic absorption is higher than emission spectrographic analysis at lower levels of concentration. Agreement between arsenic concentrations is poor and erratic e.g., 1400 g/ton by spectrographic analyses and 5900 g/ton by the confined-spot test in reference sample three. Comparisons of means for copper, tungsten, and zinc indicate the wet method is only a partial digestion. The data summarized here should not be interpreted as establishing an absolute concentration for any elements.

The agreement between spectrographic data and atomic absorption data is wholly unsatisfactory for lead. The spectrographic data are low for lead relative to atomic absorption data. For the purposes of this paper, both methods indicate homogeneity of the samples.

The spectrographic data for tungsten in the three samples in which it was detected are consistently higher than the values obtained by the colorimetric method.

SOME DETERMINANT ERRORS

During the analysis of this data certain errors were detected. We feel that it it useful to point out some of these errors, because they are types of errors common to all laboratories, and they have the effect of indicating that a sample is heterogeneous when in fact it is not.

A number of errors due to misplaced decimal points were found. These occurred in about 0.1% of the data. They appeared more frequently in the spectrographic data because of the technique used for reading and recording the data. In these data they are easily caught and corrected by reading the original spectrogram again. This type of error also occurred in transcribing data and is especially prevalent in transcribing data from source documents to the punch cards used for entering the data into a computer, where the error may reach 1%.

Errors in transcribing data are important, and their presence is difficult to establish at an early stage when processing large masses of numbers. This type of error and its effect is illustrated by a keypunching error which was present in the zirconium concentration of the nested samples for geochemical reference sample six. The reported range of the data for these samples was 50—100 g/ton with a median of 70 g/ton. One of the 128 samples was mispunched as 700 g/ton instead of 100 g/ton. The effect of this error was to change the mean from 76 to 80 g/ton, the standard deviation from 13 to 57 g/ton, and the coefficient of variation from 17 to 71.

There are several sources of error attributable to the analyst. One error made during the analysis of this data allowed us to obtain some information on the variance due to reading a spectrogram. In the case in point the spectrographer read the 22 samples on one spectrographic film for chromium. He recorded these values in the adjacent column under the cobalt heading, then read chromium again and recorded this set of readings under the correct

TABLE VIII

Variation in reading chromium analyses from the same spectrogram

Sample	Chromium (ppm)		Difference in reporting intervals
	1st reading	2nd reading	
1	30	30	0
2	30	30	0
3	20	20	0
4	50	30	−1
5	30	30	0
6	20	20	0
7	30	20	−1
8	30	30	0
9	30	50	+1
10	20	30	+1
11	30	30	0
12	30	30	0
13	30	30	0
14	30	20	−1
15	50	50	0
16	30	30	0
17	30	30	0
18	30	30	0
19	30	30	0
20	30	30	0
21	30	30	0
22	30	30	0

heading. This data is tabulated in Table VIII. The chromium line was read the same 17 of 22 times (77%) and differently 5 of 22 times (23%); when it was read differently it was only by 1/6th order, the next reporting interval. It was read an interval higher two times and an interval lower three times. Neither the mean, 30 g/ton, nor the standard deviation were significantly different for the two sets of readings. The variation in reading these 22 samples is, however, equivalent to the overall variation in the set of 50 samples.

An inadvertent error in sample control led to the detection of another systematic error. A set of subsamples was to be analyzed in duplicate for gold. The samples were analyzed once by one analyst on one instrument and the duplicates were done by another analyst on another instrument. The results are presented in Fig.1 and point up the degree of variation possible within a given laboratory.

CONCLUSIONS

The authors conclude that the overall variance determined from the data presented here is within the variance expected for the methods used for the

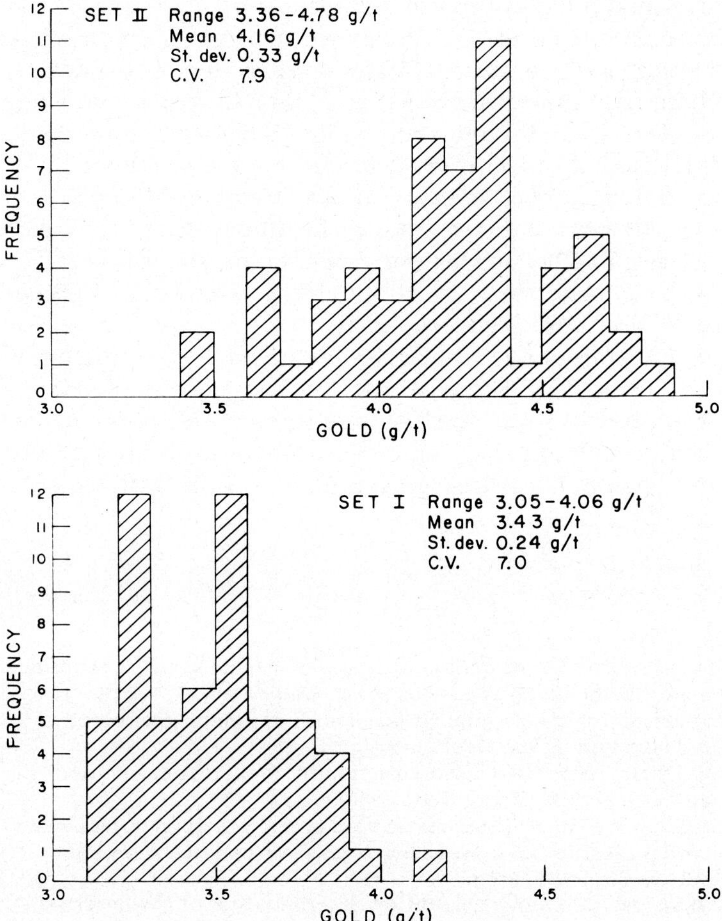

Fig.1. Comparison of duplicate analysis of gold by two analysts using different instruments.

analyses; thus the reference samples are homogeneous within these limits. In
order to determine the degree of homogeneity of the reference samples more
precisely, it will be necessary to develop more precise analytical techniques
over the ranges of concentration of interest to exploration geochemists. Many
of the analytical techniques are precise enough at present for the types of
mineral resources being sought; some are not. In the search for concealed ore
bodies where geochemical differences will be more subtle, it will surely be
necessary to improve on the present analytical methods so that we may measure the natural variations as a major component of the total variance observed.

ACKNOWLEDGEMENTS

The committee on Geochemical Analyses, appointed by J.A. Coope in 1970, was formed to study the analytical suitability and efficiency of analytical methods used in geochemical exploration. To achieve this goal, the committee implemented the collection and preparation of six samples of about 450 kg each; subsamples of these have been distributed to 176 laboratories and analyzed repeatedly by the U.S. Geological Survey laboratories to determine the homogeneity. This report is a product of the study initiated by the committee: H.W. Lakin, Chairman, Denver, Colo.; Alexei A. Beus, Member, Moscow, U.S.S.R.; Harold Bloom, Member, Denver, Colo.; I.L. Elliott, Member, Vancouver, B.C., Canada; Aslak Kvalheim, Member, Trondheim, Norway; R.H. Mazzucchelli, Member, Kalgoorlie, W.A., Australia; Mike Thompson, Member, London, England; and Norio Tono, Member, Japan.

Many individuals, in addition to the committee members, have contributed their time and talent to this project. Some of these individuals have been acknowledged in the text, but it is not possible here to credit all of the others whose contributions have been indirectly important. The authors are particularly indebted to Alfred Miesch, U.S. Geological Survey, for discussions relating to sampling methods.

REFERENCES

Almond, H., 1953. Field method for determination of traces of arsenic in soils — confined spot procedure using a modified Gutzeit apparatus. Anal. Chem., 25(11): 1766—1767

Bloom, H., 1955. A field method for the determination of ammonium citrate-soluble heavy metals in soils and alluvium. Econ. Geol., 50(5): 533—541

Canney, F.C. and Hawkins, D.B., 1958. Cold acid extraction of copper from soils and sediments — a proposed field method. Econ. Geol., 53(7): 877—886

Grimes, D.J. and Marranzino, A.P., 1968. Direct-current arc and alternating-current spark emission spectrographic field methods for the semiquantitative analysis of geologic materials, U.S. Geol. Survey Circ. 591, 6 pp.

Hubert, A.E. and Lakin, H.W., 1973. Atomic absorption determination of thallium and indium in geologic material. In: M.J. Jones (Editor), Geochemical Exploration 1972. Institution of Mining and Metallurgy, London, pp.383—387

Lakin, H.W., Curtin, G.C. and Hubert, A.E., *with chapters by* Shacklette, H.T. and Doxtader, K.G., 1974. Geochemistry of gold in the weathering cycle. U.S. Geol. Survey Bull., 1330 (in press)

Myers, A.T., Havens, R.G. and Dunton, P.J., 1961. A spectrochemical method for the semiquantitative analysis of rocks, minerals, and ores. U.S. Geol. Survey Bull., 1084-I: 207—229

Nakagawa, H.M. and Harms, T.F., 1968. Atomic absorption determination of cadmium in geologic materials. In: Geological Survey Research 1968. U.S. Geol. Survey Prof. Paper, 600-D: D207—D209

Nakagawa, H.M. and Thompson, C.E., 1968. Atomic absorption determination of tellurium. In: Geological Survey Research 1968. U.S. Geol. Survey Prof. Paper, 600-B: B123—B125

Smith, G.F., Hardy, L.V. and Gard, E.L., 1929. The segregation of analyzed samples. Ind. Eng. Chem., Anal. Ed., 1(4): 228—230

Thompson, C.E. and Lakin, H.W., 1957. A field chromatographic method for the determination of uranium in soils and rocks. U.S. Geol. Survey Bull., 1036-L: 209—220

Thompson, C.E., Nakagawa, H.M. and Van Sickle, G.H., 1968. Rapid analysis for gold in geologic materials. In: Geological Survey Research 1968. U.S. Geol. Survey Prof. Paper, 600-B: B130—B132

Vaughn, W.W. and McCarthy, Jr., J.H., 1964. An instrumental technique for the determination of submicrogram concentrations of mercury in soils, rocks, and gas. In: Geological Survey Research 1964. U.S. Geol. Survey Prof. Paper, 501-D: D123—D127
Ward, F.N., 1951a. Determination of molybdenum in soils and rocks — a geochemical semimicro field method. Anal. Chem., 23(5): 788—791
Ward, F.N., 1951b. A field method for the determination of tungsten in soils. U.S. Geol. Survey Circ. 119, 4 pp.
Ward, F.N. and Lakin, H.W., 1954. Determination of traces of antimony in soils and rocks. Anal. Chem., 26(7): 1168—1173
Ward, F.N., Lakin, H.W., Canney, F.C. et al., 1963. Analytical methods used in geochemical exploration by the U.S. Geological Survey. U.S. Geol. Survey Bull., 1152, 100 pp.
Ward, F.N., Nakagawa, H.M., Harms, T.F. and Van Sickle, G.H., 1969. Atomic absorption methods of analysis useful in geochemical exploration. U.S. Geol. Survey Bull., 1289, 45 pp.

SAMPLING AND ANALYSIS OF GEOCHEMICAL MATERIALS FOR GOLD

BRUCE W. BROWN and GORDON R. HILCHEY

Chemex Labs Limited, North Vancouver, B.C. (Canada)
Sumitomo Metal Mining Canada Limited, Vancouver, B.C. (Canada)

ABSTRACT

The material sampled, sample size, its preparation, and the size of the analytical portion
are important factors in exploration geochemistry for gold. These factors are discussed in
detail and some procedures recommended. Samples from three properties in British Columbia were investigated and a recommended laboratory procedure developed.

INTRODUCTION

This study was the result of inconsistencies observed in the results of analyses of geochemical materials for gold on a prospect in northern British
Columbia (M.G.S. property). Sample materials from this property were re-
analysed and additional samples taken from two other properties in southern
British Columbia.

With the current high price of gold, there is a great deal of interest in exploration for this metal. Unfortunately, standard analytical techniques used
are not capable of the precision and low detection limits necessary on low-
grade material now considered to be geochemically significant. Furthermore,
some of the common sampling and sample preparation techniques appear to
be inadequate.

Gold is one of the most difficult of all elements to analyse for because:
(1) extremely small concentrations of gold are significant, (2) it is usually in
discrete particles of significant size, rarely if ever uniformly distributed, and
(3) once liberated it tends to segregate easily because of its high specific
gravity relative to other minerals and rock fragments.

Most of the theoretical considerations are dealt with at length in the literature. This paper deals with the practical problems faced by the geochemist in
obtaining meaningful gold analyses.

SAMPLING PROBLEMS

There have been numerous studies of sample size and its effect on precision
of analyses (Gy, 1966; Clifton et al., 1969). The problem can be visualized
from the following example.

Assume that a sample contains 5 g Au per ton. A 200-g sample will then contain 1 mg or 1 ppm. However, a 1-mg piece of gold is a rather common size found in placer deposits. Thus, if the gold is all in a single 1-mg piece, there is no way to obtain a representative 5-g sample for chemical extraction or a 30-g sample for fire assay. Even if the gold were in forty 0.025-mg pieces, more than one-third of the 5-g samples would contain no gold (Clifton et al., 1969, fig.5).

Clifton et al. (1969) concluded that 20 pieces of gold are required in the analytical portion. For the above case, the gold should occur in pieces with an effective weight of slightly more than 1 μg or in a size range of 0.05—0.10 mm, depending on particle shape. Clearly, if the sample contains a significant amount of "coarse" gold (e.g. greater than 0.10 mm in diameter) there is a problem in obtaining a meaningful analysis.

There are three ways to solve the problem of sample size: (1) increase the size of the analytical portion, (2) reduce the size of the gold, or (3) segregate the coarse gold and treat separately.

If the size of the analytical portion is increased, we reach practical and economic limits. Until recently, the standard sample weight for geochemical analysis was 2 g. Some analysts are now using 5 g and it appears that 10 g is quite practical. Anything larger results in serious laboratory problems and unacceptable costs. The standard analytical portion for fire assaying is 1 assay ton (29.166 g), although 1/2-A.T. portions have frequently been used in North America. In South Africa the 2-A.T. portion was introduced at the Consolidated Main Reef Mines and Properties in 1958 (Coxon and Sichel, 1959). Although increasing the size of the analytical portion is a significant help in improving precision, alone it will not be satisfactory in all cases.

The size of the gold particles can be reduced by pulverizing. Since gold is very malleable, size reduction is accomplished by abrasion and smearing of gold on rock particles. This results in a more uniform distribution of gold but also has limitations. Reduction to 95% minus 150 mesh (0.104 mm) is readily accomplished, but further reduction is increasingly time-consuming and expensive.

If coarse gold remains in the sample after pulverizing, it must be segregated and treated separately. Segregation of coarse gold can be accomplished by sieving pulverized material through a screen passing 95% of the sample. If metallics are found on the screen they must be removed and analysed separately. In this case the weight of the pulverized sample and weight of the metallics must be known.

If the gold is liberated, as is often the case in sediment samples except close to their source, the coarser gold can be segregated mechanically in the first step of sample preparation. When the particles of gold are very fine and rather uniformly distributed through the rock fragments, pulverizing and rolling will of course be adequate preparation. Rolling the sample is also the simplest and probably one of the most effective techniques in overcoming the segregation problem which occurs whenever samples containing fine gold or gold-bearing sulphides are handled. Cone-and-quartering is effective on coarser material.

SAMPLING AND ANALYTICAL INVESTIGATION

Samples from three properties were subjected to various preparation and analytical procedures.

Field areas

The M.G.S. property is located in northern British Columbia at an elevation of about 1800 m in a sub-arctic climatic zone. Bedrock comprises intermediate flows and agglomerates probably of Upper Triassic age. Hydrothermal alteration is related to fracturing. Pyrite is common and numerous quartz-carbonate veins containing galena, sphalerite and chalcopyrite were observed. Gold appears to be closely related to base metals. Minute particles of gold have been panned from the soils which are residual or colluvial.

Dusty Mac property lies in the southern Okanagan Valley of British Columbia at an elevation of about 400 m in a semi-arid temperate climate. Bedrock is brecciated Tertiary andesite (Church, 1973). There is a large amount of introduced quartz in the ore zone but only small amounts of base-metal sulphides and minor pyrite. The soil is shallow and residual with a poorly developed profile.

The H.M. property is situated in a similar climatic environment to the Dusty Mac. Bedrock is granitic and mineralization is confined to narrow high-grade veinlets. Soils are shallow and residual or colluvial with poorly developed profiles.

Field sampling

Duplicate 300—400-g B-horizon samples were obtained at the Dusty Mac and H.M. properties. Samples are highly siliceous, friable and light brown. In several locations the sample was obtained immediately above bedrock. At each station duplicate bulk samples weighing between 700 g and 7 kg were also obtained. Care was taken to obtain geochemically consistent material and only large rock fragments were discarded. The bulk sample composites were obtained by combining numerous samples previously analysed for gold.

Sample preparation

Replicate samples were dried at 80° C and screened to minus 80 mesh. The plus 80-mesh material was wet-screened, dried, pulverized and rolled.

The larger bulk samples were dried, quartered and divided into approximately equal portions. Half of the bulk sample was screened to minus 80 mesh. The minus 80-mesh material was halved using a Jones sampler. Of the two minus 80-mesh fractions obtained, one was rolled 150 times, whereas the other was analysed without further physical processing. The plus 80-mesh portion was crushed, pulverized, screened to minus 150 mesh and rolled.

A series of composite samples from the M.G.S. property were obtained by combining many gold-bearing soil samples from an area about 300 m in diameter. The composites were processed to produce materials of several mesh sizes.

Sample analyses

All the digestion, preconcentration and atomic absorption methods are adequately outlined by Beamish and VanLoon (1972).

Aqua regia

The analytical procedure for the analyses of gold-bearing geochemical materials with aqua regia is modified from the method used by Tindall (1965). 10 g of prepared material is ashed in a porcelain crucible for 1 hour at 550°C. The ashed residue is transferred to a 250-ml beaker for acid digestion. 20 ml of aqua regia is added and the beaker swirled and placed on a warm hot plate. Use of a watch glass to reduce the rate of evaporation is helpful but not mandatory. The sample is digested to dryness and a second 20-ml aliquot of aqua regia added. Digestion is again taken to dryness and the residue baked for at least 1 hour on the hot plate. The residue is then dissolved and released from the beaker surface with 75 ml of 25% hydrochloric acid. A few drops of nitric acid is added to stabilize the gold in solution. If the residue is extremely fine a filtration step may be desirable to improve subsequent chemical extraction of gold. The digested sample is stable for at least 10 days in a closed container.

Solvent extraction is performed in a 125-ml Teflon or Pyrex separatory funnel into which all of the digestion solution is decanted. 3 ml hydrobromic acid is added followed by 10 ml of methylisobutyl ketone (MIBK). Gold bromide is extracted into the organic phase by vigorously shaking the mixture for 60 seconds. The organic phase is then allowed to separate from the aqueous phase which is discarded. 25 ml of an acid solution containing 2% hydrobromic and 2% hydrochloric acid is added to the MIBK extract. 10 seconds of vigorous shaking removes excess iron from the MIBK.

A set of gold standards is extracted in the same fashion and used to calibrate the instrumentation (Techtron AA5). The extraction—preconcentration of gold into an organic solvent effectively eliminates interferences observed in the analytical technique and also provides a much lower detection limit. Thus, with a 10-g sample and 5 ml MIBK it is possible to obtain a detection limit of 10 ppb.

Fire assay

Standard fire assay methods are used to obtain an Ag—Au alloy bead which is digested in aqua regia and the gold content determined by atomic absorption. A much lower detection limit is possible by this method than by weighing the precious metal beads.

Hydrobromic, HCl—NaBrO₃

A 5-g sample of unashed pulp is weighed into a 250-ml beaker. The pulp is moistened with a small amount of water and 0.5 g of sodium bromate, 10 ml hydrochloric acid and 15 ml of hydrobromic acid added. The beaker is immediately covered with a watch glass. Sample and reagents are heated gently, swirled and digested slowly at low heat to dryness. Sample residue is dissolved in 75 ml of 25% hydrochloric acid. Extraction—preconcentration of gold for atomic absorption analyses has been described.

Results

Data appear to indicate that aqua regia digestion will recover more gold than the comparative HBr digestion on unashed material and that pre-ashing of the sample improves recovery (Table I). However, fire assay preconcentration has not been shown to be significantly superior to aqua regia digestion, although results show a smaller variance because of the larger analytical portion.

Table II illustrates some of the problems of sampling and analysis for gold. Some duplicate analyses differ substantially, while others are remarkably close; gold appears to be preferentially concentrated in minus 80-mesh material on the H.M. property, but not on the Dusty Mac; rolling appears to improve reproducibility on the Dusty Mac, but not at H.M. The 300-g field sample appears to be adequate on these two properties.

Table III shows that screening M.G.S. samples concentrated gold in the finer fraction and that pulverizing reduced the relative standard deviation of the sample population. Pulverizing to minus 200 mesh with glass mortar and pestle did not reduce the relative standard deviation, compared with minus 150 mesh mechanical pulverizing. This is probably due to the superior abrasive action of the mechanical pulverizer. The lower relative standard deviation of DM 27 compared with M.G.S. COMPO A, and B shows the influence of sample character.

CONCLUSIONS

The minimum size of sample that will give an acceptable analytical precision depends on the number and size of gold particles in the sample. About 300 g of residual B soil provide sufficient material for analytical purposes. The plus 80-mesh (pulverized) and minus 80-mesh (screened) material derived from residual soils are remarkably similar in elemental content. Although the minus 80-mesh fraction appears to be satisfactory in some cases, experience indicates that pulverizing of the entire sample is desirable. Results can be improved, within limits, by pulverizing and rolling the sample, and by taking larger portions for analysis. Segregation of liberated gold during handling of the sample can cause extremely erratic results but can be partially overcome by rolling the sample before weighing. Mechanical preconcentration of the gold before

TABLE I

Sample process technique and the analyses of gold (ppb) by atomic absorption spectrophotometry

Sample No.	HBr, HCl—NaBrO$_3$ digestion (sample unashed)	Aqua regia digestion (sample unashed)	Aqua regia digestion (sample ashed)	Fire assay (Ag—Au alloy dissolved in aqua regia)
D.M. 6	350	1150	680	950
D.M. 6	375	760	375	580
D.M. 6	680	930	500	615
D.M. 10A	870	1300	1405	1645
D.M. 10A	1200	1385	1205	1510
D.M. 10A	900	1180	6080	1455
D.M. 10B	2400	3400	3600	3675
D.M. 10B	4320	2920	4800	5155
D.M. 10B	2480	2920	4520	4040
D.M. 37	50	80	50	35
D.M. 37	30	50	50	55
D.M. 37	50	50	50	55
D.M. 38	50	50	50	55
D.M. 38	50	50	30	35
D.M. 38	50	50	50	55
H.M. 12	250	280	350	410
H.M. 12	190	280	405	370
H.M. 12	220	280	375	370
H.M. 13	280	560	850	540
H.M. 13	280	560	620	540
H.M. 13	405	540	680	595

Note: all samples were pulverized to minus 150 mesh prior to multiple gold analyses. A 5-g portion of each sample was used for all acid digestions.
The fire assay preconcentrations utilized 29.166 g of sample.

analysis will be required in some cases. Pre-ashing of a 10-g sample before digestion with aqua regia followed by methylisobutyl ketone (MIBK) extraction and atomic absorption analysis is the preferred analytical procedure.

ACKNOWLEDGEMENTS

Acknowledgement is made to Mr. H.H. Bichler, Mr. H.P. Shafer and Mr. B.L. Twaites of Chemex Labs for their contribution to field and laboratory activities.

TABLE II

Multiple analyses of duplicate geochemical samples for gold content

Duplicate samples	−80-mesh fraction (sample rolled)		−80-mesh fraction (sample unrolled)		+80-mesh fraction (pulverized to −100 mesh)
D.M. 3	310	350	1230	375	310
D.M. 4	375	405	310	310	620
D.M. 18	160	220	220	310	160
D.M. 19	190	190	220	350	190
D.M. 20	160	160	190	250	280
D.M. 21	130	160	130	250	1070
D.M. 28	310	250	405	375	405
D.M. 29	470	440	540	375	440
D.M. 31	50	50	30	80	30
D.M. 32	30	30	30	80	30
D.M. 35	80	30	30	80	80
D.M. 36	80	80	80	50	80
H.M. 1	560	790	470	560	160
H.M. 2	760	500	790	620	220
H.M. 5	1930	2520	1865	2400	2400
H.M. 6	850	930	700	2000	190
H.M. 10	1020	955	1050	1070	955
H.M. 11	680	790	730	930	560
H.M. 16	80	50	80	80	< 30
H.M. 17	80	80	50	30	110
H.M. 18	30	80	50	30	< 30
H.M. 19	190	50	30	50	< 30
H.M. 26	130	130	130	80	< 30
H.M. 27	130	130	110	110	375
H.M. 30	80	50	50	50	< 30
H.M. 31	50	130	80	30	< 30

Note: gold analyses using a 5-g sample.

We also wish to gratefully acknowledge the cooperation of Sumitomo Metal Mining Canada Ltd. and to thank Dr. G.H. Laycraft, Executive Vice-President of Dusty Mac Mines Ltd., for permission to sample the property and release the results.

TABLE III

Statistical parameters

Sample No.	Number of determinations	Mean Au (ppb)	Standard deviation	2 standard deviations
M.G.S. COMPO A	20	1062	325	650
M.G.S. COMPO A$_1$	20	1096	236	472
M.G.S. COMPO B	10	760	609	1218
M.G.S. COMPO B$_1$	19	584	118	236
M.G.S. COMPO C	20	759	158	316
D.M. 27	20	1141	196	392

Notes. (1) Procedures used to prepare samples M.G.S. COMPO A$_1$, M.G.S. COMPO B$_1$ and D.M. 27: (a) sample material was mechanically pulverized, (b) pulverized material was screened to 100% minus 150 mesh, (c) screened material was rolled 200 times.
(2) Sample M.G.S. COMPO C was pulverized to 100% minus 200 mesh using a glass mortar and pestle. Each composite rolled 150 times.
(3) M.G.S. COMPO A and M.G.S. COMPO B are composed of minus 80-mesh and minus 20- plus 80-mesh material, respectively.

REFERENCES

Beamish, F.E. and VanLoon, J.C., 1972. Recent Advances in the Analytical Chemistry of the Noble Metals. Pergamon Press, Oxford, 511 pp.
Church, B.N., 1973. Geology of the White Lake basin. B.C. Dept. Mines and Pet. Res. Bull., 61: 89—92
Clifton, H.E., Hunter, R.E., Swanson, F.J. and Phillips, R.L., 1969. Sample size and meaningful gold analysis. U.S. Geol. Survey Prof. Paper 625-C: 1—17
Coxon, C.H. and Sichel, H.S., 1959. Quality control of routine mine assaying and its influence on underground evaluation. J.S. Afr. Inst. Min. Metall., 59: 489—517
Gy, P., 1966. Sampling of Materials in Bulk — Theory and Practice (Handbook in French). Société de l'Industrie Mineral., Saint-Etienne
Tindall, F.M., 1965. Silver and gold assay by atomic absorption spectrophotometry. Atomic Absorpt. Newslett., 4(9): 339—340

THE MODE OF OCCURRENCE OF TRACE ELEMENTS IN SOILS AND STREAM SEDIMENTS APPLIED TO GEOCHEMICAL EXPLORATION

ARTHUR W. ROSE

Department of Geosciences, Pennsylvania State University, University Park, Pa. (U.S.A.)

ABSTRACT

Trace elements in soils and stream sediments can occur as major elements in trace minerals, as trace constituents of unweathered minerals from the parent material, as trace constituents in minerals formed during weathering, and as ions adsorbed to colloidal particles or in the exchange layer of clays. The behavior of trace elements in these occurrences can be understood in terms of solubility of trace minerals and host minerals, ion exchange reactions, and surface chemistry of colloidal particles. Methods for selectively removing trace elements in the iron oxide, manganese oxide, organic, and exchangeable sites are reviewed in terms of chemical principles. These methods or simplified versions of them can be used to detect weak anomalies, to investigate the origin of puzzling anomalies, and to plan methods of partial extraction.

INTRODUCTION

As geochemical exploration becomes more complicated and sophisticated, there is a greater need to approach problems on the basis of chemical and physical principles rather than empiricism and past practice. In order to do this, natural materials and solutions must be understood on a more fundamental level, in order to interpret data on geochemical surveys, and in order to select and design analytical methods using partial extraction.

In the area of stream sediments and soils, we are concerned with understanding the mobility, geochemical characteristics, and chemical behavior of trace elements in a variety of modes of occurrence. Important modes of occurrence for trace elements and parameters controlling their mobility include the following:

(1) Occurrence as a major element in a trace mineral, such as copper in malachite ($Cu_2CO_3(OH)_2$), lead in anglesite ($PbSO_4$) or native gold. The behavior of trace elements in these forms will be dependent mainly on simple solubility and solution chemistry.

(2) Occurrence as a trace constituent in a mineral inherited from the pre-weathering parent material, such as zinc in magnetite, lead in K-feldspar or copper in biotite. Trace element behavior in this mode will depend on the properties of the host mineral. If and when this host mineral is destroyed,

then simple solution chemistry and solubility will come into play. Until this time, the element will behave like the host mineral, being sorted and partitioned according to its grain size, density, and chemical behavior. Minerals inherited from pre-existing rocks are likely to be unstable in the surficial environment and to slowly weather, so that the trace element will slowly be released and may then become mobile.

(3) Occurrence as a trace constituent in the lattice of a mineral formed during weathering, or occluded as a trace mineral in such a phase, or adsorbed on such a mineral and covered over by further precipitation. Examples are zinc in the octahedral sites of montmorillonite and vermiculite clays, cobalt and copper in iron or manganese oxides, and mercury in organic compounds. Controls on such materials will be the stability and properties of the host phase. Because these phases are formed during weathering, they are likely to be chemically stable in many surficial environments. However, because of the fine grain size and imperfect crystal lattices of most of these phases, elements in such sites tend to be somewhat more accessible to the surrounding solutions than those in coarser-grained, more perfect lattices of minerals formed in igneous, metamorphic or diagenetic environments.

(4) Occurrence as a trace constituent adsorbed as a counter ion on the surface of an iron or manganese oxide, colloidal clay or organic particle, or in the exchange layer of a clay mineral. Elements in these sites are controlled mainly by ion exchange equilibria (somewhat modified by effects of the environment on the substrate). These elements are very sensitive to changes in the composition of the enclosing solution.

A wide variety of partial extraction methods involving acids, buffer solutions, complexing agents, reducing agents and fluxes have been used to improve contrast of anomalies, but in a great many cases, little chemical basis has been presented for understanding the action of these solvents on natural materials. One purpose of this paper is to suggest some approaches to understanding the action of the various extraction treatments. In addition, some data obtained by a sequence of extractions on samples from several areas is presented as an example of the usefulness of the methods in determining mode of occurrence.

The major emphasis in discussing these treatments is on mineral-solution equilibria, but a second factor that must be considered is reaction rate, especially with laboratory extraction procedures. Although it is difficult to generalize on reaction rates, comments are made where relevant information is available.

STABILITY OF IMPORTANT MINERALS

Trace metals contained in the lattice of minerals composed of major elements are dependent on stability of the major phases, so such stabilities must be considered in any discussion of partial extraction procedures. The minerals formed during weathering tend to be fine grained and relatively sensitive to changes in their environment. In contrast, minerals that have resisted weathering are usually relatively resistant to chemical treatments. For this reason, the stability of clays, Fe—Mn oxides, and organic materials is of most interest in terms of the effect of partial extraction treatments on release of trace elements.

The stability of iron and manganese oxides in water as a function of Eh and pH is summarized in Fig.1. In general, the iron oxides are stable under all conditions except very acid solutions or relatively reducing conditions. Manganese oxides are stable (at $Mn^{2+} \leqslant 10^{-6}M$) only under very oxidizing neutral to basic conditions. Because of the variety of organic compounds, no single stability can be specified, but most organic compounds are truly stable only near the H_2—H_2O line near the bottom of the diagram, though meta-stability is obviously pronounced.

The clay minerals vary in their stability from one mineral to another, but dissolution of clay (as contrasted to conversion of one clay to another) will be controlled largely by solubility of aluminum. Within the pH range of 5—8, aluminum solubility is low (Parks, 1972) and effects on clay minerals are expected to be at a minimum. At more acid pH values (3 and less), clays dissolve significantly, and trace elements in their lattices will be released at a significant rate.

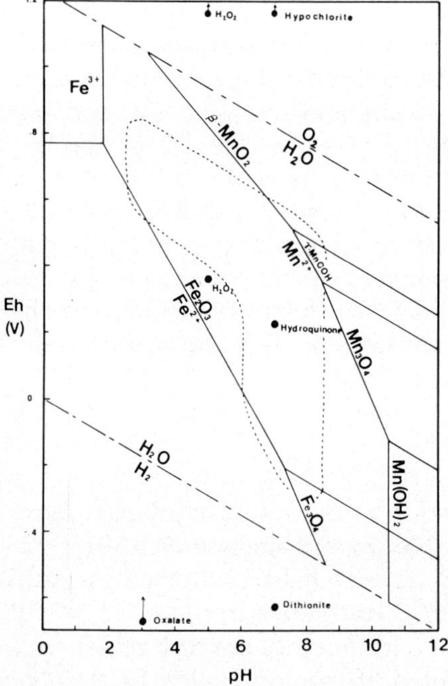

Fig.1. Eh—pH conditions of chemical treatments in relation to stability of iron and manganese oxides and oxyhydroxides. Dotted lines enclose the Eh—pH range of the natural environment. Data for iron species from Garrels and Christ (1965); manganese species from Bricker (1965). Standard oxidation potentials of reagents from Latimer (1952), calculated to Eh for reactant concentrations used in each method and product concentrations of $10^{-6}M$.

Based on solubility data of this sort, one can formulate procedures for selectively dissolving one constituent of a sediment or soil without affecting others and can estimate the effect of various treatments on the components of a sample. For instance, moderately reducing conditions at neutral pH will dissolve manganese oxides with minimal effect on clays, iron oxides, and organic matter. More strongly reducing conditions, preferably at a slightly acid pH, would be chosen to remove iron and manganese oxides without dissolving clays. Strongly acid conditions, especially if not highly oxidizing, would dissolve iron and manganese oxides, clays, and most trace minerals, but attack on coarse igneous and metamorphic grains would be slow. The general procedure is readily extended to other phases that may be present in the sample, such as sulfides or carbonates.

Once the host mineral is dissolved, the behavior of the trace metal depends on its solubility in the medium. For instance, Cu^{2+} is very insoluble at pH 7 because of formation of CuO or equivalent compounds. Therefore, a second requirement of a selective extraction procedure is the solubility of the trace metal of interest. The trace metal must either be soluble in the solution used to decompose the host, or a second treatment must be used to dissolve the trace metals released in the first step. The former alternative is clearly preferable but is not always convenient.

Similar considerations are involved in the solution of trace metals occurring as trace minerals. The behavior of copper and molybdenum as a function of Eh and pH has been discussed by Hansuld (1966), and similar considerations are useful for other minerals. In order for the minerals to dissolve in reasonable times, a considerably undersaturated solution is needed, both because of possible slow reaction of the trace mineral and because of possible difficult accessibility of the trace mineral, owing to occurrence within composite grains.

Another approach to understanding the behavior of trace elements in samples is presented by Ellis et al. (1967). These workers have examined the rate of dissolution of copper and iron from samples of soils and stream sediments reacted with a variety of extractants.

REMOVAL OF IRON OXIDES

Variations of two methods have been used in recent years to remove iron oxides selectively from soils and sediments. Mehra and Jackson (1959) describe the use of sodium dithionite ($Na_2S_2O_4$, also called sodium hyposulfite) in a buffered citrate solution of pH 7.3 at 80°C to remove iron oxides, including hematite and goethite, from soils. Magnetite and ilmenite are not attacked, and iron silicates such as nontronite and glauconite are only slightly dissolved. The dithionite acts as a strong reducing agent in basic solution (Eh = −0.7 V, observed by Mehra and Jackson, 1959). Citrate acts to complex iron and thus promotes the reaction. This method, with minor modifications, has been used to extract trace elements in iron oxides by Rose and Suhr (1971), Gibbs (1973), and others.

The other method uses ammonium oxalate to reduce and complex iron. The reducing action of oxalate ion is catalyzed by light, so can be varied from a weak attack in the dark to strong dissolution in sunlight or UV light (De Endredy, 1963; McKeague and Day, 1966). The pH of the medium is generally 3 or lower, but the reaction goes at a slower rate in solutions of higher pH. Related methods, using nascent hydrogen and H_2S as reducing agents, are referenced in Mehra and Jackson (1959). Tartrate has also been used as a complexing agent instead of citrate (Dion, 1944).

Both dithionite and oxalate dissolve some aluminium and silicon in addition to iron. The dissolved aluminium and silicon are thought to be derived from amorphous or very poorly crystalline clays and oxyhydroxides. Mehra and Jackson (1959) present data showing removal of 2—15% SiO_2 and 1.5—10% Al_2O_3 from soils with dithionite, the higher values being from allophane-rich soils. The oxalate treatment in the dark does not dissolve appreciable hematite and goethite (McKeague and Day, 1966). A comparison of dithionite- and oxalate-extractable iron and aluminium has been used to distinguish well-crystalline from poorly-crystalline iron oxides in soils (Gamble and Daniels, 1972).

In general, it appears that the dithionite method dissolves iron oxides more thoroughly, that both methods dissolve some amorphous to poorly crystalline clays, and that the effect of low pH (3) on clays with the oxalate may be offset by the effect of higher temperatures with the dithionite. Several authors discuss the effects of iron removal techniques on clay mineralogy (Dion, 1944; Harward et al., 1962).

Iron oxides are also soluble in strong acids, as suggested by Fig.1 and common digestion techniques. Concentrated ($6M$) HCl will dissolve large amounts of earthy hematite, even when cold, based on experiments by the writer, but more dilute HCl and even concentrated oxidizing acids such as HNO_3 and $HClO_4$ dissolve only small amounts (10—30 ppm) when cold. Heating to boiling for 1 hour results in solution of considerable iron (3000 ppm) by $1M$ HCl and only slightly smaller amounts (1000—3000 ppm) by $1M$ and $6M$ HNO_3, but the hot $0.1M$ acids dissolve only minor amounts (10—100 ppm). These solubilities correspond approximately with relations expected from the Eh—pH diagram (Fig.1) and the underlying solubility relationships. A solubility of $10^{-2}M$ Fe^{3+} (560 ppm) is reached only around pH 0 ($1M$ acid) under oxidizing conditions (Garrels and Christ, 1965). For thorough solution of iron oxides, acids of at least $1M$ strength, preferably combined with elevated temperatures, are needed. Under such conditions, one can expect considerable solution of clays as well.

REMOVAL OF MANGANESE OXIDES

Methods that have been used for removal of manganese oxides include the following:

(1) Hydroxylamine hydrochloride, mostly used at pH values of 1— 4

(Canney and Nowlan, 1964; Chester and Hughes, 1967; Chao, 1972; Chao and Sanzolone, 1973).

(2) Hydroquinone (0.2%) in an ammonium acetate buffer (Sherman et al., 1972; Horsnail et al., 1969) at pH 7 for 6 hours.

(3) Hydrogen peroxide at pH values of 3—5 and elevated temperatures (Jackson, 1956; Kunze, 1965; Rose and Suhr, 1971).

(4) Dithionite in acid solution (Daniels et al., 1962; Anderson and Jenne, 1970; Gibbs, 1973) usually with a citrate buffer to complex iron released along with manganese.

(5) Ammonium oxalate at pH 3.3 (Le Riche and Weir, 1963; De Endredy, 1963) will dissolve manganese oxides as well as iron oxides.

All of these reagents act as reducing agents, as indicated by the Eh values plotted on Fig.1 for the approximate conditions of the experiments. Note that the first three, which reduce manganese but not iron, are not as strong reducing agents as the latter two.

Chao (1972) has tested hydroxylamine hydrochloride at pH 1 and room temperature on a variety of iron and manganese oxides. Poorly crystallized manganese oxide dissolved instantly; well-crystallized MnO_2 (pyrolusite) dissolved in 30—60 minutes; hematite, goethite, and magnetite dissolved only in traces; amorphous $Fe(OH)_3$ dissolved to an extent of about 8%; and natural iron oxides to a lesser extent. He recommends a pH of 2, but values of 3 seem nearly as good, and even higher pH values should result in slow dissolution.

In addition to the treatments described above, manganese oxides are soluble in strong acids along the lines discussed under iron oxides.

REMOVAL OF ORGANIC MATERIAL AND SULFIDES

Two main treatments have been used for selective dissolution of organic matter: hydrogen peroxide (Kunze, 1965; Rose and Suhr, 1971) and sodium hypochlorite (Gibbs, 1973), both very strong oxidizing agents. Treatment with hydrogen peroxide does not completely dissolve some types of organic matter, such as cellulose, but probably decomposes the varieties which contain the bulk of the trace metals. It also dissolves manganese oxides and sulfides. Sodium hypochlorite does not appear to dissolve manganese oxides but is expected to decompose sulfides.

In addition to these two reagents, organic matter can be decomposed by oxidizing acids such as nitric and perchloric acids, obviously with at least partial decomposition of clays, iron oxides, sulfides, and other components. Some buffer solutions, such as sodium acetate, also dissolve part of the organic material, judging from the color of extracts.

Lynch (1971) describes a technique for selectively dissolving sulfides from rocks utilizing H_2O_2 with ascorbic acid.

EXCHANGEABLE AND ADSORBED IONS

Based on differences in behavior, several categories of occurrence falling into this group can be differentiated.

Ions in exchange sites in clay minerals

The largest ion exchange effects are observed for montmorillonite and vermiculite, which have cation exchange capacities 5—10 times as great as typical kaolinites and most illites (Carroll, 1959). In the montmorillonites and vermiculites, the basic lattice of the clay mineral has a net negative charge, which is balanced by cations such as Ca^{2+} and Na^+ held between the layers, usually with envelopes of water molecules. The exchange capacity is essentially fixed by the nature of the basic clay lattice and is relatively insensitive to changes in the surrounding solution, at least up to the point at which the clay mineral is no longer stable.

Reactions for trace constituents in exchange sites are of the type:

$$Ca\text{-clay} + Zn^{2+} = Zn\text{-clay} + Ca^{2+}$$

where Ca-clay indicates the clay with Ca^{2+} in the exchange site, Zn-clay indicates clay with Zn^{2+} in the exchange site, and Ca^{2+} and Zn^{2+} are solution species. If both solution species are present, the clay will have a mixed population of ions in the exchange site. The content of Zn in the clay is dependent on Ca^{2+} in solution as well as Zn^{2+}.

Experimental data giving distribution coefficients between solution and · clay are available for major elements and some trace elements (Carroll, 1959). In general, ions of higher valence are preferred in the exchange sites of clays over low valent ions. However, some univalent ions (K^+, NH_4^+) form strong bonds in the interlayer space, cause collapse of the interlayer space, and are no longer exchangeable. In this situation, the ions are said to be fixed. Partly for this reason, NH_4^+ is commonly used to displace other ions from exchange sites in laboratory studies of clay systems. In general, divalent ions, such as Mg^{2+} are preferable to univalent ions such as Na^+ for this purpose, but this approach does not usually seem to have been followed in setting up laboratory procedures for estimating exchangeable ions.

Reaction rates for ion exchange in clays are very rapid if the clays are dispersed, and ions in solution have ready access to the clay particles. Under these conditions, the exchange reaction goes essentially to equilibrium in seconds (Kennedy and Brown, 1966). However, in many natural clays, the particles are agglomerated into clumps or coated with iron or manganese oxides or other impediments to reaction. As a result, the system approaches equilibrium much more slowly, in 3—7 minutes, based on studies by Kennedy and Brown (1966). Such agglomeration probably accounts at least in part for the continuing slow reaction observed in many dithizone tests (Hawkes, 1963). In such cases, diffusion along grain boundaries within particles, rather than

698

simply within the interlayer space, is probably the rate-controlling factor.

For some trace elements, complex ions are known to enter the interlayer space in addition to simple cations. For instance, $ZnOH^+$, $ZnCl^+$, and $CuOH^+$ have been recognized as significant components in exchange sites (Elgabaly and Jenny, 1943; Demumbrum and Jackson, 1956a,b). In such cases, the pH or content of Cl^- in solution is an additional control on the content of the trace element in the clay.

Ions adsorbed on colloidal iron and manganese oxides

Colloidal particles of many insoluble oxides are non-stoichiometric, containing an excess of cations or anions, depending on whether the surrounding solution has a high concentration of one or the other of the ions forming the particle. Two types of adsorption are possible for such colloidal particles. Ions may be adsorbed onto the surface of the particle and become a part of the lattice, in this way contributing to determining the electrical potential of the particle and its exchange capacity (potential-determining ions). For iron and manganese oxides, the important potential-determining ions are H^+, OH^-, and various iron and manganese species in solution, especially Fe^{2+} and Mn^{2+} (Parks and De Bruyn, 1962, Morgan and Stumm, 1964). If H^+ or OH^- are potential-determining ions, as is the case with oxides and hydroxides, the pH at which the particle is uncharged is termed the zero point of charge (ZPC) or isoelectric point. Morgan and Stumm (1964) find that Zn^{2+} is strongly adsorbed by manganese oxides, and suggest that strongly hydrolyzable species such as Zn^{2+} will be more strongly sorbed than ions such as Ca^{2+} and Mg^{2+}.

A second mode of occurrence is as counter-ions held near the particle to balance the charge of the particle. The ions adsorbed as counter ions are those of charge opposite to the particle. Values of ZPC for iron oxides range from 6.7 for goethite to 8.5 for amorphous iron oxyhydroxide and 8.6 for recently precipitated hematite (Parks, 1965). In neutral and acid solution, the iron oxides are, therefore, positively charged and will adsorb anions such as molybdate and arsenate. In contrast, for manganese oxides, Murray et al. (1968) report values of ZPC of 1.5 to 7.5, depending on mineral species, with the lower values for poorly crystallized species forming directly from solution. The manganese oxides thus tend to have negative charges when newly precipitated and will adsorb cations (Posselt et al., 1968). The counter ions follow ion exchange equilibria analogous to the clay minerals; however, the variable exchange capacity caused by changes in H^+, OH^-, or other potential-determining ions makes the situation more complex than for clay minerals.

Ions adsorbed on broken surfaces of clays constitute most of the exchangeable ions in kaolinite and illite and behave analogously to the counter ions on oxides. Analyses of leachates produced by dissolution of iron oxides show that the oxides contain high concentrations of trace elements in the range of tenths of a percent. Manganese oxides show similar behavior, as indicated by the results of Canney et al. (1964) and the high content of nickel, copper, and other elements in sea-floor manganese nodules. It is likely that these cations are strongly sorbed potential-determining ions rather than counter-ions.

Counter-ions exchange rapidly if access to solution is easy. Potential-determining ions adsorbed to the lattice may be difficult or impossible to desorb, particularly if the crystal has aged and recrystallized since adsorption. Because the exchange capacity, charge, and crystallinity of oxides are easily changed by changes in the external environment, by aging, and especially by treatments such as heating and drying, the behavior of these materials in partial extraction procedures must be carefully considered. Any kind of sample treatment is likely to modify the amount of exchangeable ions found in these materials.

Ions adsorbed on surfaces of colloidal organic particles

A high exchange capacity is generally found for organic material in soils and stream sediments (Carroll, 1959). The general principles involved are similar to those discussed under iron and manganese oxides, but the complexity and range of organic materials make detailed treatment complex, and little information on this aspect of soils and sediments is yet available. Adsorbed cations will tend to be released by exchange with high concentrations of cations in the surrounding solution, and their behavior will be affected by any treatment that modifies the character of the organic material, such as heating, drying, oxidation, or natural organic reactions.

Procedures for analysis of exchangeable ions

In order to remove ions from exchange sites, the equilibrium in the exchange reaction:

Tr-clay + M(aq) = M-clay + Tr(aq)

where Tr(aq) and M(aq) are the trace and exchanging major ion in solution, and Tr-clay and M-clay indicate the same ions in exchange sites, must be shifted to the right. This may be accomplished by a large excess of the exchanging major ion or by making the activity of the trace ion very low by means of complexing or removal from solution or some combination of these two. The traditional dithizone tests and more recent procedures using EDTA (Thierweiler and Lindsay, 1969; Maynard and Fletcher, 1973) utilize the latter approach to aid in exchange. The main function of pH buffers in these procedures is to control the complexing reaction, but they also may serve the useful purpose of avoiding decomposition and dissolution of minerals because of acid pH values, as noted in the discussion of mineral stability.

APPLICATIONS

Two general areas are envisioned for application of selective solution and partition studies of the type discussed above. Exploration geochemists frequently want to know why certain samples are high in an indicator element, either because they want to understand the controls for "false" anomalies or because they want to understand the source and mode of dispersion of anomalies indicative of ore. In general, they want to have an understanding of the geochemical behavior and mobility of indicator elements.

A second general area of interest is planning the best method of partial extraction in order to optimize the contrast of ore-related anomalies relative to background. If the mode of occurrence of metal in anomalous and background samples can be specified, then an analytical method for selectively extracting the components to give high contrast can be developed.

In soils, a common question is the extent of leaching or residual concentration in forming the soil from the parent material and the controls on leaching. If the indicator element occurs mainly as an immobile primary mineral, then it has probably not been leached to any great extent and may have been concentrated by preferential leaching of other constituents. If the indicator element occurs in iron oxides, its abundance may depend on the abundance of iron in the parent material or the behavior of iron during weathering. Similar inferences can be made concerning indicator elements enriched in organic fractions or clays.

In stream sediments, the source of metal in anomalous samples is frequently in question. Study of the mode of occurrence can lead to recognition of the source as clastic, hydromorphic, or biogenic. Horsnail et al. (1969) have used this approach in investigating the source of high trace element contents in stream sediments in poorly drained areas containing abundant Fe—Mn oxides.

Three examples of the kind of data obtained from selective extraction experiments are described below. In all three cases, the method used for the partial extractions was that summarized in Rose and Suhr (1971), with modifications in the case of platinum and palladium to dissolve the platinum metals.

In soils of the Stillwater complex, Montana, palladium is found to be concentrated in clays and in the organic fraction (Fig.2), both phases formed during weathering (Fuchs, 1972; Fuchs and Rose, 1974). This distribution indicates a relatively soluble host mineral for palladium in the parent material and at least a slight degree of mobility. In contrast, platinum in some of the same samples is highest in the silt fraction, probably occurring largely within chromite. In other samples, the platinum is high in the iron oxide, organic, and exchangeable fraction; platinum is also found to be much less leached from the soils than palladium. All these features are in agreement with predictions from an Eh—pH diagram showing that platinum is expected to be less mobile than palladium. As a result, dispersion of palladium is expected to form larger halos around sources, and partial extractions may be useful for palladium but not platinum.

In the soils of the State Line chromite district, Pennsylvania, nickel and

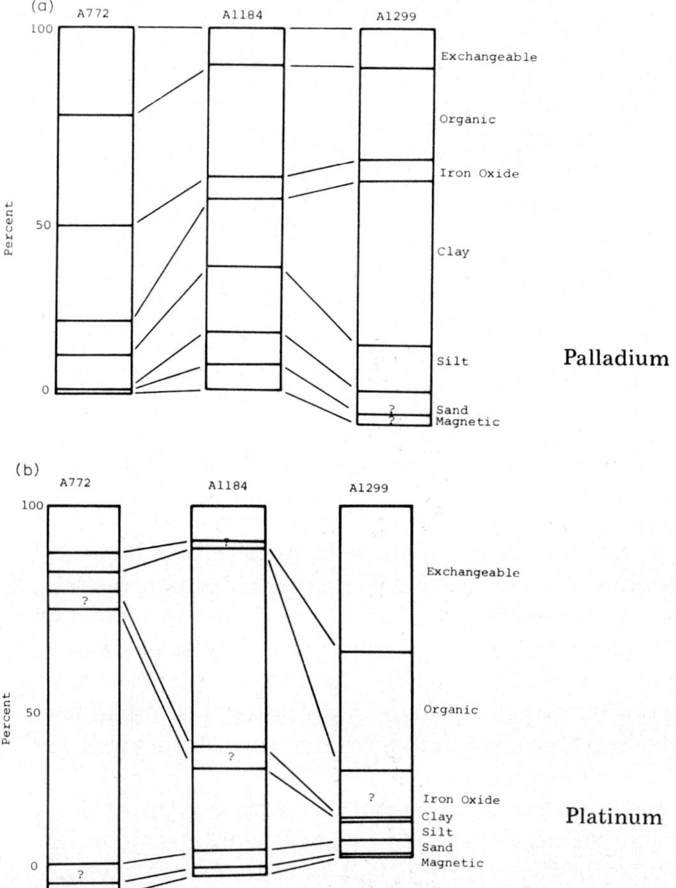

Fig.2. Contribution of palladium and platinum from phases in soils of the Stillwater Complex, Montana. Division of bar indicates percentage of total element in the sample contributed by various phases. Query indicates that amount was below the detection limit of the analytical method. (a) palladium, (b) platinum.

chromium anomalies occur over chromite ores (Pennington, 1973). As shown in Fig.3, chromium is strongly concentrated in the sand and silt fractions, indicating that residual chromite is responsible for the anomalies, a conclusion reinforced by microscopic examination of the sand and silt. In contrast, nickel occurs in large amounts in the iron-oxide fraction. The relationship of the nickel anomalies to the chromite ores is not understood but it is clear that the magnitude of nickel anomalies may be affected by variations in the behavior of iron in the soils. Profiles through the soils show that the highest nickel values occur in the B-soil horizon, as might be expected from the occurrence in iron oxides.

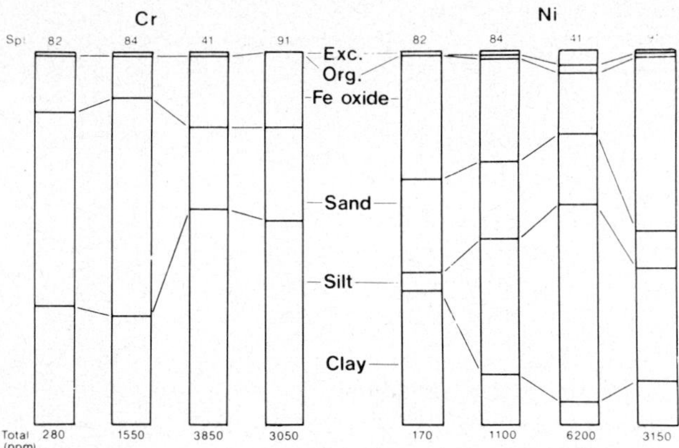

Fig.3. Contribution of chromium and nickel from phases in soils of the State Line chromite district, Pennsylvania. Division of bar indicates percentage of total element in the sample contributed by the phases.

A similar set of data on partition of elements was presented by Rose and Suhr (1971) for stream sediments. The partition of cobalt, copper, nickel, and zinc among the phases is summarized in Table I in which the average percentage of metal contributed by each phase in background samples decreases in the order shown.

Note that the clay fraction contains the highest proportion of metal for three of the four elements, and that iron oxides are the most important for cobalt and are important for all elements.

This data can be applied to estimating the results of various sample dissolution treatments. A leach with concentrated hot acid would remove the metal in iron oxides and most of the metal in clays as well as from the exchangeable and organic fractions, but would probably extract only a limited proportion of the sand and silt fractions. Weak cold acid would dissolve only a small proportion of the metal outside the organic and exchangeable fractions. The variability of amounts liberated by these treatments from sample to sample can be estimated by examination of the data for individual samples.

TABLE I

Average percentage of metals contributed by various phases in background samples

	Co (%)		Cu (%)		Ni (%)		Zn (%)
Fe oxide	45	clay	35	clay	30	clay	34
Silt	19	silt	22	Fe oxide	29	Fe oxide	22
Sand	14	Fe oxide	19	silt	24	silt	20
Clay	11	sand	16	sand	13	sand	17
Exch.	9	organ.	7	exch.	3	exch.	5
Organ.	2	exch.	1	organ.	1	organ.	2

It is clear that processes affecting precipitation and fractionation of iron oxides in the sediments are of importance for all elements, especially cobalt, and that variations of clay content, perhaps related to size sorting, can have a significant effect in background concentrations.

CONCLUSIONS

A wide variety of techniques are available for selective extraction of trace metals from various fractions in stream sediments and soils. The resulting data can be useful in interpreting the source of anomalies and in estimating the effects of various acid attacks, buffers, and other partial extractions on the samples. In many cases, the effect of a given treatment can be estimated by considering the stability of various trace metal host phases in the extractant. The mobility and general behavior of trace elements in the natural environment can similarly be estimated.

APPENDIX (from Latimer, 1952)

$$\text{Dithionite } S_2O_4{}^{2-} + 4OH^- = 2SO_3{}^{2-} + 2H_2O + 2e^- \qquad E^0 = -1.12$$
$$\text{Oxalate (approx.) } H_2C_2O_4 = 2CO_2 + 2H^+ + 2e^- \qquad E^0 = -0.49$$
$$\text{Hydrogen peroxide } H_2O_2 = 2H^+ + O_2 + 2e^- \qquad E^0 = 0.682$$
$$\text{Hydrogen peroxide } 2H_2O = H_2O_2 + 2H^+ + 2e^- \qquad E^0 = 1.77$$
$$\text{Hydroquinone } C_6H_4(OH)_2 = C_6H_4O_2 + 2H^+ + 2e^- \qquad E^0 = 0.699$$
$$\text{Hypochlorite } Cl^- + 2OH^- = ClO^- + H_2O + 2e^- \qquad E^0 = 0.94$$

REFERENCES

Anderson, B.J. and Jenne, E.A., 1970. Free iron and manganese oxide content of reference clays. Soil Sci., 109: 163—169

Bricker, O.P., 1965. Some stability relations in the system Mn—O_2—H_2O at 25° and one atmosphere total pressure. Am. Min., 50: 1296—1354

Canney, F.C., Dennen, W.H. and Post, E.B., 1964. Some observations on the evaluation of stream sediment geochemical data (abstract). Min. Eng., 16(12): 92

Canney, F.C. and Nowlan, G.A., 1964. Solvent effect of hydroxylamine hydrochloride in the citrate soluble heavy metals test. Econ. Geol., 59: 721—723

Carroll, D., 1959. Ion exchange in clays and other minerals. Bull. Geol. Soc. Am., 70: 749—780

Chao, T.T., 1972. Selective dissolution of manganese oxides from soils and sediments with acidified hydroxylamine hydrochloride. Soil Sci. Soc. Am. Proc., 36: 764—768

Chao, T.T. and Sanzolone, R.F., 1973. Atomic absorption spectrophotometric determination of microgram levels of Co, Ni, Cu, Pb, and Zn in soil and sediment extracts containing large amounts of Mn and Fe. J. Res. U.S. Geol. Surv., 1: 681—686

Chester, R. and Hughes, M.J., 1967. A chemical technique for the separation of ferro-manganese minerals, carbonate minerals and adsorbed trace elements from pelagic sediments. Chem. Geol., 2: 249—262

Daniels, R.B., Brasfield, J.F. and Kiecken, F.F., 1962. Distribution of sodium hydrosulfite extractable Mn in some Iowa soil profiles. Soil Sci. Soc. Am. Proc., 26: 75—78

De Endredy, A.S., 1963. Estimation of free iron oxides in soils and clays by a photolytic method. Clay Min. Bull., 5: 209—217

704

Demumbrum, L.E. and Jackson, M.L., 1956a. Copper and zinc exchange from dilute
   neutral solutions by soil colloidal electrolytes. Soil Sci., 81: 353—357
Demumbrum, L.E. and Jackson, M.L., 1956b. Infrared absorption evidence on exchange
   reaction mechanism of copper and zinc with layer silicate clays and peat. Soil Sci. Soc.
   Am. Proc., 20: 334—337
Dion, H.G., 1944. Iron oxide removal from clays and its influence on base exchange
   properties and X-ray diffraction properties of clays. Soil Sci., 58: 411—424
Elgabaly, M.M. and Jenny, H., 1943. Cation and anion interchange with zinc mont-
   morillonite clays. J. Phys. Chem., 47: 399—408
Ellis, A.J., Tooms, J.S., Webb, J.S. and Bicknell, J.A., 1967. Application of solution
   experiments in geochemical prospecting. Trans. Inst. Min. Metall., Sect. B, 76: B25—
   B39
Fuchs, W.A., 1972. Geochemical behavior of platinum, palladium and associated elements
   in the weathering cycle in the Stillwater Complex, Montana. M.S. Thesis, Pa. State
   Univ., University Park, Pa., 92 pp.
Fuchs, W.A. and Rose, A.W., 1974. The geochemical behavior of platinum and palladium
   in the weathering cycle in the Stillwater Complex, Montana. Econ. Geol. (in press)
Gamble, E.E. and Daniels, R.B., 1972. Iron and silica in water, acid ammonium oxalate
   and dithionite extracts of some North Carolina coastal plain soils. Soil Sci. Soc. Am.
   Proc., 36: 939—943
Garrels, R.M. and Christ, C.L., 1965. Solutions, Minerals and Equilibria. Harper and Row,
   New York, N.Y., 450 pp.
Gibbs, R.J., 1973. Mechanisms of trace metal transport in rivers. Science, 180: 71—73
Hansuld, J.A., 1966. Eh and pH in geochemical exploration. Can. Min. Metall. Bull.,
   59(647): 315—322
Harward, M.E., Theisen, A.A. and Evans, D.D., 1962. Effect of iron removal and dispersion
   methods on clay mineral identification by X-ray diffraction. Soil Sci. Soc. Am. Proc.,
   26: 535—541
Hawkes, H.E., 1963. Dithizone tests. Econ. Geol., 58: 579 586
Horsnail, R.F., Nichol, I. and Webb, J.S., 1969. Influence of variations in the secondary
   environment on the metal content of drainage sediments. Q. Colo. School of Mines,
   64(1): 307—322
Jackson, M.L., 1956. Soil Chemical Analysis, Advanced Course. Privately printed, Madison,
   Wisc. (2nd printing)
Kennedy, V.C. and Brown, T.E., 1966. Experiments with a sodium ion electrode as a
   means of studying cation exchange rates. In: W.F. Bradley and S.W. Bailey (Editors),
   Proc. 13th Natl. Conf. Clays and Clay Minerals. Pergamon Press, Oxford, pp.351—352
Kunze, G.W., 1965. Pretreatment for mineralogical analyses. In: C.A. Black (Editor),
   Methods of Soil Analysis. American Society of Agronomy, Madison, Wisc., pp.568—577
Latimer, W.M., 1952. The Oxidation States of the Elements and their Potentials in
   Aqueous Solutions. Prentice Hall, Englewood Cliffs, N.J., 392 pp.
Le Riche, H.H. and Weir, A.H., 1963. A method of studying trace elements in soil fractions.
   J. Soil Sci., 14: 225—235
Lynch, J.J., 1971. The determination of copper, nickel and cobalt in rocks by atomic
   absorption spectrometry using a cold leach. In: R.W. Boyle (Editor), Geochemical
   Exploration. Can. Inst. Min. Metall., Spec. Vol., 11: 313—314
Maynard, D.E. and Fletcher, W.K., 1973. Comparison of total and partial extractable copper
   in anomalous and background peat samples. Geochem. Explor., 2: 19—24
McKeague, J.A. and Day, J.H., 1966. Dithionite and oxalate-extractable Fe and Al as aids
   in differentiating various classes of soils. Can. J. Soil Sci., 46: 13—22
Mehra, O.P. and Jackson, M.L., 1959. Iron oxide removal from soils and clays by a
   dithionite-citrate system buffered with sodium carbonate. Proc. 5th Natl. Conf. Clays
   and Clay Min., 5: 317- 327

Morgan, J.J. and Stumm, W., 1964. Colloid chemical properties of manganese dioxide.
J. Colloid Sci., 19: 347- 359

Murray, D.J., Healy, T.W. and Fuerstenau, D.W., 1968. The adsorption of aqueous metal
on colloidal hydrous manganese oxide. In: R.F. Gould (Editor), Adsorption from
Aqueous Solution. Advances in Chemistry Series, No. 79. American Chemical Society,
Washington, D.C., pp.74—81

Parks, G.A., 1965. The isoelectric points of solid oxides, solid hydroxides and aqueous
hydroxo complex systems. Chem. Rev., 65: 177—198

Parks, G.A., 1972. Free energies of formation and aqueous solubilities of aluminum
hydroxides and oxide hydroxides at 25°C. Am. Min., 57: 1163- 1189

Parks, G.A. and De Bruyn, P.L., 1962. The zero point of charge of oxides. J. Phys. Chem.,
66: 967—973

Pennington, D.I., 1973. Chromium and nickel in soil as geochemical indicators for
chromite deposits in the State Line district, Pennsylvania. M.S. Thesis, Pa. State Univ.,
University Park, Pa., 62 pp.

Posselt, H.S., Anderson, F.J. and Weber, W.J., 1968. Cation sorption on colloidal hydrous
manganese dioxide. Environ. Sci. Technol., 2: 1087—1093

Rose, A.W. and Suhr, N.H., 1971. Major element content as a means of allowing for background variation in stream sediment geochemical exploration. In: R.W. Boyle (Editor),
Geochemical Exploration. Can. Inst. Min. Metall., Spec. Vol., 11: 587—593

Sherman, G.D., McHargue, J.S. and Hodgkiss, W.S., 1942. Determination of active
manganese in soil. Soil Sci., 54: 253- 257

Thierweiler, J.F. and Lindsay, W.L., 1969. EDTA-ammonium carbonate soil test for zinc.
Soil Sci. Soc. Am. Proc., 33: 49 54

DIRECT-READING EMISSION SPECTROSCOPY ANALYSIS OF GEOCHEM-ICAL SAMPLES

LIBERTO DE PABLO GALAN*

Consejo de Recursos Naturales No Renovables, Mexico, D.F. (Mexico)

ABSTRACT

A procedure is proposed for multi-element analysis of geochemical samples by direct-reading emission spectrometry. Rocks are classified according to their chemical composition, expressed in terms of seven major components in 13 different groups. Synthetic matrices are prepared, spiked with trace elements in concentrations from 4600 to 0.99 ppm, and sparked in an 8-A d.c. arc. Intensity ratios and concentrations are correlated by equations of the type:

$$(I_{El}/I_{Pd})_C = (I_{El}/I_{Pd})_M + \alpha_{SiO_2}(SiO_2) + \ldots + \alpha_{Al_2O_3}(Al_2O_3) + \alpha_{CaO}(CaO)$$
$$+ \alpha_{MgO}(MgO) + \alpha_{Na_2O}(Na_2O) + \alpha_{K_2O}(K_2O) + \alpha_{Fe_2O_3}(Fe_2O_3)$$

where α is a function correcting for interelement effect among major components and on trace elements. Analyses obtained for international standards are reported.

INTRODUCTION

Direct-reading emission spectroscopy has been added in recent years to the various methods used for analysis of mineral samples. Its advantages include relatively simple sample preparation, speed, and capability to simultaneously determine several elements at low concentrations. These advantages offset the high initial investment and complexity.

Interelement and matrix effects, which are generally accepted as the main cause of error, are usually diminished by use of standards comparable to sample composition, spiking, or by dilution. However, in geochemical exploration large numbers of samples of variable chemical or mineralogical composition are normal and it is not practical to employ lengthy or expensive correction procedures. In this paper, a procedure is presented for chemical analysis of geochemical samples by direct-reading emission spectrometry, corrected for interelement and matrix effects on the assumption that intensities measured for metals at low concentrations are largely affected by major components of the matrix.

---

* Also: Instituto de Geologica, Universidad Nacional A. de Mexico, Mexico, D.F.

708

TABLE I

Composition of selected matrices

Matrix	Rock	Composition* (%)						
		SiO$_2$	Al$_2$O$_3$	CaO	MgO	Na$_2$O	K$_2$O	Fe$_2$O$_3$
A	Sandstones, cherts	75 —95 85.0	2 — 6 4.0	0 — 7 3.0	0 — 5 2.0	0 —25 1.0	0 —6 2.0	0 — 6 3.0
B	Alkali granites, alkali rhyolites, granites, rhyolites, quartz monzonites, quartz latites	68 —75 71.5	12 —15 13.5	0 — 7 3.5	0 — 5 2.5	2.5— 5 3.0	0 —6 3.0	0 — 6 3.0
C	Sandstones	61 —68 71.0	6 —12 9.0	3 — 7 5.0	0 — 7 3.5	0 — 4 2.5	0 —6 3.5	3 — 8 5.5
D	Granodiorites, rhyodacites, quartzdiorites, dacites, alkali, syenites, alkali trachytes, greywackes, phylites	61 —68 64.5	13 —18.5 17.0	0 — 7 4.0	0 — 5 1.5	0.5— 8 4.5	0 —6 4.0	0 — 9 4.5
E	Syenites, trachytes, latites, monzonites, monzodiorites, latite andesites, slates, tillites, pelagic clays, shales, mica schists, kinzingites, sillimante gneisses	53 —61 57.0	15 —20 18.0	0 — 8.5 5.0	0 — 5 4.0	0.5— 8 3.0	0.5—8 5.0	5 —11 8.0
F	Nepheline syenites, phonolites	53 —61 57.0	19 —22 20.5	0 — 7 1.5	0 — 5 2.5	8 — 9.5 9.0	0 —6 5.0	0 — 6 4.5
G	Anorthosites	53 —61 57.0	25 —26 25.5	2.5—12 9.5	0 — 5 1.0	2.5— 5 4.0	0 —6 1.0	0 — 6 2.0
H	Diorites, gabbros, tholeitic basalts, olivine basalts, essexites, leucitic tephrites, andesites, shales	45 —54 49.0	13 —18 16.5	2.5—12 9.5	0.5—10 7.5	0.5— 8 3.5	0 —6 3.5	7 —13 10.5
I	Nepheline tephrites, ijolites, nephelites, olivine leucitites	36 —45 42.0	10 —19 16.0	2.5—12 9.5	13 —15 10.5	1 —10 7.0	0 —6 3.5	7 —13 11.5
J	Olivine mililitites	36 —45 40.5	7 —12 11.0	12 —18 15.0	13 —17 15.0	0.5— 8 2.5	0 —6 1.5	13 —16 14.5
K	Peridotites	36 —45 40.5	3 — 5 4.0	2.5—12 7.0	32 —36 34.0	0 — 2.5 0.5	0 —6 0.5	13 —16 13.5
L	Limestones, carbonatites	5 —15 10.5	1 — 3 2.0	39 —48 43.5	0.5—10 4.5	0 — 2.5 1.0	0 —6 1.5	0 — 6 2.0

* Limits and averages.

METHODOLOGY

Consideration of bulk chemical composition of materials likely to be encountered allows their classification into the groups presented in Table I. The average composition of each matrix was synthetically prepared, calcined to 1200°C, finely ground, and mixed with Spec Mix 1000, at a step ratio of $1/\sqrt[3]{10}$ so as to cover concentrations of elements from 4640 to 0.99 ppm. To these mixtures, in a 20/30/50 ratio, graphite and lithium borate were added, the borate being previously spiked with the equivalent of 500 ppm Pd in the final sample. Mixtures were placed in 3/16-inch necked graphite electrodes 1/4 inch deep, upper cathode plane, and excited in an 8-A d.c. arc. Following a 5-second preburn, spectra were recorded for 40 seconds in a direct reading Philips emission spectrometer set to read 28 elements at the wavelengths indicated in Table II.

Spectral intensities were read automatically from a digital voltmeter connected to the phototubes. Values recorded for each element were divided by intensity of the palladium line to obtain intensity ratios (I_{El}/I_{Pd}) which were plotted versus concentration (Figs.1 and 2). For the major components, SiO_2, Al_2O_3, CaO, MgO, Na_2O, K_2O, and Fe_2O_3, measured intensity ratios are divided by their respective concentrations to compute intensity functions which are plotted against concentrations to obtain a series of smooth, easily correlated curves (for example, Figs.3 and 4). Samples or matrices that do not fall on the curve are considered to have their intensity ratios affected by interelement effects. As an example, for Al_2O_3 in Ca-rich matrix L (Fig.3), the intensity ratio is lower than expected whereas for K_2O (Fig.4), in the same matrix, it is too high. This suggests interelement effect of CaO on Al_2O_3 and K_2O.

TABLE II

Analytical lines

Element	Line wavelength (Å)	Element	Line wavelength (Å)
Zn	2138.56	Mo	3170.35
Cd	2288.02	Cu	3247.54
Be	2348.61	Ag	3280.68
Pd	2447.91	Na	3302.32
Sb	2598.06	Co	3453.50
Mn	2798.27	Ni	3492.96
Pb	2833.07	Ti	3653.49
Si	2881.58	Mg	3838.26
W	2946.00	Al	3944.03
Pt	3064.71	K	4044.14
Bi	3067.71	U	4244.37
Fe	3083.74	Cr	4289.72
Ca	3158.87	V	4379.24
Sn	3175.02	Ba	4554.04

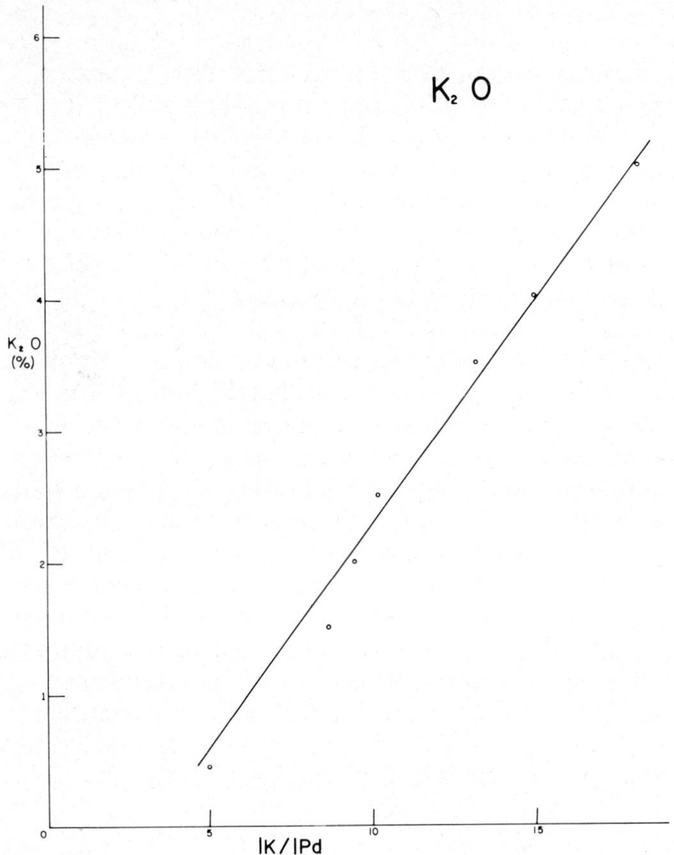

Fig.1. Plot of concentration of K$_2$O versus intensity ratio.

For a given concentration of any of the seven major components, the intensity ratio/concentrations as measured experimentally and as they should have been if unaffected by interelement effects, are read from the graphs. Let:

$$\left(\frac{I_{El}/I_{Pd}}{(El)}\right)_C = \left(\frac{I_{El}/I_{Pd}}{(El)}\right)_M + \alpha_{SiO_2}(SiO_2) + \alpha_{Al_2O_3}(Al_2O_3) + \alpha_{CaO}(CaO)$$

$$+ \alpha_{MgO}(MgO) + \alpha_{Na_2O}(Na_2O) + \alpha_{K_2O}(K_2O) + \alpha_{Fe_2O_3}(Fe_2O_3) \quad (1)$$

where $[(I_{El}/I_{Pd})/(El)]_C$ = corrected intensity function; $[(I_{El}/I_{Pd})/(El)]_M$ = measured intensity function; α = correction factor, proportional to the effect of the major component on the intensity function; (El) = concentration of measured element; and (SiO$_2$) = concentration of major component.

With eq.1 as the model, a series of simultaneous equations is estab-

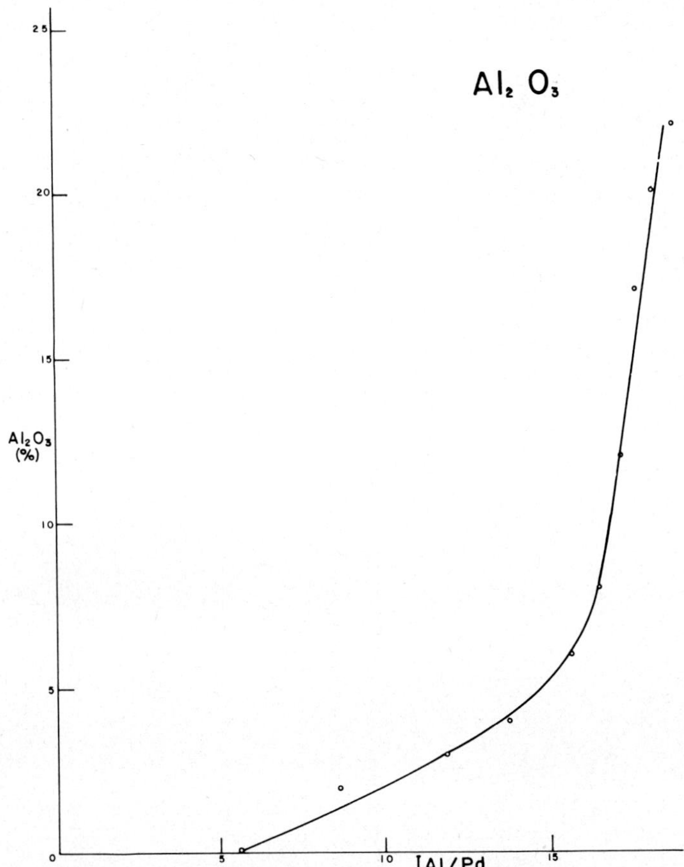

Fig.2. Plot of concentration of Al_2O_3 versus intensity ratio.

lished for each major element in the various matrices and solved for the correcting factor proportional to the effect of major components on measured intensities. As an example, such a calculation is illustrated for K_2O in Table III making reference to Fig.4. Solution of this system of six equations by Gaussian reduction allows the calculation of the six factors correcting for the effect of SiO_2, Al_2O_3, CaO, MgO, Na_2O, and Fe_2O_3 on K_2O. Calculated factors are presented in Table IV. Equations correlating intensity ratios and concentrations are now computed for each element using eq.2. Results, summarized in Table V, enable the estimation of the seven major components in unknowns:

$$(\text{El}) = A + B \left(\frac{I_{\text{El}}}{I_{\text{Pd}}}\right) + C \left(\frac{I_{\text{El}}}{I_{\text{Pd}}}\right)^2 + \ldots \tag{2}$$

To calculate concentrations in samples now becomes a relatively simple

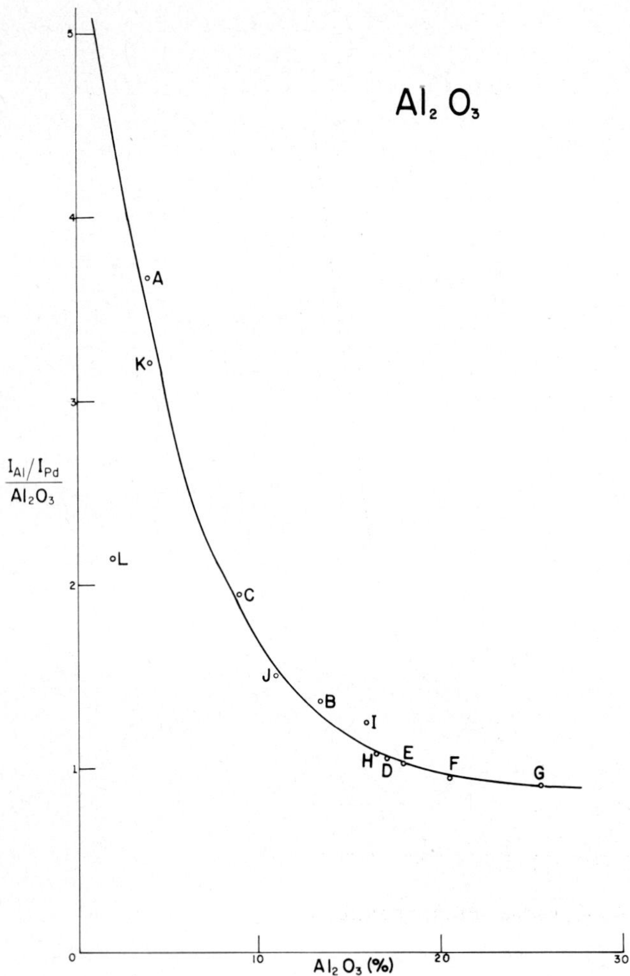

Fig.3. Plot of intensity function versus concentration of Al_2O_3.

procedure involving: (1) measurement of intensity ratios with reference to a proper internal standard (Pd in the present case), (2) calculation of concentrations with the equations in Table V, (3) computation of corrected intensity functions (Table IV), and (4) estimation of new concentrations from corrected intensities. The procedure, being one of asymptotic approximation, is repeated two or three times to obtain the final corrected concentrations.

Determination of trace elements, which range in concentration from 4600 to 0.99 ppm in each of the 13 matrices, is treated similarly. As before, intensity ratios are plotted versus concentrations. However, there is one curve for each element and matrix, to give a total of 13 curves per element. This is illustrated for lead in Fig.5, where the influence of calcium (matrix L) is readily apparent.

These curves are considered to correspond to the interelement effect of major components on trace elements. To compute this effect for a given trace element concentration, one matrix is selected as reference while the others are treated as deviations. Using eq.3, for each element at a fixed concentration, a system of equations is established equating measured intensity functions for the reference and the other matrices to correcting factors and concentrations.

$$\left(\frac{I_{El}}{I_{Pd}}\right)_R = \left(\frac{I_{El}}{I_{Pd}}\right)_M + \alpha_{SiO_2}(SiO_2) + \alpha_{Al_2O_3}(Al_2O_3) + \alpha_{CaO}(CaO)$$

$$+ \alpha_{MgO}(MgO) + \alpha_{Na_2O}(Na_2O) + \alpha_{K_2O}(K_2O) + \alpha_{Fe_2O_3}(Fe_2O_3) \qquad (3)$$

where $(I_{El}/I_{Pd})_R$ = intensity ratio for a trace element in the selected reference matrix, at a fixed concentration; $(I_{El}/I_{Pd})_M$ = intensity ratio for the same element and concentration, in a particular matrix; α = function correcting for the effect of major components on the intensity ratios of the trace elements;

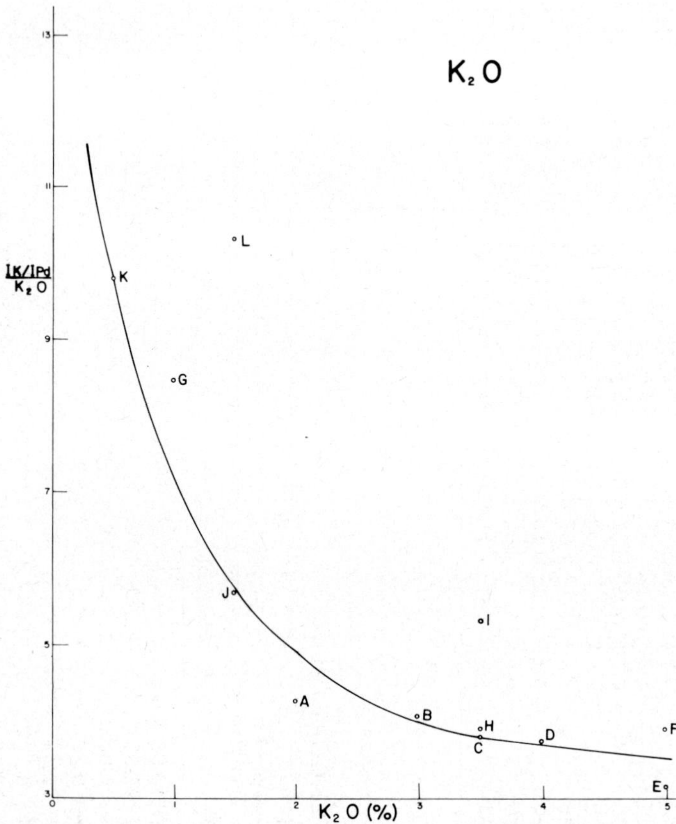

Fig.4. Plot of intensity function versus concentration of K_2O.

TABLE III

Data for calculation of the correction factor α for K_2O

Matrix	$\left(\dfrac{I_K/I_{Pd}}{(K_2O)}\right)^{\star}_C =$	$\left(\dfrac{I_K/I_{Pd}}{(K_2O)}\right)^{\star}_M$	$+\;\dfrac{\alpha_{SiO_2}}{\times\,(SiO_2)}$	$+\;\dfrac{\alpha_{Al_2O_3}}{\times\,(Al_2O_3)}$	$+\;\dfrac{\alpha_{CaO}}{\times\,(CaO)}$	$+\;\dfrac{\alpha_{MgO}}{\times\,(MgO)}$	$+\;\dfrac{\alpha_{Na_2O}}{\times\,(Na_2O)}$	$+\;\dfrac{\alpha_{K_2O}}{\times\,(K_2O)}$	$+\;\dfrac{\alpha_{Fe_2O_3}}{\times\,(Fe_2O_3)}$
E	3.63	3.04	57.0	18.0	5.0	4.0	3.0	0	8.0
F	3.63	3.92	57.0	20.5	1.5	2.5	9.0	0	4.5
D	3.75	3.75	64.5	17.0	4.0	1.5	4.5	0	4.5
C	3.80	3.80	71.0	9.0	5.0	3.5	2.5	0	5.5
I	3.80	5.37	42.0	16.0	9.5	10.5	7.0	0	11.5
B	3.82	4.13	71.5	13.5	3.5	2.5	3.0	0	3.0
K	10.30	9.80	40.5	NC	7.0	34.0	0.5	0	13.5
G	7.65	8.45	57.0	NC	9.5	1.0	4.0	0	2.0
L	5.90	10.33	10.5	NC	43.5	4.5	1.0	0	2.0
J	5.90	5.66	40.5	NC	15.0	15.0	2.5	0	14.5
A	4.72	4.27	85.0	NC	3.0	2.0	1.0	0	3.0

$\star$ From Fig.4.

TABLE IV

Values of the correction factor α

El	Limits of concentration (%)	α						
		Si	Al	Ca	Mg	Na	K	Fe
SiO_2	0—40.36	—	—	—	—	—	—	—
Al_2O_3	0—25.57	4.335×10^{-3}	—	1.166×10^{-2}	1.319		0.1105	−0.299
CaO	0—51.00	1.054×10^{-2}	-2.900×10^{-2}	—	-1.418×10^{-3}	3.410×10^{-2}	—	-7.951×10^{-3}
MgO	0—10.76	—	—	—	—	—	—	—
Na_2O	0— 4.44	—	—	—	—	—	—	—
K_2O	0— 5.100	0.427	−1.053	1.206	−4.343	—	—	0.211
Fe_2O_3	0—16.45	0.307	-7.299×10^{-2}	−0.199	−1.373	−1.102	−1.105	—

TABLE V

Equations to calculate concentration of major elements from intensity ratios

$$(El) = A + B(I_{El}/I_{Pd}) + C(I_{El}/I_{Pd})^2 + \ldots$$

El	Limits		A	B	C
	I_{El}/I_{Pd}	El (%)			
SiO_2	0 — 3.50	0 — 8.40	1.4886×10^{-6}	0.9999	0.4000
	3.50— 4.00	8.40—12.00	2.5807	4.0670	1.6616
	4.00— 5.90	12.00—40.36	100.1010	−46.7600	6.2093
Al_2O_3	5.40—14.00	0 — 3.52	−2.4434	0.4264	0
	14.00—16.00	3.52— 6.43	2.7534	−0.8407	6.6919×10^{-2}
	16.00—20.00	6.43—20.68	128.5060	−16.5488	0.5578
	20.00—20.70	20.68—25.57	−86.6696	4.1097	6.3418×10^{-2}
CaO	0.25— 4.07	0 —51.00	−3.4364	13.4475	0
MgO	0.92— 3.00	0 — 1.20	0.2816	0.3060	0
	3.00— 6.50	1.20— 5.10	1.2870	−0.6327	0.1914
	6.50— 8.50	5.10—10.76	27.6605	−8.2649	0.7384
Na_2O	1.43— 3.00	0 — 1.92	−1.7561	1.2281	0
	3.00— 4.20	1.92— 4.55	1.1366	−1.0213	0.4366
K_2O	3.45—18.30	0 — 5.00	−1.1807	0.3385	0
Fe_2O_3	0 — 3.50	0 — 2.04	0.1363	0.4154	3.7143×10^{-2}
	3.50— 7.00	2.04— 6.10	2.9833	−0.9299	0.1966
	7.00— 8.00	6.10—11.89	58.6054	−18.7495	1.6137
	8.00— 8.30	11.89—16.45	−137.5090	18.5500	0

and (SiO_2) = concentration of major component percent for lead. The procedure is illustrated in Table VI, where matrix C was selected as the reference (Fig.5).

Since this calculation is done at different concentrations, the corresponding correction factors also differ. When these are plotted, curves such as those presented for lead in Fig.6 are obtained, relating the influence of each major component to concentration of the trace element. For the trace elements, the correction factor does not have a definite value but is a function of the concentration, as expressed by these graphs. Corresponding equations are easily computed (Table VII, for Mn). It should be stressed that one basic advantage of such correlations is that matrix effects are defined in terms of interelement influence of seven major rock components thus avoiding, to an extent, the use of similar matrices for standards and samples.

Equations relating intensity ratios and concentrations for the reference matrices are then obtained (Table VIII). Estimation of unknown concentrations of trace metals then involves: (1) measuring intensity ratios, (2) compu-

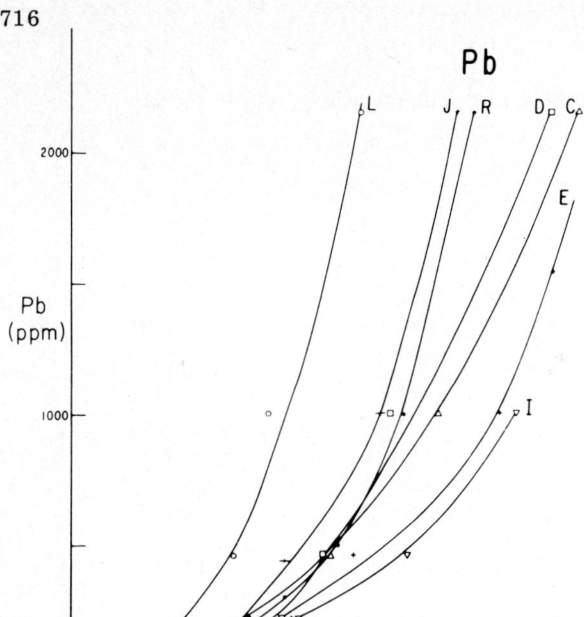

Fig. 5. Concentration of lead versus intensity ratio.

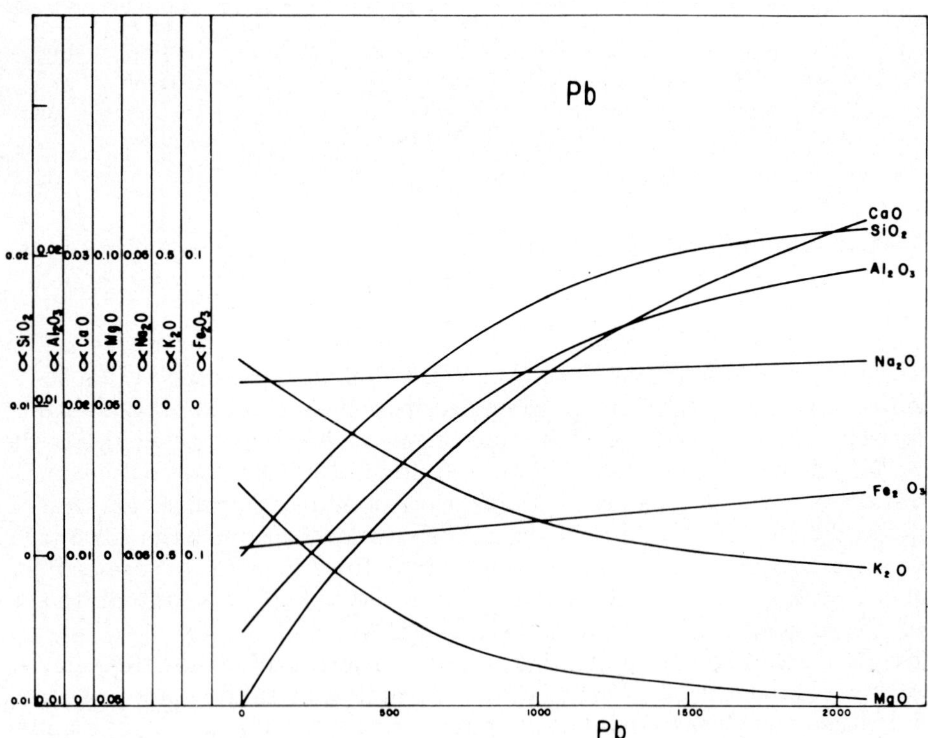

Fig. 6. Plot of the factor correcting for the influence of major components versus concentration of lead.

TABLE VI

Model calculation of the function α correcting for the effect of major components on trace elements (Pb)

Pb conc. (ppm)	Matrix	$\left(\dfrac{I_{Pb}}{I_{Pd}}\right)_R =$	$\left(\dfrac{I_{Pb}}{I_{Pd}}\right)_M$	$+\ \alpha_{SiO_2}$ $\times (SiO_2)$	$+\ \alpha_{Al_2O_3}$ $\times (Al_2O_3)$	$+\ \alpha_{CaO}$ $\times (CaO)$	$+\ \alpha_{MgO}$ $\times (MgO)$	$+\ \alpha_{Na_2O}$ $\times (Na_2O)$	$+\ \alpha_{K_2O}$ $\times (K_2O)$	$+\ \alpha_{Fe_2O_3}$ $\times (Fe_2O_3)$
0	C	0.24	0.24	71.0	9.0	5.0	3.5	2.5	3.5	5.5
	A		0.36	85.0	4.0	3.0	2.0	1.0	2.0	3.0
	B		0.21	71.5	13.5	3.5	2.5	3.0	3.0	3.0
	D		0.22	64.5	17.0	4.0	1.5	4.5	4.0	4.5
	E		0.26	57.0	18.0	5.0	4.0	3.0	5.0	8.0
	G		0.28	57.0	25.5	9.5	1.0	4.0	1.0	2.0
	J		0.31	42.0	16.0	9.5	10.5	7.0	3.5	11.5

$$\left(\frac{I_{Pb}}{I_{Pd}}\right)_C = \left(\frac{I_{Pb}}{I_{Pd}}\right)_M - 3.786 \times 10^{-3} \times (SiO_2) - 8.824 \times 10^{-3} \times (Al_2O_3) + 4.085 \times 10^{-2} \times (CaO) + 3.877 \times 10^{-2} \times (MgO) + 4.219 \times 10^{-3} \times (Na_2O) + 0.154 \times (K_2O) - 9.871 \times 10^{-2} \times (Fe_2O_3)$$

(Pb conc. = 0)

Pb conc. (ppm)	Matrix	$\left(\dfrac{I_{Pb}}{I_{Pd}}\right)_R =$	$\left(\dfrac{I_{Pb}}{I_{Pd}}\right)_M$	$\times (SiO_2)$	$\times (Al_2O_3)$	$\times (CaO)$	$\times (MgO)$	$\times (Na_2O)$	$\times (K_2O)$	$\times (Fe_2O_3)$
215	C	0.65	0.65	70.0	8.9	4.9	3.5	2.5	3.5	5.4
	B		0.54	70.5	13.5	3.5	2.5	3.0	3.0	3.0
	D		0.75	63.5	16.8	3.9	1.5	4.5	3.9	4.5
	I		0.85	41.2	15.7	9.4	10.3	6.7	3.5	11.3
	J		0.59	39.8	10.8	14.7	14.7	2.5	1.5	14.1
	K		0.95	39.8	3.9	6.7	33.2	0.5	0.5	13.2
	L		0.43	9.8	2.0	42.7	4.5	1.0	1.5	2.0

$$\left(\frac{I_{Pb}}{I_{Pd}}\right)_C = \left(\frac{I_{Pb}}{I_{Pd}}\right)_M - 1.041 \times 10^{-2} \times (SiO_2) - 4.588 \times 10^{-3} \times (Al_2O_3) + 1.226 \times 10^{-2} \times (CaO) - 2.330 \times 10^{-2} \times (MgO) + 4.595 \times 10^{-2} \times (Na_2O) - 0.237 \times (K_2O) + 6.776 \times 10^{-3} \times (Fe_2O_3)$$

(Pb conc. = 215)

TABLE VII

Equations to calculate the correction factors for trace elements

Trace element (El)	Major component	$\alpha = A + B(El) + C(El)^2 + \ldots$			
		limits of concentration El (ppm)	A	B	C
Mn	SiO_2	0—1200	6.855×10^{-2}	5.375×10^{-5}	-2.407×10^{-8}
		1200—2000	9.769×10^{-2}	3.107×10^{-6}	0
	Al_2O_3	0—1000	0.239	2.714×10^{-4}	-8.821×10^{-8}
		1000—2000	0.346	2.879×10^{-5}	0
	CaO	0— 900	4.027×10^{-2}	2.194×10^{-4}	-1.066×10^{-7}
		900—2000	0.157	1.070×10^{-5}	0
	MgO	0— 500	2.141	-4.173×10^{-3}	2.357×10^{-6}
		500—1300	1.680	-2.411×10^{-3}	5.838×10^{-7}
		1300—2000	0.846	-1.308×10^{-3}	2.328×10^{-7}
	Na_2O	0— 700	7.858×10^{-3}	-4.150×10^{-3}	1.857×10^{-6}
		700—1400	-0.419	-2.899×10^{-3}	9.108×10^{-7}
		1400—2000	-1.880	-7.030×10^{-4}	6.845×10^{-8}
	K_2O	0—2000	0.101	-1.604×10^{-3}	0
	Fe_2O_3	0—2000	-0.561	4.431×10^{-4}	0

tation of concentrations from equations in Table VIII, (3) calculation of correcting factors from major and trace components and equations typified in Table VII, (4) estimation of final concentrations from corrected intensity ratios. As for major elements, the calculation is done twice to obtain a final value.

The mathematical model and equations presented are simple and the same for both major and minor components. However, the complexity of the calculation increases with the number of elements, demanding the use of a small computer which can be connected directly to the spectrometer to produce rapid results and reduce total time for analysis to 50 seconds.

RESULTS

The final proof of reliability of the proposed procedure is presented in results obtained for standard samples. Many were considered but only data for the U.S. Geological Survey reference samples are reported here (Table IX). For the data in Table IX, major components were determined only to permit correction of intensities and concentrations for the trace metals. It is our opinion that results presented are acceptable for use in geochemical exploration.

TABLE VIII

Equations to calculate concentrations of trace elements from intensity ratios:

$$(El) = A + B(I_{El}/I_{Pd}) + C(I_{El}/I_{Pd})^2 + \ldots$$

El	Limits		A	B	C
	I_{El}/I_{Pd}	El (ppm)			
Mn	3.25—11.00	0— 150	—62.556	19.304	0
	11.00—15.00	150— 400	—407.581	49.629	0
	15.00—22.00	400—2000	—3338.430	241.904	0
Mo	1.35— 4.00	0— 700	—442.822	292.843	0
	4.00— 7.80	700—2000	—720.000	355.000	0
Pb	0 — 0.25	0— 25	-5.086×10^{-6}	100.000	0
	0.25— 1.40	25—1074	—41.485	—200.335	669.746
	1.40— 1.84	1074—2000	1583.300	2315.000	0
Cd	2.70— 4.00	0— 48	—96.957	34.783	0
	4.00— 6.00	48— 250	—1380.320	436.834	—27.569
	6.00— 6.50	250— 310	—14011.200	4275.980	—317.113
	6.50—11.00	310—2000	—2913.410	583.998	12.507
Ag	0.20— 2.50	0— 105	—6.059	32.866	2.811
	2.50— 4.20	105— 275	7.375	2.801	13.898
	4.20— 5.50	275— 570	100.185	—94.548	32.944
	5.50— 6.00	570— 895	—336.909	—7.404	32.318
	6.00— 8.00	895—2000	—2797.160	553.102	7.952
Zn	1.00— 1.60	0— 95	378.414	—712.831	333.715
	1.60— 1.90	95— 290	—923.360	634.399	0
	1.90— 2.80	290—2000	—3169.460	1843.340	0
Cu	0.50— 4.00	0— 208	—25.696	42.900	3.575
	4.00— 6.00	208— 460	243.340	—96.417	22.018
	6.00— 7.60	400—2000	19140.800	—6372.610	543.524
Sn	0.12— 1.09	0—2000	—162.505	1187.000	602.923

CONCLUSIONS

Concentration of trace metals in geochemical samples can be measured with sufficient precision by direct reading emission spectrometry by first estimating the concentration of the seven major rock components SiO_2, Al_2O_3, CaO, MgO, Na_2O, K_2O, and Fe_2O_3. Intensity ratios for trace metals are then corrected for matrix effects with a system of exponential equations that can easily be programmed for a small computer. Since matrix effects are defined as functions of major components the use of standards not closely comparable to the samples becomes feasible.

720

TABLE IX

Analyses of standard samples

Component	G-2		GSP-1		AGV-1		BCR-1		GSE	
	rep.*	det.	rep.	det.	rep.	det.	rep.	det.	rep.	det.
Major elements (%)										
SiO_2	69.19	>40.36	67.27	>40.36	58.99	>40.36	54.88	>40.36	—	>40.36
Al_2O_3	15.34	16.55	15.11	11.80	17.01	25.57	13.65	15.20	—	25.57
CaO	1.98	2.82	2.03	2.32	4.98	5.12	6.95	6.38	—	1.61
MgO	0.78	0.77	0.95	0.83	1.43	1.23	3.28	3.31	—	5.04
Na_2O	4.15	4.25	2.88	2.74	4.33	4.39	3.31	3.32	—	7.53
K_2O	4.51	4.25	5.48	4.35	2.89	2.40	1.68	2.00	—	0
Fe_2O_3	4.28	3.75	6.88	3.45	9.04	5.80	23.48	16.45	—	0
Trace elements (ppm)										
Mn	265	232	326	290	728	690	1350	1280	650	680
Mo	1.2	0	1.6	0	3.7	0	3.9	0	520	
Pb	28.7	30	52.4	45	35.4	35	18	24	580	550
Cd	0.01	0	0.02	0	0.02	0	0.07	0	460	500
Ag	0.05	0	0.07	0	0.1	0	0.04	0	380	345
Zn	74.9	80	143	100	112	105	132	160	500	540
Cu	10.7	10	35.2	30	63.7	60	22.4	20	510	545
Sn	1	0	6.5	2	3.2	0	1.4	0	450	440

* rep.: reported value; det.: value determined in this study.